Biochemistry of Drug Resistance

Sarfraz Ahmed · Suvash Chandra Ojha ·
Muhammad Najam-ul-Haq · Muhammad Younus ·
Muhammad Zaffar Hashmi
Editors

Biochemistry of Drug Resistance

Springer

Editors
Sarfraz Ahmed
Department of Basic Sciences
University of Veterinary and Animal Sciences
Narowal, Pakistan

Muhammad Najam-ul-Haq
Department of Chemistry
Bahauddin Zakariya University
Multan, Pakistan

Muhammad Zaffar Hashmi
Department of Chemistry
COMSATS University Islamabad
Islamabad, Pakistan

Suvash Chandra Ojha
Department of Infectious Diseases
The Affiliated Hospital of Southwest Medical University
Luzhou, China

Muhammad Younus
Department of Pathobiology
University of Veterinary and Animal Sciences
Narowal, Pakistan

ISBN 978-3-030-76319-0 ISBN 978-3-030-76320-6 (eBook)
https://doi.org/10.1007/978-3-030-76320-6

This Springer imprint is published by the registered company Springer Nature Switzerland AG
The registered company address is: Gewerbestrasse 11, 6330 Cham, Switzerland

Foreword

Drug resistance is used in the sense of resistance that bacteria, hosts, or living cells may have developed or have evolved, resulting in a decrease in the efficacy of a drug to treat a disease or disorder. Drug resistance is currently rising to dangerously high levels worldwide and jeopardizing our ability to cure even common infectious diseases. With the use of herbs, modern medicine started to end up with chemically based drugs. Such drugs may be used as DNA, RNA, protein, and lipid-based therapies. Drug resistance can, with high health risks, lead to socially and economically adverse effects. The general opinion is that pandemic drug use in humans and farm animals has shown to a significant driver of global drug resistance. A high degree of resistance to previous medicines contributes to creating novel treatments that may cope with these infectious and other diseases.

Depending on the type of drug and the cell type, there are myriad ways in which all populations of cells develop drug resistance. However, drug resistance literature was scarcely accessible or very dispersed through many disciplines without meaningful interpretation. Fighting the growing resistance to drugs is a race against time. Therefore, understanding the different molecular mechanisms underlying antibiotic resistance and therapeutic other medicines will help create new drugs to address increasing drug resistance.

This monograph is an extensive collection of state-of-the-art research and development information on the subject by leading experts worldwide. Its comprehensive list of chapters covers almost every issue, including drug resistance to bacterial diseases, viral diseases, parasitic diseases, fungal diseases, ENT infectious diseases, skin diseases, lung diseases, neurological disorders, kidney diseases, heart diseases, liver diseases, cancer, HIV and AIDS, reproductive diseases, ocular diseases, diabetes, and drug-resistance management. Therefore, this monograph will be a welcome addition to the subject's literature. To the best of my knowledge, the topics covered in this monograph provide deeper insights into drug resistance and encourage readers to learn more about this fascinating phenomenon. I highly appreciate and congratulate editors for their diligent work and devotion in making this valuable contribution in the field of drug resistance.

Narowal, Pakistan

Sarfraz Ahmed

Preface

Resistance to drugs continues to have a substantial global impact on public health and economic development. Apart from death and injury, chronic illness results in more extended hospital stay, the need for more expensive medications, and financial problems for those affected. Without effective antimicrobials, the advancement of modern medicine in treating diseases, including major non-infectious diseases and cancer chemotherapy, will be at greater risk. The constant need to develop new antimicrobial drugs that function against multi-drug resistant pathogens has spurred researchers on their investment in various drug development techniques, including discovering new applications for existing drugs. Albeit the process is especially promising for fast-tracking compounds in clinical trials, it nevertheless has significant barriers to success.

This book's framework comes into being as we progressively began to speak with peers from diverse backgrounds, many of whom were or had heard of drug resistance within their respective specialties. We realized that enormous information on different aspects of drug resistance was straggly available to the researchers and academicians. Also, drug resistance literature across disciplines was scarcely accessible or highly dispersed without any meaningful interpretation. Therefore, there was an immense need for a comprehensive collection of knowledge on drug resistance, biochemistry, kinetics, dynamics, and management.

This book is a compilation of the scientific work carried out by dedicated teams of scientists devoted to human well-being. This book discusses recent advances in drug resistance that can facilitate young minds to turn their ideas into discovering new medicines for the benefit of society. The central focus of this book is to cover critical issues in drug resistance biochemistry. All chapters are logically chosen and structured to provide detailed state-of-the-art information on drug resistance's broad aspects. Its detailed chapter list includes almost all issues, including drug resistance to bacterial diseases, viral diseases, parasitic diseases, fungal diseases, ENT infectious diseases, skin diseases, lung diseases, neurological disorders, kidney diseases, heart diseases, liver diseases, cancer, HIV and AIDS, reproductive diseases, ocular diseases, diabetes, and drug-resistance management. All chapters and their contents are supported by thorough citations of available literature, estimates, and conclusions based on practical facts of current research and development in this area.

We hope readers will find *Biochemistry of Drugs Resistance* enjoyable to read and useful for gaining deeper insights into drug resistance. This book provides a wealth of information based on a realistic evaluation of modern drug resistance with an emphasis on biochemistry. Many chapters in the book are written at a reasonably advanced level and should be of interest to graduate students and practitioners. It highlights the prospects of the drug mode of action, dynamics, and all possible resistance mechanisms. We hope readers will take the time to read all of the chapters, no matter which medical or scientific discipline you practice. We encourage readers to do so because, first of all, each chapter is super interesting. Secondly, because learning how drug resistance evolves may awaken you to explore its use in novel ways. We genuinely hope that one of the advantages of this book will be to encourage responsible research to resolve drug resistance and support patients by making research therapy accessible.

Narowal, Pakistan — Sarfraz Ahmed
Luzhou, China — Suvash Chandra Ojha
Multan, Pakistan — Muhammad Najam-ul-Haq
Narowal, Pakistan — Muhammad Younus
Islamabad, Pakistan — Muhammad Zaffar Hashmi

Acknowledgements Thanks to Higher Education Commission of Pakistan NRPU projects 7958 and 7964; Pakistan Science Foundation project PSF/Res/CP/C-CUI/Envr (151). Furthermore, thanks are due to Pakistan Academy of Sciences project 3-9/PAS/98 for funding.

Contents

Chapter 1
Introduction to Drugs, Drug Targets and Drug Resistance

Saima Ashraf, Nabila Bashir, Nadia Rashid, Adeel Hussain Chughtai, Khalid Mahmood Zia, Saadat Majeed, Muhammad Naeem Ashiq, Ghulam Murtaza, and Muhammad Najam-ul-Haq

Abstract Drugs are the moieties used in treating diseases. Drugs include vaccines, antidepressants, antibiotics, antiviral, anesthetics, stimulants, and inhibitors. Drugs perform functions in the living system and are associated with the change in body metabolism. Various mechanisms are linked to drug distribution and drug action. Mechanisms are specific and they target specific region or process in the body. Mechanisms for antibiotic resistance include modification of bacterial proteins which are the targets of antibiotics, enzymatic modification of antibiotics, reduction in drug uptake, efflux pump modification, plasmid modification, and inhibition of metabolic pathways. Drugs are used individually or in combinations to maximize each other's effect and minimize the risk. Factors defining drugs efficiency include nature of disease, medical history, drug dosage, treatment time, age, and environmental factors. Drug resistance is the ineffectiveness of drug against a disease after specific period. Factors associated with drug resistance are the excessive or inappropriate use of drugs and the modification in target sites including genetic mutations. Genetic mutations lead to therapeutic and economic problems as new therapies and drug combinations require time and resources. Some pathogens are resistant to multiple drugs known as multi-drug resistance. Drugs designing by considering these challenges can minimize the drug resistance.

S. Ashraf · N. Bashir · N. Rashid · A. H. Chughtai · S. Majeed · M. N. Ashiq · M. Najam-ul-Haq (✉)
Institute of Chemical Sciences, Bahauddin Zakariya University, Multan 60800, Pakistan
e-mail: najamulhaq@bzu.edu.pk

K. M. Zia
Department of Chemistry, Government College University, Faisalabad, Pakistan

G. Murtaza
Department of Pharmacy, COMSATS University Islamabad (CUI), Lahore Campus, Lahore, Pakistan

S. Ahmed et al. (eds.), *Biochemistry of Drug Resistance*,
https://doi.org/10.1007/978-3-030-76320-6_1

1.1 Drugs

A drug is a chemical substance when consumed causes change in organism's body functions both physically and psychologically (Rajesh, 2015). It can be used for the treatment, prevention, diagnosis, or cure of diseases and to enhance/effect the cellular functions. Drugs can be consumed via inhalation, injection, ingestion, smoking, absorption, and patch on skin. A drug produces pharmacological effect by interacting with specific biological sites or cell components in the body. Drug interactions occur via physical and chemical bonds like hydrogen, ionic, covalent and van der Waals forces (Sarker & Nahar, 2007). Drugs with short period of interaction form weak bonds while long actions produce strong and irreversible bonds with their targets. There are many types of target molecules showing interactions with drugs.

1.1.1 Classification of Drugs

1.1.1.1 Classification Based on Pharmacological Effect

Pharmacological effect shows how a drug effects cells of an organism. Various drugs have different pharmacological effects on the body and thus categorized on this basis.

Vaccines

Vaccination is a cost-effective measure for controlling infectious diseases. The discovery of vaccines was a critical point in the war between human and microbes. Vaccination has protected mankind from diseases like smallpox and measles etc. Till now 27 vaccines have been developed to prevent and control infections (Vartak & Sucheck, 2016). Vaccines are the biological preparations made of small amount of attenuated or dead pathogens (virus or bacterium) that elicit immunity to target pathogens (Pulendran & Ahmed, 2011). Vaccine sparks body's immune response to fight off, destroy and memorize pathogens if they ever encounter again and cause infections. By injecting vaccine, some minor infections with fever may occur. Fever is caused as immunity requires few days or sometime weeks to produce antibodies. To understand the working of vaccines, lets first look at how the body fights against illness. Pathogens enter in the body and multiply to cause infection leading to illness. Body's Defense System (Immune System) fights against such infections through white blood cells also known as immune cells present in blood. Macrophages, B-lymphocytes, and T-lymphocytes are the components of white blood cells. Macrophages swallow and digest germs, and dead or dying cells. The macrophages leave behind parts of invading germs called antigens. Body identifies antigens as danger and stimulates antigen-specific antibodies to attack them. B-lymphocytes are defensive white blood cells which produce antibodies that attack antigens. T-lymphocytes are another type of defensive white blood cell which attack the infected cells present in the body. Immune system through T-lymphocytes and B-lymphocytes remembers how to fight the same pathogen in future (Vetter et al., 2018). Vaccines have thus rapid effect, save life, and promote health.

Types of Vaccines

(i) **Live Attenuated Vaccine**
Live vaccine consists of antigenic substance containing living microorganisms cultivated under controlled conditions in culture or embryo that make them less virulent. When administered, they show protective and long-lasting immune response in host against pathogens. Viral vaccines protect against diseases like measles, rubella, varicella-zoster, oral poliovirus, chickenpox, mumps, rotavirus and yellow fever (Victoria et al., 2010). For example, the live attenuated mycobacterium vaccine containing virulently modified strains of "BCG" cures deadly and infectious Tuberculosis disease. Live attenuated vaccines have disadvantages like they are not given to immunocompromised individuals as small quantity of live pathogens can produce infections. Sometimes these vaccines mutate to virulent form and cause health problems (Sinha & Bhattacharya, 2014).

(ii) **Inactivated Vaccines**
Inactivated vaccines are created by inactivating the propagated pathogens via treating them in culture with chemicals like binary ethyleneimine, radiations or heat. Inactivated vaccines provide weaker immunity as compared to live vaccines, therefore immunological adjuvant and booster shots are required periodically after vaccination. These vaccines immune against polio, influenza, plague, cholera, typhoid and pertussis (Sanders et al., 2015).

(iii) **Toxoids**
Toxoid vaccines are inactivated toxins made by inactivating bacterial toxins (exotoxin) through treatment with heat or chemicals (formaldehyde). These vaccines induce anti-toxoid antibodies capable to bind with toxin and neutralize them. This vaccination is effective for diseases caused by secreted toxins like tetanus and diphtheria (Hermanson, 2013; Yadav et al., 2014).

(iv) **Subunit/Recombinant Vaccines**
Subunit vaccines use only a fragment of target pathogens (e.g. proteins, polysaccharides) which provoke immune response. Subunit vaccines are produced either by well-established expression systems or by conventional biochemical technologies. These vaccines are cost-effective compared to traditional vaccines (Tan & Jiang, 2017). Subunit vaccines are used against hepatitis b, human papillomavirus and pneumococcal disease (Schiller & Lowy, 2018).

(v) **Conjugate Vaccines**
Conjugate vaccines induce immune response against polysaccharides capsule present as an external layer on range of bacteria. Bacteria can synthesize variety of polysaccharides. Antibodies against polysaccharides of pathogenic bacteria like Hib, pneumococcus and meningococcus protect from diseases

(Rappuoli et al., 2019). Example of this type of vaccine is Haemophilus influenza type B (Hib) (Goldblatt, 2000) (Table 1.1).

Antidepressants

Hypertension is a medical condition where arteries have elevated pressure. Factors like blood plasma levels, hormonal activity, environmental conditions (stress) are few of the reasons of primary hypertension. Secondary hypertension can result due to diabetes, kidney disease, hyperthyroidism, pregnancy, obesity, and sleep apnea (Charles et al., 2017). There are several classes of antihypertensive drugs used to control, treat, and prevent hypertension. Diuretics, Angiotensin-converting enzyme (ACE) inhibitors and calcium channel blockers are used as the first line treatment (Khalil & Zeltser, 2020). Thiazide diuretics work by inhibiting reabsorption of salts (sodium) in the convoluted tubule of kidneys and thus promote diuresis and natriuresis. Commonly used thiazide diuretics are hydrochlorothiazide (HCTZ), indapamide and Chlorthalidone (Akbari & Khorasani-Zadeh, 2019). Calcium is the main reason of contraction of arteries and heart cells. CCBs help in lowering blood pressure by preventing calcium entry into the heart cells and arteries walls through L-type calcium channels (Ozawa et al., 2006). ACE inhibitors prevent angiotensin converting enzyme to produce angiotensin II, a strong vasoconstrictor. ACE inhibitors thus lower blood pressure and prevent contraction of cardiac myocytes and smooth muscles (Herman & Bashir, 2019). Table 1.2 shows antidepressant drugs, their mode of actions and side effects.

Antibiotics

Antibiotics slowdown the growth or destroy bacteria. Penicillin's discovery by Alexander Fleming in 1928 as first antibiotic was a breakthrough in modern medicine. Penicillin prevents cross-linking of peptidoglycan in the cell wall of bacteria by inhibiting transpeptidase enzyme and results in bursting of bacterium. Penicillin is classified with regard to the chemical substitution of side chain attached with (4 membered) beta lactam ring (Lobanovska & Pilla, 2017). Types of penicillin antibiotics include penicillin V, penicillin G, amoxicillin, and ampicillin. Another antibiotic is tetracycline used for treating infections caused by bacteria (both gram-negative and gram-positive), mycoplasmas, protozoans, chlamydiae or rickettsia. Tetracycline inhibits protein synthesis by reversibly binding with receptors on microbial RNA. They are bacteriostatic and thus prevent bacteria from multiplying. Chlortetracycline, Doxycycline and Oxytetracycline belong to this class (Chopra & Roberts, 2001). Cephalosporins are the β-lactam antibiotics working like penicillin for range of bacteria. Cephalosporins treat diseases like bacteremia, pneumonia, meningitis and soft tissue infections (Marshall & Blair, 1999). Other antibiotics include glycopeptides (Kang & Park, 2015), aminoglycosides (Krause et al., 2016), sulfonamides (Kleiner, 2018) and quinolone (Aldred et al., 2014).

Table 1.1 Types of vaccines

Vaccine type	Detail	Advantages	Disadvantages	Disease
Live attenuated	Vaccine strains are cultivated in controlled conditions of culture or embryo that make them less virulent	Induction of long-lasting immune response in host	Adverse effect in immunocompromised individuals. Mutate to virulent form	Tuberculosis, Measles, Smallpox, Oral polio vaccine, Mumps, Rotavirus, Yellow fever
Inactivated vaccines	Pathogens killed in culture through radiations, chemical treatment, or heat	Cannot replicate	Shorter length response. Booster shots are required for long term immunity	Influenza, Plague, Cholera, Typhoid, Whole cell Pertussis
Toxoid vaccine	Inactivated toxins. They generate strong immune response against toxins	Excellent stability, Highly efficient	Not highly immunogenic	Tetanus toxoid, Diphteria toxoid
Subunit/recombinant vaccine	Purified antigens. Fragments (polysaccharides, proteins) of virus or bacteria	Safe and non-replicating vaccine	Repeated vaccination or booster are required	Smallpox, Hepatitis B, Haemophilus influenza type B, Pneumococcal (PCV-7, PCV-10, and PCV-13)
Conjugate vaccine	Antigens enclosed in polysaccharides' coating	Improved immune response, longer protection, herd immunity	The lack of protective immune response in young children	Haemophilusinfluenzae type B (Hib)

Table 1.2 Antidepressant drugs

Drugs	Mechanism of action	Side effects	Examples
Thiazide diuretics	Inhibit Na/Cl symporter	Dizziness, blurred vision, stomach upset, weakness	Chlorothiazide, Hydrochlorothiazide, Chlorthalidone, Indapamide
Calcium channel blockers	Prevent Ca^{2+} entry in cell	Constipation, fatigue, headache, nausea, fast heartbeat	Amlodipine, Isradipine, Nifedipine
ACE inhibitors	Inhibit conversion of angiotensin I to angiotensin II	Cough, fatigue, hypotension, azotemia	Benazepril, Perindopril, Lisinopril, Captopril
Beta blockers	Blocking effect of epinephrine hormone	Bradycardia, depression, fatigue, constipation, sexual dysfunction	Atenolol, Nadolol, Propranolol, Metoprolol
Loop diuretics	Inhibit Na^{+}, Ca^{2+}, and K^{+} reabsorption	Electrolyte imbalance, dehydration, hyperlipidemia, hyperuricemia	Bumetanide, Furosemide, Torsemide, Bumetanide
Potassium sparing diuretics	Na^{+} channel inhibitor, Reduced K^{+} secretion	Muscles cramp, skin rashes, aches and pain, headache	Amiloride, Triamterene, Eplerenone
Alpha-blockers	Blockage of alpha-1 receptor	Orthostatic hypotension and Tachycardia	Prazosin, Doxazosin
Angiotensin II Receptor Blockers (ARBs)	Blocking of type 1 angiotensin II receptors on arteries	Angioedema, renal failure	Losartan, Irbesarten, Telmisartan

Drugs for Specific Disease Manifestations

This class of drugs target the specific cells hampered because of any disorder in the body. These drugs alter the mechanism of disease either by inhibition of diseased cells or regulation of healthy cells, for example, antipsychotics also referred as neuroleptics (Finkel et al., 2009). These drugs target the mental disorders including psychosis, primarily in bipolar disorder and schizophrenia (Lally & MacCabe, 2015). Their long-term use may produce adverse effects like gynecomastia, weight gain, metabolic syndrome, and impotence. In 1940s, typical antipsychotics were discovered as first-generation while the second generation of atypical antipsychotics like Clozapine was developed in 1960s (Hippius, 1989). Both generations block the dopamine receptors however some atypical medications also operate via serotonin receptors. These medications are employed in schizoaffective disorders with mood stabilizer or antidepressant. Atypical medications are prescribed in psychotic depression in combination with an antidepression. These medications work by blocking dopamine D_2 receptors to reduce its effect in brain dopaminergic pathways, as psychotic effects are produced due to excess dopamine release. These medications impact the activity of neurotransmitters in the brain.

Drugs for Non-specific Diseases

These drugs treat disorders like runny nose, nasal congestion, sneezing, fever, influenza, and sinus. These medications do not contain cough relieving ingredients. This class includes analgesics also referred as pain relievers or painkillers. They act without inducing sleep or unconsciousness. Analgesics function on both central and peripheral nervous systems (CNS and PNS). Analgesics are different from anesthetics which temporarily or sometimes permanently affect the sensation. Analgesics include drugs like NSAID's (non-steroidal anti-inflammatory drugs), paracetamol and opioids as oxycodone and morphine. The extent of pain suggests the use of analgesics. Traditional drugs are not effective in neuropathic pain and benefit can come from drugs like anticonvulsants and tricyclic anti-depressants which are not the normal analgesics (Dworkin et al., 2003). Analgesics work by blocking the pain signals of brain or disrupt the brain's interpretation of those signals. In some cases, they terminate the synthesis of prostaglandins by enzyme inhibition. They also stimulate opioid receptors which relieve the chronic pain (Smith et al., 2014).

Antihistamine is a group of drugs for relieving allergies and allergic rhinitis. They are inexpensive, over the counter, and generic medicines that treat skin rashes, sneezing and nasal congestion. Antihistamines are for short term processes whereas chronic allergies like sinusitis and asthma cannot be treated. They function by opposing the histamine receptors and thus categorized as e.g. H_1-antihistamine and H_2-antihistamine. H_1-antihistamine acts by the attachment of H_1 receptors in cells to cure problems associated with nose, inner ear, sneezing, motion sickness and insomnia. H_2 receptors bind cells in gastrointestinal tract for treating ulcers and gastric acid issues (Panula et al., 2014).

Anesthetics, Muscle Relaxants and Diagnostic Agents

(a) **Anesthetics**
They make humans unconscious in controlled manner. Anesthetics are diversified compounds which perform via neuronal passages. They cause reduced consciousness, immobility, decreased responsiveness and amnesia (Chou et al., 2002). They are administered through inhalation and injection. Sometimes, anesthetics are given in combination with other drug to avoid side effects or uncontrolled consciousness. Anesthetics agonize GABA receptors for their effect, e.g. benzodiazepines, isoflurane, etomidate and propofol (Louis, 2011).

(b) **Muscle Relaxants**
They deal with skeletal muscles and reduce muscle tone. Spasmolytic and neuromuscular blockers are the muscle relaxants. Neuromuscular blockers hinder the transmission at NMJ (Neuromuscular junction) and cause no activity

in CNS and induce temporary paralysis in intensive care and surgical operations. Spasmolytics as "centrally acting" muscle relaxants alleviate musculoskeletal pain and decrease spasticity in neurological conditions (Bowman, 2006). Baclofen, metaxalone, methocarbamol and cyclobenzaprine are the examples of muscle relaxants.

(c) **Diagnostic Agents**

Diagnostic agents help in diagnosing diseases and their causes. They are termed as "helping aids" in detecting abnormalities in cells, tissues, or organs. Their dose depends on the type of test and can be administered via injection, mouth, eye etc. In radiopharmaceuticals, the radioactive isotope is employed to determine the disorder. For example, I-131 detects goiter in thyroid gland, P-32 and Sr-90 diagnose skin and other cancers, N-13, Rb-82 and Tl-201 detect heart disorders. Their composition and activity differ because of their working nature, e.g. sulphobromophthalein sodium detects liver disorder whereas p-amino hippuric acid, phenosulphophthalein and inulin determine kidney dysfunction (Sano et al., 2001). Rose Bengal targets serum albumin for finding abnormalities. Protirelin targets thyrotropin releasing hormone receptor, i.e. GPCR which activates intracellular signaling molecules and respond regarding the function of cells.

Enlon and tensilon containing Edrophonium, are acetylcholinesterase inhibitors. They detect muscle disorder referred as myasthenia gravis. Some medications in surgical operations cause muscle contractions which can be reversed by edrophonium. Acetylcholine in the body inhibits acetylcholinesterase by stimulating the muscarinic and nicotinic receptors (Caliandro et al., 2009).

1.1.1.2 Classification Based on Drug Action

Immune Response

The immune system is activated against pathogens to avoid disorders in the body. Longer presence of pathogens causes more damage. Immune response can be innate and adaptive. The innate part responds immediately to every pathogen and is non-specific. Adaptive part includes cells like antibodies, B cells and T cells that are specific to the pathogen.

Some drugs are immuno suppressant and can be advantageous in organ transplantation where they reduce the probability of organ rejection. Immune responsive drugs include cortisone, hydrocortisone, prednisone and triamcinolone (Demirkiran et al., 2009). Some drugs are immunostimulant which activate the immune system to work against pathogens (Rich et al., 2012). Immunosuppressants work by consuming body healthy tissues and weaken the immune system while immuno stimulants induce pathogen-linked patterns recognized by the receptors for quick recovery. Cytokines (signaling channels of immune system) can also activate the immune system. Example of immune response is the histamine which reacts differently to an injury or infection. Mast cells and basophils produce histamine in response to infection. Histamine enhances capillaries permeability to platelets and restrict some proteins to pathogens in the infected part. In injury, mast cells in injured part release histamine which broadens nearby blood vessels for enhanced permeability.

Protease Inhibitors

Protease inhibitors inhibit proteins cleavage to small peptide chains and thus are antiviral. The antiviral drugs block virus replication instead of destroying or deactivating the viruses by inhibiting the cleavage of proteins into peptides (Daniels & Nicoll, 2011). These proteins also contain virus, and the cleavage inhibition stops their replication. Some of the protease inhibitor antiviral drugs are atazanir, indivanir, ritonavir and fosamprevanir (Madruga et al., 2007).

Dihydropteroate Synthetase (DPHS) Inhibitors

DHPS enzyme assists the synthesis of dihydropteroate in bacteria. The anti-diabetic drug (sulphonamide) binds with DHPS and inhibits this enzyme. DHPS enzyme is essential for the folate synthesis in bacteria. Inhibition of folate synthesis inhibits the amino acids synthesis which are crucial for bacterial functioning. Sulphonamide thus inhibits the reproduction of bacteria rather than killing them (Bauman, 2015).

1.1.1.3 Classification Based on Chemical Structure

Opioids

They include illegal drugs like heroin, synthetic opioids like fentanyl (50–100% more potent than morphine and given in extreme pain conditions), and legally prescribed pain relievers such as oxycodone, hydrocodone, codeine, and morphine (Strømgaard et al., 2017). These drugs act on opioid receptors to produce morphine like effects and relieve pain. Sometimes body becomes opioid tolerant due to internalization, downregulation, and desensitization of opioid receptors (Gudin, 2012). Opioid addiction must not be confused with opioid tolerance as both phenomena are different. Patients using opioids for non-cancer or cancer pain may become tolerant (Adesoye & Duncan, 2017). The homeostatic mechanisms of body counteract the effect of drug resulting in opioid tolerance. Opioids also include some abuse deterrent formulations which are given in Table 1.3.

Table 1.3 Abuse deterrent Opioids (2015)

Brand name	Generic name	Abuse deterrent mechanism
Oxaydo	Oxycodone	Crush/extraction resistant
Zohydro ER	Hydrocodone	Crush/extraction resistant
Emdeda ER	Morphine-naltrexone	Agonist-antagonist
Troxyca ER	Oxycodone-naltrexone	Agonist-antagonist
Morpha bond	Morphine	Crush/extraction resistant
Xtampza ER	Oxycodone	Crush/extraction resistant
Hysingla ER	Hydrocodone	Crush/extraction resistant
Arymo ER	Morphine	Crush/extraction resistant
Opana ER	Oxymorphone HCl	Crush/extraction resistant

Er = Extended release, HCl = Hydrogen chloride

Alcohol

This class of drugs are antiseptic and used in many situations. Alcohol in the form of wine was initially prescribed to aid digestion, for the stimulation of appetite and as general tonic (Williams, 1980). However, it was banned in many countries due to harmful effects (O'Connell et al., 2003). Epidemiological researches show that alcohol is beneficial in moderate levels and prevents heart disease (Corrao et al., 2000), diabetes (Koppes et al., 2005) and stroke (Reynolds et al., 2003). Evidence from red wine suggests that antioxidants are responsible for its beneficial effects (St Leger, 1979). Later studies proved that alcohol modifies the fat content and blood coagulation factors (Rimm et al., 1996). The use of alcoholic medications varies with age as risk factor must be considered in aged people because of physiological changes occurring due to the prescribed and non-prescribed medicines (Aira et al., 2005).

Alcohol interacts with medicines and changes their effects. Even the moderate alcohol levels interfere the activity of a medicine (Tapert et al., 2002). Drug-alcohol interactions can be explained as pharmacokinetic and pharmacodynamic interactions. The medication metabolism is harmed by the interference of alcohol and is termed as pharmacokinetic interaction. These interactions are observed in liver as the metabolism of both alcohol and drug occurs in liver by the same enzymes. Alcohol enhances the medication effects in some cases, specifically in CNS i.e. sedation (Carrière et al., 1996). Negative effects are observed when herbal medicines interact with alcohols. The drugs interacting with alcohols are e.g. muscle relaxants, barbiturates, antidepressants, histamine H_2 receptor antagonists, antihistamines, anti-inflammatory agents, opioids, and antibiotics (Weathermon & Crabb, 1999).

Benzodiazepines and Barbiturates

Benzodiazepines consist of fused diazepine and benzene rings. Benzodiazepines are psychoactive drugs also termed as 'benzos' (Shorter, 2005). Benzodiazepines were the most prescribed medications in 1977 (Washton & Zweben, 2006). They belong to a group of medicines known as minor tranquilizer (Haverkampf, 2017). They work by increasing the gamma-aminobutyric acid (GABA, a neurotransmitter) effect at $GABA_A$ receptor which results in anxiolytic, hypnotic/sleep inducing, and muscle relaxant properties (Page, 2002). So, seizures, agitation, insomnia, muscle spasms, anxiety and the issues of alcohol withdrawal can be resolved by using benzodiazepines. They can be the potential pre-medicines in medical and dental procedures (Olkkola & Ahonen, 2008). Benzodiazepines drugs are alprazolam, clorazepate, diazepam, lorazepam (Ativan), and chlordiazepoxide (Librium).

Barbiturates act as depressant in Central Nervous System and are active as hypnotics, anxiolytics, and anticonvulsants. Barbiturates are prescribed in cluster headache/acute migraines, treatment of epilepsy and as general anesthesia (Vaux & Golder, 2003). Phenobarbital (a barbiturate medication) treats epilepsy, anxiety, and hypnotics. Sodium thiopental is a general anesthetic and primidone an anticonvulsant (Johns, 1975). Barbiturates bind to GABA receptor as agonist and response is produced. GABA is the inhibitory neuro receptor in CNS of mammalian cell. Barbiturates inhibit CNS by stimulating the inhibitory GABA receptor and cause depression in CNS (Löscher & Rogawski, 2012). GABA is a chloride containing channel. When GABA binds to chloride, it alters the voltage in brain cells by increasing the number of negative charges. Brain cells become resistant to nerve impulses and depress them. Sometimes barbiturates resistance is due to the enhanced p450 cytochrome content which is followed by reduction in sleep (Chiara et al., 2013).

Cocaine and Other Stimulants

Cocaine is a stimulant employed as recreational drug. It enters the body by injection, snorting and smoking (Pomara et al., 2012). It exerts physical and mental disorders. Physical effects include large pupils, sweating and fast heart rate while mental symptoms include loss of agitation and restlessness (Harper & Jones, 2006). High blood pressure and high body temperature is observed with the large dose. It medically diminishes bleeding in nasal treatment and in numbing. Cocaine increases the amount of three neurotransmitters in brain cells. They are dopamine, serotonin, and norepinephrine. Dopamine activates brain cells for muscle mobility and the mind to work faster (Kalivas & McFarland, 2003). Cocaine can cross by breaking the blood–brain barrier. It stops the uptake of these neurotransmitters and thus hampers CNS (Karch, 2009).

Sulfonamides

Sulfonamides known as sulfa drugs contain sulfonamide functional group. They are antibacterial (Henry, 1943) with antimalarial and antifungal properties and treat cough and allergies. Sulfonamide drugs include sulfadimethoxine (long-acting), sulfadimidine, sulfacetamide, sulfisomidine, sulfafurazole (short-acting), sulfanitran, sulfamoxole, and sulfamethoxazole (intermediate-acting). Sulfonamides also contain a group of sulfonylyreas (anti-diabetic agents), diuretics, antiretrovirals, and dermatologicals (Pelet & Matheux, 2011). Sulfa drugs bind with DHPS and inhibit this enzyme which is essential for folate synthesis in bacteria. Side effects of sulfa drugs are dizziness, headache, rash, folate deficiency, gastrointestinal disorders such as diarrhea, stomach upset, vomiting and fatigue (Chung, 2016). Sulfa drugs can also accelerate DHPS synthesis leading to the resistance of sulfonamides. This mechanism is mediated by the mutations in DHPS genes which can be isolated to explore this mechanism (Zhang et al., 2009).

Amphetamines

Amphetamine is branded as Evekeo and used as a stimulant in CNS. The chemical name of amphetamine is α-methylphenethylamine (Klee, 1997). It treats ADHD (attention deficit hyperactivity disorder), obesity and narcolepsy (Rasmussen, 2006). Amphetamine therapeutic effect in the form of euphoria and weakness. It improves muscle strength, fatigue resistance and reaction time (Carvalho et al., 2012). Side effects include nausea, increase in heart rate, blood pressure, abdominal cramps and vomiting (Adler et al., 2008). Primary mechanism for amphetamines is the release of dopamine, serotonin and norepinephrine from nerve terminals (Biaggioni & Robertson, 2009). Abuse or unprescribed use of amphetamines disturb the reuptake of neurotransmitters leading to amphetamine resistance.

1.1.1.4 Classification Based on Molecular Targets

Enzymes

Enzymes also termed as biological catalysts which speed up the biochemical reactions. They are proteins acting on substrate for conversion into products. Every metabolic action requires enzyme for the fast completion (Buchner). Enzymatic study is referred as enzymology and pseudoenzyme analysis emphasizes enzymes which have lost activity in the evolutionary processes (Murphy et al., 2017). Their catalytic activity has been affected because of change in their amino acid sequence (Murphy et al., 2014).

(a) **Protein Kinases**
These enzymes cause phosphorylation by chemically adding phosphate groups to proteins. Phosphorylation affects the enzyme activity which in turn alters the function of target protein. About 500 protein kinases are found in human genome (Manning et al., 2002). Plants and bacteria also contain kinases. Kinases may modify 30% of human proteins and regulate cellular pathways, especially the signal transduction. They remove phosphate moiety form ATP followed by its attachment to one of the three amino acids (Serine, Tyrosine and Threonine). Majority of kinases operate on threonine and serine while others work on tyrosine. Some kinases work on all three amino acids (Dhanasekaran & Reddy, 1998). Histidine residues are phosphorylated by histidine kinase (Besant et al., 2003).

(b) **Proteases**
Protease also termed as proteinase or peptidase, catalyzes the breakdown of large proteins into small peptide chains or in single amino acids. They perform cleavage through hydrolysis (King et al., 2014). Proteases also have functions like ingested proteins digestion, cell signaling and proteins catabolism (Shen et al., 2006). Protease activates a nucleophile that attacks peptide carbon followed by the migration of electrons of C–O bond to oxygen. This produces high energy intermediate and protease stabilizes it by splitting into small chain peptides (Berg et al., 2012).

(c) **Esterases**
Esterase is a hydrolase enzyme that converts ester into alcohol and acid via hydrolysis. Esterases have activity based on the nature of substrate, biological action, and protein structure.

(d) **Phosphatase**
Phosphatase detaches phosphate group or performs opposite to that of kinase. It splits phosphoric acid into alcohol and phosphate ion via hydrolysis. As phosphatase catalyzes hydrolysis, it can be categorized as hydrolase. Phosphorylation (by kinases) followed by dephosphorylation (by phosphatases) have role in signaling and cellular regulation (Liberti et al., 2013). Phosphatases are different from phosphorylases which are responsible for phosphate group transfer (from hydrogen phosphate) to acceptor.

Structural Proteins

These proteins develop protective capsid or capsule. They constitute cell shape, compose vertebral bone and cartilage in connective tissues. For example, tubulin via polymerization forms long chains and assists the skeletal system. Tubulin after certain structural changes forms microtubules that regulate transport in the cell (Jenkins et al., 2002).

Membrane Transport Proteins

These proteins involve the movement of molecules, ions, macromolecules, and micro molecules across the membrane. They are found in the cell and perform the function of transport. Diffusion is also supported by these proteins. The binding of transport proteins to solutes bring them to one side of membrane and their release to other end is accomplished by alteration in conformation of transport proteins (Medicine, 2000).

Nucleic Acids

Macro biomolecules or biopolymers are referred as DNA or RNA. They consist of nucleotides formed by the combination of 5-carbon sugar, nitrogenous base, and a phosphate group. DNA is formed in the presence of deoxyribose sugar and RNA in the presence of ribose. DNA serves as the "life genetic code" for information storage while RNA is involved in proteins synthesis. Nucleic acids have the potential of genetic level cure to the diseases (Elson, 1965).

1.2 Receptors (Drug Targets)

Receptors are the protein molecules within or on the cell that recognize and bind with ligands (e.g. hormones, chemical messenger substances or particular drugs) and regulate the cellular process (Currie, 2018). Drug-receptor interaction is important in biological functions such as drug delivery, cell proliferation and immune response (Ma et al., 2018). Drug molecules upon interaction with receptor may activate or inactivate a receptor. Activation of receptor may enhance or diminish a biological function. Few drugs show specific binding with receptors and thus are selective. Selectivity is the ability of a drug to bind with specific receptor.

Potency of a drug is determined by its affinity and efficacy. The term affinity is the ability of drug to affect a receptor and intrinsic efficacy is the potential of drug-receptor complex to initiate a cellular process (Lambert, 2004). Drug may act as agonist or antagonist based on the difference in efficacy. An agonist is a drug which activates receptor and produces functional response. The neurotransmitters (e.g. histamine, acetylcholine, and norepinephrine), hormones and drugs (e.g. phenylephrine, isoproterenol, and morphine) act as agonist. Antagonist is a drug which prevents receptor activation. Neutral antagonist prevents the action of an agonist. There are reversible and irreversible types of antagonists. Reversible forms weak bond and dissociates from receptor while irreversible makes stable chemical bond with receptors. Antagonism can be competitive or non-competitive. In competitive antagonism, antagonist competes with agonist for the same binding site. For example, a competitive antagonist for all opioid receptors like morphine is naloxone.

In non-competitive antagonism, simultaneous binding of both agonist and antagonist reduce the effect of agonist.

Alfred Joseph Clark proposed receptor theory which informs about drug receptor interaction and cellular functions altered thereof. Rate of formation of drug-receptor complex is proportional to the dose of drug (Zhu, 1993).

$$\text{Drug} + \text{Receptor} \rightleftharpoons \text{Drug-Receptor complex}$$

Above equation is derived from Langmuir isotherm which shows effect of drug receptor interaction on forward (k1) and backward (k2) reaction constants.

1.2.1 Types of Receptors

There are four types of receptors based on molecular structures.

1.2.1.1 Ion Channels

(a) **Ligand Gated and Voltage Gated**
Ligand gated ion channels are the pore like transmembrane proteins which bind with ligands and change the permeability of cell membrane to provide gateway for selective ions to move across the membrane (Abdul Kadir, 2018). These are proteins spread around the membrane and are responsible for ions channels which can be categorized based on the cause of channel opening and closing. So, ion channels are either ligand-gated or voltage-gated. Opening and closing of voltage gated channels correspond to variation in membrane potential. Activation of this channel occurs in the presence of ligand and no other biological intermediates are involved. Sometimes amino acids side chains may block the pores. Binding of receptors to the extracellular part of channel causes conformational changes and opens the pore for flow of ions like calcium, sodium, chlorine, magnesium and hydrogen (Hanukoglu, 2017). Receptors activation unlocks the transmembrane ion channel and potential of the cell changes due to flow of ions across the membrane and cause activation or prevention of electrical impulses. Examples of this channel include acetylcholine, GABAA receptor, glycine, Glutamate and serotonin (Suvarna, 2011) (Fig. 1.1).

(b) **G-protein Linked Receptors**
G protein-linked receptor also termed as heptahelical receptor binds with extracellular substances and activates intracellular molecule called G-protein (guanine nucleotide-binding protein), also referred as 7TM receptors (α-helices transmembrane receptors). There are various types of GPCRs—human genome has 1000 types of encoded receptors. They mediate responses to signal molecules including amines, amino acids, neurotransmitters, lipids, hormones,

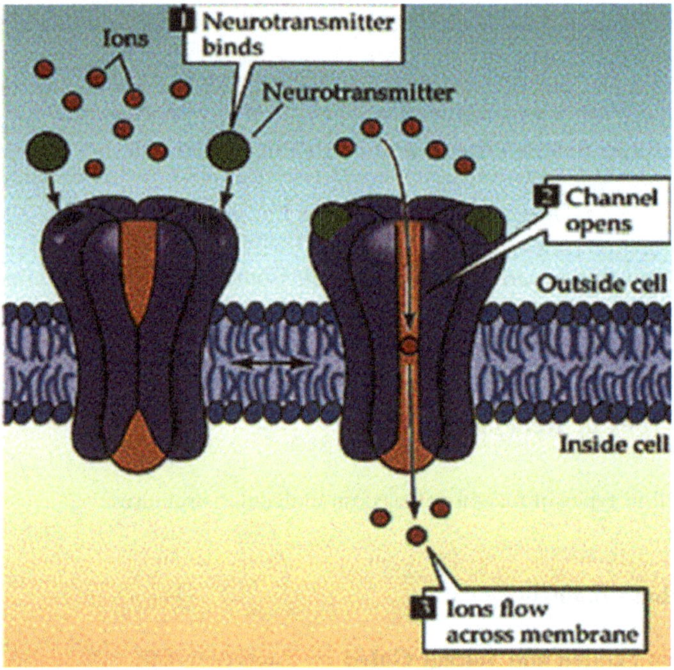

Fig. 1.1 Ligand gated ion channel (Pacheco, 2007)

proteins, local mediators, and also to the derivatives of fatty acids and amino acids (Gether, 2000). GPCRs are titled because of their binding ability and regulation of G protein activity. They also mediate the activities like kidney function, immune response, visual control, tumorigenesis, and inflammation. In 2007, β2 adrenergic receptor was analyzed as first human GPCR. Its crystal structure is similar to bovine rhodopsin (mammalian GPCR) discovered in year 2000 (Strader et al., 1994). Many receptors are activated by the same ligand like adrenaline activates 9 GPCRs, acetylcholine 5 receptors and serotonin 15 receptors (Alberts et al., 2002). For example, epinephrine binds with beta-adrenergic receptors of GPCRs, rhodopsin and prostaglandins which are inflammatory substances binding with prostaglandin E_2 receptors.

GPCRs are involved in two signal transduction mechanisms (Gilman, 1987). First is the Adenosine 3',5'-cyclic monophosphate (cAMP), a nucleotide that works as 2nd messenger and controls cellular actions like cell differentiation and growth, gene transcription followed by protein expression. In second mechanism, extracellular signal binds with receptor converting phosphatidylinositol-(4, 5)-bisphosphate (PIP2) to two messengers (diacylglycerol), DAG and IP3 (inositol triphosphate) which produce the intracellular response from extracellular signal.

GPCRs are important drug targets and 50% of USFDA approved drugs are from this class (Nuez Veulens and Rodríguez, 2009). GPCRs work by GTP-GDP exchange where GPCRs bind GDP in inactive form on plasma membrane inside the cell. It binds GTP upon becoming active. Signal molecule attachment with GPCRs changes its shape and receptor gets activated for binding with inactive G protein. Replacement of GDP with GTP activates G protein (Brass, 2003). GPCRs also control homeostasis. Brain depends on GPCRs for their practical expression. Synaptic transmission mechanisms including major senses are affected by the activation of relevant GPCRs (Oliveira et al., 1993). Newly approved drugs for psychiatric, metabolic, oncologic, neurodegenerative, and cardiovascular diseases are based on GPCR.

(c) **Receptor Tyrosine Kinases (RTKs)**
Biochemical signalling pathways initiate through ligand binding with receptor. Insulin and other peptide growth factors adhere with receptor proteins having C-terminal domains of tyrosine kinase activity. Receptor tyrosine kinases have single transmembrane segment and are monomers without ligand. Binding of ligand to extracellular domain induces conformational change in intracellular domain. Mechanism for activation of RTKs seems dimerization of receptor proteins by ligand binding as with the growth hormone receptor. Insulin receptor however exists in dimer state without ligand and ligand binding induces structural change. Upon RTK dimer formation, its cytoplasmic protein tyrosine kinase (PTK) domains cross-phosphorylate on specific Tyr residues. Autophosphorylation triggers PTK to phosphorylate other protein substrates (Voet et al., 2013).

(d) **Nuclear Receptors**
Nuclear receptors are also known as steroid receptors. Ligand like estrogen binds to receptor influencing the rate of transcription and translation of specific genes to form cellular proteins. Steroid hormones are ligand activated for the regulation of gene expression. They bind to targeted DNA comprised of specific genes sequences for controlling functions like skin differentiation and development, behavioral brain centers, bone development and reproductive tissues regulation (Kosztin et al., 1999).

Steroid hormones (such as cortisol, estrogen, and progesterone) are hydrophobic and do not readily dissolve in blood. They are transported through specific carrier proteins to their target site from point of synthesis. In target cells, these hormones cross plasma membranes through diffusion and bind to receptor proteins in the nucleus inducing conformational changes in receptor proteins. The receptors then interact with specific regulatory sequences in DNA called hormone response elements (HREs). Consequently, gene expression is altered. The receptor-hormone complex either suppresses or enhances the expression of specific genes. The regulators take effect in hours or days resulting in alterations of RNA synthesis and consequent protein synthesis is evident from altered metabolism (Nelson & Cox, 2009).

Nuclear receptors (NRs) also categorized as ligand-regulated transcription factors are another class of therapeutic drug receptor target. The nuclear receptors act for biological ligands like lipids, fatty acids, and steroid hormones. Approved drugs for nuclear receptors include steroidal contraceptives, anti-inflammatory glucocorticoids, lipid lowering agents, hormone replacement therapies and drugs related to estrogen and thyroid receptors (Kojetin & Burris, 2014). Until now 48 types of NRs had been recognized in humans (Kronenberger et al., 2015).

1.2.1.2 Functional Macromolecules

Drugs combine with targets like proteins and enzymes other than the specific receptors. Proteins function in signaling pathways, subcellular transport, structural stability, and translation. Proteins are the targets for drug action where they bind with lower molecular weight drug molecules (Makley & Gestwicki, 2013). Drugs inhibit enzymes by interacting receptors like Donepezil for inhibiting acetylcholinesterase which inactivates acetylcholine neurotransmitter for delaying the removal of amyloid plague and also cause symptomatic effect in Alzheimer's disease (Colovic et al., 2013). Transport proteins transport endogenous compounds like drugs across epithelial barriers for therapeutic functions at the site of action. These proteins are the pharmacological targets in increasing drug absorption at target site. Transport proteins like P-glycoprotein, apical sodium-dependent transporter and the peptide transporter pepT1 enhance bioactivity and the drug absorption (Swaan, 2003).

1.2.1.3 Four Phases of Pharmacokinetics

Pharmacokinetics is the study of drug in body which include four phases of drug absorption, distribution, metabolic changes, and excretion (ADME). Drug administration is categorized into two groups.

(a) **Enteral Route**
Enteral administration delivers medication via gastrointestinal (GI) tract where both ends, mouth and the anus can be used. Enteral route involves esophagus, stomach, and gastrointestinal tract. Enteral administration can be oral (through mouth), sublingual (under tongue), buccal (between gums and teeth) and rectal (into the rectum).

(b) **Parental Route**
Parental route delivers compounds (food & drug) via intravenous (through veins), subcutaneous (under the skin), intramuscular (through muscles), pulmonary (inhaled via lungs) and percutaneous (absorption via intact skin) (Kapalka, 2009).

Drug Absorption

Absorption is the process of transferring drug from its site of administration to blood stream (Devadasu et al., 2018). Oral route is convenient and the common path of administering medication. Oral absorption involves two processes, (a) drug dissolves in gastrointestinal fluid, and (b) dissolved drug goes into bloodstream through intestinal walls. Absorption of oral drug encounters GI secretions like degrading enzymes and low pH which decrease the bioavailability of drug. Peptide drug (i.e. insulin) is thus not given orally as it is susceptible to degradation. The process is controlled by physiochemical properties of drugs like permeability and solubility. Oral absorption is a lengthy process and its efficacy depends on various steps (Lin & Wong, 2017). Other methods of drug absorption (inhalation, intravenous, intramuscular) take few seconds to minutes. Absorption further depends on drug characteristics like molecular size, particle shape, physical form, concentration, chemical nature, lipid solubility, concentration gradient and degree of ionization (Chillistone & Hardman, 2017).

Drug Distribution

Drug distributes to body tissues from blood stream. Drug distribution is controlled by physiological factors which effect degree and rate of distribution. The rate is affected by blood perfusion and degree by plasma protein binding, lipid solubility, pH-pKa of drug and tissue protein binding (Maheshwari et al., 2018). Cell membrane composed of phospholipids is the barrier in drug distribution across the cell. Hydrophobic drugs (lipid soluble) cross the barrier whereas movement of hydrophilic drugs depends on the molecular size of drugs (Kelley et al., 2017).

Metabolism

Chemical alteration of pharmaceutical substances in the body is drug metabolism. Most drugs metabolize in liver as enzymes catalyze reactions including deamination, N-dealkylation, alkyl chain oxidation, azo-link cleavage, glucuronide formation and nitro-group reduction. These reactions help in drug excretion (Conney & Bubns, 1962). Rate of drug metabolism differs among patients because of variations in drug interaction, genetic factors, and coexisting disorders. In 20–95% patients, genetic factors are responsible for variation in drug responses. Cytochrome P450 2C9 (CYP2C9) enzyme metabolizes drugs like glipizide, losartan, warfarin, phenytoin and tolbutamide (Belle & Singh, 2008).

Elimination

Controlled removal of drug either as metabolite or as intact drug from the body is referred as drug elimination. Excretion paths include bile, urine, sweat, tears, and saliva. Excretion is mostly carried out by liver and kidney. Hydrophilic or polar drugs excrete by kidneys while hydrophobic ones undergo biotransformation (primary site of biotransformation is liver) before excretion (Garza et al., 2019). Factors affecting

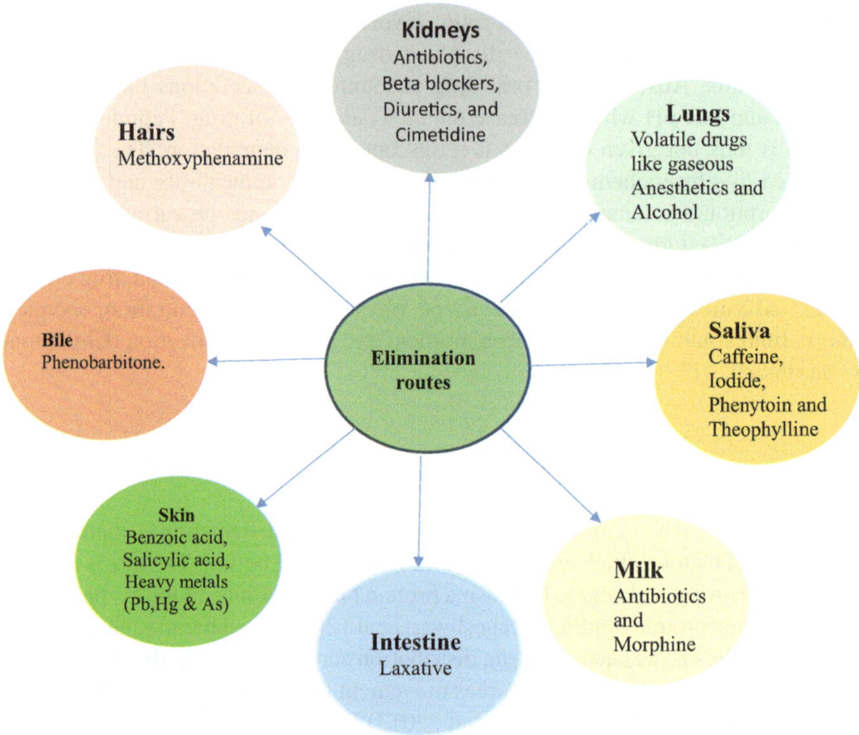

Fig. 1.2 Excretion routes

excretion through kidneys involve kidney function, glomerular filtration rate, urine pH, protein binding and urine flow (Fig. 1.2).

1.2.1.4 Drug-Drug Interactions (DDIs)

Drug interaction is defined as change in pharmacodynamic effect of one drug due to interaction with another. Two or more drugs taken together change the efficiency or potency of one drug modified due to the presence of another drug. These interactions can enhance or diminish the effect of either drug. Adverse drug interactions cause patient morbidity, increased malpractice claims and medical costs (Goldberg et al., 1996). Concomitant intake of drugs increases the chance of drugs interaction. Factors contributing in drug interactions are multiple drug therapy, poor patient compliances, patient's age, drug related factors, multiple diseases, multiple prescribers and some social factors (alcohol intake) (Alomar, 2014). Favorable interactions enhance drug effectiveness allowing the use of small doses and reducing the chance of severe action. Unfavorable interactions produce toxic effects on body (Baxter & Preston,

2010). DDIs are divided into two classes based on mechanism of interaction i.e. pharmacodynamics and pharmacokinetics. In pharmacodynamics, one drug changes the pharmacodynamics properties of another drug either by enhancing its activity (synergistic effect) or by decreasing its efficiency (antagonistic effects). Pharmacokinetic interaction relates to the change in drug concentration in various body parts. About 3–5% adverse drug reactions are caused by DDIs (Farooqui et al., 2018).

Many antidepressant drugs used concurrently with other medicines cause adverse drug interactions. Antidepressant drugs are the substrate of CYP 450 enzyme responsible for pharmacokinetics interactions. Tranylcypromine and Phenelzine are the common antidepressant drugs leading to adverse drug interaction. Adverse interactions cause death, hypertensive emergencies, suicide attempts, organ failure, transplant rejection, delirium, psychosis, adverse bleeding and hemorrhage (Nieuwstraten et al., 2006).

1.3 Drug Resistance

Drug resistance reduces drug efficiency in treating a disease. Antimicrobial resistance is the inability of antimicrobial drug to kill or inhibit growth of microbes such as viruses, bacteria, fungi, or parasites. Factors speeding up the drug resistance include misuse and overuse of antimicrobials, patient non-compliance with the antimicrobials and sub-therapeutic dosing. Mechanisms for antibiotic resistance include modification of bacterial proteins which are the targets of antibiotics, enzymatic modification of antibiotics, reduction in drug uptake, efflux pump modification, plasmid modification and inhibition of metabolic pathways. Antibiotics can also be inactivated by hydrolysis and transfer of chemical groups like phosphate, acyl, ribitoyl and nucleotidyl by bacterial enzymes preventing antibiotic to bind its target site (Loke & Hanafi, 2019).

1.3.1 Mechanisms of Drug Resistance

Mechanisms for drug resistance are summarized in Fig. 1.3 which include enzymatic modifications of drugs, prevention of antibiotic penetration or efflux pump and target modification.

1.3.1.1 Drug Modifications

Bacteria survive against antibiotics by producing enzymes that deactivate drug via adding specific functional groups to the compound. Antibiotics get inactivated or destroyed through hydrolysis. Resistance in both gram-negative and gram-positive bacteria is caused by enzymes which introduce chemical moieties to antimicrobial

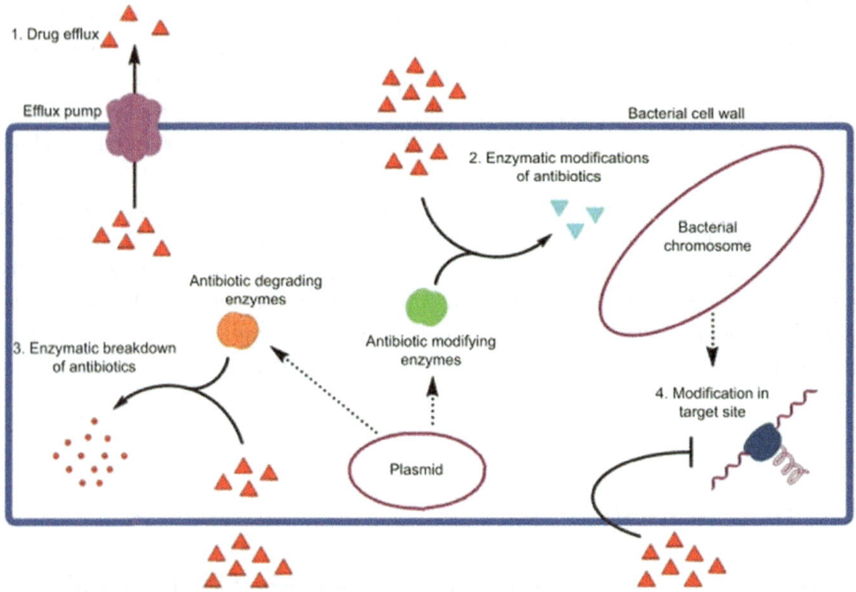

Fig. 1.3 Resistance mechanisms in antibiotics (Aslam et al., 2018)

molecules. Antibiotics affected due to enzymatic modification act by preventing protein production at ribosome. Modifying enzymes cause biochemical reactions like (1) acetylation (chloramphenicol, aminoglycosides, streptogramins), (2) phosphorylation (chloramphenicol, aminoglycosides), and (3) adenylation (lincosamides, aminoglycosides) (Wilson, 2014). For example, aminoglycoside resistance can occur through aminoglycoside modifying enzymes (AMEs) that cause covalent modification of amino or hydroxyl group of drug, weakening the drug affinity to its bacterial target (Ramirez & Tolmasky, 2010). Another example of enzymatic modification of an antibiotic is the adaptation of chloramphenicol, an antibacterial that prevents protein synthesis by linking with peptidyl-transfer component of 50S ribosome subunit (Schwarz et al., 2004). β-lactam resistance is caused by β-lactamases which destroy antibiotics. These enzymes hydrolyze amide bond within β-lactam ring, making antimicrobial ineffective (Munita & Arias, 2016). Rifampin phosphotransferases, member of an antibiotic-resistance protein inactivates rifampin through phosphorylation (Qi et al., 2016).

1.3.1.2 Prevention of Antibiotic Penetration or Efflux

Tetracycline resistance in E. coli depends on energy-efflux. Active efflux system causes antimicrobial resistance against structurally different antifungal, antibacterial and antiprotozoal drugs (Renau et al., 1998). Antibiotics have intracellular bacterial targets, e.g. located in inner membrane as in gram-negative bacteria. Drug penetrates through cytoplasmic membrane to reach target site. Bacteria may inhibit antibiotics from reaching their intracellular targets by reducing the uptake of antimicrobial drugs. This happens in gram negative pathogens which limit uptake by changing lipid membrane composition, porin channel concentration and porin channel selectivity. Hydrophilic molecules such as tetracycline, β-lactams and few fluoroquinolones are affected by modifications in outer membrane permeability as they use porins (water filled diffusion channel) to cross this barrier (Pagès et al., 2008).

Modifications of porins occur by three processes, (i) variation in porin expression level, (ii) change in the nature of porins expressed, and (iii) weakening of porin function (Nikaido, 2003). The abnormal production of OprD in *Pseudomonasaeruginosa* leads to porin-mediated resistance which otherwise facilitates the uptake of antibiotic, small peptides, carbapenems and amino acids (Terzi et al., 2015). Antimicrobial resistance can also occur because of the compounds being excreted by the bacterial cells. First efflux system developed in 1980s, pumps tetracycline out of the cytoplasm of E.coli (Poole, 2005). This efflux system of resistance affects antimicrobial classes like fluoroquinolones, β-lactams, protein synthesis inhibitors, polymyxins and carbapenems.

1.3.1.3 Target Modification

Common mechanism of resistance is the modification in target sites. Antimicrobial drugs have definite targets, and the structural modification of targets can stop drug binding and making drug ineffective. Strains resistance is known for all classes of antibiotics. Target gets modified through mutations in bacterial genome developing resistance against antibiotics, e.g. mutations in DNA gyrase and RNA polymerase results in resistance for quinolones and rifamycins (Lambert, 2005). Penicillin binding proteins (PBPs) having structural similarity with β-lactamases prevent binding of β-lactam drugs by genetic modification. This kind of resistance is common among strains of pneumococci which genetically modifies PBPs (Smith et al., 2005). Strains of *Staphylococcus aureus* become resistant to β-lactam antibiotic methicillin and others in this class through low affinity PBP and PBP2a (Stapleton & Taylor, 2002).

1.3.2 Multidrug Resistance

The resistance of microbes to antimicrobial drugs via new resistance mechanisms has complicated the treatment of infectious diseases resulting in prolonged illness, higher expenditures, and greater risk of death. Super bugs like fungi, bacteria, parasites and viruses have developed multidrug resistance with higher risk of mortality and morbidity (Tanwar et al., 2014).

Multidrug resistance in bacteria is caused by two mechanisms. In the first, multiple genes may accumulate on resistance plasmids in bacteria and each gene code resists a single drug inside single cell. In second mechanism, resistance occurs due to overexpression of genes which code for multidrug efflux pump, forcing out drugs. Some pathogens are resistant to available drugs like methicillin-resistant *Staphylococcus aureus* (MRSA) which not only resist methicillin but also macrolides, tetracycline, aminoglycosides, lincosamides and chloramphenicol (De Lencastre et al., 2007). Bacteria such as *Klebsiella pneumoniae* resist carbapenems and cephalosporin, *Escherichia coli* resists fluoroquinolones and cephalosporin, Nontyphoidal *Salmonella* resists fluoroquinolones, and *Mycobacterium tuberculosis* resists isoniazid, fluoroquinolone and rifampicin causing common infections (Nikaido, 2009).

Few antifungal drugs are available for curing fungal infections. Resistance to drugs like azole derivatives (fluconazole, ketoconazole, etc.), polyene macrolides, DNA and RNA synthesis inhibitors exist in strains of *Trichosporonbeigelii, Cryptococcus neoformans, Scopulariopsis* spp., *Aspergillus* spp., and *Pseudallescheriaboydii* (Loeffler & Stevens, 2003). Tuberculosis (TB) is worsened because of multidrug resistant (MDR) *Mycobacterium tuberculosis* and drug resistant *Mycobacterium tuberculosis* (XDR-TB). MDR-TB strains resist TB drugs like isoniazid and rifampin. XDR-TB strains resist drugs used as second line of treatment like fluoroquinolone, kanamycin, amikacin or capreomycin (Palomino & Martin, 2014).

1.3.3 Drug Resistance in Tumor

The malignant tumor chemotherapy is unsuccessful due to drug resistance. Mechanisms for anticancer drug resistance include drug inactivation, drug efflux, binding sites mutation, cell death inhibition, DNA damage repair, inhibition of tumor protein production and cell heterogeneity in malignant tumor. In vivo drug activation involves modification of drug through interaction with proteins. Anticancer drug undergoes metabolic activation for clinical efficiency. Malignant cell develops resistance to these treatments through reduced cell activation (Housman et al., 2014).

1.3.4 Economic Impact of Drug Resistance

In last two decades, there has been increase in the incidence and frequency of antimicrobial drug resistant microorganisms. It is a concern due to its impact on patients, physicians, health care administrators, public, national policy makers, and pharmaceutical producers. Economic impact of infections and colonization of these pathogens is incomplete and imprecise (McGowan 2001). Drug resistance due to gram-positive and gram-negative bacteria is associated with increase in medical cost upto $29,069 per patient. Infections due to resistant gram-negative bacteria are nearly 23% of all infections. These infections are linked to higher hospital costs, higher severity of illness, longer hospital stays and antibiotic costs (Li & Sperling, 2013).

References

Abdul Kadir, L., Stacey, M., & Barrett-Jolley, R. (2018). Emerging roles of the membrane potential: Action beyond the action potential. *Frontiers in Physiology, 9*, 1661.

Adesoye, A., & Duncan, N. (2017). Acute pain management in patients with opioid tolerance. *US Pharm, 42*, 28–32.

Adler, L. A., Goodman, D. W., Kollins, S. H., Weisler, R. H., Krishnan, S., Zhang, Y., & Biederman, J. (2008). Double-blind, placebo-controlled study of the efficacy and safety of lisdexamfetamine dimesylate in adults with attention-deficit/hyperactivity disorder. *Journal of Clinical Psychiatry, 69*, 1364.

Aira, M., Hartikainen, S., & Sulkava, R. (2005). Community prevalence of alcohol use and concomitant use of medication—a source of possible risk in the elderly aged 75 and older? *International Journal of Geriatric Psychiatry: A Journal of the Psychiatry of Late Life and Allied Sciences, 20*, 680–685.

Akbari, P., & Khorasani-Zadeh, A. (2019). Thiazide diuretics. *StatPearls [Internet].* StatPearls Publishing.

Alberts, B., Johnson, A., Lewis, J., Raff, M., Roberts, K., & Walter, P. (2002). Signaling through G-protein-linked cell-surface receptors. *Molecular Biology of the Cell. 4th edition.* Garland Science.

Aldred, K. J., Kerns, R. J., & Osheroff, N. (2014). Mechanism of quinolone action and resistance. *Biochemistry, 53*, 1565–1574.

Alomar, M. J. (2014). Factors affecting the development of adverse drug reactions. *Saudi Pharmaceutical Journal, 22*, 83–94.

Aslam, B., Wang, W., Arshad, M. I., Khurshid, M., Muzammil, S., Rasool, M. H., Nisar, M. A., Alvi, R. F., Aslam, M. A., & Qamar, M. U. (2018). Antibiotic resistance: A rundown of a global crisis. *Infection and Drug Resistance, 11*, 1645.

Bauman, R. W. (2015). *Microbiology: With diseases by body system* (4th ed.).. Pearson Education Inc.

Baxter, K., & Preston, C. L. (2010). *Stockley's drug interactions.* Pharmaceutical Press London.

Belle, D. J., & Singh, H. (2008). Genetic factors in drug metabolism. *American Family Physician, 77*, 1553–1560.

Berg, J. M., Tymoczko, J. L., & Stryer, L. (2012). *Biochemistry/Jeremy M. Berg, John L. Tymoczko, Lubert Stryer; with Gregory J. Gatto, Jr.* WH Freeman.

Besant, P. G., Tan, E., & Attwood, P. V. (2003). Mammalian protein histidine kinases. *The International Journal of Biochemistry & Cell Biology, 35*, 297–309.

Biaggioni, I., & Robertson, D. (2009). Adrenoceptor agonists & sympathomimetic drugs. *Basic & Clinical Pharmacology, 12*, 129–149.

Bowman, W. (2006). Neuromuscular block. *British Journal of Pharmacology, 147*, S277–S286.

Brass, L. F. (2003). Thrombin and platelet activation. *Chest, 124*, 18S-25S.

Buchner, E. " Biocatalyst" redirects here. For the use of natural catalysts in organic chemistry, see Biocatalysis.

Caliandro, P., Evoli, A., Stålberg, E., Granata, G., Tonali, P., & Padua, L. (2009). The difficulty in confirming clinical diagnosis of myasthenia gravis in a seronegative patient: A possible neurophysiological approach. *Neuromuscular Disorders, 19*, 825–827.

Carrière, V., Berthou, F., Baird, S., Belloc, C., & Beaune, P. (1996). Human cytochrome P450 2E1 (CYP2E1): From genotype to phenotype. *Pharmacogenetics, 6*, 203–211.

Carvalho, M., Carmo, H., Costa, V. M., Capela, J. P., Pontes, H., Remião, F., Carvalho, F., & de Lourdes, B. M. (2012). Toxicity of amphetamines: An update. *Archives of Toxicology, 86*, 1167–1231.

Charles, L., Triscott, J., & Dobbs, B. (2017). Secondary hypertension: Discovering the underlying cause. *American Family Physician, 96*, 453–461.

Chiara, D. C., Jayakar, S. S., Zhou, X., Zhang, X., Savechenkov, P. Y., Bruzik, K. S., Miller, K. W., & Cohen, J. B. (2013). Specificity of intersubunit general anesthetic-binding sites in the transmembrane domain of the human $\alpha1\beta3\gamma2$ γ-aminobutyric acid type A (GABAA) receptor. *Journal of Biological Chemistry, 288*, 19343–19357.

Chillistone, S., & Hardman, J. G. (2017). Factors affecting drug absorption and distribution. *Anaesthesia & Intensive Care Medicine, 18*, 335–339.

Chopra, I., & Roberts, M. (2001). Tetracycline antibiotics: Mode of action, applications, molecular biology, and epidemiology of bacterial resistance. *Microbiology and Molecular Biology Reviews, 65*, 232–260.

Chou, T. C., Bjorkum, A. A., Gaus, S. E., Lu, J., Scammell, T. E., & Saper, C. B. (2002). Afferents to the ventrolateral preoptic nucleus. *Journal of Neuroscience, 22*, 977–990.

CHUNG, K.-T. . (2016). Azo dyes and human health: A review. *Journal of Environmental Science and Health, Part C, 34*, 233–261.

Colovic, M. B., Krstic, D. Z., Lazarevic-Pasti, T. D., Bondzic, A. M., & Vasic, V. M. (2013). Acetylcholinesterase inhibitors: Pharmacology and toxicology. *Current Neuropharmacology, 11*, 315–335.

Conney, A., & Bubns, J. (1962). Factors influencing drug metabolism. In Advances in pharmacology. Elsevier.

Corrao, G., Rubbiati, L., Bagnardi, V., Zambon, A., & Poikolainen, K. (2000). Alcohol and coronary heart disease: A meta-analysis. *Addiction, 95*, 1505–1523.

Currie, G. M. (2018). Pharmacology, part 1: Introduction to pharmacology and pharmacodynamics. *Journal of Nuclear Medicine Technology, 46*, 81–86.

Daniels, R., & Nicoll, L. H. (2011). *Contemporary medical-surgical nursing*. Cengage Learning.

de Lencastre, H., Oliveira, D., & Tomasz, A. (2007). Antibiotic resistant Staphylococcus aureus: A paradigm of adaptive power. *Current Opinion in Microbiology, 10*, 428–435.

Demirkiran, A., Sewgobind, V. D., van der Weijde, J., Kok, A., Baan, C. C., Kwekkeboom, J., Tilanus, H. W., Metselaar, H. J., & van der Laan, L. J. (2009). Conversion from calcineurin inhibitor to mycophenolate mofetil-based immunosuppression changes the frequency and phenotype of CD4+ FOXP3+ regulatory T cells. *Transplantation, 87*, 1062–1068.

Devadasu, V. R., Deb, P. K., Maheshwari, R., Sharma, P., & Tekade, R. K. (2018). *Physicochemical, pharmaceutical, and biological considerations in GIT absorption of drugs*. Elsevier.

Dhanasekaran, N., & Reddy, E. P. (1998). Signaling by dual specificity kinases. *Oncogene, 17*, 1447–1455.

Dworkin, R. H., Backonja, M., Rowbotham, M. C., Allen, R. R., Argoff, C. R., Bennett, G. J., Bushnell, M. C., Farrar, J. T., Galer, B. S., & Haythornthwaite, J. A. (2003). Neurological reviews-advances in neuropathic pain: Diagnosis, mechanisms, and treatment recommendations. *Archives of Neurology, 60*, 1524–1536.

Elson, D. (1965). Metabolism of nucleic acids (macromolecular DNA and RNA). *Annual Review of Biochemistry, 34*, 449–486.

Farooqui, R., Hoor, T., Karim, N., & Muneer, M. (2018). Potential drug-drug interactions among patients prescriptions collected from medicine out-patient setting. *Pakistan Journal of Medical Sciences, 34*, 144.

Finkel, R., Clark, M. A., & Cubeddu, L. X. (2009). *Pharmacology: Lippincott's illustrated reviews*. Lippincott Williams & Wilkins.

Garza, A. Z., Park, S. B., & Kocz, R. (2019). Drug elimination. *StatPearls [Internet]*. StatPearls Publishing.

Gether, U. (2000). Uncovering molecular mechanisms involved in activation of G protein-coupled receptors. *Endocrine Reviews, 21*, 90–113.

Giluman, A. G. (1987). G proteins: Transducers of receptor-generated signals. *Annual Review of Biochemistry, 56*, 615–649.

Goldberg, R. M., Mabee, J., Chan, L., & Wong, S. (1996). Drug-drug and drug-disease interactions in the ED: Analysis of a high-risk population. *The American Journal of Emergency Medicine, 14*, 447–450.

Goldblatt, D. (2000). Conjugate vaccines. *Clinical & Experimental Immunology, 119*, 1–3.

Gudin, J. (2012). Risk Evaluation and Mitigation Strategies (REMS) for extended-release and long-acting opioid analgesics: Considerations for palliative care practice. *Journal of Pain & Palliative Care Pharmacotherapy, 26*, 136–143.

Hanukoglu, I. (2017). ASIC and ENaC type sodium channels: Conformational states and the structures of the ion selectivity filters. *The FEBS Journal, 284*, 525–545.

Harper, S., & Jones, N. (2006). Cocaine: What role does it have in current ENT practice? A review of the current literature. *The Journal of Laryngology & Otology, 120*, 808–811.

Haverkampf, J. (2017). *Benzodiazepines Z-drugs and dependence.*

Henry, R. J. (1943). The mode of action of sulfonamides. *Bacteriological Reviews, 7*, 175.

Herman, L. L., & Bashir, K. (2019). Angiotensin converting enzyme inhibitors (ACEI). *StatPearls [Internet].* StatPearls Publishing.

Hermanson, G. T. (2013). *Bioconjugate techniques.* Academic press.

Hippius, H. (1989). The history of clozapine. *Psychopharmacology.*

Housman, G., Byler, S., Heerboth, S., Lapinska, K., Longacre, M., Snyder, N., & Sarkar, S. (2014). Drug resistance in cancer: An overview. *Cancers, 6*, 1769–1792.

Jenkins, C., Samudrala, R., Anderson, I., Hedlund, B. P., Petroni, G., Michailova, N., Pinel, N., Overbeek, R., Rosati, G., & Staley, J. T. (2002). Genes for the cytoskeletal protein tubulin in the bacterial genus Prosthecobacter. *Proceedings of the National Academy of Sciences, 99*, 17049–17054.

Johns, M. (1975). Sleep and hypnotic drugs. *Drugs, 9*, 448–478.

Kalivas, P. W., & McFarland, K. (2003). Brain circuitry and the reinstatement of cocaine-seeking behavior. *Psychopharmacology (berl), 168*, 44–56.

Kang, H.-K., & Park, Y. (2015). Glycopeptide antibiotics: Structure and mechanisms of action. *Journal of Bacteriology and Virology, 45*, 67–78.

Kapalka, G. M. (2009). *Nutritional and herbal therapies for children and adolescents: A handbook for mental health clinicians.* Academic Press.

Karch, S. (2009). *Karch's pathology of drug abuse* (pp. 635–648). Taylor & Francis Group.

Kelley, A. G., Ling-Hsiu, L., & Berenbaum, M. (2017). *You are what you eat: Food-drug interaction in honey bees (Apis mellifera).*

Khalil, H., & Zeltser, R. (2020). Antihypertensive medications. *StatPearls [Internet].* StatPearls Publishing.

King, J. V., Liang, W. G., Scherpelz, K. P., Schilling, A. B., Meredith, S. C., & Tang, W.-J. (2014). Molecular basis of substrate recognition and degradation by human presequence protease. *Structure, 22*, 996–1007.

Klee, H. (1997). *Amphetamine misuse: International perspectives on current trends.* CRC Press.

Kleiner, D. E. (2018). *Drugs and toxins.* Elsevier.

Kojetin, D. J., & Burris, T. P. (2014). REV-ERB and ROR nuclear receptors as drug targets. *Nature Reviews Drug Discovery, 13*, 197–216.

Koppes, L. L., Dekker, J. M., Hendriks, H. F., Bouter, L. M., & Heine, R. J. (2005). Moderate alcohol consumption lowers the risk of type 2 diabetes: a meta-analysis of prospective observational studies. *Diabetes Care, 28*, 719–725.

Kosztin, D., Izrailev, S., & Schulten, K. (1999). Unbinding of retinoic acid from its receptor studied by steered molecular dynamics. *Biophysical Journal, 76*, 188–197.

Krause, K. M., Serio, A. W., Kane, T. R., & Connolly, L. E. (2016). Aminoglycosides: An overview. *Cold Spring Harbor Perspectives in Medicine, 6*, a027029.

Kronenberger, T., Keminer, O., Wrenger, C., & Windshügel, B. (2015). Nuclear receptor modulators—current approaches and future perspectives. *Drug Discovery and Development–From Molecules to Medicine*. InTech.

Lally, J., & Maccabe, J. H. (2015). Antipsychotic medication in schizophrenia: A review. *British Medical Bulletin, 114*, 169–179.

Lambert, D. (2004). Drugs and receptors. *Continuing Education in Anaesthesia Critical Care & Pain, 4*, 181–184.

Lambert, P. A. (2005). Bacterial resistance to antibiotics: Modified target sites. *Advanced Drug Delivery Reviews, 57*, 1471–1485.

Li, B., & Sperling, A. (2013). Drug resistance. In *Encyclopedia of genetics* (pp. 418–420).

Liberti, S., Sacco, F., Calderone, A., Perfetto, L., Iannuccelli, M., Panni, S., Santonico, E., Palma, A., Nardozza, A. P., & Castagnoli, L. (2013). HuPho: The human phosphatase portal. *The FEBS Journal, 280*, 379–387.

Lin, L., & Wong, H. (2017). Predicting oral drug absorption: mini review on physiologically-based pharmacokinetic models. *Pharmaceutics, 9*, 41.

Lobanovska, M., & Pilla, G. (2017). Focus: Drug development: Penicillin's discovery and antibiotic resistance: Lessons for the future? *The Yale Journal of Biology and Medicine, 90*, 135.

Loeffler, J., & Stevens, D. A. (2003). Antifungal drug resistance. *Clinical Infectious Diseases, 36*, S31–S41.

Loke, M. F., & Hanafi, A. (2019). *Molecular mechanisms responsible for drug resistance.*

Löscher, W., & Rogawski, M. A. (2012). How theories evolved concerning the mechanism of action of barbiturates. *Epilepsia, 53*, 12–25.

Louis, S. (2011). *Goodman & Gilman's pharmacological basis of therapeutics.*

Ma, W., Yang, L., & He, L. (2018). Overview of the detection methods for equilibrium dissociation constant KD of drug-receptor interaction. *Journal of Pharmaceutical Analysis, 8*, 147–152.

Madruga, J. V., Berger, D., McMurchie, M., Suter, F., Banhegyi, D., Ruxrungtham, K., Norris, D., Lefebvre, E., de Béthune, M.-P., & Tomaka, F. (2007). Efficacy and safety of darunavir-ritonavir compared with that of lopinavir-ritonavir at 48 weeks in treatment-experienced, HIV-infected patients in TITAN: a randomised controlled phase III trial. *The Lancet, 370*, 49–58.

Maheshwari, R., Sharma, P., Seth, A., Taneja, N., Tekade, M., & Tekade, R. K. (2018). *Drug disposition considerations in pharmaceutical product*. Elsevier.

Makley, L. N., & Gestwicki, J. E. (2013). Expanding the number of 'druggable' targets: Non-enzymes and protein–protein interactions. *Chemical Biology & Drug Design, 81*, 22–32.

Manning, G., Whyte, D., Martinez, R., Hunter, T., & Sudarsanam, S. (2002). cAMP-dependent protein kinase from Plasmodium falciparum: An update. *Trends in Biochemical Sciences, 27*, 514–520.

Marshall, W. F., & Blair, J. E. (1999). The cephalosporins. In *Mayo Clinic Proceedings* (pp. 187–195). Elsevier.

Mcgowan, J. R., & J. E. (2001). Economic impact of antimicrobial resistance. *Emerging Infectious Diseases, 7*, 286.

Medicine, N. L. O. (2000). *Medical subject headings*. US Department of Health and Human Services, Public Health Service, National ….

Munita, J. M., & Arias, C. A. (2016). Mechanisms of antibiotic resistance. In *Virulence mechanisms of bacterial pathogens* (pp. 481–511).

Murphy, J. M., Farhan, H., & Eyers, P. A. (2017). Bio-Zombie: the rise of pseudoenzymes in biology. *Biochemical Society Transactions, 45*, 537–544.

Murphy, J. M., Zhang, Q., Young, S. N., Reese, M. L., Bailey, F. P., Eyers, P. A., Ungureanu, D., Hammaren, H., Silvennoinen, O., & Varghese, L. N. (2014). A robust methodology to subclassify pseudokinases based on their nucleotide-binding properties. *Biochemical Journal, 457*, 323–334.

Nelson, D. L., & Cox, M. M. (2009). *Lehninger principles of biochemistry.* WH Freeman.

Nieuwstraten, C., Labiris, N. R., & Holbrook, A. (2006). Systematic overview of drug interactions with antidepressant medications. *The Canadian Journal of Psychiatry, 51*, 300–316.

Nikaido, H. (2003). Molecular basis of bacterial outer membrane permeability revisited. *Microbiology and Molecular Biology Reviews, 67*, 593–656.

Nikaido, H. (2009). Multidrug resistance in bacteria. *Annual Review of Biochemistry, 78*, 119–146.

Nuez Veulens, A. D. L., & Rodríguez, R. (2009). G Protein-coupled receptors as targets for drug design. *Biotecnología Aplicada, 26*, 24–33.

O'Connell, H., Chin, A.-V., Cunningham, C., & Lawlor, B. (2003). Alcohol use disorders in elderly people–redefining an age old problem in old age. *BMJ, 327*, 664–667.

Oliveira, L., Paiva, A. C. D. M., & Vriend, G. (1993). A common motif in G-protein-coupled seven transmembrane helix receptors. *Journal of Computer-Aided Molecular Design, 7*, 649–658.

Olkkola, K. T., & Ahonen, J. (2008). *Midazolam and other benzodiazepines*. Springer.

Opioids, A.-D. (2015). *Evaluation and labeling guidance for industry*. US Department of Health and Human Services Food and Drug Administration Center for Drug Evaluation and Research (CDER) Clinical Medical.

Ozawa, Y., Hayashi, K., & Kobori, H. (2006). New generation calcium channel blockers in hypertensive treatment. *Current Hypertension Reviews, 2*, 103–111.

Pacheco, M. A. (2007). *Receptor types and subtypes*.

Page, C. P. (2002). *Integrated pharmacology*.

Pagès, J. M., James, C. E., & Winterhalter, M. (2008). The porin and the permeating antibiotic: a selective diffusion barrier in Gram-negative bacteria. *Nature Reviews Microbiology, 6*, 893–903.

Palomino, J. C., & Martin, A. (2014). Drug resistance mechanisms in mycobacterium tuberculosis. *Antibiotics, 3*, 317–340.

Panula, P., Chazot, P., Cowart, M., Gutzmer, R., Leurs, R., Liu, W., Stark, H., Thurmond, R., & Haas, H. (2014). International Union of Basic and< tep-common: author-query> AQ9: Please check whether the updated information in ref 1 is correct.</tepcommon: author-query> Clinical Pharmacology XCVIII. Histamine Receptors. *Pharmacological Reviews, 67*, 601–655.

Pelet, T., & Matheux, F. (2011). *Methods for inhibiting gram-postive bacteria using non-phospholipid lipid vesicules*. Google Patents.

Pomara, C., Cassano, T., D'Errico, S., Bello, S., Romano, A. D., Riezzo, I., & Serviddio, G. (2012). Data available on the extent of cocaine use and dependence: Biochemistry, pharmacologic effects and global burden of disease of cocaine abusers. *Current Medicinal Chemistry, 19*, 5647–5657.

Poole, K. (2005). Efflux-mediated antimicrobial resistance. *Journal of Antimicrobial Chemotherapy, 56*, 20–51.

Pulendran, B., & Ahmed, R. (2011). Immunological mechanisms of vaccination. *Nature Immunology, 12*, 509.

Qi, X., Lin, W., Ma, M., Wang, C., He, Y., He, N., Gao, J., Zhou, H., Xiao, Y., & Wang, Y. (2016). Structural basis of rifampin inactivation by rifampin phosphotransferase. *Proceedings of the National Academy of Sciences, 113*, 3803–3808.

Rajesh, D. (2015). The need of a uniform drug classification in text books of pharmacology. *Indian Journal of Pharmacology, 47*.

Ramirez, M. S., & Tolmasky, M. E. (2010). Aminoglycoside modifying enzymes. *Drug Resistance Updates, 13*, 151–171.

Rappuoli, R., de Gregorio, E., & Costantino, P. (2019). On the mechanisms of conjugate vaccines. *Proceedings of the National Academy of Sciences, 116*, 14–16.

Rasmussen, N. (2006). Making the first anti-depressant: Amphetamine in American medicine, 1929–1950. *Journal of the History of Medicine and Allied Sciences, 61*, 288–323.

Renau, T. E., Hecker, S. J., & Lee, V. J. (1998). Antimicrobial potentiation approaches: Targets and. In *Annual reports in medicinal chemistry* (p. 121).

Reynolds, K., Lewis, B., Nolen, J. D. L., Kinney, G. L., Sathya, B., & He, J. (2003). Alcohol consumption and risk of stroke: A meta-analysis. *JAMA, 289*, 579–588.

Rich, R. R., Fleisher, T. A., Shearer, W. T., Schroeder JR, H. W., Frew, A. J., & Weyand, C. M. (2012). *Clinical immunology E-Book: Principles and practice*. Elsevier Health Sciences.

Rimm, E. B., Klatsky, A., Grobbee, D., & Stampfer, M. J. (1996). Review of moderate alcohol consumption and reduced risk of coronary heart disease: is the effect due to beer, wine, or spirits? *BMJ, 312*, 731–736.
Sanders, B., Koldijk, M., & Schuitemaker, H. (2015). Inactivated viral vaccines. In *Vaccine analysis: Strategies, principles, and control*. Springer.
Sano, K., Totsuka, Y., Ikegami, Y., & Uesugi, T. (2001). Metabolism of sulphobromophthalein I: Positional isomers of sulphobromophthalein monoglutathione conjugate. *Journal of Pharmacy and Pharmacology, 53*, 1015–1020.
Sarker, S. D., & Nahar, L. (2007). *Chemistry for pharmacy students* (pp. 322–324). Willcy & Sons Ltd.
Schiller, J., & Lowy, D. (2018). Explanations for the high potency of HPV prophylactic vaccines. *Vaccine, 36*, 4768–4773.
Schwarz, S., Kehrenberg, C., Doublet, B., & Cloeckaert, A. (2004). Molecular basis of bacterial resistance to chloramphenicol and florfenicol. *FEMS Microbiology Reviews, 28*, 519–542.
Shen, Y., Joachimiak, A., Rosner, M. R., & Tang, W.-J. (2006). Structures of human insulin-degrading enzyme reveal a new substrate recognition mechanism. *Nature, 443*, 870–874.
Shorter, E. (2005). *A historical dictionary of psychiatry*, Oxford University Press.
Sinha, J., & Bhattacharya, S. (2014). *A text book of immunology (Google book preview)*. Academic Publishers.
Smith, A. M., Feldman, C., Massidda, O., McCarthy, K., Ndiweni, D., & Klugman, K. P. (2005). Altered PBP 2A and its role in the development of penicillin, cefotaxime, and ceftriaxone resistance in a clinical isolate of Streptococcus pneumoniae. *Antimicrobial Agents and Chemotherapy, 49*, 2002–2007.
Smith, H. S., Raffa, R. B., Pergolizzi, J. V., Taylor, R., & Tallarida, R. J. (2014). Combining opioid and adrenergic mechanisms for chronic pain. *Postgraduate Medicine, 126*, 98–114.
St Leger, A. S., Cochrane, A. L., & Moore, F. (1979). Factors associated with cardiac mortality in developed countries with particular reference to the consumption of wine. *The Lancet, 313*(8124), 1017–1020.
Stapleton, P. D., & Taylor, P. W. (2002). Methicillin resistance in Staphylococcus aureus: Mechanisms and modulation. *Science Progress, 85*, 57–72.
Strader, C. D., Fong, T. M., Tota, M. R., Underwood, D., & Dixon, R. A. (1994). Structure and function of G protein-coupled receptors. *Annual Review of Biochemistry, 63*, 101–132.
Strømgaard, K., Krogsgaard-Larsen, P., & Madsen, U. (2017). *Textbook of drug design and discovery*. CRC press.
Suvarna, B. (2011). Drug-receptor interactions. *Kathmandu University Medical Journal, 9*, 203–207.
Swaan, P. W. (2003). Membrane transport proteins and drug transport. *Burger's Medicinal Chemistry and Drug Discovery, 2*, 249–293.
Tan, M., & Jiang, X. (2017). Recent advancements in combination subunit vaccine development. *Human Vaccines & Immunotherapeutics, 13*, 180–185.
Tanwar, J., Das, S., Fatima, Z., & Hameed, S. (2014). Multidrug resistance: An emerging crisis. *Interdisciplinary Perspectives on Infectious Diseases*.
Tapert, S. F., Tate, S. R., & Brown, S. A. (2002). *Substance abuse: An overview*. Springer.
Terzi, H. A., Kulah, C., Atasoy, A. R., & Ciftci, I. H. (2015). Investigation of OprD porin protein levels in carbapenem-resistant Pseudomonas aeruginosa isolates. *Jundishapur Journal of Microbiology, 8*.
Vartak, A., & Sucheck, S. J. (2016). Recent advances in subunit vaccine carriers. *Vaccines, 4*, 12.
Vaux, B., & Golder, S. (2003). *The Harvard dialect survey*. Harvard University Linguistics Department.
Vetter, V., Denizer, G., Friedland, L. R., Krishnan, J., & Shapiro, M. (2018). Understanding modern-day vaccines: What you need to know. *Annals of Medicine, 50*, 110–120.

Victoria, J. G., Wang, C., Jones, M. S., Jaing, C., McLoughlin, K., Gardner, S., & Delwart, E. L. (2010). Viral nucleic acids in live-attenuated vaccines: Detection of minority variants and an adventitious virus. *Journal of Virology, 84*, 6033–6040.

Voet, D., Voet, J. G. & Pratt, C. W. (2013). *Fundamentals of biochemistry: Life at the molecular level.*

Washton, A. M., & Zweben, J. E. (2006). *Treating alcohol and drug problems in psychotherapy practice: Doing what works*. Guilford Press.

Weathermon, R., & Crabb, D. W. (1999). Alcohol and medication interactions. *Alcohol Research & Health, 23*, 40.

Williams, S. E. (1980). The use of beverage alcohol as medicine, 1790–1860. *Journal of Studies on Alcohol, 41*, 543–566.

Wilson, D. N. . (2014). Ribosome-targeting antibiotics and mechanisms of bacterial resistance. *Nature Reviews Microbiology, 12*, 35–48.

Yadav, D. K., Yadav, N., & Khurana, S. M. P. (2014). *Vaccines: Present status and applications.* Elsevier.

Zhang, X.-X., Zhang, T., & Fang, H. H. (2009). Antibiotic resistance genes in water environment. *Applied Microbiology and Biotechnology, 82*, 397–414.

Zhu, B. T. (1993). The competitive and noncompetitive antagonism of receptor-mediated drug actions in the presence of spare receptors. *Journal of Pharmacological and Toxicological Methods,* 29, 85–91.

Chapter 2
Drugs Resistance in Bacterial Diseases

Muhammad Zeeshan Ahmed, Tazeen Rao, Waqas Ahmad, Shahzeb Hameed, Samrana Anayat, Asma Altaq, Wafa Iqbal, Muhammad Younus, Ali Ahmad, Suvash Chandra Ojha, Atif Liaqat, Rahat Naseer, Shehla Munir, Muhammad Ibrahim, and Sarfraz Ahmed

Abstract The development of resistance in micro-organisms against the antimicrobial agents is known as drug resistance, and this causes morbidity and mortality in humans and animals. Major contributing factors for the development of drug

M. Z. Ahmed · S. Hameed
Institute of Biochemistry and Biotechnology, University of the Punjab, Quaid-i-Azam campus, LahorePunjab 54590, Pakistan

T. Rao · S. Munir · M. Ibrahim
Department of Biochemistry, Bahauddin Zakariya University, MultanPunjab 60800, Pakistan
e-mail: ibrahimjam@bzu.edu.pk

W. Ahmad
Department of Clinical Sciences, University College of Veterinary and Animal Sciences, Narowal 51600, Pakistan
e-mail: waqas.hussain@uvas.edu.pk

S. Anayat
School of Biological Sciences, University of the Punjab, Quaid-e-Azam Campus, Lahore 54590, Pakistan

A. Altaq · W. Iqbal
Department of Biochemistry and Biotechnology, The Women University, Multan 60800, Pakistan

M. Younus · A. Ahmad
Department of Pathobiology, University of Veterinary and Animal Sciences Lahore, Narowal 51600, Pakistan
e-mail: younusrana@uvas.edu.pk

A. Ahmad
e-mail: aliahmad@uvas.edu.pk

S. C. Ojha
Department of Infectious Diseases, The Affiliated Hospital of Southwest Medical University, Luzhou 646000, China
e-mail: suvash_ojha@swmu.edu.cn

A. Liaqat
Department of Food Science and Technology, Khwaja Freed University of Engineering and Technology, Rahim Yar Khan 64200, Pakistan
e-mail: atif.liaqat@kfueit.edu.pk

S. Ahmed et al. (eds.), *Biochemistry of Drug Resistance*,
https://doi.org/10.1007/978-3-030-76320-6_2

resistance are the overuse of antibiotics, inaccurate implementation, extended hospitalization, and unintended prescriptions. In this chapter, the authors described the functions of the most frequently prescribed and used drugs against bacteria and reported the mechanism leading to the development of resistance. Most common drugs include aminoglycosides, amphenicols, antifolates, β-lactams, glycopeptides, rifamycins, macrolides, quinolones, and tetracyclines which are extensively used against bacterial infections in humans and animals. Health authorities may manage and control the development of drug resistance through enforced legislation, strict ethical considerations, systematic management, reliable and efficient screening activities in any country and even at international levels. Up-to-date awareness, regarding the misuse of antibiotics and resulting in serious health concerns, among health workers and communities can also reduce the burdensome of drug resistance globally. With new information, technologies, and expertise, the challenges of drug-resistance can be handled urgently, and these resources can be the better ways for the development of next-generation drugs.

Keywords Bacteria · Infections · Antibiotic · Drugs · Mode of action · Resistance mechanism

2.1 Introduction

Antibiotics are the drugs that are well-known to treat infections caused by bacteria. Antibiotics resistance occurs when bacteria show a changed response against the corresponding antibiotics or drugs. Bacteria are resistant to some antibiotics, but not to animals or humans (Ventola, 2015). These bacteria can cause infections to humans and animals which are more severe to cure than the infections caused by non-resistant bacteria. Antibiotic resistance results in increased medical costs, prolonged stay in the hospital, and a higher mortality rate. There is an urgent need in the world to change the way antibiotics are prescribed and used. Antibiotic resistance remains a major public health threat despite the existing new drugs. Behavioral changes should include measures to lessen the blowout of infections through proper handwashing, healthy food, hygiene control, safe sex, and vaccination (Haque, 2020).

Throughout the world, antibiotic resistance increases to deleteriously high stages. The new mechanisms of resistance are developing and scattering around the world, threatening our competence to treat very common infectious ailments. With the

R. Naseer
Institute of Biochemistry & Biotechnology, University of Veterinary and Animal Sciences Lahore, Lahore 54000, Pakistan
e-mail: rahat.naseer@uvas.edu.pk

S. Ahmed (✉)
Department of Basic Sciences, University of Veterinary and Animal Sciences Lahore, Narowal 51600, Pakistan
e-mail: sarfaz.ahmed@uvas.edu.pk

increasing gradient of infections, like sepsis, tuberculosis, gonorrhea, pneumonia, and other food-borne infections, the corresponding treatment strategies become more difficult and occasionally impossible when antibiotics are ineffective. Lack of standards for life longevity and care indicates an excessive risk for the spread and emergence of resistance when antibiotics are purchased for humans and animals even without the recommendations of doctor of veterinarians respectively. Without taking immediate actions, we will put ourselves in the post-antibiotic period, where even non-serious infectious diseases and minor wounds can be fatal (Ayukekbong et al. 2017).

2.2 Prevention and Control

Various factors regarding antibiotics such as inappropriate use, abusive consumption, poor control, and poor prevention of infections can accelerate antibiotic resistance, but it is not infection control. Steps have been taken at every stage of society to decrease influence and bound the blowout of resistance. For the prevention and spread control of antibiotic resistance, people should; (1) use antibiotics if these are only prescribed by a healthcare professional, (2) not demand antibiotics if there is no need according to the doctor, (3) follow the prescription to consume antibiotics dosage, (4) not use or share surplus antibiotics randomly (Llor & Bjerrum, 2014). Infections can be prevented by frequent hand washing, hygienic food, safe sexual ways, vaccination, and avoiding interaction with frail people. "World Health Organization (WHO) 5-keys to Safer Food" can be used to make hygienic food. These include keeping ourselves clean, detaching raw material from cooked food items, complete cooking at the desired temperature, maintaining the temperature needed to preserve food, filtered water for washing the food items. Select food that contains no antibiotic residues which are also required to make hygienic food (Omari et al. 2014).

For the prevention of antibiotic resistance, policy creators should make sure that a concrete national action plan is in place to combat antibiotic resistance. These plans may enforce advance monitoring of antibiotic-resistant infections to promote the correct disposal of antibiotics and remove eminence drugs to reinforce the policies for active control actions (Ayukekbong et al. 2017). Healthcare professionals should avoid infections by cleaning their hands, tools, and the surrounding environment. They need to prescribe and distribute antibiotics according to patient needs keeping in view the current trends and recommendations of drug regulatory authorities. Moreover, they also need to advise their patients about proper antibiotic use, antibiotic resistance, and misuse hazards to avoid the transmission of infections via hand-washing, sneezing, sexual intercourse, and vaccination (Knobler et al. 2003). Additionally, the health industry should also participate in the investigation and progress of novel antibiotics, diagnostic tools, and vaccines. Similarly, the agriculture segment can also regulate the extra-label use of antibiotics in the livestock sector to encourage growth or prevent disease in healthy animals. The concepts of good agricultural practices, farm to fork approach and grass to glass must be ensured with antibiotic free processing from beginning till end. These approaches are the valuable chains of food processing obtained from agricultural crops and food animals which are highly influenced by the antibiotics (Manyi-Loh et al. 2018).

2.3 Recent Developments in the Prevention of Antibiotic Resistance

Some novel antibiotics are being developed which are least effective against most of the harmful drug-resistant bacteria. Due to the ease and frequency through which people travel nowadays, antibiotic resistance has become a global health issue. When antibiotics are ineffective, expensive drugs must be used. Severe illness, long-term treatment, frequent visits to the hospitals has raised the health charges and economic load on families, and the community. Antibiotic resistance has also weakened the success of modern medicine. Surgery, such as organ transplant, chemotherapy, and other operations like cesarean sections is very risky without the use of active antibiotics in preventing and treating diseases (Redfield, 2019).

Handling antibiotics is a WHO priority. A global plan of action for immune deficiency, involving antibiotic resistance, was adopted in May 2015 by the World Health Assembly. The goals of this action plan were to confirm the anticipation and cure of infectious ailments through safe and effective drugs. There are five objectives of the "Global Action Plan on Antimicrobial Resistance" including (a) improve understanding and awareness of antimicrobial resistance, (b) reinforce observations and investigations, (c) lessen the rate of infection, (d) improve the usage of antimicrobial drugs, and (e) approve the bearable expenditure in opposing antimicrobial resistance (WHO, 2015a).

In September 2016, a policy approved by the leaders of the United Nations (UN) General Assembly in New York demonstrated a global commitment to a broadly coordinated method to tackle the sources of antimicrobial resistance in many sectors such as agriculture, human and animal health. The WHO supports the Member States in developing national antimicrobial resistance action plans created based on the Global Action Plan. The WHO has also led several initiatives to combat antimicrobial resistance (United Nations,2016).

Since 2015, the World Antimicrobial Awareness Week (WAAW) has been held from November 18–24 every year. WAAW is a universal program that promotes the alertness of antimicrobial resistance and boosts the finest practices in the overall public health professionals, and policymakers at the global level to prevent the additional prevalence of drug-resistant infections. Antibiotics are unique medicines to help in the fight against human, plant, and animal diseases. It consists of antivirals, antibacterial, antiprotozoal, and antifungals. In 2020, the motto "Antibiotics: Handle with Care" has now been replaced by "Antimicrobials: Handle with Care" to show the prevalence of drug-resistant infections (World Health Organization, 2020).

The Global Antimicrobial Resistance Surveillance System (GLASS) supported by the WHO is responsible for gathering, analyzing and distributing data associated with antimicrobial resistance globally to update policymaking, national and regional actions (WHO, 2015b). The Global Antibiotic Research and Development Partnership (GARDP) is the combined initiative of the Drugs for Neglected Diseases initiative (DNDI) and WHO, while the GARDP promotes research and advancement in collaboration with the government and independent organizations (Balasegaram & Piddock, 2020). By 2023, the partnership aims to develop and implement 4 new treatments by improving the current antibiotics and accelerating the flow of new antibiotics. The Secretary-General of the UN has recognized the Interagency Coordination Group on Antimicrobial Resistance (IACG) to monitor progress and ensure the management among global organizations and active global action to combat this risk to the health sector. The IACG is managed by UN Secretary-General and the WHO Director-General. It consists of representatives of related UN agencies, experts in several fields, and various international organizations (WHO, 2018).

2.4 Antibiotic Discovery and Supremacy

In ancient times, well-recognized methods were present to control microbial infection in many parts of the world. In 1928, due to the discovery of penicillin, many doors were unlocked to the recent period of medicine and protected many lives (Gaynes, 2017). From 1939 to 1945, the period of World War II, penicillin was known as an excellent choice because it was the very first antibiotic recognized in the world and cured many people against microbial infections. After a short time, microbes developed resistance against this drug, and again scientists started to work out to handle this alarming situation. That's why the progressions completed in the 1940s were helpless for the next onward years (American Chemical Society, 2018). Nevertheless, soon after this situation, scientists developed new methods to overcome this resistance so that clinical approaches to treat microbial infections using antibiotics can be preserved. These efforts of scientists resulted in the discovery of innovative β-lactam antibiotics. When novel antibiotics were successfully used in clinical treatments during 1962 and 1968, in the United Kingdom and the United States, the very first case of methicillin-resistant Staphylococcus aureus (MRSA) was informed (Raygada & Levine, 2009). Unfortunately, soon after a short time, nearly all the antibiotics discovered during the same decade developed resistance. So, far ahead in 1972, scientists initiated to make more efforts and designed a drug 'vancomycin' which was effective against MRSA. However, still, clinicians were doubtful about the development of vancomycin resistance. Later on, during the 1970s and early 1980s, few cases of coagulase-negative staphylococci were reported that were resistant to vancomycin. Up-till the 1980s, new strategies were made to determine the problem of resistance, and from time to time more progress was done in the area of the pharmaceutical industry. Even so, the advancements made in the development of resistance-free antibiotics were stunned and made famous the limited novel antibiotics available at that time (Raygada & Levine, 2009). Subsequently, after almost 70 years of first antibiotic treatment of a patient, once again microbial infection turned out to be a serious hazard to natural life. Even though, in recent times, the risk of resistance development against antibiotics publicizes worldwide. In the meanwhile, antibiotics have offered many revolutions in clinical settings for humanity, exclusively in surgery and medicine. These discoveries resulted in the treatment of several bacterial and other microbial infections that may arise during or after chemotherapy and main surgeries like cardiac, organ transplant or joint replacement, etc. The results of these magical shots altered the outcome of microbial infections and also prolonged the average life expectancy. During those times, in the United States, 56.4 years was the average life span but nowadays, it has increased to almost 80 years. Poor public health services of the developing countries have resulted in a low rate of morbidity and mortality caused by bacterial infections (Ventola, 2015).

2.5 Causes of Antibiotic Resistance

Recently, different factors are playing their role to develop resistance against antibiotics. Some major causes of resistance are inaccuracies in usage, insufficient instructions, carelessness, the excessive or inexpert practice of drug dosage, and online marketing. All these factors and many others together make the low-grade antibiotics very easily reachable. Furthermore, antibiotics are also used as growth promoters instead of infection control for livestock and poultry which results in antibiotic resistance (Kırmusaoğlu et al., 2019).

Mainly, the key reason for resistance development is the excessive usage of antibiotics. Sir Alexander Fleming already advised that the "public will demand [the drug and] then will begin an era … of abuses". Delicate bacteria are more susceptible to antibiotics, but at the same time permit the pathogenic bacteria to survive. These pathogens then reproduce and prosper through natural choice. Even though misuse of antibiotics is powerfully dejected, but still their relics of over-prescription have spread across the world. One of the studies has exposed the fact that in 30 to 50 percent of cases, the choice of agent, treatment suggestions, and the duration of antibiotic therapy remained unsuitable (Leekha et al., 2011). Now antibiotics are applied as a growth promoter in the live-stock world widely. One of the surveys revealed that in the United States, approximately 80 percent of the antibiotics are prescribed as a growth enhancement and also to overcome the microbial infections in animals. In addition to this study, a universal map of 228 countries showed the usage of antibiotics in livestock. In 2010, one estimation disclosed that about 63,151 tons of antibiotics were consumed and Van Boeckel et al. have also expected a 67 percent increase in antibiotic usage by 2030. In speedily developing and extremely populated countries of the world like Russia, India, Brazil, South Africa, this percentage would be roughly doubled (Boeckel et al., 2015).

2.6 Common Antibiotics Used Against Bacteria

1. Aminoglycosides
2. Amphenicols
3. Antifolates
4. β-lactams
5. Glycopeptides
6. Rifamycins
7. Macrolides
8. Quinolones
9. Tetracyclines.

2.6.1 *Aminoglycosides*

Aminoglycosides are broad-spectrum antibiotics. It is potent that inhibits the synthesis of protein. This has been the basis of antibacterial chemotherapy subsequently streptomycin antibiotics were firstly extracted from*Streptomycesgriseus* and then went toward experimental practice in 1944 (Krause et al., 2016).

2.6.1.1 Mode of Action

Aminoglycosides bind with greater affinity on 16S ribosomal RNA at its A-site of the 30S ribosome to prevent the synthesis of protein as shown in Fig. 2.1 (Mehta & Champney, 2002), while all other members of aminoglycosides exhibit diverse specificity on the A-site for different regions. Consequently, the amino-acyl transfer RNA codon misreading is caused by antibiotics that endorse mistranslation. As a result of this, the accumulation of improper amino acids occurs into the growing polypeptide chain that causes cell membrane impairment. Several aminoglycosides can also affect the synthesis of protein by directly inhibiting initiation or by blocking elongation (Krause et al., 2016).

The consequent downstream effects and the exact mechanism of binding differ by chemical structure. All aminoglycosides naturally produced the prolonged post-antibiotic effect (PAE) because these are bactericidal. The bacteria take to regain from the protein synthesis inhibition because PAE directly links to the length of time. It is assumed that PAE is associated with the subsequent disassociation of the antibiotic from its mark and secrete out from the cell. These are categorized through a central structure of amino sugars linked through the glycosidic bond to a dibasic aminocyclitol (Behera & Kulkarni, 2018). These are generally categorized into four subclasses based on the individuality of the aminocyclitol moiety (1) a mono-substituted deoxystreptamine ring (e.g., apramycin), (2) no deoxystreptamine (e.g., streptomycin), (3) 4,6-di-substituted deoxystreptamine ring (e.g., amikacin,

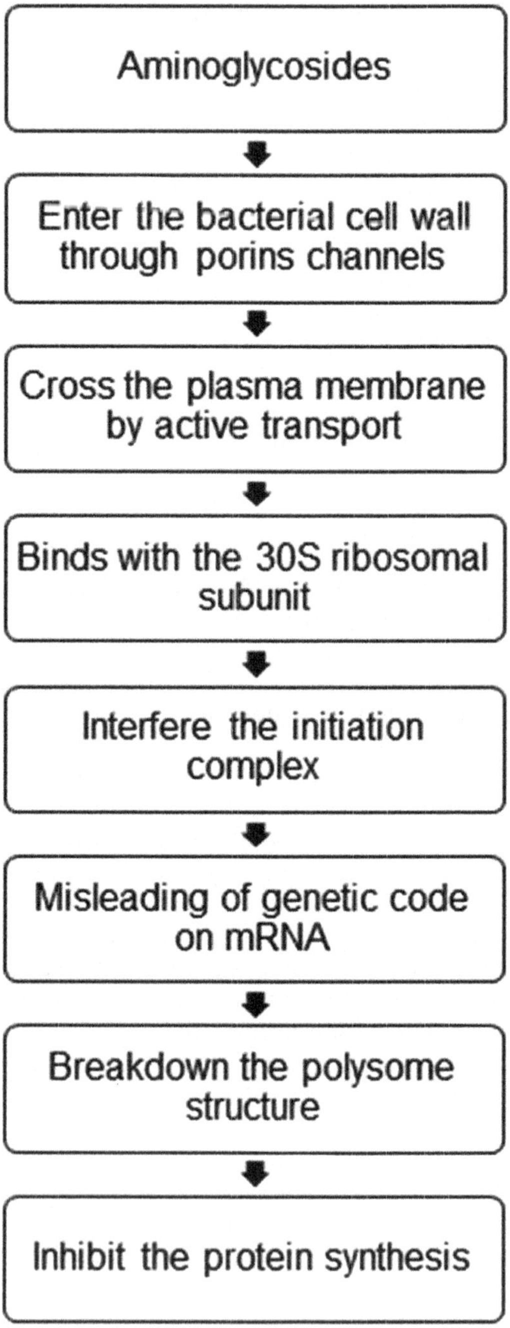

Fig. 2.1 Showing the mode of action of Aminoglycosides

gentamicin, plazomicin, and tobramycin), and (4) 4,5-di-substituted deoxystreptamine ring (e.g., ribostamycin, neomycin) (EMEA and Veterinary Medicines and Inspections, 2005).

The central structure is ornated by a diversity of hydroxyl and amino groups that affect the mode of act and vulnerability to several aminoglycoside-modifying enzymes (AMEs) linked through each of the aminoglycosides (Labby & Garneau-Tsodikova, 2013). The aminoglycoside that entered into bacteria is composed of three different stages, the first stage enhances the permeability of the bacterial membrane, while the second and third stages are reliant on energy. In the first stage, the polycationic aminoglycoside electro-statically bind to the negative charge carrying components of the membrane of bacteria like lipopolysaccharide (LPS) and phospholipids as well as teichoic acids and phospholipids of gram-negative and gram-positive organisms respectively (Krause et al., 2016). The cross bridges and the lipid components of the membrane of bacteria are stabilized by these cations. Their exclusion leads to distraction of the outer membrane, initiation of aminoglycoside uptake, and greater permeability. This approach helps intake into the cytoplasm through an energy-dependent, slow, electron-transport chain. The mistranslation of protein and protein synthesis inhibition occurs when aminoglycoside approaches cytoplasm. Cytoplasmic membrane damage caused by mistranslated proteins facilitate entry of aminoglycoside. This leads to the quick intake of supplementary aminoglycoside into the cytoplasm that improved protein synthesis inhibition, mistranslation, and enhanced cell death (Mingeot-Leclercq et al., 1999).

2.6.1.2 Mechanism of Resistance

The resistance to the aminoglycoside takes many forms, including an alteration to the target site by chromosomal mutation or a chromosomal enzyme (Fig. 2.2). All these mechanisms differ from one class to another and often involve multiple mechanisms in a provided resistant isolate. Resistance to aminoglycosides due to all prokaryotes, except Borrelia spp., has not been observed for target-site mutations. Although modern large-scale research programs include consideration of resistance of phenotypic aminoglycoside amongst significant pathogens, this research has

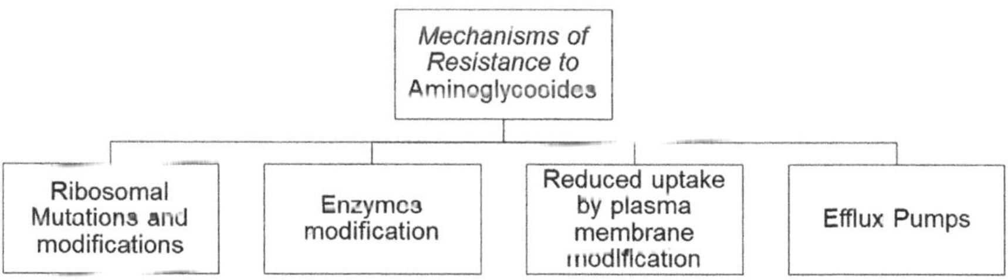

Fig. 2.2 Showing the mechanisms of resistance to Aminoglycosides

usually not emphasized the epidemiology of certain mechanisms of resistance (Labby & Garneau-Tsodikova, 2013; Ramirez & Tolmasky, 2010).

Aminoglycosides attach to the 16S rRNA aminoacyl-tRNA site of recognition (A-site) that makes up the 30S ribosomal subunit, prohibition of polypeptide synthesis, and the consecutive death of cells. Depending on several factors, resistance to aminoglycosides can occur at aminoglycoside acetyltransferases, nucleotidyltransferases, or phosphotransferase that facilitate enzymatic modification and inactivation of aminoglycoside. These are usually studied through gram-negative and gram-positive bacteria, enhanced efflux, decreasing permeability, and alteration in 30S ribosomal subunit that interferes with aminoglycoside binding (Khan et al., 2020).

For the latter, aminoglycoside resistance has been associated with both mutations' post-transcriptional modifications and nucleotide replacement. In Mycobacterium tuberculosis, streptomycin resistance is produced by point mutations in the 16S ribosomal RNA and the rpsL gene encoding the S12 protein. However, mutations within the 30S ribosomal subunit do not appear to be a usual mechanism of aminoglycoside resistance among fast-growing pathogenic bacteria in general. However, among aminoglycoside-producing Actinobacteria, including Micromonosporaspp and Streptomyces spp., which are impulsively immune to these metabolites of their own, post-transcriptional alteration of 16S rRNA is usually noticed. Post-transcriptional methylation of the N-1 position of nucleotide A1408 or N-7 position of nucleotide G1405 on 16S rRNA by numerous 16S-RMTases is mediated by this mechanism. Intrinsic N7-G1405 16S-RMTases are present in both Micromonospora spp. and Streptomyces spp. while intrinsic N1-A1408 16S-RMTases have been reported only in Streptomyces spp. N7-G1405 16S-RMTases, but not in 4,5-disubstituted and monosubstituted DOS agents, giving resistance to 4,6-disubstituted DOS agents for example tobramycin, gentamicin and amikacin. N1-A1408 16S-RMTases, in comparison, can confer resistance to both of these classes (Plattner et al., 2020).

2.6.2 *Amphenicols*

Chloramphenicol and thiamphenicol are mostly used in amphenicol (chloramphenicol) antibiotics. Some countries are currently researching the fluorine-containing compound to obtain drugs with a greater ability to penetrate the membrane, low toxicity, and better effectiveness in killing microorganisms. Chloramphenicol has been removed from the Streptomycesvenezuelae culture and is now being mass-produced by chemical synthesis. This product exists in the form of an L-body. Chloramphenicol antibiotics can behave on a broad-spectrum antibacterial antibiotic's 50S subunit of the bacterial ribosome, interfering with protein synthesis (PubChem, 2019).

The main cellular elements for protein synthesis are the 70S ribosome of bacterial cells. It is composed of two 50S and 30S subunits. Chloramphenicol inhibits the action of peptidyl transferase by reversibly binding to the 50S subunits, interfering with the mechanism of binding the amino acid containing-aminoacyl-tRNA to the 50S subunits, thus preventing the formation of the new peptide chain, causing protein synthesis to be inhibited. Chloramphenicol can also associate with the ribosome of human mitochondria the 70S and thus inhibits the synthesis of mitochondrial protein in the human body and produce toxicity to the human body. Since the binding between the 70S ribosome and chloramphenicol is reversible, bacteriostatic antibiotics are considered. However, it also has a bactericidal effect on some types of bacteria at high drug concentrations with a strong bactericidal effect on influenza bacilli at even lower concentrations (Scholar, 2007).

2.6.2.1 Mode of Action

By inhibiting protein synthesis, chloramphenicol is bacteriostatic. It prohibits elongation of the protein chain by reducing the formation of the bacterial ribosome in peptidyl transferases. The chloramphenicol interferes directly with substrates binding in the ribosome contrast with macrolides which block sterically the development of the cultivated peptide and the chloramphenicol directly implies the production of the peptide in the ribosome (Fig. 2.3) (Henras et al., 2017).

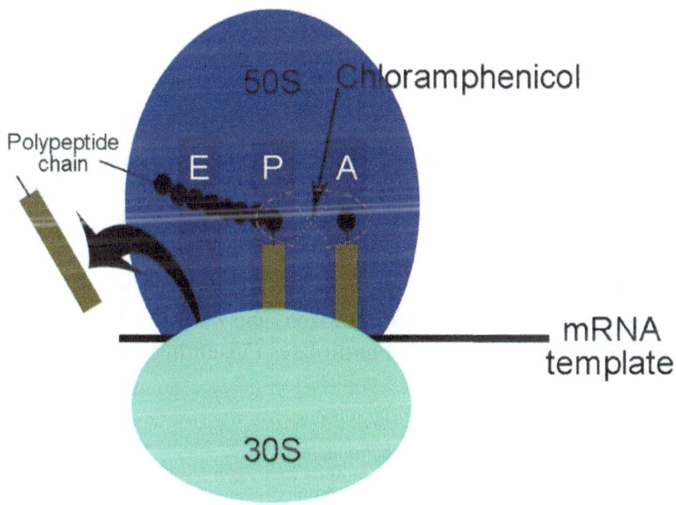

Fig. 2.3 Showing the mode of action of chloramphenicol

2.6.2.2 Mechanism of Resistance

The three chloramphenicol resistance mechanisms are known: reduced membrane permeability, 50S ribosomal subunit mutation, and acetyltransferase chloramphenicol growth as shown in Fig. 2.4 (Fernández et al., 2012; Schwarz et al., 2004). The serial passage of bacteria makes it simple to pick in vitro a diminished membrane permeability to chloramphenicol and is the most frequent low chloramphenicol resistance mechanism. The cat-gene confers a high-level resistance, and this enzyme gene codes chloramphenicol acetyltransferases to inactivate the enzyme by binding one or two acetyl groups of acetyl S-coenzyme A to hydroxyl groups on chloramphenicol-molecule covalently. Through acetylation, chloramphenicol does not bind to the ribosome. Mutations in the 50S ribosomal subunit are unusual in terms of resistance-conferring. A plasmid that also accounts for resistance to other medicines may contain chloramphenicol resistor. Examples include ACCoT (A = ampicillin, C = chloramphenicol, Co = co-trimoxazole, and T = tetracycline), a plasmid that mediates several drug resistances to typhoid (also known as R factors). Since 2014, some strains of Enterococcus Faecium and Pseudomonas Aeruginosa are chloramphenicol resistant. There are few Veillonella spp. resistance to varying levels of Chloramphenicol was also established with Staphylococcus capitis strains (Morita et al., 2013).

2.6.3 *Antifolates*

Folic acidifies the primary role of folic acid in the body as the cofactor of various methyltransferases involved in thymidine, methionine, serine, and purine biosynthesis. Folic acid is a cofactor in different methyltransferases in the body, which are also associated with serine, thymidine, and purine. Antifolates thus prohibit cell divisions, synthesis of DNA/RNA, and repair and synthesis of proteins. Some of them selectively inhibit folate action, such as pyrimethamine, proguanil, and trimethoprim, in microbial species like fungi, protozoa, and bacteria. Inhibiting the DHFR (dihydrologalreductase) is the bulk of antifolates (Visentin et al., 2012).

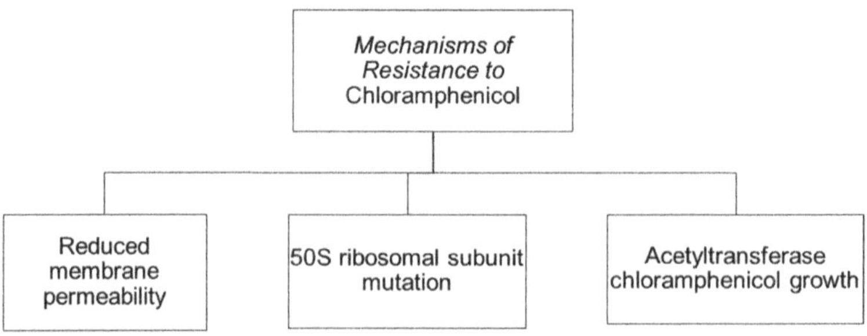

Fig. 2.4 Showing the three mechanisms of resistance of chloramphenicol

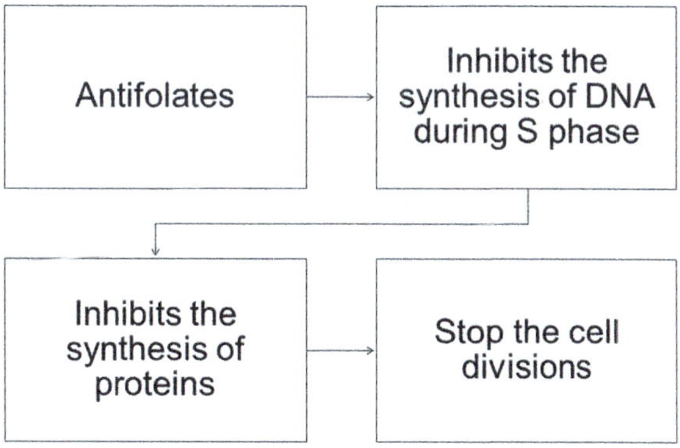

Fig. 2.5 Showing the mode of action of Antifolates

2.6.3.1 Mode of Action

Many inhibitors are mainly DHFR, but raltitrexed is a synthase inhibitor and a pemetrexed enzyme. During the synthesis of DNA and RNA specifically, antifolates become cytotoxic during the S-phase of the cell cycle (Fig. 2.5). They are, therefore, more toxic to rapidly splitting cells—such as myeloid and malignant cells, and GI and oral molecules which more often duplicate their DNA and thus inhibit their growth and proliferation (Raimondi et al., 2019; Visentin et al., 2012).

2.6.3.2 Mechanism of Resistance

Although the role of folate as a treatment for cancer is well known, the cellular response reduces its long-term effectiveness. The cell starts to transcribe more DHF reductase, which decreases DHF to THF, in response to decreased tetrahydrofolate (THF). As methotrexate is an efficient inhibitor of DHF reductase, higher DHF reductase levels can resolve the activation of the drug. Many new medicines have been produced to reduce the resistance of antifolates (Fig. 2.6) (Gangjee et al., 2007; Kordus & Baughn, 2019).

2.6.4 β-lactams

Beta-lactams are antibiotics that have the nucleus of a beta-lactam ring. Monobactams, Carbapenems, Cephalosporins, cephamycins (cephems), Clavams, and Penicillin are all subclasses of β-lactams The bacterial cell wall synthesis demands that all beta-lactams bind and inactivate enzymes (Page, 2012).

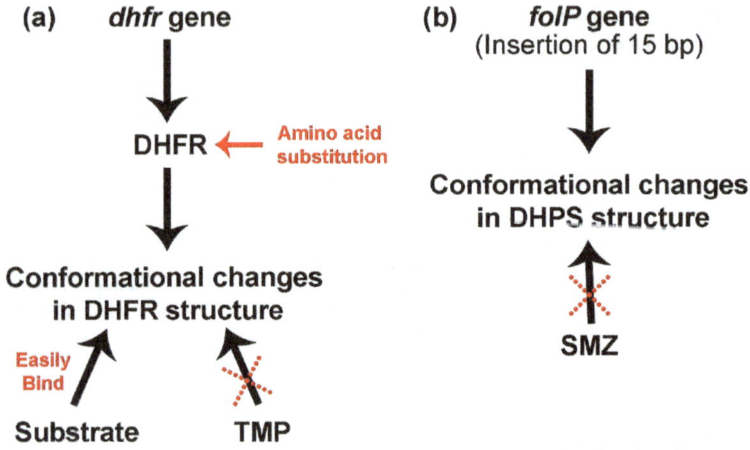

Fig. 2.6 Mechanisms of antifolate resistance in bacterial cells (Kordus & Baughn, 2019)

2.6.4.1 Mode of Action

The β-lactam antibiotic exerts its antimicrobial impact by preventing the formation of the layer of peptidoglycan in the cell wall of bacteria. The β-lactam permanently binds to the receptor, penicillin-binding proteins (PBPs), on the cell wall. This binding inhibits transpeptidase activity that further interferes with peptidoglycan formation. This disruption causes a targeted microbe to die as shown in Fig. 2.7 (Page, 2012).

Fig. 2.7 Showing the mode of action of β-lactam antibiotic

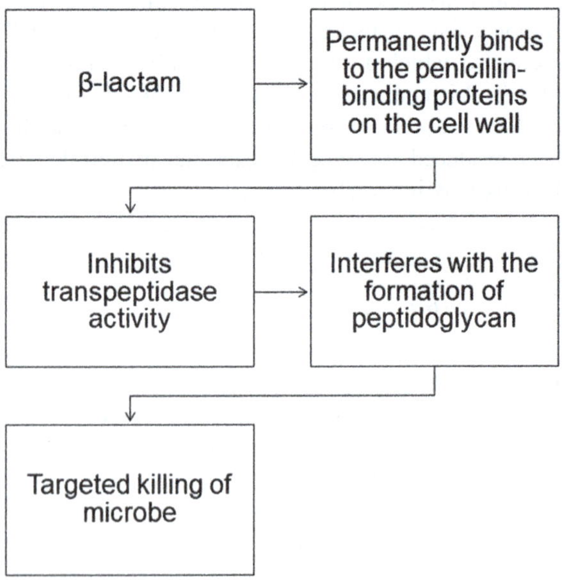

2.6.4.2 Drug Resistance and Its Mechanism

Extensive and unnecessary prescription of antibiotics cause serious issues such as drug resistance and a viable population of pathogens in the respiratory tract can be propagated. Acute rhinosinusitis is primarily caused by H. Moraxella and Influenza. Resistance to β-lactam antibiotics is demonstrated by around one-third of these bacterial strains. Resistance is demonstrated by two mechanisms, by generating β-lactamase and by resistance mediated by non-β-lactamase as shown in Fig. 2.8 (Jacobs, 2017; Page, 2012).

Plasmid gene in bacteria encodes β-lactamase, and this gene is similar in structure to the penicillin-binding protein (PBP) possessing high β-lactam affinity. Inactivation of antibiotics can be caused by the interaction between bacterial enzymes and β-lactam, whereas β-lactamase alters the structure of the antibody making it unable to bind any receptor. The binding of β-lactamase to the antibiotic is a reversible mechanism for the recovery of the enzyme to attack the next antibiotic molecule. A non-β-lactamase derived resistance mechanism, such as the change in the structure of PBPs, can also cause antibiotic resistance. The binding of antibiotics to the receptor can be inhibited by a mutation in the ftsI gene encoding PBP. Three conserved motifs called SXXK, SSN, and KTG form the active site of transpeptidase. Mutations in these motifs can cause conformational changes in the receptor protein's active site and render it inaccessible for antibiotic attachment. It is possible to overcome antibiotic resistance using β-lactamase inhibitors. These inhibitors bind irreversibly to resistance-causing enzymes that are inactive (Jacobs, 2017; Page, 2012).

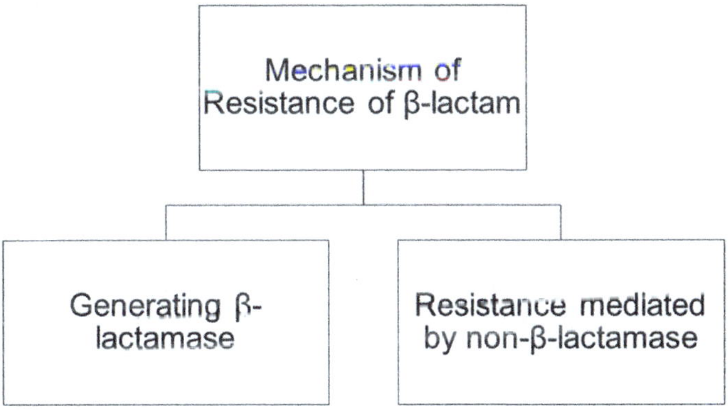

Fig. 2.8 Mechanism of resistance of β-lactam

2.6.5 Glycopeptides

Glycopeptide antibiotics consist of glycosylated polycyclic or cyclic non-ribosomal peptides. These antibiotics form a class of drugs that have a microbial origin. Remarkable glycopeptides antibiotics are comprised of anti-infective antibiotics decaplanin, vancomycin, complestatin, teicoplanin, corbomycin, ramoplanin, telavacin, and anti-tumor antibiotic bleomycin. In MRSA (methicillin-resistant Staphylococcus aureus) infection, the use of vancomycin is proved to be beneficial. Some members in this class can restrict cell wall synthesis in susceptible microorganisms by inhibiting the synthesis of peptidoglycan. They also inhibit the addition of new amino acid units to peptidoglycan through attachment with the amino acids present in the cell wall, particularly acyl-D-alanyl-D-alanine. Moreover, many members of this class can also prevent the function of the enzyme glycosyltransferases, which polymerizes building blocks of sugar/amino acid into peptidoglycan (Zeng et al., 2016).

2.6.5.1 Mode of Action

Earlier research on glycopeptide activity revealed the vancomycin-mediated intracellular aggregation of metabolic precursors involving lipid carrier linked MurNAc-pentapeptide (N-acetyl-muramic pentapeptide; lipid II), and the restriction of amino acid attachment to the bacterial peptidoglycan. This proposed intrusion occurs in the transglycosylation step during the synthesis of peptidoglycan. Distribution studies of iodinated vancomycin have exhibited the surface-localized binding in bacterial cells. Moreover, it was discovered that vancomycin can make complexes with lipid II, but it is unable to inhibit the incorporation of a distinct tetrapeptide-linked precursor (MurNAc-tetrapeptide) into peptidoglycan. Therefore, in Gaffkyiahomari (which uses MurNAc-tetrapeptide) peptidoglycan synthesis cannot be restricted. Consequently, it was assumed that vancomycin terminates pentapeptide moiety (which belongs to lipid II) by binding to d-alanyl-d-alanine. Thus, vancomycin inhibits the process of transglycosylation via corresponding to steric hindrance and the transpeptidation through substrate sequestration. Hydrogen bonding is responsible for the binding of vancomycin to d-alanyl-d-alanine leading to the termination of pentapeptide (Zeng et al., 2016).

2.6.5.2 Mechanism of Resistance

The presence of the outer membrane in gram-negative bacilli provides them intrinsic resistance against vancomycin and lipoglycopeptides. Outer membrane inhibits these complex and large molecules from entering into the bacterial cells leading to the target site. In comparison with members of non-glycopeptide antibiotics, gram-positive bacteria have also developed resistance against vancomycin (glycopeptides) after a long period. In 1987, the first clinically isolated vancomycin resistance strain was documented following the first introduction of vancomycin. Today, greater than 30% of intensive care units show resistance against vancomycin in E. faecalis (Enterococcus faecalis) infections, and these resistant strains are collectively known as VRE (vancomycin-resistant Enterococcus). After subsequent development of resistance against vancomycin in Enterococcus, abrupt resistance has been evolved in S. aureus (VRSA) due to horizontal transfer of a gene from resistant strains of Enterococci. The first strain of S. aureus with decreased susceptibility to teicoplanin and vancomycin was clinically isolated from a patient in Japan. However, in 2002, the first completely resistant VRSA strain (teicoplanin MIC (minimum inhibitory concentration) = 128 μg/ml; vancomycin MIC > 256 μg/ml) was reported from the United States. Recently, some workers from Brazil have also reported similar resistant strains of staphylococcus. As vancomycin forms complexes with peptidoglycan precursors, its binding affinity with peptidoglycan precursor determines its activity. The presence of certain enzymes causes the production of lower-affinity binding precursors, which is the main reason behind the mechanism of resistance to vancomycin. In these precursors, the d-alanine amino acid at carboxy-terminal is replaced by d-lactate, as a result, hydrogen-bond interaction is lost and binding interaction decreases 1000-fold. Alternatively, the d-alanine amino acid may also be substituted with d-serine residue, causing six-fold depletion in binding affinity likely produced by steric hindrance (Zeng et al., 2016).

2.6.6 *Rifamycins*

Rifamycins are a special group of antibiotics and belong to a subclass of ansamycins. These are naturally produced by a bacterium "Amycolatopsisrifamycinica" or prepared through artificial methods as well. As these antibiotics are remarkably effective against mycobacteria, therefore, these are usually used to treat leprosy, MAC (mycobacterium avium complex), and tuberculosis infections. Rifamycin group comprises of classic rifamycin drugs together with their derivatives rifalazil, rifabutin, rifaximin, rifapentine, and rifampicin (also known as rifampin). In the United States, Rifamycin is approved for the cure of travelers' diarrhea in some conditions.

Rifamycins are powerful bactericidal agents as they restrict bacterial DNA-dependent RNA-polymerase. Rifampin analogs can kill intracellular &semi-dormant state mycobacteria and other actively proliferating extra-cellular organisms. For tuberculosis, the addition of rifampin to therapeutic regimens can minimize the time of treatment from 12 to 6 months in case of active disease, while the duration of treatment can be reduced from 9 months to 2–3 months for latent infections. Owing to their potencies and powerful sterilizing activities, rifamycins are found to have an effective role in the treatment of latent Mycobacterium tuberculosis infections (Mahalaxmi et al., 2008).

2.6.6.1 Mode of Action

Rifamycins have a great affinity for prokaryotic RNA-polymerase, and this affinity allows them to inhibit DNA-dependent synthesis of RNA. It also provides them effective antibacterial properties. Rifamycins are highly selective for prokaryotic RNA-polymerase and show very little affinity for eukaryotic analogous enzymes. Crystal structure data showed that binding of rifamycin with RNA-polymerase causes extreme steric clashes with the growing chain of the oligonucleotide, also known as a steric-occlusion mechanism. After the initiation of the chain extension process, rifamycin binding to RNA-polymerase cannot inhibit the biosynthesis process. A single change in the amino acid residue of bacterial DNA dependent RNA-polymerase can lead to high-level single step resistance in rifamycins (Maslow & Portal-Celhay, 2014).

2.6.6.2 Mechanism of Resistance

Recently, the development of resistance to rifampin has given rise to a significant problem in TB control. Rifamycins resistance arises when a mutation occurs in rpoB (β-subunit of the ribosomal polymerase gene). During rifamycin monotherapy, extemporaneous mutations occur in 81 base pair region of rpoB. It is notable that resistance to rifamycin occurs along the course of therapy against active TB disease, and does not occur during therapy against latent TB infection. Nearly all rifampin-resistant TB clinical isolates also show resistance to isoniazid. Isolates showing resistance to rifamycins have reduced susceptibility to rifabutin, rifapentine, and the other rifamycins (Chaisson, 2003).

2.6.7 *Macrolides*

Macrolides, a polyketide class of natural products, are made up of large macrocyclic lactone ring to which at least one or more deoxy sugars, generally desosamine and cladinose, may be anchored. These lactone rings are generally 14–16 membered. Macrolides are used as pharmaceutical agents because of their antifungal and antibacterial activities. These are found to be bacteriostatic as they only inhibit or suppress bacterial growth instead of neutralizing them completely (Dhakal et al., 2019).

2.6.7.1 Mode of Action

Macrolidesinhibit the bio-synthesis of protein by preventing the attachment of peptidyltransferase to the growing peptide bound for the transfer of RNA to the next amino acid (identically to chloramphenicol) and also block the translation of bacterial ribosomal. Another potential method is premature detachment of the peptidyl transfer RNA from the ribosome. Macrolides do so by attaching reversibly to the P-site on the small (50S) subunit of the bacterial ribosome, and this is considered a bacteriostatic act. Macrolides are actively centralized within leukocytes, and hence these are immediately transferred during an infection (Vázquez-Laslop & Mankin, 2018).

2.6.7.2 Mechanism of Resistance

As several methods of resistance have been monitored, hence the resistance for macrolides is not limited to a single method. Particularly, methods to (i) reduce the intracellular assembly of macrolides, (ii) modify the ribosome (target), (iii) protect the ribosome (target) (iv) and chemical alteration of antibiotics are monitored in clinical isolates. The primary mode of bacterial resistance toward macrolides is caused by post-transcriptional methylation on the 23S ribosomal RNA of bacteria. This attained resistance may be either chromosomal or plasmid-mediated, i.e., through mutation, and causes cross-resistance to lincosamides, macrolides & streptogramins (an MLS resistant phenotype).Two other rarely seen classes of acquired resistance include the manufacturing of drug-deactivating enzymes (kinases or esterases), and also the production of efflux proteins (active ATP dependent), involved in the transfer of the drug to the exterior of the cell. In penicillin-sensitive patients, strep throat (Group A streptococcal (GAS) disease induced by Streptococcus pyogenes) has been treated using Azithromycin. Anyhow, macrolide-resistant types of GAS are not unusual. Another alternative for these patients is a cephalosporin (Kirst, 2014).

2.6.8 *Quinolones*

A group of antibiotics having a bicyclic interior structure homologous to the 4-quinolone compound is known as quinolones.

2.6.8.1 Mode of Action

Fluoroquinolone/quinolone is a bactericidal and broad-spectrum antimicrobial agent. Quinolones cease the bacterial cell division by blocking the bacterial DNA interfering with transcriptional, recombinant, and replicating enzymes. During replication of DNA in bacteria, ofloxacin binds & blocks DNA gyrase enzyme and ceases unwinding of DNA double helix. It also impedes the bacterial type II & IV topoisomerases. In this way, cell division in bacteria is halted. Quinolones, attached at the active site of splitting ligation with DNA-enzyme, form cleavage complexes with stable concentration. This disturbs the replication process of DNA & colloids cleavage complexes resulting in a chromosomal disintegration. Due to this destruction, DNA rectifying processes are activated because of the activation of SOS systems. The SOS systems which are activated in turn are the following: cell filaments become elongated because of the expression of LexA suppressor & activation of antitoxin-toxin modules causes apoptosis (Hooper & Jacoby, 2016).

2.6.8.2 Mechanism of Resistance

Ofloxacin belongs to the 4-quinolone group & contains the quality of plasmid-borne resistance in bacteria. It is confirmed that ofloxacin-resistance in bacteria is developed by a mutation in the chromosome. Resistance developed against ofloxacin was tested by using the KL16 genus of Escherichia coli, as a model system. In several in vitro studies concerning the development of resistance to new 4-quinolone, a low frequency of mutation [1 × 10(−10) to 1 × 10(−12)] was observed in bacteria that were cultured under regular circumstances of the laboratory. Auxotrophic, temperature-sensitive, and slow-growing organisms were obtained. The unstable mutations were noticed, & mutants were reversed rapidly to the sensitivity of ofloxacin, while the selection was done in the unavailability of ofloxacin. The spontaneous mutations in undergrowth states, involving the lumen of the gut with restricted oxygen supply, were observed in studies. Similar instability of the mutants, in addition to the same phenotype, was also noticed under these conditions. The physiology of ofloxacin-resistant mutants exhibited metabolic defects associated with environmental changes (Hooper & Jacoby, 2016).

Aldred et al. (2014) described three resistance mechanisms against quinolones, firstly, the plasmid-mediated, secondly, the chromosome mediated, and thirdly target mediated resistance. In the third category, a mutation in gyrase and topoisomerase IV decreases the quinolone-enzyme interaction. The Opr protein is involved in the protection of the enzyme–DNA complex, and joining of Opr protein is reduced in plasmid-mediated resistance. In the case of resistance mediated by chromosomes, under-expression of porins reduces the intake of medicine in gram-negative bacteria (Aldred et al., 2014).

2.6.9 *Tetracyclines*

A tetracycline is a broad-spectrum group of antibiotics showing activity against a vast variety of gram-negative aerobic, anaerobic, and gram-positive bacteria, chlamydiae, cell wall lacking mycoplasmas, mycobacterium, helicobacter, listeria, rickettsia & protozoan parasites like Giardia lamblia, Plasmodium falciparum, and Entamoeba histolytica. They have been applied widely for therapy in humans for urogenital tract and respiratory diseases like Lyme, rickettsial, and periodontal diseases (Chopra & Roberts, 2001).

2.6.9.1 Mode of Action

Tetracycline works by inhibiting protein synthesis, and translation through sticking with the 30S ribosomal subunit at the initial stage in various ways. The binding of the messenger RNA translation complex to the aminoacyl transfer RNA is inhibited by tetracycline. It is mentioned in a few studies that tetracycline can bind with both 23S and 16S rRNAs. Besides, tetracyclines have been observed to repress matrix metalloproteinases. This method does not contribute to their antimicrobial outcomes but promotes considerable research on chemically modified tetracyclines (like incyclinide) for the cure of acne, diabetes rosacea & several kinds of neoplasms. It has been disclosed that tetracyclines are not just active antagonistic towards broad-spectrum of bacteria, but these are also efficient against viruses, few non-infectious conditions and protozoa. The binding of tetracycline to double-stranded RNA (dsRNA) could be an indication of their wide variety of effects. It may also be ascribed to the inherent feature of ribosomal protein-synthesizing routes among bacteria. In September 2007, incyclinide was proclaimed to be unproductive for rosacea. Various trials have investigated unmodified and modified tetracycline to cure human cancers, while very favorable outcomes were obtained with chemically modified tetracyclines for people suffering from Kaposi Sarcoma (Chopra & Roberts, 2001).

2.6.9.2 Mechanism of Resistance

Cells can get resistance to tetracycline by efflux, enzymatic inactivation of tetracycline, protection of ribosomes, ribosome mutation, and decreased permeability. Inactivation is the unusual kind of resistance, in which a group of antibiotic destructase known as NADPH dependent oxidoreductase, inactivates tetracycline. For instance, the oxidoreductase alters the C11, which is a site of oxytetracycline. Ribosome binding and chelation of $sMg2^{+}$, both are needed for the biological action of oxytetracycline & the alteration attenuates the binding, resulting in deactivation of the oxytetracycline (Chopra & Roberts, 2001).

In the most popular method of the reaction, efflux, several resistance genes code for a membrane protein that causes active pumping of tetracycline to the exterior of the cell by replacing a proton for antibiotic tetracycline cation complex. This exchange results in a decreased concentration of tetracycline in the cytoplasm. In the ribosomal protection method, a resistance gene codes for a protein that may have various effects, based on which gene is shifted. Twelve groups of ribosomal protection proteins/genes have been identified. These protective proteins exhibit the following possible actions: (1) blocking tetracycline so that it cannot bind with the ribosome, (2) sticking to the ribosome but twisting the structure to permit transfer RNA binding, whilst tetracycline is bound, 93) binding to the ribosome & removing tetracycline (Chopra & Roberts, 2001).

2.7 Conclusion

Drug resistance can be stopped through hand-washing, consumption of hygienic and properly cooked food, vaccinations, and taking antibiotics only in direly needed situations. Avoiding these simple directions has put animals and human health at greater risk. The development of antibiotic resistance should be monitored and prevented at an early stage during drug preparation, as this will help the drug industries and research institutions to collaborate to improve perceptions about the possible mechanism of resistance against drugs. The early identification can also help to control drug resistance more effectively compared to the preventive methods adopted in clinical grounds during practice. Today, we have relatively innovative techniques to understand the emergence of resistance and resultant responses, but future challenges for the rapid and reliable identification of drug resistance include the use of the best surveillance methods, technologies, and expertise to enhance food security, global health, and well being.

References

Aldred, K. J., Kerns, R. J., & Osheroff, N. (2014). Mechanism of quinolone action and resistance. *Biochemistry, 53*, 1565–1574.

American Chemical Society. (2018). *Alexander Fleming Discovery and Development of Penicillin - Landmark*. American Chemical Society. Am. Chem. Soc.

Ayukekbong, J. A., Ntemgwa, M., & Atabe, A. N. (2017). The threat of antimicrobial resistance in developing countries: Causes and control strategies. *Antimicrobial Resistance and Infection Control, 6*, 47.

Balasegaram, M., & Piddock, L. J. V. (2020). The global antibiotic research and development partnership (GARDP) not-for-profit model of antibiotic development. *ACS Infect. Dis, 6*, 1295–1298.

Behera, A., & Kulkarni, S. S. (2018). Chemical synthesis of rare, deoxy-amino sugars containing bacterial glycoconjugates as potential vaccine candidates. *Molecules, 23*, 1997.

Chaisson, R. E. (2003). Treatment of chronic infections with rifamycins: Is resistance likely to follow? Antimicrob. *Agents and Chemotherapy, 47*, 3037–3039.

Chopra, I., & Roberts, M. (2001). Tetracycline antibiotics: mode of action, applications, molecular biology, and epidemiology of bacterial resistance. *Microbiology and Molecular Biology Reviews, 65*, 232–260.

Dhakal, D., Sohng, J. K., & Pandey, R. P. (2019). Engineering actinomycetes for biosynthesis of macrolactonepolyketides. *Microbial Cell Factories, 18*, 137.

EMEA, and Veterinary Medicines and Inspections. (2005). *Committee for medicinal products for veterinary use phenoxymethypenicillin.*

Fernández, M., Conde, S., De La Torre, J., Molina-Santiago, C., Ramos, J. L., & Duque, E. (2012). Mechanisms of resistance to chloramphenicol in Pseudomonas putida KT2440. *Antimicrobial Agents and Chemotherapy, 56*, 1001–1009.

Gangjee, A., Kurup, S., & Namjoshi, O. (2007). Dihydrofolate reductase as a target for chemotherapy in parasites. *Current Pharmaceutical Design, 13*, 609–639.

Gaynes, R. (2017). The discovery of penicillin—new insights after more than 75 years of clinical use. *Emerging Infectious Diseases, 23*, 849–853.

Haque, M. (2020). Antibiotic resistance: A complete trouble maker for healthcare system especially among lower and middle-income countries. *Advances in Medical, Dental and Health Sciences, 2*, 42–43.

Henras, A. K., Plisson-Chastang, C., Humbert, O., Romeo, Y., & Henry, Y. (2017). Synthesis, function, and heterogeneity of snoRNA-guided posttranscriptional nucleoside modifications in eukaryotic ribosomal RNAs. In *Enzymes* (pp. 169–213). Academic Press.

Hooper, D. C., & Jacoby, G. A. (2016). Topoisomerase inhibitors: Fluoroquinolone mechanisms of action and resistance. *Cold Spring Harbor Perspectives in Medicine, 6*, a025320.

Jacobs, M. R. (2017). Mechanisms of resistance in haemophilusinfluenzae and moraxella catarrhalis. In *Antimicrobial Drug Resistance* (pp. 867–888). Springer International Publishing.

Khan, F., Pham, D. T. N., & Kim, Y. M. (2020). Alternative strategies for the application of aminoglycoside antibiotics against the biofilm-forming human pathogenic bacteria. *Applied Microbiology and Biotechnology, 104*, 1955–1976.

Kırmusaoğlu, S., Gareayaghi, N., & Kocazeybek, B. S. (2019). Introductory chapter: The action mechanisms of antibiotics and antibiotic resistance. In *Antimicrobials, antibiotic resistance, antibiofilm strategies and activity methods*. IntechOpen.

Kirst, H. A. (2014). Macrolide antibiotics. In W. Schönfeld, & H. A. Kirst (Eds.) *Antimicrobials: New and old molecules in the fight against multi-resistant bacteria* (pp. 211–230). Birkhäuser Basel.

Knobler, S. L., Lemon, S. M., Najafi, M., & Burroughs, T. (2003). *Strategies to contain the development and consequences of resistance.*

Kordus, S. L., & Baughn, A. D. (2019). Revitalizing antifolates through understanding mechanisms that govern susceptibility and resistance. *MedChemComm, 10*, 880–895.

Krause, K. M., Serio, A. W., Kane, T. R., & Connolly, L. E. (2016). Aminoglycosides: An overview. *Cold Spring Harbor Perspectives in Medicine, 6*, a027029.

Labby, K. J., & Garneau-Tsodikova, S. (2013). Strategies to overcome the action of aminoglycoside-modifying enzymes for treating resistant bacterial infections. *Future Medicinal Chemistry, 5*, 1285–1309.

Leekha, S., Terrell, C. L., & Edson, R. S. (2011). General principles of antimicrobial therapy. In *Mayo clinic proceedings* (pp. 156–167). Elsevier Ltd.

Llor, C., & Bjerrum, L. (2014). Antimicrobial resistance: Risk associated with antibiotic overuse and initiatives to reduce the problem. *Therapeutic Advances in Drug Safety, 5*, 229–241.

Mahalaxmi, Y., Rao, C. S., Suvarnalaxmi, G., Satish, T., Sudhakar, P., & Prakasham, R. S. (2008). Rifamycin B production pattern in nocardia RSP-3 strain and influence of barbital on antibiotic production. *Current Trends in Biotechnology and Pharmacy, 2*, 173–181.

Manyi-Loh, C., Mamphweli, S., Meyer, E., & Okoh, A. (2018). Antibiotic use in agriculture and its consequential resistance in environmental sources: Potential public health implications. *Molecules, 23*, 795.

Maslow, M. J., & Portal-Celhay, C. (2014). Rifamycins. In *Mandell, Douglas, and Bennett's principles and practice of infectious diseases* (pp. 339–349.e3). Elsevier Inc..

Mehta, R., & Champney, W. S. (2002). 30S ribosomal subunit assembly is a target for inhibition by aminoglycosides in Escherichia coli. *Antimicrobial Agents and Chemotherapy, 46*, 1546–1549.

Mingeot-Leclercq, M. P., Glupczynski, Y., & Tulkens, P. M. (1999). Aminoglycosides: Activity and resistance. *Antimicrobial Agents and Chemotherapy, 43*, 727–737.

Morita, Y., Tomida, J., & Kawamura, Y. (2013). Responses of Pseudomonas aeruginosa to antimicrobials. *Frontiers in Microbiology, 4*, 422.

Omari, R., Quorantsen, E. K., Omari, P., Oppey, D., & Asigbee, M. (2014). Prevailing food safety practices and barriers to the adoption of the WHO 5-keys to safer food messages in rural cocoa-producing communities in Ghana. *Internet Journal of Food Safety, 16*, 36–44.

Page, M. G. P. (2012). Beta lactam antibiotics. In *Antibiotic discovery and development* (pp. 79–117). Springer US.

Plattner, M., Gysin, M., Haldimann, K., Becker, K., & Hobbie, S. N. (2020). Epidemiologic, phenotypic, and structural characterization of aminoglycoside-resistance gene Aac(3)-IV. *International Journal of Molecular Sciences, 21*, 1–11.
PubChem. (2019). *Chloramphenicol | C11H12Cl2N2O5*. PubChem.
Raimondi, M. V., Randazzo, O., Franca, M. L., Barone, G., Vignoni, E., Rossi, D., & Collina, S. (2019). DHFR inhibitors: Reading the past for discovering novel anticancer agents. *Molecules, 24*, 1140.
Ramirez, M. S., & Tolmasky, M. E. (2010). Aminoglycoside modifying enzymes. *Drug Resist. Updat, 13*, 151–171.
Raygada, J. L., & Levine, D. P. (2009). Methicillin-resistant Staphylococcus aureus: A growing risk in the hospital and in the community. *American Health & Drug Benefits, 2*, 86–95.
Redfield, R. R. (2019). Antibiotic resistance threats in the United States. *Centers for Disease Control and Prevention, 148*, 938–941.
Scholar, E. (2007). Chloramphenicol. In *XPharm: The comprehensive pharmacology reference* (pp. 1–7). Elsevier Inc.
Schwarz, S., Kehrenberg, C., Doublet, B., & Cloeckaert, A. (2004). Molecular basis of bacterial resistance to chloramphenicol and florfenicol. *FEMS Microbiology Reviews, 28*, 519–542.
United Nations (2016). *At UN, global leaders commit to act on antimicrobial resistance*. UN News Cent.
Van Boeckel, T. P., Brower, C., Gilbert, M., Grenfell, B. T., Levin, S. A., Robinson, T. P., Teillant, A., & Laxminarayan, R. (2015). Global trends in antimicrobial use in food animals. *Proceedings of the National Academy of Sciences U. S. A., 112*, 5649–5654.
Vázquez-Laslop, N., & Mankin, A. S. (2018). How macrolide antibiotics work. *Trends in Biochemical Sciences, 43*, 668–684.
Ventola, C. L. (2015). The antibiotic resistance crisis: Causes and threats. *P T J, 40*, 277–283.
Visentin, M., Zhao, R., & Goldman, I. D. (2012). The antifolates. *Hematology/oncology Clinics of North America, 26*, 629–648.
WHO. (2018). *Future Global Governance for Antimicrobial Resistance IACG Discussion Paper July 2018* (pp. 1–86).
WHO. (2015b). *Global antimicrobial resistance surveillance system*. World Health Organization.
WHO. (2015a). *WHO library cataloguing-in-publication data global action plan on antimicrobial resistance*.
World Health Organization. (2020). *World antimicrobial awareness week*.
Zeng, D., Debabov, D., Hartsell, T. L., Cano, R. J., Adams, S., Schuyler, J. A., McMillan, R., & Pace, J. L. (2016). Approved glycopeptide antibacterial drugs: Mechanism of action and resistance. *Cold Spring Harbor Perspectives in Medicine, 6*, a026989.

Chapter 3
Drugs Resistance Against Viral Diseases

Atif Liaqat, Muhammad Farhan Jahangir Chughtai, Adnan Khaliq, Tariq Mehmood, Samreen Ahsan, Kanza Saeed, Syed Junaid Ur Rahman, Nimra Sameed, Shoaib Aziz, Rabia Iqba, Waqas Ahmad, and Saeed Ur Rahman

Abstract Viruses can cause different types of infections including respiratory, skin, foodborne and sexually transmitted ones. Worldwide trades of food and food by products and international travelling have resulted in the transmission of various viruses from developing countries to the America and European region. Polymerase chain reaction has been used for the confirmatory diagnosis of several viral infections due to its high specificity and sensitivity. Besides preventive approaches, drugs have also been developed to manage viral infections, but genomic mutations in viruses result in drug resistance. Hence, viruses replicate despite the presence of anti viral drugs. Several molecular and cellular factors including viral attachment to host cell receptors, virus entry, replication rate, genetic barrier, drug resistant viruses etc. lead to high mutation rate. Pharmacological models have also been derived to understand viral dynamics in antiviral treated patients. RNA viruses are more susceptible to mutations and develop drug resistance which can possibly deplete the effects of previously consumed safe drugs in a very short time, making the chances of drug resistance more inevitable. For this purpose, novel strategies are required for the development of antiviral drugs which may combat the challenge of drug-resistance in challenging community of global health.

A. Liaqat (✉) · M. F. J. Chughtai · A. Khaliq · T. Mehmood · S. Ahsan · K. Saeed · S. J. U. Rahman · N. Sameed · S. Aziz
Department of Food Science and Technology, Khwaja Freed University of Engineering and Technology, Rahim Yar Khan 64200, Pakistan
e-mail: atif.liaqat@kfueit.edu.pk

R. Iqba
Department of Food Science and Technology, Government College Women University, Faisalabad 38000, Pakistan

W. Ahmad
Department of Clinical Sciences, University College of Veterinary and Animal Sciences, Narowal 51600, Pakistan

S. U. Rahman
Interdisciplinary Research Center in Biomedical Materials (IRCBM), COMSATS University Islamabad, LahoreCampus, Punjab 54000, Pakistan

Department of Oral Biology, Institute of Basic Medical Sciences (IBMS), Khyber Medical University, Peshawar, Pakistan

S. Ahmed et al. (eds.), *Biochemistry of Drug Resistance*,
https://doi.org/10.1007/978-3-030-76320-6_3

3.1 Introduction

Viruses are intricate protein us particles that require a living body for their replication and survival causing infections which are commonly known as viral infection. After cellular attachments, viruses insert genetic material into the host cell to hijack molecular pathways of the host cells and deploy their own mechanism to make further viral copies. In case of active infections, viral multiplication results into death of primary host cells followed by movement of virion towards other healthy cells. Viral entries also represent propagation presenting symptoms of tissue damage, cellular destructions and immune disorders (Sattentau, 2008).

Drug resistance is mainly the reduction of susceptibility to a drug which is known as the phenotype. This phenotype is regulated mainly by specific mutations of the viral genome that alters the viral target protein or drug activator. The mutations basically cause resistance that can change the shapes and chemical characteristics of viral proteins to either stop drug binding or permit the virus to get entry into the cell even in the presence of drug (Neagu et al., 2018).

In immuno-compromised patients, trends of antiviral drug resistances are continuously escalating and supposed as an evolutionary process. Primary reason of resistance is mutation in genetic makeup that directly interferes with the drug molecule inhibition potential against viral target. For example, antivirals designed to halt antiviral block fusion between viral particle and target cell membrane, such mutations usually alter a viral protein physical and chemical characteristic which ultimately effect drug binding sites or enables the virus to pass barrier even in the presence of controlling drug (Neagu et al., 2018).

3.2 Types of Viral Infections

There are various types of viral infections which are dependent upon type of host, environment, host cell receptors, pathogenesis and clinical profile among patients. The most common infections are described below. The partial inhalation of virus affects throat, lungs and nose of host body causing pneumonia and related infections of lungs. Such viruses include Rhinovirus, seasonal influenza and respiratory syncytial viruses (Tregoning & Schwarze, 2010). The causative agents include *Molluscum contagiosum*, Herpes simplex virus-1 and Varicella-zoster virus. In these infections, appearance of rashes and blisters on host skin surface are the most common symptoms (Tan & Goh, 2006).These infections originate from infected food and food by products such as water, milk and meat of small and large food animals. Typically, these infections usually cause food poisoning, but mostly the signs and symptoms are related to the gastrointestinal tract and stomach. For example, Norovirus and Rotavirus cause gastrointestinal illness and Hepatitis A virus directly infects the liver (Jaykus et al., 2012). These are transferred from one person to others through direct contact via body fluids and blood. The genital herpes, hepatitis B, human

Papillomavirus and human immunodeficiency viruses are included in list of sexually transmitted infections. Beside these infections, different viruses are also recognized for other viral infections. For example, Epstein-barr virus cause herpes, West Nile is transmitted due to mosquito bites and present symptoms of fever and inflammation in brain and spinal cord linings (Caswell & Manavi, 2020).

3.3 History

Exotic infections like Ebola and Dengue fever were once restricted to developing countries only, however, with the increasing trade and travelling across the globe these infections, made their way to the American and European world. For human and zoonotic infections, travelers and international traders are supposed as possible carriers of human immune deficiency virus, SARS corona and West Nile virus. About 70% of the total zoonotic infections are transferred to humans because of domestic and wild animals (Cleaveland et al., 2007) and result in anthropogenic change by bridging the cross species viral jumps through cross-contamination and spill over. Evolution in viral genetics results in the emergence of novel viruses which are considered as huge threat to the health and wellbeing of human and animal population. Such genomic manipulations sometimes occured liberately in order to use genetically modified viral entity as biohazard agents (Lipkin, 2013).

3.4 Clinical Diagnostics

In some cases, virus induces disease in the primary organ and the disease spreads from its predilection site to other parts of the body. In contrast to this, same disease is being induced by different viral strains due to difference in their antigenic structures. Therefore, the incidence and prevalence of a typical disease varies depending upon the type of a viral strain. Cerebrospinal fluid is used as selected sample to conduct polymerase chain reaction to determine the sequence of causative agent for Herpes Encephalitis and to further develop therapeutic interventions (Klapper et al., 1993), whereas in West Nile encephalitis experiments, the same biological fluid is less reliable to diagnose the disease through polymerase chain reaction because IgM antibodies may not be detected (Macdonald et al., 2005) In some circumstances, viral footprints found to be undetected in primary target locations as well as on its adjacent sites and shift to other body organs as noted in enterovirus PCR based detection from fecal samples. Despite of these successful examples, these assays may be failed to detect an etiological agent as reported in viral encephalitis and pneumonia cases. These problems arise due to the following reasons; selection of inappropriate biological samples, and intermittent shedding of viruses into body fluids (Witsø et al., 2007).

3.5 Pathogenesis

The viral pathogenic cycle consists of four steps; (1) viral cell implantation on entry portal of host cell-viruses at this stage implant themselves on the host cell receptors through numerous modes like skin penetration, gastrointestinal track, or through genital or respiratory routes. The ultimate infection outcomes depend on the dose as well as viral attachment site, infectivity and virulence, (2) replication in host body, (3) propagation and dissemination at target organ, and (4) shedding of adult virions in conducive environment. The incubation period of the virus depends upon the following conditions as virus gets accessibility to cellular tissues to initiate its multiplication as well as if viral bodies demonstrate sensitivity against host defensive systems.

The viral pathogenesis may be exhibited at cellular or specified target body tissue level. Cellular destruction and death may occur due to diversion of cell energy demand. The synthesis of macromolecules is halted in host cell to insist viral mRNA for ribosomes and finally viral transcription promoters allow RNA polymerases to hinder interferon defense systems. Receptor sites for viral cell and transcription factors are responsible for recognition of viral enhancer sequence and promoters to stimulate replication potential in tissue-based pathogenesis (tropism) (Nagai & Yoshida, 1984).

3.6 Management

The disease control strategy is usually designed by manipulating biological, physical and environmental nature of virus that connects with molecular pathways involving pathogenesis of the virion. Preventive schemes such as drug therapy, chemoprophylaxis and immunization are employed to suppress and eradicate vector population. At environmental level, proper sewage disposal and creation of negative pressure ventilation setup are practices for the management and control of viral diseases. In behavioral interventions, quarantine measures of infected individuals and use of personal protection equipments can also play pivotal role in effective prevention (Fraise et al., 2013).

3.6.1 Therapy and Chemoprophylaxis

Chemoprophylaxis of viral infections are riveted on viral replication cycle and host metabolic pathway. Novel therapies include in-silico computer programs to generate exquisite models of precise protein–protein interactions to study their effect on host cell. These advance techniques exhibit their own limitations like poor correlation between in-silico and in-vitro models, contingent acute toxicity and strain difference etc. (Razonable, 2011).

Numerous drugs are being tested (in-vitro and in-vivo studies) and marketed with an improved antiviral efficacy. Major drug classes contain methanocarbathymidine, cyclopropavir, cidofovir, favipiravir and pleconaril for the treatment of Herpes, Cytomegalo, Adeno, Influenza and Hantaviruses respectively. These antiviral drugs target kinase pathways and initiate apoptosis by binding viral RNA (De Clercq, 2013).

3.6.2 *Immunization*

Immunization is a process of developing resistance against infectious diseases by developing strong immune defense mechanism (Plotkin, 2014). The active and passive immunization are of two types for production of antibodies on vaccine administration. Active vaccines are employed in order to provide the protection for longer period and with minimum associated side effects by enhancing production of antibodies. These may include dead virus or live weakened virus, but the latter one has more strong and instantaneous protective effect as compared to the killed vaccines specifically when administered through the natural portal in order to boost immunity. Contrary to this rule, sometimes the live virus preparations are completely successful like nasal spray vaccine against influenza virus and herpes zoster vaccine; however, some of the non-replicating types like HPV and HBV showed unpredictably increased efficiency against viral diseases. Both live and killed active viruses are highly efficient against hepatitis A; even though live virus provides long-lasting immunity (Irving et al., 2012).

There are number of disadvantages associated with the administration of live vaccines, as few viral strains are weak. Therefore, there is possibility of their reversion to virulence or reactivation as noted against herpes viruses in laboratory trials. In comparison to active vaccine, passive immunization contains high level of antibodies to produce the resistance against the causative agent and is being consumed soon after the exposure to provide short-term prevention. It is obtained from the person just recovered from the disease and the one having high antibody titers. This immunization is limited for well-defined exposures to rabies virus or in immuno-compromised patients.

3.7 Viral Drug Therapy (Principles and Their Kinetics)

Antiviral drugs are small organic molecules which are used to inhibit the growth and proliferation of viruses within host. These are scarce as compared to antibiotics which are being used to kill bacteria and fungi. As viruses rely on host for their growth and propagation, a small molecule of antiviral drug can cause toxic effects in host cells due to structural or mechanistic similarity of inhibition, resultantly viral drug development becomes cumbersome due to imparted toxicity. Another reason of antiviral drugs being obsolete is the development of mutation in viruses and drug resistance. The drugs can target and attack on the following parts of viruses.

3.7.1 Viral Proteins

The significant role of viral proteins is evident in the attachment, pathogenesis, replication and other biological processes of viruses. Functional domains in viral proteins and their structures can be mapped and imaged can be acquired that can help us out in designing a drug of choice. For example, in HIV many efforts have been done in developing drugs to exploit virus proteins. Most of HIV recommended drugs target the viral all or few enzymes such as protease, reverse transcriptase or integrase (Richman & Nathanson, 2016).

These drugs exhibit potential to block the activity of reverse transcriptase. For inactivation of enzyme, the reverse transcriptase inhibitors are nucleoside and non-nucleoside in nature. Nucleosides are incorporated in newly synthesized DNA and tend to inhibit its elongation while, non-nucleosides can bind directly to the enzyme and inhibit its function and bioavailability. The second drug target is proteases that can bind catalytic site of this enzyme and degrade the viral proteins and generate mature proteins. A fusion inhibitor of HIV-1, enfuvirtide binds at gp41 and inhibits the formation of pores required for viral entry to target cells (Hartman & Buckheit, 2012). It has been reported that enfuvirtidemolecule (T20) bind to heptad repeat and prevents the formation of hairpin to finally block the entry of HIV.

3.7.2 RNA Interference

In this phase, RNA molecule tends to inhibit the expression of gene by neutralization of targeted mRNA molecule. This phenomenon is used widely in clinical trials of HIV to silence the genes.

3.7.3 Viral Mutagens

Host immune responses towards antivirals and vice versa are factors that determine the viral survival. Mutations in viruses select the fittest mutants among all genetic variants and these variants pose adaptation to drug treatments and host immune system. *i.e.* an antiviral drug, ribavirin has been known to act through the mechanism in which mutagens are used to increase the non-sense mutations.

3.7.4 *Cell Targets*

In cell-based targets, the therapeutics inhibits the viral infection by preventing the attachment of virus to CCR5 which is also known as chemokine receptor and hinders the fusion of virus and host cellular membranes (Hartman & Buckheit, 2012). Viruses that cause persistent illness are considered as appealing targets for antiviral drugs and such viruses are also diagnosed properly before starting a therapy. Hepatitis B and C viruses are such examples that require long time therapeutic interventions for their complete clearance or suppression. In immunologically deficient mice, cellular immunity and antibodies in case of West Nile virus play important role in infectious outcomes whereas passive antibodies exhibit potential to increase the survival chances in mice.

3.7.5 *Interferons*

Type 1 interferon (IFN-α and–β) has significant inherent immune response. IFN induces pleotropic response to inhibit the viral genome replication through development of antigen based immune response.

3.7.6 *Maturation Inhibitors*

These inhibitors interfere with the last stage of HIV replication as the viral proteins are packed and freed from host cell membrane to make fresh viral particles. Bevirimat is a betulinic acid like compound used to inhibit the HVI maturation (Nguyen et al., 2012).

3.8 Drug Resistance as Adaptive Process in Viral Infections (Possible Factors)

The possible factors in affected populations are prolonged drug exposure, viral replication and induced resistance. Diagnosis of specific drug resistance depends on viral mutation linked to a specific drug and determined on the basis of phenotypic assays. Health disorders and their consequences due to viral drugs evolution are continuously being investigated within the framework of genetic population, and ultimately lead to a continuous improvement in the field of monitoring. In the field of experimental evolution complex factors, like selection, population size and nucleotide diversity can be monitored, and be precisely measured among patients at frequent intervals and can be used as base of treatment decision (Renzette et al., 201). Fine scale development enabled the researchers to observe new mutations leading to antiviral drug resistance development. Drug resistance management therapies include drug delivery optimization along with alternative therapies designed on the basis of resistance mechanism and development of new antiviral drugs. As a consequence of antiviral drug resistance, patients often develop inherent toxicity which leads to severe disease development followed by death with no effective alternative therapies (Wiesch et al., 2010).

3.9 Evolution and Mechanism of Drug Resistance in Viral Infected Patients

Viruses have formed various resistance mechanisms that make them able to elude the effect of antivirals. As an outcome of this, numerous viruses have developed resistance to almost all current treatment of their effects. This issue although is not new, but it is becoming acute and if we have to gain new insights into ways to combat this problem then there is ultimate need to understand the resistance mechanisms of viruses (McKeegan et al., 2002).

Robust destructive therapies against viral fitness, halts viral genome replications. Insufficient treatment designs lead to partial destruction of genome replication mechanism, generating a selective pressure condition. Genome replicates under such stressed conditions lead to a rapid resistance adaptation. Due to this reason, drug development systems require innovation and improvement either through developing new drugs or new combinations of existing drugs (Irwin et al., 2016).

3.9.1 Mutation in Virus Physiology and Biology

Viral replication process is not the same for all viruses and it drives the productions of new drugs to target different virus life cycle at their different stages and parts. It is necessary to know these critical differences to develop novel and effective drugs that will also counter for evolution towards resistance. The extent and type of substitution (which is a precise type of mutation that causes alteration in encoded amino acids and single base within a coding sequence) needed to confer resistance is different on the basis of target and class of antiviral drug which is frequently enumerated in the form of genetic hurdle for resistance. Many old antiviral drugs have been known to have less genetic obstructions and most recurrent viruses usually need only one or sometimes two substitutions to attain resistance. More chances of drug resistance may occur during the pathway that needs two or more substitution and the difference in these pathways can also be variable on the basis of virus genotype. The resistance does not arise so readily and the main objective for a single or cocktail of drug synthesis is to enhance this genetic barrier (Irwin et al., 2016).

Variation in viral replication mechanisms leads to the urge of new drugs development, capable of targeting various stages of viral life. Viruses entities usually exist as quasi species; a population structure that involves very large number of variant genomes, termed mutant spectra, mutant swarms or mutant clouds. These genomes are powered by high mutation rates, mutants arise continually, and they change in relative frequency as viral replication proceeds. Understanding these changes is crucial for predicting and coping resistance evolution (Sanjuán et al., 2004).

This resistance occurs at a cost of viral genome health, this deficit is further enhanced in the process of compensatory mutations development. It happens, either to restore normal protein structures and functions or to re-attain potentially lost original state. Among these conditions, situation selection depends upon population size, rate of mutation of the locus, drug environment or genomic stability of individuals' carriers of compensatory mutation (Tanaka & Valckenborgh, 2011).

In an uncertain fluctuating environment, these could be neutral, while in extreme cases when target cell is absent, it could be deleterious and mostly results in the removal of primary resistance mutation. This primary resistance is antagonistically epistatic with compensatory mutations and significantly observed by population (Karkhah et al., 2018; Khudyakov, 2010).

There is need for promising efforts to prevent the most futile nature of drug resistance as well as to target mutational processes instead of replicative processes. When mutation rate of population elevates, specifically in case of novel mutations, it tends to generate influx with numerous effects. Muller's Ratchet, is one of these processes that results in losing most of appropriate class through genetic drifts, resulting in operations that reduce fitness over time. It is mostly observed in non-recombining viruses (Muller, 1964). The second effect is Hill-Robertson interference which includes the processes related to linkage, fixation probabilities of mutations, and with chances to avoid deleterious mutations (Hill and Robeterson, 1966).

Both effects usually synchronize over time, thus reducing fitness and leads to population extinction or mutational meltdown. In case of RNA viruses like IAV and HCV, in which replication process usually takes place either at error threshold condition or in environment where rate of error is far beyond meltdown (Holmes, 2003). Novel drug designs are usually made by taking benefit of this, and elevating rate of base mutation so that population exceed these threshold levels (Pauly & Lauring, 2015).

3.9.2 Adherence and Non-adherence of Antiviral Agent

In clinical terms, occurrence of interruption and its frequency are regarded as adherence whereas complete treatment interruption of dosage as well as timing is known as non-adherence which ultimately leads to treatment failures and formation of new resistant mutants (Fox et al., 2008).

Antiviral drug regimen disturbance is a great risk to human health, as such interruptions repopulate virus in host in case of incomplete eradication (Chotiyaputta et al., 2012). Initial treatments against viruses act as a bottleneck and hinder their survival in drug resistant individuals. In such individuals, if drug dosage patterns are disturbed, it may result in dramatic increase in viral population and developing new population is far more resistant against drugs in comparison to the previous generations rendering antiviral therapy more cumbersome (Wang et al., 2011). Diversity in intrahost concentrations assists in the judgment of variation magnitude and to know about nature of fluctuating environments (Pennings et al., 2014).

The fluctuating selection coefficients were previously estimated using novel polymorphism-based data collection techniques, from phenotypic data collected at the expense of both labor and time (Miura et al., 2013). The traditional tests are in continuous improvements to accommodate estimations (Gossmann et al., 2014). So, these novels techniques can track fluctuation in drug adherence from being beneficial to neutral, and from neutral to deleterious resistant mutant generations.

3.9.3 Life Cycle Synchronization

This drug resistance mainly occurs due to genetic mutation in viral genome which enables the efficient viral replication even in the presence of drugs designed to control the viral population and resultantly mutant strains ultimately outcompetes old genome. Antiviral therapies with initial good prognosis against viral load, eventually becomes less effective in individual or entire affected populations due to generation of new drug resistant strains. As the viral life cycle is mutable characteristic, it alters in accordance with successive dose duration and interval during treatment. The treatment period synchronizes with viral population and drug levels. In this way, virus adapt sustaining patterns even in absence of mechanism for counteracting the drug effects by hiding in high drug concentration periods and highly recommended for practice.

The concept of cryptic resistance, an unobserved effect that prevents viral suppression by antiviral drugs, through a mechanism known as drug binding resistance or life cycle completion of viral cell in the presence of drugs. Continuous fluctuating drug levels in the patient's body provided the foundation of idea that presence of drug is also a time varying factor. Antivirals are usually administered in a defined size, discrete dosage and regular time intervals. Immediately after administration, the drug levels are high causing maximum possible damage to relevant viral life cycle stage. However, their levels are optimized to support viral suppression amid dose intervals. This drug level fluctuations are assumed as periodic and virus adapt the growing conditions followed by complete target life cycle phase in the period of low drug concentrations in order to avert drug effects.

3.10 Effect of Drug Resistance on Viral Replication and Pathogenesis

The viral infections significantly affect human mortality and morbidity. The scientists have developed vaccines to avoid viral morbidity rate more than a century ago, but it has been observed in previous decades that drugs have developed to directly target viral replication. These extremely potent drugs are accessible to treat human immunodeficiency virus (HIV), influenza A and B viruses, hepatitis B and C viruses, varicella zoster virus (chickenpox virus), cytomegalovirus, Epstein Barr virus and herpes simplex viruses (Neagu et al., 2018). Drugs block the explicit phase of the virus life cycle and binds to protein coat of viruses or else interfere with a precarious step in viral replication that enables to decrease the growth rate of virus. These antiviral drugs bind to cell membrane of target virus and halt the transcription of viral genome. Resultantly, the virus integrates in the genome of host cell or viral proteins may undergo post translational cleavages. Antiviral actions may also be rendered ineffective, in few or large number of people, by the development of drug resistant strains (Campos et al., 2016; Piret & Boivin, 2016).

To define the significance of drug resistance, the efficiency of genetic barrier plays an important role other than host and virus factors. The genetic barrier represents the type and extent of mutations that should be accrued for the virus to develop significant drug resistance while keeping drivable growth. Phenotypic methods estimate susceptibility of drug by making serial dilutions of drug and cultural growth of viral inoculums and concentration of drug noted require inhibiting viral growth by 50 (IC50) or 90% (IC90) from the level noticed devoid of drug (Strasfeld & Chou, 2010).

3.10.1 *Molecular Mechanisms of Mutation Rates*

The high mutation rate can be defined as introduction of error-prone polymerases base substitution due to deficient of 3' to 5' exonuclease action and it will increase the chances of resistance. High mutation rate depends upon several molecular factors *i.e.* (1) replication rate of virus, (2) viral load, (3) viral fitness, (4) genetic barrier and (5) ability of viruses to progress substitutions under selective pressure of antiviral drugs. One harmony among all viruses is their replication that involves RNA, and RNA replication needs an RNA-dependent RNA or DNA polymerase encoded by the virus. These polymerases deficient a 3'–5' endonuclease activity which played a role in addition to remove mistakenly incorporated nucleosides. The classic mutation level for retroviruses and RNA viruses such as HIV is 1×10^{-3} or HCV1×10^{-4} errors/base/replication cycle. Its mean that in 10,000 nitrogenous bases of a genome of virus only one error is introduced during each replication cycle; equivalent rates for dsDNA viruses that would typically be in between 1×10^{-6} and 1×10^{-8} errors/base/replication cycle. The other factor for high mutation is the rapid viral replication rate that can be stated as high viral load due to faster rate of replication which will increase number of mutant viruses within a swarm and it will increase chances of resistance. The high and rapid level of viral replication is usually augmented by a high turnover of the virus, with very short viral life. This tendency with error prone polymerase, allows the formation of viral swarm soon after infection. These viral swarms usually composed of lineages and usually deviate from the initiator virus. The extent of deviation usually depends on replication cycles and a swarm with extensive diversity that can be established as if higher replication rounds occur then it will be prior to treatment and it usually happened in HCV or HIV infections. If the antivirals are absent then as consequences of rapid viral replication rate, the higher viral load in RNA-dependent viruses is well-known at equilibrium. Viral fitness can be described as capability of virus to replicate within its giving environment. Resistance substitution often impairs fitness. With a reduction of fitness, the chances of resistance also decrease. The substitutions of amino acids will manifest the reduced fit of virus because of changes occurring at functional critical residue. These substitution changes may do not compete with remaining population endure a slight variant that can be diagnosed within the viral swarm, or it may be misplaced (Mason et al., 2018). It is very simplified method to quantify number of genetic changes *e.g.*in transition phase nitrogenous bases change as purine to purine base turnover occur more speedily as 75 times in comparison to transversions where purine change in pyrimidine or pyrimidine changed as purine base (Powdrill et al., 2011). Hence, substitutions of amino acid that need transversion require more genetic barrier as compared to those that need transition (Mason et al., 2018).

Every so often, a virus can modify resistance towards antiviral drugs by attaining mutations and when this stage occur it becomes difficult for a drug to bind with virus and inhibit its growth. These above stated four mutation causing factors (error-prone polymerases, rapid and high-level viral replication, viral fitness and genetic barrier) can cause antiviral resistance and they should be customized for every drug.

3.11 Pharmacological Model for Viral Dynamics in Antiviral Treated Patients

An antiviral drug exhibits significant effect on infection and its tendency to reduce infectivity is dependent on amount of drug used $[D\ (t)]$ for specific period. Its reduction level is calculated from Hill dose–response curve, where IC_{50} indicates the concentration inhibiting 50% of population, M quantifies the steepness of the decay curve and β_0 is the infectivity in the absence of treatment. If the drug is taken consistently, drug levels $D(t)$ peak immediately at $C_{\max}$ following each dose and then decay exponentially with half-life *th*, before the next dose. The amount of drug which is used can be estimated from following Eq. (3.1).

$$D(t) = C_{max} 2 - t/th \qquad (3.1)$$

$$\beta(t) = \frac{\beta_0}{1 + \left[\frac{D(t)}{IC_{50}}\right]M} \tag{3.2}$$

If a drug is taken after interval of T days and reduces infection for a time. This time is recorded as fT and assumed as *on window* and $\beta\ (t) = 0$ for that time. For the remaining time between doses, the drug is assumed as to be completely inactive, such that $\beta\ (t) = \beta 0$ for that time (Eq. 3.3). The efficacy of the drug is denoted by f and perfect drug therapy would have efficacy equal to 1. It is noted that drug efficacy can be varied in model by changing $C_{\max}$, *th*, *IC*50, or *M* (Fig. 3.1c):

$$\beta(t) = \beta_0(1 - D(t)) \tag{3.3}$$

The viral cell completes its life cycle stages to produce new virions. These stages comprise; removal of coating from viral cell, viral genome transcription and translation, production of multiple copies of viral genomics and assembling of virus protein matter. The sequence of these steps may vary viral cells but delay of time between viral production and infection of a cell is supposed as same. This intracellular time delay is a prerequisite to assess drug tolerance in infected cells (Herz et al., 1996). For incorporation of delays into a viral dynamic model, commonly two models are

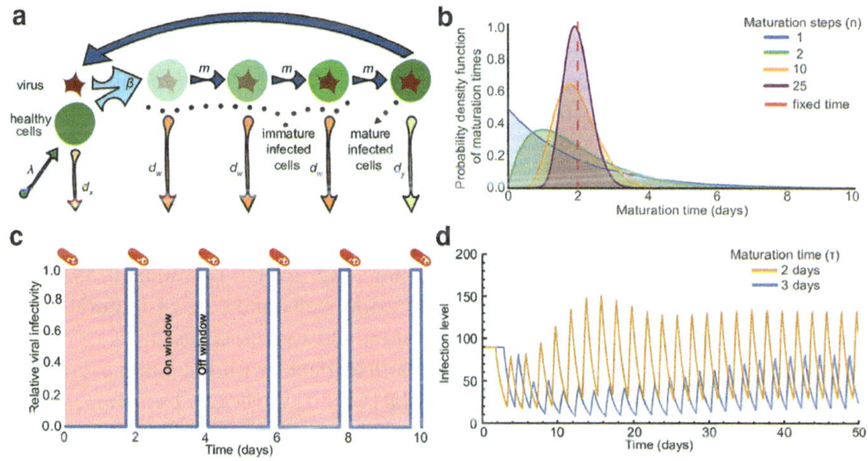

Fig. 3.1 Graphical display of viral dynamic model in a patient (available in: Neagu et al., 2018). **a** Diagram for incorporation the infected cell maturation process. **b** Probability distribution functions of maturation time. **c** The infectivity of virus $\beta(t)$ in relation with infectivity of virus in the absence of the drug $\beta 0$ alters in response to drug levels (blue line). In the simple on–off model (a step function), drug levels are ªonº for a fraction f of the time between doses (red shading), reducing viral fitness to zero, and ªoffº for the rest of the interval (viral fitness returns to baseline). **d** Time course of infection levels when the maturation time is fixed to $\tau = 2$ days (yellow line) and $\tau = 3$ days (blue line), and the drug dosage is modeled as a periodic step function. The synchronized strain (maturation time of 2 days, yellow line) reaches higher time-averaged infection level than the unsynchronized strain. In these examples, drug period $T = 2$ days and drug efficacy $f = 0.85$

employed. For the first method (Fig. 3.1a), a series of n immature phases wi, are introduced as described earlier (Mittler et al., 1998). The population of cells is comprised of healthy cells x, one or more stages of immature infected cells w, and mature (virus-producing) infected cells y. The maturation time is the time it takes for an infected cell to pass through all the maturation phases. The equation for model is derived as below.

3.11.1 Derivation of Models

3.11.1.1 Model 1

$$
\begin{aligned}
x'(t) &= \lambda - \beta(t)x(t)y(t) - d_x x(t) \\
w_1'(t) &= \beta(t)x(t)y(t) - (m_1 + d_w)w_1(t) \\
w_i'(t) &= m_{i1}w_{i1}(t) - (m_i + d_w)w_i(t) \quad \forall 2 \le i \le n \\
y'(t) &= m_n w_n(t) - d_y y(t)
\end{aligned}
\qquad \text{(Model 1)}
$$

These equations depict that infection is mainly occurring in mature and healthy cells and its rate is dependent on drug concentration and time during which drug treatment is given. The time of maturation τ is the time required to become fully mature in state y from a newly infected cell in the first immature state w_1. The Gamma (Erlang) distribution is reported for probability of maturation times and noted that each step of maturation at same rate and can be written as $mn = mi\,for$ all phases of immaturity (Floyd et al., 2010).

$$
p(\tau) = m_n^n \tau^n 1_{(l)} m_n^{\tau} / (n-1)! \qquad (3.4)
$$

Mathematically, it is assumed that maturation rate, mn, is a linear function of n (number of maturation steps) and mn = nm. The inverse relationship is found between square root of the number of maturation steps and rate of maturation, if standard deviation of maturation times is taken and expressed as $\sigma_{st} = 1/(\sqrt[m]{n})$ (Fig. 3.1b). The maturation times are gamma-distributed with the same average of 2 days, for different numbers of maturation steps ($n = 1, 2, 10, 25$).Therefore, strains with greater maturation rates have more deviation in their time to mature than strains with lower expected times in a number of intermediate phases. In case of greater number of phases, the maturation time is equal to 1/m and is fixed and these steps allow viral cell to grow in better way to achieve its suitable size.

3.11.1.2 Model 2

The second method for modeling intracellular delay uses delay differential equations with a fixed delay of $\tau = 1/m$, between infection and start of virion production (Dixit & Perelson, 2004; Nelson & Perelson, 2002). According to this method the equations for model are given as;

$$x'(t) = \lambda - \beta(t)x(t)y(t) - +x(t)$$
$$y'(t) = \beta(t-\tau)x(t-\tau)y(t-\tau)e^{d}w^{\tau} - d_y y(t) \quad \text{(Model 2)}$$

From these two models, the formula for basic reproductive ratio, R_0 can be derived by putting $\beta(t) = \beta$ as constant in both models (Eq. 3.5). For the multi-stage model, R_0 is calculated as;

$$R_0 = \frac{\lambda\beta}{d_x d_y}(nm|nm + d_w)n \quad (3.5)$$

Similarly, R_0 is given for the fixed-delay model.

$$R_0 = \frac{\lambda\beta}{d_x d_y} e^{d_x}/m \quad (3.6)$$

It is important to keep in mind for the limit of large n, the model can be described by two equivalent equations (Model 1 and 2) and R_0 are given by Eqs. (3.5 and 3.6). The variables used in models and R_0 are explained in Table 3.1.

In case of $dw > 0$ means, if there is possibility for an immature cell to die prior to produce viral body then reproductive ratio value is increased and ultimately infected cells mature as possible as they can and vice versa. In an ideal and basic model, reproductive ratio value (R_0), is kept sufficient to determine the outcome of the system in qualitative manners. It is reported that infection persists till an equilibrium stage for system, if $R_0 > 1$, whereas if $R_0 < 1$, then infection ultimately declines and go extinction in the system (Mittler et al., 1998). In addition, R_0 value can be derived for competition between multi-strain exhibiting different average maturation times as well as stochastic versions of each single- and multi-strain model can be created

Table 3.1 Variables used for viral dynamic model

Variables	Description
λ	Production rate of target cells
x	Healthy target cells population size
d_x	Per capita die rate
y	Population of mature infected cells
w_i	Population of immature infected cells
B(t)	Rate of infection to healthy cells
Subscript i	No of immature phases if these exist more than one
m	Maturation rate of immature cells
d_y	Death rate of mature cells
R_0	Basic reproductive ratio
d_w	Death rate of immature cells

(Nelson et al., 2000).The description of variables used in the pharmacological viral dynamics and drug treatment models is presented.

Conclusively, a common assumption for simplification of virus dynamic model and reduction in computation burden in stimulation is used. Generally, free viral cell dynamics is considered as faster than infected cells. It is observed that viral cells tend to reproduce in large numbers and quickly reach to quasi-steady state as compared to the level of infected cells. Consequently, with this assumption, the abundance of infected cells is proportional to level of free viral cell population. In this case, infectivity parameter $\beta(t)$ is equal to $b(t)k/c$, where k is the rate constant to produce free virus by mature infected cells, $b(t)$ is the infectivity of a free virion, and c is the rate constant for clearance of free virus. It is also reported that infected immature cells are unable to produce new viral cells, therefore they do not directly experience cytotoxic effect of viral cell release and production and this ultimately triggers the cytolytic responses of immunity system. So, death rate of healthy cells is comparable to immature infected cells (Neagu et al., 2018).

3.12 Antiviral Drug Resistance Testing (In-Vitro Evaluation) in Viral Infected Patients

There are two significant uses of drug resistance testing: to assess drugs and to oversee patients. There are different measures for medicate opposition, and their utility requires normalization, quality affirmation and approval for each of these implementations. The achievement of antiviral therapy is to lower the virus replication; if levels of virus are not fully repressed, drug resistance develops quickly (Borroto-Esoda et al., 2006).

3.12.1 Principal of Drug Resistance Testing

The ubiquity of resistance drug can be described as "the quantity of resistance drug presented in a delaminating population in a given time period". After some time, there has been an expanded assortment and accessibility of antiretroviral drugs. Be that as it may, suboptimal treatment responses can evolve to antiretroviral drug resistance development (Santoro et al., 2008).

3.12.1.1 HIV (Human Immunodeficiency Virus)

First report of human immunodeficiency syndrome was examined in 1981. According to reports it was measured that, HIV was most pervasive among patients involved in sexually transmitted diseases, clinics, intravenous users of drugs and hemophiliacs

(Reeves & Doms, 2002). Immediately after the recognition of AIDS in human beings, outbursts of decay and dreadful infections were examined in captive colonies of rhesus Asian macaques in USA anthropoid centers and this feature was called as simian AID (Acquired Immune Deficiency Syndrome) (Kumarasamy et al., 2003). It was evaluated that worldwide approximately 90% of patients infected with HIV were residents of developing countries. There are hardly few reports regarding survival history people having AIDS from poor resource settings and developing countries (Newmann et al., 2000).

There are 2 types of HIV, contamination with HIV type 1 (HIV-1) normally happens across mucosal surfaces or by direct immunization. HIV-1 relate closely to simian immunodeficiency virus (SIV) accumulated from chimpanzees (SIV-CPZ) (De Cock et al., 1993). The infection first experiences dendritic cells (DCs), which therefore encourage spread of HIV-1 to CD4+ T lymphocytes. DC-SIGN, a HIV-septic DC receptor, ties HIV-1 at its 120 space -gp without needing direct disease of the cell and transports HIV-1 to lymphoid tissue. Indications commonly happen 2–6 weeks after exposure to HIV-1 and frequently include fever, exhaustion, pharyngitis, weight reduction, night sweats, lymphadenopathy, myalgias, migraine, queasiness, and looseness of the bowels (Kassutto & Rosenberg, 2004).

HIV type 2 (HIV-2) relate closely by SIV from sooty or collard mangabeys (SIV-SM). Subtypes of HIV-2 are fundamentally confined to West Africa and can be arranged as scourge subtypes (A & B) and non-epidemic subtypes (C–G) (Reeves & Doms, 2002).Contrasts in viral burden might be dependable: detailed consequences of infection culture and polymerase chain response examination propose that at high CD4 + T-lymphocyte, viral burden is lower in HIV-2 as compared to HIV-1 infected people (De Cock et al., 1993).

In-Vitro Testing Method in HIV

Resistance mutations occur less likely with increasing trough plasma concentration of protease inhibitors. Secondly, decreased HIV-1 RNA level makes drug failure less likely to occur (Molla et al., 1996). There are several factors contributing towards antiretroviral drugs failure that may develop resistance in patients. These factors lack adherence to regular medications, decreased drug potency, variable pharmacokinetics in individuals or host-immune response and reduced drug activation (Potter et al., 2004). RNA viral replications may result in drug resistance and failure (Fig. 3.2).

There are eight reverse transcriptase inhibitor (NRTIs) and four co-plans of NRTIs affirmed for treatment in HIV-1 infected patients. The decision of NRTI blend to start treatment is driven by intensity, dosing recurrence/pill trouble, poisonous, obstruction profile, pharmacokinetic collaborations and individual patient concerns. Tenofovir disoproxil fumarate (TDF) and emtricitabine (FTC) are NRTIs with once-day by day dosing affirmed for use in HIV-1 contamination. A fixed-portion blend of TDF and FTC is likewise accessible. Results from clinical examinations with the blend of TDF and FTC as a component of a completely once-day by day routine with either

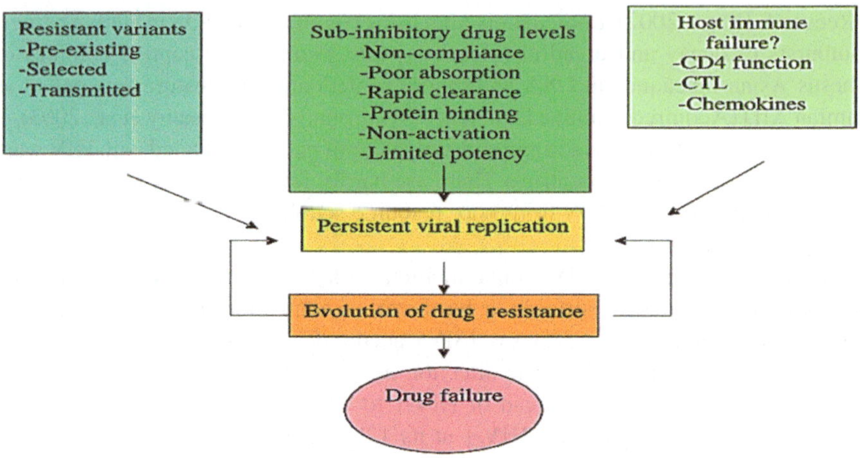

Fig. 3.2 Factors involved in HIV drug resistance in viral infected patients. Available in: Potter et al. (2004)

a PI or a NNRTI have exhibited noteworthy virological viability and insignificant improvement of obstruction (Borroto-Esoda et al., 2006).

In a study conducted in Malawi, NRTI failure has been studied. In this study, patient's demographic details were collected along antiretroviral therapy history and CD4 cell count. Before initiation of second line treatment, patients' blood samples for viral load, plasma storage and CD4 cell count were obtained at initial assessment for treatment failure. On samples more than one NRTI mutations, phenotyping was performed while genotyping was done on all samples. Results of study indicated that NRTI with elevated activity in therapy is zidovudine, lamivudine, tenofovir following lamivudine, tenofovir and finally abacavir, didanosine (Hosseinipour et al., 2009). NNRTI therapy suggested that about 90% patients experienced suppression of HIV RNA by 1 year after antiretroviral therapy (Hosseinipour et al., 2013). HIV strains can be divided into subtypes, groups (M, N, O) and subgroups(A–D, F–H, J & K) and recombinant forms (CRFs, CRF01–CRF16) (Robertson et al., 2000).

Efficacy of antiviral drugs in-vivo and in-vitro has been studied in subtype B HIV-1, which predominates in Africa and North America, representing the limited proportion of people so, work is required on non-B types. In another study, HIV non-B types were selected having mutation in protease gene with atypical substitutions at positions destined with antiretroviral drug resistance in reverse transcriptase and protease. *In-vitro* susceptibility to these antiretroviral drugs and influence of these mutations was determined. Phenotypic study showed the resistance against antiretroviral drugs: ritonavir, saquinavir, nelfinavir, zidovudine, didanosine, stavudine; delavirdine, nevirapine and efavirenz. Some mutations were associated with in vitro decreased susceptibility; 1 in 3 strains with M46I/L mutations in protease, 1/2 A98S, K101N, V108I, V179I, and P236L as in reverse transcriptase. In protease, presence of minor mutations were not linked to decreased susceptibility of protease inhibitors (Vergne et al., 2006).

Several studies have shown that polymorphism in reverse transcriptase and protease have no influence on highly active antiretroviral therapy (HAART) in HIV-1 infected patients (Bocket et al., 2005). Moreover, study is required to investigate the effect on non-B subtypes.

Concurrent cytotoxity examines were performed utilizing the medication grids at concentration levels for TFV, FTC and 3TC were 50, 5 and 10 μM, respectively. Resultantly twofold of serial dilutions were then executed over the plate. Uninfected MT-2 cells were plated at a last convergence of 3×10^4 cells per well on 96-well plates containing different concentrations of compound test followed by incubation in 5% CO_2 for 5 days. Viability of cells was then examined using XTT based assay (Yeni et al., 2004).

Results from the antiviral tests were broken down utilizing the *isobologram strategy* and the *Mac-Synergy program* to decide expected synergistic as well as opposing exercises. The *isobologram assessment* depends on the Lowe additivity model. Portion reaction bends are created for each medication alone and in blend and used to decide EC50 values for each medication alone or within the sight of the fixed grouping of the subsequent medication. Regular points were plotted on graph to create the isobologram. The power of the communication can be estimated and the factual contrast from portion insightful additivity tried by a one-followed t-test.

The *Mac-Synergy program* computes hypothetical added substance esteem for each medication mix dependent on the qualities created by the medications alone utilizing a free impacts model. The hypothetical added substance esteems are deducted from the trial esteems produced by each medication mix to give an estimation of collaboration (positive worth) or hostility (negative worth). These collaborations and additionally threat esteems are plotted on a three-dimensional chart with their comparing drug blends. These collaboration and threat volumes are then factually assessed utilizing the 95% certainty level and are communicated in μM2% (Gathe et al., 2004).

It is concluded that blend of TFV and FTC demonstrated as synergistic enemy of HIV movement *in-vitro* and this action corresponded with the degrees of intracellular phosphorylation noted for each compound. These outcomes bolster the utilization of these medications as a double NRTI spine in mix treatment for the treatment of HIV-1 sickness (Borroto-Esoda et al., 2006).

3.12.1.2 HSV (Herpes Simplex Virus)

This disease is supposed as ordinary infection and self-limited and ends in 7 days. It is characterized by the appearance of blisters on mucosal membrane. This is generally just an annoyance, despite the fact that they can be substantially more serious in the host that is immune compromised. HSV intermittent infection is ascribed to the foundation of inactivity inside neurons of the comparing sensory peripheral ganglia. For whatever length of time that viral DNA stays flawless inside cores of in contaminated tactile neurons, the infection can use cell proteins to start viral quality translation, which can come full circle in the union of viral proteins and get together and results in irresistible viral particles release (Millhouso & Wigdahl, 2000). Herpes virus type-2 (HSV-2) is the main causative agent for most genital herpes and is pervasive mostly through sexual transmission worldwide (Xu et al., 2006).

In-Vitro Testing Method in HSV

Some immunological and clinical factors are responsible for developing resistance in herpes simplex virus against antiviral therapy. The factors that determine the development of resistance in HSV are; lack of HSV specific cell mediated immunity that results in delayed clearing of HSV from mucocutanious lesions and prolonged usage of antiviral therapy. The third factor can be minimal adherence to therapy by some HIV patients due to large burden of medications. Acyclovir is used as first effective drug against HSV. Valacyclovir, an aminoacyl prodrug of acyclovir that is converted to acyclovir by host cells in liver and intestinal walls (Levin et al., 2004).

Acyclovir triphosphate, active metabolite of acyclovir and valacyclovir prevents further replication of HSV (Fig. 3.3). It works by inhibiting DNA polymerase enzyme of virus and terminating viral DNA strand (Elion et al., 1999). Penciclovir an active ingredient of famciclovir have the same action mechanism as valacyclovir (Hodge & Perkins, 1989). HSV-1 and HSV-2 in-vitro resistance to acyclovir results due to mutation in viral genes encoding DNA polymerase or Thymidine kinase (TK) (Coen & Schaffer, 1980).

As recommended treatment is acyclovir (ACV) which is a nucleoside sample of guanosine that must be phosphorylated multiple times. The first phosphorylation is accomplished by the viral encoded thymidine kinase (TK) protein which permits ACV to get dynamic during infection contaminated cells. Number second and third phosphorylation is accomplished by cell thymidylate kinases. ACV triphosphate acts by serious restraint of viral DNA polymerase and it is a DNA chain eliminator. This across the board utilization of ACV has led to the rise of HSV strains resistant to ACV (Morfin & Thouvenot, 2003).

The peptides can be based on polyamide-dynamic ester science utilizing a Milligen 9050 pep synthesizer. For experiment, the chemicals as 4-(Hydroxyl-methyl) Pheno-acetic corrosive connected polyamide/kieselguhr gum (pepsin KA), Fmoc-amino corrosive pentauoro phenyl (Pfp), and 3-hydroxy-2,3-dehydro-4-oxo-benzotriazin esters were used. Cleavage of peptidyl resin and side chain can be done at ambient temperature in 2 h by using 5 mg of peptidyl-tar in 1 mL of blend containing tri-choloro acetic acid, ethyl methyl sulfide, water and thioanisol. The crude peptides undergo purification after extraction of ether and resins through HPLC, Sephadex gel filtration or ion exchange chromatographic techniques. The purified peptides were kept under frozen storage as a stock solution as concentration of 1 mg per mL in double distillated water at temperature of −20 °C whereas infection stocks were set up on monolayers of Hep-2 cells in support medium. It is suggested that tainted cells should be gathered 48–72 h prior to disease testing. After collection of infected

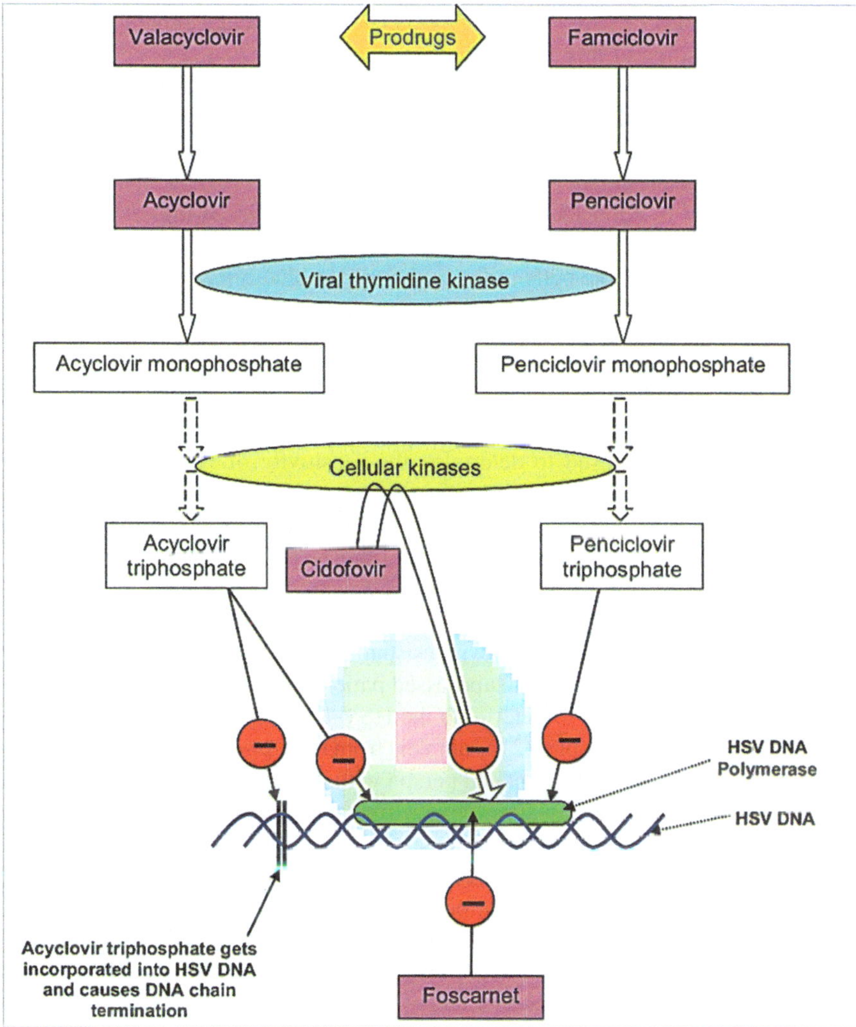

Fig. 3.3 Mechanism of drug against HSV. Available in: Nath and Thappa (2009)

samples, these are solidified followed by defrost multiple times in order to centrifuge at 1000 rpm (10 min). The infection containing supernatants are stored at temperature of −80 °C until utilized. After that, the virus cell goes under titration by the assay of cytopathic effect by utilizing the method of Karber (Belaid et al., 2002).

The direct immuno fluorescence assay is also suggested for detection of HSV. In this method, trypsin separated Hep-2 cell-monolayers are washed 3 times with PBS and spread in the well in the form of slides. After sliding, it undergoes to drying and subjected to cold storage at temperature of −20 °C. Furthermore, it is recolored with a fluorescent conjugated monoclonal counter acting agent in order to separate

HSV-1 polypeptides of 150,000 and 85,000 Da. This molecular separation range is assumed as limitation of this assay (Caviness et al., 2010).

For the estimation of the impact of every peptide on cell suitability, cytotoxicity assay can be suggested. For experiment performance, 104 Hep-2 cells in 0.2 mL culture medium are seeded into each well of 96-well plates. The cells were permitted to develop confluence at that point, and the prepared cultures are incubated at 37 °C for 24, 48 and 72 h. After incubation, medium was transferred into detached cells using 0.25% trypsin solution (Chentoufi et al., 2008).

The quantity of feasible cells utilizing the trypan blue-rejection strategy can be calculated, and level of cytotoxicity can be determined by the formula;

$$\%\ \text{cytotoxicity} = \frac{100 \times \text{ no. of viable cells incubated with peptide}}{\text{No. of viable cells incubate without peptide}}$$

Phenotypic assay is used to determine the sensitivity of HSV towards antivirals including viral DNA-inhibition assay, dye-uptake assay and assessing plaque reduction (Levin et al., 2004). Among all of these methods most commonly used is plaque reduction assay for determining viral drug resistance due to an established correlation between in-vitro susceptibility of acyclovir determined by this assay and a clinical response obtained from 243 HSV samples from 115 HSV infected patients (Safrin et al., 1994). Acyclovir resistant isolates have been collected from immuno-mpromised and immuno-suppressed patients and they vary greatly.

A study was carried out to determine the HSV-1 and HSV-2 resistance towards antiviral drugs. Clinical samples (211 samples) from HSV positive patients (Immuno-compromised, immuno-competent and person with unknown immune system) were collected. The samples were assayed for acyclovir and foscarnet drugs susceptibility. In first step genotypic testing was done consisting of UL30 (Pol) gene sequencing and Thymidine kinase (gene UL23). In second step phenyotypic plaque reduction assay was performed on detection of mutations. Results showed antiviral drug resistance was more prevalent in HSV-2 isolates comparatively HSV-1. Gene UL23 frame shift reading, TK mutations resulted in ACV drug resistance (Burrel et al., 2013).

Another study conducted in Netherland by Stránská et al. (2005) showed the ACV resistance in immunosuppressed patients and least resistance in immuno-competent patients. New mutations were determined in Thyamine Kinase genes in these isolates. Results suggested the need to access the antiviral drug susceptibility in immunosuppressed patients. It is estimated that reduced susceptibility of HSV towards ACV is high in immuno-compromised patients about 14% and may go higher to 36% in hematopoietic stem cell transplant patients (Piret & Boivin, 2011).

3.12.1.3 VZV (Varicella-Zoster Virus)

The varicella virus (VZV) is recognized as human herpes virus 3 (HHV-3).It is composed of double stranded DNA genome and omnipresent alpha virus in nature. Naturally, VZV only cause infection in human beings and have no reservoir with animals. It mainly targets epithelial cells, T-lymphocytes and ganglia. When it is in dormant stage in ganglionic neurons, it causes chicken pox which is a primary infection of this virus. It also causes serious ocular and neurological diseases with persistent pain known as post herpetic neuralgia (PHN). Vaccines and anti-viral drugs are available for both zoster and varicella-based infections (Gershon et al., 2015).

In-Vitro Testing Method in VZV

For a long time, with the exception of in immuno-compromised patients, infections of VZV were treated with medications and acetaminophen to overcome itching. Today, antiviral drugs are available with significant results and are being used frequently. Antiviral therapy for VZV became assessable in the mid-1980s in the form of the analog acyclovir nucleoside (ACV), DNA polymerase inhibitor and DNA chain terminator. ACV is injected intravenously if high drug levels are needed or it may be given as orally to patients (Gershon & Gershon, 2013).

Varicella vaccination was presented in 1995 in United States, in the infant immune schedule. Presently, most other nations are focusing on mass Varicella vaccination. Mass inoculation has two threats: it could build the quantity of Varicella cases in grown-ups, where seriousness is more noteworthy, and increment instances of zoster. Three methods were used to inoculate vaccine: (1) Routine vaccinating at the age of 1-year (infant strategy); (2) Vaccination on 2- 11-year-old kids in the program of first year (catch up method); last (3) Routine vaccinating at the age of 12 years (Brisson et al., 2003; Edmunds & Brisson, 2002). This vaccination appears to show excellent protection around 95–100% opposed to severe disease (Edmunds & Brisson, 2002).

In clinical measurements, it is noted that acyclovir (ACV) are linked to mutations in thymidine kinase enzyme. Some resistant virus for ACV causes nucleotides deletions that lead to thymidine kinase codon. Genotypic and phenotypic assay for detection of ACV Resistant VZV with mutations in the thymidine kinase are frequently used to measure resistance of VZV drugs (Piret & Boivin, 2016).

The first discovered drug against varicella-zoster virus was acyclovir (ACV). Long time usage to ACV in immunocompromised patients leads to development of drug resistance since the ACV resistance in immunocompetent patients with herpetic keratitis. The drugs used for treatment of varicella zoster virus directly act on viral DNA polymerase. Acyclovir and penciclovir as well as prodrugs valaciclovir and famciclovir are used as first line drugs in the treatment of VZV. Resistance of VZV towards ACV is not described in immunocompetent individuals except in one patient with ACV resistant Varicella zoster virus keratitis. In clinical samples of VZV isolates resistance towards ACV is conferred due to mutations in enzyme thymidine kinase (ORF36 gene) and DNA polymerase Pol (ORF28 gene).

Genotypic testing is most widely used method in assessing drug resistant VZV samples (Piret & Boivin, 2016). Viral drug resistance in VZV can be determined by plaque reduction assay by utilizing fibroblastic cells. Clinical efficiency of method is low due to low isolation rate and slow growth. The enzyme linked immunosorbent assay, real time PCR, DNA hybridization and flow cytometry can be used to improve the antigen expression and objectivity of readout (Thi et al., 2006). Other phenotypic assay includes cell viability determination with chem-iluminescent substrate, transgenic HSV inducible reporter cell line expressing beta-galactosidase. At enzymatic level VZV resistance to ACV drug is related to different mechanisms including; reduced activity of viral enzyme thymidine kinase, low production of thymidine kinase and a protein viral thymidine kinase altered.

The enzyme can phosphorylate thymidine but does not phosphorylate ACV. Genotypic by DNA sequencing is an altered method used to phenotypic assay. The whole thymidine kinase gene as well as conserved region of VZV is sequenced. The development of efficient method for the detection of viral mutant is by next generation sequencing method. In order to characterize the unknown mutations molecular biology-based assay can be used to generate VZV recombinant viruses such as transfection of overlapping cosmids and plasmids allow only rapid site directed mutagenesis in gene of interest.

The phenotypic strategy includes PCR implication of the VZV and TK followed by the examination of phenotype of its encoded protein in microscopic organisms which are in any case insufficient in endogenous TK action (Sahli et al., 2000).Some recombinant technologies have been implemented for the manipulation of herpes virus genome. For the generation of single substitute nucleotide in phosphotransferase genes, *en passant* mutagenesis approach is used. Every mutant of VZV was then differentiated with its fluorescent red counterpart. This application is treasure not only for unclear resistance mutations but also for novel drug candidate's assessment against herpes virus infections (Brunnemann et al., 2015).

To monitor susceptibility of VZV resistance to antiviral drugs genotypic assay includes extraction of viral DNA from viral stocks and clinic samples with an instrument known as Magna pure compact, stored at temperature of 20 °C for evaluation. DNA polymerase (ORF28) and TK (ORF36) gene complete length amplifications (1st and 2nd round annealing primers for PCR stored at temperature of 52 and 59 °C for TK and DNA polymerase, and ordering were carried out utilizing primers, with the help of expand high-fidelity proof-read enzyme (Perrier et al., 2016).

In genotypic studies, plaque reduction assay (PRA) can also be performed in melanoma cell line (Mewo) culture for the quantification of ACV EC50 and FOS. Mewo cells are fatten to convergence in 24 well plates of culture and then inoculated with isolates of VZV. After incubating for 1 h, media contain mixers of 10% formaldehyde and antiviral are added. Lastly, this mixture stained with 1% concentration of crystal violet and 70% ethanol. Total number of plaques viral was computed and EC50s were determined by Karber technique. It is reported that EC50 of 30 μM for ACV and 160 μM for FOS are suggested for VZV infected patients (Frobert et al., 2007).

3.12.1.4 CMV (Cytomegalovirus)

Cytomegalovirus (HCMV), also known as human herpesvirus (HHV-5), is a ubiquitous virus that belongs to the virid herpes family, a subfamily of the beta herpesvirus subfamily (Landolfo et al., 2003). These are considered as prototypes of beta-herpesviruses and share the structure of the virion, genome of double-stranded DNA, dynamics of expression of viral genes, survival of the host after primary infection and the reactivation of latency with other herpesviruses (Jarvis & Nelson, 2002). The HCMV genome is the largest of all herpesviruses, with almost 230,000 base pairs, a single origin of replication, the DNA polymerase gene (UL54) and the complete set of genes required for replicating 'viral DNA (Britt & Boppana, 2004). The laboratory strain AD169 was the first HCMV strain to be fully followed with a shorter genome than other clinical isolates due to the deletion of the 15 kbp segment (Cunningham et al., 2010). Viral genes are numbered according to their location in the genome. For example, UL54 is the 54th gene in a single long span of the HCMV genome. HCMV is a common virus and its epidemiology varies in different parts of the world depending on social, economic and age groups. HCMV infection is most prevalent among developing countries and lower socio-economic individuals. These are almost asymptomatic. Indeed, the virus is maintained with a clinically undetectable delay or a low rate of spread. However, HCMV is an important concern for certain high-risk special groups, and its delayed primary infection or reactivation is very important, especially in immuno-compromised patients, including HIV-infected and transplant patients (Cannon et al., 2010).

In-Vitro Testing Method in HCMV

Presently, three approved antiviral drugs are used to prevent and to treat HCV infections and disorders: (h) ganciclovir, foscarnet (FOS) and cidofovir (CDV). Despite low bioavailability (except valganciclovir (VGCV)), clinical resistance and associated toxicity, these agents have been used to treat various forms of HCMV in immuno-compromised patients (Cvetkovic & Wellington, 2005).

The incidence of resistance to HCMC varies widely depending on the type of patient, and many other risk factors have been identified like; (1) Factors related to the disease such as patient and SOT, underlying disease, type, degree of immuno-suppression of the host and incidence of HCMV disease (Baldanti et al., 2004). Long-term antiviral therapy, antiviral concentration, or poor absorption and limited bioavailability are factors affecting treatment (Schreiber et al., 2009).In fact, putative drug resistance is suspected if a persistent or increased viral load or overt disease of HCMV develops after at least 2 weeks of treatment (Kleiboeker et al., 2014). The antiviral resistance of human cytomegalovirus (HCMV) is due to the UL97 gene (coding). Besides this, viral phosphotransferase p UL97) and UL54 genes also exhibit resistance. So, mutations in HCMV are particularly important in order to targets pUL97 and pUL54 for current therapeutic strategies against recommended GCV, FOS and CDV (Razonable & Emery, 2004).

(i) **CMV QNAT**

This assay detects and quantifies CMV nucleic acid Results reported in IU/mL, various assays including Roche (COBAS 6800 and 8800), Qiagen (artus), Abbott, and many laboratory-developed tests (home-brew assays) are available. It is used for rapid and sensitive method for diagnosis of CMV infection, surveillance for preemptive therapy, monitoring of antiviral response, prognosticator of risk of CMV disease but there are certain limitations of this test *i.e.* it lacks consensus of viral load threshold that is applicable across patients and clinical settings. Patients with different risk profiles will likely have different viral thresholds (Dioverti et al., 2017).

(ii) **Antigenemia**

This assay uses monoclonal antibody to detect CMV pp65 antigen expressed in leukocytes during the early period of CMV replication, reported as number of pp65-positive cells per number of leukocytes counted. Its clinical use includes sensitive diagnosis of CMV infection, surveillance for preemptive therapy, monitoring of antiviral response, prognosticator of risk of CMV disease. Limitations include lack of assay standardization, need for samples to have enough leukocytes (limited in HSCT and neutropenia) and intensively lacking automation subjective interpretation (Ross et al., 2011).

(iii) **CMV-specific T cells**

This assay detects INF-γ in blood or cells after stimulation ex vivo with CMV antigens. Various assays include QuantiFERON-CMV, T-spot.CMV, T-track.CMV. There is still lack of prospective studies and interventional trials that utilize these assays for disease management (McGoldrick et al., 2013).

3.12.1.5 IAV (Influenza A Virus)

This virus is one of the most persistent and unpredictable human pathogens. Globally, influenza causes unpredictable pandemics, consistent seasonal epidemics, and regular fatal zoonotic diseases. The virus spreads by aerosol and results in acute febrile respiratory disease in humans, generally known as "influenza." This is especially severe for young children, the elderly and immuno-compromised individuals. Influenza viruses are a burden on the world's population and economies (Viboud et al., 2016). The World Health Organization estimates that influenza viruses cause 1 billion cases of HIV each year. Influenza kills 500,000 people with 3–5 million serious illnesses around the globe.

The alternating annual influenza vaccination programs for the northern and southern hemispheres are the primary means of preventing or controlling seasonal influenza pandemics. Two groups of antiviral drugs: (1) M2 ion channel blockers (amantadine and rimantadine) neuraminidase inhibitors (NA) (NAI; oseltamivir, peramivir, zanamivir,andlaninamivir) have been approved for the treatment of influenza. However, as with antibiotic resistance, the development of antiviral resistance in the influenza virus is a matter of serious concern. Therefore, INI is currently the only type of influenza drug currently in use, as most circulating influenza viruses have gained resistance to M2 channel blockers (Webster & Govorkova, 2014).

In-Vitro Testing Method in IAV

The determination of adamantine as M2 ion channel inhibitor in clinical experiments is supposed as indicator for resistance in IAV patients. Adamantane, amantadine (symmethyl), and rimantadine (flumazine) are the first approved classes of anti-AVI drugs. The membrane protein and the unique trans-membrane domain of M2 are integrated type III tetramers that form four helical radii that act as a pH-sensitive trigger ion channel within the IAV envelope (Ma et al., 2009). IAV enters the host cell by receptor-mediated endocytosis and is found by HA-mediated fusion of the viral envelope and endosomal membrane. Prior to fusion, the low pH of the endosome is activated. This event leads to the dissociation of the M1 protein from the vRNP nucleus and its subsequent release into the host cell. The cytoplasm initiating adamantane replication of IAV binds to the pore of the M2 channel and directly or allosterically blocks proton conductivity, thereby inhibiting vRNP release and IA replication (Schnell & Chou, 2008). Propeller tetramer (L26, S31, and L38 amino acids), tend to increase the pore size of the helix-helix assembly and hydrophilicity as well as destabilization of the channel (Thomaston & DeGrado, 2016). IAV resistance to adamantane was first time reported during the 1980 epidemic. The U.S. Centers for Disease Control and Prevention provides advice on the use of adamantane in the treatment of IAV infections (WHO, 2005).

3.12.1.6 HBV (Hepatitis B Virus)

HBV, a DNA virus which replicates through RNA intermediates. The viral polymerase enzyme, that also has reverse transcriptase (RT) activity, has a high error rate and lacks the corrective function necessary to repair improperly integrated bases (Lai et al., 2003). The high replication rate of HBV produces 10^{12} virions per day. This is combined with 1010-5 substitutions per base and mutation rate in the replication cycle (Fung et al., 2005).

In individuals with ongoing viral replication, almost 10^{10-11} point mutations occur daily in the whole viral genome. Some mutations result in severe loss of function, compromising the ability of the virus to stop the mutation from replicating. Other variants show varying degrees of fitness. Therefore, in infected hosts, HBV is present in different forms in different strains that are subtypes of the virus. Some variants are always predominant under the influence of regulators such as immune-and drug-mediated selection pressures, while others exist only as rare low-frequency mutations. HBV mutations that result in reduced drug sensitivity occur naturally at the target of antiviral therapy and circulate in viral subspecies prior to the introduction of treatment. Given the overall functional cost, in the absence of drug pressure, drug-resistant mutants of HBV circulate at a low frequency, usually in the form of a single mutant, allowing escape detection in a routine and sensitive manner. When drug replication continues under viral selective pressure, the variant gains selective advantage and gradually becomes the dominant species. Due to continued replication of the virus under drug pressure, individual mutants genetically develop and acquire additional primary and secondary mutations, including compensatory changes that restore viral fitness and enhance pathogenicity (Tacke et al., 2004).

Over the past 20 years, significant progress has been made in the prevention and treatment of HBV infection through antiviral resistance management strategies (Locarnini et al., 2015). Currently, several oral inhibitors of interferon alpha-2b and pegylated interferon alpha (pegIFN-alpha) as well as the HBV polymerase enzyme-based techniques have been approved for the treatment (Terrault et al., 2016).

Polymerase inhibitors containing both nucleosides and nucleotide analogs (NAs) are generally more potent and have fewer side effects than IFN-based therapies. Current treatment options for de novo treatment of chronic hepatitis B (CHB) have a low risk of drug resistance selection, but long-term, indefinite treatment is usually required to maintain viral pressure. The need for better treatment is increasing because many patients accumulate to drug resistance and cross-resistance to HBV through previous fewer effective treatments and there are limited treatment options. Even more people are at risk of drug resistance because of the limited access to effective antiviral compounds in low- and middle-income backgrounds. Several new compounds are under trial phase to pave the way for the development of a finite therapeutic system and potentially eradicate HBV (Tong & Revill, 2016).

HBV tends to lose its replication ability if treatment interruption caused by the appropriate drug-sensitive ("wild-type") strain and inhibit this virus to proliferate. In patients who discontinue treatment for more than few weeks, resistant variants cannot usually be detected by conventional techniques. However, the loss of resistance is clear. Therapeutic variants remain active and tend to develop rapidly when suboptimal treatment is continued. The HBV also exhibits potential to create a genetic mutation in covalently closed circular DNA (cccDNA) to remain persistant in infected liver cells (Herbers et al., 2013).

HBV cccDNA is prepared from the gentle circular DNA genome of the incoming virion and persists episomally in the nucleus of liver cells for a long time (Chang et al., 2010). Although NA effectively inhibits HBV replication, NA has limited efficacy in reducing cccDNA reservoirs. The aim of the present study is to target the reservoir and cure HBV infection. On the other hand, HBV cccDNA retains the long-term memory of selected resistant strains from which new replicating strains may reappear (Buti et al., 2015).

3.13 Strategies for Improvement in Antiviral Agents Efficiency

Antiviral agent's affectivity can be adopted as important measure for improving drug efficiency among patients. Particularly, people living with AIDS and HIV patients must be recruited against viral infections. According to era of modern medicines, naturalistic use of lithium treatment prevents reoccurrence of viral disorders and enhances antiviral agents' affectivity. The antiviral potential of lithium favors to reduce and stop recurrence of labial-herpetic cells. In addition to this treatment, perpetual monitoring of blood serum level is mandatory.

For preventing long term toxicity and other side effects, the utilization of antiviral agents must be minimized. Antiviral drugs come in action when symptoms of illness are noted. After becoming susceptible to viral infection, the symptoms start to diminish and ultimately shorten the time of infection from further prolongation.

Designing of effective and safe antiviral drug is a rather difficult task because virus uses host cell material for replication. This replication process make that process difficult rather being in confront with virus. Due to structural and functional variations in viral taxonomy it is difficult to develop antiviral drugs. The development of antiviral drugs is a time-consuming process due to absence of knowledge of mode of action of targeted virus.

In existing era, it is not an efficient output for discovering effective antiviral drugs with little bit side effects. When genetic sequence of viral DNA began to reveal then researchers came to know about the reproductive and multiplication cycle of viruses. The antiviral drugs are designed to target viral proteins and parts of proteins by reducing the likelihood of side effects. But researchers are rather confused about the different species of proteins and these species-oriented variants results in different structure of proteins. So, that's why many antiviral drugs results in failure of antiviral drugs in broad spectrum. Hence, these targeted proteins can be resynthesized in sophisticated laboratories by bacterial cells in order to avoid from development of resistant strain (Paintsil & Cheng, 2009).

3.14 Future of Antiviral Drug Resistance

In last four decades, development of several antiviral drugs has been seen with therapeutic effects for the treatment of debilitating and life-threatening disorders. Some of these diseases caused by HIV, hepatitis B virus, herpes viruses and influenza virus. Virus cultivation in laboratories, viral enzymes identification and their molecular study are the major technical breakthroughs supported recent advances in viral drug development. Emerging viruses and biohazard agents are potential threat for humanity in current era. Modern antiviral drug therapies are restricted to a number of viral diseases. In the development of antiviral therapies, there is a major challenge of drug resistant viral mutants and this challenge is compromising the treatment efficiency of viral infectious diseases. When conventional specified antiviral drugs are used for the treatment of their particular viral disease, initially brings good results effectively but the continuous use of these drugs enhances the selective pressure for resistant mutation in that virus this process will support the emergence of drug resistant viruses.

RNA viruses are more prone towards mutation and results in emergence of drug resistance, which can turn previously effective drugs useless in a very short time. All this process brings economic losses for researchers or donors. New antiviral drug development is multi-million dollars investment with a long-time requirement. If we see in the case of Covid-19 which caused highly destructive global pandemic, the world is unable to produce the anti Covid-19 drug or vaccine even after more than half million deaths. What if! The drug introduces in market and the virus develops resistance. The whole investment will be a loss along with a big number of human lives.

Innovative strategies are needed in the development of antiviral drugs that can combat the challenge of drug-resistance and could be effective in broad-spectrum. For the function, spread, reproduction and infections viruses require functions of host. So, there should be development of antiviral drugs which have capacity to target host-encoded pathways or functions, along with this limit the drug resistance mutation potential. In future of antiviral drugs efforts must accelerate the development of broad-spectrum antiviral drugs by the discovery of host targets and this job should be on top priotized tasks of scientists and medical experts (Cho & Glenn, 2020).

3.15 Conclusion

Viral infections include different types such as respiratory, skin, foodborne and sexually transmitted ones. Viruses and ultimately infections may spread through travelling. For diagnostic purpose of infections, polymerase chain reaction has been widely used. Prevention of contamination and use of drugs are some ways for the management of viral infections. However, genomic mutation develops drug resistance among them, resulting in ineffectiveness of drugs. Virus load, replication rate, genetic barrier, ability of viruses to grow in the presence of antiviral drugs etc. lead to high mutation rate. There are pharmacological models derived for exhibition of viral dynamics in antiviral treated patients. Research on viral mutation continues to be carried out. For example, HIV non-B types having mutation in protease gene with atypical substitutions at positions destined for antiretroviral drug resistance have been studied. Efficacy of antiviral drugs, both in-vivo and in-vitro has also been carried out in subtype B HIV-1, predominating in Africa and North America are presenting the limited proportion of people so work is required on non-B types. A phenotypic study on HIV non-B types has shown resistance against antiretroviral drugs:

ritonavir, saquinavir, nelfinavir, zidovudine, didanosine, stavudine; delavirdine, nevirapine and efavirenz. Novel strategies are obligatory for the development of antiviral drugs which may combat the challenge of drug-resistance in a broad-spectrum way.

References

Baldanti, F., Lurain, N., & Gerna, G. (2004). Clinical and biologic aspects of human cytomegalovirus resistance to antiviral drugs. *Human Immunology, 65*(5), 403–409.

Belaid, A., Aouni, M., Khelifa, R., Trabelsi, A., Jemmali, M., & Hani, K. (2002). In vitro antiviral activity of dermaseptins against herpes simplex virus type 1. *Journal of Medical Virology, 66*(2), 229–234.

Bocket, L., Cheret, A., Deuffic-Burban, S., Choisy, P., Gerard, Y., de la Tribonnière, X., Viget, N., Ajana, F., Goffard, A., Barin, F., & Mouton, Y. (2005). Impact of human immunodeficiency virus type 1 subtype on first-line antiretroviral therapy effectiveness. *Antiviral Therapy, 10*(2), 247–254.

Borroto-Esoda, K., Vela, J. E., Myrick, F., Ray, A. S., & Miller, M. D. (2006). In vitro evaluation of the anti-HIV activity and metabolic interactions of tenofovir and emtricitabine. *Antiviral Therapy, 11*(3), 377.

Brisson, M., Edmunds, W. J., & Gay, N. J. (2003). Varicella vaccination: Impact of vaccine efficacy on the epidemiology of VZV. *Journal of Medical Virology, 70*(S1), S31–S37.

Britt, W. J., & Boppana, S. (2004). Human cytomegalovirus virion proteins. *Human Immunology, 65*(5), 395–402.

Brunnemann, A. K., Bohn-Wippert, K., Zell, R., Henke, A., Walther, M., Braum, O., Maschkowitz, G., Fickenscher, H., Sauerbrei, A., & Krumbholz, A. (2015). Drug resistance of clinical varicella-zoster virus strains confirmed by recombinant thymidine kinase expression and by targeted resistance mutagenesis of a cloned wild-type isolate. *Antimicrobial Agents and Chemotherapy, 59*(5), 2726–2734.

Burrel, S., Aime, C., Hermet, L., Ait-Arkoub, Z., Agut, H., & Boutolleau, D. (2013). Surveillance of herpes simplex virus resistance to antivirals: A 4-year survey. *Antiviral Research, 100*(2), 365–372.

Buti, M., Tsai, N., Petersen, J., Flisiak, R., Gurel, S., Krastev, Z., Schall, R. A., Flaherty, J. F., Martins, E. B., Charuworn, P., & Kitrinos, K. M. (2015). Seven-year efficacy and safety of treatment with tenofovir disoproxil fumarate for chronic hepatitis B virus infection. *Digestive Diseases and Sciences, 60*(5), 1457–1464.

Campos, A. B., Ribeiro, J., Boutolleau, D., & Sousa, H. (2016). Human cytomegalovirus antiviral drug resistance in hematopoietic stem cell transplantation: Current state of the art. *Reviews in Medical Virology, 26*(3), 161–182.

Cannon, M. J., Schmid, D. S., & Hyde, T. B. (2010). Review of cytomegalovirus seroprevalence and demographic characteristics associated with infection. *Reviews in Medical Virology, 20*(4), 202–213.

Caswell, R. J., & Manavi, K. (2020). Emerging sexually transmitted viral infections: Review of human T-lymphotropic virus-1 disease. *International Journal of STD & AIDS, 31*(2), 92–99.

Caviness, A. C., Oelze, L. L., Saz, U. E., Greer, J. M., & Demmler-Harrison, G. J. (2010). Direct immunofluorescence assay compared to cell culture for the diagnosis of mucocutaneous herpes simplex virus infections in children. *Journal of Clinical Virology, 49*(1), 58–60.

Chang, T. T., Lai, C. L., Kew Yoon, S., Lee, S. S., Coelho, H. S. M., Carrilho, F. J., Poordad, F., Halota, W., Horsmans, Y., Tsai, N., & Zhang, H. (2010). Entecavir treatment for up to 5 years in patients with hepatitis B e antigen–positive chronic hepatitis B. *Hepatology, 51*(2), 422–430.

Chentoufi, A. A., Binder, N. R., Berka, N., Durand, G., Nguyen, A., Bettahi, I., Maillère, B., & BenMohamed, L. (2008). Asymptomatic human CD4+ cytotoxic T-cell epitopes identified from herpes simplex virus glycoprotein B. *Journal of Virology, 82*(23), 11792–11802.

Chotiyaputta, W., Hongthanakorn, C., Oberhelman, K., Fontana, R. J., Licari, T., & Lok, A. S. F. (2012). Adherence to nucleos (t) ide analogues for chronic hepatitis B in clinical practice and correlation with virological breakthroughs. *Journal of Viral Hepatitis, 19*(3), 205–212.

Cho, N. J., & Glenn, J. S. (2020). Materials science approaches in the development of broad-spectrum antiviral therapies. *Nature Materials, 19*, 813–816.

Cleaveland, S., Haydon, D. T., & Taylor, L. (2007) Overviews of pathogen emergence: Which pathogens emerge, when and why?. In *Wildlife and emerging zoonotic diseases: The biology, circumstances and consequences of cross-species transmission* (pp. 85–111). Springer.

Coen, D. M., & Schaffer, P. A. (1980). Two distinct loci confer resistance to acycloguanosine in herpes simplex virus type 1. *Proceedings of the National Academy of Sciences, 77*(4), 2265–2269.

Cunningham, C., Gatherer, D., Hilfrich, B., Baluchova, K., Dargan, D. J., Thomson, M., Griffiths, P. D., Wilkinson, G. W., Schulz, T. F., & Davison, A. J. (2010). Sequences of complete human cytomegalovirus genomes from infected cell cultures and clinical specimens. *The Journal of General Virology, 91*(Pt 3), 605.

Cvetkovic, R. S., & Wellington, K. (2005). Valganciclovir. *Drugs, 65*(6), 859–878.

De Clercq, E. (2013). A cutting-edge view on the current state of antiviral drug development. *Medicinal Research Reviews, 33*(6), 1249–1277.

De Cock, K. M., Adjorlolo, G., Ekpini, E., Sibailly, T., Kouadio, J., Maran, M., Brattegaard, K., Vetter, K. M., Doorly, R., & Gayle, H. D. (1993). Epidemiology and transmission of HIV-2: Why there is no HIV-2 pandemic. *JAMA, 270*(17), 2083–2086.

Dixit, N. M., & Perelson, A. S. (2004). Complex patterns of viral load decay under antiretroviral therapy: Influence of pharmacokinetics and intracellular delay. *Journal of Theoretical Biology, 226*(1), 95–109.

Dioverti, M. V., Lahr, B. D., Germer, J. J., Joseph, D. Y., Gartner, M. L., & Razonable, R. R. (2017). Comparison of Standardized Cytomegalovirus (CMV) viral load thresholds in whole blood and plasma of solid organ and hematopoietic stem cell transplant recipients with CMV infection and disease. In *Open forum infectious diseases* (Vol. 4, no. 3, pp. 143). 2017 Summer.

Edmunds, W. J., & Brisson, M. (2002). The effect of vaccination on the epidemiology of varicella zoster virus. *Journal of Infection, 44*(4), 211–219.

Elion, G. B., Furman, P. A., Fyfe, J. A., de Miranda, P., Beauchamp, L., & Schaeffer, H. J. (1999). The selectivity of action of an antiherpetic agent, 9-(2-hydroxyethoxymethyl) guanine. *Reviews in Medical Virology, 9*(3), 147.

Floyd, D. L., Harrison, S. C., & Van Oijen, A. M. (2010). Analysis of kinetic intermediates in single-particle dwell-time distributions. *Biophysical Journal, 99*(2), 360–366.

Fox, Z., Phillips, A., Cohen, C., Neuhaus, J., Baxter, J., Emery, S., Hirschel, B., Hullsiek, K.H., Stephan, C., Lundgren, J. and SMART Study Group. (2008). Viral re-suppression and detection of drug resistance following interruption of a suppressive NNRTI-based regimen. *AIDS (london, England), 22*(17), 2279.

Fraise, A. P., Maillard, J. Y., & Sattar, S. A. (2013). *Hugo and Ayliffe's principles and practice of disinfection, preservation and sterilization* (5th ed.). . Wiley-Blackwell.

Frobert, E., Ooka, T., Cortay, J. C., Lina, B., Thouvenot, D., & Morfin, F. (2007). Resistance of herpes simplex virus type 1 to acyclovir: Thymidine kinase gene mutagenesis study. *Antiviral Research, 73*(2), 147–150.

Fung, S. K., Andreone, P., Han, S. H., Reddy, K. R., Regev, A., Keeffe, E. B., Hussain, M., Cursaro, C., Richtmyer, P., Marrero, J. A., & Lok, A. S. (2005). Adefovir-resistant hepatitis B can be associated with viral rebound and hepatic decompensation. *Journal of Hepatology, 43*(6), 937–943.

Gathe, J., Podzamczer, D., Johnson, M., Schwartz, R., Yeh, V., Travers, N., Luff, K., King, M., Tressler, R., & Brun, S. (2004). Once-daily vs. twice-daily lopinavir/ritonavir in antiretroviral-naïve patients: 48-week results. *BID, 75*(61), 55.

Gershon, A. A., & Gershon, M. D. (2013). Pathogenesis and current approaches to control of varicella-zoster virus infections. *Clinical Microbiology Reviews, 26*(4), 728–743.

Gershon, A. A., Breuer, J., Cohen, J. I., Cohrs, R. J., Gershon, M. D., Gilden, D., Grose, C., Hambleton, S., Kennedy, P. G., Oxman, M. N., & Seward, J. F. (2015). Varicella zoster virus infection. *Nature Reviews Disease Primers, 1*(1), 1–18.

Gossmann, T. I., Waxman, D., & Eyre-Walker, A. (2014). Fluctuating selection models and McDonald-Kreitman type analyses. *PLoS ONE, 9*(1), e84540.

Hartman, T. L., & Buckheit, R. W. (2012). The continuing evolution of HIV-1 therapy: Identification and development of novel antiretroviral agents targeting viral and cellular targets. Molecular Biology International.

Herbers, U., Amini-Bavil-Olyaee, S., Mueller, A., Luedde, T., Trautwein, C., & Tacke, F. (2013). Hepatitis B e antigen-suppressing mutations enhance the replication efficiency of adefovir-resistant hepatitis B virus strains. *Journal of Viral Hepatitis, 20*(2), 141–148.

Herz, A. V., Bonhoeffer, S., Anderson, R. M., May, R. M., & Nowak, M. A. (1996). Viral dynamics in vivo: Limitations on estimates of intracellular delay and virus decay. *Proceedings of the National Academy of Sciences, 93*(14), 7247–7251.

Hill, W. G., & Robertson, A. (1966). The effect of linkage on limits to artificial selection. *Genetics Research, 8*(3), 269–294.

Hodge, R. A., & Perkins, R. M. (1989). Mode of action of 9-(4-hydroxy-3-hydroxymethylbut-1-yl) guanine (BRL 39123) against herpes simplex virus in MRC-5 cells. *Antimicrobial Agents and Chemotherapy, 33*(2), 223–229.

Holmes, E. C. (2003). Error thresholds and the constraints to RNA virus evolution. *Trends in Microbiology, 11*(12), 543–546.

Hosseinipour, M. C., Gupta, R. K., Van Zyl, G., Eron, J. J., & Nachega, J. B. (2013). Emergence of HIV drug resistance during first-and second-line antiretroviral therapy in resource-limited settings. *The Journal of Infectious Diseases, 207*(suppl_2), S49–S56.

Hosseinipour, M. C., van Oosterhout, J. J., Weigel, R., Phiri, S., Kamwendo, D., Parkin, N., Fiscus, S. A., Nelson, J. A., Eron, J. J., & Kumwenda, J. (2009). The public health approach to identify antiretroviral therapy failure: High-level nucleoside reverse transcriptase inhibitor resistance among Malawians failing first-line antiretroviral therapy. *AIDS (london, England), 23*(9), 1127.

Irving, S. A., Patel, D. C., Kieke, B. A., Donahue, J. G., Vandermause, M. F., Shay, D. K., & Belongia, E. A. (2012). Comparison of clinical features and outcomes of medically attended influenza A and influenza B in a defined population over four seasons: 2004–2005 through 2007–2008. *Influenza and Other Respiratory Viruses, 6*(1), 37–43.

Irwin, K. K., Renzette, N., Kowalik, T. F., & Jensen, J. D. (2016). Antiviral drug resistance as an adaptive process. *Virus Evolution*, 2(1).

Jarvis, M. A., & Nelson, J. A. (2002). Human cytomegalovirus persistence and latency in endothelial cells and macrophages. *Current Opinion in Microbiology, 5*(4), 403–407.

Jaykus, L. A., D'Souza, D. H., & Moe, C. L. (2012). Foodborne viral pathogens. *Food microbiology: Fundamentals and frontiers* (pp. 619–649).

Karkhah, A., Javanian, M., & Ebrahimpour, S. (2018). The role of regulatory T cells in immunopathogenesis and immunotherapy of viral infections. *Infection, Genetics and Evolution, 59*, 32–37.

Kassutto, S., & Rosenberg, E. S. (2004). Primary HIV type 1 infection. *Clinical Infectious Diseases, 38*(10), 1447–1453.

Khudyakov, Y. (2010). Coevolution and HBV drug resistance. *Antiviral Therapy, 15*(3), 505.

Klapper, P. E., Cleator, G. M., Tan, S. V., Guiloff, R. J., Scaravilli, F., Ciardi, M., Aurelius, E., & Forsgren, M. (1993). Diagnosis of herpes simplex encephalitis with PCR. *The Lancet, 341*(8846), 691.

Kleiboeker, S., Nutt, J., Schindel, B., Dannehl, J., & Hester, J. (2014). Cytomegalovirus antiviral resistance: Characterization of results from clinical specimens. *Transplant Infectious Disease, 16*(4), 561–567.

Kumarasamy, N., Solomon, S., Flanigan, T. P., Hemalatha, R., Thyagarajan, S. P., & Mayer, K. H. (2003). Natural history of human immunodeficiency virus disease in southern India. *Clinical Infectious Diseases, 36*(1), 79–85.

Lai, C. L., Dienstag, J., Schiff, E., Leung, N. W., Atkins, M., Hunt, C., Brown, N., Woessner, M., Boehme, R., & Condreay, L. (2003). Prevalence and clinical correlates of YMDD variants during lamivudine therapy for patients with chronic hepatitis B. *Clinical Infectious Diseases, 36*(6), 687–696.

Landolfo, S., Gariglio, M., Gribaudo, G., & Lembo, D. (2003). The human cytomegalovirus. *Pharmacology and Therapeutics, 98*(3), 269–297.

Levin, M. J., Bacon, T. H., & Leary, J. J., (2004). Resistance of herpes simplex virus infections to nucleoside analogues in HIV-infected patients. *Clinical Infectious Diseases, 39*(Supplement_5), S248–S257.

Lipkin, W. I. (2013). The changing face of pathogen discovery and surveillance. *Nature Reviews Microbiology, 11*(2), 133–141.

Locarnini, S., Hatzakis, A., Chen, D. S., & Lok, A. (2015). Strategies to control hepatitis B: Public policy, epidemiology, vaccine and drugs. *Journal of Hepatology, 62*(1), S76–S86.

Ma, C., Polishchuk, A. L., Ohigashi, Y., Stouffer, A. L., Schön, A., Magavern, E., Jing, X., Lear, J. D., Freire, E., Lamb, R. A., & DeGrado, W. F. (2009). Identification of the functional core of the influenza A virus A/M2 proton-selective ion channel. *Proceedings of the National Academy of Sciences, 106*(30), 12283–12288.

Macdonald, J., Tonry, J., Hall, R. A., Williams, B., Palacios, G., Ashok, M. S., Jabado, O., Clark, D., Tesh, R. B., Briese, T., & Lipkin, W. I. (2005). NS1 protein secretion during the acute phase of West Nile virus infection. *Journal of Virology, 79*(22), 13924–13933.

Mason, S., Devincenzo, J. P., Toovey, S., Wu, J. Z., & Whitley, R. J. (2018). Comparison of antiviral resistance across acute and chronic viral infections. *Antiviral Research, 158*, 103–112.

McGoldrick, S. M., Bleakley, M. E., Guerrero, A., Turtle, C. J., Yamamoto, T. N., Pereira, S. E., Delaney, C. S., & Riddell, S. R. (2013). Cytomegalovirus-specific T cells are primed early after cord blood transplant but fail to control virus in vivo. *Blood, The Journal of the American Society of Hematology, 121*(14), 2796–2803.

McKeegan, K. S., Borges-Walmsley, M. I., & Walmsley, A. R. (2002). Microbial and viral drug resistance mechanisms. *Trends in Microbiology, 10*(10), s8–s14.

Millhouso, S., & Wigdahl, B. (2000). Molecular circuitry regulating herpes simplex virus type 1 latency in neurons. *Journal of Neurovirology, 6*(1), 6–24.

Mittler, J. E., Sulzer, B., Neumann, A. U., & Perelson, A. S. (1998). Influence of delayed viral production on viral dynamics in HIV-1 infected patients. *Mathematical Biosciences, 152*(2), 143–163.

Miura, S., Zhang, Z., & Nei, M. (2013). Random fluctuation of selection coefficients and the extent of nucleotide variation in human populations. *Proceedings of the National Academy of Sciences, 110*(26), 10676–10681.

Molla, A., Korneyeva, M., Gao, Q., Vasavanonda, S., Schipper, P. J., Mo, H. M., Markowitz, M., Chernyavskiy, T., Niu, P., Lyons, N., & Hsu, A. (1996). Ordered accumulation of mutations in HIV protease confers resistance to ritonavir. *Nature Medicine, 2*(7), 760–766.

Morfin, F., & Thouvenot, D. (2003). Herpes simplex virus resistance to antiviral drugs. *Journal of Clinical Virology, 26*(1), 29–37.

Muller, H. J. (1964). The relation of recombination to mutational advance. *Mutation Research/fundamental and Molecular Mechanisms of Mutagenesis, 1*(1), 2–9.

Nagai, Y. O., & Yoshida, T. E. (1984). Viral pathogenesis: Mechanism of acute and persistent infections with paramyxoviruses. *Nagoya Journal of Medical Sciences, 46*, 1–7.

Nath, A. K., & Thappa, D. M. (2009). Newer trends in the management of genital herpes. *Indian Journal of Dermatology, Venereology, and Leprology, 75*(6), 566.

Neagu, I. A., Olejarz, J., Freeman, M., Rosenbloom, D. I., Nowak, M. A., & Hill, A. L. (2018). Life cycle synchronization is a viral drug resistance mechanism. *PLoS Computational Biology, 14*(2), e1005947.

Nelson, P. W., & Perelson, A. S. (2002). Mathematical analysis of delay differential equation models of HIV-1 infection. *Mathematical Biosciences, 179*(1), 73–94.

Nelson, P. W., Murray, J. D., & Perelson, A. S. (2000). A model of HIV-1 pathogenesis that includes an intracellular delay. *Mathematical Biosciences, 163*(2), 201–215.

Newmann, S., Sarin, P., Kumarasamy, N., Amalraj, E., Rogers, M., Madhivanan, P., Flanigan, T., Cu-Uvin, S., McGarvey, S., Mayer, K., & Solomon, S. (2000). Marriage, monogamy and HIV: A profile of HIV-infected women in south India. *International Journal of STD and AIDS, 11*(4), 250–253.

Nguyen, H. T., Trujillo, A. A., Sheu, T. G., Levine, M., Mishin, V. P., Shaw, M., Ades, E. W., Klimov, A. I., Fry, A. M., & Gubareva, L. V. (2012). Analysis of influenza viruses from patients clinically suspected of infection with an oseltamivir resistant virus during the 2009 pandemic in the United States. *Antiviral Research, 93*(3), 381–386.

Paintsil, E., & Cheng, Y. C. (2009). Antiviral agents. In *Encyclopedia of microbiology* (pp. 223–257).

Pauly, M. D., & Lauring, A. S. (2015). Effective lethal mutagenesis of influenza virus by three nucleoside analogs. *Journal of Virology, 89*(7), 3584–3597.

Pennings, P. S., Kryazhimskiy, S., & Wakeley, J. (2014). Loss and recovery of genetic diversity in adapting populations of HIV. *PLoS Genetics, 10*(1), e1004000.

Perrier, M., Désiré, N., Deback, C., Agut, H., Boutolleau, D., & Burrel, S. (2016). Complementary assays for monitoring susceptibility of varicella-zoster virus resistance to antivirals. *Journal of Virological Methods, 233*, 10–14.

Piret, J., & Boivin, G. (2011). Resistance of herpes simplex viruses to nucleoside analogues: Mechanisms, prevalence, and management. *Antimicrobial Agents and Chemotherapy, 55*(2), 459–472.

Piret, J., & Boivin, G. (2016). Antiviral resistance in herpes simplex virus and varicella-zoster virus infections: Diagnosis and management. *Current Opinion in Infectious Diseases, 29*(6), 654–662.

Plotkin, S. (2014). History of vaccination. *Proceedings of the National Academy of Sciences, 111*(34), 12283–12287.

Potter, S. J., Chew, C. B., Steain, M., Dwyer, D. E., & Saksena, N. K. (2004). Obstacles to successful antiretroviral treatment of HIV-1 infection: Problems & perspectives. *Indian Journal of Medical Research, 119*, 217–237.

Powdrill, M. H. , Tchesnokov, E. P., Kozak, R. A., Russell, R. S., Martin, R., Svarovskaia, E. S., Mo, H., Kouyos, R. D., & Götte, M. (2011). Contribution of a mutational bias in hepatitis C virus replication to the genetic barrier in the development of drug resistance. *Proceedings of the National Academy of Sciences, 108*(51), 20509–20513.

Razonable, R. R., & Emery, V. C. (2004). Management of CMV infection and disease in transplant patients. *Herpes: The Journal of the IHMF, 11*(3), 77–86.

Razonable, R. R. (2011, October). Antiviral drugs for viruses other than human immunodeficiency virus. In *Mayo clinic proceedings* (Vol. 86, No. 10, pp. 1009–1026). Elsevier.

Reeves, J. D., & Doms, R. W. (2002). Human immunodeficiency virus type 2. *Journal of General Virology, 83*(6), 1253–1265.

Renzette, N., Bhattacharjee, B., Jensen, J. D., Gibson, L., & Kowalik, T. F. (2011). Extensive genome-wide variability of human cytomegalovirus in congenitally infected infants. *PLoSPathogenetics, 7*(5), e1001344.

Richman DD, Nathanson N. Antiviral Therapy. In *Viral pathogenesis* (pp. 271–287). Academic Press.

Robertson, D. L., Anderson, J. P., Bradac, J. A., Carr, J. K., Foley, B., Funkhouser, R. K., Gao, F., Hahn, B. H., Kalish, M. L., Kuiken, C., & Learn, G. H. (2000). HIV-1 nomenclature proposal. *Science, 288*(5463), 55–55.

Ross, S. A., Novak, Z., Pati, S., & Boppana, S. B. (2011). Diagnosis of cytomegalovirus infections. *Infectious Disorders-Drug Targets, 11*(5), 466–474.

Safrin, S., Elbeik, T., Phan, L., Robinson, D., Rush, J., Elbaggari, A., & Mills, J. (1994). Correlation between response to acyclovir and foscarnet therapy and in vitro susceptibility result for isolates of herpes simplex virus from human immunodeficiency virus-infected patients. *Antimicrobial Agents and Chemotherapy, 38*(6), 1246–1250.

Sahli, R., Andrei, G., Estrade, C., Snoeck, R., & Meylan, P. R. (2000). A rapid phenotypic assay for detection of acyclovir-resistant varicella-zoster virus with mutations in the thymidine kinase open reading frame. *Antimicrobial Agents and Chemotherapy, 44*(4), 873–878.

Sanjuán, R., Moya, A., & Elena, S. F. (2004). The distribution of fitness effects caused by single-nucleotide substitutions in an RNA virus. *Proceedings of the National Academy of Sciences, 101*(22), 8396–8401.

Santoro, M. M., Ciccozzi, M., Alteri, C., Montieri, S., Alexiev, I., Dimova, I., Ceccherini-Silberstein, F., Beshkov, D., Rezza, G., & Perno, C. F. (2008). Characterization of drug-resistance mutations in HIV type 1 isolates from drug-naive and ARV-treated patients in Bulgaria. *AIDS Research and Human Retroviruses, 24*(9), 1133–1138.

Sattentau, Q. (2008). Avoiding the void: Cell-to-cell spread of human viruses. *Nature Reviews Microbiology, 6*(11), 815–826.

Schnell, J. R., & Chou, J. J. (2008). Structure and mechanism of the M2 proton channel of influenza A virus. *Nature, 451*(7178), 591–595.

Schreiber, A., Härter, G., Schubert, A., Bunjes, D., Mertens, T., & Michel, D. (2009). Antiviral treatment of cytomegalovirus infection and resistant strains. *Expert Opinion on Pharmacotherapy, 10*(2), 191–209.

Stránská, R., Schuurman, R., Nienhuis, E., Goedegebuure, I. W., Polman, M., Weel, J. F., Wertheim-Van Dillen, P. M., Berkhout, R. J., & van Loon, A. M. (2005). Survey of acyclovir-resistant herpes simplex virus in the Netherlands: Prevalence and characterization. *Journal of Clinical Virology, 32*(1), 7–18.

Strasfeld, L., & Chou, S. (2010). Antiviral drug resistance: Mechanisms and clinical implications. *Infectious Disease Clinics, 24*(3), 809–833.

Tacke, F., Gehrke, C., Luedde, T., Heim, A., Manns, M. P., & Trautwein, C. (2004). Basal core promoter and precore mutations in the hepatitis B virus genome enhance replication efficacy of Lamivudine-resistant mutants. *Journal of Virology, 78*(16), 8524–8535.

Tan, H. H., & Goh, C. L. (2006). Viral infections affecting the skin in organ transplant recipients. *American Journal of Clinical Dermatology, 7*(1), 13–29.

Tanaka, M. M., & Valckenborgh, F. (2011). Escaping an evolutionary lobster trap: Drug resistance and compensatory mutation in a fluctuating environment. *Evolution: International Journal of Organic Evolution, 65*(5), 1376–1387.

Terrault, N. A., Bzowej, N. H., Chang, K. M., Hwang, J. P., Jonas, M. M., & Murad, M. H. (2016). AASLD guidelines for treatment of chronic hepatitis B. *Hepatology (Baltimore, Md.), 63*(1), 261.

Thi, T. N., Deback, C., Malet, I., Bonnafous, P., Ait-Arkoub, Z., & Agut, H. (2006). Rapid determination of antiviral drug susceptibility of herpes simplex virus types 1 and 2 by real-time PCR. *Antiviral Research, 69*(3), 152–157.

Thomaston, J. L., & DeGrado, W. F. (2016). Crystal structure of the drug-resistant S31N influenza M2 proton channel. *Protein Science, 25*(8), 1551–1554.

Tong, S., & Revill, P. (2016). Overview of hepatitis B viral replication and genetic variability. *Journal of Hepatology, 64*(1), S4–S16.

Tregoning, J. S., & Schwarze, J. (2010). Respiratory viral infections in infants: Causes, clinical symptoms, virology, and immunology. *Clinical Microbiology Reviews, 23*(1), 74–98.

Vergne, L., Stuyver, L., Van Houtte, M., Butel, C., Delaporte, E., & Peeters, M. (2006). Natural polymorphism in protease and reverse transcriptase genes and in vitro antiretroviral drug susceptibilities of non-B HIV-1 strains from treatment-naive patients. *Journal of Clinical Virology, 36*(1), 43–49.

Viboud, C., Simonsen, L., Fuentes, R., Flores, J., Miller, M. A., & Chowell, G. (2016). Global mortality impact of the 1957–1959 influenza pandemic. *The Journal of Infectious Diseases, 213*(5), 738–745.

Wang, D., Hicks, C. B., Goswami, N. D., Tafoya, E., Ribeiro, R. M., Cai, F., Perelson, A. S., & Gao, F. (2011). Evolution of drug-resistant viral populations during interruption of antiretroviral therapy. *Journal of Virology, 85*(13), 6403–6415.

Webster, R. G., & Govorkova, E. A. (2014). Continuing challenges in influenza. *Annals of the New York Academy of Sciences, 1323*(1), 115.

Witsø, E., Palacios, G., Rønningen, K. S., Cinek, O., Janowitz, D., Rewers, M., Grinde, B., & Lipkin, W. I. (2007). Asymptomatic circulation of HEV71 in Norway. *Virus Research, 123*(1), 19–29.

World Health Organization Global Influenza Program Surveillance Network. (2005). Evolution of H5N1 avian influenza viruses in Asia. *Emerging Infectious Diseases, 11*(10), 1515.

Xu, F., Sternberg, M. R., Kottiri, B. J., McQuillan, G. M., Lee, F. K., Nahmias, A. J., Berman, S. M., & Markowitz, L. E. (2006). Trends in herpes simplex virus type 1 and type 2 seroprevalence in the United States. *JAMA, 296*(8), 964–973.

Yeni, P. G., Hammer, S. M., Hirsch, M. S., Saag, M. S., Schechter, M., Carpenter, C. C., Fischl, M. A., Gatell, J. M., Gazzard, B. G., Jacobsen, D. M., & Katzenstein, D. A. (2004). Treatment for adult HIV infection: 2004 recommendations of the International AIDS Society-USA Panel. *JAMA, 292*(2), 251–265.

ZurWiesch, P. S., Engelstädter, J., & Bonhoeffer, S. (2010). Compensation of fitness costs and reversibility of antibiotic resistance mutations. *Antimicrobial Agents and Chemotherapy, 54*(5), 2085–2095.

Chapter 4
Drug Resistance in Protozoal Infections

Hafiz Muhammad Rizwan, Haider Abbas, Muhammad Sohail Sajid, Mahvish Maqbool, Malcolm K. Jones, Muhammad Irfan Ullah, and Nabeel Ijaz

Abstract Parasitic infections have enormous health, social, and economic influence worldwide, predominantly in tropical and subtropical areas. Diseases caused by protozoan parasites, like malaria, leishmaniasis, trypanosomiasis, toxoplasmosis, giardiasis, etc. have major contributions in the global death rate due to infections. The burden of these parasitic protozoans has increased around the globe due to the lack of effective, commercially-availablevaccines, and the production of less efficacious drugs with poor pharmacokinetics. Unfortunately, in areas with endemic protozoan infections, antiparasitic drugs use and efficacy have been compromised by drug resistance. The development of resistance has been found to be associated with decreased drug uptake, the export of drugs from parasites, genetic modifications, loss of drug activity, and alteration of the drug target. Recently, the isolation and characterization of resistance-related genes and proteins have increased the knowledge about drug resistance greatly and provided a way forward for the identification of new drug candidates. In this chapter, our focus will be on the mode of action and mechanism of resistance development of those drugs used to treat protozoal

H. M. Rizwan (✉) · H. Abbas
Department of Pathobiology (Parasitology Section), KBCMA College of Veterinary and Animal Sciences, Narowal, Sub-campus University of Veterinary and Animal Sciences, Lahore, Pakistan
e-mail: hm.rizwan@uvas.edu.pk

M. S. Sajid · M. Maqbool
Department of Parasitology, Faculty of Veterinary Sciences, University of Agriculture, Faisalabad 38040, Pakistan

M. S. Sajid
One Health Laboratory, Centre for Advanced Studies in Agriculture and Food Security, University of Agriculture, Faisalabad, Pakistan

M. K. Jones
School of Veterinary Science, The University of Queensland, Gatton Queensland 4343, Australia

M. I. Ullah
Department of Pathobiology, Faculty of Veterinary Sciences, Bahauddin Zakariya University, Multan, Pakistan

N. Ijaz
Department of Clincial Sciences, Faculty of Veterinary Sciecnes, Bahauddin Zakariya University, Multan, Pakistan

S. Ahmed et al. (eds.), *Biochemistry of Drug Resistance*,
https://doi.org/10.1007/978-3-030-76320-6_4

infections in human including malaria, toxoplasmosis, trypanosomiasis, giardiasis, cystoisosporiasis, babesiosis, and leishmaniasis.

4.1 Introduction

Protozoa are microscopic eukaryotic organisms. About 65,000 protozoans, or protistan, species have been identified. Most of them are free-living and are recognized due to their complex internal structures and metabolic activities (Shanan et al., 2015). The trophozoite, or feeding stage, is the basic developmental stage for protozoa, either for intracellular or extracellular species. Different transmission routes used by these protozoa to reach the host cell include direct, vector-borne, predatory-prey, and fecal–oral route. For many enteric protozoans, dormant cysts are formed which can survive in harsh environment without nutrients, oxygen, and water for a long period (Antonovics et al., 2017).

Protozoan diseases are considered an important section of infectious diseases around the globe and have a significant effect on health, social life, and economy. Malaria, amoebic dysentery, cutaneous and visceral leishmaniasis, toxoplasmosis, cystoisosporiasis, African sleeping sickness, Chagas' disease, babesiosis, and giardiasis, are important parasitic diseases affecting one-sixth of world population (Fletcher et al., 2012). Tropical and subtropical areas were considered as hubs for parasitic diseases, but recently, the changes in climate and human interventions have changed the natural distribution of these diseases so that they have spread to high-income countries (Capela et al., 2019).

The armamentarium of drugs related to protozoa is inadequate and resistance development is a major hindrance in drug effectiveness. The useful and active drugs like chloroquine in malaria, sulfonamide in toxoplasmosis, diloxanide in amebiasis, and metronidazole in anaerobic parasites have become less effective due to resistance of previously susceptible parasites (Aspinall et al., 2002). A list of drugs used against protozoan diseases is given (Table 4.1) along with mode of action and mechanism of resistance. The lack of commercially available and effective vaccines against human protozoan parasites and the development of resistance against drugs are the major challenges in protozoan control. Moreover, most of the research related to drug resistance is done on laboratory animals which are unable to mimic the natural host-parasite interaction (Monzote & Siddiq, 2011).

The host immune system also plays an important role in the disease pathogensis. The individuals who contain the protozoal infection in their body but cannot eliminate it, become carriers and serve as a source of transmission. In those regions, where high frequency of well-tolerated protozoal diseasesis reported, population was never treated in a fashion to eliminate these diseases. Because the elimination of infection in such areas may lower the immunity of individuals which may lead to a high rate of reinfection (Barrett, 2018; Capela et al., 2019). Most of the impact of human protozoans is due to lack of vaccine, development of resistance, unavailability of safe and affordable drugs. Therefore, new antiprotozoal drugs with better efficacy are the need

Table 4.1 Mode of action and mechanism of resistance of common antiprotozoal drugs

Sr. No.	Disease	Drug	Mechanism of action	Mechanisms of resistance	References
1.	Malaria	Chloroquine	It acts by • Interfering with the elimination of the hematin, toxic byproduct of hemoglobin degradation, by the parasite	Resistance develop due to • Due to mutations at *Pfcrt* and (*Pfmdr*) gene	Fidock et al. (2000), Valderramos and Fidock (2006), Ebstie et al. (2016)
		Primaquine	It acts by • Targeting the electron transport chain and mitochondria of the parasite	Resistance develop due to • Mutations at CYP2D6 gene	Battle et al. (2014), Marcsisin et al. (2016), Talapko et al. (2019)
		Quinine	Unknown	Resistance develop due to • Single nucleotide polymorphisms in pfmdr1, pfcrt, pfmrp1 and amplification of pfmdr1 gene	Cowman et al. (1994), Achan et al. (2011), Cui et al. (2015)
		Lumefantrine	Unknown	Resistance develop due to • Mutation at Pfmdr N86	Sisowath et al., (2005), Achan et al. (2011)
		Mefloquine	Still not cleared however, it may be due to • Damage to parasite membrane	Resistance develop due to • Pvmdr1 gene amplification	Suwanarusk et al. (2008), Imwong et al. (2008), Achan et al. (2011), Talapko et al. (2019)
		Sulfadoxine and pyrimethamine	Act as • Folic acid antagonist	Resistance develop due to • Mutations in DHFR and DHPS	Wongsrichanalai et al. (2002), Lumb et al. (2011), Nankoberanyi et al. (2014)

(continued)

Table 4.1 (continued)

Sr. No.	Disease	Drug	Mechanism of action	Mechanisms of resistance	References
		Atovaquone	It acts by • Blocking mitochondrial electron transport by acting as cytochrome *bc*11 complex inhibitor	Resistance develop due to • Mutations in DHFR and DHPS	Fry and Pudney (1992), Peterson et al. (1988)
		Proguanil	It acts by • Blocking deoxythymidylate production by acting as DHFR inhibitor in the form of cycloguanil	Resistance develop due to • Mutations in DHFR and DHPS	Peterson et al., (1988), Srivastava and Vaidya (1999)
		Artemisinin	Still debatable however, it may be due to • Production of free radicals, on activation by haem, to interrupt the proteins of parasites • Haem and *Pf* ATP6 (Ca^{2+} transporter)	Resistance develop due to • PfATP6 S769N molecular marker	Jambou et al. (2005), Shandilya et al. (2013), Tilley et al. (2016)
2.	Toxoplasmosis	Pyrimethamine	It acts by • Blocking the biosynthesis of purines and pyrimidines by inhibition of DHPS (dihydropteroate synthase) and DHFR (dihydrofolate reductase)	Resistance develop due to • Change in targets like DHPS and DHFR	Reynolds et al. (2001), Alday and Doggett (2017)

(continued)

Table 4.1 (continued)

Sr. No.	Disease	Drug	Mechanism of action	Mechanisms of resistance	References
		Atovaquone	It acts by • Inhibition of electron transport chain of mitochondria which indirectly inhibit the enzymes linked with mitochondria	Resistance develop due to Its binding at oxidizes ubiquinol domain of cytochrome b	McFadden et al. (2001), Lawres et al. (2016)
3.	Babesiosis	Diminazene aceturate	It acts by • Binding with kinetoplastid DNA (kDNA) and causes disturbance of kinetoplast replication and function • It also inhibits mitochondrial topoisomerase II	Resistance develop due to • Change in mitochondrial DNA (mtDNA) i.e. • Mutation cytochrome b (CYTb), • Cytochrome c oxidase subunit I (COXI), • Cytochrome c oxidase subunit III (COXIII) genes	James and Giles (1985), Peregrine (1994), Matsuu et al. (2006), Sakuma et al. (2009)
		Atovaquone	It acts by • Inhibition of the rate of oxygen consumption • Also acts as ubiquinone (mitochondrial protein involved in electron transport chain) analog for protozoa	Resistance develop due to • Its binding at oxidizes ubiquinol domain of cytochrome b	Murphy and Lang-Unnasch (1999), Baggish and Hill (2002)

(continued)

Table 4.1 (continued)

Sr. No.	Disease	Drug	Mechanism of action	Mechanisms of resistance	References
4.	Cystoisosporiasis	Trimethoprim-sulfamethoxazole	Act as • Folic acid antagonist	Exact mechanism of this resistance in *C. belli* is unknown, however, • *Trypanosoma cruzi* also has DHFR and resistance to trimethoprim develop due to DHFR insensitivity to this drug • Single mutation to thymine auxotrophy	Then (1982), Wongsrichanalai et al. (2002), Lumb et al. (2011), Mayers et al. (2017)
5.	Trypanosomiasis	Pentamidine	It acts by • Destruction of kinetoplast DNA • Modulation of membrane potential of mitochondria	Resistance develop due to • Loss of function of P2 aminopurine transporter	Franco et al. (2018)
		Suramin	It acts by • Inhibiting glycolytic enzymes	• No resistance is reported in humans • Resistant strains are developed in laboratory conditions • It was hypothesis that variant surface glycoprotein (VSG) might be involved in the resistance mechanism	Wiedemar et al. (2018)

(continued)

Table 4.1 (continued)

Sr. No.	Disease	Drug	Mechanism of action	Mechanisms of resistance	References
		Melarsoprol	• The mode of action is still not clearly understood • It shows an affinity for vicinal thiols in proteins particularly and for sulfhydryl groups • Inhibit ATP synthesis	The resistance develop due to the • Loss of drug uptake due to mutation in transporters AQP2, AQP3, and P2	Graf et al. (2015), Fairlamb and Horn (2018), Thomas et al. (2018)
6.	Leishmaniasis	Sodium Stibogluconate	It acts by • Inhibition of Trypanothione reductase which is the basic requirement of parasite for its survival in the host	Resistance is contributed due to • High intracellular thiols levels • MRP1 (multidrug resistance associated protein 1) • Overexpression of TXN/TXNPx (tryparedoxin/ tryparedoxin peroxidase system), • ATP-binding cassette (ABC) transporters	Ponte-Sucre et al. (2017), Turcano et al. (2018)

(continued)

Table 4.1 (continued)

Sr. No.	Disease	Drug	Mechanism of action	Mechanisms of resistance	References
		Pentamidine	Still not cleared however, it may be due to • *Inhibition of ribonucleic acid* (RNA) polymerase, • Interaction with kinetoplast DNA (kDNAs), • Interference with polyamine synthesis, • Apoptotic death, • Inhibition of topoisomerase II (TOPII)	Resistance is reported and may be due to • *Aquaglyceroporin* 2 mutation • Overexpression of pentamidine resistance protein 1 (PRP1)	Yang et al. (2016), Jeacock et al. (2017)
		Paromomycin	Still not clear but it may be due to • Alteration in membrane fluidity, • Protein synthesis inhibition, • Respiratory dysfunction, • Decreasing potential of mitochondrial membrane, • Lipid metabolism	No resistance is described until now, however, various hypotheses have been described based on laboratory-derived resistant strains	Fernandez et al. (2011), Hendrickx et al. (2017)
7.	Giardiasis	Metronidazole	It acts by • Damages/breakage of the double-strand DNA • Forming covalent adducts with cysteines in proteins • Targets thioredoxin reductase by inhibiting its function as a disulfide reductase • Degradation of elongation factor 1-γ enzyme	• Exact mechanism of resistance is still not clear • Reduced pyruvate ferredoxin oxidoreductase (PFOR) activity and knock-down of PFOR expression was observed in Metronidazole resistant strain	Leitsch et al. (2011), Uzlikova and Nohynkova (2014)

(continued)

Table 4.1 (continued)

Sr. No.	Disease	Drug	Mechanism of action	Mechanisms of resistance	References
		Albendazole	It acts by • Binding to the β-tubulin and inhibit cytoskeleton • Polymerization resulting in severe structural defects	Resistance develop due to • Changed expression of various genes including a VSP-like protein, • Cytoskeletal aberrations, • Chromosomal rearrangements	MacDonald et al. (2004), Argüello-García et al. (2009), Paz-Maldonado et al. (2013)

of time. In this chapter, we will focus on drug resistance against common protozoal infections of public health significance i.e., malaria, cutaneous and visceral leishmaniasis, toxoplasmosis, cystoisosporiasis, African sleeping sickness, babesiosis, and giardiasis.

4.2 Malaria

Agents of human malaria belong to the genus *Plasmodium,* a member of the order Haemospororida, and family Plasmodiidae (Robert et al., 2014; Walker et al., 2017). Charles Louis Alphonse Laveran, a French army doctor, was the first person to identify the malaria parasite in 1880 (Talapko et al., 2019). *Plasmodium (P.) malariae, P. knowlesi, P. falciparum* and *P. ovale* are among the *Plasmodium* spp. prevalent in sub-tropics and tropics of the globe (Ashley et al., 2018; Walker et al., 2017; WHO, 2018). In low- and middle-income countries, malaria has higher public health significance (Dhiman, 2019; Mohammadi et al., 2011). Among 435,000 reported deaths due to malaria in 2017, about 61% were children having less than five-year age, and mostly werefrom Sub-Saharan Africa (WHO, 2018). Malaria is prevalent around the globe, but most specifically occurs in Eastern Mediterranean, Western Pacific, South-East Asia, Africa and Americas (WHO, 2018). Among *Plasmodium* spp., only *P. knowlesi* is considered as zoonotic and many cases are reported in Southeast Asia. *Macacafascicularis*, the long-tailed macaque, serves as the natural host for *P. knowlesi* (Ashley et al., 2018; Millar & Cox-Singh, 2015).

The life cycle of *Plasmodium*species commences in humans with the injection of sporozoites during feeding by the female mosquito of the genus *Anopheles.* These sporozoites then move to the liver where non-sexual multiplication takes place, producing by abundant merozoites through intra-hepatocellular schizogony. The life cycle of different species of *Plasmodium* differs. For example, in *P. ovale* and *P. vivax*, a few sporozoites enter a new phase, called hypnozoites, before asexual replication and serve as a source of recrudescence of the infection (Jones & Good, 2006). After replication, the hepatic merozoites escape the liver to infect erythrocytes, for multiple rounds of asexual reproduction, before some go on to become sexual phases, the macro-and micro-gametocytes, which are taken up by mosquitoes during a blood meal. The sexual phase occurs in the mosquitoes leading to sporozoites formation (Pavitra et al., 2016).

Out of five species of human *Plasmodium, P. vivax* is responsible for morbidity, while most mortality is due to *P. falciparum.* Commonly, *P. vivax* is found in areas where transmission intensity is very low. This indicates that more disease incidence in these areas might be due to relapses occurring due to the hepatic dormant stages of *P. vivax* (Guerra et al., 2010). Benign type malaria is caused by *P. ovale* and *P. malariae.* In addition, *P. ovale* also has dormant hepatic hypnozoites participating in relapses of disease, as with *P. vivax. Plasmodium malariae* has very low parasitemia and can persist for longer times in the human host than other malarias. *Plasmodium knowlesi* has very short reproduction cycle among the human malaria parasites, and is responsible for fatal and rapidly progressive infection (Lee et al., 2011). The resistant strains of *P. falciparum* to various antimalarials have emerged (Corey et al., 2016).

There are several mechanisms by which *Plasmodium* evade immune mechanisms of the host. The most important is intracellular parasitisim. Free sporozoites are exposed to the antibodies but sporozoites or merozoites when enter the RBCs of the host escape the recognition of $CD8^+$ T cells which normally work with MHC molecules. Infected RBCs when adhere to endothelial cells also protect them to enter liver and spleen. Antigenic polymorphism also restricts the recongnition of antigenically different epitopes by antibodies and T cells. This also limit the possibilities of development of anti-malarial vaccines (Plebanski et al., 1999). There are several other mechanisms on host side which also aid the malarial parasites to evade immune mechanisms (Hisaeda et al., 2005).

4.2.1 Drugs Related to Malaria (Mode of Action and Mechanism of Resistance)

Malaria therapy is carried out through drugs. There are different types of drugs for this purpose like chloroquine, primaquine, antifolates, arylamino alcohols and artemisinin-based combination therapy which are typically efficacious against the symptomatic asexual blood stages. Almost 14 drugs have been listed for the treatment of malaria and only 6 drugs for prophylaxis in the database of WHO either in the form of monotherapy or in combination therapies (WHO, 2019a). Malaria therapy faces challenges due to the broad range of *Plasmodium* species, the stages of theirlife cycles and lack of standardized assays to screen drugs.

Mostly drugs available for malaria treatment are effective against the clinical asexual blood forms of the parasite because the early hepatic infection is asymptomatic. The major screening tool for the anti-malarials is the growth inhibition assay for asexual forms (Burrows et al., 2011; Wells et al., 2009). The primaquine can be effective against the dormant liver forms of the parasites and gametocyte stages because identification and testing is not easy for such drugs. Although, primaquine is effective for sexual stages and hypnozoites of the *Plasmodium* spp., it is contraindicated in pregnant women or patients deficient in glucose-6-phosphate dehydrogenase (G6PD) (Alonso et al., 2011; WHO, 2009).

The lack of standardized assays is a major challenge faced for the development of drugs against the dormant stages of the parasites in the liver (Wells et al., 2010). Rodent models do not harbour dormant hepatic stages of *Plasmodium* spp., so these can only be used to screen drugs which are effective against blood stages (Langhorne et al., 2011). The only models available to screen hepatic stage drugs like primaquine is the non-human primates but these have their own limitations in the form of high cost, less availability at many research facilities and low throughput. There was no success attained with in vitro models for drug screening against hepatic stages either due to low infection rates or problematic sporozoites procurement besides hepatic cells retain in vivo properties (Silvie et al., 2003). However, it is also difficult to differentiate between the dormant hepatic stages and early hepatic pre erythrocytic stages of the parasite but still research is continuing to develop hypnozoite assays (Chattopadhyay et al., 2010; Dembele et al., 2011).

4.2.2 Resistance to Chloroquine

Among the 4-aminoquinolones, including piperaquine, pyronaridine, aminophenols, amodiaquine and dimeric 4-aminoquinoline, perhaps chloroquine has been the most dominant compound for treatment of uncomplicated malaria. The use of chloroquine as an antimalarial for *P. falciparum* infection is limited due to the development of resistance, but it is still being used for vivax malaria despitereports on therapy failure (Commons et al., 2018). Amodiaquine is used in chloroquine—resistant cases of parasites due to its better potency, although it is chloroquine—like in its propensity for developing resistance. The mode of action (MoA) of chloroquine is related to its accumulation in the digestive vacuole of the parasite, and limiting the elimination of hematin, the toxic by product of hemoglobin degradation (Combrinck et al., 2013; Ebstie et al., 2016; Vennerstrom et al., 1992).

An effective concentration of chloroquine does not properly accumulate within the food vacuole of chloroquine-resistant parasites. The *P. falciparum* chloroquine resistance transporter (*Pfcrt*) was found to be associated with the development of this resistance on genetic crosses (Fidock et al., 2000) as mutations of*Pfcrt* leads to areduction in the concentration of chloroquine accumulated within the food vacuole. Both at in vivo and in vitro levels, the development of resistance against chloroquine is related to multiple genetic polymorphisms in the gene but the substitution of lysine with threonine (at residue 76) is the major mutation (Djimde et al., 2001; Lakshmanan et al., 2005). Chloroquine resistance is not only caused due through mutation of *Pfcrt* but also in *P. falciparum* multidrug resistance (*Pfmdr*) gene. The degree of resistance caused due to mutations in these genes is also affected by the genetic properties of the strain (Cooper et al., 2005; Valderramos & Fidock, 2006).

4.2.3 Resistance to Primaquine

Pamaquine, primaquine and tafenoquine are 8-aminoquinolones. Primaquine has some efficacy for erythrocytic stages and is the only licensed antimalarial drug for the hypnozoite form and gametocytes of *Plasmodium* spp. but it is contraindicated in pregnant women or G6PD deficit patients (Battle et al., 2014; Hobbs & Duffy, 2011; Howes et al., 2016; WHO, 2009; Wells et al., 2010). Primaquine acts by targeting the electron transport chain and mitochondria of the parasite (Talapko et al., 2019). In addition to other factors lowering its efficacy, resistance to primaquine in parasites has also been reported (Battle et al., 2014; Marcsisin et al., 2016).

The reason for development of resistance to primaquine has been reported to be mutations at *CYP2D6* gene, as the CYP2D6 enzyme is involved in the hydroxylation of primaquine to convert it into active form within liver cells (Battle et al., 2014; Dijanic et al., 2018; Marcsisin et al., 2016).

4.2.4 Resistance to Arylamino Alcohols

Arylamino alcohols including lumefantrine, halofantrine, mefloquine and quinine are also used as antimalarial drugs (Martins et al. 2006; Quiliano et al., 2016). Quinine was the first drug used to cure malaria (Cowman et al., 2016; Hobbs & Duffy, 2011). Parenteral artesunate proved more efficacious in recent trials while parenteral quinine is considered as a better choice for severe malaria (WHO, 2009). The MoA of arylamino alcohols, including quinine, as antimalarials is unknown (Achan et al., 2011).

The resistance to quinine was first reported in 1965 (Thompson et al., 1965). Single nucleotide polymorphisms in *pfmdr1*, *pfcrt*, *P. falciparum* multidrug resistance protein-1 (*pfmrp1*) and amplification of *pfmdr1* gene have been reported in quinine resistance (Cowman et al., 1994; Cui et al., 2015). The development of resistance to lumefantrine was proposed due to the *Pfmdr N86*, molecular marker (Sisowath et al., 2005). In vitro assessment showed that chloroquine susceptible-allele, 86Y was also found to be hypersensitive to mefloquine (Woodrow & Krishna, 2006) and Pvmdr1 gene amplification can result in resistance to mefloquine in *P. vivax* (Imwong et al., 2008; Suwanarusk et al., 2008). The efficacy of antimalarial drugs is indirectly proportional to copy number of *pfmdr* which means therapy failure with mefloquine can happen if *pfmdr* copy number is increased (Alker et al., 2007; Price et al., 2006). By reducing the copy number of mRNAs for *pfmdr*, Sidhu et al. (2006) reported the increased susceptibility to drugs like quinine, halofantrine, and mefloquine.

4.2.5 Resistance to Antifolate Drugs

Sulfadoxine and pyrimethamine work synergistically when used in combination. Pyrimethamine interferes with the enzyme dihydrofolate reductase (DHFR), while sulfadoxineinterferes with dihydropteroate synthase (DHPS) of the folate synthesis pathway, ultimately inhibiting the synthesis of tetrahydrofolate. Parasite survival depends on the folate synthesis pathway and these drugs are categorized as antifolates (Lumb et al., 2011; Wongsrichanalai et al., 2002). These enzyme-blocking drugs were found efficacious in malaria treatment in the 1930s. Moreover, these sulfonamide compounds, having better pharmacokinetic properties, were initially used during the 1940s and 50 s. After the more efficacious quinine derivatives became available, the usage of these antifolates was reduced (Schlitzer, 2007). In 1971, a sulfadoxine and pyrimethamine combination was incorporated as a second-line of treatment for malaria, but resistance developed against these drugs in a shorter period of time than resistance to chloroquine developed (Baird, 2005; Wells et al., 2009).

One of the first antifolates used to treat malaria was Proguanil in 1945 (Curd et al., 1945). In 1991, atovaquone was used against protozoans (Hudson et al., 1991). Proguanil and atovaquone when used in combination, were provenas synergistically-effective antimalarials. The MoA of atovaquone involves blockage of electron transport in the mitochondria by acting as a cytochrome *bc*11 complex inhibitor (Fry & Pudney, 1992). Proguanil, when used as a monotherapy for malaria, blocks deoxythymidylate production by acting as a DHFR inhibitor in the form of cycloguanil, but this does not happen when used in combination with atovaquone, instead it reduces atovaquone concentration necessary for cure of malaria (Srivastava & Vaidya, 1999). This proguanil-atovaquone combination is still used in the case of artemisinin-based therapy failure or chloroquine-resistant patients for therapy of malaria, but its usage in endemic areas is limited due to rapid development of resistance (Talapko et al., 2019).

Mutations in DHFR and DHPS lead to the development of resistance against the antifolate antimalarial medications. The extent of this resistance is increased due to stepwise modifications in the nonsynonymous substitution (Alday & Doggett, 2017; Nankoberanyi et al., 2014; Peterson et al., 1988; Plowe, 1997). The variation in the enzymatic activity is correlated with the order in which mutations occur, although clear data is not available (Gregson & Plowe, 2005). In South America and Southeast Asia, the main source of resistance is the mutation at DHFR (polymorphism at residue 164) while the drug is being extensively used in sub-Saharan Africa and DHFR (I164L) was not found in this region (Sibley et al., 2001). Most African malaria-naïve children were unable to benefit from sulfadoxine and pyrimethamine as antimalarial therapy but in contrast, the combination was found efficacious in curing pregnant women from malaria (Ter Kuile et al., 2007). The reason for this discrepancy might be the presence of mutations (polymorphisms) in DHPS and DHFR in the African region, leading towards lower degree of resistance (Nzila et al., 2005). The efficacy of this antimalarial agent is inversely proportional to gene copy number in the *Plasmodium* genome. The areas with high antifolate medication pressure lead to increase in copy number of GTP-cyclohydrolase I, the first gene in de novo folate synthesis. Increased copy number of this gene is linked to the parasites having mutations in DHPS and DHFR, leading to the development of resistance against antifolates and this copy number is related to the compensation for the fluctuations in the enzyme activity responsible for resistance (Nair et al., 2008).

4.2.6 *Resistance to Artemisinin-Based Combination Therapy (ACT)*

Commonly used monotherapies for *Plasmodium* infection have been replaced more recently by artemisinin-based combination therapy (ACT) and have proven successful in complicated cases of *P. falciparum* malaria (Askling et al., 2012; Muñoz et al., 2015; Pousibet-Puerto et al., 2016; WHO, 2015). ACT has reduced the global malaria burden (WHO, 2019b). The artesunate-mefloquine, artesunate with sulfadoxine-pyrimethamine, artemether-lumefantrine, dihydroartemisinin-piperaquine, and artesunate-amodiaquine are the combinations of antimalarial agents used as ACTs (Ashley et al., 2018; Kremsner & Krishna, 2004; Talapko et al., 2019). For complete malarial therapy through ACT, a second drug having longer half-life is required to act with artemisinin orits derivatives, owing to their short half-lives. This combination will impede the development of resistance as parasites are already becoming less sensitive to artemisinin (Eastman & Fidock, 2009; Woodrow & White, 2017). These drugs have also been found to minimize the carriage of gametocytes but they have no role in prevention of transmission of mature gametocytes that had developed at the onset of therapy (Bousema et al., 2006; Smithuis et al., 2010). Triple ACTs with artemether–lumefantrine plus amodiaquine and dihydroartemisinin–piperaquine plus mefloquine have been successful in treating uncomplicated cases of malaria due to *P. falciparum* (van der Pluijm et al., 2020).

The MoA of artemisinin is still uncertain (O'Neill et al., 2010). Some theories suggest that artemisinin leads to the production of free radicals, on activation by haem, to interrupt the proteins that help in parasite survival (Tilley et al., 2016). Haem and *Pf*ATP6 (Ca^{2+} transporter) were also found to be involved in these drugs actions (Shandilya et al., 2013). Artemisinin was reported to inhibit the phosphatidylinositol-3-kinase of *P. falciparum* (Mbengueet al., 2015). Another study showed that artemisinin catalyzed the unfolded protein response pathways, ultimately reducing the chances of parasite survival (Mok et al., 2015).

Some 30 reports of resistance to artemisinin have been reported to 2018, particularly that of dihydroartemisinin and piperaquine, which is due to single line of multidrug-resistant *Plasmodium* (Amato et al., 2018; Imwong et al., 2017). In addition to resistance against artemisinin, resistance has also developed to partner drugs (sulfadoxine pyrimethamine, piperaquine, amodiaquine and mefloquine) in ACT (van der Pluijm et al., 2019). In this regard, WHO initiated a Global Plan to contain the spread of this resistance (WHO, 2011). Artemesinin resistance might develop due to mutations in the various genes and the knowledge of the molecular basis of this provides a way forward to contain the spread (Imwong et al. 2010). The transporter, PfATP6 is influenced by artemisinins and this transporter is an ortholog of mammalian sarcoendoplasmic reticulum Ca2 ATPase. Thus, the polymorphism in the gene may affect the susceptibility of this drug (Eckstein-Ludwig et al., 2003). The PfATP6 S769N polymorphism is said to be the molecular marker responsible for the development of resistance to artemisinin (Jambou et al., 2005). This artemisinin resistance was identified in French Guiana, while absent in Africa and China (Ferreira et al., 2007; Zhang et al., 2008). The efficacy of antimalarial drugs like artesunate-mefloquine and artemether-lumefantrine is indirectly proportional to copy number of *Pfmdr* which means therapy failure with these drugscan happen if *pfmdr* copy number is increased (Alker et al., 2007).

4.3 Toxoplasmosis

Toxoplasma gondii belongs to phylum Apicomplexa. Approximately 1/3rd of the human population is affected by this parasite (Montoya & Liesenfeld, 2004; Tenter et al., 2000). Among food-borne parasitic infections, toxoplasmosis is the third most common disease in which hospitalization is required (Vaillant et al., 2005). There are two possible ways for *Toxoplasma* infection to occur in humans: (1) cyst ingestion through raw or uncooked meat or (2) oocyte consumption through contaminated water, food, fruits, and vegetables, etc. It can also be transmitted through the congenital route from mother to fetus (Sepúlveda-Arias et al., 2014). Infection is asymptomatic in immune-competent individuals while in immunocompromised, it may be severe (Wang et al., 2017). A higher rate of ocular toxoplasmosis has been recorded in South America as compared to other areas of the world (Glasner et al., 1992).

Three main clonal lineages are present for *T. gondii* (i) Type I contain a virulent strain i.e. RH (ii) Type II contain avirulent strains, PRU, and ME49 and (iii) Type III contain avirulent strains named NED (Howe & Sibley, 1995). Type II clonal lineage is most common in America and Europe and infects both the animals and the human population. In South America, non-clonal strains of *Toxoplasma* were isolated from infected humans and this strain causes more clinical symptoms as compared to clonal strain present in Europe (Shwab et al., 2014).

A combination of sulfadiazine (SDZ) and pyrimethamine (PYR) is recommended for the treatment of toxoplasmosis. The combination of these drugs showed severe side effects including thrombocytopenia, hypersensitivity, neutropenia, leucopenia, and a severe decrease in platelet count (Montazeri et al., 2015). In addition, this drug combinationcan also cause some uncommon reactions i.e., toxic epidermal necrolysis, agranulocytosis, hepatic necrosis, and Stevens-Johnson syndrome (McLeod et al., 2006). Other drugs used to treat toxoplasmosis are cotrimoxazole (trimethoprim-sulfamethoxazole), azithromycin, spiramycin, dapsone, clarithromycin, and atovaquone but these drugs have no effect on bradyzoite form and are poorly tolerated (Montazeri et al., 2018).

The use of antifolate drugs to treat long term congenital toxoplasmosis and chorioretinitis showed no effect, especially in AIDS patients (Villena et al., 1998). It became a controversy that either this failure is due to change in pharmacological parameters (malabsorption of drug, drug intolerance) or due to lower susceptibility of *Toxoplasma* strain or the development of resistant strain (Meneceur et al., 2008). Resistance mechanism in *Toxoplasma* is not completely understood (Kropf et al., 2012).

Toxoplasma gondii has also developed certain mechanisms to bypass host immune response to establish long term survival in the infected host. Effector proteins released from secretory granules of *Toxoplasma gondii* cause changes in the host pathogen interaction through antimicrobial molecule production, control of immuneregulatory cytokine production, gene transcription of host, apoptosis and dysregulation in signaling pathways of host (Lima & Lodoen., 2019).

4.3.1 Drugs Related to Toxoplasmosis (Mode of Action and Mechanism of Resistance)

Pyrimethamine (PYR) is a folic acid antagonist used to treat *Toxoplasma* infection along with sulfonamide. Pyrimethamine blocks the biosynthesis of purines and pyrimidines (required for multiplication of cell and synthesis of DNA) by inhibition of DHPSand DHFR. Pyrimethamine resistance develop due to change in targets like DHPS and DHFR (Alday & Doggett, 2017; Reynolds et al., 2001).

Atovaquone, an analog of ubiquinone, is a hydroxynaphthoquinone that has anti-*Pneumocystis*, antimicrobial and antiprotozoal activity. It acts by inhibiting electron transport chain of mitochondria which indirectly inhibit the enzymes linked with mitochondria. Inhibition of such metabolic activities may lead to blockade of ATP and nucleic acid synthesis. The resistance of atovaquone develops due to its binding at oxidizes ubiquinoldomain of cytochrome b (Lawres et al., 2016; McFadden et al., 2001).

4.3.2 Resistance to Pyrimethamine

A combination of SDZ and PYR is recommended to treat acute toxoplasmosis, which can inhibit the major enzymes i.e., DHPS and DHFR. These are the main enzymes involved in pyrimidine biosynthesis in the parasite. Treatment of toxoplasmosis by blocking DHFR in humans may cause folate deficiency and results in embryopathies and hematological side effects. To avoid this folate deficiency, these drugs are administrated along with folic acid (Alday & Doggett, 2017; Rajapakse et al., 2013).

Pyrimidine and purine analogs, adenosine arabinoside, 5-fluorodeoxyuridine, and 6-thioxanthine have found viable against resistant strains of *Toxoplasma* in an in vitro study, despite the defects in nucleotide synthesis enzymes (Pfefferkorn & Borotz, 1994). Resistance hasoccurred in RH strains as compared to type I and type II strains in an in vitro experiment through single point mutation in *Toxoplasma* DHFR-thymidylate synthase e.g., L98S, W25R, and L134H (Reynolds et al., 2001). By transgenic and site-directed mutagenesis experiments, a number of mutations were induced in the DHFR-thymidylate synthase gene for PYR resistance and mutation in T83N was found to be responsible for PYR resistance (Donald & Roos, 1993). Combined mutations at sites T83N, S36R, and F245S resulted in an enhanced degree of resistance (Reynolds et al., 2001). Pyrimethamine resistance variesin different *Toxoplasma* strains, therefore, it provides a way forward to explore the sources of treatment failure and possible mechanisms of drug resistance.

4.3.3 Resistance to Atovaquone

Atovaquone (ATV) is used for the treatment of clinical toxoplasmosis. The drug is successfully used for the treatment of *T. gondii, Babesia microti*, and *P. falciparum* infection. ATV is effective against the chronic bradyzoite stage of the *Toxoplasma* and acts by inhibiting the electron transport chain of mitochondria (Tomavo & Boothroyd, 1995). ATV is not effective against toxoplasmosis in AIDS patients and in recipients of hematopoietic cell transplant (Gajurel et al., 2016). This drug acts by targeting the membrane-bound enzyme cytochrome bc1 complex (CYT bc1), present in the inner mitochondrial membrane. This complex plays a significant role in the respiratory electron transport chain. There are two active sites, one that oxidizes ubiquinol (bc1 Qo site) and other that reduces ubiquinone (bc1 Qi site) (Crofts, 2004; Lawres et al., 2016). By transferring protons to the intermembrane space, CYT bc1 generates an electrochemical gradient, reduces cytochrome c, and for pyrimidine biosynthesis creates ubiquinone.

It is shown that ATV targets the CYT bc1 enzyme of *Toxoplasma* by binding the Qo domain of cytochrome b. The resistance toATV in *T. gondii* is attributed to major mutational changes in M129L and I254L, although another study conducted on these mutations have not confirmed the role of these mutations in resistance (McFadden et al., 2001; Meneceur et al., 2008). Endochin-like quinolones (ELQs) showed an active response in the case of ATV- resistant *Babesia* and *Plasmodium* (Lawres et al., 2016). However, in an in vivo study, GSK932121, GW844520, and 4(1H)-pyridone compounds were found effective in malaria treatment (Capper et al., 2015). The mode of action of these ELQs is the same as that of ATV but they inhibit the Qi site of CYT bc1. Moreover, low doses of ELQ-316 and ELQ-271 showed effective results in the treatment of latent and acute *Toxoplasma* infection (Doggett et al., 2012). The 4(1H)-pyridone and ELQs act at the Qi site of cytochrome b in *T. gondii,* therefore, these drugs are considered efficacious against ATV resistant *T. gondii.*

4.4 Babesiosis

Babesia spp. are known as the most common haemoprotozoan of the world. These parasites cause tick-borne fever in domestic and wild animals, along with humans, and are agents of the disease known as babesiosis (Hamšíkováet al., 2016). To date, only hard ticks (Family Ixodidae) are known to be responsible for transmission of

Babesia species. As trophozoites in erythrocytes of mammals, the *Babesia* parasites are pyriform in shape, a feature that gives rise to their common name, the piroplasms. *Babesia* species are commonly transmittedthrough ticks belonging to the genera *Hyalomma, Rhipicephalus*, and *Dermacentor* (Soulsby, 1982; Telford et al., 1993). More than 100 *Babesia* species have been identifiedincluding Babesia *(B.) ovis, B. canis, B. bigemina, B. microti, B. divergens, B. crassa, B. bovis, B. duncani, B. venatorum* prevalent in Italy, in North America, Europe, Asia and Austria (Soulsby, 1982; Najm et al., 2014).

Both vertebrate and invertebrate hosts are involved in the completion of its life-cycle. Sexual and asexual development take place during its life cycle. Sexual development (gametogony) occurs in the midgut of vector tick, while a form of asexual development (sporogony) occurs in tick salivary glands. In erythrocytesof the vertebrate host, the parasites multiply by by binary fission (Becker et al., 2013). The parasite enters into the vertebrate bloodstream through the bite of the tick vector, undergoes rapid multiplication, targets RBCs and causes their lysis and leads to the spread of infection in the bloodstream (Urquhart et al., 1996).

Its clinical manifestations are variable, and depend on the the host immune system, age, concurrent infections, and genetic factors. Chronic cases are mostly asymptomatic, while in acute cases, infection is mainly characterized by anemia, fever, icterus mucous, hemoglobinuria, anorexia, and low packed cell volume (Smith et al., 2013). In case of humans, symptoms vary from 1 to 6 weeks and in some cases may take 3 months (Benach & Habicht, 1981). Malaria like symptoms including chills, fatigue, fever, myalgia, and malaise reported in human. In some cases, nausea, weight loss, emesis, and hematuria were also reported. In the case of high parasitic burden, hepatomegaly and splenomegaly were also reported (Benach & Habicht, 1981; Persing & Conard, 1995).

A well explained and established phenomenon of immune evasion in *Babesia bovis*is antigenic variation. The mechanism is almost similar as described earlier in malarial parasite in which infected RBCs adhere themselves to the venous endothelium through their ligands which develop in these pathogens strategies to evade host immune system protection. There is not much work established other than the mentioned one to evade host immunity in babesiosis (Allred and Al-Khedery 2006).

4.4.1 Drugs Related to Babesiosis (Mode of Action and Mechanism of Resistance)

Atovaquone (ATV) is an antiprotozoal compound and used against apicomplexan parasites including *Babesia* spp., *T. gondii*, and *Plasmodium* spp. (Baggish & Hill, 2002; Kessl et al., 2003). ATV is very effective in *B. divergens* infection (Vial & Gorenflot, 2006). Atovaquone along with azithromycin (AZ) is more effective as compared to quinine and clindamycin used for the treatment and prevention of babesiosis in hamsters (Wittner et al., 1996). A combined regimen of ATV and

AZis found more efficient in human babesiosis especially in immunocompetent adult patients (Krause et al., 2000). This regimen effectively inhibits *B. divergens* growth. It works by inhibiting the rate of oxygen consumption (Murphy & Lang-Unnasch, 1999) and acts as a ubiquinone (mitochondrial protein involved in electron transport chain) analog for Protozoa (Baggish & Hill, 2002).

A surfen-derived aromatic compound diminazene aceturate (DA) is a conventionally used drug for babesiosis. The compound is marketed as 45% diaceturate salt, along with 55% antipyrine, which is added as stabilizer due to short stability of diminazene in water (Fairclough, 1962). It is the most common anti-trypanosomal agent and also found effective in babesiosis. DA binds irreversibly to dsDNA in grooves present between complementary strands with the help of specific binding with sites rich in adenine and thymine base pairs (Peregrine, 1994). This binding has a high affinity for kinetoplastid DNA (kDNA) and causes disturbance of kinetoplast replication and function (James & Giles, 1985). Diminazene also inhibits mitochondrial topoisomerase II (Peregrine, 1994). DA is highly effective against *B. bigemina* as compared to *B. divergens* and *B. bovis* (Kuttler, 1981).

4.4.2 Resistance to Diminazene Aceturate

The mechanism of resistance to DA is unclear but there are different theories to explain this resistance. The mitochondrial membrane appears to be a primary target of drugs that are effective against apicomplexan parasites. It is suggested that some modification of mitochondrial molecules are involved in DA resistance (De Souza et al., 2009; Mi-Ichi et al., 2005). These changes include, changes in mitochondrial DNA (mtDNA), particularly terough mutation of cytochrome b (CYTb), cytochrome c oxidase subunit I (COXI) or cytochrome c oxidase subunit III (COXIII) (Matsuu et al., 2006; Sakuma et al., 2009). However, a study of DA-resistant *B. gibsoni*, suggested that there is neither change in amino acid and nucleotides of these genes, norin their transcription changes so, these are not involved in the DA resistance mechanism (Rajapakshage et al., 2012a). During normal physiological conditions, most protozoan parasites including *Plasmodium* spp. (Mogi & Kita, 2009), *B. gibsoni* (Yamasaki et al., 2003) fulfill their energy requirements through their glycolytic system. Adaption in the glycolytic pathway during resistance is reported in different parasites. It is reported that DA-resistant strains showed a higher level of ATP and glucose as compared to wild type (susceptible) parasites (Rajapakshage et al., 2012b).

Heat shock proteins (HSPs) are reported to play an important role in stress resistance development and help in adaption in the environment. The higher level of HSPs enhances the tolerance against the stress, along that HSP27 and HSP90 involved in drug resistance mechanisms (Lindquist, 1986). The HSP70 of D- resistant strains in *B. gibsoni* is found to be expressed at lower levels during long term selection pressure with DA (Hwang et al., 2010).

4.4.3 Resistance to Atovaquone

A combination of atovaquone (ATV) and azithromycin (AZ) is used for babesiosis and it is found more effective than a combination therapy of clindamycin and quinine. However, resistance to ATV-AZ is reported in some cases and some different possible mechanisms have been reported for these cases of resistance. The mechanism of action of ATV is through targeting the electron transport chain in mitochondria by binding with cytochrome b protein (CYTb) while in the case of AZ, it is through inhibition of protein translation in the apicoplast (Birth et al., 2014; Sidhu et al., 2007). Resistance to the ATV–AZ combination is due to mutational changes in binding regions of its target (Lemieux et al., 2016) and involve point mutations in the CYTb and RPL4 genes. Two significant point mutations in CYTb include an adenine to guanine shift at position 815 and a tyrosine to cysteine change at position 272. In the RPL4 gene, there appears to bea substitution from arginine to cysteine at position 86 that gives rise to the insensitivity of that molecule. (Simon et al., 2017). This mutation, along with other mutations in CYTb ubiquinol-binding pocket (Qo domain), have been shown to be associated with susceptibility of ATV in *T. gondii* (McFadden et al., 2000), and *Plasmodium* spp. (Birth et al., 2014). In the case of resistance to AZ, corresponding mutation near the R86 conserved regions of the RPL4 gene is associated with AZ resistance in *Plasmodium* and various bacteria (Sidhu et al., 2007).

4.5 Cystoisosporiasis

The coccidian protozoan, *Isospora* (*I.*) infects humans and companion animals like cats and dogs and it has been known for2 centuries (Levine, 1973). It was considered until the 1970's that *Isospora* species has more than one host (Dubey, 2018). *Isospora belli* infects the gastrointestinal system of humans. Previously, the parasite was misdiagnosed as *Isospora hominis* but Wenyon (1923) gave the name, based on the distinctive morphology of the oocysts of *I. belli* (the large, bell-like oocyst that are 25–30 × 12-15 μm in size). After the discovery of tissue cysts in 1987, *I. belli* named as *Cystoisospora belli* (*C. belli*) (Barta et al., 2005). The recent review on the significance of *C. belli* has been documented by Dubey and Almeria (2019). This disease is generally asymptomatic or causes mild disease in immunocompetent humans but occurs with severe and prolonged symptoms in HIV patients or non-HIV patients receiving immunosuppressive drugs due to tumors (e.g., human T-cell leukemia virus 1 associated T-cell lymphoma) or after organ transplantation (Akateh et al., 2018; de Hovitz et al., 1986; Ud Din et al., 2012; Marathe & Parikh, 2013; Sanches et al., 2015).

The unsporulated oocyst are excreted in the stool of human. In the environment, sporulation occurs resulting in sporulated oocyst usually containing 2 sporocysts and 4 sporozoites in each sporocyst (Jongwutiwes et al., 2007). Sometimes sporulation occurs in the intestine and partially or fully sporulated oocysts are passed in the feces (Nateghi et al., 2014). There is direct fecal-oral route of transmission with no involvement of paratenic host. In immunosuppressed patients, tissue cysts develop in gut lamina propria, spleen, lymph nodes (mesenteric, mediastinal and trachea-bronchial), and liver which are extra-intestinal organs (Lindsay et al., 2014; Swanson et al., 2018). This parasite (*C. belli*) is worldwide in distribution but disease is prevalent mostly in tropical and subtropical regions of the globe including America, South-East Asia, Caribbean, and Africa (Dubey & Almeria, 2019).

4.5.1 Drugs Related to Cystoisosporiasis (Mode of Action and Mechanism of Resistance)

The trimethoprim-sulfamethoxazole combined therapy is recommended for cystoisosporiasis among the anticoccidial drugs tested against *C. belli* (Comin & Santucci, 1994; Haider et al., 2013; Verdier et al., 2000; Woon et al., 2016). The pyrimethamine and sulfadiazine or sulfadoxinemay act as the alternative for patients hypersensitive to trimethoprim-sulfamethoxazole combination therapy (Weiss et al., 1988) and other option is ciprofloxacin in this scenario (Esvan et al., 2018). Other drugs, which have been reported to be successful in treating cystoisosporiasis in humans, include nitazoxanide (Marathe & Parikh, 2013), nitrofurantoin (French et al., 1964), albendazole (Dionisio et al., 1996), and roxitrhomycin (Musey et al., 1988).

4.5.2 Resistance to Trimethoprim-Sulfamethoxazole

Although, trimethoprim-sulfamethoxazole combination provides rapid relief, relapses of clinical infection may occur most often in immunocompromised humans (Boyles et al., 2012). In case of relapses, prolonged therapy or increased daily doses may require to eliminate the disease (Hasan et al., 2017). The side effects of this combination of drugs have been faced in the form of allergic reactions resulting in Lyell syndrome, decrease in number of white blood cells and skin rash (Esvan et al., 2018). In addition, development of resistance or failure of therapy through trimethoprim-sulfamethoxazole combination were reported in some cases of cystoisosporiasis in humans (Boyles et al., 2012; Mudholkar & Namey, 2010).

This resistance to combination is considered as resistance to both drugs and there would be numerous possibilities for that but the exact mechanism of this resistance in *C. belli* is unknown. However, it has been reported that *T. cruzi* also has DHFR and resistance to trimethoprim develop due to DHFR insensitivity to this drug. The single mutation to thymine auxotrophy can cause resistance to the sulfonamides, trimethoprim, and their combinations (Mayers et al., 2017; Then, 1982).

4.6 Trypanosomiasis

Human African Trypanosomiasis (HAT), is caused by *Trypanosoma brucei* (*T.b.*) in Africa. HAT is transmitted to animals and humans by the bite of the tsetse fly (*Glossina* species). A slow-progressing form caused by *T.b. gambiense* is present in central and west Africa and the faster-progressing form caused by *T.b. rhodesiense* is present in southern and eastern Africa. An increase in HAT cases has been reported from 1980 to 2006 after which there was significant reduction. Recently, with the help of control measures and eradication strategies adopted by WHO and other organizations, there has been a reduction in disease burden (Barrett, 2018; Kennedy & Rodgers, 2019). If infection remains untreated, it becomes fatal in humans. The life cycle consists of two phases, one in human being and second in the insect vector. The tsetse fly is hematophagous and possesses vector capabilities for trypanosomiasis. After ingesting the trypanosomes from skin or blood of mammals, tsetse flies become infected. In the insect midgut, these trypanosomes undergo a series of changes by multiplying before entering salivary glands and human infective metacyclic forms develop. In the next phase of the lifecycle, these infected flies bite a new host. These trypanosomes can reside in the host for a long period and can originate disease even after treatment. So, routine analysis of body fluids, and follow-up after 24 months of treatment are the necessary precautions (Buscher et al., 2017).

Two major disease states of HAT occur in effected humans. The first is "acute bloodstream-form" which can be lethal to the host, and can affect spleen, heart, and liver. The second form in which trypanosomes can cross the blood–brain barrier and can enter the central nervous system is called the "encephalitic stage". In the second form, severe damage takes place in meninges resulting in meningoencephalitis (Buscher et al., 2017; Franco et al., 2018).

These trypanosomal protozoans are versatile organisms and are capable to adapt to drug pressure. Drug resistance may be due to (a) change in the molecular target of the drug (b) change in influx/efflux ratio (c) change in general repair and defense mechanisms. In trypanosomiasis, resistance rate is lower than the other protozoal diseases because drugs used for its treatment are multitargeted, except eflornithine. Due to the misuse of similar drugs to treat the animal and human trypanosomiasis endemic and rural areas showed higher levels of drug resistance (Franco et al., 2018).

Trypanosoma normally revolves among salivary glands and alimentary tract of Tsetse fly and blood or tissue of the final host i.e., mammals. During life, this pathogen experiences several morphological froms incuding procyclic and metacyclic forms and epimestigote forms. All these stages have different mechnisms of inhabitation within specific site of the host. Several biochemical and metabolic changes have been adopted by different life cycle stages supported by changes in the expression of genes. These variations help trypanosoma to evade host immune responses. At vector end, Tsetse fly after a long evolutionary survival, has developed the ability in their saliva to create an immune tolerant environment suitable for the establishment and proliferation of parasite (Bernard et al., 2014). Moreover, immune evasion mechanisms in terms of morphological, structural and metabolic alterations are also established by Trypanosoma to fulfill its nutrient requirement in the host (of the parasite also acticate immune evasion mechanisms (Barry & Emery, 1984).

4.6.1 Drugs Related to Trypanosomiasis (Mode of Action and Mechanism of Resistance)

Pentamidine (PMD) is used to treat the 1ststage of disease caused by *T. b. gambiense* (g-HAT) and *T. b. rhodesiense* (r-HAT). Pentamidine inhibits the synthesis of RNA and DNA by interfering with nuclear mechanisms. Resistance against pentamidine in HAT is reported and is due to loss of P2 aminopurine transporter function (Franco et al., 2018).

Suramin is also used against the 1st stage of r-HAT and g-HAT. It acts by inhibiting glycolytic enzymes. Laboratory derived resistant strain is reported in the case of suramin (Wiedemar et al., 2018).

Melarsoprol is used for the treatment of the second form of the disease caused by g-HAT and r-HAT. The mechanism of action of melarsoprol is still not understood, fully.

4.6.2 Resistance to Pentamidine

PMD is a drug of choice for leishmaniasis but also used to treat trypanosomiasis. It is still used as a therapeutic and tolerated by the patients with minor side effects. PMD is found more effective for 1st stage disease in g-HAT than comparable stage in r-HAT. Transporters that are involved in PMD uptake and internalization include P2 adenosine/adenine and aquaglyceroporin 2 (AQP2) (Franco et al., 2018). With the help of endocytosis and active transport, PMD is accumulated in the nucleus and mitochondria. Pentamidine is a DNA-binding-drug, causing destruction of kinetoplast DNA and modulation in mitochondrial membrane potential (Baker & Welburn, 2018; Thomas et al., 2018).

Resistance in PMD is due to loss of function of P2 aminopurine transporter, which helps in the uptake of PMD in *T. brucei*. The cross-resistance of PMD or melarsoprol is a major concern for the treatment of HAT (Franco et al., 2018). Along with P2 transporter/AT1, surface protein like AQP2 (helps in osmoregulation) are involved in cross-resistance of PMD/melarsoprol (Quintana et al., 2018).

4.6.3 Resistance to Suramin

Suramin is a drug of choice to treat 1st stage r-HAT. It is also used in synergism with stage second treatment drugs i.e., melarsoprol, eflornithine, and nifurtimox while in case of pentamidine, it has inhibition effect (Babokhov et al., 2013). This drug mainly targets the glycolysis process along with some other pathways like inhibition of an enzyme (6-phosphogluconate dehydrogenase) which belongs to the pentose phosphate pathway. The exact mechanism of action is not well understood but there are some hypotheses according to which several enzymes are involved for the development of resistance in humans (Babokhov et al., 2013; Wiedemar et al., 2018). Suramin can also inhibit cytokinesis. Recently, due to continuous mitosis and defect in cytokinesis, two or more nuceli were observed in individual *T. brucei* trophozoites after suramin treatment (Thomas et al., 2018).

Suramin has been used for around a hundred years, and the compound has a 44–54 days half-life in humans. There are no reports as yet of suramin resistance in HAT in humans. Resistant strains can be developed in laboratory conditions. It was confirmed that trypanosomes of the bloodstream mask themselves with variant surface glycoprotein (VSG) which protects them from the immune defense of mammalian hosts and might be involved in the resistance mechanism. The VSGSur gene expression was found correlated with suramin resistance just after exposure to high concentrations of suramin (Wiedemar et al., 2018).

4.6.4 Resistance to Melarsoprol

Melarsoprol (Mel B) is an inorganic arsenical compound which is used for trypanosomiasis treatment. Mel B was first used in 1858 and is used presently for intravenous administration to treat both forms of HAT. The Mel B is approved for the first and second stage of r-HAT and g-HAT, respectively. Mel B showed a number of severe side effects, the most severe one is a reactive encephalopathy. Due to its toxicity, Mel B has been replaced with nifurtimox-eflornithine combination therapy (NECT) in West Africa to treat g-HAT but in East Africa, Mel B is still in use to treat the advanced stage of r-HAT (Thomas et al., 2018). Melarsoprol and NECT are the only drugs which cancross the blood–brain barrier, therefore, these drugs are used to treat the 2nd stage of trypanosomiasis.

The mode of action of this trivalent arsenical compound is not clearly understood. It shows an affinity for vicinal thiols in proteins particularly and for sulfhydryl groups. The particular intake of Mel B is through AQP2, Adenosine transporter 1 (AT1 gene), and P2 transporters of the parasite. After entering the cell, a complex is formed when the drug reacts with the dithiol group and inhibits trypanothione reductase. However, cell death due to this mechanism is not reported but in contrast, glycolysis may cause cell lysis. It was described recently that Mel B is involved in disrupting mitosis

(Fairlamb & Horn, 2018; Thomas et al., 2018). The resistance mechanism is due to the loss of drug uptake due to mutation in transporters AQP2, AQP3, and P2 (Graf et al., 2015).

4.7 Leishmaniasis

It is a complex disease, mostprevalent in tropical/sub-tropical areas. More than 50 species in the genus *Leishmania*have been discovered and of these, 20 are pathogenic to humans. Leishmaniasis is endemic in 98 countries and of one million annual cases, about 26,000–65,000 deaths have been reported (Perez-Cabezas et al., 2019). Leishmaniasis exists in four different forms, visceral leishmaniasis (VL) or kala-azar (a very severe form leading to fatal outcomesif persists), post kala-azar dermal leishmaniasis (PKDL; this form occursafter 6–12 months of possible cure of VL), cutaneous leishmaniasis (CL; a commonform of the disease causing skin lesions, ulcers, and scars) and mucocutaneous leishmaniasis (MCL; causing mucus membrane destruction of nose, throat, and mouth) (Burza et al., 2018).

The disease is transmitted by 90 species of the phlebotomine sand fly. The life cycle of the parasite consists of promastigotes forms, present in the sand fly gut, and the amastigotes form, present in the macrophages of the host. The life cycle starts when an infected sand fly vector takes a blood meal from the human host and injects the metacyclic promastigote stage. These promastigotes undergo phagocytosis by macrophages then they convert into non-flagellated amastigotes. The parasites then transfer to new cells, invading adjacent macrophages of the reticuloendothelial system, followed by those of the bone marrow, spleen and liver. The reproduction of parasites occurs through binary fission (Alemayehu & Alemayehu, 2017).

In low-income countries, therapeutic measures display success rate, due to resistance, toxicity, and the high cost of drugs. Due to the emergence of co-infection of HIV-VL, there has been a delay in the eradication of the disease in some countries (Diro et al., 2014). Several resistance mechanisms were identifiedin the last few years. To control drug resistance, it is necessary to know the biochemistry and molecular characteristics of the drugs (Capela et al., 2019).

Several mechnaisms of immune evasions Leishmania are studies extensively in the world including alteration in signaling pathoways of host cells, changes in the activity of phagocytic cells, modulation of profiles of effector cells including cytokines and chemokines and trafficking at infection site (Gupta et al., 2013). Complement system-based destruction of amastigotes of *Leismania tropica* are more susceptible as compared to the amastigotes of *Leishmania donovani*. This might be due to more resistant nature of the later pathogen (Hoover et al., 1984). Componenets of innate immunity including neutrophils and natural killer cells are actively involved in the destruction of Leishmania infection by producing several antimicrobial substances including nitric oxide, platelet activating factors, neutrophil extracellular traps and neutrophil elastase but still destruction depends on the strain of Leishmania (Riberio-Gomes et al., 2007). Metacyclic promastigotes which are infective restrict the entry of complement by showing resistance to lysis through increased expression of protein kinases and deactivation of classical pathway (Puentes et al., 1990). *Leishmania major* activate the suppressors to reduce the cytokine activation proteins to limit the cytokine induction through production of Toll like receptors-2 (de Veer et al., 2003).

4.7.1 Drugs Related to Leishmaniasis (Mode of Action and Mechanism of Resistance)

Sodium stibogluconate is a generally recommended drug used for all four forms of leishmaniasis. It can be used in combination with paromomycin. This drug inhibits Trypanothione reductase which is the basic requirement of parasite for its survival in the host. Resistance against sodium stibogluconate has beenreported and this resistance is contributed to by high intracellular thiols levels, MRP1 (multidrug resistance associated protein 1), overexpression of TXN/TXNPx (tryparedoxin/tryparedoxin peroxidase system), or due to ATP-binding cassette (ABC) transporters (Ponte-Sucre et al., 2017; Turcano et al., 2018).

Another drug used for leishmaniasis is pentamidine. It is effective in VL and CL and HIV co-infection cases. Its mode of action is still not clear; however, it may be due to inhibition of ribonucleic acid (RNA) polymerase, interaction with kinetoplast DNA (kDNAs), interference with polyamine synthesis, apoptotic death, and inhibition of topoisomerase II (TOPII). Pentamidine resistance is reported and may be due to aquaglyceroporin-2 mutation and overexpression of pentamidine resistance protein 1 (PRP1) (Jeacock et al., 2017; Yang et al., 2016).

Paromomycin is also used for the treatment of PKDL and CL form of leishmaniasis. Its mode of action is not clear but it may be due to alteration in membrane fluidity, protein synthesis inhibition, respiratory dysfunction, decreased potential of mitochondrial membrane, and lipid metabolism. No resistance against paromomycin is described until now, however, various hypotheses have been described based on laboratory-derived resistant strains (Fernandez et al., 2011; Hendrickx et al., 2017).

4.7.2 Resistance to Sodium Stibogluconate

The only effective drug of choice for leishmaniasis treatment is sodium stibogluconate (SSG), a drug with satisfactory results for all forms of leishmaniasis. Side effects encountered with this drug are cardiac and renal toxicity and pancreatitis. No oral preparation is available, only injectable form is present (Berg et al., 2013). The MoA of SSG includes inhibition of trypanothione reductase (TR) enzyme, which helps in reducing trypanothione, neutralization of reactive oxygen species presented by macrophages with the help of *Leishmania* TXN/TXNPx system. Trypanosomatid parasites use trypanothione as a detoxifying system in response to oxidative damage while mammals use glutathione (Turcano et al., 2018).

Emergence of resistance may be due to certain factors including (i) decrease in concentration of drug in parasite either through low intake of drug or through increase efflux of a drug; (ii) inhibition of drug activation (iii) and gene amplification (Maltezou, 2010). Over expression of the TXNP × system and high levels of intracellular thiol are the reasons for the high degree of resistance to Sb(III). The activation of inhibition of Sb(V) and decreased intake of Sb(III) active form by amastigotes (during metabolism of thiol) have been verified in in vivo antimonial resistance. The lower expression of genes, glutamylcysteine synthetase (involved in the metabolism of glutathione), ornithine decarboxylase (involved in the metabolism of the trypanothione), and aquaglyceroporin 1 which involved in the intake of Sb(III), have been observed.

The resistance of SSG has also been reported due to the overexpression of membrane bound ABC transporters on the *Leishmania* surface. Resistance developed when ABC transporters changes in efflux as well as in accumulation of the intracellular drug (Ponte-Sucre et al., 2017). The SSG resistance is also developed due to permeability of glycoprotein (P-gp) and MRP1 in the host cells through up-regulation by Sb-resistant parasite, inhibition of intracellular drug accumulation and decrease in antimony influx (Garcia-Salcedo et al., 2016; Maltezou, 2010).

4.7.3 Resistance to Pentamidine

Pentamidine (PMD) was first time used in 1937 to treat sleeping sickness. The first report of its use for leishmaniasis was reported in 1949 from India for the treatment of VL. This drug is used for CL caused by *L. panamensis* and *L. guyanensis* (WHO, 2010). Side effects observed for this drug include myocarditis, hypotension, hypoglycemia, fever, and renal toxicity.

Due to these side effects, it was banned in India in the 1990s (Nagle et al., 2014).

The exact mode of action is not clear but it is assumed that PMD disturbs the inner membrane potential of parasite mitochondria. The mitochondrial accumulation of this drug induces apoptotic death of *L. donovani* (bloodstream form) through an increase of cytosolic Ca2 + , inhibition of respiratory chain complexes (I, II, and III), and generation of reactive oxygen species (ROS) thus increasing the cytotoxicity of the drug (Chakravarty & Sundar, 2010; Yang et al., 2016). This drug also targets DNA topoisomerases (TOPs), which are a basic requirement during transcription, replication, repair, and recombination of DNA topology. Topoisomerases enzymes, TOPI and TOPII of *Leishmania* parasites, are different from that of humans and perform an important role in organizing the kinetoplast DNA network. It is selective for TOPII of *Leishmania* (Cortazar et al., 2007).

In *L. donovani*, PMD is also found to be susceptible for competitive inhibition of arginine transport, whereas in *L. mexicana*and *L. infantum*, PMD is affectedby a non-competitive inhibition of spermidine transport and putrescine. In *L. mexicana*, PMD can enter into amastigotes and promastigotes through a carrier-mediated process. In amastigotes, the velocity of PMD is higher than it is in promastigotes (Bray et al., 2003).

In *Leishmania*, mutations of various transporters are responsible for PMD resistance e.g. the ABC transporters identified from many *Leishmania* species are involved in resistance (Pramanik et al., 2019). The PMD resistance protein 1 (PRP1) and p-glycoprotein (PGP) are members of ABCC7 (a superfamily of ABC transporters) (Coelho et al., 2007). Aquaglyceroporin 2 (AQP2), belonging to a surface channel protein family, is responsible for passive transport of water across the membrane, and is involved in resistance process against higher concentrations of PMD in trypanosomes (Baker et al., 2012). It is found that the same AQP2 protein mutation is also responsible for PMD resistance in leishmaniasis (Jeacock et al., 2017). Along with this P-glycoprotein inhibitor, verapamil and $Ca2^{+}$ channel blockers are also involved in the inhibition of PMD efflux (Basselin et al., 2002).

To overcome the resistance, some flavonoid dimers were synthesized which remarkably reversed the resistance process through increased accumulation of drugs in mitochondria of *L. enriettii*. The flavonoid dimers, along with quinacrine, showed a synergistic effect on reversing the resistance of PMD in *Leishmania* (Wong et al., 2009).

4.7.4 Resistance to Paromomycin

Paromomycin (PMM) was first time used in 2006 to treat VL. PMM can be administered through oral and intramuscular injections having no secondary effects (Wiwanitkit, 2012). PMM interacts with ribosomal subunits and inhibits protein synthesis during bacterial infections. In *E.coli,* this drug binds to 16S rRNA on the major groove of the A-site and causes misreading of mRNA. The mode of action of PMM against *Leishmania* is still not well understood, but it is thought that it causes protein synthesis inhibition by binding and changing conformationally at the A-site of 16S rRNA. During translation, the modification happens at the N1 position of A1492 and A1493. Other mechanisms of action include (i) decrease in mitochondrial membrane potential (ii) change in membrane fluidity and (iii) change in lipid metabolism (Chawla et al., 2011; Fernandez et al., 2011). Endocytosis plays role in uptake of PMM and facilitate its binding with surface proteins of the parasite such as P-type H^{+} ATPase, prohibitin, paraflagellar rod (PFR) 1D and 2D (Légaré and Ouellette, 2017).

The advantages of the use of PMM as an anti-*Leishmania* drug include easy availability, economics, safe profile, and short duration of administration. Its physiochemical properties suggest that low concentration should be used at the infection site. Administration of drug with some delivery systems such as solid lipid nanoparticles enhances its penetration into macrophages and boosts immune responses (Heidari-Kharaji et al., 2016). A formulation uses albumin microspheres as a delivery system for PMM to target *Leishmania* in macrophages for VL treatment (Khan et al., 2013).

Conjugation of PMM with stearylamine (SA)-bearing phosphatidylcholine (PC) liposomes has also been used, which results in increase in immuneresponses, zerotoxicity, and effective anti-Leishmanial activity (Banerjee et al., 2011).

The resistance of PMM in prokaryotes may be due to enzymatic inactivation of the drug, decreased drug accumulation, and mutation in ribosomal binding sites. In the case of eukaryotic *Leishmania*, the resistance mechanism is not clear but it may be due to decreased drug uptake (Bhandari et al., 2014; Hendrickx, et al., 2017). Some studies suggest that the resistance in *L. donovani* is due to decrease in the binding of drugs with the cell surface and reduction in drug accumulation (Jhingran et al., 2009). Vesicles and vacuoles are involved in vesicular trafficking are associated with resistant strains as compared to others along with a high level of glycolytic enzymes and augmented basal level of stress proteins from the HSP70 family (Ouelletteet al., 2004). A high level of glycolytic enzymes shows that resistant strains depend upon glycolysis for their energy requirements.

4.8 Giardiasis

Giardia is a gastro-intestinal parasite with cosmopolitan distribution. The species responsible for human disease is common called *Giardia (G.) lamblia* (syn. *intestinalis, duodenalis*) (Buret, 2008). *Giardia* infection is mostly self-limiting and present without symptoms. In case of severe giardiasis, clinical symptoms like diarrhea, malabsorption, bloating, and nausea may occur. Lactose intolerance and irritable bowel syndrome are post-infection symptoms (Farthing, 1997; Halliez & Buret, 2013). In developing countries, about 33% of the population carries the parasite, with high prevalence of giardiasis.

Giardia is one of the most common protozoal parasites affecting the gastrointestinal (GI) tract of humans. This protist settles in the small intestine and completes its lifecycle in two stages. The first stage, the cystic form, is metabolically inactive, acid-resistant, and enter the stomach with water or food. The stage transits the stomach to the small intestine, where it excysts, releasing trophozoites, the active feeding stage. This stage either attaches to the mucosa of the intestine or freely moves in the lumen. A small number of trophozoites can undergo developmental changes and convert into cysts that pass with feces and contaminate water and food after its excretion. When a host ingests this cyst a new cycle of infection starts in the host (Leitsch, 2015).

No prophylaxis is available against *Giardia* infections, so chemotherapy is the only option used for the treatment. A large range of drugs is available against giardiasis. The most common drugs used for giardiasis treatment are 5-nitroimidazoles metronidazole, benzimidazole, albendazole, and tinidazole (Tejman-Yarden et al., 2013). Metronidazole and albendazole alone are effective in about 90% of cases but there exist recalcitrant giardiasis, in which drug resistance is suspected (Tejman-Yarden & Eckman, 2011). In such drug-resistance cases, metronidazole and albendazole or quinacrine combination therapy is used (Mørch et al., 2008). It is reported that drug resistance is more frequent in *Giardia*than is appreciated, a matter of high attention because it is a widely distributed and prevalent parasite.

4.8.1 Drugs Related to Giardiasis (Mode of Action and Mechanism of Resistance)

The MoA of metronidazole is to cause breakage of the double-strand DNA in anaerobes and microaerophiles (Uzlikova & Nohynkova, 2014). Metronidazole also acts by forming covalent adducts with cysteines in proteins. The compound particularly targets thioredoxin reductase, a redox enzyme, in *G. lamblia* by inhibiting its function as a disulfide reductase. These mechanisms cause severe oxidative stress. It is observed that when cysteine is added in growth medium it protects *Giardia* from the effects of metronidazole. Metronidazole also causes degradation/breakdown of elongation factor 1-γ enzyme required for protein translation in *Giardia* (Leitsch et al., 2012).

Albendazoletargets β-tubulin (Edlind et al., 1990), which is the major component of the *G. lamblia* cytoskeleton (Holberton & Ward, 1981). Mebendazole and albendazole bind to β-tubulin (MacDonald et al., 2004) and cause inhibition of microtubule polymerization resulting in severe structural defects (Chavez et al., 1992).

4.8.2 Resistance to Metronidazole

Metronidazole (5-nitroimidazole) was developed against *Trichomonas* in the late 1950's (Cosar & Julou, 1959). After that, it was also found useful against *Giardia* and currently, it is listed as the most important drug for giardiasis by WHO. To become toxic to the parasite, metronidazole (a prodrug) needs to be reduced at its nitro group (Upcroft & Upcroft, 2001). Microaerophilic and anaerobic organisms possess strong reducing physiology, while the redox potential of metronidazole is very low (Samuelson, 1999; Yarlett et al., 1985). Resistance in the case of metronidazole is reported in anaerobic and microaerophilic pathogens as well as in *Giardia*. Resistant *Giardia* strains were isolated frequently from refractory giardiasis victims (Lemée et al., 2000). In laboratory conditions, it is very easy to produce resistance in *Giardia* against metronidazole and other 5-nitroimidazoles by exposure to UV light (Argüello-García et al., 2009; Müller et al., 2007).

The mechanism of resistance to metronidazole has been studied for the last 40 years but the exact mechanism is still unclear. Two basic factors are described. These include the activation of pyruvate ferredoxin oxidoreductase (PFOR) and 5-nitroimidazoles present in anaerobic and microaerophilic organisms being central for reduction (Narikawa, 1986). The PFOR reduced the ferredoxin and transfer an electron to the metronidazole nitro group (Liu et al., 2000; Yarlett et al., 1985). In laboratory investigations, reduced PFOR activity was observed in metronidazole-resistant *Giardia* strains (Leitsch et al., 2011). Knock-down of PFOR expression with the help of hammerhead ribozymes produced metronidazole resistant strains of *Giardia* (Dan et al., 2000). The cell lines of *Giardia* were found to be resistant to second-generation 5-nitroimidazole (Dunn et al., 2010), possessing fully functional PFOR and ferredoxin, and were shown to be highly cross-resistant to metronidazole (Leitsch et al., 2011). I Activation of the ferredoxin pathway or PFOR is not the basic requirement for metronidazole resistance. Some studies have indicated that the metabolism of metronidazole is also controlled by nitroreductase 1 and 2. The activation of metronidazole to reactive metabolites can be controlled by nitroreductase 1 (Nillius et al., 2011). While, nitroreductase 2 provides protective effect by reducing the nitro group with six electrons to non-toxic amino group and overexpression of this nitroreductase 2 made *Giardia* more tolerant to metronidazole (Müller et al., 2013). In metronidazole-resistant cell lines of *Giardia*, the thioredoxin reductase (TrxR) factor was not found to be effective in the reduction of 5-nitroimidazoles (Leitsch et al., 2009).

4.8.3 Resistance to Albendazole

Albendazole was first used to treat helminth infections in 1983 was later found to be effective against *G. lamblia* (Reynoldson et al., 1991). At the dose rate of 400/day for 5 days, it was found as effective as metronidazole. However, due to its poor absorption in the human gut, albendazole showed some side effects (Solaymani-Mohammadi et al., 2010). So, it is considered as alternative therapy tometronidazole for giardiasis. Other drugs like mebendazole and benzimidazole were also found useful in giardiasis treatment (Almirall et al., 2011; Edlind et al., 1990). The teratogenic effect in rats was recorded for both albendazole and mebendazole (Escobedo & Cimerman, 2007).

Albendazole resistance is reported from several clinical and laboratory cases (Leméeet al., 2000; Paz-Maldonado et al., 2013). Some *Giardia*isolates have IC50's to albendazole that are 1000-fold higher that that of susceptible isolates (Paz-Maldonado et al., 2013). Albendazole resistance in helminths is due to the mutation of amino acid at positions 167 and 200 of the β-tubulin gene (Beech et al., 2011). For *Giardia,* conflicting observations exist for potential sites that confer resistance. In another study, an altered sequence of amino acid of β-tubulin was identified in the prolonged exposure of albendazole with increasing concentration (Jiménez-Cardoso et al., 2009), whereas others found no such alterations in the sequence. In some studies, modified expression of various genes (including a VSP-like protein), cytoskeletal aberrations, chromosomal rearrangements (Argüello-García et al., 2009), several enzymes (i.e. triosephosphate isomerase and NADH oxidase), and constituents of the ventral disk (Paz-Maldonado et al., 2013) were found in cell lines resistant to albendazole. Surprisingly, PFOR was also found to be up regulated in cell lines resistant to albendazole (Argüello-García et al., 2009).

So far, these are the assumptions but it is described that benzimidazole targets more regions in *G. lamblia* as compared to helminths.

4.9 Conclusion

Resistance to drugs in eukaryotic microorganisms is an unavoidable global issue that hampers the advances in human health which have been made in the course of the last 50 years. The same trend regarding resistance development is followed by bacteria and viruses but this condition is considered acute in case of eukaryotic organisms due to their abundant population in the underdeveloped countries of globe. These underdeveloped countries have minimal resources and facilities to contain the drug resistance through the advancement of new medications, antibodies, vaccines and diagnostics, while developed nations require funds to carry out the research. However, this threat can be overcome by exploring the biological aspects of these eukaryotic pathogens through research. Moreover, most of vector-borne disease are due to eukaryotic pathogens therefore control of vector or disease transmission routes can help in containment of pathogens spread with the exception of opportunistic organisms. This can be done through transmission-blocking immunogens or medications focusing on *Plasmodium*, or the use of vector control strategies, for example, tsetse traps, spray of insecticides in peri-residential and indoors, application of insecticide coated bed nets and better housing development. One of the vector population control strategies is the sterile insect release limiting the disease transmission. Like few bacteria, eukaryotic pathogens have also been reported to represent mutual and social interactions with their hosts or vectors to survive. These social interactions can help in the control of parasite population or the development of life cycle stages through blocking or copying signals for correspondence or their transduction pathways, providing way forward for the novel and alternative parasite control strategies which might not lead to rapid development of resistance.

Regardless of new targets or approaches can be recognized, there is a genuine need to improve the delivery and organization of medicines. Quality, cost, and distribution of drugs are very important factors. Moreover, the evolutionary theory and epidemiological modelling both are crucial in extending the drugs life over longest period thereby increasing the yield on substantial cost related to development and introduction of a new drug. Targeted treatment, instead of implementing mass medication, is critical to restrict the development of resistance. In this regard, the implementation of supply chain control, study of disease and accurate diagnostic tools are required. In addition, on development of resistance, funds should be deployed for the discovery and availability of new and alternate drugs. The successful containment of eukaryotic microbes depends on slowing the process of resistance emergence and boosting the process of identification of new drug candidates.

References

Achan, J., Talisuna, A. O., Erhart, A., Yeka, A., Tibenderana, J. K., Baliraine, F. N., Rosenthal, P. J., & D'Alessandro, U. (2011). Quinine, an old anti-malarial drug in a modern world: Role in the treatment of malaria. *Malaria Journal, 10*(1), 144–155.

Akateh, C., Arnold, C. A., Benissan-Messan, D., Michaels, A., & Black, S. M. (2018). *Cystoisospora belli* gallbladder infection in a liver transplant donor. *Case Report in Infectious Diseases, 2018*, 1–5.

Alday, P. H., & Doggett, J. S. (2017). Drugs in development for toxoplasmosis: advances, challenges, and current status. *Drug Desigen, Development and Therapy, 11*, 273–293.

Alemayehu, B., & Alemayehu, M. (2017). Leishmaniasis: A review on parasite, vector and reservoir host. *Health Science Journal, 11*, 1–6.

Alker, A. P., Lim, P., Sem, R., Shah, N. K., Yi, P., Bouth, D. M., Tsuyuoka, R., Maguire, J. D., Fandeur, T., Ariey, F., Wongsrichanalai, C., & Meshnick, S. R. (2007). Pfmdr1 and in vivo resistance to artesunate-mefloquine in *falciparum* malaria on the cambodian-Thai border. *The American Journal of Tropical Medicine and Hygiene, 76*, 641–647.

Allred, D. R., & Al-Khedery, B. (2006). Antigenic variation as an exploitable weakness of babesial parasites. *Veterinary Parasitology, 138*(1–2), 50–60.

Almirall, P., Escobedo, A. A., Ayala, I., Alfonso, M., Salazar, Y., Canete, R., Cimerman, S., Galloso, M., Olivero, I., Robaina, M., & Tornés, K. (2011). Mebendazole compared with secnidazole in the treatment of adult giardiasis: A randomised, no-inferiority, open clinical trial. *Journal of Parasitology Research, 2011*, 1–6.

Alonso, P. L., Brown, G., Arevalo-Herrera, M., Binka, F., Chitnis, C., Collins, F., Doumbo, O. K., Greenwood, B., Hall, B. F., Levine, M. M., & Mendis, K. (2011). A research agenda to underpin malaria eradication. *PLoS Medicine, 8*(1), 1–8.

Amato, R., Pearson, R. D., Almagro-Garcia, J., Amaratunga, C., Lim, P., Suon, S., Sreng, S., Drury, E., Stalker, J., Miotto, O., & Fairhurst, R. M. (2018). Origins of the current outbreak of multidrug-resistant malaria in southeast Asia: A retrospective genetic study. *The Lancet Infectious Diseases, 18*(3), 337–345.

Antonovics, J., Wilson, A. J., Forbes, M. R., Haue, H. C., Kallio, E. R., Leggett, H. C., Longdon, B., Okamura, B., Sait, S. M., & Webster, J. P. (2017). The evolution of transmission mode. *Philosophical Transactions of the Royal Society B-Biological Sciences, 372*, 1–12.

Argüello-García, R., Cruz-Soto, M., Romero-Montoya, L., & Ortega- Pierres, G. (2009). In vitro resistance to 5-nitroimidazoles and benzimidazoles in *Giardia* duodenalis: Variability and variation in gene expression. *Infection, Genetics and Evolution, 9*, 1057–1064.

Ashley, E. A., Pyae Phyo, A., & Woodrow, C. J. (2018). Malaria. *The Lancet, 391*(10130), 1608–1621.

Askling, H. H., Bruneel, F., Burchard, G., Castelli, F., Chiodini, P. L., Grobusch, M. P., Lopez-Vélez, R., Paul, M., Petersen, E., Popescu, C., & Ramharter, M. (2012). Management of imported malaria in Europe. *Malaria Journal, 11*(1), 328–342.

Aspinall, T. V., Joynson, D. H. M., Guy, E., Hyde, J. E., & Sims, P. F. G. (2002). The molecular basis of sulfonamide resistance in *Toxoplasma gondii* and implications for the clinical management of toxoplasmosis. *Journal of Infectious Diseases, 185*, 1637–1643.

Babokhov, P., Sanyaolu, A. O., Oyibo, W. A., Fagbenro-Beyioku, A. F., & Iriemenam, N. C. (2013). A current analysis of chemotherapy strategies for the treatment of human African trypanosomiasis. *Pathogens and Global Health, 107*, 242–252.

Baggish, A. L., & Hill, D. R. (2002). Antiparasitic agent atovaquone. *Antimicrobiol Agents and Chemotherapy, 46*(5), 1163–1173.

Baird, J. K. (2005). Effectiveness of antimalarial drugs. *New England Journal of Medicine, 352*(15), 1565–1577.

Baker, C. H., & Welburn, S. C. (2018). The Long Wait for a New Drug for Human African Trypanosomiasis. *Trends in Parasitology, 34*, 818–827.

Baker, N., Glover, L., Munday, J. C., Andres, D. A., Barrett, M. P., de Koning, H. P., & Horn, D. (2012). Aquaglyceroporin 2 controls susceptibility to melarsoprol and pentamidine in African trypanosomes. *Proceeding of the Natlional Acadamy of Sciences USA, 109*, 10996–11001.

Banerjee, A., De, M., & Ali, N. (2011). Combination therapy with paromomycin-associated stearylamine-bearing liposomes cures experimental visceral leishmaniasis through Th1-biased immunomodulation. *Antimicrobiol Agents and Chemotherapy, 55*, 1661–1670.

Barrett, M. P. (2018). The elimination of human African trypanosomiasis is in sight: Report from the third WHO stakeholders meeting on elimination of gambiense human African trypanosomiasis. *PLoS Neglected Tropical Diseases, 12*, 1–4.

Barry, J. D., & Emery, D. L. (1984). Parasite development and host responses during the establishment of Trypanosoma brucei infection transmitted by tsetse fly. *Parasitology, 88*(1), 67–84.

Barta, J. R., Schrenzel, M. D., Carreno, R., & Rideout, B. A. (2005). The genus *Atoxoplasma* (Garnham 1950) as a junior objective synonym of the genus *Isospora* (Schneider 1881) species infecting birds and resurrection of *Cystoisospora* (Frenkel 1977) as the correct genus for *Isospora* species infecting mammals. *Journal of Parasitology, 91*, 726–727.

Basselin, M., Denise, H., Coombs, G. H., & Barrett, M. P. (2002). Resistance to pentamidine in *Leishmania mexicana* involves exclusion of the drug from the mitochondrion. *Antimicrobiol Agents and Chemotherapy, 46*, 3731–3738.

Battle, K. E., Karhunen, M. S., Bhatt, S., Gething, P. W., Howes, R. E., Golding, N., Van Boeckel, T. P., Messina, J. P., Shanks, G. D., Smith, D. L., & Baird, J. K. (2014). Geographical variation in *Plasmodium vivax* relapse. *Malaria Journal, 13*(1), 1–16.

Becker, C. A. M., Malandrin, L., Larcher, T., Chauvin, A., Bischoff, E., & Bonnet, S. I. (2013). Validation of BdCCp2 as a marker for Babesia divergens sexual stages in ticks. *Experimental Parasitology, 133*, 51–56.

Beech, R. N., Skuce, P., Bartley, D. J., Martin, R. J., Prichard, R. K., & Gilleard, J. S. (2011). Anthelmintic resistance: markers for resistance, or susceptibility? *Parasitology, 138*, 160–174.

Benach, J. L., & Habicht, G. S. (1981). Clinical characteristics of human babesiosis. *Journal of Infectious Diseases, 144*, 1481.

Berg, M., Mannaert, A., Vanaerschot, M., Van der Auwera, G., & Dujardin, J. C. (2013). (Post-) Genomic approaches to tackle drug resistance in *Leishmania*. *Parasitology, 140*, 1492–1505.

Bernard, Q., Jaulhac, B., & Boulanger, N. (2014). Smuggling across the border: how arthropod-borne pathogens evade and exploit the host defense system of the skin. *Journal of Investigative Dermatology, 134*(5), 1211–1219.

Bhandari, V., Sundar, S., Dujardin, J. C., & Salotra, P. (2014). Elucidation of cellular mechanisms involved in experimental paromomycin resistance in *Leishmania donovani*. *Antimicrobial Agents and Chemotherapy, 58*, 2580–2585.

Birth, D., Kao, W. C., & Hunte, C. (2014). Structural analysis of atovaquone-inhibited cytochrome bc1 complex reveals the molecular basis of antimalarial drug action. *Nature Communications, 5*, 1–11.

Bousema, J. T., Schneider, P., Gouagna, L. C., Drakeley, C. J., Tostmann, A., Houben, R., Githure, J. I., Ord, R., Sutherland, C. J., Omar, S. A., & Sauerwein, R. W. (2006). Moderate effect of artemisinin-based combination therapy on transmission of *Plasmodium falciparum*. *Journal of Infectious Diseases, 193*, 1151–1159.

Boyles, T. H., Black, J., Meintjes, G., & Mendelson, M. (2012). Failure to eradicate Isospora belli diarrhoea despite immune reconstitution in adults with HIV–a case series. *PLoS ONE, 7*(8), 1–4.

Bray, P. G., Barrett, M. P., Ward, S. A., & de Koning, H. P. (2003). Pentamidine uptake and resistance in pathogenic protozoa: Past, present and future. *Trends in Parasitology, 19*, 232–239.

Buret, A. G. (2008). Pathophysiology of enteric infections with Giardia duodenalis. *Parasite, 15*, 261–265.

Burrows, J. N., Chibale, K., & Wells, T. N. (2011). The state of the art in anti-malarial drug discovery and development. *Current Topics in Medicinal Chemistry, 11*, 1226–1254.

Burza, S., Croft, S. L., & Boelaert, M. (2018). Leishmaniasis. *Lancet, 392*, 951–970.

Buscher, P., Cecchi, G., Jamonneau, V., & Priotto, G. (2017). Human African trypanosomiasis. *Lancet, 390*, 2397–2409.

Capela, R., Moreira, R., & Lopes, F. (2019). An overview of drug resistance in protozoal diseases. *International Journal of Molecular Science, 20*, 5748–5778.

Capper, M. J., O'Neill, P. M., Fisher, N., Strange, R. W., Moss, D., Ward, S. A., Berry, N. G., Lawrenson, A. S., Hasnain, S. S., Biagini, G. A., & Antonyuk, S. V. (2015). Antimalarial 4 (1H)-pyridones bind to the Qi site of cytochrome bc1. *Proceedings of the National Academy of Sciences, 112*(3), 755–760.

Chakravarty, J., & Sundar, S. (2010). Drug resistance in Leishmaniasis. *Journal of Global Infectious Diseases, 2*, 167–176.

Chattopadhyay, R., Velmurugan, S., Chakiath, C., Andrews Donkor, L., Milhous, W., Barnwell, J. W., Collins, W. E., & Hoffman, S. L. (2010). Establishment of an in vitro assay for assessing the effects of drugs on the liver stages of *Plasmodium vivax* malaria. *PLoS ONE, 5*(12), 1–8.

Chavez, B., Cedillo-Rivera, R., & Martinez-Palomo, A. (1992). *Giardia lamblia*: Ultrastructural study of the in vitro effect of benzimidazoles. *Journal of Protozoology, 39*, 510–515.

Chawla, B., Jhingran, A., Panigrahi, A., Stuart, K. D., & Madhubala, R. (2011). Paromomycin affects translation and vesicle-mediated trafficking as revealed by proteomics of paromomycin-susceptible-resistant *Leishmania donovani*. *PLoS ONE, 6*(10), 1–12.

Coelho, A. C., Messier, N., Ouellette, M., & Cotrim, P. C. (2007). Role of the ABC transporter PRP1 (ABCC7) in pentamidine resistance in Leishmania amastigotes. *Antimicrobial Agents and Chemotherapy, 51*, 3030–3032.

Combrinck, J. M., Mabotha, T. E., Ncokazi, K. K., Ambele, M. A., Taylor, D., Smith, P. J., Hoppe, H. C., & Egan, T. J. (2013). Insights into the role of heme in the mechanism of action of antimalarials. *ACS Chemical Biology, 8*(1), 133–137.

Comin, C. E., & Santucci, M. (1994). Submicroscopic profile of Isospora belli enteritis in a patient with acquired immune deficiency syndrome. *Ultrastructural Pathology, 18*, 473–482.

Commons, R. J., Simpson, J. A., Thriemer, K., Humphreys, G. S., Abreha, T., & Alemu, S. G. (2018). The effect of chloroquine dose and primaquine on *Plasmodium vivax* recurrence: A worldwide antimalarial resistance network systematic review and individual patient pooled meta-analysis. *Lancet Infectious Diseases, 18*, 1025–1034.

Cooper, R. A., Hartwig, C. L., & Ferdig, M. T. (2005). pfcrt is more than the *Plasmodium falciparum* chloroquine resistance gene: A functional and evolutionary perspective. *Acta Tropica, 94*, 170–180.

Corey, V. C., Lukens, A. K., Istvan, E. S., Lee, M. C. S., Franco, V., Magistrado, P., Coburn-Flynn, O., Sakata-Kato, T., Fuchs, O., Gnädig, N. F., Goldgof, G., Linares, M., Gomez-Lorenzo, M. G., De Cózar, C., Lafuente-Monasterio, M. J., Prats, S., Meister, S., Tanaseichuk, O., Wree, M., … Winzeler, E. A. (2016). A broad analysis of resistance development in the malaria parasite. *Natural Communication, 7*(1), 1–9.

Cortazar, T. M., Coombs, G. H., & Walker, J. (2007). *Leishmania panamensis*: Comparative inhibition of nuclear DNA topoisomerase II enzymes from promastigotes and human macrophages reveals anti-parasite selectivity of fluoroquinolones, flavonoids and pentamidine. *Experimental Parasitology, 116*, 475–482.

Cosar, C., & Julou, L. (1959). The activity of 1-(2-hydroxyethyl)-2-methyl-5-nitroimidazole (R. P. 8823) against experimental *Trichomonas vaginalis* infections. *Annales De L'institut Pasteur (paris), 96*, 238–421.

Cowman, A. F., Galatis, D., & Thompson, J. K. (1994). Selection for mefloquine resistance in *Plasmodium falciparum* is linked to amplification of the pfmdr1 gene and cross-resistance to halofantrine and quinine. *Proceeding of National Acadamy of Science USA, 91*(3), 1143–1147.

Cowman, A. F., Healer, J., Marapana, D., & Marsh, K. (2016). Malaria: Biology and disease. *Cell, 167*(3), 610–624.

Crofts, A. R. (2004). The cytochrome bc1 complex: Function in the context of structure. *Annual Review of Physiology, 66*, 689–733.

Cui, L., Mharakurwa, S., Ndiaye, D., Rathod, P. K., & Rosenthal, P. J. (2015). Antimalarial drug resistance: Literature review and activities and findings of the ICEMR network. *American Journal of Tropical Medicine and Hygiene, 93*(3), 57–68.

Curd, F. H. S., Davey, D. G., & Rose, F. L. (1945). Studies on synthetic antimalarial drugs. *Annual Tropical Medicine Parasitology, 39*, 208–216.

Dan, M., Wang, A. L., & Wang, C. C. (2000). Inhibition of pyruvate-ferredoxin oxidoreductase gene expression in *Giardia lamblia* by a virusmediated hammerhead ribozyme. *Molecular Microbiology, 36*, 447–456.

De Hovitz, J. A., Pape, J. W., Boncy, M., & Johnson, W. D. (1986). Clinical manifestations and therapy of *Isospora belli* infection in patients with the acquired immunodeficiency syndrome. *New England Journal of Medicine, 315*(2), 87–90.

De Souza, W., Attias, M., & Rodrigues, J. C. F. (2009). Particularities of mitochondrial structure in parasitic protists (Apicomplexa and Kinetoplastida). *International Journal of Biochemistry and Cell Biology, 41*, 2069–2080.

de Veer, M. J., Curtis, J. M., Baldwin, T. M., DiDonato, J. A., Sexton, A., McConville, M. J., et al. (2003) MyD88 isessential for clearance of Leishmania major: Possible role for lipophosphoglycan and toll-likereceptor 2 signaling. *European Journal of Immunology, 33*, 2822–2831. [PubMed:14515266].

Dembele, L., Geg, A., Zeeman, A. M., Franetich, J. F., Silvie, O., Rametti, A., Le Grand, R., Dereuddre-Bosquet, N., Sauerwein, R., van Gemert, G. J., Vaillant, J. C., Thomas, A. W., Snounou, G., Kocken, C. H., & Mazier, D. (2011). Towards an in vitro model of *Plasmodium hypnozoites* suitable for drug discovery. *PLoS ONE, 6*, 1–7.

Dhiman, S. (2019). Are malaria elimination efforts on right track? An analysis of gains achieved and challenges ahead. *Infectious Diseases of Poverty, 8*(1), 1–19.

Dijanic, C., Nickerson, J., Shakya, S., Dijanic, A., & Fabbri, M. (2018). Relapsing malaria: A case report of primaquine resistance. *Case Reports in Infectious Diseases, 2018*, 1–3.

Dionisio, D., Sterrantino, G., Meli, M., Leoncini, F., Orsi, A., & Nicoletti, P. (1996). Treatment of isosporiasis with combined albendazole and ornidazole in patients with AIDS. *AIDS, 10*(11), 1301–1302.

Diro, E., Lynen, L., Ritmeijer, K., Boelaert, M., Hailu, A., & van Griensven, J. (2014). Visceral Leishmaniasis and HIV coinfection in East Africa. *PLoS Neglectd Tropical Diseases, 8*, 10.

Djimde, A., Doumbo, O. K., Cortese, J. F., Kayentao, K., Doumbo, S., Diourte, Y., Dicko, A., Su, X. Z., Nomura, T., Fidock, D. A., Wellems, T. E., Plowe, C. V., & Coulibaly, D. (2001). A molecular marker for chloroquine-resistant falciparum malaria. *New England Journal of Medicine, 344*, 257–263.

Doggett, J. S., Nilsen, A., Forquer, I., Wegmann, K. W., Jones-Brando, L., Yolken, R. H., Bordón, C., Charman, S. A., Katneni, K., Schultz, T., & Burrows, J. N. (2012). Endochin-like quinolones are highly efficacious against acute and latent experimental toxoplasmosis. *Proceedings of the National Academy of Sciences, 109*(39), 15936–15941.

Donald, R., & Roos, D. S. (1993). Stable molecular transformation of *Toxoplasma gondii*: A selectable dihydrofolate reductase-thymidylate synthase marker based on drug-resistance mutations in malaria. *Proceding of the National Acadamy of Sciences USA, 90*, 11703–11707.

Dubey, J. P. (2018). A review of *Cystoisospora felis* and *C. revolta*-induced coccidiosis in cats and public health. *Veterinary Parasitology, 263*, 34–48.

Dubey, J., & Almeria, S. (2019). *Cystoisospora belli* infections in humans: The past 100 years. *Parasitology, 146*(12), 1490–1527.

Dunn, L. A., Burgess, A. G., Krauer, K. G., Eckmann, L., Vanelle, P., Crozet, M. D., Gillin, F. D., Upcroft, P., & Upcroft, J. A. (2010) A new-generation 5-nitroimidazole can induce highly metronidazole-resistant Giardia lamblia in vitro. *International Journal of Antimicrobial Agents, 36*(1), 37–42.

Eastman, R. T., & Fidock, D. A. (2009). Artemisinin-based combination therapies: A vital tool in efforts to eliminate malaria. *Nature Reviews Microbiology, 7*, 864–874.

Ebstie, Y. A., Abay, S. M., Tadesse, W. T., & Ejigu, D. A. (2016). Tafenoquine and its potential in the treatment and relapse prevention of *Plasmodium vivax* malaria: The evidence to date. *Drug Design, Development and Therapy, 10*, 2387–2399.

Eckstein-Ludwig, U., Webb, R. J., Van Goethem, I. D., East, J. M., Lee, A. G., Kimura, M., O'Neill, P. M., Bray, P. G., Ward, S. A., & Krishna, S. (2003). Artemisinins target the SERCA of *Plasmodium falciparum. Nature, 424*, 957–961.

Edlind, T. D., Hang, T. L., & Chakraborty, P. R. (1990). Activity of the anthelmintic benzimidazoles against *Giardia lamblia* in vitro. *Journal of Infectious Diseases, 162*, 1408–1411.

Escobedo, A. A., & Cimerman, S. (2007). Giardiasis: A pharmacotherapy review. *Expert Opinion on Pharmacotherapy, 8*, 1885–1902.

Esvan, R., Suleková, L. F., Gabrielli, S., Biliotti, E., Palazzo, D., Spaziante, M., & Taliani, G. (2018). Severe diarrhoea due to *Cystoisospora belli* infection in a Good syndrome patient. *Parasitology International, 67*, 413–414.

Fairclough, R. (1962). In *Proceedings 9th meeting of the international scientific council for trypanosomiasis research*; commision for technical cooperation in Africa, Lagos, CCTA Pub. No. 88: Conakrt, Guinera, (pp 81–86).

Fairlamb, A. H., & Horn, D. (2018). Melarsoprol resistance in African trypanosomiasis. *Trends in Parasitology, 34*, 481–492.

Farthing, M. J. (1997). Giardiasis. *Journal of Pediatric Gastroenterology and Nutrition, 24*, 79–88.

Fernandez, M. M., Malchiodi, E. L., & Algranati, I. D. (2011). Differential effects of paromomycin on ribosomes of *Leishmania mexicana* and mammalian cells. *Antimicrobioal Agents and Chemotherapy, 55*, 86–93.

Ferreira, I. D., Lopes, D., Martinelli, A., Ferreira, C., Do Rosario, V. E., & Cravo, P. (2007). In vitro assessment of artesunate, artemether and amodiaquine susceptibility and molecular analysis of putative resistance-associated mutations of *Plasmodium falciparum* from Sao Tome and principe. *Tropical Medicine and International Health, 12*, 353–362.

Fidock, D. A., Nomura, T., Talley, A. K., Cooper, R. A., Dzekunov, S. M., Ferdig, M. T., Ursos, L. M., Sidhu, A. B., Naude, B., Deitsch, K. W., Su, X. Z., Wootton, J. C., Roepe, P. D., & Wellems, T. E. (2000). Mutations in the *P. falciparum* digestive vacuole transmembrane protein PfCRT and evidence for their role in chloroquine resistance. *Molecular Cell, 6*, 861–871.

Fletcher, S. M., Stark, D., Harkness, J., & Ellisa, J. (2012). Enteric protozoa in the developed world: A public health perspective. *Clinical Microbiology Review, 25*, 420–449.

Franco, J., Scarone, L., & Comini, M. A. (2018). Drugs and drug resistance in African and American trypanosomiasis. In M. Botta, (Eds.), *Neglected diseases: Extensive space for modern drug discovery*, (pp. 97–133). Academic Press Inc.

French, J. M., Whitby, J. L., & Whitfield, A. G. W. (1964). Steatorrhea in a man infected with coccidiosis (*Isospora belli*). *Gastroenterology, 47*, 642–648.

Fry, M., & Pudney, M. (1992). Site of action of the antimalarial hydroxynaphthoquinone, 2-[trans-4-(4'-chlorophenyl) cyclohexyl]-3-hydroxy-1,4-naphthoquinone (566C80). *Biochemical Pharmacology, 43*, 1545–1553.

Gajurel, K., Gomez, C., Dhakal, R., Vogel, H., & Montoya, J. (2016). Failure of primary atovaquone prophylaxis for prevention of toxoplasmosis in hematopoietic cell transplant recipients. *Transplant Infectious Diseses, 18*, 446–452.

Garcia-Salcedo, J. A., Unciti-Broceta, J. D., Valverde-Pozo, J., & Soriano, M. (2016). New approaches to overcome transport related drug resistance in trypanosomatid parasites. *Frontiers in Pharmacology, 7*, 351–364.

Glasner, P. D., Silveira, C., Kruszon-Moran, D., Martins, M. C., Burnier, M., Jr., Silveira, S., Camargo, M. E., Nussenblatt, R. B., Kaslow, R. A., & Belfort, R., Jr. (1992). An unusually high prevalence of ocular toxoplasmosis in southern Brazil. *American Journal of Ophthalmology, 114*(2), 136–144.

Graf, F. E., Baker, N., Munday, J. C., De Koning, H. P., Horn, D., & Mäser, P. (2015). Chimerization at the AQP2–AQP3 locus is the genetic basis of melarsoprol–pentamidine cross-resistance in

clinical *Trypanosoma brucei gambiense* isolates. *International Journal for Parasitology: Drugs and Drug Resistance, 5*(2), 65–68.
Gregson, A., & Plowe, C. V. (2005). Mechanisms of resistance of malaria parasites to antifolates. *Pharmacology Review, 57*, 117–145.
Guerra, C. A., Howes, R. E., Patil, A. P., Gething, P. W., Van Boeckel, T. P., Temperley, W. H., Kabaria, C. W., Tatem, A. J., Manh, B. H., Elyazar, I. R., Baird, J. K., Snow, R. W., & Hay, S. I. (2010). The international limits and population at risk of *Plasmodium vivax* transmission in 2009. *PLoS Neglected Tropical Diseases, 4*(8), 1–11.
Gupta, G., Oghumu, S., & Satoskar, A. R. (2013). Mechanisms of immune evasion in leishmaniasis. In *Advances in applied microbiology* (Vol. 82, pp. 155–184). Academic Press.
Haider, M., Sharma, A., Dogra, V., & Mishra, B. (2013). Isospora spp. in chronic diarrhea: A case report. *Journal of Pharmacy and Bioallied Scienes, 5*(4), 327–328.
Halliez, M. C., & Buret, A. G. (2013). Extra-intestinal and long-term consequences of *Giardia duodenalis* infections. *World Journal of Gastroenterology, 19*, 8974–8985.
Hamšíková, Z., Kazimírová, M., Haruštiaková, D., Mahríková, L., Slovák, M., Berthová, L., Kocianová, E., & Schnittger, L. (2016). Babesia spp. in ticks and wildlife in different habitat types of Slovakia. *Parasites and Vectors, 9*, 292.
Hasan, M., Bishburg, E., & Nagarakanti, S. (2017). Recurrent *Cystoisospora belli* in a patient with HTLV-1 infection. *Idcases, 10*, 122.
Heidari-Kharaji, M., Taheri, T., Doroud, D., Habibzadeh, S., Badirzadeh, A., & Rafati, S. (2016). Enhanced paromomycin efficacy by solid lipid nanoparticle formulation against *Leishmania* in mice model. *Parasite Immunology, 38*, 599–608.
Hendrickx, S., Van den Kerkhof, M., Mabille, D., Cos, P., Delputte, P., Maes, L., & Caljon, G. (2017). Combined treatment of miltefosine and paromomycin delays the onset of experimental drug resistance in *Leishmania infantum. PLoS Neglected Tropical Disease, 11*, 1–10.
Hisaeda, H., Yasutomo, K., & Himeno, K. (2005). Malaria: Immune evasion by parasites. *The International Journal of Biochemistry and Cell Biology, 37*(4), 700–706.
Hobbs, C., & Duffy, P. (2011). Drugs for malaria: something old, something new, something borrowed. F1000 *Biology Reports, 3*, 1–9.
Holberton, D. V., & Ward, A. P. (1981). Isolation of the cytoskeleton from Giardia. Tubulin and a low-molecular-weight protein associated with microribbon structures. *Journal of Cellular Sciences, 47*, 139–166.
Hoover, D. L., Berger, M., Nacy, C. A., Hockmeyer, W. T., & Meltzer, M. S. (1984). Killing of Leishmania tropica amastigotes by factors in normal human serum. *The Journal of Immunology, 132*(2), 893–897.
Howe, D. K., & Sibley, L. D. (1995). *Toxoplasma gondii* comprises three clonal lineages: Correlation of parasite genotype with human disease. *Journal of Infectious Diseases, 172*, 1561–1566.
Howes, R. E., Battle, K. E., Mendis, K. N., Smith, D. L., Cibulskis, R. E., Baird, J. K., & Hay, S. I. (2016). Global epidemiology of *Plasmodium vivax. The American Journal of Tropical Medicine and Hygiene, 95*(6), 15–34.
Hudson, A. T., Dickins, M., Ginger, C. D., Gutteridge, W. E., Holdich, T., Hutchinson, D. B. A., Pudney, M., Randall, A. W., & Latter, V. S. (1991). 566C80: A potent broad spectrum anti-infective agent with activity against malaria and oppor- tunistic infections in AIDS patients. *Drugs in Experimental Clinical Research, 17*, 427–435.
Hwang, S. J., Yamasaki, M., Nakamura, K., Sasaki, N., Murakami, M., Rajapakshage, B. K. W., Ohta, H., Maede, Y., & Takiguchi, M. (2010). Reduced transcript levels of the heat shock protein 70 gene in diminazene aceturate-resistant *Babesia gibsoni* variants under low concentrations of diminazene aceturate. *Japanese Journal of Veterinary Research, 58*(4), 155–164.
Imwong, M., Dondorp, A. M., Nosten, F., Yi, P., Mungthin, M., Hanchana, S., Das, D., Phyo, A. P., Lwin, K. M., Pukrittayakamee, S., Lee, S. J., Saisung, S., Koecharoen, K., Nguon, C., Day, N. P., Socheat, D., & White, N. J. (2010). Exploring the contribution of candidate genes to artemisinin resistance in *Plasmodium falciparum. Antimicrobial Agents and Chemotherapy, 54*(7), 2886–2892.

Imwong, M., Pukrittayakamee, S., Pongtavornpinyo, W., Nakeesathit, S., Nair, S., Newton, P., Nosten, F., Anderson, T. J., Dondorp, A., Day, N. P., & White, N. J. (2008). Gene amplification of the multidrug resistance 1 gene of *Plasmodium vivax* isolates from Thailand, Laos, and Myanmar. *Antimicrobial Agents and Chemotherapy, 52*(7), 2657–2659.

Imwong, M., Suwannasin, K., Kunasol, C., Sutawong, K., Mayxay, M., Rekol, H., Smithuis, F. M., Hlaing, T. M., Tun, K. M., van der Pluijm, R. W., & Tripura, R. (2017). The spread of artemisinin-resistant *Plasmodium falciparum* in the Greater Mekong subregion: A molecular epidemiology observational study. *The Lancet Infectious Diseases, 17*(5), 491–497.

Jambou, R., Legrand, E., Niang, M., Khim, N., Lim, P., Volney, B., Ekala, M. T., Bouchier, C., Esterre, P., Fandeur, T., & Mercereau-Puijalon, O. (2005). Resistance of *Plasmodium falciparum* field isolates to in vitro artemether and point mutations of the SERCAtype PfATPase6. *Lancet, 366*, 1960–1963.

James, D. M., & Giles, H. M. (1985). Pharmacology and usage. In J. W. Sons (Ed.).

Jeacock, L., Baker, N., Wiedemar, N., Maser, P., & Horn, D. (2017). Aquaglyceroporin-null trypanosomes display glycerol transport defects and respiratory-inhibitor sensitivity. *PLoS Pathology, 13*(3), 1–16.

Jhingran, A., Chawla, B., Saxena, S., Barrett, M. P., & Madhubala, R. (2009). Paromomycin: Uptake and resistance in *Leishmania donovani. Molecular Biochemistry and Parasitology, 164*, 111–117.

Jiménez-Cardoso, E., Eligio-García, L., Cortés-Campos, A., Flores- Luna, A., Valencia-Mayoral, P., & Lozada-Chávez, I. (2009). Changes in betagiardin sequence of *Giardia intestinalis* sensitive and resistant to albendazole strains. *Parasitology Research, 105*, 25–33.

Jones, M. K., & Good, M. F. (2006). Malaria parasites up close. *Natural Medicine, 2*, 170–171.

Jongwutiwes, S., Putaporntip, C., Charoenkorn, M., Iwasaki, T., & Endo, T. (2007). Morphologic and molecular characterization of *Isospora belli* oocysts from patients in Thailand. *American Journal of Tropical Medicine and Hygiene, 77*, 107–112.

Kennedy, P. G. E., & Rodgers, J. (2019). Clinical and neuropathogenetic aspects of human African trypanosomiasis. *Frontier Immunology, 10*, 11.

Kessl, J. J., Lange, B. B., Merbitz-Zahradnik, T., Zwicker, K., Hill, P., Meunier, B., Palsdottir, H., Hunte, C., Meshnick, S., & Trumpower, B. L. (2003). Molecular basis for atovaquone binding to the cytochrome bc1 complex. *Journal of Biological Chemistry, 278*(33), 31312–31318.

Khan, W., Sharma, S. S., & Kumar, N. (2013). Bioanalytical method development, pharmacokinetics, and toxicity studies of paromomycin and paromomycin loaded in albumin microspheres. *Drug Test and Analysis, 5*, 453–460.

Krause, P. J., Lepore, T., Sikand, V. K., Gadbaw, J., Jr., Burke, G., Telford, S. R., Brassard, P., Pearl, D., Azlanzadeh, J., Christianson, D., McGrath, D., & Spielman, A. (2000). Atovaquone and azithromycin for the treatment of babesiosis. *The New England Journal of Medicine, 343*(20), 1454–1458.

Kremsner, P. G., & Krishna, S. (2004). Antimalarial combinations. *Lancet, 364*(9430), 285–294.

Kropf, C., Debache, K., Rampa, C., Barna, F., Schorer, M., Stephens, C. E., Ismail, M. A., Boykin, D. W., & Hemphill, A. (2012). The adaptive potential of a survival artist: Characterization of the in vitro interactions of *Toxoplasma gondii* tachyzoites with di-cationic compounds in human fibroblast cell cultures. *Parasitology, 139*(2), 208–220.

Kuttler, K. L. (1981). Babesiosis. In M. Ristic, J. P. Keire, (Eds.), (p. 25). Academic Press.

Lakshmanan, V., Bray, P. G., Verdier-Pinard, D., Johnson, D. J., Horrocks, P., Muhle, R. A., Alakpa, G. E., Hughes, R. H., Ward, S. A., Krogstad, D. J., Sidhu, A. B., & Fidock, D. A. (2005). A critical role for PfCRT K76T in *Plasmodium falciparum* verapamil-reversible chloroquine resistance. *The EMBO JOurnal, 24*, 2294–2305.

Langhorne, J., Buffet, P., Galinski, M., Good, M., Harty, J., Leroy, D., Mota, M. M., Pasini, E., Renia, L., Riley, E., Stins, M., & Duffy, P. (2011). The relevance of non-human primate and rodent malaria models for humans. *Malaria Journal, 10*(1), 1–4.

Lawres, L. A., Garg, A., Kumar, V., Bruzual, I., Forquer, I. P., Renard, I., Virji, A. Z., Boulard, P., Rodriguez, E. X., Allen, A. J., & Pou, S. (2016). Radical cure of experimental babesiosis in immunodeficient mice using a combination of an endochin-like quinolone and atovaquone. *Journal of Experimental Medicine, 213*(7), 1307–1318.

Lee, K. S., Divis, P. C., Zakaria, S. K., Matusop, A., Julin, R. A., Conway, D. J., Cox-Singh, J., & Singh, B. (2011). *Plasmodium knowlesi*: Reservoir hosts and tracking the emergence in humans and macaques. *PLoS Pathogens, 7*(4), 1–11.

Légaré, D., & Ouellette, M. (2017). Drug resistance in leishmania. In M. B. Gotte, A. Matlashewski, G. Wainberg, A. Mark & S. Donald, (Eds.), *Handbook of antimicrobial resistance*, (pp. 313–341). Springer.

Leitsch, D. (2015). Drug resistance in the microaerophilic parasite *Giardia lamblia. Current Tropical Medicinal Report, 2*, 128–135.

Leitsch, D., Burgess, A. G., Dunn, L. A., Krauer, K. G., Tan, K., Duchêne, M., Upcroft, P., Eckmann, L., & Upcroft, J. A. (2011). Pyruvate: ferredoxin oxidoreductase and thioredoxin reductase are involved in 5-nitroimidazole activation while flavin metabolism is linked to 5-nitroimidazole resistance in *Giardia lamblia. Journal of Antimicrobial Chemotherapy, 66*(8), 1756–1765.

Leitsch, D., Kolarich, D., Binder, M., Stadlmann, J., Altmann, F., & Duchêne, M. (2009). *Trichomonas vaginalis*: metronidazole and other nitroimidazole drugs are reduced by the flavin enzyme thioredoxin reductase and disrupt the cellular redox system. Implications for nitroimidazole toxicity and resistance. *Molecular Microbiology, 72*, 518–536.

Leitsch, D., Schlosser, S., Burgess, A., & Duchêne, M. (2012). Nitroimidazole drugs vary in their mode of action in the human parasite *Giardia lamblia. Internattional Journal of Parasitolgical Drugs and Drug Resistance, 2*, 166–170.

Lemée, V., Zaharia, I., Nevez, G., Rabodonirina, M., Brasseur, P., Ballet, J. J., & Favennec, L. (2000). Metronidazole and albendazole susceptibility of 11 clinical isolates of *Giardia duodenalis* from France. *Journal of Antimicrobial Chemotherapy, 46*(5), 819–821.

Lemieux, J. E., Tran, A. D., Freimark, L., Schaffner, S. F., Goethert, H., Andersen, K. G., Bazner, S., Li, A., McGrath, G., Sloan, L., & Vannier, E. (2016). A global map of genetic diversity in *Babesia microti* reveals strong population structure and identifies variants associated with clinical relapse. *Nature Microbiology, 1*(7), 1–7.

Levine, N.D. 1973. *Protozoan parasites of domestic animals and of man*, (2nd edn, pp. 1–406). Burgess.

Lima, T. S., & Lodoen, M. B. (2019). Mechanisms of human innate immune evasion by Toxoplasma gondii. *Frontiers in Cellular and Infection Microbiology, 9*, 103.

Lindquist, S. (1986). The heat-shock response. *Annual Review of Biochemistry, 55*, 1151–1191.

Lindsay, D. S., Houk, A. E., Mitchell, S. M., & Dubey, J. P. (2014). Developmental biology of *Cystoisospora* (Apicomplexa: Sarcocystidae) monozoic tissue cysts. *Journal of Parasitology, 100*, 392–398.

Liu, S. M., Brown, D. M., O'Donoghue, P., Upcroft, P., & Upcroft, J. A. (2000). Ferredoxin involvement in metronidazole resistance of *Giardia duodenalis. Molecular Biochemistry and Parasitology, 108*, 137–140.

Lumb, V., Das, M. K., Singh, N., Dev, V., Khan, W., & Sharma, Y. D. (2011). Multiple origins of *Plasmodium falciparum* dihydropteroate synthetase mutant alleles associated with sulfadoxine resistance in India. *Antimicrobial Agents and Chemotherapy, 55*, 2813–2817.

MacDonald, L. M., Armson, A., Thompson, A. R., & Reynoldson, J. A. (2004). Characterisation of benzimidazole binding with recombinant tubulin from *Giardia duodenalis, Encephalitozoon intestinalis*, and *Cryptosporidium parvum. Molecular Biochemistry and Parasitology, 138*, 89–96.

Maltezou, H. C. (2010). Drug resistance in visceral Leishmaniasis. *Journal of Biomedicine and Biotechnology, 2010*, 1–8.

Marathe, A., & Parikh, K. (2013). Severe diarrhoea due to *Cystoisospora belli* in renal transplant patient on immunosuppressive drugs. *Indian Journal of Medicine and Microbiology, 31*, 185–187.

Marcsisin, S. R., Reichard, G., & Pybus, B. S. (2016). Primaquine pharmacology in the context of CYP 2D6 pharmacogenomics: current state of the art. *Pharmacology and Therapeutics, 161*, 1–10.

Martins, T. M., Domingos, A., Berry, C., & Wyatt, D. M. (2006). The activity and inhibition of the food vacuole plasmepsin from the rodent malaria parasite *Plasmodium chabaudi*. *Acta Tropica, 97*(2), 212–218.

Matsuu, A., Miyamoto, K., Ikadai, H., Okano, S., & Higuchi, S. (2006). Short report: cloning of the *Babesia gibsoni* cytochrome b gene and isolation of three single nucleotide polymorphisms from parasites present after atovaquone treatment. *American Journal of Tropical Medicine and Hygiene, 74*, 593–597.

Mayers, D. L., Sobel, J. D., Ouellette, M., Kaye, K. S., & Marchaim, D. (2017). Bacterial genotyping drug resistance assays. In A. Huletsky, & M. G. Bergeron, (Eds.), *Antimicrobial drug resistance: Clinical and epidemiological aspects*, (2nd Edn, p. 85) Springer International Switzerland.

Mbengue, A., Bhattacharjee, S., Pandharkar, T., Liu, H., Estiu, G., Stahelin, R. V., Rizk, S. S., Njimoh, D. L., Ryan, Y., Chotivanich, K., & Nguon, C. (2015). A molecular mechanism of artemisinin resistance in *Plasmodium falciparum* malaria. *Nature, 520*(7549), 683–687.

McFadden, D. C., Camps, M., & Boothroyd, J. C. (2001). Resistance as a tool in the study of old and new drug targets in Toxoplasma. *Drug Resisttance Updates, 4*, 79–84.

McFadden, D. C., Tomavo, S., Berry, E. A., & Boothroyd, J. C. (2000). Characterization of cytochrome b from *Toxoplasma gondii* and Qo domain mutations as a mechanism of atovaquone-resistance. *Molecual Biochemistry and Parasitology, 108*, 1–12.

McLeod, R., Boyer, K., Karrison, T., Kasza, K., Swisher, C., Roizen, N., Jalbrzikowski, J., Remington, J., Heydemann, P., Noble, A. G., & Mets, M. (2006). Outcome of treatment for congenital toxoplasmosis, 1981–2004: The national collaborative Chicago-based, congenital toxoplasmosis study. *Clinical Infectious Diseases, 42*(10), 1383–1394.

Meneceur, P., Bouldouyre, M. A., Aubert, D., Villena, I., Menotti, J., Sauvage, V., Garin, J. F., & Derouin, F. (2008). In vitro susceptibility of various genotypic strains of *Toxoplasma gondii* to pyrimethamine, sulfadiazine, and atovaquone. *Antimicrobial Agents and Chemotherapy, 52*(4), 1269–1277.

Mi-Ichi, F., Miyadera, H., Kobayashi, T., Takamiya, S., Waki, S., Iwata, S., Shibata, S., & Kita, K. (2005). Parasite mitochondria as a target of chemotherapy: Inhibitory effect of licochalcone A on the *Plasmodium falciparum* respiratory chain. *Annals of the New York Academy of Sciences, 1056*, 46–54.

Millar, S. B., & Cox-Singh, J. (2015). Human infections with *Plasmodium knowlesi*-zoonotic malaria. *Clinical Microbiological Infections, 7*, 640–648.

Mogi, T., & Kita, K. (2009). Identification of mitochondrial complex II subunits SDH3 and SDH4 and ATP synthase subunits a and b in Plasmodium spp. *Mitochondrion, 9*(6), 443–453.

Mohammadi, M., Ansari-Moghaddam, A., Raiesi, A., Rakhshani, F., Nikpour, F., & Haghdost, A. (2011). Baseline results of the first malaria indicator survey in Iran at household level. *Malaria Journal, 10*, 1–6.

Mok, S., Ashley, E. A., Ferreira, P. E., Zhu, L., Lin, Z., Yeo, T., Chotivanich, K., Imwong, M., Pukrittayakamee, S., Dhorda, M., & Nguon, C. (2015). Population transcriptomics of human malaria parasites reveals the mechanism of artemisinin resistance. *Science, 347*(6220), 431–435.

Montazeri, M., Daryani, A., Ebrahimzadeh, M., Ahmadpour, E., Sharif, M., & Sarvi, S. (2015). Effect of propranolol alone and in combination with pyrimethamine on acute murine toxoplasmosis. *Jundishapur Joirnal of Microbiology, 8*, 1–5.

Montazeri, M., Mehrzadi, S., Sharif, M., Sarvi, S., Shahdin, S., & Daryani, A. (2018). Activities of anti-Toxoplasma drugs and compounds against tissue cysts in the last three decades (1987 to 2017), a systematic review. *Parasitology Research, 117*, 3045–3057.

Montoya, J. G., & Liesenfeld, O. (2004). Toxoplasmosis. *Lancet, 363*, 1965–1976.

Monzote, L., & Siddiq, A. (2011). Drug development to protozoan diseases. *Open Medicine and Chemistry Journal, 5*, 1–3.

Mørch, K., Hanevik, K., Robertson, L. J., Strand, E. A., & Langeland, N. (2008). Treatment-ladder and genetic characterisation of parasites in refractory giardiasis after an outbreak in Norway. *Journal of Infectious Diseases, 56*, 268–273.

Mudholkar, V. G., & Namey, R. D. (2010). Heavy infestation of *Isospora belli* causing severe watery diarrhea. *Indian Journal of Pathology and Microbiology, 53*(4), 824–825.

Müller, J., Schildknecht, P., & Müller, N. (2013). Metabolism of nitro drugs metronidazole and nitazoxanide in *Giardia lamblia*: characterization of a novel nitroreductase (GlNR2). *Journal of Antimicrobres and Chemotherapy, 68*, 1781–1789.

Müller, J., Wastling, J., Sanderson, S., Müller, N., & Hemphill, A. (2007). A novel *Giardia lamblia* nitroreductase, GlNR1, interacts with nitazoxanide and other thiazolides. *Antimicrobial Agents and Chemotherapy, 51*, 1979–1986.

Muñoz, J., Rojo-Marcos, G., Ramírez-Olivencia, G., Salas-Coronas, J., Treviño, B., Arellano, J. L. P., Torrús, D., Vilches, M. J. M., Ramos, J. M., Alegría, I., & López-Vélez, R. (2015). Diagnóstico y tratamiento de la malaria importada en España: Recomendaciones del grupo de trabajo de malaria de la sociedad Española de medicina tropical y salud internacional (SEMTSI). *Enfermedades Infecciosas y Microbiología Clínica, 33*(6), 1–13.

Murphy, A. D., & Lang-Unnasch, N. (1999). Alternative oxidase inhibitors potentiate the activity of atovaquone against Plasmodium falciparum. *Antimicrobial Agents and Chemotherapy, 43*(3), 651–654.

Musey, K. L., Chidiac, C., Beaucaire, G., Houriez, S., & Fourrier, A. (1988). Effectiveness of roxithromycin for treating *Isospora belli* infection. *Journal of Infectious Diseases, 158*(3), 646.

Nagle, A. S., Khare, S., Kumar, A. B., Supek, F., Buchynskyy, A., Mathison, C. J., Chennamaneni, N. K., Pendem, N., Buckner, F. S., Gelb, M. H., & Molteni, V. (2014). Recent developments in drug discovery for leishmaniasis and human African trypanosomiasis. *Chemical Reviews, 114*(22), 11305–11347.

Nair, S., Miller, B., Barends, M., Jaidee, A., Patel, J., Mayxay, M., Newton, P., Nosten, F., Ferdig, M. T., & Anderson, T. J. (2008). Adaptive copy number evolution in malaria parasites. *PLoS Geneicst, 4*, 1–10.

Najm, N. D., Kayser, E. M., Hoffman, L., Herb, I., Fensterer, V., Pfister, K., & Silaghi, C. (2014). A molecular survey of Babesia spp. and Theileria spp. in red foxes (*Vulpes vulpes*) and their ticks from Thuringia Germany. *Ticks and Tick Borne Disases, 4*, 386–391.

Nankoberanyi, S., Mbogo, G. W., LeClair, N. P., Conrad, M. D., Tumwebaze, P., Tukwasibwe, S., Kamya, M. R., Tappero, J., Nsobya, S. L., & Rosenthal, P. J. (2014). Validation of the ligase detection reaction fluorescent microsphere assay for the detection of *Plasmodium falciparum* resistance mediating polymorphisms in Uganda. *Malaria Journal, 13*, 1–5.

Narikawa, S. (1986). Distribution of metronidazole susceptibility factors in obligate anaerobes. *Journal of Antimirobes and Chemotherapy, 18*, 565–574.

Nateghi, R. M., Nikmanesh, B., Haghi-Ashtiani, M. T., Monajemzadeh, M., Douraghi, M., Ghalavand, Z., & Kashi, L. (2014). *Isospora belli* associated recurrent diarrhea in a child with AIDS. *Journal of Parasitic Diseases, 38*, 444–446.

Nillius, D., Müller, J., & Müller, N. (2011). Nitroreductase (GlNR1) increases susceptibility of *Giardia lamblia* and *E. coli* to nitro drugs. *Journal of Antimicrobes and Chemotherapy, 66*, 1029–1035.

Nzila, A., Ochong, E., Nduati, E., Gilbert, K., Winstanley, P., Ward, S., & Marsh, K. (2005). Why has the dihydrofolate reductase 164 mutation not consistently been found in Africa yet? *Transaction of Royal Society of Tropical Medicine and Hygiene, 99*, 341–346.

O'Neill, P. M., Barton, V. E., & Ward, S. A. (2010). The molecular mechanism of action of artemisinin-the debate continues. *Molecules, 15*, 1705–1721.

Ouellette, M., Drummelsmith, J., & Papadopoulou, B. (2004). Leishmaniasis: Drugs in the clinic, resistance and new developments. *Drug Resistance, 7*, 257–266.

Pavitra, N. R., Jorge, M. S., Arnab, P., Thomas, J. T., & Gunnar, R. M. (2016). Translational repression of the cpw-wpc gene family in the malaria parasite Plasmodium. *Parasitology International, 65*, 463–471.

Paz-Maldonado, M. T., Argüello-García, R., Cruz-Soto, M., Mendoza- Hernández, G., & Ortega-Pierres, G. (2013). Proteomic and transcriptional analyses of genes differentially expressed in *Giardia duodenalis* clones resistant to albendazole. *Infection, Genetic and Evoloution, 15*, 10–17.

Peregrine, A. S. (1994). Chemotherapy and delivery systems: Haemoparasites. *Veterinary Parasitology, 54*(1–3), 223–248.

Perez-Cabezas, B., Cecilio, P., Gaspar, T. B., Gartner, F., Vasconcellos, R., & Cordeiro-da-Silva, A. (2019). Understanding resistance versus susceptibility in visceral Leishmaniasis using mouse models of Leishmania infantum infection. *Frontiers in Cellular and Infection Microbiology, 9*, 1–30.

Persing, D. H., & Conrad, P. A. (1995). Babesiosis: New insights from phylogenetic analysis. *Infectious Agents of Diseases, 4*, 182 195.

Peterson, D. S., Walliker, D., & Wellems, T. E. (1988). Evidence that a point mutation in dihydrofolate reductase-thymidylate synthase confers resistance to pyrimethamine in falciparum malaria. *Proceeding of National Acadamy of Sciences USA, 85*(23), 9114–9118.

Pfefferkorn, E., & Borotz, S. E. (1994). *Toxoplasma gondii*: characterization of a mutant resistant to 6-thioxanthine. *Experimental Parasitology, 79*, 374–382.

Plebanski, M., Flanagan, K. L., Lee, E. A. M., Reece, W. H. H., Hart, K., Gelder, C., et al. (1999). Interleukin 10-mediated immunosuppression by a variant CD4 T cell epitope ofPlasmodiumfalciparum. *Immunity, 10*, 651–660.

Plowe, C. V., Cortese, J. F., Djimde, A., Nwanyanwu, O. C., Watkins, W. M., Winstanley, P. A., Estrada-Franco, J. G., Mollinedo, R. E., Avila, J. C., Cespedes, J. L., Carter, D. & Doumbo, O. K. (1997). Mutations in *Plasmodium falciparum* dihydrofolate reductase and dihydropteroate synthase and epidemiologic patterns of pyrimethamine-sulfadoxine use and resistance. *Journal of Infectious Diseases, 176*, 1590–1596

Ponte-Sucre, A., Gamarro, F., Dujardin, J. C., Barrett, M. P., Lopez-Velez, R., Garcia-Hernandez, R., Pountain, A. W., Mwenechanya, R., & Papadopoulou, B. (2017). Drug resistance and treatment failure in leishmaniasis: A 21st century challenge. *PLoS Neglected Tropical Diseases, 11*, 1–24.

Pousibet-Puerto, J., Salas-Coronas, J., Sánchez-Crespo, A., Molina-Arrebola, M. A., Soriano-Pérez, M. J., Giménez-López, M. J., Vázquez-Villegas, J., & Cabezas-Fernández, M. T. (2016). Impact of using artemisinin-based combination therapy (ACT) in the treatment of uncomplicated malaria from *Plasmodium falciparum* in a non-endemic zone. *Malaria Journal, 15*, 339.

Pramanik, P. K., Alam, M. N., Roy Chowdhury, D., & Chakraborti, T. (2019). Drug resistance in protozoan parasites: An incessant wrestle for survival. *Journal of Global Antimicrobial Research, 18*, 1–11.

Price, R. N., Uhlemann, A. C., van Vugt, M., Brockman, A., Hutagalung, R., Nair, S., Nash, D., Singhasivanon, P., Anderson, T. J., Krishna, S., White, N. J., & Nosten, F. (2006). Molecular and pharmacological determinants of the therapeutic response to artemetherlumefantrine in multidrug-resistant *Plasmodium falciparum* malaria. *Clinical Infectious Diseases, 42*, 1570–1577.

Puentes, S. M., Da Silva, R. P., Sacks, D. L., Hammer, C. H., & Joiner, K. A. (1990). Serum resistance of metacyclic stage Leishmania majorpromastigotes is due to release of C5b–9. *Journal of Immunology., 145*, 4311–4316.

Quiliano, M., Mendoza, A., Fong, K. Y., Pabón, A., Goldfarb, N. E., Fabing, I., Vettorazzi, A., López de Cerain, A., Dunn, B. M., Garavito, G., Wright, D. W., Deharo, E., Pérez-Silanes, S., Aldana, I., & Galiano, S. (2016). Exploring the scope of new arylamino alcohol derivatives: Synthesis, antimalarial evaluation, toxicological studies, and target exploration. *International Journal for Parasitology Drugs and Drug Resistance, 6*(3), 184–198.

Quintana, J. F., Del Pino, R. C., Yamada, K., Zhang, N., & Field, M. C. (2018). Adaptation and therapeutic exploitation of the plasma membrane of African trypanosomes. *Genes, 9*(7), 1–16.

Rajapakse, S., Chrishan Shivanthan, M., Samaranayake, N., Rodrigo, C., & Deepika Fernando, S. (2013). Antibiotics for human toxoplasmosis: A systematic review of randomized trials. *Pathogens and Global Health, 107*, 162–169.

Rajapakshage, B. K. W., Yamasaki, M., Hwang, S. J., Sasaki, N., Murakami, M., Tamura, Y., Lim, S. Y., Nakamura, K., Ohta, H., & Takiguchi, M. (2012a). Involvement of mitochondrial genes of *Babesia gibsoni* in resistance to diminazene aceturate. *Journal of Veterinary Medical Science, 5*, 1–45.

Rajapakshage, B. K. W., Yamasaki, M., Murakami, M., Tamura, Y., Lim, S. Y., Osuga, T., Sasaki, N., Nakamura, K., Ohta, H., & Takiguchi, M. (2012b). Analysis of energy generation and glycolysis pathway in diminazene aceturate-resistant *Babesia gibsoni* isolate in vitro. *Japanese Journal of Veterinary Research, 60*(2 & 3), 51–61.

Reynolds, M. G., Oh, J., & Roos, D. S. (2001). In vitro generation of novel pyrimethamine resistance mutations in the *Toxoplasma gondii* dihydrofolate reductase. *Antimicrobial Agents and Chemotherapy, 45*, 1271–1277.

Reynoldson, J. A., Thompson, R. C., & Meloni, B. P. (1991). In vivo efficacy of albendazole against *Giardia duodenalis* in mice. *Parasitology Research, 77*, 325–328.

Ribeiro-Gomes, F. L., Moniz-de-Souza, M. C., Alexandre-Moreira, M. S., Dias, W. B., Lopes, M. F., Nunes, M. P., et al. (2007). Neutrophils activate macrophages for intracellular killing of Leishmania major through recruitment of TLR4 by neutrophil elastase. *Journal of Immunology., 179*, 3988–3994.

Robert, M. A., Okamoto, K. W., Gould, F., & Lloyd, A. L. (2014). Antipathogen genes and the replacement of disease-vectoring mosquito populations: A model-based evaluation. *Evoloutionary Applied, 10*, 1238–1251.

Sakuma, M., Setoguchi, A., & Endo, Y. (2009). Possible emergence of drug-resistant variants of Babesia gibsoni in clinical cases treated with atovaquone and azithromycin. *Journal of Veterinary International Medicine, 23*, 493–498.

Samuelson, J. (1999). Why metronidazole is active against both bacteria and parasites. *Antimicrobial Agents and Chemotherapy, 43*, 1533–1541.

Sanches, B. F., Morgado, J., Carvalho, N., & Anjos, R. (2015). Multiple parasitic infections in a cardiac transplant recipient. *BMJ Case Reports, 2015*, 2014207033.

Schlitzer, M. (2007). Malaria chemotherapeutics part I: History of antimalarial drug development, currently used therapeutics, and drugs in clinical development. *Chemical Medical Chemistry, 2*(7), 944–986.

Sepúlveda-Arias, J. C., Gómez-Marin, J. E., Bobić, B., Naranjo-Galvis, C. A., & Djurković-Djaković, O. (2014). Toxoplasmosis as a travel risk. *Traveler Medicine and Infectious Diseases, 12*, 592–601.

Shanan, S., Abd, H., Bayoumi, M., Saeed, A., & Sandstrom, G. (2015). Prevalence of protozoa species in drinking and environmental water sources in Sudan. *Biomedicine Research International, 5*, 1–5.

Shandilya, A., Chacko, S., Jayaram, B., & Ghosh, I. (2013). A plausible mechanism for the antimalarial activity of artemisinin: A computational approach. *Scientific Reports, 3*(1), 1–7.

Shwab, E. K., Zhu, X. Q., Majumdar, D., Pena, H. F., Gennari, S. M., Dubey, J. P., & Su, C. (2014). Geographical patterns of *Toxoplasma gondii* genetic diversity revealed by multilocus PCR-RFLP genotyping. *Parasitology, 141*(4), 453–461.

Sibley, C. H., Hyde, J. E., Sims, P. F., Plowe, C. V., Kublin, J. G., Mberu, E. K., Cowman, A. F., Winstanley, P. A., Watkins, W. M., & Nzila, A. M. (2001). Pyrimethamine-sulfadoxine resistance in *Plasmodium falciparum*: What next? *Trends in Parasitology, 17*, 582–588.

Sidhu, A. B., Sun, Q., Nkrumah, L. J., Dunne, M. W., Sacchettini, J. C., & Fidock, D. A. (2007). In vitro efficacy, resistance selection, and structural modeling studies implicate the malarial parasite apicoplast as the target of azithromycin. *Journal of Biology and Chemistry, 282*, 2494–2504.

Sidhu, A. B., Uhlemann, A. C., Valderramos, S. G., Valderramos, J. C., Krishna, S., & Fidock, D. A. (2006). Decreasing pfmdr1 copy number in *Plasmodium falciparum* malaria heightens susceptibility to mefloquine, lumefantrine, halofantrine, quinine, and artemisinin. *Journal of Infectious Diseases, 194*, 528–535.

Silvie, O., Rubinstein, E., Franetich, J. F., Prenant, M., Belnoue, E., Renia, L., Hannoun, L., Eling, W., Levy, S., Boucheix, C., & Mazier, D. (2003). Hepatocyte CD81 is required for *Plasmodium falciparum* and *Plasmodium yoelii* sporozoite infectivity. *Natural Medicine, 9*, 93–96.

Simon, M. S., Westblade, L. F., Dziedziech, A., Visone, J. E., Furman, R. R., Jenkins, S. G., Schuetz, A. N., & Kirkman, L. A. (2017). Clinical and molecular evidence of atovaquone and

azithromycin resistance in relapsed Babesia microti infection associated with rituximab and chronic lymphocytic leukemia. *Clinical Infectious Diseases, 65*(7), 1222–1225.
Sisowath, C., Stromberg, J., Martensson, A., Msellem, M., Obondo, C., Bjorkman, A., & Gil, J. P. (2005). In vivo selection of *Plasmodium falciparum* pfmdr1 86N coding alleles by artemether-lumefantrine (Coartem). *Journal of Infectious Diseases, 191*, 1014–1017.
Smith, F. D., Ellse, L., & Wall, R. (2013). Prevalence of babesia and anaplasma in ticks infesting dogs in Great Britain. *Veterinary Parasitology, 198*, 18–23.
Smithuis, F., Kyaw, M. K., Phe, O., Win, T., Aung, P. P., Oo, A. P., Naing, A. L., Nyo, M. Y., Myint, N. Z., Imwong, M., Ashley, E., Lee, S. J., & White, N. J. (2010). Effectiveness of five artemisinin combination regimens with or without primaquine in uncomplicated falciparum malaria: An open-label randomised trial. *Lancet Infectious Diseases, 10*, 673–681.
Solaymani-Mohammadi, S., Genkinger, J. M., Loffredo, C. A., & Singer, S. M. (2010). A meta-analysis of the effectiveness of albendazole compared with metronidazole as treatments for infections with *Giardia duodenalis. PLoS Neglected Tropical Diseases, 4*(5), 1–10.
Soulsby, E. J. L. (1982). *Helminths, arthropods and protozoa of domestic animals* (7th Edn.), Bailliere Tindall.
Srivastava, I. K., & Vaidya, A. B. (1999). A mechanism for the synergistic antimalarial action of atovaquone and proguanil. *Antimicrobial Agents and Chemotherapy, 43*, 1334–1339.
Suwanarusk, R., Chavchich, M., Russell, B., Jaidee, A., Chalfein, F., Barends, M., Prasetyorini, B., Kenangalem, E., Piera, K. A., Lek-Uthai, U., Anstey, N. M., Tjitra, E., Nosten, F., Cheng, Q., & Price, R. N. (2008). Amplification of pvmdr1 associated with multidrug-resistant *Plasmodium vivax. Journal of Infectious Diseases, 198*(10), 1558–1564.
Swanson, E. A., March, J. K., Clayton, F., Couturier, M. R., Arcega, R., Smith, R., & Evason, K. J. (2018). Epithelial Inclusions in gallbladder specimens mimic parasite infection: Histologic and molecular examination of reported *Cystoisospora belli* infection in gallbladders of Immunocompetent patients. *American Journal of Surgical Pathology, 42*, 1346–1352.
Talapko, J., Škrlec, I., Alebić, T., Jukić, M., & Včev, A. (2019). Malaria: The past and the present. *Microorganisms, 7*(6), 179–196.
Tejman-Yarden, N., & Eckman, L. (2011). New approaches to the treatment of giardiasis. *Current Opinion in InfectiousDiseases, 24*, 451–456.
Tejman-Yarden, N., Miyamoto, Y., Leitsch, D., Santini, J., Debnath, A., Gut, J., McKerrow, J. H., Reed, S. L., & Eckmann, L. (2013). A reprofiled drug, auranofin, is effective against metronidazole-resistant *Giardia lamblia. Antimicrobial Agents and Chemotherapy, 57*(5), 2029–2035.
Telford, S. R., III., Gorenflot, A., Brasseur, P., & Spielman, A. (1993). Babesial infections in humans and wildlife. In J. P. Kreier (Ed.), *Parasitic protozoa* (pp. 1–47). Academic Press.
Tenter, A. M., Heckeroth, A. R., & Weiss, L. M. (2000). *Toxoplasma gondii*: from animals to humans. *International Journal of Parasitology, 30*, 1217–1258.
Ter Kuile, F. O., van Eijk, A. M., & Filler, S. J. (2007). Effect of sulfadoxine-pyrimethamine resistance on the efficacy of intermittent preventive therapy for malaria control during pregnancy: A systematic review. *JAMA, 297*, 2603–2616.
Then, R. L. (1982). Mechanisms of resistance to trimethoprim, the sulfonamides, and trimethoprim-sulfamethoxazole. *Clinical Infectious Diseases, 4*(2), 261–269.
Thomas, J., Baker, N., Hutchinson, S., Dominicus, C., Trenaman, A., Glover, L., Alsford, S., & Horn, D. (2018). Insights into antitrypanosomal drug mode-of-action from cytology-based profiling. *PLoS Neglcted Tropical Diseases, 12*, 1–19.
Thompson, P. E., Bayles, A., Olszewski, B., & Waitz, J. A. (1965). Quinine-resistant *Plasmodium berghei* in mice. *Science, 148*(3674), 1240–1241.
Tilley, L., Straimer, J., Gnädig, N. F., Ralph, S. A., & Fidock, D. A. (2016). Artemisinin action and resistance in *Plasmodium falciparum. Trends in Parasitology, 32*, 682–696.
Tomavo, S., & Boothroyd, J. C. (1995). Interconnection between organellar functions, development and drug resistance in the protozoan parasite, *Toxoplasma gondii. International Journal of Parasitology, 25*, 1293–1299.

Turcano, L., Torrente, E., Missineo, A., Andreini, M., Gramiccia, M., Di Muccio, T., Genovese, I., Fiorillo, A., Harper, S., Bresciani, A., & Colotti, G. (2018). Identification and binding mode of a novel Leishmania Trypanothione reductase inhibitor from high throughput screening. *PLoS Neglected Tropical Diseases, 12*(11), 1–21.

Ud Din, N., Torka, P., Hutchison, R. E., Riddell, S. W., Wright, J., & Gajra, A. (2012). Severe *Isospora* (*Cystoisospora*) *belli* diarrhea preceding the diagnosis of human T-cell-leukemia-virus-1-associated T-cell Lymphoma. *Case Reports in Infectious Diseases, 2012*, 1–4.

Upcroft, P., & Upcroft, J. A. (2001). Drug targets and mechanisms of resistance in the anaerobic protozoa. *Clinical Microbiology Review, 14*, 150–164.

Urquhart, G. M., Armour, J., Duncan, J. L., Dunn, A. M., & Jennings, F. W. (1996). *Veterinary parasitology* (2nd ed., p. 307). Blackwell Scince limited.

Uzlikova, M., & Nohynkova, E. (2014). The effect of metronidazole on the cell cycle and DNA in metronidazole-susceptible and -resistant Giardia cell lines. *Molecular Biochemsity and Parasitology, 198*, 75–81.

Vaillant, V., Valk, H. D., Baron, E., Ancelle, T., Colin, P., Delmas, M. C., Dufour, B., Pouillot, R., Strat, Y. L., Weinbreck, P., & Jougla, E. (2005). Foodborne infections in France. *Foodbourne Pathogens and Disease, 2*(3), 221–232.

Valderramos, S. G., & Fidock, D. A. (2006). Transporters involved in resistance to antimalarial drugs. *Trends in Pharmacological Sciences, 27*, 594–601.

van der Pluijm, R. W., Imwong, M., Chau, N. H., Hoa, N. T., Thuy-Nhien, N. T., Thanh, N. V., Jittamala, P., Hanboonkunupakarn, B., Chutasmit, K., Saelow, C., & Runjarern, R. (2019). Determinants of dihydroartemisinin-piperaquine treatmet failure in *Plasmodium falciparum* malaria in Cambodia, Thailand, and Vietnam: A prospective clinical, pharmacological, and genetic study. *The Lancet Infectious Diseases, 19*(9), 952–961.

van der Pluijm, R. W., Tripura, R., Hoglund, R. M., Phyo, A. P., Lek, D., Ul Islam, A., Anvikar, A. R., Satpathi, P., Satpathi, S., Behera, P. K., & Tripura, A. (2020). Triple artemisinin-based combination therapies versus artemisinin-based combination therapies for uncomplicated *Plasmodium falciparum* malaria: a multicentre, open-label, randomised clinical trial. *The Lancet, 2020*, 1345–1360.

Vennerstrom, J. L., Ellis, W. Y., Ager Jr, A. L., Andersen, S. L., Gerena, L., & Milhous, W. K. (1992). Bisquinolines. 1. N, N-bis (7-chloroquinolin-4-yl) alkanediamines with potential against chloroquine-resistant malaria. *Journal of Medicinal Chemistry, 35*(11), 2129–2134.

Verdier, R. I., Fitzgerald, D. W., Johnson, W. D., Jr., & Pape, J. W. (2000). Trimethoprim–sulfamethoxazole compared with ciprofloxacin for treatment and prophylaxis of *Isospora belli* and *Cyclospora cayetanensis* infection in HIV-infected patients: a randomized, controlled trial. *Annals of Internal Medicine, 132*(11), 885–888.

Vial, H. J., & Gorenflot, A. (2006). Chemotherapy against babesiosis. *Veterinary Parasitology, 138*(1–2), 147–160.

Villena, I., Aubert, D., Leroux, B., Dupouy, D., Talmud, M., Chemla, C., Trenque, T., Schmit, G., Quereux, C., Guenounou, M., & Pluot, M. (1998). Pyrimethamine-sulfadoxine treatment of congenital toxoplasmosis. Follow-up of 78 cases between 1980 and 1997. *Scandinavian Journal of Infectious Diseases, 30*(3), 295–300.

Walker, N., Nadjm, B., & Whitty, C. (2017). Malaria. *Medicine, 42*, 52–58.

Wang, Z. D., Liu, H. H., Ma, Z. X., Ma, H. Y., Li, Z. Y., Yang, Z. B., Zhu, X. Q., Xu, B., Wei, F., & Liu, Q. (2017). *Toxoplasma gondii* infection in immunocompromised patients: A systematic review and meta-analysis. *Frontiers in Microbiology, 8*, 1–12.

Weiss, L. M., Perlman, D. C., Sherman, J., Tanowitz, H., & Wittner, M. (1988). *Isospora belli* infection: Treatment with pyrimethamine. *Annals of Internal Medicine, 109*, 474–475.

Wells, T. N., Alonso, P. L., & Gutteridge, W. E. (2009). New medicines to improve control and contribute to the eradication of malaria. *Nature Reviews Drug Discovery, 8*(11), 879–891.

Wells, T. N., Burrows, J. N., & Baird, J. K. (2010). Targeting the hypnozoite reservoir of *Plasmodium vivax*: The hidden obstacle to malaria elimination. *Trends in Parasitology, 26*(3), 145–151.

Wenyon, C. M. (1923). Coccidiosis of cats and dogs and the status of the Isospora of man. *Annals of Tropical Medicine and Parasitology, 17*, 231–288.
WHO. (2009). Guidelines for the treatment of malaria world health organization. http://whqlibdoc.who.int/publications/2010/9789241547925_eng.pdf
WHO. (2010) Control of the Leishmaniases; World Health Organization.
WHO. (2011). Global plan for Artemisinin Resistance Containment World Health Organization. http://www.who.int/malaria/publications/atoz/9789241500838/en/index.html.
WHO. (2015). *Guidelines for the treatment of malaria* (3rd ed., p. 2015). World Health Organization.
WHO. (2018). World health organization. World Malaria Report 2018. WHO. 2018. Accessed July 5, 2020. https://apps.who.int/iris/bitstream/handle/10665/275867/9789241565653-eng.pdf?ua=1
WHO. (2019a). *Model list of essential medicines*. http://www.who.int/medicines/publications/essentialmedicines/en/. Accessed June 24, 2020.
WHO. (2019b). World malaria report 2019b. World Health Organization.
Wiedemar, N., Graf, F. E., Zwyer, M., Ndomba, E., Kunz Renggli, C., Cal, M., Schmidt, R. S., Wenzler, T., & Mäser, P. (2018). Beyond immune escape: A variant surface glycoprotein causes suramin resistance in *Trypanosoma brucei*. *Molecular Microbiology, 107*(1), 57–67.
Wittner, M., Lederman, J., Tanowitz, H. B., Rosenbaum, G. S., & Weiss, L. M. (1996). Atovaquone in the treatment of *Babesia microti* infections in hamsters. *The American Journal of Tropical Medicine and Hygiene, 55*(2), 219–222.
Wiwanitkit, V. (2012). Interest in paromomycin for the treatment of visceral leishmaniasis (kala-azar). *Therapeutics and Clinical Risk Management, 8*, 323–328.
Wong, I. L., Chan, K. F., Zhao, Y., Chan, T. H., & Chow, L. M. (2009). Quinacrine and a novel apigenin dimer can synergistically increase the pentamidine susceptibility of the protozoan parasite Leishmania. *Journal of Antimicrobial Chemotherapy, 63*(6), 1179–1190.
Wongsrichanalai, C., Pickard, A. L., Wernsdorfer, W. H., & Meshnick, S. R. (2002). Epidemiology of drug-resistant malaria. *The Lancet Infectious Diseases, 2*(4), 209–218.
Woodrow, C. J., & Krishna, S. (2006). Antimalarial drugs: recent advances in molecular determinants of resistance and their clinical significance. *Cellular and Molecular Life Sciences CMLS, 63*(14), 1586–1596.
Woodrow, C. J., & White, N. J. (2017). The clinical impact of artemisinin resistance in Southeast Asia and the potential for future spread. *FEMS Microbiology Reviews, 41*(1), 34–48.
Woon, S. A., Yang, R., Ryan, U., Boan, P., & Prentice, D. (2016). Chronic Cystoisospora belli infection in an immunocompetent Myanmar refugee–microscopy is not sensitive enough. *BMC Infectious Diseases, 16*(1), 221.
Yamasaki, M., Hossain, M. A., Jeong, J. R., Chang, H. S., Satoh, H., Yamato, O., & Maede, Y. (2003). *Babesia gibsoni*–specific isoenzymes related to energy metabolism of the parasite in infected erythrocytes. *Journal of Parasitology, 89*(6), 1142–1146.
Yang, G., Choi, G., & No, J. H. (2016). Antileishmanial mechanism of diamidines involves targeting kinetoplasts. *Antimicrobial Agents and Chemotherapy, 60*(11), 6828–6836.
Yarlett, N., Gorrell, T. E., Marczak, R., & Müller, M. (1985). Reduction of nitroimidazole derivatives by hydrogenosomal extracts of *Trichomonas vaginalis*. *Molecular and Biochemical Parasitology, 14*(1), 29–40.
Zhang, G., Guan, Y., Zheng, B., Wu, S., & Tang, L. (2008). No PfATPase6 S769N mutation found in *Plasmodium falciparum* isolates from China. *Malaria Journal, 7*(1), 122–141.

Chapter 5
Antifungal Drugs: Mechanism of Action and Resistance

Muhammad Zeeshan Ahmed, Tazeen Rao, Ali Saeed, Zeeshan Mutahir, Shahzeb Hameed, Samrana Inayat, Haseeba Shahzad, Najeeb Ullah, Muhammad Abaid-Ullah, Muhammad Ibrahim, Shehla Munir, Asghar Javed, Jallat Khan, Muhammad Khalid, and Sarfraz Ahmed

Abstract Overuse of antifungal drugs in recent years has developed resistance against these drugs. The significant clinical influence of resistance has resulted in increased interest in the study of antimicrobial resistance from all possible angles. The mechanisms of resistance, better methods of resistance detection, alternative methods to treat the infections of resistant organisms, and all the possible strategies to control and prevent the occurrence of resistance should be addressed. In this

M. Z. Ahmed · Z. Mutahir · S. Hameed · H. Shahzad
Institute of Biochemistry and Biotechnology, University of the Punjab, Quaid-i-Azam campus, Lahore 54590, Punjab, Pakistan
e-mail: zeeshan.ibb@pu.edu.pk

T. Rao
Department of Biochemistry, Bahauddin Zakariya University, Multan 60800, Punjab, Pakistan

A. Saeed
Institute of Molecular Biology & Biotechnology, Bahauddin Zakariya University Multan, Multan 60800, Punjab, Pakistan
e-mail: alisaeed@bzu.edu.pk

S. Inayat
School of Biological Sciences, University of the Punjab, Quaid-e-Azam Campus, Lahore 54590, Pakistan

N. Ullah
Laboratory for Structural Biology of Infection and Inflammation C/O DESTY, Build. 22 A NotkestraBe, 8522607 Hamburg, Germany
e-mail: najeeb.ullah@chemie.uni-hamburg

M. Abaid-Ullah
Department of Life Sciences, Khwaja Fareed University of Engineering & Information Technology, Rahim Yar Khan 64200, Pakistan
e-mail: m.abaidullah@kfueit.edu.pk

M. Ibrahim · S. Munir
Department of Biochemistry, Bahauddin Zakariya University, Multan 60800, Pakistan
e-mail: ibrahimjam@bzu.edu.pk

A. Javed
Department of Pathology, Nishtar Medical University, Multan 60800, Pakistan
e-mail: sarfraz.ahmed@uvas.edu.pk

S. Ahmed et al. (eds.), *Biochemistry of Drug Resistance*,
https://doi.org/10.1007/978-3-030-76320-6_5

chapter, common drugs used against fungi, their mode of action, and mechanism of resistance are discussed. Allylamines and azole inhibit the ergosterol synthesis, echinocandin binds with glucan synthase to inhibit the synthesis of the cell wall, 5-fluorocytosine (5-FC) inhibits the synthesis of macromolecules, and polyenes physiochemically interact with sterols in the plasma membrane. While the alteration in the biosynthesis of sterols, overexpression, reduction, and alteration in concentration or structure of target enzymes could contribute to the development of resistance against the antifungal drugs. All the ongoing efforts to identify the resistance and its emergence is used for a safe future.

Keywords Antifungal drug · Fungi · Infections · Mode of action · Resistance mechanism

5.1 Introduction

The world has witnessed a troubling change over the past few decades, where traditional antimicrobial agents are becoming more and more inefficient in combating contagious diseases. There is still a significant need for human health for the production of novel therapeutics to combat pathogenic microbes, and also a need to anticipate and avoid the detention development to our present antimicrobial armamentarium, with the current situation widely defined as a post-antibiotic age (Cheesman et al., 2017). In the sense of fungal pathogens, this problem of infectious diseases is of particular concern considering the lack of various groups of fungicide for the medicaments of bacterial pneumonia and the appearance and growth of fungal pathogens that are multidrug-resistant (Konopka et al., 2019).

They remain a fairly undervalued contributor to human illness and morbidity, considering the destructive effects that pathogenic fungi have on human health. Fungi attack billions and annually kill approximately 1.5 million people, which is equivalent to the infant mortality of more commonly known ailments like tuberculosis or malaria. In recent decades, the incidence of severe fungal diseases has increased due to the rising number of immunocompetent patients, as well as cancer patients, recipients of organ transplants, respectively (World Health Organization, 2015).

A substantial rise in the prevalence of resistance to antibacterial and antifungal agents has been observed in the past decade. In U.S. hospitals, as well as in the environment, exposure to antimicrobial agents has significant consequences for mortality,

J. Khan · M. Khalid
Department of Chemistry, Khwaja Fareed University of Engineering & Information Technology, Rahim Yar Khan 64200, Pakistan

S. Ahmed (✉)
Department of Basic Sciences, University of Veterinary and Animal Sciences Lahore, Narowal 51600, Pakistan
e-mail: sarfraz.ahmed@uvas.edu.pk

illness, and health care costs. Significant attention has therefore been paid to developing a more comprehensive understanding of antimicrobial resistance pathways, improved methods for identifying resistance whenever it arises, effective antiviral options for treating infections caused by drug-resistant species, and, in the first place, strategies for preventing the growth and spread of resistance (Clarke, 2006; World Health Organization, 2020).

All of this focus has been dedicated to the analysis of bacterial resistance for a variety of reasons: (1) infectious diseases are responsible for the bulk of group and nosocomial diseases; (2) a broader range of resistance mechanisms are available for study in the significant and growing number of antibacterial classes; and (3) the ability to transfer bacterial resistance determinants (Fair & Tor, 2014). For many factors, the research of fungicide resistance has underperformed behind other antibiotic resistance. Maybe most important, fungal diseases have not been viewed as significant infections until recently. The annual mortality rate due to infections, for instance, was steady between 1950 and around 1970. In conjunction with many developments in medical treatment, this rate has risen dramatically since 1970, including further extensive use of drugs that weaken the immune function, the regular and frequently systematic use of diverse disinfectants, the widespread use of injectable devices in residents, and the prevalence of chronic anticoagulant bacterial diseases like AIDS. These advances and the resulting rise in fungal infections strengthened the discovery of new, better, and safer agents to battle dermatomycosis (World Health Organization, 2020).

5.2 Problems with Comparing Antifungal and Antibacterial Resistance

While we assume that a comparative analysis among antibacterial and antifungal resistance mechanisms is a valuable way to gain a viewpoint on nosocomial infections in the two kingdoms, many factors inherently restrict the contrast. First, there are very considerable differences in the structures of bacteria and fungi (such as the multicellular presence of most fungi and the prolonged development period of fungi relative to bacteria), and the antifungal and antibacterial agents available attack the function and structure which are most important to the species to be prevented. Many bacterial pathogens, for example, prevent steps that are necessary for phospholipids, the basic component of the bacterial cell membrane (Frey-Klett et al., 2011).

Most fungicides, on the other hand, attack either the production or action of an enzyme, an essential component of the membrane of fungal cells. There are major similarities, however, between the methods by which fungus confer antibodies to inhibitors of ethylene biosynthesis and microbes show antibodies to anti-cell wall compounds. A comparative analysis is restricted with some other forms of resistant bacteria by the possibility that there are no antiviral analogs of certain groups of antibacterial agents (protein synthesis receptors like macrolides, tetracyclines, and clarithromycin; targeted therapies such as fluoroquinolones; and metabolomic agents including trimethoprim-sulfamethoxazole) (Langdon et al., 2016). Contrary, among currently accessible antimicrobial agents, antifungal nucleoside analogs like 5-FC have no relatives (while they are expressed among the artifact compounds). The ability of fungi to develop ribosomal opposition or angiotensin mutants, as well as the ability of bacteria to build immunity to cytotoxic agents, is still uncertain. Surprisingly, if used in conjunction with amphotericin B, the antimicrobial RNA polymerase blocker rifampin, which exhibits no probability and impact of risks towards pathogens, proves very effective against many microbial strains (Ghannoum & Rice, 1999). As a result of the action of amphotericin B on the fungal membrane, this synergistic behavior was due to increased absorption of rifampin further into the fungal cell. Comparable synergism was shown by Polak et al., utilizing porcine candidiasis models, between amphotericin B and 5-FC. Most researchers have asserted that the method for this synergism is to increase 5-FC uptake due to membrane dysfunctionality caused by the interaction of amphotericin B-ergosterol (Polak, 1978).

This synergism is close to the observed bacteriostatic synergism mechanism among cell wall active products and enterococcal aminoglycosides where only enhanced intracellular oxytetracycline concentrations are observable when streptomycin is paired with in vitro penicillin targeting *Enterococcus faecalis* (Krause et al., 2016).In comparison to the suggestion that amphotericin B enhances 5-FC absorption, data collected and indicated that such dual agents work consecutively towards *Candida albicans* (*C. albicans*) to impact synergy rather than in conjunction.

Here the second drawback of the constant comparison between the mechanisms of antibacterial and antifungal resistance was that these general classes of resistance mechanisms in fungi have still not been established. Antibacterial agent resistance results from antibiotics alteration, antimicrobial target modification, decreased access to the network, or any confluence of such processes. Antimicrobial alteration is certainly the most fundamental resistance mechanism for the antibacterial groups of β-lactam (β-lactamases) and an aminoglycoside (penicillinase enzymes). On the other hand, while there has been a specific, unnamed source of dermatophytic fungal depletion of nystatin, there is no evidence to indicate that antibiotic modulation is an effective antifungal restraining factor. But on the other side, considerable evidence indicates that significant resistance mechanisms to antifungals are both objective modifications and decreased exposure to targets (sometimes in combination). In resistant bacteria, certain pathways have essential comparisons (Kapoor et al., 2017).

The third drawback of the analogy is that our understanding of the transmission of genetic exchanges is much more developed than our awareness of fungal exchanges, whether they exist. To promote the transfer of resistance and toxicity determinants among as well as between organisms, bacteria appoint the majority of activities of transposons, plasmids, and bacterial spores. As a consequence, there is an abundance of prospects for the rapid growth of advanced resistance and the capacity for resistance emergence and dissemination, though in the absence of specific choice by particular antimicrobial pressures. In comparison, the analgesic resistance mentioned to date typically includes the appearance of spontaneous pathogens (as in the growing importance of *Candidakrusei* in areas of widespread use in some health centers) or the slow, gradual modification of individual cells or structures to prevent the action of an antifungal to which widespread access has occurred (Monteiro & Santos, 2019). The last major drawback in the comparison of antibacterial and antifungal resistant strains is the existence of standard strains of bacteria and plasmid DNA to be used in the microbial antibiotic-resistant analysis. This accessibility enables resistance determinants to be isolated in very well environments, such that it is possible to determine the precise influence of individual resistant strains. In contrast to bacterial antibiotic resistance, the existence of really well strains and methods for DNA distribution enables an even more systematic contribution to the subject of the physiology of microbial mechanisms of resistance (Fair & Tor, 2014). For instance, moving the plasmid through a good strain, usually, *Escherichia*, is the first step in studying plasmid-mediated β-lactamases in the microbiota. Through this manner, needlessly complicated processes like membrane modification can be regulated and it is possible to make fair comparisons of the resistance levels imposed by various β-lactamases. The fact that identical standardized schemes are not accessible in fungi suggests that resistant experiments almost always take place in the pathogenic isolates itself, making it practically impossible to assess the precise significance of various resistance mechanisms to the phenotypic of susceptibility (Davies, 1996).

5.3 Assessing Resistance Factors

Medical resistance corresponds to treatment failures in which, following application at a normal dosage, a person tends to react to an antibacterial drug. The growth of resistant pathogens is largely dependent on various host and bacterial variables. Host's immune strength is an essential aspect, and to contain and kill a disease, fungicidal drugs should work synergistically. Sufferers with extreme immune dysfunction are much more unsuccessful treatment because, without the help of an immune reaction, the antibacterial drug could fight the infection (Knight, 1975). The existence of indwelling catheters, artificial valves for the heart, and other surgical instruments may also lead to refractory infections, as infectious substances bind to these items and build biofilms that resist drug action. Suitable therapy needs that at a concentration necessary for the action of antibiotics, each drug enters the site of infection. The pharmacokinetics of many drugs are understood, but at all sites of inflammation, we do not have a solid knowledge of drug absorption. Thus, at inadequate amounts, certain microbes are subjected to medicines. Both of these things influence resistant bacteria, which relates to the choice of species that, following exposure to medicinal anti-inflammatory levels, will propagate. During treatment, such conditions lead greatly to medication errors. Microorganism resistance includes both main resistance genes that are generally less responsive to a given antifungal and secondary pathogenic organisms that, despite clinical manifestations, develop a resistance particular attribute in a strain that is otherwise resistant (Cavalheiro et al.,

2018). In main strains that are immune, the molecular mechanisms involved in resistant strains are sometimes showed at different stages and these would be addressed in-depth in this chapter.

5.4 Common Drugs Used Against Fungal Infections

1. Allylamines
2. Azoles
3. Echinocandin
4. Flucytosine
5. Polyenes.

5.4.1 Allylamines

During a chemical development program for the synthesis of new central nervous system drugs, the allylamine class of antifungal agents was uncovered by chance. The substance with a novel chemical composition was formed by an unexpected chemical process. This substance, later called naftifine, was reported to have better antifungal efficacy when tested in vitro in the regular biologic bacteriological analysis. This feature has been verified both in scientific research of animal bacterial genomes and in human testing (Petranyi et al., 1984). The label of this new chemical class of antifungals is associated with the existence of a tertiary feature of allylamine, that is, a nitrogen atom with a neighboring double bond, in its chemical composition. All active members of the sequence of allylamines express this structural role. Two allyl amines-naftifine, the official member of this group, and the analog terbinafine were studied clinically. Naftifine is purely therapeutically effective. Terbinafine synthesis resulted from research undertaken to improve the allylamines' pro-file antifungal effectiveness by altering their chemical composition. Terbinafineincludes two major structural shifts, a triple bond, and corresponding alkyl side chain branching. The tertbutyl acetylene group's substitution of the phenyl ring in the naftifine molecule offers better antifungal effectiveness, nasal potency, and a wider range of activities, like *Aspergillus fumigatus* and *C. albicans* as well as *dermatophytes* (Krishnan-Natesan, 2009). In vitro data shows that terbinafine is especially active in the treatment of *dermatophytes*, molds, and dimorphic fungi. It was observed that mean minimum inhibitory concentrations (MICs) of terbinafine range from 0.001 to 0.01~g/m 1.4 for *Trichophyton spp.*, *Epidermophytonfloccosum*, and *Microsporummay*. Similarly, MIC values for *Aspergillus spp* and *Sporothrix schenckii* range from 0.63 and 0.2~g/ml, respectively. *Terbinafine's* yeast activity has been found to be lower and differs between various species and strains. MIC values against C, for instance, are ten-fold to 100-fold lower than against *C. parapsilosis* and *albicans*. The in vivo toxicity of terbinafine toward *dermatophytes* is around one to 100 times bigger than that of naftifine, 2–30% bigger than that of ketoconazole, and roughly 10 times higher than those of itraconazole, the triazole relative (Cuenca-Estrella et al., 2008; Public Health Agency of Canada, 2011).

5.4.1.1 Mode of Action

There is a novel class of ergosterol biosynthetic inhibitors that is allylamines include terbinafine and naftifine. These inhibitors are chemically as well as functionally different from other main classes of antifungal representatives. The major allylamines are terbinafine which is highly active against *dermatophytes* both in vitro as well as in vivo. The National Committee for Clinical Laboratory Standards (NCCLS) done the latest study about terbinafine by using the M27 process and revealed the fact that its geometric mean MIC in contrast to 179 clinical separates of *C. albicans* was 1.2 μg/ml. In addition to these findings, earliest confirmations from different groups of researchers and scientists point out that terbinafine is also effective against some *C. albicans* strains which are resistant to azole as well as against *Cryptococcus neoformans* by using a similar assay technique. Moreover, experiments using animal models exploring the usefulness of this representative against dispersed candidiasis are in the process (Ryder et al., 1998).

Allyl amines best perform their actions by inhibiting the initial steps of ergosterol biosynthesis. As a result of this inhibition, the sterol precursor squalene accumulates, and due to the absence of other sterol intermediates signifying that when squalene epoxidated, allylamine inhibition of sterol synthesis takes place. Squalene epoxidation is a reaction catalyzed by squalene epoxidase (Fig. 5.1). Further studies with isolated squalene epoxidase specify that it is the target for allyl amine action. Accretion of squalene instead of ergosterol insufficiency leads to fungal cell death by increasing the permeability of the membrane which resulted in interruption of cellular organization (Jordá & Puig, 2020).

5.4.1.2 Mechanism of Resistance

Even though terbinafine was given to patients, medically failed to perform its function but in the case of human pathogenic fungi, no resistance to allylamines such as terbinafine and naftifine has been observed. On the other hand, one can expect resistance to these agents due to frequent usage. Berkow and Lockhart (2017) have informed a *C. glabrata* strain that developed fluconazole resistance and stated cross-resistance to allylamines such as terbinafine. Further studies on these agents showed that some fungal species already present a piece of machinery that develops resistance to allylamines and CDR1 can use terbinafine as substrate (Berkow & Lockhart, 2017).

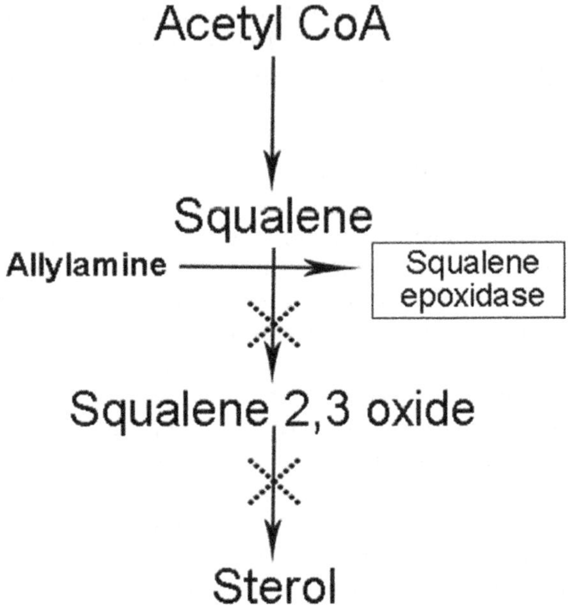

Fig. 5.1 Showing the mode of action of allylamine by inhibiting the synthesis of sterol

5.4.1.3 Correlation with Antibacterial Resistance

Subsequently, allylamine-resistant fungi are so far not sound defined, links of resistance mechanisms are controversial. Still, it is worth stating that different antibacterial agents that are involved in successive steps of cell wall synthesis bear a resemblance to numerous azoles, allylamines, and polyenes action sites. These antibacterial agents include; phosphomycin is a phosphoenolpyruvate analog (act at an early step), penicillin (act at an intermediate step in peptidoglycans synthesis), and vancomycin (acts at the final step in cross-linking). Further studies of bacterial cell wall synthesis showed that the various mechanisms of action of the antifungal agent have been clarified by considering the increase of particular precursors after antibiotic exposure. As all the antibiotics effective at different stages of a similar process, in the meantime it is not amazing that particular mutations to many compounds will have resulted in cross-resistance (Lee & Lee, 2018).

5.4.2 Azoles

To treat invasive and superficial fungal infections, antifungal azoles are commonly used. These antifungals comprise the three generations of triazole and imidazole-containing medications.

a. First generation includes bifonazole, clotrimazole, ketoconazole and econazole

b. Second generation includes fluconazole and itraconazole
c. Third generation includes voriconazole and posaconazole (Mast et al., 2013).

Imidazole present in the ring system of first-generation compounds and due to their poor water stability and restricted oral bioavailability, these compounds are generally used topically. These drugs are not the best choice for orally administered due to their undesirable side effects including skin irritation, hepatotoxicity, gastrointestinal and endocrine disturbances which are drug dependent (Verma et al., 2013). Due to the development of less toxic and more effective systemic azoles, the first-generation azole, Ketoconazole is no longer the drug of choice as it could be used systemically. Triazole ring is present in the second and third generation antifungal azoles instead of imidazole structure and clinically exhibit a broader spectrum of activity than imidazole containing azoles. Triazole-containing antifungals have improved oral bioavailability as well as pharmacokinetic/pharmacodynamics properties in contrast to antifungal imidazole (Mast et al., 2013). Furthermore, due to their enhanced affinities to the target enzymes in the invading microorganisms, these agents have noticeably upgraded safety profiles. Recently, isavuconazole, albaconazole, and ravuconazole were also included in third-generation azoles and are under study in medical trials. These agents are still not appropriate for clinical practice in the United States.

5.4.2.1 Mode of Action

Imidazole and triazoles are the two main classes of antifungals azole. Presently existing imidazole are butoconazole, miconazole, clotrimazole, econazole, and ketoconazole. The triazoles used extensively are terconazole, itraconazole, and fluconazole. Both classes share the same mechanisms of action and also are fungistatic representatives. By hindering cytochrome P450-dependent 14 alpha-sterol demethylase, the azoles restrict the permeability and the synthesis of fungal cell membranes (Fig. 5.2). The main sterol of most fungal cell membranes is ergosterol which is synthesized by this enzyme. Due to the inhibition of this enzyme, fungal cell membrane integrity damage as a result of cell division is inhibited. Because of cell division and cell membrane integrity damage, intracellular swelling occurs which interferes with cellular adhesion and epithelial penetration. Furthermore, all this phenomenon resulted in the protrusion of cytoplasm through the cell membrane (Bardal et al., 2011).

As mentioned earlier, ketoconazole, one of the main imidazole achieved extensive acceptance because of its usage as the only accessible oral agent for clinical treatment of systemic fungal infections. Due to the discovery of new triazoles, ketoconazole's usefulness has been surpassed. But still, ketoconazole remains active as a second-line agent in contrary to certain Paracoccidioidomycosis, blast mycosis, candida species, coccidioidomycosis, and histoplasmosis. In immunocompromised patients having systemic fungal infections, itraconazole and fluconazole proved to be more useful.

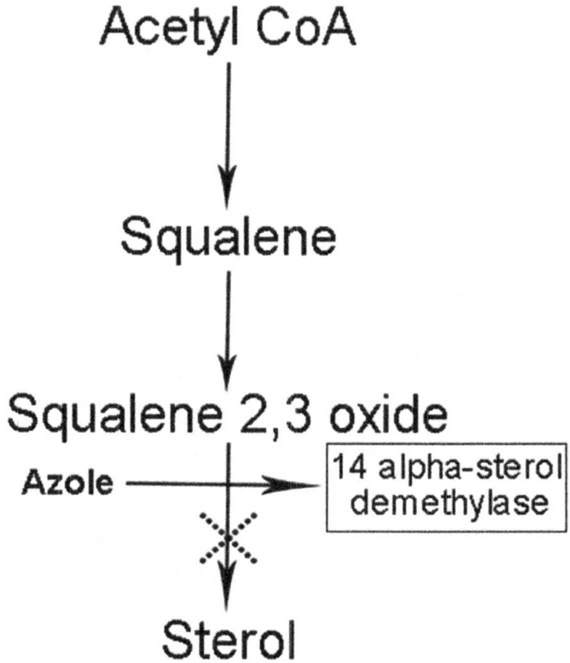

Fig. 5.2 Showing the mode of action of azole by inhibiting the synthesis of sterol

Moreover, these compounds have low drug-related toxicity (Sinawe & Casadesus, 2020).

5.4.2.2 Mechanism of Resistance

In candida species, there are three main resistance mechanisms to azoles that have been defined as shown in Fig. 5.3 (Berger et al., 2017). In any known fungal strain, more than one mechanism can be operative with additive effects. At the site of action, active efflux pumps developed which results in a low level of drug concentration. There are two main gene families of transporters that encodes the efflux pumps in candida species. These include; *CDR* genes (ATP-binding cassette superfamily) and *MDR* genes (major facilitators class). In the case of azole-resistant *C. albicans*, *CDR1*, *CDR2*, and *MDR1* genes seem to be up-regulated. In addition to these transporters' genes, *PDH1* and *CgCDR1* genes have been noticed in *C. glabrata* candida species as well as *CdMDR1* and *CdCDR1* in *Candida dubliniensis*. However, nearly all azoles are resistant to up-regulation of the *CDR* gene but efflux pumps encoded by *MDR* genes have contracted spectrum particular for fluconazole. As it was found earlier that the *ERG11* gene encodes for the target enzyme lanosterol C14α-demethylase, a mutation in this gene leads to the inhibition of azoles binding to the enzymatic site. Likewise, in *C. krusei* isolates, the intrinsic resistance to fluconazole has been recognized to the decreased affinity of *ERG11* to the drug. More than 80 amino acid

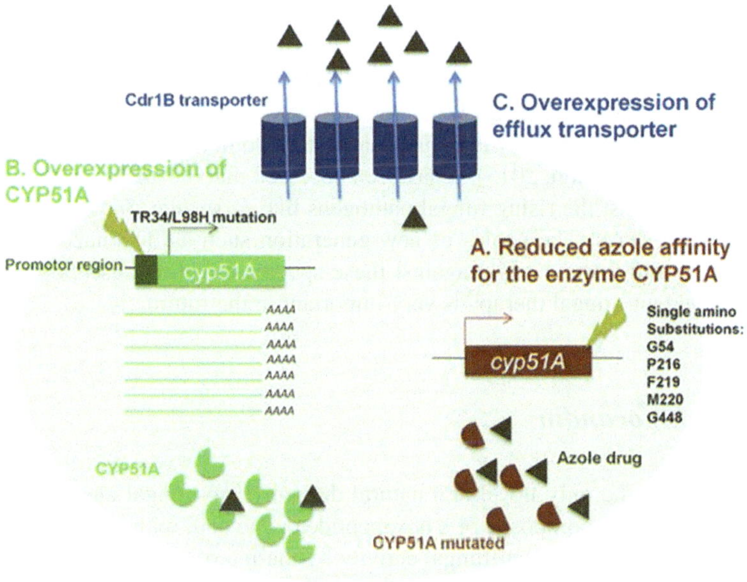

Fig. 5.3 Showing the three reported resistance mechanisms against azole drug (Berger et al., 2017)

substitutions in *ERG11p* have been reported and in similar genes, different mutations can co-occur with additive effects (Cowen et al., 2015).

Azole-susceptible strains have less intracellular *ERG11p* concentrations as compare to some candida strains which are less susceptible to azoles. Consequently, ergosterol synthesis can't be effectively inhibited by routine therapeutic concentrations as the antifungal agent is overwhelmed. Upregulation of target enzyme can be accomplished with reduced degradation of the gene product or high rate of transcription and also with gene amplification. On the other hand, these phenomena are believed to add a minute to the general resistance load in Candida species due to an uncertain increase in enzyme level (Ghannoum & Rice, 1999). Ergosterol is the fungal membrane that becomes depleted with the exposure of azole with the aggregation of 14α-methyl-3,6-diol, a toxic product that stops the growth of fungi. While the mutation in the *ERG3* gene will stop the conversion of 14 α-methyl fecosterol to 14α-methyl-3,6-diol. Substitution of ergosterol with the 14 α-methy lfecosterol helps to make the membrane functional and stop the azole action on the biosynthesis of ergosterol. The mutations in the *ERG3* gene made the Candida strains polyenes resistant due to the deficiency of ergosterol in the plasma membrane (Martel et al., 2010).

Although the azole resistance in the *Aspergillus* species is well known. Efflux pump expression caused the reduction of intracellular itraconazole concentration is the first mechanism. Other mechanisms are more common that is due to the alteration in the 14α-sterol demethylase enzyme encoded by the genes *cyp51A* and *cyp51B*. The substitution at position G54 caused the cross-resistance toposaconazole and itraconazole, while the substitution at position M220 led to the phenotypic resistance to all the azole with the elevation of MICs. Recently, another new mechanism of resistance in *Aspergillus fumigatus* against azole has been reported. In which overexpression of the proteins has been observed by the mutations in the promoter region of gene *cyp51A* (Wiederhold & Patterson, 2015). It has been observed that the azole has shown variable activity against the rising fungal pathogens like *Fusarium*, *Scedosporium*, and *zygomycetes* species. The azoles of new-generation such as fluconazole have not shown any antimycotic activity against these species therefore susceptible testing for the guided antifungal therapy is very important in the future.

5.4.3 Echinocandin

Echinocandin is the only novel and natural derivative antifungal class of drugs in medical history. It is composed of a hexapeptide cyclic core with a side chain of N-linked fatty-acyl. It shows antifungal activity by the interruption in the structure of the cell wall that is not found in mammalian cells. Echinocandin non-competitively binds to the active site of the (1,3)-β-D-glucan synthase enzymes. In *C.glabrata*, this enzyme is encoded by *FKS1* and *FKS2* while in *C. albicans*, encoded by only *FKS1*. After the blockage in the synthesis of (1,3)-β-D-glucan, the structure of the cell wall is disrupted and osmotic imbalance in the cell leads to the fungicidal effect. At this time, echinocandins are considered as a first-line of treatment for candidiasis to inhibit the activity of Candida species. While this class of drug is considered ineffective against *Cryptococcus*. The major advantages for the use of echinocandins are the higher fungal selectivity, low potential in drug to drug interaction, and excellent safety profile. Similar to polyenes, echinocandins are not good to use as orally and that is why the clinical use of echinocandins is limited to administration via intravenous. Recently, an inhibitor of 1,3-β-D-glucan synthase such as Ibrexafungerp with advanced structural changes that are available in both intravenous and oral formulations. This is now in phase II and III clinical trials. With the widespread clinical use of Echinocandin, resistance against this drug is also reported in recent years (Perlin, 2011).

5.4.3.1 Mode of Action

Echinocandins bind non-competitively with the β-(1,3)-glucan synthase enzymes to inhibit the catalytic process of uridine diphosphate-glucose (UDP-glucose) polymerization into β-(1,3)-glucan that is important in the structure of fungal cell wall as shown in Fig. 5.4. Lesser the β-(1,3)-glucan results in the decrease of rigidity and integrity of the cell and swallowing of the cell leads to the burst of the cell. Glucan synthase consists of two subunits: one is the activating subunit (RHO) and the other is the catalytic subunit (FKS) which is bound with the plasma membrane. This enzyme

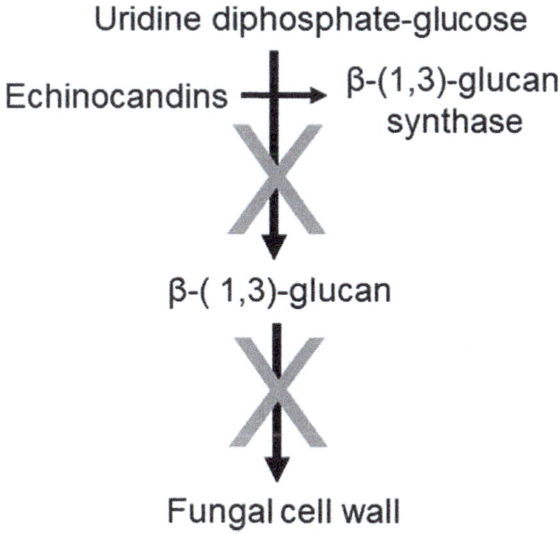

Fig. 5.4 Showing the mode of action of echinocandins

is a UDP-glucosyl-transferase. The activity of guanosine triphosphatase at activating subunit activates the catalytic subunit to form the glucan fibrils by the polymerization of UDP-glucose. The polymer extrudes towards the plasma membrane from where the polymer is incorporated into the cell wall of fungi in the periplasmic space. Still, the clear mechanism of echinocandins to inhibit the synthesis of the cell wall of fungi is not known. *Cryptococcus neoformans* have a single homolog copy of catalytic subunit i.e., FKS1 that's expression is essential for the viability of cells. While FKS in *C. neoformans* is found under the control of transcriptional signal i.e., calcineurin. It was found that the caspofungin exposure results in a decrease in the number of β-1,6 glucan in the cell wall with the possibility of inhibition in the synthesis of β-1,3 glucan in the cell wall of *C. neoformans* (Douglas, 2001).

5.4.3.2 Mechanism of Resistance

In some organisms such as *C. neoformans*, their cell wall integrity does not depend upon the β-(1-3) linked glucan. So those organisms are less responsive to echinocandins. *Zygomycetes*, *Sporothrixschenckii*, *Rhizopussp*, *Histoplasma capsulatum*, *Fusarium sp.*, and *Blastomyces dermatitidis* have not or very less been affected by Echinocandins in vitro. Caspofungin resistance has not been reported because of the limited clinical reports to date. However, in vitro studies of exposure of caspofungin to *C. albicans* in small concentrations with an increase in one-fold from 0.06 to 0.125 μg/mL in MIC was observed. *GNS1* gene was deleted while the mutations in the *FKS1* and *FKS2* genes were correlated with the low-levels and high-levels of resistance against echinocandins. The low level of resistance against echinocandins resulted in the overexpression of Golgi protein, *Sbe2P* that is associated with

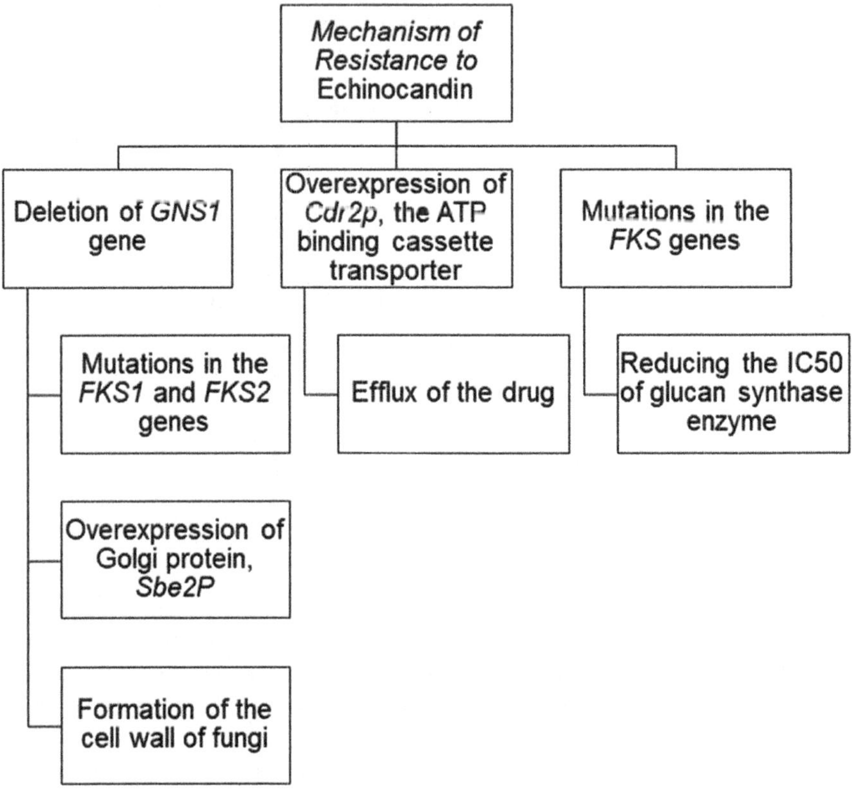

Fig. 5.5 Showing the mechanism of resistance in echinocandin

the formation of the cell wall (Fig. 5.5). While in *Aspergillus sp.*, and some other pathogenic fungi the in vitro resistance against the echinocandins has not yet been reported (Shields et al., 2011).

The resistance mechanism associated with efflux pumps in the *Aspergillus* and *Candida* species against azole is important while their role against the echinocandins has not been reported and understood. Previous studies had reported that echinocandins are not the substrate of efflux pumps but recent studies have reported the resistance against the caspofungin due to the overexpression of *Cdr2p*, the ATP binding cassette transporter in *C. albicans*. This is just the efflux of the drug, not the cell wall or membrane changes that are still unknown as shown in Fig. 5.5 (Cannon et al., 2009). In a U.S. Hospital, the rate of resistance against the echinocandins was reported to be increased from 4.9 to 12.3% in *C. glabrata* in the last 10-years of the period. The resistance of echinocandin in *C. auris* has also been reported.

In Candida species, the resistance against the echinocandin is mainly associated with the mutations in the *FKS* genes. In the *FKS1* gene, the mutations in the regions of hot-spot of amino acids 641–649 and 1357–1364 confirmed the resistance against echinocandin. These mutations led to reducing the IC50 of glucan synthase enzyme with the elevation of MIC values that results in the cross-resistance with various echinocandins as discussed in Fig. 5.5. The high frequency of substitution of serine 645 in the hot-region results in prominent phenotypical resistance. The resistance against echinocandin in *C. albicans* related to *FKS2* and *FKS3* genes is not so much known; however, the deletion of the genes increased the transcription level of *FKS1* led to lesser susceptibility of echinocandin (Garcia-Effron et al., 2009).

In *C. glabrata*, mutations at similar regions as in *C. albicans* in both genes *FKS1* and *FKS2* confirmed the high resistance against echinocandin. The *FKS2* is more expressed and shows higher mutations as compared to *FKS1*. The clinically isolated echinocandin-resistant has shown higher substitutions in *FKS2* at serine 663 residue that is similar to the *C. albicans* serine 645. The substitution of phenylalanine at serine 639 in the sequence of the *FKS1* gene in clinical isolates of *C. auris* led to the echinocandins resistance. Moreover, the substitution of tyrosine and proline at serine 639 residue in the clinical isolates reported resistance against echinocandin in *C. auris* (Hou et al., 2019). Few efforts have been done to prevent microbial resistance against echinocandins. However, in an old study, it is reported that mutations in the signaling pathway of calcineurin were related to the hypersensitivity to echinocandins. While the administration of inhibitors of calcineurin decreased the resistance.

5.4.4 *Flucytosine*

5-FC is one of the oldest cytosine analog fluorinated antifungal agents. In 1957, 5-FC was first synthesized as an anti-tumor drug but it was not found effective against tumors. After four years it was proved to be effective against cryptococcosis and candidosis in mice and then in 1968, it was used in humans. It was also reported as an effective drug against the fungi that cause chromoblastomycosis. Recent the interest in the use of 5-FC has been developed due to two recent advancements. One is the role in the treatment of colorectal carcinoma and the other is the increased use in combination with azole antifungal drugs such as fluconazole, itraconazole, and ketoconazole (Vermes et al., 2000).

5.4.4.1 Mode of Action

The antifungal activity of the 5-FC is not reported while its conversion into 5-FU in the susceptible fungal cells results in antimycotic activity. These cells absorbed the 5-FC by the enzyme cytosine permease, which is also the system of transport for adenine, cytosine, and hypoxanthine. The following compounds compete with the uptake of 5-FC by utilizing the energy that is coupled to the gradient of protons. After entering and converting the 5-FC to 5-FU by the process of deamination with the cytosine deaminase enzyme, the specificity of this conversion will become limited of the 5-FC to the antifungal spectrum. The fungi are not sensitive to 5-FC because of the lack of active metabolite converting enzyme cytosine deaminase. Whereas 5-FU is highly toxic to the mammalian cells and very poor uptake by the fungal cell, it is not recommended to be used as an antifungal drug. The 5-FU conversion in the cells of fungal after entering in the form of 5-FC, two mechanisms become active to work as an antifungal drug (Polak, 1983).

The first mechanism is the conversion of 5-FU into 5-fluorouridine-tri phosphate (FUTP) via 5-fluorouridine monophosphate (FUMP) and 5-fluorouridine diphosphate (FUDP). This FUTP becomes incorporated in place of uridylic acid in the RNA of fungi to alter the amino-acylation of tRNA to inhibit the synthesis of protein after disturbing the amino acids pool. While the second mechanism is the synthesis of 5-fluorodeoxyuridine monophosphate (FdUMP) from 5-FU by uridine monophosphate pyrophosphorylase. FdUMP is the inhibitor of the enzyme, thymidylate synthetase, which is important for the biosynthesis of DNA because thymidylate synthetase is the main source of thymidine. FdUMP inhibits the synthesis of fungal DNA (Fig. 5.6) (Quinto-Alemany et al., 2012).

However, it is yet not clear that the antimycotic effect of 5-FC is only due to the following two pathways or not. In a study, it was confirmed that the use of 5-FC in the *C. albicans*, inhibit the synthesis of DNA and RNA, 5-FU incorporation in the RNA, synthesis of ribosomal protein was inhibited with the increase in the level of FdUMP. While in another study, 5-FU incorporation and a decrease in the concentrations of FdUMP was not responsible for the antifungal activity of 5-FC. Due to these findings, it is controversial that the 5-FC is responsible for the inhibition of the synthesis of fungal DNA (Ghannoum & Rice, 1999).

5.4.4.2 Mechanism of Resistance

The resistance against the 5-FC has widely been discussed in many studies. **Two main mechanisms of resistance** have been identified. The first one is the **mutations that cause the deficiency of the enzymes to uptake and transport the 5-FC**, the second one is the **enhanced production of pyrimidines** that encounter with the antimetabolites of the 5-FC leads to a decline in its antifungal activity as shown in Fig. 5.7 (Vermes et al., 2000).

Another 5-FC resistance acquired by the fungal cells is the synthesis of defective uridine monophosphate pyrophosphorylase as discussed in Fig. 5.7. Two phenotypic different 5-FC resistant strains have been reported: the phenotype class 1 resistant restrains have shown no effect at a high concentration of 5-FC while phenotype class 2 have acquired the resistance after long exposure of 5-FC at low concentrations. The resistance development is due to the selection of non-susceptible mutants and this resistance is developed in the secondary population. The occurrence of 5-FC resistance has variations in different species. 7–8% resistant strains in *C. albicans*, *unspeciated candida*, and *Torulopsis glabrata*, 1–2% in *C. neoformans*, and about 22% in *Candida spp.* except *C. albicans* were found. Whereas the resistance of 5-FC in the primary population has not been found (Bhattacharya et al., 2020; Spampinato & Leonardi, 2013).

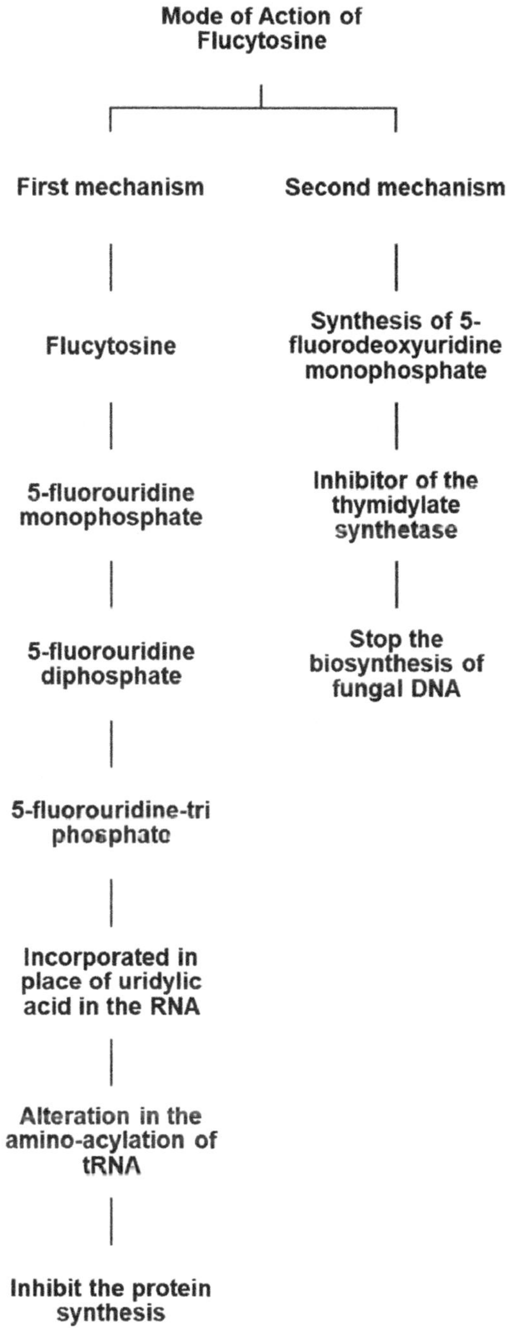

Fig. 5.6 Showing the two modes of action of flucytosine

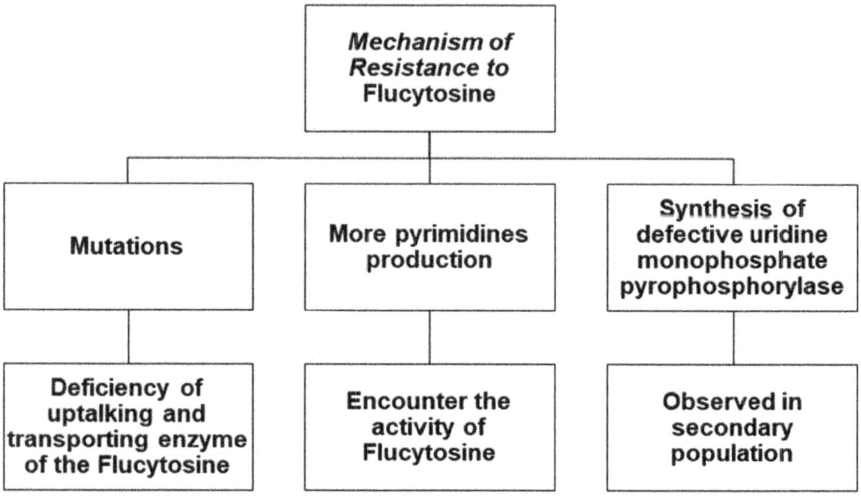

Fig. 5.7 Showing the mechanism of resistance to flucytosine

5.4.5 Polyenes

Polyenes are the antimycotic drugs that include the amphotericin B, nystatin, and pimaricin that induce the physical changes in the cell membrane after binding with the sterols of the cell membranes to form a complex of fungal cells. The molecular size and number of conjugated bonds influenced the affinity of particular polyene macrolide with ergosterol in the cell membrane of fungi. Amphotericin B has higher ergosterol and cholesterol that are present in the cell membrane of fungi and eukaryotes respectively (Kristanc et al., 2019).

5.4.5.1 Mode of Action

Lengthy structures of polyenes make the channels in the cell membrane of fungi that cause the loss of permeability of the membrane leads to the loss of important molecules. The hydrogen ion influx and potassium ion efflux from the cell cause cellular acidification and stop the enzymatic functions. Moreover, the amino acids and sugars also leaked from the cells. Fungistatic action can be obtained by using different concentrations of polyenes while fungicidal actions can be generated by varying the pH and using a high concentration of drugs (Ghannoum & Rice, 1999).

Before the discovery of azoles, polyene such as amphotericin B were used as standard therapy as an antifungal agent. There is a relationship between the presence of sterol in the cell membrane of the cells and the susceptibility of polyene. All organisms except resistant are susceptible to polyenes due to the presence of sterols in the plasma membrane. Early studies also supported the activity of polyene on the membrane sterols, while the fungi can be protected by using sterol to the growth medium and the inhibitor of polyenes. This effect was due to the physicochemical interactions between polyenes and sterols that inhibit the binding of drugs with the cellular sterols. This interaction was further confirmed by the direct spectrophotometric analysis where UV absorbance was significantly decreased with the addition of sterols in the solutions of a polyene (Tevyashova et al., 2020).

Amphotericin B is the larger polyenes that interact with the sterol on the membrane to produce aqueous pores that containing an annulus for the eight molecules of amphotericin B hydrophobically linked with the membrane sterols. This conformation produces a pore with polyene hydroxyl groups facing inward that disturb the permeability of the cell, leaking the main components from the cells that eventually cause the death of the organism. While polyenes produced oxidative damage to kill *C. albicans* (Tevyashova et al., 2020).

Amphotericin B is found to be an effective antimycotic drug but its limited therapeutic index leads to less utilization. The amphotericin B has been used in liposomes to transfer its higher doses with less cytotoxicity to the mammals. Many liposomal amphotericin B such as Amphoteck, ABELCET, and AmBisome and liposomal nystatin such as Nyotran are under clinical and preclinical evaluation. It is speculated that once the liposomal amphotericin B has been approved, it will be used in a selective transfer mechanism to transfer the drug from donor liposome to target position in the fungal cell membrane by using phospholipases (Hamill, 2013).

5.4.5.2 Mechanism of Resistance

Polyene has been clinically used as an antibiotic for the last 30 years, but resistance is still very rare in it. Candida species such as *C. glabrata*, *C. guilliermondii*, and *C. lusitaniae* have reported few resistant isolates. Similarly, in a study, it was reported that the resistance development is based on natural selection in resistant cells of small numbers in the population. A lower affinity of binding nystatin to sterols after natural modification in the resistant cells is reported. In the presence of nystatin, the growth rate depends on the ability of affinity of the drug to bind with the membrane sterols. Greater the sterol-nystatin affinity, the greater the damage to the membrane. It was deduced that the daughter strain will be more resistant as compared to the parent that was susceptible to the nystatin. It is a fact that the resistance to polyene gradually disappears after passing through the media in which nystatin is not available, which probably represents the culture population through the cells that produce sterols highly related to nystatin. The basic molecular genetics of these shifts do not work well in sterols (O'Shaughnessy et al., 2009).

While in other studies mutation was considered as the cause of resistance as compared to selection. All the information related to the mechanism of resistance in fungi species against polyenes has been obtained from the different studies related to the increasing number of antifungal drugs in the growing cells, cells exposed to the concentration gradients, and using mutagenic agents to create mutant strain by single-step mutation. A biochemical hypothesis was proposed by Hamilton-Miller in 1973 that the resistance is generated due to qualitative or quantitative changes in the content of sterols in the plasma membrane of the cell (Hamilton Miller, 1973). The binding affinity of polyene with resistant cells is very less as compared to susceptible cells. In mutant *C. albicans*, the reduced binding affinity of polyenes could be associated with the substitution of sterol that could bind some or all polyenes, without the change in the composition of sterol, a reduction in the total content of ergosterol in the cell, and readjustment of ergosterol to make thermodynamically or sterically less binding of polyenes.

Scientists have found evidence related to all these possibilities. In a study, it was reported that the inducible resistance was developed due to a reduction in the content of ergosterol by an adaptation mechanism in the *C. albicans* (Čapek & Brůna, 1976). The less content of ergosterol was due to inhibition of synthesis of ergosterol, not of the enzymatic degradation. In another study, 27 polyene-resistant strains of *C. albicans* were studied and reported a 74 to 85% decrease in the content of ergosterol. Hence the less susceptibility of polyenes was due to less content of ergosterol. The substitution of other sterols in the cell membrane leads to a change in the fluidity of the membrane and less affinity to amphotericin B. This confirmed the resistance mechanism of amphotericin B in *Leishmania donovani* (Ksiezopolska & Gabaldón, 2018).

In some studies, it was also checked that the role of cell wall components in the polyene's interaction with their active site, the plasma membrane. The susceptibility of polyenes with the stationary- and exponential-phases in the candida cells. It was reported that the stationary phase cells were found more resistant as compared to the exponential-phase. After this, it was considered that in the exponential phase cells can disintegrate and resynthesize the cell wall at a high rate that helps to access the polyene to the plasma membrane easily. While in stationary-phase cells, the disintegration and synthesis of the cell wall were at less rate. In a study of *dermatophytic* fungi, an induced enzyme system was reported that degrade the nystatin. But still, no other study reported and confirmed this verdict (O'Shaughnessy et al., 2009).

5.5 Conclusion

The development of resistance to antimicrobial agents is a coherent and unavoidable consequence while treating human infections by using these agents. The availability of molecular genetic tools has rapidly enhanced our understanding of the mechanisms of development of antifungal resistance and, to a large extent, and efforts to prosper novel and adequate compounds to use in the future. With the increasing use and easy availability of antifungal agents, it is expected that we will find a variety of resistant fungal strains to these agents. Ongoing efforts to study the development of antifungal resistance mechanisms and experimental systems (like those available in bacteria), to report and limit the resistance against these agents. It will be an important part of the development strategy to make safer and stronger compounds for the future.

References

Bardal, S. K., Waechter, J. E., & Martin, D. S. (2011). Infectious diseases. In *Applied pharmacology* (pp. 233–291). Elsevier.

Berger, S., El Chazli, Y., Babu, A. F., & Coste, A. T. (2017). Azole resistance in *Aspergillus fumigatus*: A consequence of antifungal use in agriculture? *Frontiers in Microbiology, 8*, 1024.

Berkow, E. L., & Lockhart, S. R. (2017). Fluconazole resistance in candida species: A current perspective. *Infection and Drug Resistance, 10*, 237.

Bhattacharya, S., Sae-Tia, S., & Fries, B. C. (2020). Candidiasis and mechanisms of antifungal resistance. *Antibiotics, 9*, 1–19.

Cannon, R. D., Lamping, E., Holmes, A. R., Niimi, K., Baret, P. V., Keniya, M. V., Tanabe, K., Niimi, M., Goffeau, A., & Monk, B. C. (2009). Efflux-mediated antifungal drug resistance. *Clinical Microbiology Reviews, 22*, 291–321.

Čapek, A., & Brůna, L. (1976). Antimicrobial agents—XXVI. Antifungal activity of 2-metliylthio-5-(3-iodopropargyioxy)pyrimidine toward pathogenic yeast-like microorganisms. *Folia Microbiologica, 21*, 155–156.

Cavalheiro, M., Pais, P., Galocha, M., & Teixeira, M. C. (2018). Host-pathogen interactions mediated by MDR transporters in fungi: As pleiotropic as it gets! *Genes, 9*, 332.

Cheesman, M. J., Ilanko, A., Blonk, B., & Cock, I. E. (2017). Developing new antimicrobial therapies: Are synergistic combinations of plant extracts/compounds with conventional antibiotics the solution? *Pharmacognosy Reviews, 11*, 57–72.

Clarke, C. R. (2006). Antimicrobial resistance. *Veterinary Clinics of North America: Small Animal Practice., 36*, 987–1001.

Cowen, L. E., Sanglard, D., Howard, S. J., Rogers, P. D., & Perlin, D. S. (2015). Mechanisms of antifungal drug resistance. *Cold Spring Harbor Perspectives in Medicine., 5*, a019752.

Davies, J. (1996). Origins and evolution of antibiotic resistance. *Microbiología, 12*, 9–16.

de Andrade Monteiro, C., & dos Santos, J. R. A. (2019). Phytochemicals and their antifungal potential against pathogenic yeasts. In *Phytochemicals in human health*. IntechOpen.

Douglas, C. M. (2001). Fungal β(1,3)-D-glucan synthesis. *Medical Mycology, 39*, 55–66.

Fair, R. J., & Tor, Y. (2014). Antibiotics and bacterial resistance in the 21st century. *Perspectives in Medicinal Chemistry, 6*, 25–64.

Frey-Klett, P., Burlinson, P., Deveau, A., Barret, M., Tarkka, M., & Sarniguet, A. (2011). Bacterial-fungal interactions: Hyphens between agricultural, clinical, environmental, and food microbiologists. *Microbiology and Molecular Biology Reviews, 75*, 583–609.

Garcia-Effron, G., Park, S., & Perlin, D. S. (2009). Correlating echinocandin MIC and kinetic inhibition of fks1 mutant glucan synthases for *Candida albicans*: Implications for interpretive breakpoints. *Antimicrobial Agents and Chemotherapy, 53*, 112–122.

Ghannoum, M. A., & Rice, L. B. (1999). Antifungal agents: Mode of action, mechanisms of resistance, and correlation of these mechanisms with bacterial resistance. *Clinical Microbiology Reviews, 12*, 501–517.

Gomez-Lopez, A., Buitrago, M. J., Mellado, E., & Rodriguez-Tudela, J. L. (2008). In vitro activities of 35 double combinations of antifungal agents against *Scedosporium apiospermum* and *Scedosporium prolificans*. *Antimicrobial Agents and Chemotherapy, 52*, 1136–1139.

Hamill, R. J. (2013). Amphotericin B formulations: A comparative review of efficacy and toxicity. *Drugs, 73*, 919–934.

Hamilton Miller, J. M, T. (1973) Chemistry and biology of the polyene macrolide antibiotics. *Microbiological Reviews, 37*, 166–196.

Hou, X., Healey, K. R., Shor, E., Kordalewska, M., Ortigosa, C. J., Paderu, P., Xiao, M., Wang, H., Zhao, Y., Lin, L. Y., & Zhang, Y. H. (2019). Novel *FKS1* and *FKS2* modifications in a high-level echinocandin resistant clinical isolate of *Candida glabrata*. *Emerging Microbes and Infections, 8*, 1619–1625.

Jordá, T., & Puig, S. (2020). Regulation of ergosterol biosynthesis in *Saccharomyces cerevisiae*. *Genes, 11*, 1–18.

Kapoor, G., Saigal, S., & Elongavan, A. (2017). Action and resistance mechanisms of antibiotics: A guide for clinicians. *Journal of Anaesthesiology Clinical Pharmacology, 33*, 300–305.
Knight, J. F. (1975). Antibiotic resistance. *Medical Journal of Australia, 1*, 730–731.
Konopka, J. B., Casadevall, A., Taylor, J. W., Heitman, J., & Cowen, L. (2019). One health: Fungal pathogens of humans, animals, and plants.
Krause, K. M., Serio, A. W., Kane, T. R., & Connolly, L. E. (2016). Aminoglycosides: An overview. *Cold Spring Harbor Perspectives in Medicine, 6*, a027029.
Krishnan-Natesan, S. (2009). Terbinafine: A pharmacological and clinical review. *Expert Opinion on Pharmacotherapy, 10*, 2723–2733.
Kristanc, L., Božič, B., Jokhadar, ŠZ., Dolenc, M. S., & Gomiščček, G. (2019). The pore-forming action of polyenes: From model membranes to living organisms. *Biochimica Et Biophysica Acta—Biomembranes., 1861*, 418–430.
Ksiezopolska, E., & Gabaldón, T. (2018). Evolutionary emergence of drug resistance in candida opportunistic pathogens. *Genes, 9*, 461.
Langdon, A., Crook, N., & Dantas, G. (2016). The effects of antibiotics on the microbiome throughout development and alternative approaches for therapeutic modulation. *Genome Medicine., 8*, 39.
Lee, H., & Lee, D. G. (2018). Novel approaches for efficient antifungal drug action. *Journal of Microbiology and Biotechnology, 28*, 1771–1781.
Martel, C. M., Parker, J. E., Bader, O., Weig, M., Gross, U., Warrilow, A. G. S., Rolley, N., Kelly, D. E., & Kelly, S. L. (2010). Identification and characterization of four azole-resistant erg3 mutants of *Candida albicans. Antimicrobial Agents and Chemotherapy, 54*, 4527–4533.
Mast, N., Zheng, W., Stout, C. D., & Pikuleva, I. A. (2013). Antifungal azoles: Structural insights into undesired tight binding to cholesterol-metabolizing cyp46a1s. *Molecular Pharmacology, 84*, 86–94.
Public Health Agency of Canada. (2011). Epidermophyton floccosum, microsporum SPP., Trichophyton SPP.
O'Shaughnessy, E. M., Lyman, C. A., & Walsh, T. J. (2009). Amphotericin B: Polyene resistance mechanisms. In *Antimicrobial Drug Resistance* (pp. 295–305).
Perlin, D. S. (2011). Current perspectives on echinocandin class drugs. *Future Microbiology, 6*, 441–457.
Petranyi, G., Ryder, N. S., & Stütz, A. (1984). Allylamine derivatives: New class of synthetic antifungal agents inhibiting fungal squalene epoxidase. *Science, 224*, 1239–1241.
Polak, A. (1978). Synergism of polyene antibiotics with 5-fluorocytosine. *Chemotherapy, 24*, 2–16.
Polak, A. (1983). Mode of action of 5-fluorocytosine and 5-fluorouracil in *dematiaceous* fungi. *Medical Mycology, 21*, 15–25.
Quinto-Alemany, D., Canerina-Amaro, A., Hernández-Abad, L. G., Machín, F., Romesberg, F. E., & Gil-Lamaignere, C. (2012). Yeasts acquire resistance secondary to antifungal drug treatment by adaptive mutagenesis. *PLoS ONE, 7*, 42279.
Ryder, N. S., Wagner, S., & Leitner, I. (1998). In vitro activities of terbinafine against cutaneous isolates of *Candida albicans* and other pathogenic yeasts. *Antimicrobial Agents and Chemotherapy, 42*, 1057–1061.
Shields, R. K., Nguyen, M. H., Du, C., Press, E., Cheng, S., & Clancy, C. J. (2011). Paradoxical effect of caspofungin against candida bloodstream isolates is mediated by multiple pathways but eliminated in human serum. *Antimicrobial Agents and Chemotherapy, 55*, 2641–2647.
Sinawe, H., & Casadesus, D. (2020). Ketoconazole. *StatPearls*.
Spampinato, C., & Leonardi, D. (2013). Candida infections, causes, targets, and resistance mechanisms: Traditional and alternative antifungal agents. *BioMed Research International*.
Tevyashova, A. N., Bychkova, E. N., Solovieva, S. E., Zatonsky, G. V., Grammatikova, N. E., Isakova, E. B., Mirchink, E. P., Treshchalin, I. D., Pereverzeva, E. R., Bykov, E. E., & Efimova, S. S. (2020). Discovery of amphamide, a drug candidate for the second generation of polyene antibiotics. *ACS Infectious Diseases, 6*, 2029–2044.

Verma, A., Joshi, S., & Singh, D. (2013). Imidazole: Having versatile biological activities. *Journal of Chemistry*.

Vermes, A., Guchelaar, H. J., & Dankert, J. (2000). Flucytosine: A review of its pharmacology, clinical indications, pharmacokinetics, toxicity and drug interactions. *Journal of Antimicrobial Chemotherapy, 46*, 171–179.

Wiederhold, N. P., & Patterson, T. F. (2015). Emergence of azole resistance in *Aspergillus*. *Seminars in Respiratory and Critical Care Medicine, 36*, 673–680.

World Health Organization. (2015). *Tuberculosis mortality nearly halved since 1990.*

World Health Organization. (2020). *Antimicrobial resistance.*

Chapter 6
Drug Resistance in Ear, Nose, and Throat Infections

Muhammad Zeeshan Ahmed, Zeeshan Mutahir, Tazeen Rao, Arshad Islam, Nayyab Hameed, Saifullah Shakeel, Haseeba Shahzad, Mazhar Ali, Shahzeb Hameed, Syed Hizbullah, and Saeed Ur Rahman

Abstract Microbial infections of the ear, nose, and throat are the most common problems encountered by human beings in routine life, and their treatment with antimicrobial drugs is frequently prescribed in clinics. The frequent use of prescribed antimicrobial drugs as well those used by the patients without the prescription of a physician, mostly in the under-developing regions, has triggered drug resistance in microorganisms. The resistance of pathogenic microorganisms towards the drugs available in clinics has now become a major hindrance in the treatment of infectious diseases. The management of drug resistance is the main concern and challenging task related to the use of drugs against different microorganisms. This chapter comprehensively discusses the microbiomes of the ear, nose, and throat, pathogenic microorganisms, and related major infections as well as the mode of the action of antimicrobial drugs that are used to treat those infections. Moreover, the mechanism of resistance developed by the microorganisms has been described.

Keywords Ear · Nose · Throat · Infections · Antimicrobial drug · Mode of action · Resistance mechanism

---

M. Z. Ahmed (✉) · Z. Mutahir · N. Hameed · H. Shahzad · S. Hameed · S. Hizbullah
Institute of Biochemistry and Biotechnology, University of the Punjab, Quaid-i-Azam campus, Lahore 54590, Punjab, Pakistan

Z. Mutahir
e-mail: zeeshan.ibb@pu.edu.pk

T. Rao · S. Shakeel · M. Ali
Department of Biochemistry, Bahauddin Zakariya University, Multan 60800, Punjab, Pakistan

A. Islam
Department of Plastic Surgery, Lady Reading Hospital Peshawar, Peshawar, Pakistan

S. U. Rahman
Interdisciplinary Research Center in Biomedical Materials (IRCBM), COMSATS University Islamabad, Lahore Campus, Lahore 54000, Punjab, Pakistan

Department of Oral Biology, Institute of Basic Medical Sciences (IBMS), Khyber Medical University, Peshawar, Pakistan

S. Ahmed et al. (eds.), *Biochemistry of Drug Resistance*,
https://doi.org/10.1007/978-3-030-76320-6_6

6.1 Introduction

Since the last few years, the world is facing a very serious issue of the development of resistance by microbes against antimicrobials. The under-developing world has been enormously contributing to the emergence of antimicrobial resistance due to the frequent, prescribed, over-dosage as well as over-use of antimicrobials against infectious diseases (Hart & Kariuki, 1998). Consequently, the existing antimicrobial agents are becoming less or non-effective. The development of novel drug discovery strategies to develop novel therapeutics has, therefore, become the focus of scientists throughout the world to not only combat pathogenic microbes but also predict and avoid the changes in the development of resistance (Lee et al., 2020). No doubt, antimicrobials are the wonders in medical history that significantly contributed to the survival of human civilization to combat infectious diseases. Antimicrobials are derived from natural, synthetic or semisynthetic sources that help to kill microbes or stop their growth. The term antimicrobial is generally used for pharmaceutical agents that kill or inhibit the growth of bacteria (antibiotics), fungi (antifungal), protozoa (antiprotozoal), or viruses (antiviral) (Das & Patra, 2017). Antimicrobials act as cytotoxic or cytostatic to microbial, mainly inhibit their deoxyribonucleic acid (DNA), ribonucleic acid (RNA), or protein synthesis, disorganizing the cell, inhibiting the cell membrane and cell wall synthesis, or other definitive actions (Zaman et al., 2017).

In the last six decades, a huge number of antimicrobials have been developed and used against different microorganisms to treat the related disease. The excessive use of antimicrobials has resulted in the emergence of a significantly high number of resistant strains, which, unfortunately, has led to the inefficacy of numerous clinically available antimicrobials. In the past, the production or development of new antimicrobial was directly proportional to the emergence of the resistant strain. But now scientists focus mainly on the alteration or modification of existing antimicrobials to fight with the currently prevailing resistance (Davies, 1996). Also, with the increase in cases of antimicrobial resistance by different strains, the development of new antimicrobial is becoming hard-hitting. One of the major reasons behind the emergence of resistance in the developing world is the ready availability of all antimicrobials and their use without any medical prescription (Zaman et al., 2017). According to the World Health Organization (WHO), resistance in micro-organisms helps them to fight against antimicrobial drugs, which made treatment ineffective and helps the increased spread of infection. WHO also reported the highest level of incidence of resistance in bacteria than other microbes (viruses and fungi). *Escherichia coli* is reportedly resistant to cephalosporin and fluoroquinolones, *Staphylococcus aureus* to methicillin, *Klebsiella pneumoniae* to cephalosporin, and carbapenems, *Streptococcus pneumoniae* to penicillin, and so on (Tanwar et al., 2014).

Antimicrobial drug resistance has been classified into two main categories i.e., primary resistance and secondary resistance. In primary resistance, microorganisms have never confronted the drug in the host. If a person or host has never been treated with any type of drug for a specific disease and then treated with a drug for the first time for example anti-tuberculosis (anti-TB) treatment evolves multi-drug-resistant tuberculosis (MDR-TB) for the first time against anti-TB drugs is considered as primary drug resistance. Whereas in secondary resistance, resistance is developed after the drug exposure. If a host develops resistance after the first line MDR-TB and anti-TB drugs (primary resistance) history to another drug is known as secondary drug resistance. Antimicrobial drug resistance is also categorized into three other classes: intrinsic resistance, extensive resistance, and clinical resistance. Clinical evidence-based on a single strain of micro-organism insensitivity used to treat illness to first-line drugs is called intrinsic resistance. In extensive resistance, micro-organism withstand ability helps against one or two most efficient drugs. Clinical resistance in the pathogens is also developed with the use of a high dosage of antibiotics (Tanwar et al., 2014).

Through years of research and clinical practice in the last decade, it has been found that microorganisms causing ear, nose, and throat (ENT) infections are becoming resistant to more and more antimicrobials as evidenced in clinical practices. But it is the severe and chronic nature of some of the infections that make the situation extremely difficult to manage and diagnose the disease. Sometimes, excessive administration of drugs results in the unsuccessful treatment of disease. This situation of overuse of antimicrobials drugs in our clinics has increased the drug resistance ability and enhance the challenge to handle the infection (Vlastarakos et al., 2007).

In this chapter, the development and mechanism of resistance by the microbes that cause infection in the ear, nose, and throat, and the consequent complications are briefly discussed.

6.2 Ear Infection

Ear infections are caused by viruses, bacteria, and sometimes fungus that can be painful due to fluid and inflammation. Ear infections can be acute or chronic. Acute infections are short in time but painful whereas chronic infections remain for long durations. Sometimes recurrence causes permanent hearing loss.

The human ear comprises three parts: the outer, middle, and inner ear. The outer ear is exposed to the environment and most of the aerobic bacterial flora, viruses, and fungi are present here. Gram-positive bacteria are frequently found in the external auditory canal than gram-negative bacteria. *Bacillus, coryneform, enterococci, micrococci, staphylococci*, and *streptococci* are the main Gram-positive bacteria found in the ear. Pre-dominant species of the Gram-positive bacteria are *Alloiococcus otitis, E. coli, Micrococcus luteus, S. capitis, S. epidermidis, S. warnen, Staphylococcus auricularis*, and *Turicellaotitidis*. *Moraxellaosloensis* and *Pseudomonas aeruginosa* are the Gram-negative bacteria found in the outer ear toa lesser extent. *C. Albicans, Candida parapsilosis*, and *Penicillium* species are the fungal species found toa lesser extent in the outer ear of humans. It is not an easy task for bacteria to travel the middle ear, but still the alpha-hemolytic *streptococci* bacteria are present in the healthy middle ear. Contrary to the outer and middle ear, the inner ear is safer from microbial entry due to the bony structure and strong immune system.

6.2.1 Otitis Media (Middle Ear Infection)

Otitis media is the common middle ear infection with three forms including acute otitis media (AOM), otitis media with effusion (OME), and chronic suppurative otitis media (CSOM). It can occur at any age but is most common in children less than 2 years of age. Bacteria, viruses, or sometimes both can cause otitis media infection but *Streptococcus pneumoniae* bacteria with non-typeable *Haemophilusinfluenzae* (NTHi), and *Moraxellacatarrhalis* are the main cause of this infection. Adenoviruses, coronaviruses, human metapneumovirus, influenza viruses, picornaviruses, and respiratory syncytial viruses (RSV) may also include in the cause of otitis media (Danishyar & Ashurst, 2020).

6.2.1.1 Treatment

Amoxicillin is used as a first-line drug uncomplicated AOM but there is a cross-reaction risk of this drug with other beta-lactam drugs like cephalosporins. Cefaclor, cefixime, cefuroxime axetil, erythromycin-sulfisoxazole, and trimethoprim-sulfamethoxazole are second-line effective drugs used against resistant bacterial strains (Antibiotics et al., 1998).

6.2.1.2 Amoxicillin

Amoxicillin is amino penicillin created against resistant strains of bacteria and included in the most used antibiotics. Amoxicillin is mostly used against gram-positive bacteria with some gram-negative as well. Federal Drug Administration (FDA) has approved Amoxicillin for the treatment of ear, nose, throat, lower respiratory, *Helicobacterpylori*, tonsillitis, genitourinary tract, and skin infections. It is used against *Actinomyces*, *Corynebacteria*, *Clostridial* species, *Enterococcus*, some *Escherichia coli*, *Haemophilusinfluenzae*, *Salmonella*, and most *Streptococcus* species (Kwon, 2016).

(a) Mechanism of action

Amoxicillin is a β-lactam antibiotic that exerts its antimicrobial effect by preventing the formation of the peptidoglycan layer in the bacterial cell wall (Table 6.1). β-lactam antibodies show their mechanism by irreversible binding to the receptor, penicillin-binding proteins (PBPs), on the cell wall. This binding inhibits the activity of transpeptidase, hence disturb the formation of peptidoglycan (Fig. 6.1). This disturbance leads to the death of a targeted microbe (Jacobs, 2017).

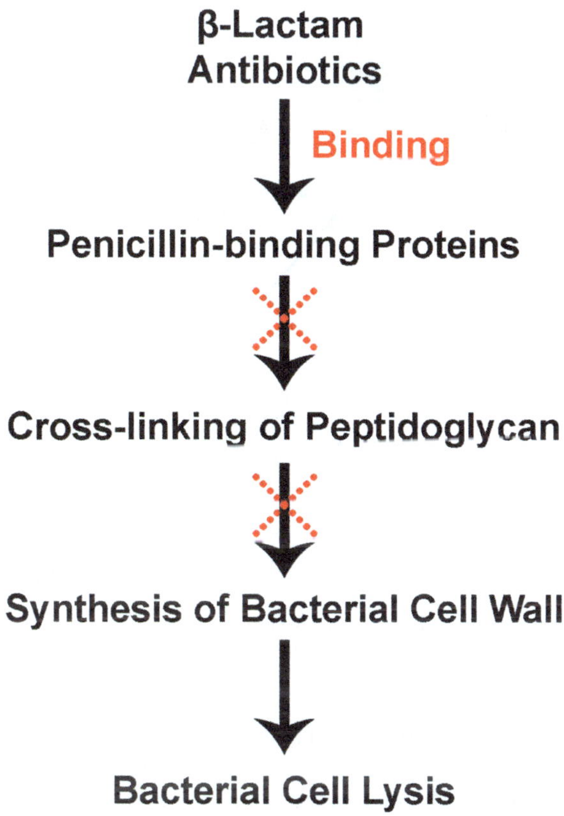

Fig. 6.1 Action mechanism of β-lactam antibiotics on bacterial cell wall

(b) Drug resistance and its mechanism

Extensive prescription of antibiotics can induce serious problems such as drug-resistance and a higher population of pathogens in the respiratory tract. Acute rhinosinusitis is mainly caused by *H. Infuenzae* and *Moraxella*. About one-third of these bacterial strains show resistance against β-lactam antibiotics. They show resistance through two mechanisms i.e., (i) by producing β lactamase, and (ii) by non-β-lactamase mediated resistance (Table 6.1).

Bacteria have a plasmid gene that encodes for β-lactamase. This protein has a structural resemblance to the penicillin-binding protein (PBP) and has a high affinity for the β lactam. This interaction between bacterial enzymes and β-lactam can cause the inactivation of antibiotics. β-lactamase alters the structure of the antibody by the permanent opening or deformation of the molecule and makes it unable to bind its receptor (Fig. 6.2) (Massova & Mobashery, 1998). The binding of β-lactamase to the antibiotic is a reversible process so that the enzyme can be recovered to target the next molecule of antibiotics (Jacobs, 2017). Antibiotic resistance can also be induced by a non-β-lactamase derived resistance mechanism. This type of mechanism is carried out by changing the structure of PBPs. Mutation in the gene ftsI that encode for PBP

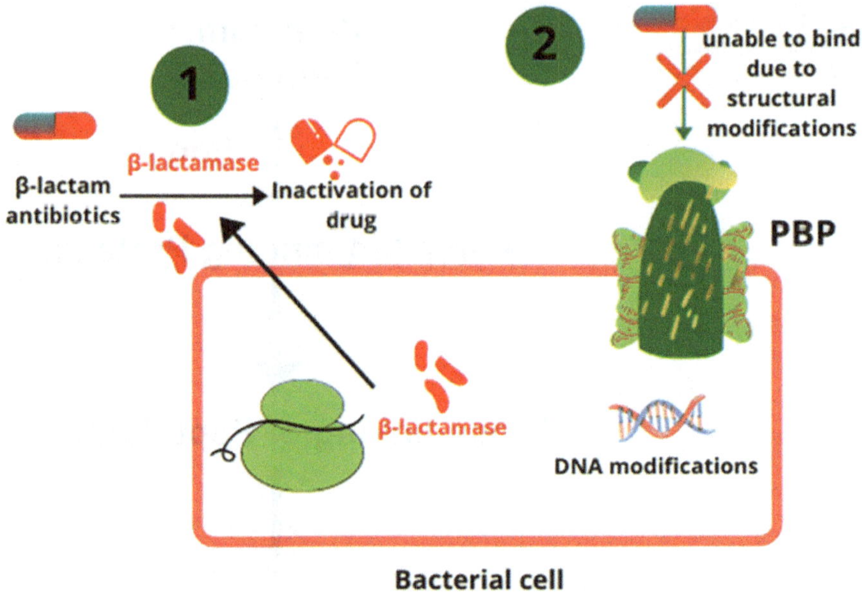

Fig. 6.2 Mechanism of bacterial resistance against β-lactam antibiotics by producing β-lactamase and changing the structure of PBP

can inhibit the binding of antibiotics to the receptor (Tristram et al., 2007). The active site of transpeptidase is formed by the three conserved motifs named SXXK, SSN, and KTG. Mutation in these motifs can cause conformational changes in the active site of receptor protein and make it unavailable for the attachment of antibiotics. Antibiotic resistance can be overcome by using the inhibitors of β-lactamase. These inhibitors bind irreversibly to inactive the enzymes that cause resistance.

6.2.1.3 Ceftriaxone

Ceftriaxone is a third-generation cephalosporin antibiotic used for bone infection, endocarditis, intra-abdominal infection, meningitis, and middle ear bacterial infections. It is not recommended for those who have had anaphylaxis to penicillin.

(a) **Mode of action**

Ceftriaxone is a β-lactam antibiotic that works to inhibit the synthesis of the cell wall (Fig. 6.1). β-lactam antibiotics bind themselves to the PBPs that are transpeptidases, also known as transamidases, and catalyze the peptidoglycan cross-linkages help in synthesizing the cell wall of bacteria. The ceftriaxone structure resembles the moiety of D-alanyl-D-alanine and if the normal substrate D-alanyl-D-alanine is present then PBP tackles the ring of beta-lactam in ceftriaxone. The PBPs cross-linking peptidoglycan activity helps to maintain and repair the cell wall of bacteria so that the PBPs inhibition destruct and damage the cell wall leading to cell lysis (Rawls, 2014). Their resistance mechanism of β-lactam antibiotic (Table 6.1) is already discussed above and shown in Fig. 6.2.

6.2.1.4 Trimethoprim-Sulfamethoxazole

Co-trimoxazole is another name of trimethoprim-sulfamethoxazole, which is also an antibiotic used against many bacterial infections (Smilack, 1999).

(a) **Mechanism of action of Trimethoprim-sulfamethoxazole**

Trimethoprim-sulfamethoxazole shows its antimicrobial effect by inhibiting the Dihydrofolate Reductase (DHFR) (Table 6.1). These antibiotics interfere with cellular metabolism and prevent the conversion of dihydrofolate to tetrahydrofolate. Tetrahydrofolate is very important for the formation of nitrogenous bases. It serves as a co-factor to form purines, serine, methionine, thymidylate, and other compounds. These antibiotics cause microbial death by preventing DNA replication. Trimethoprim is a competitive inhibitor of dihydrofolate, and sulfamethoxazole is a substrate analog to the para-aminobenzoic acid which is a precursor to the formation of dihydrofolate. So, these compounds in a combined relation inhibit the catalytic activity of DHFR and Dihydropteroate Synthetase (DHPS). So, the formation of dihydrofolate and its conversion into tetrahydrofolate is inhibited using antibiotics (Fig. 6.3) (Jacobs, 2017).

(b) **Drug resistance mechanism of Trimethoprim-sulfamethoxazole:**

The resistance mechanism against Trimethoprim-sulfamethoxazole (TMP-SMZ) is associated with the mutation in a bacterial gene that codes for the receptor enzymes (Table 6.1). In the case of *S. aureus*, the *dhfr* gene is responsible to encode the DHFR. Single amino acid substitution can cause conformational changes in the structure of DHFR and make it unavailable for the binding of antibiotics. Modification in the gene sequence can change the affinity of the enzyme for the antibiotic. Hence, TMP failed to bind with DHFR after mutation, but the enzyme hasthe same affinity for the original substrate. Resistance against SMZ can be achieved by altering the gene that codes for DHPS. Insertion of 15 bp in the gene folP can change the structure of DHPS. This mutation in the bacterial gene can inhibit the binding of sulfamethoxazole to the enzyme DHPS. Microbes show higher resistance when the trimethoprim and sulfamethoxazole are used separately, but the combined applications of TMP and SMZ can reduce the effect of resistance (Fig. 6.4) (Kordus & Baughn, 2019).

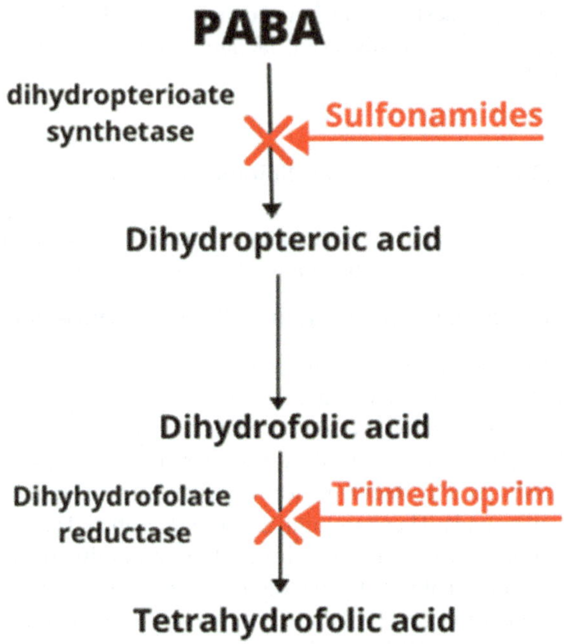

Fig. 6.3 Antibiotics inhibition of tetrahydrofolate (THF) and Deoxynucleoside triphosphates (dNTP) synthesis

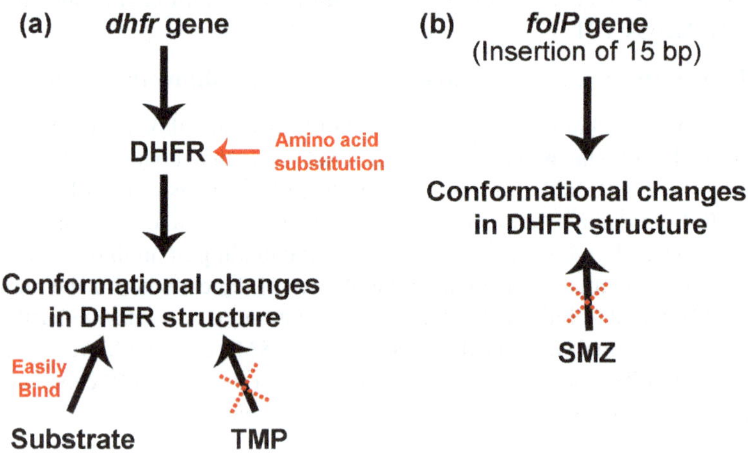

Fig. 6.4 Resistance mechanism against TMP-SMZ through mutation in the gene that codes for DHFR and DHPS

6.2.2 *Swimmer's Ear (Outer Ear Infection)*

Swimmer's ear is an outer ear infection and known as otitis externa. The main reason for this infection is the moisture and water in the ear that help to grow bacteria and damage the skin line of the ear canal. Bacteria are the main cause of this infection but sometimes fungus and viral infections are also involved. *Pseudomonas*, *Staphylococcus*, and *Streptococcus* are common bacteria that cause this infection (Lewin & Smith, 1988).

6.2.2.1 Ofloxacin

Ofloxacin is the common drug used to inhibit bacterial growth in the swimmer's ear infection. It is a quinolone antibiotic and used only against bacterial infections (Lewin & Smith, 1988).

(a) **Mode of action of Ofloxacin**

Ofloxacin is a broad-spectrum, quinolone/fluoroquinolone and bactericidal antimicrobial agent. It stops the bacterial cell division by inhibiting the bacterial DNA replicating, repairing transcriptional and recombinant enzymes (Table 6.1). Ofloxacin binds and inhibits the DNA gyrase enzyme and stops untwisting the DNA double helix in the DNA replication of bacteria. It also inhibits the bacterial type II topoisomerases and topoisomerase IV. In this way, bacterial cell division is stopped (Lewin & Smith, 1988). Pham et al. (2019) reported an intracellular mechanism of action of quinolones that bind at the active site of cleavage ligation with enzyme-DNA cleavage complex and make cleavage complexes with steady-state concentration. This disturbs the process of replication of DNA and colloid the cleavage complexes that result in a chromosomal breakdown. As a result of this damage, DNA repairing mechanisms become activated due to the SOS systems activation. Following are the SOS systems that become activated: activation of toxin-antitoxin modules leads to apoptosis and due to expression of LexA repressor cell filaments become extended (Fig. 6.5).

(b) **Drug resistance and its mechanism**

Ofloxacin is a member of the 4-quinolone group and has the feature of plasmid-borne resistance in bacteria. In a study, it is confirmed that mutation in chromosome developed the resistance in bacteria against ofloxacin. KL16 strain of *Escherichia coli* was used as a model system to test the resistance development against ofloxacin. In various in vitro studies regarding resistance development to new 4-quinolones, it has been noticed that the frequency of mutation was very low ($1 \times 10(-10)$ to $1 \times 10(-12)$) for bacteria that were grown under regular conditions of the laboratory. The organisms obtained were temperature-sensitive, auxotrophic, and slow-growing. The unstable mutations were observed, and mutants were reverted readily to the sensitivity of ofloxacin when the selection was carried out in the absence of ofloxacin. In studies, under growth conditions, the spontaneous mutations were found closely related to in vivo situations including lumen of gut along with restriction of oxygen supply which showed the frequency of mutations in a range of $1 \times 10(-8)$. The same phenotype, as well as equal instability of the mutants, were observed under these

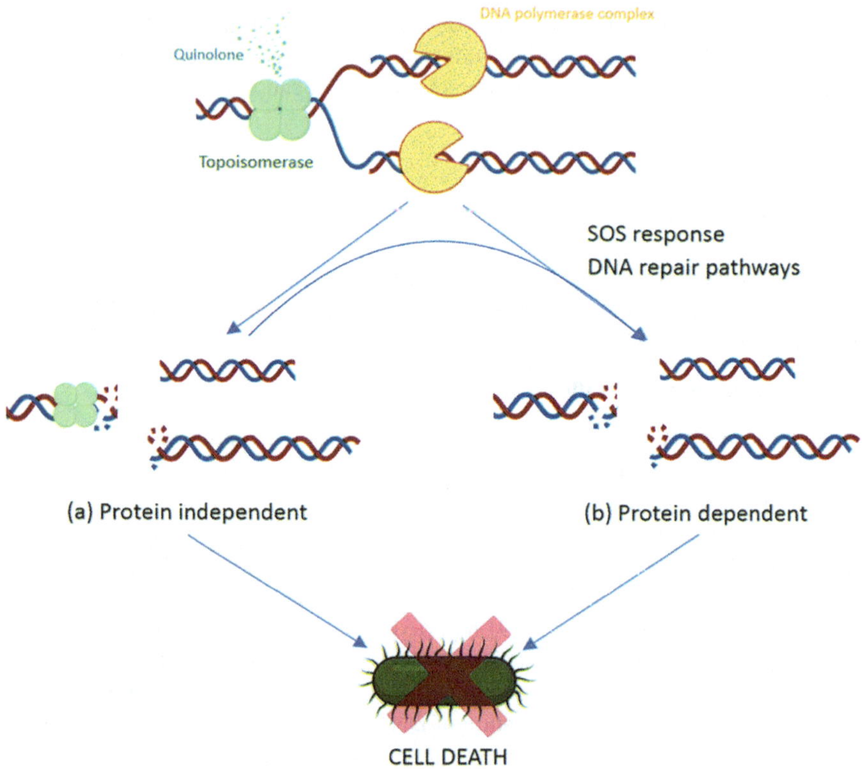

Fig. 6.5 Intracellular mechanism action of quinolones (Pham et al., 2019)

conditions as were found previously. The physiology of mutants that were ofloxacin-resistant displayed metabolic defects related to being able to adjust to environmental changes (Cattoir et al., 2006).

Aldred et al. (2014) reported three resistance mechanisms to target mediated, plasmid-mediated and chromosome-mediated resistance found against the quinolone (Fig. 6.6). In target-mediated resistance, the mutation in topoisomerase IV and gyrase weakens the enzyme-quinolone interaction. In plasmid-mediated resistance, the binding of the Onr protein is decreased that protects the DNA-enzyme complex from the quinolones drug. In chromosome-mediated resistance, porins under-expression decrease the uptake of drug in gram-negative bacteria and decrease the retention of drug in the cell by the chromosome-encoded efflux pumps overexpression (Table 6.1).

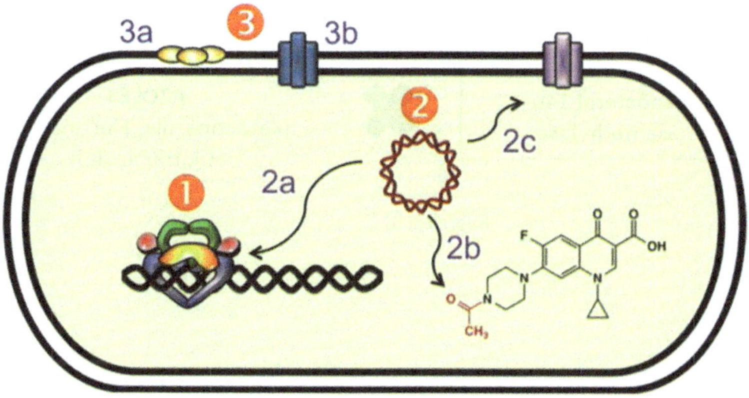

Fig. 6.6 Mechanisms of quinolone resistance (Aldred et al., 2014)

6.2.3 *Otomycosis (Outer Ear Infection)*

Otomycosis is another outer ear superficial mycotic infection that affects one or both ears. It is most common in people who live in high temperatures or tropical areas. The main cause of this infection is a fungus. About 60 different fungal species are found in this infection but the two most common species are *Aspergillus* and *Candida*. Sometimes, the infection becomes more complicated when bacterial infection combines with the fungal infection. *Candida* infection is more complicated in clinical detection because of the similar appearance to the *Aspergillus* (Anwar & Gohar, 2014).

6.2.3.1 Clotrimazole

For this infection, topical clotrimazole is the preferred imidazole antifungal drug because of its efficacy against both fungal species (Anwar & Gohar, 2014).

(a) **Mode of action**

Clotrimazole inhibits the lanosterol 14 demethylase enzyme involved in the ergosterol synthesis (Fig. 6.7 and Table 6.1). With the inhibition of ergosterol synthesis, the fungal cell membrane is not able to stabilize itself, fungal growth is reduced, and fungicidal action takes place. This drug not only inhibits the synthesis of ergosterol but also shows pharmacological effects. Calcium-dependent potassium channel blockage, intracellular calcium depletion, sarcoplasmic reticulum Ca^{2+}-ATPase inhibition, and voltage-dependent calcium channels blockage is also done with Clotrimazole for the fungicidal action. A high dose of this drug exhibits a mild inhibitory effect while its high dose completely blocks the synthesis of ergosterol (Haller, 1985).

(b) **Drug resistance and its mechanism**

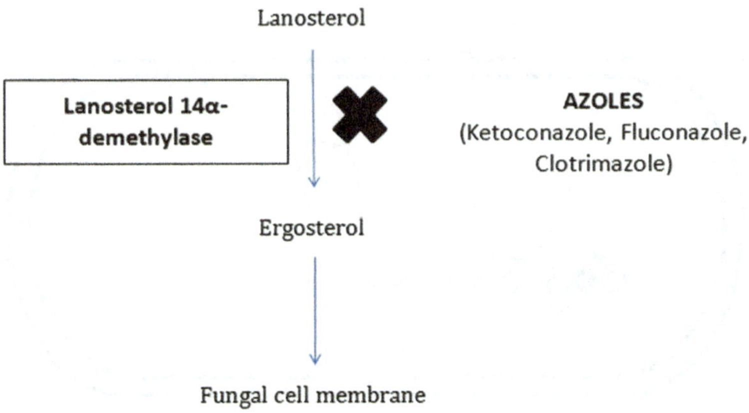

Fig. 6.7 Flowchart depicting the mechanism of action of azole (Chowdhry, 2017)

In *Candida*spp, three main mechanisms of resistance are reported (Table 6.1). First is the increase in the production of Erg11 enzyme targeted by azole, second is the point mutation in the Erg11 enzyme and cause hindrance in binding and third is the multidrug efflux pumps Cdr1 and Cdr2 mediate the drug efflux belong to the superfamily of ATP-binding cassette transporter. Expression of ATP-Binding Cassette (ABC) multidrug transporters is associated with the *Candida* species drug resistance. Recently it was found that drug efflux pumps of the Drug: H^+Antiporter (DHA) family are also associated with drug resistance. Till now only Mdr1 transporter of the *Candidaalbicans* is found relevant to the antifungal drug resistance. Costa et al. (2016) reported that C. Glabrata DHA transporters CgAqr1, CgQdr2, CgTpo1_1, and CgTpo3 are involved in the medical attainment of drug resistance of clotrimazole.

Three azole drug resistance mechanisms were reported by Berger et al. (2017). The first one is the mutation in the structure of the CYP51A enzyme that reduces the binding affinity of the azole. Second is the overexpression of the gene cyp51A gene because of the insertion of 34-bp in the promotor region that helps to combine with the leucine to histidine substitution at codon 98 which results in the increase of CYP51A at the cellular level. The third is the efflux transporter genes overexpression results in a decrease in azole drug accumulation intracellular (Fig. 6.8).

6.3 Nose Infection

Microbiomes are the complex communities of microorganism which use the human body as their microhabitat where they adopt the symbiotic relationship with the host for a living (Kumpitsch et al., 2019). According to the specificity of their different composition and function, these microbes' niche different body sites such as airways and gastrointestinal tract. These microbial colonies interact with internal

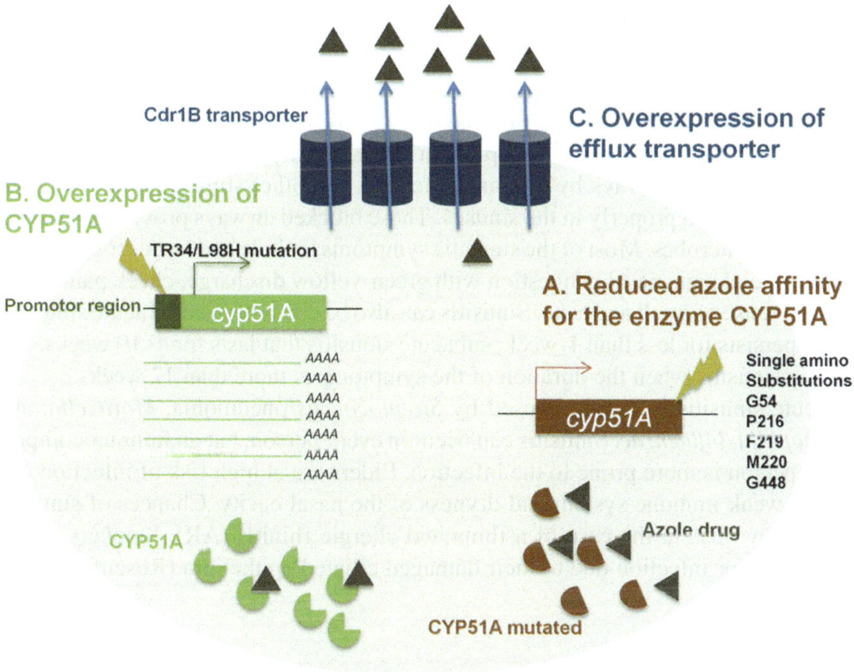

Fig. 6.8 The three mechanisms of resistance of azole drugs (Berger et al., 2017)

and external factors that are involved in aging, immune response, and pathogenesis of the diseases. Most of the microbial components of the nasal cavity composed of different bacteria, including the species from phyla *Firmicutes*, *Actinobacteria*, *Bacteriodetes*, and *Proteobacteria* (Shilts et al., 2016). Other components of the microbiome include viruses, fungi, and archaea. All these components provide the first line of defense by filtering, humidifying, and warming the inhaled air (Yan et al., 2013).

At different life stages, different microbiota niches the nasal cavity. *Moraxella*, *Staphylococcus*, *Streptococcus*, *Dolosigranulum*, *Haemophilus*, and *Corynebacterium* present in surplus in nares and nasopharynx at the time of birth and remains until the age of 24 months (Frayman et al., 2017). In adults, *Actinobacteria* and *Firmicutes* are present in anterior nares. The bacterial colonies change mostly in middle age and the anterior nares are dominated by *Cutibacterium*, *Propionibacterium*, *Corynebacterium*, and *Staphylococcus* in elders (>60 years old) (Schenck et al., 2016). These microbiomes are very important for diagnostic purposes and can serve as targets for therapies. Human health is associated with the complex interaction between microbiota and host (Bassis et al., 2014). *Dysbacteriosis* is an imbalance of the commensal bacteria which protect their host from pathogenic bacteria. The loss of beneficial bacteria can result in the occurrence of multiple types of diseases (Coburn et al., 2015).

6.3.1 Acute Rhinosinusitis

Sinusitis is an infection of the upper respiratory tract (URT) which is characterized by the inflammation and swelling of maxillary sinuses. This is also known as rhinosinusitis and affects millions of people worldwide each year. Sinusitis is due to the blockage of nasal airways by the viral infection and other stimuli in which mucus cannot be drained properly in the sinuses. These blocked airways provide the habitat to grow the microbes. Most of the sinusitis symptoms include headache, cough, fever, fatigue, facial pain, nasal congestion with green-yellow discharge, cheek pain, and a reduced sense of smell and taste. Sinusitis can also be characterized by acute sinusitis, which persists for less than 1-week, subacute sinusitis that lasts for 4–10 weeks, and chronic sinusitis when the duration of the symptoms is more than 12 weeks.

Acute sinusitis is mainly caused by *Streptococcus* pneumonia, *Moraxella*, and *HaemophilusInfluenzae*. Sinusitis can occur in every person, but an immunocompromised person is more prone to the infection. Elders are at high risk of infection due to their weak immune system and dryness of the nasal cavity. Chances of sinusitis increase by 30% in the case of asthma and allergic rhinitis (AR). Smokers are an easy target for infection due to their damaged ciliated epithelium (Rosenfeld et al., 2015).

6.3.1.1 Treatment

Non-drug therapies are helpful for the treatment of acute sinusitis which includes hydration, steaming, and sinus irrigation. These therapies help to restore the moisture of mucosal epithelium and aid to drain the mucus from sinuses. Saline nasal irrigation helps to expel the bacterial debris from the sinuses. These therapies can help to resolve the sinusitis in the patients with the symptoms for at least 7–10 days. However, their therapies have no well-defined efficacy rates. Certain side effects of drugs can be avoided by using these conventional methods. The main side effect is the disturbance of gastrointestinal flora. The patients who have less severe symptoms of maxillary acute sinusitis and do not get improved after 10 days can be treated with antibiotics. The use of antibiotics depends upon the condition of patients and the duration of symptoms. Antibiotic resistance must be taken into account while prescribing antibiotics (Jacobs, 2017).

6.3.1.2 Antibiotics

Amoxicillin serves as the first line of a pharmaceutical agent to treat acute rhinosinusitis. The patients with allergic effects against penicillin and amoxicillin are recommended to use other antibiotics. Doxycycline has a high cure rate than amoxicillin. Trimethoprim-sulfamethoxazole is a good antibiotic choice for patients who have resistance to penicillin (Jacobs, 2017). The mode of action and resistance mechanism (Table 6.1) is discussed above (Figs. 6.3 and 6.4).

6.3.2 *Chronic Rhinosinusitis*

There is a complex relation between URT and microbiota that these commensal microbes protect the respiratory tract from the pathogenic microbes. Dysbiosis of microbiota can lead to serious diseases like acute rhinosinusitis, chronic rhinosinusitis (CRS), and AR in which commensal bacteria play a pathogenic role and cause the infection (Miyake & Bleier, 2019). Different stimuli, such as toxins, allergens, and systemic immune system components bring the compositional and structural changes in commensal bacteria that can cause CRS (De Boeck et al., 2019; Sivasubramaniam & Douglas, 2018). CRS is mainly characterized by the inflammation of the nasal cavity or the physical blockage in the form of nasal polyps, tumors, and abscesses. There are two phenotypes of CRS including (i) chronic rhinosinusitis with nasal polyps (CRSwNP), and (ii) chronic rhinosinusitis without nasal polyps (CRSsNP). CRSwNP is class-I-inflammation, mainly induced the production of the cytokines IL-4, IL-5, and IL-13. These cytokines then induce inflammation in the presence of other stimulants like antigens. These cytokines are not present under normal conditions, but in the presence of antigens activated immune system secrete these cytokines to induce inflammation. CRSsNP is class-II-inflammation which is caused by the production of TNF-α. TNF-α is not directly involved in inflammation but this is a leading agent that triggers the cascade mechanism to induce the inflammation. Even TNF-α stimulates the secretion of interleukins that lead to autoimmune diseases (Bachert & Akdis, 2016).

Staphylococcus, *Corynebacterium*, *Streptococcus*, *Propionibacterium*, *P. acnes*, and *S. aureus* are frequently present in URT of CRS patients (Wos-Oxley et al., 2016). Some exceptions are there for *S. aureus* which is also present in a healthy person. Studies have shown that a low concentration of *S. aureus* can induce IL-10, which is an important anti-inflammatory cytokine. While the higher concentration of *S. aureus* can reduce the IL-10 and promote inflammation (Schwartz et al., 2016). 80% of CRS patients have been diagnosed with bacterial biofilms in their sinuses. These biofilms are the specific communities of microorganisms surrounded by an extracellular matrix. Biofilms are not directly involved in inflammation, but they promote the pro-inflammatory response.

6.3.2.1 Treatment

Treatment for CRS is provided by saline irrigation, corticosteroids, and antibiotics. FDA did not approve any antibiotic for the treatment of CRS, but the experts believe that antibiotics can reduce the load of pathogenic microbes involved in sinuses and decrease their diversity. The use of antibiotics may exert both positive and negative effects on a patient's health. Long courses of antibiotics may alter the gastrointestinal microbes unfavorably for up to two years. CRS patients require a long duration of treatment and poorly respond to conventional antibiotics. amoxicillin, clindamycin, levofloxacin, ciprofloxacin, and trimethoprim-sulfamethoxazole are commonly used antibiotics against CRS (Maxfield et al., 2017). The mode of action and resistance mechanism of amoxicillin (Figs. 6.1 and 6.2) and trimethoprim-sulfamethoxazole (Figs. 6.3 and 6.4) are already discussed above (Table 6.1).

6.3.3 Allergic Rhinitis

AR is a pandemic disease that affects about 10–25% of the population. AR is a common airway disease in which exposure to different allergens causes the hypersecretion of Ig-E. This disease is characterized by hyperactivation, hypersecretion, and hypersensitive nasal mucosa. AR is mainly caused by aeroallergens, oxidative pollutants, and domestic allergens. These allergens may include house dust mites, animal dander, pollens, molds, insects, nitric oxide, ozone, and sulfur oxides. There are different symptom levels of AR that depend upon the duration of allergen exposure and patient sensitivity. The main symptoms include runny nose, sneezing, itchy nose, red watery eyes, loss of sense of smell, sleeping disturbance, and affect the quality of life as patients cannot perform their routine work properly. AR was previously categorized as seasonal, perennial, and occupational that was based upon the exposure of allergens. The recent classification of AR-based upon Allergic Rhinitis and its Impact on Asthma (ARIA). Intermittent or persistence classification is based on duration and mild or moderate-severe based on symptoms (Varshney & Varshney, 2015).

6.3.3.1 Treatment

Treatment of AR depends upon the symptom's severity. H1 antihistamine and intranasal corticosteroids (INCS) are two main drugs that may be used to treat AR. When patients experience mild symptoms, doctors may prescribe oral antihistamine and prescribe the nasal antihistamine under severe conditions. INCS serve as first-line therapy for the treatment of moderate-to-severe symptoms. Leukotriene receptor inhibitors are used when the patients are suffering from both AR and Asthma. All these types of treatments can provide short or long-term relief, but none of these can completely cure the disease (Varshney & Varshney, 2015).

(a) **Mechanism of action of Glucocorticoid**

Glucocorticoid (GC) is naturally synthesized by the adrenal cortex under the stress condition. The body uses GC to reduce immune activity and to treat the diseases that are mainly caused by a hyperactive immune system. The pharmacological effect of GC expresses at a molecular level through the mechanism of transactivation and transrepression.

GC crosses the cell membrane and binds to the glucocorticoid receptor (GR) in the cytoplasm (Table 6.1). GR is present in the inactivated form before the steroid-binding and attached to the stabilizing proteins. These proteins are heat shock protein 90 (hsp90) and immunophilins. Binding of steroid to the receptor in cytoplasm cause to release of the immunophilins. This dissociation causes the binding of the GR-steroid complex to the Dynein motor protein. This activated complex then translocates into the nucleus and affects the transcription. The GR-steroid complex then dimerizes and binds to the regulatory region of the glucocorticoid responsive element (GRE) gene. GC shows their anti-inflammatory effect by inhibiting COX-2 and inflammatory cytokines like IL-4, IL-5, IL-13, IL-18, and TNF-α (Oakley & Cidlowski, 2013). The non-genomic anti-inflammatory effect of GC is exerted by the upregulation of Annexin A1 which is directly involved in the production of the anti-inflammatory cytokine IL-10. At a cellular level, GC inhibits the recruitment of inflammatory cytokines into the nasal mucosa and promotes the apoptosis of eosinophil (Table 6.1). GC inhibits the maturation of mast cells and the release of histamine as shown in Fig. 6.9 (Girol et al., 2013).

(b) **Drug resistance and its mechanism**

GC resistance is mainly caused by the elevated level of GR-β and IL-17 in neutrophils. The gene that encodes for human GR contains 9 exons. GR-α is obtained when the

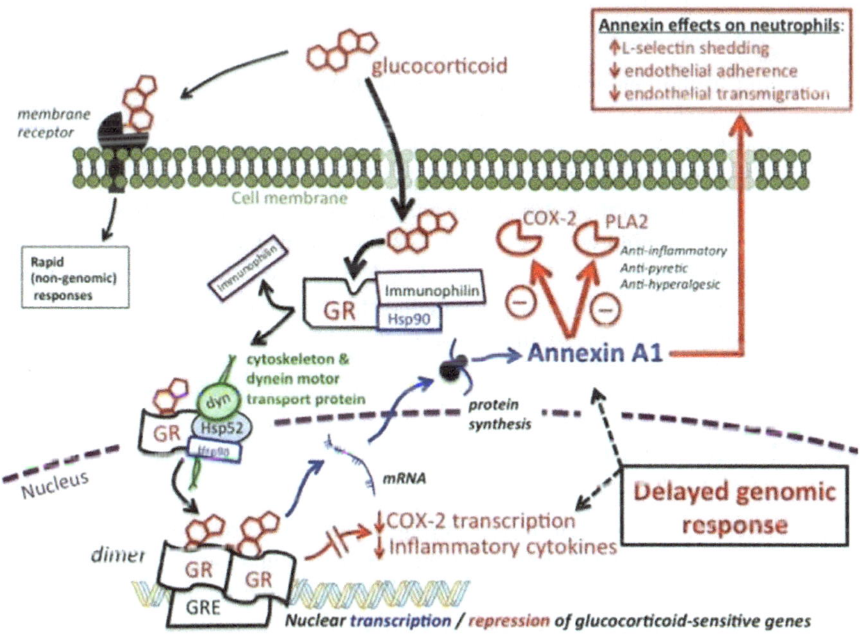

Fig. 6.9 Action mechanism of GCs against inflammatory diseases (Girol et al., 2013)

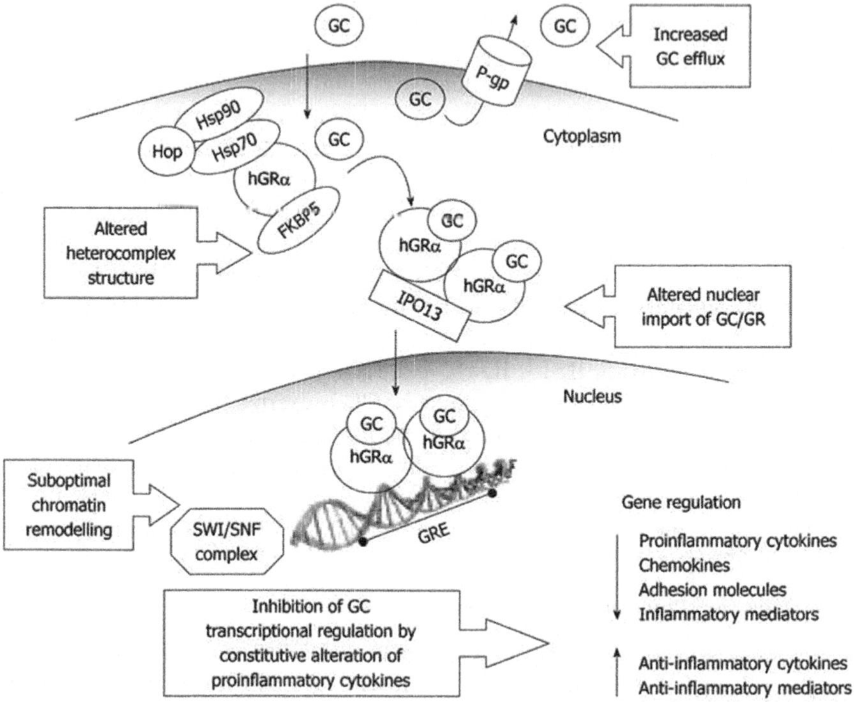

Fig. 6.10 Multiple resistance mechanisms against GCs (de Iudicibus et al., 2011)

exon 8 bind to the beginning region of exon 9. While the alternative version of this receptor is obtained when exon 8 joins to the downstream of exon 9. These two proteins have the same no of amino acids 727, but they don't have the same function(Kino et al., 2009). GR-β lacks the binding ability of the GCs. An increased level of GR-β can form the heterodimer with GR-α and block the active site of this required receptor to bind with GCs. Thus GR-β serves as the inhibitor of GR-α and masts the anti-inflammatory effect of GCs (Fig. 6.10) (de Iudicibus et al., 2011). Other mechanisms of GC resistance may include the efflux system, mutation in GR-α, and improper conjugation of GC with other molecules (Table 6.1).

6.4 Throat Infection

Throat infection, also known as sore throat, is the most common infection. It is an agonizing, dry, and harsh feeling in the throat. About 13 million patients visit the doctor due to sore throat each year. Sore throats have different types depending on the part of the throat affected: laryngitis that affect the larynx, pharyngitis that affect the pharynx's and tonsillitis affect the tonsils of the throat. There are many causes of a throat infection in which eight are the most common include: Allergies, bacterial infections, chemicals, colds, dry air, the flu, gastroesophageal reflux disease (GERD), irritants, injury, smoke, strep throat, tumor, and viral infections. In humans, the oral, throat, and respiratory tract are connected and ecologically colonized with numerous microbes. *Bacteroidetes*, *Firmicutes*, and *Fusobacteria* in the oral cavity while *Achromobacterxylosoxidans*, *Burkholderiacepacian* complex, and *Pseudomonasaeruginosa* are the dominant microbes in the lungs. *Bordetella pertussis, Chlamydia pneumoniae, Haemophilus influenza, Mycoplasma pneumoniae*, and *Streptococcus pneumoniae* are the bacteria that cause respiratory infections in humans (Gong et al., 2013, 2014).

6.4.1 Colds and Flu

The most common cold is chest and head cold, as well as all types of seasonal flu, which are caused by viruses. It is very important to know about the difference between the symptoms of cold and flu. As compare to flu, a cold is considered a milder respiratory illness. Colds cause discomfort for a few days while flu from days to weeks. Serious health problems like hospitalizations and *pneumonia* can take place as a result of flu. More than 200 types of viruses have been observed serologically to be responsible for causing human UTR infection, and rhinoviruses are the most common (Eccles, 2005).

Both cold and flu are syndromes that are caused by UTR viral infection. Defining the syndromes is a very difficult task due to variations in duration, severity, and types of symptoms. Among all the colds, 30–50% occur due to rhinoviruses and 10–15% due to coronaviruses that are considered the second most common cold-causing agents. There is a 5–15% participation of the influenza virus and other cold virus-like RSV causes much illness that is flue like (Eccles, 2005).

6.4.1.1 Treatment

Cold and flu cannot be recovered by taking antibiotics. Antibiotics are used for the treatment of bacterial infections like urinary tract, skin, strep throat, and ear infections. Using antibiotics for infections that are not treatable with them reduces the effectiveness of that antibiotic against infection they are assumed for treatment. This situation is known as antibiotic resistance. Antibiotics should be never be used for the treatment of flu and cold. Oseltamivir (Tamiflu®), rimantadine (Flumadine®), amantadine (Symmetrel®), and zanamivir are some of the anti-flu medicine available in clinics.

6.4.1.2 Amantadine

Amantadine is an agent that is used against the virus. It also shows an antiparkinsonian activity. In early 2000, amantadine was used for the treatment of *Influenza A*. In 2006, a meta-analysis showed the decreased symptoms of influenza by one day and reduced the severity of fever by the drug. However, a high level of resistance has emerged recently from North America and Asia. As of 2011, due to high resistance, the Center for disease control (CDC) has barred the treatment of *influenza A* (Lyons & Pahwa, 2012).

(a) **Mechanism of action**

The mode of action of amantadine is not clear. Its anti-viral activity is associated with interference in viral DNA replication. It damages the transmembrane domain of the M2 protein of the virus which escapes the entry of nucleic acid (infectious) of the virus in the cells of the host. During replication, it also inhibits the assembly of isolates of influenza A virus (IAV) from each subtype that are H1N1, H2N2, and H3N2 (Fig. 6.11 and Table 6.1). On the other hand Influenza, B contains M2 protein with a different structure which makes it ineffective against amantadine (Lyons & Pahwa, 2012).

(b) **Drug resistance and its mechanism**

During the years 2003–2004, the resistance of adamantine against IAV was enhanced rapidly. The percentage of adamantane-resistant isolates of IAV that was submitted

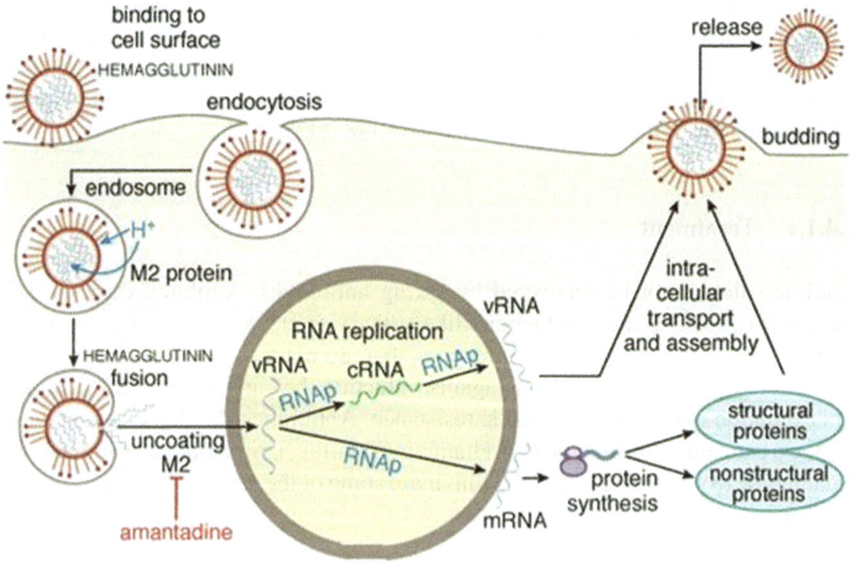

Fig. 6.11 Mode of action of amantadine (*Source* Antibiotic drugs, information, description on amantadine. "www.antibiotics-info.org/amantadine.html")

from the whole world to the WHO collaborating center for epidemiology, surveillance, and control of influenza was 0.4–12.3% during 1994–1995 and 2003–2004 respectively. During the influenza season of 2005–06, CDC determined that 92% of IAV taken from patients, from 26 states showed changes at amino acid 31 of the M2 gene that caused resistance to adamantanes. Due to high resistance to adamantanes, rimantadine, and amantadine are not preferred for the treatment of viral infections and chemoprophylaxis of strains of IAV (Hussain et al., 2017).

To treat the IAV infections, two types of drugs i.e. neuraminidase (NA) and adamantanes inhibitors which target the NA and M2 ion channel of the virus have been approved. By creating mutations in viral components, IAV swiftly acquires resistance to both these drugs. The H1N1 and H3N2 are the most common subtypes of IAV that have created resistance to adamantanes. Up to 100%, IAV subtype H1N1 was found resistant to oseltamivir (most common NA Inhibitor) until 2009. However, the IAV H1N1 subtype of the 2009 pandemic that was sensitive to NA inhibitors is now the most dominant influenza virus strains of the season and has replaced the H1N1 variants that were oseltamivir-resistant pre-2009 (Table 6.1) (Hussain et al., 2017).

6.4.2 *Pharyngitis*

Pharyngitis is the disorder in which inflammation of the throat takes place due to viral or bacterial infection. The most common symptoms including are cough, body aches, fever, swollen lymph nodes in the neck, and congestion. The severity of symptoms can be enhanced from a mild range. Pharyngitis is caused by both viral infections and bacterial infections. Viral infections including are mononucleosis and influenza and bacterial infections like an allergic reaction, strep throat, and gastric reflux. Pharyngitis can also occur due to any environmental irritant. Pharyngitis occurs due to bacterial and viral infection is contagious. People with a weak immune system like in HIV infection are at more risk of suffering from pharyngitis. *Streptococcus pyogenes* (GAS infection) is the most common pharyngitis causing bacterial agents in both children and adults. Some viruses that contribute to the development of pharyngitis are Epstein-Barr virus, rhinovirus, coronavirus, influenza virus, and Cold adenovirus (Bower, 2012).

For the treatment of Group A beta-hemolytic streptococcal (GABHS) pharyngitis, oral penicillin is being used currently. Alternatively, amoxicillin is also reliable and has benefits in terms of more palatability and easier dosing (Bower, 2012).

6.5 Conclusion

Many anti-microbial drugs are produced by micro-organisms and those micro-organisms are resistant to those drugs. Antibiotics were discovered in the twentieth century and now there are millennia of antibiotics used against bacteria. Meanwhile, their excess and unbalanced use of drugs develop resistance in micro-organisms. This makes it more challenging to treat infectious diseases. The ENT infection is the most common so there is a need for management of the use of drugs against micro-organisms. Physicians must implement an accurate clinic practice to adjust resistance development. The pharmaceutical companies must try to aware of the community about the resistance to advertisements and counseling.

Table 6.1 Mechanism of action and resistance of drugs used against ENT diseases

Type of infections	Disease	Drug	Mechanism of action	Mechanism of drug resistance	References
Ear infection	Otitis media (middle ear infection)	Ceftriaxone	Inhibit the synthesis of bacterial cell wall	• By producing β-lactamase • By non-β-lactamase mediated resistance	(Jacobs, 2017; Rawls, 2014)
		Amoxicillin	Preventing the formation of the peptidoglycan layer in the bacterial cell wall	• By producing β-lactamase • By non-β-lactamase mediated resistance	(Jacobs, 2017)
		Trimethoprim-sulfamethoxazole	Inhibit the dihydrofolate reductase (DHFR)	Mutation in the dhfr gene leads to single amino acid substitution in the DHFR cause change in the structure that makes antibiotics unable to bind it	(Jacobs, 2017; Kordus & Baughn, 2019)
	Swimmer's ear (outer ear infection)	Ofloxacin	Inhibit the bacterial cell division by inhibiting the bacterial DNA replicating, repairing transcriptional and recombinant enzymes	• Mutation in topoisomerase IV and gyrase • Binding of the One protein is decreased that protect the DNA-enzyme complex • Porins under-expression decrease the uptake of the drug • Decrease the retention of drug in the cell by the chromosome-encoded efflux pumps overexpression	(Aldredet al., 2014; Pham et al., 2019)

(continued)

Table 6.1 (continued)

Type of infections	Disease	Drug	Mechanism of action	Mechanism of drug resistance	References
	Otomycyosis (outer ear infection)	Clotrimazole	Inhibits the lanosterol 14-demethylase enzyme involved in the ergosterol synthesis	• Mutation in the structure of the CYP51A enzyme that reduces the binding affinity of the azole • Overexpression of the gene cyp51A gene • Overexpression of efflux transporter genes	(Berger et al., 2017; Haller, 1985)
Nose infection	Acute rhinosinusitis	Amoxicillin	Preventing the formation of the peptidoglycan layer in the bacterial cell wall	• By producing β-lactamase • By Non-β-lactamase mediated resistance	(Jacobs, 2017)
	Chronic rhinosinusitis	Trimethoprim-sulfamethoxazole	Inhibit the dihydrofolate reductase (DHFR)	Mutation in the dhfr gene leads to single amino acid substitution in the DHFR cause change in the structure that makes antibiotics unable to bind it	(Jacobs, 2017; Kordus & Baughn, 2019)

(continued)

Table 6.1 (continued)

Type of infections	Disease	Drug	Mechanism of action	Mechanism of drug resistance	References
	Allergic rhinitis	Glucocorticoid (GC)	• GC binds to the intracellular GC receptors and causes changes in the gene transcription • GC also inhibits the selection of inflammatory cytokines which leads to eosinophil apoptosis at the cellular level	• Elevated of GR-β and IL-17 in neutrophil • Efflux system • Mutation in GR-α • Improper conjugation	(de Iudicibus et al., 2011; Girol et al., 2013)
Throat infection	Colds and flu	Amantadine	• Damages the transmembrane domain of the M2 protein of the virus which escapes the entry of nucleic acid (infectious) of the virus in the cells of the host • Inhibits the assembly of isolates of influenza A virus from each subtype that are H1N1, H2N2, and H3N2	H1N1 and H3N2 are the most common subtypes of IAV that have created resistance to adamantanes but there is still unknown	(Hussain et al., 2017; Lyons & Pahwa, 2012)

(continued)

Table 6.1 (continued)

Type of infections	Disease	Drug	Mechanism of action	Mechanism of drug resistance	References
	Pharyngitis	Penicillin	Preventing the formation of the peptidoglycan layer in the bacterial cell wall	• By producing β-lactamase • By non-β-lactamase mediated resistance	(Jacobs, 2017)

References

Aldred, K. J., Kerns, R. J., & Osheroff, N. (2014). Mechanism of quinolone action and resistance. *Biochemistry, 53*, 1565–1574.

Antibiotics, S., Used, B. E., Treat, T. O., Antibiotic, W., The, I. S., Of, D., & For, C. (1998). Antibiotic management of acute otitis media. *Paediatrics & Child Health, 3*, 265–267.

Anwar, K., & Gohar, M. S. (2014). Otomycosis; clinical features, predisposing factors, and treatment implications. *Pakistan Journal of Medical Sciences, 30*, 2–5.

Bachert, C., & Akdis, C. A. (2016). Phenotypes and emerging endotypes of chronic rhinosinusitis. *The Journal of Allergy and Clinical Immunology: in Practice, 4*, 621–628.

Bassis, C. M., Tang, A. L., Young, V. B., & Pynnonen, M. A. (2014). The nasal cavity microbiota of healthy adults. *Microbiome, 2*, 1–5.

Berger, S., Chazli, Y. E., Babu, A. F., & Coste, A. T. (2017). Azole resistance in *Aspergillusfumigatus*: A consequence of antifungal use in agriculture? *Frontiers in Microbiology, 8*, 1024.

Bower, J. R. (2012). Pharyngitis. *Netter's Infectious Disease*, 177–182.

Cattoir, V., Lesprit, P., Lascols, C., Denamur, E., Legrand, P., Soussy, C. J., & Cambau, E. (2006). In vivo selection during ofloxacin therapy of *Escherichia coli* with combined topoisomerase mutations that confer high resistance to ofloxacin but susceptibility to nalidixic acid. *Journal of Antimicrobial Chemotherapy, 58*, 1054–1057.

Chowdhry, S. (2017). Topical antifungals used for treatment of seborrheic dermatitis. *Journal of Bacteriology & Mycology: Open Access, 4*, 76.

Coburn, B., Wang, P. W., Caballero, J. D., Clark, S. T., Brahma, V., Donaldson, S., Zhang, Y., Surendra, A., Gong, Y., Tullis, D. E., et al. (2015). Lung microbiota across age and disease stage in cystic fibrosis. *Scientific Reports, 5*, 10241.

Costa, C., Ribeiro, J., Miranda, I. M., Silva-Dias, A., Cavalheiro, M., Costa-de-Oliveira, S., Rodrigues, A. G., & Teixeira, M. C. (2016). Clotrimazole drug resistance in *Candida glabrata* clinical isolates correlates with increased expression of the drug: H^+ antiporters CgAqr1, CgTpo1_1, CgTpo3, and CgQdr2. *Frontiers in Microbiology, 7*, 526.

Danishyar, A., & Ashurst, J. V. (2020). *Acute otitis media*. StatPearls Publishing.

Das, B., & Patra, S. (2017). Antimicrobials: Meeting the challenges of antibiotic resistance through nanotechnology. In *Nanostructures for antimicrobial therapy* (pp. 1–22)

Davies, J. (1996). Origins and evolution of antibiotic resistance. *Microbiología, 12*, 9–16.

De Boeck, I., Wittouck, S., Martens, K., Claes, J., Jorissen, M., Steelant, B., van den Broek, M. F. L., Seys, S. F., Hellings, P. W., Vanderveken, O. M., et al. (2019). Anterior nares diversity and pathobionts represent sinus microbiome in chronic rhinosinusitis. *MSphere, 4*.

de Iudicibus, S., Franca, R., Martelossi, S., Ventura, A., & Decorti, G. (2011). Molecular mechanism of glucocorticoid resistance in inflammatory bowel disease. *World Journal of Gastroenterology, 17*, 1095–1108.

Eccles, R. (2005). Understanding the symptoms of the common cold and influenza. *The Lancet Infectious Diseases, 5*, 718–725.

El-Din, M. B. (2018). Validated analytical methods for the determination of some antimicrobial drugs. CU Theses.

Frayman, K. B., Armstrong, D. S., Grimwood, K., & Ranganathan, S. C. (2017). The airway microbiota in early cystic fibrosis lung disease. *Pediatric Pulmonology, 52*, 1384–1404.

Girol, A. P., Mimura, K. K. O., Drewes, C. C., Bolonheis, S. M., Solito, E., Farsky, S. H. P., Gil, C. D., & Oliani, S. M. (2013). Anti-inflammatory mechanisms of the annexin A1 protein and its mimetic peptide Ac2-26 in models of ocular inflammation in vivo and in vitro. *The Journal of Immunology, 190*, 5689–5701.

Gong, H.-L., Shi, Y., Zhou, L., Wu, C.-P., Cao, P.-Y., Tao, L., Xu, C., Hou, D.-S., & Wang, Y.-Z. (2013). The composition of microbiome in larynx and the throat biodiversity between laryngeal squamous cell carcinoma patients and control population. *PLoS ONE, 8*, e66476.

Gong, H., Shi, Y., Zhou, X., Wu, C., Cao, P., Xu, C., Hou, D., Wang, Y., & Zhou, L. (2014). Microbiota in the throat and risk factors for laryngeal carcinoma. *Applied and Environment Microbiology, 80*, 7356–7363.

Haller, I. (1985). Mode of action of clotrimazole: Implications for therapy. *American Journal of Obstetrics and Gynecology, 152*, 939–944.

Hart, C. A., & Kariuki, S. (1998). Antimicrobial resistance in developing countries. *BMJ, 317*, 647–650.

Hussain, M., Galvin, H. D., Haw, T. Y., Nutsford, A. N., & Husain, M. (2017). Drug resistance in *influenza A* virus: The epidemiology and management. *Infection and Drug Resistance, 10*, 121–134.

Jacobs, M. R. (2017). Mechanisms of resistance in *Haemophilus influenzae* and *Moraxella catarrhalis*. In *Antimicrobial drug resistance* (pp. 867–888).

Kino, T., Su, Y. A., & Chrousos, G. P. (2009). Human glucocorticoid receptor isoform β: Recent understanding of its potential implications in physiology and pathophysiology. *Cellular and Molecular Life Sciences, 66*, 3435–3448.

Kordus, S. L., & Baughn, A. D. (2019). Revitalizing antifolates through understanding mechanisms that govern susceptibility and resistance. *Medicinal Chemistry Communications, 10*, 880–895.

Kumpitsch, C., Koskinen, K., Schöpf, V., & Moissl-Eichinger, C. (2019). The microbiome of the upper respiratory tract in health and disease. *BMC Biology, 17*, 1–20.

Kwon, Y. H. (2016). Amoxicillin. In *Helicobacter Pylori* (pp. 387–396). Springer.

Lee, Y., Puumala, E., Robbins, N., & Cowen, L. E. (2020). Antifungal drug resistance: Molecular mechanisms in *Candida albicans* and beyond. *Chemical Reviews*.

Lewin, S. C., & Smith, J. T. (1988). Bactericidal mechanisms of ofloxacin. *Journal of Antimicrobial Chemotherapy, 22*, 1–8.

Lyons, K. E., & Pahwa, R. (2012). Amantadine. In *Parkinson's Disease* (2nd edn., pp. 819–824). CRC Press.

Massova, I., & Mobashery, S. (1998). Kinship and diversification of bacterial penicillin-binding proteins and β-lactamases. *Antimicrobial Agents and Chemotherapy, 42*, 1–17.

Maxfield, A. Z., Korkmaz, H., Gregorio, L. L., Busaba, N. Y., Gray, S. T., Holbrook, E. H., Guo, R., & Bleier, B. S. (2017). General antibiotic exposure is associated with increased risk of developing chronic rhinosinusitis. *The Laryngoscope, 127*, 296–302.

Miyake, M. M., & Bleier, B. S. (2019). Future topical medications in chronic rhinosinusitis. *International Forum of Allergy & Rhinology, 9*, S32–S46.

Oakley, R. H., & Cidlowski, J. A. (2013). The biology of the glucocorticoid receptor: New signaling mechanisms in health and disease. *The Journal of Allergy and Clinical Immunology, 132*, 1033–1044.

Pham, T. D. M., Ziora, Z. M., & Blaskovich, M. A. T. (2019). Quinolone antibiotics. *Medicinal Chemistry Communications, 10*, 1719–1739.

Rawls, S. M. (2014). Antibiotics, β-Lactam. In *Encyclopedia of the neurological sciences* (pp. 207–209). Elsevier.

Rosenfeld, R. M., Piccirillo, J. F., Chandrasekhar, S. S., Brook, I., Ashok Kumar, K., Kramper, M., Orlandi, R. R., Palmer, J. N., Patel, Z. M., Peters, A., et al. (2015). Clinical practice guideline (update): Adult sinusitis. *Otolaryngology–head and Neck Surgery, 152*, S1–S39.

Schenck, L. P., Surette, M. G., & Bowdish, D. M. E. (2016). Composition and immunological significance of the upper respiratory tract microbiota. *FEBS Letters, 590*, 3705–3720.

Schwartz, J. S., Peres, A. G., Endam, L. M., Cousineau, B., Madrenas, J., & Desrosiers, M. (2016). Topical probiotics as a therapeutic alternative for chronic rhinosinusitis: A preclinical proof of concept. *American Journal of Rhinology & Allergy, 30*, e202–e205.

Shilts, M. H., Rosas-Salazar, C., Tovchigrechko, A., Larkin, E. K., Torralba, M., Akopov, A., Halpin, R., Peebles, R. S., Moore, M. L., Anderson, L. J., et al. (2016). Minimally invasive sampling method identifies differences in taxonomic richness of nasal microbiomes in young infants associated with mode of delivery. *Microbial Ecology, 71*, 233–242.

Sivasubramaniam, R., & Douglas, R. (2018). The microbiome and chronic rhinosinusitis. *World Journal of Otolaryngology-Head and Neck Surgery, 4*, 216–221.

Smilack, J. D. (1999). Trimethoprim-sulfamethoxazole. In *Mayo clinic proceedings* (pp. 730–734). Elsevier.

Tanwar, J., Das, S., Fatima, Z., & Hameed, S. (2014). Multidrug resistance: An emerging crisis. *Interdisciplinary Perspectives on Infectious Diseases*, 1–7.

Tristram, S., Jacobs, M. R., & Appelbaum, P. C. (2007). Antimicrobial resistance in *Haemophilus influenzae. Clinical Microbiology Reviews, 20*, 368–389.

Varshney, J., & Varshney, H. (2015). Allergic rhinitis: An overview. *Indian Journal of Otolaryngology-Head and Neck Surgery, 67*, 143–149.

Venter, H. (2019). Reversing resistance to counter antimicrobial resistance in the World Health Organisation's critical priority of most dangerous pathogens. *Bioscience Reports*, 39.

Vlastarakos, P. V., Nikolopoulos, T. P., Maragoudakis, P., Tzagaroulakis, A., & Ferekidis, E. (2007). Biofilms in ear, nose, and throat infections: How important are they? *The Laryngoscope, 117*, 668–673.

Wos-Oxley, M. L., Chaves-Moreno, D., Jáuregui, R., Oxley, A. P. A., Kaspar, U., Plumeier, I., Kahl, S., Rudack, C., Becker, K., & Pieper, D. H. (2016). Exploring the bacterial assemblages along the human nasal passage. *Environmental Microbiology, 18*, 2259–2271.

Yan, M., Pamp, S. J., Fukuyama, J., Hwang, P. H., Cho, D. Y., Holmes, S., & Relman, D. A. (2013). Nasal microenvironments and interspecific interactions influence nasal microbiota complexity and *S. aureus* carriage. *Cell Host and Microbe, 14*, 631–640.

Zaman, S. B., Hussain, M. A., Nye, R., Mehta, V., Mamun, K. T., & Hossain, N. (2017). A review on antibiotic resistance: Alarm bells are ringing. *Cureus, 9*, e1403.

Chapter 7
Drug Resistance in Skin Diseases

Humaira Shah and Sher Zaman Safi

Abstract Skin is the largest organ of the body and the biggest barrier against pathogens. Skin diseases have become one of the most challenging medical problems in clinical practices, and a tremendous burden to the healthcare system in terms of cost and consumption of institutional resources. The emergence of drug resistance is a threat and concern in various clinical implications of skin diseases, particularly those that have secondary intervention such as chemotherapy, as the main treatment option. Multidrug resistant organisms include bacteria and other microorganisms that have grown resistance to antimicrobial drugs. Resistance to antimicrobial drugs reduces the effectiveness of treatment and increases the severity of dermatologic diseases. Methicillin-resistant *Staphylococcus aureus* (MRSA), Vancomycin-Resistant Enterococci (VRE), and multidrug-resistant gram-negative rods (MDR GNR) bacteria are the common multidrug-resistant organisms (MDROs). Skin cancers have poor prognosis and treatment options, due to resistance to conventional chemotherapies. The exact mechanisms of resistance in skin cancers are not known, however most tumors likely resist through reduced apoptosis. Many molecular pathways, enzymes and genes are implicated in the process of reduced "apoptosis induction". Similarly, acyclovir (ACV) and related analogues are successful drugs against HSV infections, however the emergence of drug resistance to ACV has become an obstacle in the treatment of HSV infections, especially in patients with low immunity. In this chapter, we have attempted to discuss major skin diseases and relevant factors that contribute to drug resistance in many dermatologic conditions.

H. Shah
University of Management and Technology (UMT), Lahore, Pakistan

S. Z. Safi (✉)
Interdisciplinary Research Centre in Biomedical Materials (IRCBM),
COMSATS University Islamabad, Lahore Campus, Lahore, Pakistan
e-mail: dr.szsafi@cuilahore.edu.pk

S. Ahmed et al. (eds.), *Biochemistry of Drug Resistance*,
https://doi.org/10.1007/978-3-030-76320-6_7

7.1 Introduction

Global Burden of Disease (GBD) 2013, reports cutaneous and subcutaneous diseases as the 4th leading cause of disability, with a total 1.79% global burden of disease worldwide. These infections are estimated to be 18th leading global disability-adjusted life years (DALYs) (Karimkhani et al., 2017). Skin diseases pose significant threat to a patient's mental health, well-being, ability to function and their social participation (Seth et al., 2017). Skin diseases are diverse, and there have been attempts to categorize them into several categories such as viral skin diseases, fungal skin diseases, dermatitis, keratinocyte carcinoma, malignant skin melanoma, psoriasis, pyoderma, dermatitis, acne vulgaris, scabies and other skin diseases (Karimkhani et al., 2017). The burden of skin diseases exhibits age and geographic related variations. Dermatitis has the highest DALY rate in Sub-Saharan Africa while melanoma causes the greatest burden in resource-rich countries such as North America and Australia. Similarly, the mortality due to melanoma was 4.7 times higher in resource-rich regions as compared to resource-poor countries in 2010 (Boyers et al., 2014).

In 2011, WHO declared "combat drug resistance: no action today, no cure tomorrow." (Sharma, 2011). Antimicrobial resistance (AMR) in skin diseases is a serious global threat for effective treatment and control, despite multiple efforts to combat the situation (Fleming-Dutra et al., 2016; Velez & Sloand, 2016). Antibiotics prescription in dermatological diseases is very high and prolonged than any other specialty (Barbieri et al., 2019). This overuse of antibiotics has helped the microorganisms to resist every antibiotic (Fig. 7.1) introduced clinically, agriculturally or otherwise, by utilizing every possibility of horizontal gene transfer (Davies & Davies, 2010). Emergence of multidrug resistance in bacteria have made treatment difficult or untreatable with conventional drugs. In many settings, the health care professionals rely on broad-spectrum antibiotics due to non-availability of antimicrobial susceptibility patterns or identification of causative microorganisms. Dramatic increases in emerging resistance occurs and, when coupled with poor infection control practices, resistant bacteria can easily be disseminated to the other patients and the environment (Frieri et al., 2017).

Multidrug resistance (MDR) is on the rise, in all parts of the world. New mechanisms of resistance are developing and outspreading globally, threatening our ability to treat common skin infections (Morales et al., 2012). In common clinical settings, skin issues present a variety of forms, ranging from superficial infections that could be controlled with topical antibiotics to severe forms of infections that could lead to death if the infection is not properly treated (Arias & Murray, 2009). Common multidrug-resistant organisms (MDROs) include methicillin-resistant *Staphylococcus aureus* (MRSA), Vancomycin-Resistant Enterococci (VRE), and multidrug-resistant gram-negative rods (MDR GNR) bacteria such as *Enterobacter* species, E.coli, *Klebsiella pneumoniae, Acinetobacter baumannii,* and *Pseudomonas aeruginosa. Staphylococcus aureus* and *Streptococcus* bacterial strains are the most frequent causes of skin infections. Methicillin-resistant *Staphylococcus aureus* (MRSA) is a strain of *S.*

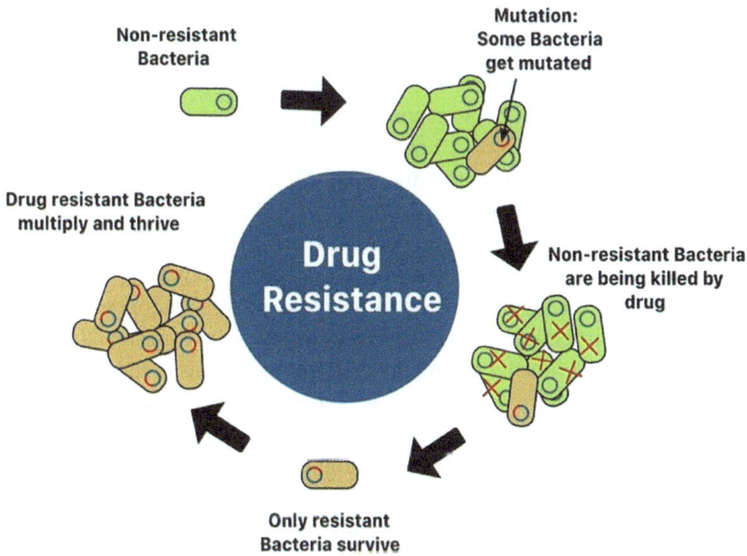

Fig. 7.1 Flow chart of the process through which bacteria become resistant to drugs

aureus that has acquired resistance to β-lactam antibiotics, which include penicillin and cephalosporins (Magiorakos et al., 2012).

Skin cancer has emerged as one of the most common malignancies in the world, where 1 in every 3 newly diagnosed malignancies is of skin (Leiter & Garbe, 2008). According to estimates of the World Health Organization(WHO) global occurrence of melanoma skin cancer is about 132,000 and non-melanoma is 2 to 3 million cases every year (GCO, 2020). Constant exposure to sunlight, change in climate, ozone depletion that leads to more UV radiations and changes in lifestyle at social and individual levels are major causes of skin cancer. Exposure to radiations causes mutations in DNA, consequently inactivating apoptosis in healthy cells (Ekwueme et al., 2011). The melanoma skin cancer is far more severe in terms of general characteristics and progression than non- melanoma skin cancer (basal cell and squamous cell carcinomas). Despite high occurrence of non-melanoma skin cancers, these malignancies have slow progression and are easier to treat. On the contrary, melanoma skin cancer, although comprising less than 1% of cases, is very aggressive, highly metastatic, drug resistant and has a high mortality rate (Laikova et al., 2019).

The poor prognosis and treatment of skin cancers stem from resistance to conventional chemotherapies, induced either by intrinsic or extrinsic pathways (Holohan et al., 2013). The exact mechanisms conferring this intrinsic drug resistance in melanoma cells are poorly understood, however a range of genes associated with apoptosis, cell signaling, inflammation and oxidative stress, are either activated or deactivated, consequently pose a poor response to treatment (Casas et al., 2011). The key genes and genetic pathways, involved in the skin cancer drug resistance are nuclear factor K (NF-κB), protein kinase B (PKB/Akt), mitogen-activated protein kinase (MAPK) and phosphoinositide-3 kinase (PI3K) (Muthusamy & Piva, 2010; Casas et al., 2011). Most drugs in cancer treatment ultimately act through the induction of apoptosis in cancerous cells, however activation of anti-apoptotic agents in skin cancer treatment might be a reason for relatively low apoptosis (Grossman & Altieri, 2001).

Herpes simplex virus (HSV), a member of the Herpesviridae family, causes a range of infectious diseases called herpes, from mild to severe mucocutaneous lesions in the oral cavity or genital infections (Jiang et al., 2016; Whitle & Roizman, 2001). HSV are worldwide in distribution and 60–95% of the adult population is affected (Jiang et al., 2016). Acyclovir (ACV) and other related nucleoside analogues are successful drugs against HSV infections, however the emergence of drug resistance to ACV has become an obstacle in the treatment of HSV infections, especially in patients with low immunity (Jiang et al., 2016). ACV resistance generally ranges from 3.5 to 10% (Piret & Bovine, 2011), however some clinical surveys have reported ACV-resistant HSV isolates of up to 36% (Langston et al., 2002). The primary target of anti-HSV drugs is the DNA polymerase gene of HSV. A minor mutation in the DNA polymerase gene of HSV may build resistance to many anti-HSV drugs including ACV (Bacon et al., 2003). Other ACV resistance mechanisms include decreased production of HSV encoded thymidine kinase (TK) and altered substrate specificity of HSV DNA polymerase (Gilbert et al., 2002).

Here we attempt to discuss the principal skin diseases and dermatological conditions which have been linked to drug resistance. We have reviewed the relevant host and pathogen-related factors that contribute to drug resistance in skin diseases. Other aspects such as mechanisms of drug resistance, alterations in drug targets, drug-induced DNA damage, various enzyme systems, regulatory aspects of apoptosis, multidrug-resistant bacterial strains and associated molecular pathways are discussed. We have also covered various chemotherapeutic options for skin diseases with a detailed mechanistic review of drugs and resistance.

7.2 Acne Vulgaris

Acne vulgaris is a common multifactorial chronic inflammatory disease of pilosebaceous units. It typically appears at puberty, when the development and secretory activity of sebaceous glands increases under the influence of androgen (Tan & Tan, 2005). Globally, acne ranks 8th in overall disease prevalence (Hay et al., 2014) and 2nd among global burden of skin diseases (Karimkhani et al., 2017). Acne is typically categorized according to age but higher prevalence is reported in the second decade of life (Hay et al., 2014). Acne may not be a life-threatening infection, yet its social and psychological effects are profound. Situation becomes grave with development of severe symptoms and scarring.

The pathogenetics of the disease can be attributed to four identified mechanisms. First is the high sebum excretion rate due to influence of androgens. Second key factor is follicular hyperkeratinization, when follicle cells become cohesive and normal shedding does not occur. This paves way for development of microcomedone, invisible small bumps on acne prone skin which finally leads to characteristic lesions of acne. Third important element in development of acne is colonization of skin by an anaerobic diphtheroid bacteria, *Propionibacterium acnes (P.acness)* now named as Cutibacterium. Acnes (C. acnes). Last contributing factor is the various products of inflammation (Gollnick et al., 1991). Although the source of follicular hyperkeratinization is not fully known but a few reports suggest follicular deficiency of linoleic acid to be a causative factor. Fourth is the expression of inflammatory cytokine interleukin-1 (IL-1) by follicular keratinocytes. This stimulates the comedogenesis (Guy, 1996; Morello et al., 1976). Comedogenesis is abnormal differentiation of the follicular infundibulum with excessive cornification thus forming a horny plug called microcomedo. Recent reports suggest that the regional differences in production of dihydrotestosterone, a potent androgen may also contribute to the pathogenesis of acne vulgaris. Recently the role of neurogenic factors such as neuropeptides, innervation, neurotrophic factors and neuropeptide-degrading enzymes are also reported to play part in the inflammatory process of acne (Tan & Tan, 2005). There are many factors, but our focus will be on *P.acnes,* the resistant strain which has developed resistance against a wide range of antimicrobial agents.

7.2.1 Propionibacterium Acnes/Cutibacterium Acnes

Propionibacterium acnes is a facultative, rod shaped anaerobic, gram-positive bacteria which along with *Staphylococcus, Streptococcus, Pseudomonas spp and Corynebacterium* is a major colonizer of normal skin (Cogen et al., 2008). These propionibacteria play an important role in balance and support of the natural microbiome of skin. But under certain set of conditions, they can trigger cutaneous diseases. Among these *P.acness/C.acnes* are the most predominant opportunists, causing a wide range of cutaneous infections such as acne vulgarisand several other soft tissue infections (McLaughlin et al., 2019). In light of the fact that *P.acnes/C.acnes*constitutes part of the normalskin microbiome, Koch's postulates remain unsatisfied in relation to the establishment of a causal effect. Therefore, its role in the pathogenesis of inflammatory acne is circumstantial. The evidence is primarily based on the effectiveness of antimicrobial agents in improving the symptoms. Moreover, the presence of antibiotic resistant strains is usually linked to nonresponse to antibiotic therapy thus opposing the very argument that the effect of antibiotics is primarily the result of anti-inflammatory activity (EADY et al., 1989).Another evidence of this association is the high antibody titer against *C. acnes* in severe acne patients in comparison to normal controls (Puhvel et al., 1964).

Sebaceous glands secrete lipid rich sebum which function as growth medium for the bacteria within the follicles. This serum triglyceride is hydrolyzed into glycerols and free fatty acids by action of an extracellular lipase. The glycerol serves as a growth medium for *P.acnes* whereas the triglyceride has comedogenic and pro-inflammatory properties (Brown & Shalita, 1998). Furthermore, Vowels et al. report that chemotactic factors produced by P.acnes attract neutrophils and lymphocytes (Vowels et al., 1995). Another study by Kim et al. suggests that in acne this bacterium triggers inflammatory cytokines by stimulation of toll-like receptor 2 (TLR2) present on monocytes (Kim et al., 2002). Recent reports suggest that proliferation of P.acnes does not trigger the disease. Instead, the activation of the innate immunity and the loss of the cutaneous diversity of might lead to the chronic inflammatory condition. In light of these findings P.acnes was renamed to *Cutibacterium acnes* (*C. acnes*) (Dréno et al., 2018).

7.2.2 Treatment Modalities of Acne

Two primary lesions common in acne are typically distributed in oil gland rich areas of face, neck, chest, shoulders and upper back. Comedones are common lesions which could be both open and closed. Second type of lesion is inflammatory lesions such as papules, pustules and nodules. The influence of post inflammatory hyperpigmentation, scarring or erythema are also important considerations to take into account during treatment. Treatment of acne depends on the type, severity and distribution of lesions. Before commencement of treatment it is important to review the skin care routine, diet and lifestyle of the patient. Different treatment regimens are available for acne. They can be broadly divided into topical such as retinoid, benzoyl peroxides and non-topical or systemic therapy which includes oral medications such as various hormones and antibiotics. (Rathi, 2011). Physical treatments such as laser or surgical removal of lesions are also used but we will focus only on chemo-treatments and resistance in *P. acnes.*

i. **Topical Treatment of Acne**

Retinoids such as tretinoin, have been in use as anti-acne agents, alone or in combinations as the first-line therapy for more than 30 years. They target the micro-comedo–precursor lesion of acne and so are preferred in maintenance therapy as well. (Rathi, 2011). Topical retinoids work for treatment of both inflammatory and non-inflammatory lesions of acne by reducing the follicular plugging and micro-comedones (Kang, 2005; Krishnan, 1976; Leyden et al., 2005; Shalita et al., 1996). Other topical retinoids such as tazarotene and salicylic acid are also reported to be effective and less irritant (Wolf, 2001; Zaenglein, 2018). Adapalene is a synthetic derivative of naphthoic acid with retinol activity. It is lipophilic and chemically more stable (Piskin & Uzunali, 2007). Besides topical retinoids, systemic retinoids such as isotretinoin is highly efficacious for treating refractory nodulocystic acne. To overcome the resistance against all trans-retinoic acid resistance, agents such as liarozole have been developed, which function as retinoic acid metabolism blocking agents (Piskin & Uzunali, 2007).

Topical benzoyl peroxide is an alternative to antibiotic treatment and is integral to first line acne therapy. It effectively reduces concentrations of C. acnes by releasing free oxygen radicals without allowing microbial resistance (Zaenglein, 2018). They are recommended to be used as key components in combination therapy to stop induction of bacterial resistance (Eady et al., 1996; Sagransky et al., 2009). Other topical agents include dapsone gel with proven clinical efficiency in patients with inflammatory acne (Del Rosso et al., 2015; Lynde & Andriessen, 2014). Azelaic acid is used for its ability to reduce post-inflammatory hyperpigmentation in mild inflammation and comedonal acne (Zaenglein, 2018).

ii. **Hormonal Therapies**

To control inflammatory lesions in adult women with acne, the use of combined oral medicine, containing progestin and estrogen are shown to have efficacy similar to that of antibiotics, however it takes longer for patients to have clinical improvement. Currently, Estrostep (ethinyl estradiol–norethindrone), Ortho tri-cyclen (ethinyl estradiol–norgestimate) and Yaz or Beyaz (ethinyl estradiol–drospirenone) are Food and Drug Administration (FDA)approved medicine for the treatment of acne vulgaris, although most other third- and fourth-generation progestin-containing combined oral contraceptives are also efficacious (Varghese, 2020; Zaenglein, 2018).

iii. **Topical and Non-Topical Antimicrobial Drugs**

Topical and non-topical (systemic oral) antibiotics form the backbone of acne treatment and have been used for treatment of acne lesions for more than 40 years (Dréno, 2016). Topical antibiotics such as erythromycin and clindamycin are the most common agents used in acne management (Rathi, 2011). But widespread resistance to antimicrobial agents is now considered a public health issue. The first report of resistance to topical use of clindamycin and erythromycin in acne was reported in the 1970s and is on rise worldwide (Dréno et al., 2018).

- **Tetracyclines**

Tertracyclin is a group of broad spectra, bacteriostatic antibiotics introduced in 1948. This group and second generation derivatives of tetracycline such as doxycycline, minocycline, lymecycline and third generation derivatives such as glycylcyclines and tigecycline have been proven to be effective in inflammatory acne treatment (Eady et al., 1993). The antibacterial features of tetracycline confer strong antibacterial properties (Mitscher, 1978). Tetracycline molecules consist of a linear fused tetracyclic nucleus with diverse functional groups. These fused rings are designated A, B, C, and D. The pharmacophore tetracycline with detectable antibacterial activity is 6-deoxy-6-demethyl tetracycline (Chopra & Roberts, 2001; Rogalski, 1985). Tetracycline enters bacterial cells through membrane porin channels by passive diffusion (Thanassi et al., 1995). Once inside, they inhibit synthesis of bacterial protein by preventing the association of charged aminoacyl-tRNA with the"A" site of the bacterial ribosome. The 30S and 50S subunits of microbial ribosomes serve as binding sites for tetracycline (Chopra, 1992).

- **Erythromycin**

Erythromycin is equally effective in treating inflammatory acne, however, compared to tetracycline, resistance to erythromycin is much more common. Moreover, substantial propagation of cross-resistant strains of propionibacteria have been recorded due to the extensive usage of topical preparations of erythromycin and clindamycin (Tan & Tan, 2005). The bacteriostatic activity of erythromycin at higher concentrations is not fully known but like other antibiotics, it might bind to the 50S subunit of bacterial ribosomal RNA complex. Mainly, it disturbs the aminoacyl translocation by preventing tRNA transfer at "A" site to "P" site of ribosomal complex. The "A" site remains occupied and results in inhibition of amino acid addition to the growing polypeptide chain. This alters the protein synthesis, consequently producing functionally useless protein. This alters the associated structural and functional activities which are critical for bacterial replication (Katzung, 2017).

7.2.3 Antibiotic Resistance in Acne Vulgaris

The antibiotic resistance in *P. acnes* to topical and non -topical antibiotics is rising each year. Reports from various countries suggest 50% of *P. acnes* strains show resistance against topical macrolides, a large group of bacteriostatic antibiotics (Walsh et al., 2016). The therapeutic failure or recurrence of acne is concurrent with heavy dependence on antibiotics in management of acne, the prolonged duration of usage and poor compliance which exerts a selection pressure on P. acnes (Eady, 1998). This antibiotic resistance can be evident as a reduced response, no response, or a relapse (Thiboutot et al., 2009). A systematic review by Cooper, reported an increase in overall incidence of antibiotic resistance in *P. acnes* from 20% in 1978 to 62% in 1996 (Cooper, 1998). Although the resistance varied for different antibiotics such as clindamycin, erythromycin, tetracycline, doxycycline and trimethoprim. Resistance to minocycline was rare (Cooper, 1998; Eady et al., 1989). Due to increased resistance, the use of erythromycin is becoming rare (Sardana & Garg, 2014).

Chromosomal point mutations are reported to be instrumental in *P. acnes* resistance. Their continued use despite no response, exerts selection pressure on non-target bacteria which could have potentially serious consequences (Walsh et al., 2016). Long-term treatment with oral antibiotics is a threat in emergence of resistant strains of *Staphylococcus aureus* in the nares, streptococci in the oral cavity and coagulase negative staphylococci on the skin. There is a significant association between occurrence of upper respiratory tract infection due to antibiotics used in acne (Margolis et al., 2005). Besides their antimicrobial activity, antibiotics also act through anti-inflammatory and immodulatory mechanisms, which include modification of complement pathways, diminishing chemotaxis of polymorphonuclear leukocytes and inhibiting lipase production of *P.acnes* (Meynadier & Alirezai, 1998). There isevidence for development of opportunistic infections by *P. acnes* in weak, immunocompromised patients such as mediastinitis after cardiac surgery and endocarditis (Cooper, 1998; Margolis et al., 2005).

7.2.4 Mechanism of Resistance to Tetracycline

Tetracyclines is one of the essential drugs in treating a wide range of skin infections caused by bacterial pathogens, however the widespread resistance to this drug has become challenging (Chopra & Roberts, 2001). So far, there are four possible molecular mechanisms through which bacteria acquire resistance against tetracycline. These include resistance efflux, ribosome protection proteins reduced permeability, ribosome mutation, and enzymatic inactivation of tetracycline (Nguyen et al., 2014). This acquired resistance is usually conferred by horizontal gene transfer that either codes for ribosomal protection proteins or efflux pumps. The efflux pumps prevent buildup of toxic levels of tetracycline in cytoplasm by actively pumping it out (Chopra & Roberts, 2001). Seven groups of efflux pumps, falling into the major facilitator superfamily are known that give resistant properties to bacteria (Guillaume et al., 2004).

Ribosomal protectionembodies a vital tactic for promoting tetracycline resistance in both gram-negative and gram-positive bacteria. Ribosomal protection proteinswork byinteracting with the ribosome to remove the tetracycline molecules from the binding site on the ribosome thus allowing for translation to continue. These essentially are GTPases with similarity to elongation factors that bind the ribosome analogously to elongation factors and chase bound tetracycline from the 30S ribosomal subunit (Chopra & Roberts, 2001; Nguyen et al., 2014). Ribosomal mutations are rare in clinical resistance to tetracyclines, still some clinical isolates of P. *acnes* carry point mutations in the 16S ribosome that probably reduces the tetracycline binding affinity (Ross et al., 1998). The next generation tetracyclines such as tigecycline, eravacycline and omadacycline overcome the efflux and ribosome protection mediated resistance but emergence of enzymatic inactivation of these drugs has raised new concerns. The enzymatic inactivation decreases the intracellular and extracellular concentration of the drug. This eliminates the tetracycline antibiotic challenge permanently. The clinical implications of enzymatic antibiotic inactivation can be catastrophic, as documented by the worldwide spread of broad-spectrum beta-lactamases (Markley & Wencewicz, 2018; Zaenglein, 2018).

7.3 Drug Resistance in *Staphylococcus aureus*

Staphylococci are abundant colonizers of human skin. The *Staphylococcus aureus* and *Staphylococcus epidermidis* are the dominant opportunists which cause many important skin diseases (Coates et al., 2014) (Fig. 7.2). *S. aureus* causes bacterial infections of skin, bone, soft tissues and risk of increased morbidity and mortality in health care related bacteremia (Loomba et al., 2010). *S. aureus* is a major cause of infection and delayed wound healing in chronic and acute wounds (Loomba et al., 2010). Invasive infections caused by *S. aureus* often leads to sepsis (Cardona & Wilson, 2015). Another remarkable clinical feature of *S. aureus* is its ability to elude

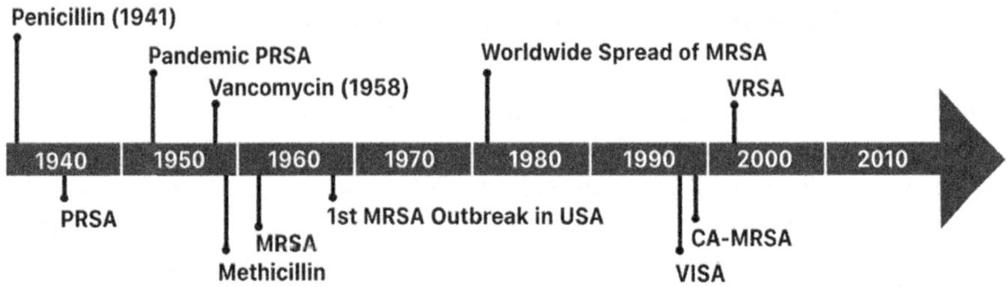

Fig. 7.2 Drug Resistance in *Staphylococcus aureus* and the epidemics

antibiotic chemotherapy by developing multidrug resistance (MDR). It can resist almost all the antibiotics developed since 1940. The use and misuse of various antibiotics over the years has paved the way to emergence of multidrug-resistant strains of *S. aureus* (Livermore, 2000). The resistance against methicillin and vancomycin is most remarkable (Hiramatsu et al., 2014). Historically, *S. aureus* developed resistance within two years of penicillin discovery when in 1942, the first penicillin-resistant strain of *S. aureus* was identified (Lakhundi & Zhang, 2018). The health-care associated methicillin-resistant *S. aureus* (MRSA) is a global epidemic, associated with high rates of mortality and morbidity (Loomba et al., 2010; Oztoprak et al., 2006; Whitby et al., 2001). According to a survey conducted by Diekema et al., in hospital settings, the MRSA accounts for at least 25 to 50% of *S. aureus* infections (Diekema et al., 2001).

7.3.1 Mechanism of Action of Methicillin

Methicillin is a semisynthetic derivative of penicillin, a β-lactam antibiotic which was developed in late 1950s. Unlike penicillin, it is not a substrate of beta-lactamase enzyme but its target is same as penicillin. (Autiero et al., 2009). The structural modification during chemical synthesis made methicillin resistant to beta lactamase. Beta lactamase is an enzyme produced by bacteria that breaks the structure of antibiotics by catalyzing the hydrolyses of betalactum ring, thereby making it ineffective (Chambers, 1997). Beta lactum antibiotics produce their action mainly by two mechanisms. First, by inhibiting the penicillin binding proteins (PBS) produced by bacteria. Peptidoglycan is 20–40 nm thick mesh-like structure that surrounds the cell of staphylococcus. It is made up of a series of short glycan chains of approximately 20 alternating *N*-acetylmuramic acid and β-1–4-*N*-acetylglucosamine residues. This *N*-acetylmuramic acid residue is attached to a pentapeptide chain called stem peptide. The glycan chains in peptidoglycan are cross-linked through the last glycine residue of a pentaglycine cross-bridge the last glycine residue of a pentaglycine cross-bridge attached to the l-lys residue (position 3) on one stem peptide and the d-alanyl-d-alanine (position 4) on another. This transpeptidation reaction is catalyzed by penicillin-binding proteins (PBPs) on exterior of cytoplasmic membrane. PBP1,

PBP2, PBP3, and PBP4 are four identified PBSs in S. aureus which are involved in transpeptidation and transglycosylation. The β-lactam antibiotics, are structurally similar to the terminal d-alanyl-d-alanine bond of the stem peptide. They interfere the cross-linking by inhibiting the transpeptidation and carboxypeptidase domains of PBS. The cell wall becomes mechanically weak without this crosslinking (Autiero et al., 2009; Stapleton & Taylor, 2002) First, the inhibition disrupts the cell wall by blocking mucin synthesis causing bacterial expansion and lysis. Secondly, it triggers the autolytic enzyme activity of the bacteria which leads to autolysis and ultimately death (Matono et al., 2018). Shortly after its clinical administration, S. aureus acquired methicillin resistance in the 1960s (Jevons, 1961).

7.3.2 *Mechanism of Methicillin Resistance in* S. aureus

S. aureus that are resistant to the isoxazoyl penicillins such as oxacillin, methicillin, and flucloxacillin are defined as Methicillin resistant S. aureus (MRSA) They are cross-resistant to all presently licensed β-lactam antibiotics (Chambers, 1997). Since the discovery of Methicillin resistant *S. aureus in* 1961, more than half of clinical strains of S. aureus have developed methicillin resistance (Hiramatsu et al., 2014). MRSA is classified into community-acquired MRSA and hospital-acquired MRSA depending on its source of origin (Hiramatsu et al., 2014). The methicillin-susceptible *S. aureus* (MSSA) transforms into notorious hospital pathogen Methicillin resistant *S. aureus* (MRSA) by acquiring the methicillin resistance gene mecA and apenicillin binding protein PBP2a. This transfer is mediated by a mobile genetic element staphylococcal cassette chromosome (SCC), a site specific transposon, through horizontal gene transfer (Chambers, 1997; Ito et al., 1999).

SCCmec is considered a new class of mobile genetic elements comprised of two mec and ccr genes complexes. The mec complex encodes methicillin resistance and ccr complex codes for recombinases which movement specific. Moreover, these elements also carry numerous genes responsible for resistance to non-β-lactam antibiotics. After accusation of mec element, numerous mutational events evolve MRSA into notorious hospital pathogens which are resistant to all available classes of antibiotics. The evolution of MRSA as community pathogen is also attributed the SCCmec element (Hiramatsu et al., 2001).

7.3.3 *Active Efflux System and Organization of Bacterial Cell Wall*

Recent researches have shed light on the role of cell wall organization and active drug efflux in resistance against various antibiotics.

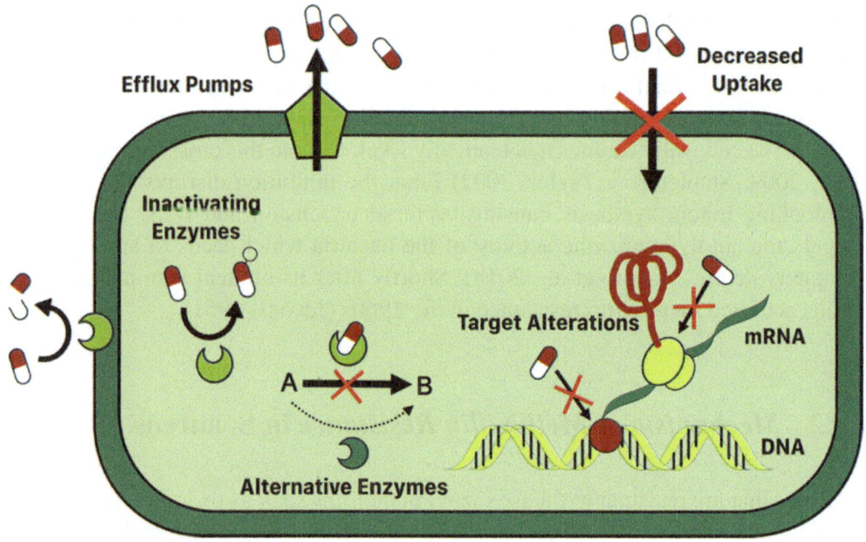

Fig. 7.3 Bacterial cell wall organization, decreased uptake and drug resistance

i. **Cell Wall Organization (Decrease/Limiting Drug Uptake)**

Restriction to drug access is not common in gram positive bacteria because they do not process an outer membrane. However, *Staphylococcus aureus*, a gram-positive bacterium, has recently developed resistance to vancomycin. The mechanism of this resistance is not fully understood. The bacteria secrete a thick cell wall which makes drug uptake difficult for the cell (Fig. 7.3). These strains are classified as VISA strains (Reygaert, 2018). This lowered permeability affects the energy metabolism of bacteria, consequently decreases the overall energy metabolism of drugs (Anuj et al., 2019; Li et al., 2013). Multiple studies suggest that lowered membrane permeability confers *S.aureus* resistance to aminoglycosides by decreasing drug uptake (Andrade et al., 2014; Hori & Hiramatsu, 1994).

ii. **Active Efflux Pump Systems**

Efflux pumps transport substances across bacterial walls. These pumps are regulated by chromosomally encoded genes in both antibiotic resistant as well as susceptible bacteria. The primary role of these pumps is to get rid of toxic substances. A large variety of chemicals are transported by multidrug resistant bacteria. They may transport a specific substance or a range of structurally different substances. The latter gives bacteria the ability to transport multiple compounds including antibiotics of multiple classes leading to multiple drug resistance (MDR). The genes which regulate these pumps are either overexpressed or expressed constitutively in the presence of suitable substrate or stimuli. Mutations in these genes modify the transport channel which leads to bacterial resistance to drugs (Blair et al., 2014; Villagra et al., 2012) When an environmental stimulus such as a substrate remains in a cell for a long time, the efflux regulating genes are expressed. This greatly enhances the ability to efflux drugs leading to resistance. Carbon source influences the efflux ability of many of these pumps (Zárate Villarroel et al., 2019).

A group of transporter proteins in a cell wall of bacteria which are resistant to antibiotics, pumps out a variety of substrates. These are termed as MDR (Multi Drug Resistance) systems. These MDR systems usually have downregulated expression under In-vitro conditions. Only in wild type strains, the MDR system is expressed which signifies its role in intrinsic resistance to antibiotics. On the contrary, their mutations in regulatory elements or induction and subsequent overexpression may serve as an important element in acquired antimicrobial resistance (Sánchez, 2003). QacA, NorA, and Smr are three categories of multidrug-pumping proteins, present on the cell membrane of *S. aureus.* QacA is considered as a significant player in MRSA. The multidrug pumping protein system are all kinases so they don't depend on hydrolysis of ATP for energy; instead electrochemical gradients are used for transport. This reversible gradient is set by H^+ on both sides of the cell membrane (Guo et al., 2020).

7.3.4 Resistance to the β-Lactam Drugs

The class of drugs with a β-lactam ring are called β-lactam drugs which include the famous penicillin and cephalosporin. A class of enzymes called β-lactamases inactivates theβ-lactam drugs.Penicillinases and cephalosporinases are two well-known β-lactamases. They break the β-lactam ring by hydrolyzing particular sites in the β-lactam drugs. The loss of ring structures inactivates the enzyme by destroying their ability to bind with the targeted penicillin binding proteins. There are many β-lactamases capable of inactivating any of the β-lactam drugs used currently in medicinal practice. Resistance via β-lactamases against penicillin and cephalosporin drugs is the most important mechanism in mediating resistance against these drugs (Bush & Jacoby, 2010).

7.3.4.1 Mechanism of Resistance Against β-Lactam Drugs

Three principle recognised mechanisms govern the resistance against β-lactam drugs. The first mechanism serves to prevent the interaction between the β-lactam drug and the targeted Penicillin binding protein. Second mechanism is hydrolysis of β-lactam drugs by β-lactamases. The third system to mediate resistance function via efflux pumps which pumps out the β-lactam drugs (Reygaert, 2018). The genes that encode for β-lactamase are innately found on chromosomes or it is acquired by plasmids. The S. aureus has plasmid carried genes of β-lactamase. In the case of MRSA, two mechanisms are responsible for mediation of resistance to β-lactam drugs (Chancey et al., 2012).

In The first mechanism the β-lactamases hydrolyzes the β-lactam drugs thereby inactivating the antibiotic. The second mechanism is pinching which prevents the antibiotic from reaching the target sites. In this method, a large amount of β-lactamase is produced that swiftly and tightly binds to the extracellular drug. This prevents the drug from reaching the intracellular spaces and thus the targets which confer MRSA its resistance against these antibiotics (Guo et al., 2020). The next class of antibiotics is tetracycline. This drug is hydrolysed by activation of tetX gene or by inactivation via transfer of a chemical group (Blair et al., 2015).

7.3.4.2 Drug Target Modification

Antimicrobial agents target multiple sites on bacterial cell walls. On the other hand, the bacteria also have the ability to modify multiple target sites to mediate resistance against those drugs. One exclusive mechanism used by gram positive bacteria against β-lactam drugs is the changes in number or structure of penicillin-binding proteins. These are transpeptidase proteins in nature which havea key role in synthesizing peptidoglycan in cell walls of bacteria. The amount of drug binding to a target is greatly affected by the change in number of penicillin-binding proteins. An increase in penicillin-binding proteins will decrease the drug binding ability. Likewise, any change in structure will reduce the binding ability of the drug to the protein or will cause complete inhibition of binding. In *S. aureus* the acquisition of mecA gene changes the structure of penicillin-binding proteins 2a, and inhibits the ability of drugs to bind to the protein (Beceiro et al., 2013; Reygaert, 2009).

Gram positive bacteria shows resistance against drugs that target ribosomal subunits. This resistance is mediated by mechanisms. One way is methylation of ribosomal subunits which is reported against a aminoglycoside and macrolides class of antibiotics. Another mechanism against tetracyclines is through ribosomal protection. Ribosomal mutation is another active mechanism harvested by bacteria against aminoglycosides and oxazolidinones (Reygaert, 2009).

7.3.5 Resistance of S. aureus to Vancomycin

Vancomycin resistance in *S. aureus* which is already resistant to methicillin (MRSA), has emerged as the most feared genetic adaptation. This is due to the fact that treatment of MRSA patients relies on vancomycin. Resistance to vancomycin shed light on how both core and accessory genomic elements influence the acquisition of resistance to antibiotics. Two forms of vancomycin resistance have been reported in *S. aureus*. The intermediate vancomycin resistant *S.aureus* (VISA) strains and high-level vancomycin-resistant *S. aureus*. The long duration of treatment and repeated courses of vancomycin led to emergence of VISA strains. The structure of peptidoglycan precursors is changed by acquisition of vancomycin resistance genes (*van* genes) which decreases the ability of vancomycin to bind to the target. Multiple mutations at different levels give different degrees of resistance to vancomycin. This particular trait is termed as heteroresistance. The strains with multiple mutations and higher tolerance to minimum inhibitory concentrations (MICs) of vancomycin ssurvives the time and selective pressure.

The high-level vancomycin-resistant *S. aureus* is due to vanA gene and operan by plasmid transfer from vancomycin-resistant *Enterococcus faecalis.* The structure of peptidoglycan precursors is changed by vancomycin resistance genes (*van* genes) which decreases the ability of vancomycin to bind to the target (McGuinness et al, 2017; Miller et al, 2014).

7.4 Skin Cancer

The skin cancers have emerged as one of the most common malignancies in the world where one in every 3 newly diagnosed malignancies is of skin (Leiter & Garbe, 2008). Although it affects men and women of all ages and ethnicities, the burden of skin cancer is carried by white population of the world. Repeated exposure to sunlight, change in climate, ozone depletion leading to more UV radiations and changes in lifestyle at social and individual level are major causes of skin cancer. The increased exposure to radiations causes mutations in healthy cells that inactivates apoptosis, a process crucial to avoid conversion of normal cells to malignant (Ekwueme et al., 2011). According to estimates of the World Health Organization (WHO) global occurrence of melanoma skin cancer is about 132,000 and non-melanoma is 2 to 3 million cases every year (GCO, 2020).

The melanoma skin cancer is far more severe in terms of general characteristics and progression than non- melanoma skin cancer. The basal cell and squamous cell carcinomas comprise the bulk of non-melanoma skin cancers. The basal cell carcinoma comprises 80% of non-melanoma skin cancers. Despite high occurrence, these malignancies have slow progression and are easier to treat. On the contrary, melanoma skin cancer, although comprising less than 1% of cases, is very aggressive, highly metastatic, drug resistant and has a high mortality rate. About 80% of deaths due to skin cancers is credited to melanoma skin cancer (Laikova et al., 2019).

Development of resistance to chemotherapy is becoming a major impeding factor in treatment of melanoma and non-melanoma skin cancer. New chemotherapeutic and targeted agents are constantly developed to alter tumor growth. Unfortunately, in most cases, the effect of anticancer therapies is not long-lasting. This treatment failure results in a high mortality rate particularly in advanced cancer patients (Olszanski, 2014). In this section we will briefly discuss the drug resistance and mechanism of resistance in both melanoma and non- melanoma skin cancers but the primary focus will be on melanoma skin cancer.

7.4.1 Melanoma Skin Cancer Treatment

Melanoma is a cancer that stems from malignant transformation in melanocyte, the melanin producing cells in the skin. Besides its high mortality rate, the metastatic melanoma is a deadly, heterogeneous and highly mutated form of cancer. Diagnosis at an early stage is mostly curable but metastatic stages are hard to treat (Mandalà & Romano, 2019). The past 50 years have witnessed no significant improvement in efficiency of melanoma treatment with less than 5% survival rate. The non- metastasized melanoma can be treated by surgery, chemotherapy and radiotherapy but the treatment choices become very narrow for metastasized melanoma. Until last decade, chemotherapy with high dose interleukin-2 (IL-2) and alkylating dacarbazine (DTIC) were only options for treatment. Interferon α-2b was the only approved adjuvant therapy for patients at high risk of progression (Mandalà & Romano, 2019). More recently the advances in immunotherapies such as the immune checkpoint inhibitors and targeted therapies such as MAPK pathway inhibitors have brought revolution in the treatment of melanoma (Kozar et al., 2019). However, due to rapid development of resistance, these therapies are deemed highly unsuccessful and unreliable means of treatment for malignant melanoma.

i. **Dacarbazine and Temozolomide**

Dacarbazine (DTIC) also known as imidazole carboxamide, is an alkylating chemotherapeutic agent and only FDA approved drug to treat melanoma. Despite its poor efficacy, the standard and combination treatment of metastatic melanoma is still dacarbazine. So far, no other treatment has shown efficient benefit over dacarbazine (Mishra et al., 2018). Temozolomide is an orally administered analog of dacarbazine. (Hofmann et al., 2006; Middleton et al., 2000; Patel et al., 2011) Conventional chemotherapies kill tumors by targeting the cell division or DNA replication. (Link, 2019). When these drugs are metabolically activated, they attack bases of DNA and alkylate them. This prevents the division of rapidly growing cancer cells (Sanada et al., 2004). Bioactivation of dacarbazine to MTIC and then to diazomethane occurs in the liver. Diazomethane is an alkylating agent which primarily targets DNA. It methylates guanine base at the N-7 and O-6 positions (Kewitz et al., 2014). Guanine is one of the four building blocks of DNA called nucleotides. The alkylated strands of DNA stick together in a manner making cell division impossible. This slowing down of cell division affects cancer more than normal cells because cancer cells divide rapidly (Kewitz et al., 2014).

ii. **Immunotherapy**

The intrinsic resistance of melanoma to chemotherapy is circumvented by different approaches. One such approach is immunotherapy, in which different immune-stimulators are administered to enhance the immune response against tumor in melanoma patients (Mishra et al., 2018).

Cancer immunotherapies could be classified as active immunotherapy and passive immunotherapy. Active immunotherapy stimulates intrinsic factors of the immune system of patients. Passive immunotherapy such as adoptive cell transfer and tumor-specific mAbs, carries already active immune factors (Papaioannou et al., 2016). Interleukin-2 (IL-2), monoclonal antibodies (ipilimumab), thymosin α1 and interferon α (IFN-α) are commonly used immunostimulators for melanoma treatment (Chen et al., 2013). Although these treatments show substantial improvement in about 30% of patients with advanced stage diseases, most patients either develop side effects or do not respond to immunotherapie, which make termination of treatment essential (Abdul-Karim & Cowey, 2017).

7.4.2 *Mechanism of Resistance to Chemotherapy in Melanoma*

The drug resistance develops either through primary or intrinsic resistance at the onset or during cytostatic drug application (Kalal et al., 2017) (Figs. 7.4 and 7.5).

i. **Resistance Mediated by Alteration in Enzyme Activation**

The alteration or inactivation of drugs is one of the mechanisms through which drug resistance occurs. Drugs function by binding to their intracellular targets. Once inactivated, they become unavailable for binding with the targets. This inactivation is mostly catalyzed by enzymes. For many alkylating agents used as anticancer drugs,

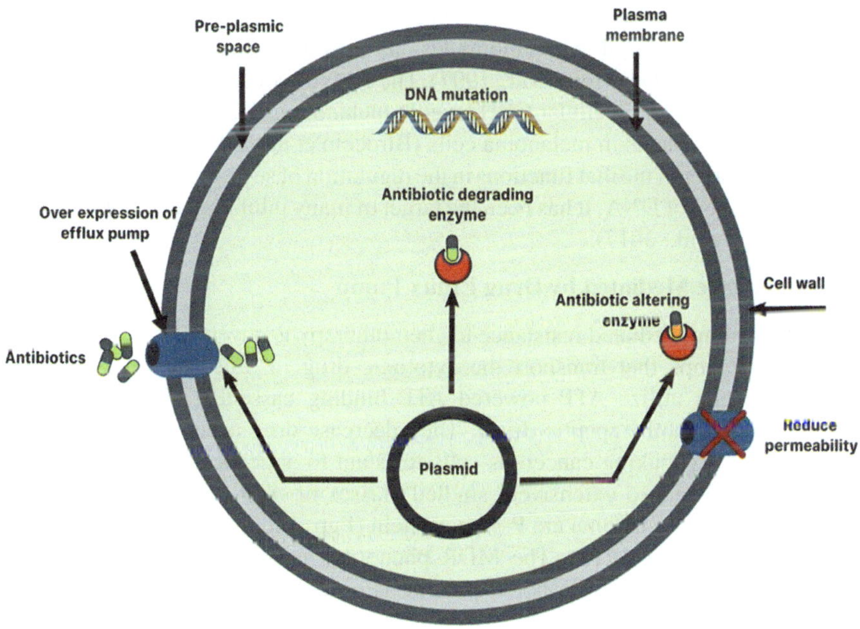

Fig. 7.4 Mechanism of drug resistance in melanoma

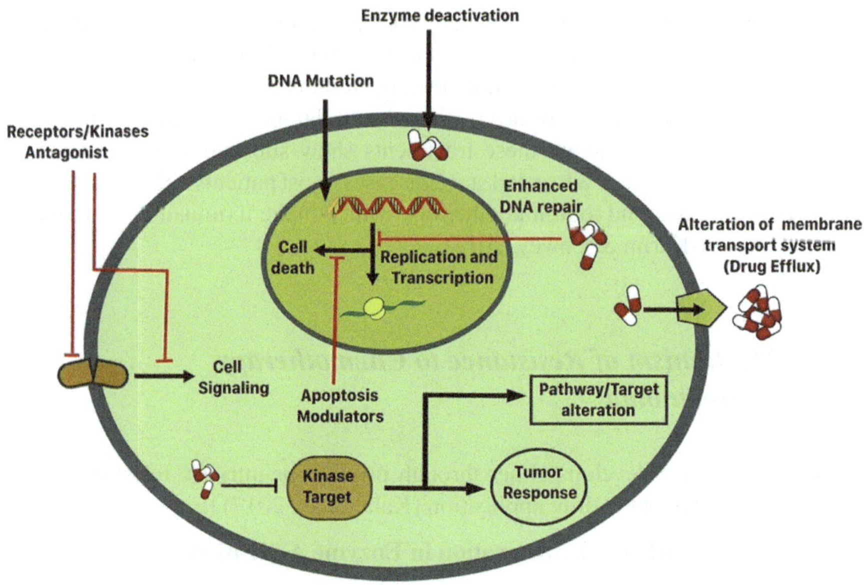

Fig. 7.5 Mechanism of drug resistance in melanoma

the enzyme glutathione-S-transferase (GST) catalyze the conjugation of drug with glutathione (GSH) which is a powerful antioxidant (La, 2009; Townsend & Tew, 2003). GSH inhibits oxidative stress which can damage DNA and RNA. Higher levels of GST is reported in melanoma lesions as compared to benign melanocytes (Mannervik et al., 1987; Moral et al., 1997). The c-Myc-induced apoptosis was having dependence on the intracellular GSH level in melanoma. The down-regulation of c-Myc induces apoptosis in melanoma cells (Biroccio et al., 2002). Topoisomerase is another nuclear enzyme that functions in the regulation of segregation, recombination and transcription of DNA. It has been the target of many inhibitory chemotherapeutic agents (Kalal et al., 2017).

ii. **Resistance Mediated by Drug Efflux Pump**

The efflux-pump mediated resistance to chemotherapy is governed by energy mediated efflux pumps that transport the cytotoxic drug in drug resistant cell lines. Cancerous cells utilize ATP powered ATP binding cassette (ABC) transporters to force out chemotherapeutic drugs. They decrease drug accumulation by mediating drug efflux making cancerous cells resistant to anticancer agents (Lowrence et al., 2019). The two extensively studied classes of ATP-dependent drug transporter proteins in melanoma are P-glycoprotein (Pgp) and the multidrug resistance-associated proteins (MRPs). The MDR phenotype is often associated with over expression of these ABC transporter proteins (Pgp, MRPs) (Hee Choi & Yu, 2014). There are 48 known ABC transporters in humans. The melanoma cells express the key transporters of drug resistance such as ABCC1 (MRP1), ABCB1 (MDR1),

ABCC2 (MRP2),ABCB5, ABCA9, ABCD1, ABCB8 and ABCG2 (MXR). There are 48 recognized human ABC transporters. The melanoma cells also express the well studied targeted mediators of drug resistance such as ABCC1 (MRP1), ABCB1 (MDR1), ABCC2 (MRP2), ABCB5, ABCB8, ABCA9, ABCD1 and ABCG2 (MXR) (Kalal et al., 2017).

iii. **Resistance Mediated by DNA Repair**

DNA repair is one of the established mechanisms of drug resistance. The anticancer drugs damage DNA either directly or/and indirectly. There are mechanisms that can repair the damage of DNA (Borst et al., 2000)f. The chemotherapeutic drugs which are alkylating agents, induce cytotoxic O^6-chloroethylguanine DNA lesions through adducts formation. This results in inter-strand crosslinks in DNA which results in inhibition of replication or transcription. This ultimately arrests the cell cycle in the G2 phase. This repair of damaged DNA mediates a major resistance pathway in tumor cells by impairing the cytotoxic effect of drugs (Fu et al., 2012). Melanoma cells resistant to drugs show an increased DNA repair (Rünger et al., 2000). O^6 methylguanine-DNA methyltransferase (MGMT) is a DNA damage repairing enzyme found in high concentrations in melanoma cells (Middleton et al., 1998).

iv. **Resistance by Modulation of the Apoptotic Pathway**

Apoptosis is a multifactorialbiological process in which many chemical and molecular pathways are involved. There are two well-characterized caspase-activating cascades that regulate caspase-mediated cell death. The extrinsic apoptotic pathway involves initiator caspases (caspase-8/caspase-10), resulting in a proteolytic cascade. The extrinsic pathway is although a primary mediator of cellular immunity, but can also be implicated in the response of tumor cells to drug therapy. The Intrinsic apoptosis involves Apaf-1/Casp9 mitochondrial pathway (Hussein et al., 2003b). The normal melanocytes depend on anti-apoptotic Bcl-2 family members for survival. However, the role of Bcl-2 as a melanoma gene remains controversial. A few reports associate poor survival of metastatic melanoma patients to high Bcl-2 level, however others show Bcl-2 downregulation during progression of melanoma and eliminate Bcl-2 to be used as a prognostic factor (Soengas & Lowe, 2003).

The acquired defects in apoptotic machinery needs extensive research to find the progression of melanoma and associated complications. The mutation of p53, a tumor suppressor gene is known to play a role in drug induced apoptosis and has been linked to chemoresistance. However, melanomas show a low frequency of p53 mutations even though they exhibit extreme chemoresistance (Hussein et al., 2003a; Soengas & Lowe, 2003). Possible explanations could be that p53 does not influence treatment response in melanoma, or loss of p53 increases drug sensitivity or its activity could be disabled by lesions, disrupting other components of the pathway (Soengas & Lowe, 2003).

7.4.3 Resistance to Immunotherapy and Targeted Drug Therapy in Melanoma

In more than 50% of melanoma patients, the MAPK pathway activation via mutations in the serine/threonine kinase BRAF resulted in beginning of targeted therapy with small molecule inhibitors acting on mutated BRAF with overall better results than chemotherapy (Ascierto et al., 2012). The treatment landscape for advanced stage melanoma was revolutionized with these targeted therapies in 2011 with the approval of targeted BRAF inhibitors (BRAFi) vemurafenib and ipilimumab (Olszanski, 2014). Later on, MEK inhibitors (MEKi) were developed. Trametinib was the first MEKi to be approved by FDA in 2013. The combined therapy of MEKi and BRAFi has shown better results than monotherapy (Eroglu & Ribas, 2016). Combining the immune checkpoint therapies and MAPKi therapies in melanoma patients for durable response proved rather insufficient as cancerous cells appear rather unresponsive to immunotherapy. As a result, the benefits of treatment with both these regimes of cancer therapy are under investigation (Kozar et al., 2019). At first all these inhibitors showed outstanding response with significant reduction of tumor burden, but the development of drug resistance jeopardized the long-term success of these therapies.

Although BRAFi and MEKi delays cell proliferation in cells mutated for the BRAF and MAPK pathway by reducing ERK activation through efficiently inhibiting the MAPK pathway. In up to 80% of BRAFi-resistant tumours, MAPK pathway reactivation occurs which represent the high dependence of tumour cells on the MAPK pathway as they quickly acclimate to its inhibition (Moriceau et al., 2015). Another commonly activated pathway in melanoma drug resistance is PI3K-mTOR pathway (Kozar et al., 2019). The tumor cells can also adapt to drug –induced stress. In 74% patients treatment with BRAFi monotherapy or in combination with a MEKi showed an increase in autophagy (Levy et al., 2017). Another interesting feature exhibited by melanoma cells, to avoid drugs, is their ability to switch in an epithelial to mesenchymal transition.The most common changes in phenotype are related to the RTK AXL and master transcription factor MITF. These changes involve changes in rate of proliferation, metabolic rewiring and differentiation/dedifferentiation (Levy et al., 2006).

In past decades various anticancer drugs were developed against melanoma but these therapies merely gave encouraging outcomes. Although a myriad of resistance mechanisms has been identified yet extensive work is required to deeply understand the molecular, genetic and chemical nature of resistance in melanoma therapy to provide a durable treatment outcome.

7.4.4 *Basal Cell Carcinoma*

Non-melanoma skin cancers (NMSC) mainly consist of basal cell carcinoma (BCC) and squamous cell carcinoma (SCC). Besides Merkel cell carcinoma, adnexal tumors, dermatofibrosarcoma protuberans, angiosarcoma, and cutaneous lymphoma are also considered NMSC (Amaral & Garbe, 2017). Basal cell carcinoma (BCC) is one of the most common skin cancers, which is rising worldwide (Mohan & Chang, 2014). Surgery or radiotherapy are commonly used management strategies for BCC. Although these treatment options cure ≈1–10% of BCCs but recurrence is reported in mismanaged treatment, aggressive carcinomas or carcinoma refractory to standard treatments. In such cases it can become locally advanced or occasionally metastasize. This leads to high morbidity and mortality (Mohan & Chang, 2014; Puig & Berrocal, 2015).

7.4.4.1 Treatment Modalities and Resistance

NMSC can be treated with wide range of available options such asingenol mebutate, 5-fluorouracil, imiquimod, methotrexate, diclofenac, cetuximab, vismodegib, photodynamic therapy and radiotherapy. Although all of these treatments show a high histological and clinical response rate but some tumours do not respond to these drugs due to primary or acquired resistance (Gracia-Cazana et al., 2016). Photodynamic Therapy (PDT) is also used to treat Basal cell carcinoma (BCC). Variable outcome has been with improved results after pretreatment and repeated PDT cycles but recurrence of BCC is also reported (Griffin & Lear, 2016; Lucena et al., 2019; Mascaraque et al., 2020). The understanding of molecular pathogenesis of BCC has paved the way for the exploration of treatment options for advanced BCC by targeting the Hedgehog (Hh) signalling pathway inhibitors. Sonidegib(Odomzo) and Vismodegib (Erivedge) are targeted drugs which are used to treat a few recurrent or advanced BCC in skin (Cowey, 2013; Danhof et al., 2018).

Resistance whether primary/intrinsic (does not respond to treatment) or secondary/acquired (regrowth of tumor), is a knownhurdle inthe targeted anticancer therapies. One mechanism of resistance against PDT treatment is alteration in expression of proteins associated with cell death such as P53. The activation of WNT/β-catenin pathway, presence of cancer stem cells and epithelial to mesenchymal transition are other active mechanisms involved in acquisition of resistance against PTD therapy (Lucena et al., 2019).

7.5 Herpes Simplex Virus (HSV)

Herpes simplex virus (HSV) is an important viral pathogen of humans of the herpesviridae family. It causes a variety of infections commonly known as herpes, from mild oral cavity related mucocutaneous lesions, or genital infections to life threatening complications (Whitle & Roizman, 2001). HSV are worldwide in distribution and 60–95% of the adult population is affected. This family of viruses includes eight members which are grouped into three subfamilies (α, β, and γ) based on genomic and biological similarities. The α-herpes viruses of humans include herpes simplex viruses (HSV-1, HSV-2) and varicella zoster virus (VZV) (Jiang et al, 2016). The virus enters neuroglia causing persistent and latent infection from which it can reactivate.

Infections with herpes viruses are mostly subclinical. Common symptoms include ulcerative lesions and rarely does it lead to other complications such as encephalitis, blindness, keratitis or meningitis. This happens mostly in immunocompromised people (Whitley & Roizman, 2001). In oral herpes, the transmission of HSV-1 occurs primarily via oral to oral contact causing infection in or around the mouth (WHO). Although it can also be transmitted via apparently normal skin or oral surface, the risk of transmission is higher from active sores. HSV-1 infections remain life long and are usually acquired during childhood. HSV-1 could transmit from apparently normal skin or oral surfaces, even without any symptoms. This becomes further risky when there are active sores. Most HSV-1 infections are acquired during childhood, and infection is lifelong (Who-herpes). HSV infection is reported to add to the burden of human immunodeficiency virus HIV by increasing the risk for developing HIV infection. Cytomegalovirus, HHV-6, and HHV-7 are common human β-herpes viruses. Kaposi's sarcoma and Epstein–Barr virus (EBV) are common γ-herpes viruses (Jiang et al, 2016).For detection of HSV-1 and HSV-2, serological tests centered on use of type-specific immunoassays are used that detect antibodies to these viruses. These tests help in detection of both symptomatic and asymptomatic infections (Pebody et al. 2004). For treatment of HSV infections, the acyclovir (ACV) and associated drugs proved very successful (De Clercq, 2013). The prolonged use of these drugs has resulted in drug resistant strains (Piret & Boivin, 2011; Toriyama et al, 2014).

7.5.1 Treatment Modalities of HSV

The licensed drugs for HSV infections can be divided into 3 broad classes. The acyclic guanosine analogues (Acyclovir, Ganciclovir, Penciclovir, Valaciclovir, Valganciclovir, Famciclovir) (Bedel et al, 2012; Cattamanchi et al, 2011; Drake et al, 2012; Janoly-Dumenil et al, 2012; Ogungbenro et al, 2009), acyclic nucleotide analogues (Cidofovir, Adefovir dipivoxil) (Gokulgandhi et al, 2012; Momper et al, 2013; Vossen et al, 2014), and pyrophosphate analogues (Foscarnet) (Gilbert et al, 2011). All the major licensed anti-HSV drugs share a common mode of action. While each drug has a different pathway to its target they all act to inhibit the action of viral DNA polymerase by preventing its synthesis (James & Prichard, 2014). First line treatment of HSV infection includes a series of anti-HSV nucleoside drugs, such as ganciclovir, valacyclovir (VCV) and famciclovir. These drugs target viral DNA replication (Jiang et al, 2016).

The second line therapies include pyrophosphate analog, the acyclic nucleoside phosphonate analog, cidofovir and foscarnet (James & Prichard, 2014).

7.5.2 Acyclovir—A Nucleoside Analog

Acyclovir (ACV) is the standard antiviraltreatment of HSV infections. ACV is a highly effective acyclic purine nucleoside analog based antiviral drug against HSV-1 HSV-2 and varicella zoster virus. It has very low cytotoxicity against normal host cells (Elion, 1982).

7.5.2.1 Mechanism of Action (Nucleoside Analog)

ACV is a deoxyguanosine analog. To function as the viral DNA polymerase inhibitor, itundergoes phosphorylation (monophosphate) inside the HSV-infected cell (Elion, 1982) (Fig. 7.6). The first step of phosphorylation of ACV within infected cells occurs by action of HSV thymidine kinase (HSV-TK) which phosphorylates to the 5′-monophosphate. The ACV has a greater affinity (~200 times) to human thymidine kinases than viral thymidine kinases. This property of ACV makes it remarkably safe in HSV treatment (De Clercq & Holý, 2005). The second and third steps of phosphorylation occur via cellular kinases. First there is progressive conversion of monophosphate to the 5′-diphosphate and finally into an active form of ACV triphosphate derivatives (ACV-TP) (Hornig & Mecgregor, 2014; Reardon & Spector, 1989). This active form of drug competes with naturally occurring nucleoside triphosphates. As it replicates, the active triphosphate is incorporated into the growing DNA chain because of the absence of 3′ prime hydroxyl, which results in chain termination. Valacyclovir is an oral prodrug of ACV with improved bioavailability (De Clercq & Holý, 2005; Hornig and Mecgregor, 2014; Piret et al., 2011). As the antiviral action of ACV needs two viral enzymes, the TK and HSV polymerase, inhibition or failure of any of the two processes or both may lead to resistance against the drug (Jiang et al, 2016).

Penciclovir is another acyclic guanosine analog. It also acts via a TK dependent pathway of phosphorylation. The active penciclovir-triphosphate is different from ACV triphosphate as it allows a limited amount of continued chain elongation. It has a 30 hydroxyl group on acyclic side chains which allow elongation to be continued at a limited pace (Hornig and Mecgregor, 2014; Piret et al., 2011).

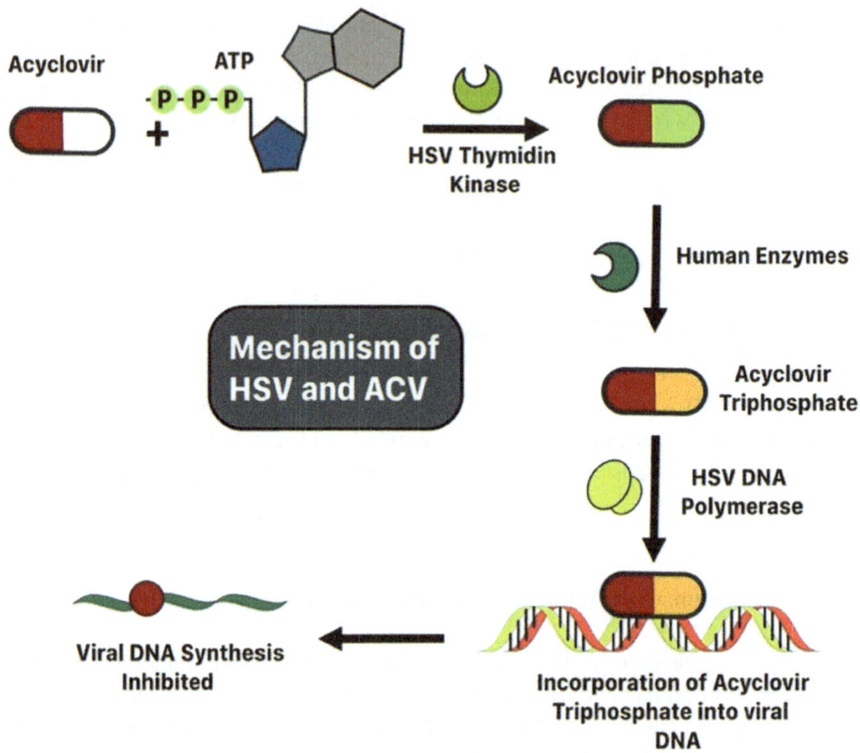

Fig. 7.6 Mechanism of Acyclovir in HSV

7.5.3 Mechanism of Viral Resistance

A prolonged treatment with ACV and its derivatives may lead to resistance against the drug (Jiang et al, 2016) (Fig. 7.7). Graft recipients and cancer patients face recurring and persistent infection by many classes of herpes viruses such as human cytomegalovirus, herpes simplex viruses varicella-zoster virus (Jenkins et al. 2003; Yenikomshian et al. 2015). A higher prevalence of (4–10%) ACV-resistant HSV infections has been recorded in immunocompromised patients. Also the transplant recipients of hematopoietic stem cells sustain a higher risk of developing resistance against anti HSV drugs (Piret et al., 2011). There are three reported active mechanisms of resistance against ACV by HSV. First one occurs via decreased production of viral TK. The second works by complete absence in activity of viral TK activity. The third mechanism is governed by change in substrate specificity of viral TK protein and DNA polymerase.

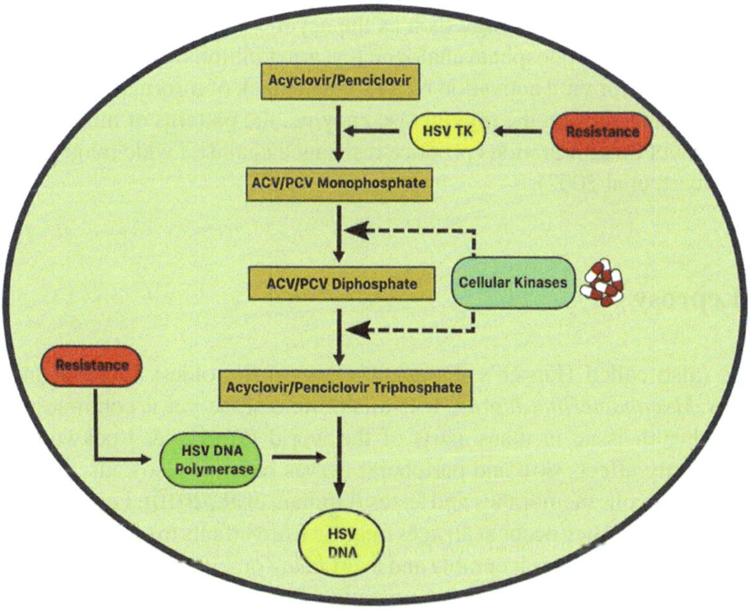

Fig. 7.7 Mechanism of drug resistance of HSV drugs

7.5.3.1 Resistance Due to Altered or Deficient Viral TK

The primary mechanism for acquisition of resistance to ACV and other nucleoside analogues by HSV is associated with changes in viral TK which prevents the drug from undergoing the preliminary phosphorylation which is required for activation of antiviral activity of the drug (Burrel et al, 2012; Piret & Boivin, 2014; Schubert et al, 2014). This usually is the result of mutant strains of HSV which have a deficiency of viral TK enzyme or no functional enzyme at all. The deficiency in production of enzyme is due to a chain terminating mutation which produces truncated polypeptide chains of viral TK (Field & Biron, 1994). The activating/phosphorylating genes responsible for conferring resistance are TK and *UL23* kinase and the gene for viral DNA polymerase is *UL30*. TK deficiency has been reported in 95% of clinical isolates of ACV- resistant HSV (Morfin & Thouvenot, 2003; Pan & Coen, 2012; Watson-Jones et al., 2010).

7.5.3.2 Altered HSV DNA Polymerase Mediated Resistance

The drug-resistant mutants of HSV polymerase were used to study structure, function and relationship of domains within polymerase gene and possible pathogenicity of virus in animals till 1989 (Collins et al., 1989; Field & Coen, 1986). Presently used second line anti HSV drugs such as the acyclic nucleoside, phosphate derivative cidofovir and pyrophosphate analogue foscarnet inhibits viral DNA polymerase, independent of prior viral activation by TK. Due to lack of information on the particular drug binding sites on the polymerase enzyme, the patterns of mutations are not predictable, yet these mutations produce resistance against a wide range of antiviral drugs (Gilbert, et al 2002).

7.6 Leprosy

Leprosy, (also called Hansen's disease), is one of the oldest bacterial infections caused by *Mycobacterium leprae.* It is a curable disease yet it continues to be an important health issue in many parts of the world (Walker & Lockwood, 2006). Leprosy mainly affects skin and peripheral nerves but may also cause infections in bones, eyes, mucous membranes and testes (Graham et al., 2010). Leprosy infections are not age specific, they occur at all ages ranging from infants to old age people. Early treatment of leprosy makes it curable and avert many other associated complications.

In the skin, *M. leprae* has an empathy for macrophages, histiocytes and keratinocytes and plays a vital role in the release of M. leprae responding antimicrobial peptide beta-defensin (Cogen et al., 2012; Polycarpou et al., 2013). Leprosy is one of the diseases, whose development is attributed to the genetic predisposition of the host. Studies suggest that host (human) genetic factors play a role in the acquisition of leprosy, pathogenesis and its clinical course (Alter et al., 2008).

Upon the entry of *M. leprae*, Schwann cells start to differentiate into immature cells via activation of the Erk1/2 signaling pathway. This alteration produces a favourable environment for bacterial proliferation. This dedifferentiation somehow reprograms the Schawn cells into cells with "stem cells" like property conferring a wide range of possible dedifferentiation capabilities such as dedifferentiation into mesenchymal cells. The skin infected with M. leprae causes many dermatological manifestations of leprosy and the infected nerve endings generates axonaldemyelination and dysfunction. This leads to loss of sensation and consequently to deformities and disabilities (White & Franco-Paredes, 2015).

7.6.1 Treatment Modalities

Dapsone, rifampin, and clofazimine are anti-leprosy drugs included in Multi Drug Treatment MDT in the standard recommended by WHO. Other treatment regimens include fluoroquinolones, macrolides and tetracyclines.

7.6.1.1 Dapsone

Dapsone is a sulfonate drug that acts as a competitive inhibitor of p- aminobenzoic acid (PBA) by attacking a crucial enzyme in the bacterial folate biosynthesis pathway called the dihydropteroate synthase (DHPS) (Scollard et al. 2006). The PABA is an important substrate for biosynthesis of folate.This inhibits biosynthesis of folate. This causes depletion of guanosine, thymidine and adenosine pools (Kulkarni & Seydel, 1983; Williams et al., 2000).

7.6.1.2 Rifampin

Rifampin, is actually a derivative of rifamycin, (3-(4-methyl-1-piperazinyl)-imino-methylrifamycin) and a very important drug of WHO recommended MDT for leprosy. At 37 C, it specifically inhibits RANA polymerase (bacterial), that is a key enzyme for the transcription of DNA. With a binding constant of 10–9 M, it forms a very stable enzyme-drug complex.

Resistance to rifampin occurs through mutations, which leads to structural changes in the RNA polymerase β subunit. A number of RNA polymerases exist which have been found sensitive to rifampin at various degrees. Most of the newly identified drugs against Mycobacterium leprae, belong to the macrolide, fluoroquinolone and cycline drug families. Another newly identified drug moxifloxacin, that is rifapentine, fluoroquinolone and rifamycin derivative, has shown promising anti-leprosy effects in mouse model (Grosset, 2001).

7.6.2 Mechanism of Action of Dapsone

Dapsone stops the conversionof 7,8-dyhydro-6-hydroxymethylpterin-pyrophosphate and para-aminobenzoate (PABA) into 7,8-dihydrofolic acid. In this process, the key enzyme is dihydropteroate synthase (DHPS) which is encoded by the folP gene. This dapsone-induced inhibition stops the condensation and reduce the production tetrahydrofolate, that is one of the important components of nucleic acid biosynthesis in *M. leprae*. Unfortunately, there is very little or no antibacterial activity, due to dapsone's strong bacteriostasis (Holmes & Hilson., 1972). Over the period, dapsone became an effective treatment of leprosy, however within few decades, emergence of antibiotics resistance was seen among the leprosy patients. Resistance to this drug resulted mounting relapse cases, mainly due to inappropriate monotherapy (secondary resistance). Consequently, dapsone-resistant leprosy got transmitted to susceptible persons and caused new cases of dapsone-resistant leprosy (primary resistance) (Lowe, 1950; Pearson et al., 1975).

7.6.3 Drug Resistance in M. leprae

From the 1950s to 1970, dapsone was used as standard therapy against *M. leprae.* In 1970, the long term monotherapy with dapsone led to evolution of *M. leprae* strains highly resistant to the drug impeding the control programs of leprosy (Shepard et al., 1986; Vaishnavi et al., 1985). During 1960, new drugs for leprosy such as rifampin and clofazimine were already introduced which were successfully administered as antimicrobial drugs. But sadly, the over use of rifampin in combination with dapsone or alone led to development of resistance against once very successful treatment (William & Gillis, 2004).

To overcome this worldwide resistance against existing therapies, in the 1980s WHO recommended multidrug therapy (MDT). Since then administration of formations of dapsone, rifampin, clofazimine, ofloxacin and minocycline for six (paucibacillary leprosy) to twelve months (multibacillary leprosy) has proved successful in combating leprosy and the prevalence rate has fallen all around the world (Carsaladeet al., 1997; WHO, Leprosy elimination) yet the resistance is still prevalent (Lavania et al., 2018; Reibel et al., 2015). Recently, studies reported emergence of new-quinolone resistant *M. leprae* strains (Kim et al., 2011).

7.6.4 Mechanism of Resistance

Most of the knowledge of *M. leprae* is based on studies conducted on Mycobacterium tuberculosis. The drug resistance mechanism and transmission can be monitored by genotyping and molecular characterization. The *M. Laprae* acquires resistance via adaptive evolution. The major mechanisms through which the bacteria acquire resistance include spontaneous mutation due to error in replication of DNA which leads to chromosomal mutations of genes encoding target drugs. The misuse of the drugs/therapy further augments these mutations (William & Gillis, 2004). *M. leprae* has not has not been cultured in-vitro (Gutierrez et al., 2009).

7.7 Conclusion

Skin diseases are one of the leading causes of disability and contribute significantly to the global burden of diseases. The chemotherapies have always been a mainstay agent for treatment of all bacterial, and viral diseases and cancers. The rapid emergence of drug resistance to almost all once effective treatments in skin diseases is on rise either due to long- term use of these drugs or misuse such as discontinued therapy. These prolonged therapies are affecting the normal human micro-biome, thus extending the scale of damage. They alter the genetic components of either the host or the pathogen by producing mutations. The drug resistance in human skin cancer, the multidrug resistant bacteria such as *S. aureus, Propionobacteriu acnes, Mycobacterium leprae,* and the drug resistant viruses such as *Herpes viruses* are only few pearls from the vast ocean of resistant organisms in various diseases.

Although new drugs have been regularly introduced, there is an urgent need to devise potential strategies and explore new drugs. The search for new treatment modalities could be successful only if we have a complete understanding of the underlying mechanism of resistance. Also, there is an urgent need to identify new drug target sites in resistant organisms. The spread of existing resistance could be controlled by undertaking extensive research and understanding the underlying molecular mechanism of resistance.

References

Abdul-Karim, R. M., & Cowey, C. L. (2017). Challenging the standard of care in advanced melanoma: Focus on pembrolizumab. *Cancer Management and Research, 9*, 433.

Alter, A., Alcaïs, A., Abel, L., & Schurr, E. (2008). Leprosy as a genetic model for susceptibility to common infectious diseases. *Human Genetics, 123*, 227–235.

Amaral, T., & Garbe, C. (2017). Non-melanoma skin cancer: new and future synthetic drug treatments. *Expert Opinion on Pharmacotherapy, 18*, 689–699.

Andrade, J. C., Morais-Braga, M. F. B., Guedes, G. M., Tintino, S. R., Freitas, M. A., Menezes, I. R., & Coutinho, H. D. (2014). Enhancement of the antibiotic activity of aminoglycosides by alpha-tocopherol and other cholesterol derivates. *Biomedicine & Pharmacotherapy, 68*, 1065–1069.

Anuj, S. A., Gajera, H. P., Hirpara, D. G., & Golakiya, B. A. (2019). Interruption in membrane permeability of drug-resistant Staphylococcus aureus with cationic particles of nano-silver. *European Journal of Pharmaceutical Sciences, 127*, 208–216.

Arias, C. A., & Murray, B. E. (2009). Antibiotic-resistant bugs in the 21st century—a clinical super-challenge. *New England Journal of Medicine, 360*, 439–443.

Ascierto, P. A., Kirkwood, J. M., Grob, J.-J., Simeone, E., Grimaldi, A. M., Maio, M., Palmieri, G., Testori, A., Marincola, F. M., & Mozzillo, N. (2012). The role of BRAF V600 mutation in melanoma. *Journal of Translational Medicine, 10*, 1–9.

Autiero, I., Costantini, S., & Colonna, G. (2009). Modeling of the bacterial mechanism of methicillin-resistance by a systems biology approach. *PLoS ONE, 4*, e6226.

Bacon, T. H. et al. (2003). Herpes simplex virus resistance to acyclovir and penciclovir after two decades of antiviral therapy. *Clinical Microbiology Reviews, 16*(1), 114–128. https://doi.org/10.1128/CMR.16.1.114-128.2003

Barbieri, J. S., Bhate, K., Hartnett, K. P., Fleming-Dutra, K. E., & Margolis, D. J. (2019). Trends in oral antibiotic prescription in dermatology, 2008 to 2016. *JAMA Dermatology, 155*, 290–297.

Beceiro, A., Tomás, M., & Bou, G. (2013). Antimicrobial resistance and virulence: a successful or deleterious association in the bacterial world? *Clinical Microbiology Reviews, 26*, 185–230.

Bedel, A. N., Hemmelgarn, T. S., & Kohli, R. (2012). Retrospective review of the incidence of cytomegalovirus infection and disease after liver transplantation in pediatric patients: Comparison of prophylactic oral ganciclovir and oral valganciclovir. *Liver Transplantation, 18*, 347–354.

Biroccio, A., Benassi, B., Filomeni, G., Amodei, S., Marchini, S., Chiorino, G., Rotilio, G., Zupi, G., & Ciriolo, M. R. (2002). Glutathione influences c-Myc-induced apoptosis in M14 human melanoma cells. *Journal of Biological Chemistry, 277*, 43763–43770.

Blair, J. M., Richmond, G. E., & Piddock, L. J. (2014). Multidrug efflux pumps in Gram-negative bacteria and their role in antibiotic resistance. *Future Microbiology, 9*, 1165–1177.

Blair, J. M., Webber, M. A., Baylay, A. J., Ogbolu, D. O., & Piddock, L. J. (2015). Molecular mechanisms of antibiotic resistance. *Nature reviews microbiology, 13*(1), 42–51.
Borst, P., Evers, R., Kool, M., & Wijnholds, J. (2000). A family of drug transporters: the multidrug resistance-associated proteins. *Journal of the National Cancer Institute, 92*, 1295–1302.
Boyers, L. N., Karimkhani, C., Naghavi, M., Sherwood, D., Margolis, D. J., Hay, R. J., Williams, H. C., Naldi, L., Coffeng, L. E., & Weinstock, M. A. (2014). Global mortality from conditions with skin manifestations. *Journal of the American Academy of Dermatology, 71*(1137–1143), e17.
Brown, S. K., & Shalita, A. R. (1998). Acne vulgaris. *The Lancet, 351*, 1871–1876.
Burrel, S., Bonnafous, P., Hubacek, P., Agut, H., & Boutolleau, D. (2012). Impact of novel mutations of herpes simplex virus 1 and 2 thymidine kinases on acyclovir phosphorylation activity. *Antiviral Research, 96*, 386–390.
Bush, K., & Jacoby, G. A. (2010). Updated functional classification of β-lactamases. *Antimicrobial Agents and Chemotherapy, 54*, 969–976.
Cardona, A. F., & Wilson, S. E. (2015). Skin and soft-tissue infections: a critical review and the role of telavancin in their treatment. *Clinical Infectious Diseases, 61*, S69–S78.
Casas, A., di Venosa, G., Hasan, T., & Batlle, A. (2011). Mechanisms of resistance to photodynamic therapy. *Current Medicinal Chemistry, 18*, 2486–2515.
Cattamanchi, A., Saracino, M., Selke, S., Huang, M. L., Magaret, A., Celum, C., Corey, L., Wald, A., & Casper, C. (2011). Treatment with valacyclovir, famciclovir, or antiretrovirals reduces human herpesvirus-8 replication in HIV-1 seropositive men. *Journal of Medical Virology, 83*, 1696–1703.
Chambers, H. F. (1997). Methicillin resistance in staphylococci: molecular and biochemical basis and clinical implications. *Clinical Microbiology Reviews, 10*, 781–791.
Chancey, S. T., Zähner, D., & Stephens, D. S. (2012). Acquired inducible antimicrobial resistance in Gram-positive bacteria. *Future Microbiology, 7*, 959–978.
Chen, J., Shao, R., Zhang, X. D., & Chen, C. (2013). Applications of nanotechnology for melanoma treatment, diagnosis, and theranostics. *International Journal of Nanomedicine, 8*, 2677.
Chopra, I. et al. (1992). Tetracyclines, molecular and clinical aspects. *The Journal of Antimicrobial Chemotherapy, 29*(3), 245–277. https://doi.org/10.1093/jac/29.3.245.
Chopra, I., & Roberts, M. (2001). Tetracycline antibiotics: Mode of action, applications, molecular biology, and epidemiology of bacterial resistance. *Microbiology and Molecular Biology Reviews, 65*, 232–260.
Coates, R., Moran, J., & Horsburgh, M. J. (2014). Staphylococci: colonizers and pathogens of human skin. *Future Microbiology, 9*, 75–91.
Cogen, A., Nizet, V., & Gallo, R. (2008). Skin microbiota: A source of disease or defence? *British Journal of Dermatology, 158*, 442–455.
Cogen, A. L., Walker, S. L., Roberts, C. H., Hagge, D. A., Neupane, K. D., Khadge, S., & Lockwood, D. N. (2012). Human beta-defensin 3 is up-regulated in cutaneous leprosy type 1 reactions. *PLoS Neglected Tropical Diseases, 6*, e1869.
Collins, P., Larder, B., Oliver, N., Kemp, S., Smith, I., & Darby, G. (1989). Characterization of a DNA polymerase mutant of herpes simplex virus from a severely immunocompromised patient receiving acyclovir. *Journal of General Virology, 70*, 375–382.
Cooper, A. J. (1998). Systematic review of Propionibacterium acnes resistance to systemic antibiotics. *Medical Journal of Australia, 169*, 259–261. https://doi.org/10.5694/j.1326-5377.1998.tb140250.x.
Cowey, C. L. (2013). Targeted therapy for advanced Basal-cell carcinoma: vismodegib and beyond. *Dermatology and Therapy, 3*, 17–31.
Danhof, R., Lewis, K., & Brown, M. (2018). Small molecule inhibitors of the hedgehog pathway in the treatment of basal cell carcinoma of the skin. *American Journal of Clinical Dermatology, 19*, 195–207.
Davies, J., & Davies, D. (2010). Resistance origins and evolution of antibiotic. *Microbiology and Molecular Biology Reviews, 74*, 417–433.

de Clercq, E., & Holý, A. (2005). Acyclic nucleoside phosphonates: a key class of antiviral drugs. *Nature Reviews Drug Discovery, 4*, 928–940.
de Carsalade, G.-Y., Wallach, D., Spindler, E., & Pennec, J. (1997). Daily multidrug therapy for leprosy; results of a fourteen-year experience. *International Journal of Leprosy and Other Mycobacterial Diseases, 65*, 37.
De Clercq, E. (2013). Antivirals: past, present and future. *Biochemical Pharmacology, 85*, 727–744.
del Rosso, J. Q., Kircik, L., & Gallagher, C. J. (2015). Comparative efficacy and tolerability of dapsone 5% gel in adult versus adolescent females with acne vulgaris. *The Journal of Clinical and Aesthetic Dermatology, 8*, 31.
Diekema, D., Pfaller, M., Schmitz, F., Smayevsky, J., Bell, J., Jones, R., Beach, M., & Group, S. P. (2001). Survey of infections due to Staphylococcus species: frequency of occurrence and antimicrobial susceptibility of isolates collected in the United States, Canada, Latin America, Europe, and the Western Pacific region for the SENTRY Antimicrobial Surveillance Program, 1997–1999. *Clinical Infectious Diseases, 32*, S114–S132.
Drake, A. L., Roxby, A. C., Ongecha-Owuor, F., Kiarie, J., John-Stewart, G., Wald, A., Richardson, B. A., Hitti, J., Overbaugh, J., & Emery, S. (2012). Valacyclovir suppressive therapy reduces plasma and breast milk HIV-1 RNA levels during pregnancy and postpartum: a randomized trial. *Journal of Infectious Diseases, 205*, 366–375.
Dréno, B., Pécastaings, S., Corvec, S., Veraldi, S., Khammari, A., & Roques, C. (2018). Cutibacterium acnes (Propionibacterium acnes) and acne vulgaris: a brief look at the latest updates. *Journal of the European Academy of Dermatology and Venereology, 32*, 5–14.
DRÉNO, B. (2016). Bacteriological resistance in acne: a call to action. *European Journal of Dermatology*, 26, 127–132.
Eady, E. A. et al. (1989). Erythromycin resistant propionibacteria in antibiotic treated acne patients: association with therapeutic failure. *The British Journal of Dermatology, 121*(1), 51–57. https://doi.org/10.1111/j.1365-2133.1989.tb01399.x.
Eady, E. A. (1998). Bacterial resistance in acne. *Dermatology, 196*, 59–66.
Eady, E. A., Cove, J., Holland, K., & Cunliffe, W. (1989). Erythromycin resistant propionibacteria in antibiotic treated acne patients: association with therapeutic failure. *British Journal of Dermatology, 121*, 51–57.
Eady, E., Jones, C., Gardner, K., Taylor, J., Cove, J., & Cunliffe, W. (1993). Tetracycline-resistant propionibacteria from acne patients are cross-resistant to doxycycline, but sensitive to minocycline. *British Journal of Dermatology, 128*, 556–560.
Eady, E., Bojar, R., Jones, C., Cove, J., Holland, K., & Cunliffe, W. (1996). The effects of acne treatment with a combination of benzoyl peroxide and erythromycin on skin carriage of erythromycin resistant propionibacteria. *British Journal of Dermatology, 134*, 107–113.
Ekwueme, D. U., Guy JR, G. P., Li, C., Rim, S. H., Parelkar, P., & Chen, S. C. (2011). The health burden and economic costs of cutaneous melanoma mortality by race/ethnicity–United States, 2000 to 2006. *Journal of the American Academy of Dermatology, 65*, S133. e1-S133. e12.
Elion, G. B. (1982). Mechanism of action and selectivity of acyclovir. *The American journal of medicine, 73*, 7–13.
Eroglu, Z., & Ribas, A. (2016). Combination therapy with BRAF and MEK inhibitors for melanoma: latest evidence and place in therapy. *Therapeutic Advances in Medical Oncology, 8*, 48–56.
Field, A. K., & Biron, K. K. (1994). The end of innocence" revisited: Resistance of herpesviruses to antiviral drugs. *Clinical Microbiology Reviews, 7*, 1–13.
Field, H., & Coen, D. M. (1986). Pathogenicity of herpes simplex virus mutants containing drug resistance mutations in the viral DNA polymerase gene. *Journal of Virology, 60*, 286–289.
Fleming-Dutra, K. E., Hersh, A. L., Shapiro, D. J., Bartoces, M., Enns, E. A., File, T. M., Finkelstein, J. A., Gerber, J. S., Hyun, D. Y., & Linder, J. A. (2016). Prevalence of inappropriate antibiotic prescriptions among US ambulatory care visits, 2010–2011. *JAMA, 315*, 1864–1873.
Frieri, M., Kumar, K., & Boutin, A. (2017). Antibiotic resistance. *Journal of Infection and Public Health, 10*, 369–378.

Fu, D., Calvo, J. A., & Samson, L. D. (2012). Balancing repair and tolerance of DNA damage caused by alkylating agents. *Nature Reviews Cancer, 12*, 104–120.

GCO. (2020). (Global Cancer Obsevatory, https://gco.iarc.fr/today/data/factsheets/cancers/17-Non-melanoma-skin-cancer-fact-sheet.pdf).

Gilbert, C., Bestman-Smith, J., & Boivin, G. (2002). Resistance of herpesviruses to antiviral drugs: clinical impacts and molecular mechanisms. *Drug Resistance Updates, 5*, 88–114.

Gilbert, C., Azzi, A., Goyette, N., Lin, S.-X., & Boivin, G. (2011). Recombinant phenotyping of cytomegalovirus UL54 mutations that emerged during cell passages in the presence of either ganciclovir or foscarnet. *Antimicrobial Agents and Chemotherapy, 55*, 4019–4027.

Gokulgandhi, M. R., Barot, M., Bagui, M., Pal, D., & Mitra, A. K. (2012). Transporter-targeted lipid prodrugs of cyclic cidofovir: A potential approach for the treatment of cytomegalovirus retinitis. *Journal of Pharmaceutical Sciences, 101*, 3249–3263.

Gollnick, H. P., Zouboulis, C. C., Akamatsu, H., Kurokawa, I., & Schulte, A. (1991). Pathogenesis and pathogenesis related treatment of acne. *The Journal of Dermatology, 18*, 489–499.

Gracia-Cazana, T., Gonzalez, S., & Gilaberte, Y. (2016). Resistance of nonmelanoma skin cancer to nonsurgical treatments. *Part I: Topical Treatments. Actas Dermo-Sifiliográficas (english Edition), 107*, 730–739.

Graham, A., Furlong, S., Margoles, L. M., Owusu, K., & Franco-Paredes, C. (2010). Clinical management of leprosy reactions. *Infectious Diseases in Clinical Practice, 18*, 235–238.

Griffin, L. L., & Lear, J. T. (2016). Photodynamic therapy and non-melanoma skin cancer. *Cancers, 8*, 98.

Grosset, J. H. (2001). Newer drugs in leprosy. International Journal of Leprosy and Other Mycobacterial Diseases, *69*, S14–S18.

Grossman, D., & Altieri, D. C. (2001). Drug resistance in melanoma: mechanisms, apoptosis, and new potential therapeutic targets. *Cancer and Metastasis Reviews, 20*, 3–11.

Guillaume, G., Ledent, V., Moens, W., & Collard, J.-M. (2004). Phylogeny of efflux-mediated tetracycline resistance genes and related proteins revisited. *Microbial Drug Resistance, 10*, 11–26.

Guo, Y., Song, G., Sun, M., Wang, J., & Wang, Y. (2020). Prevalence and Therapies of Antibiotic-Resistance in Staphylococcus aureus. *Frontiers in Cellular and Infection Microbiology, 10*, 107.

Gutierrez, M., Supply, P., & Brosch, R. (2009). *Pathogenomics of mycobacteria*. Karger Publishers.

Guy, R. et al. (1996). Modeling acne in vitro. *The Journal of Investigative Dermatology, 106*(1), 176–182. https://doi.org/10.1111/1523-1747.ep12329907.

Hay, R. J., Johns, N. E., Williams, H. C., Bolliger, I. W., Dellavalle, R. P., Margolis, D. J., Marks, R., Naldi, L., Weinstock, M. A., & Wulf, S. K. (2014). The global burden of skin disease in 2010: an analysis of the prevalence and impact of skin conditions. *Journal of Investigative Dermatology, 134*, 1527–1534.

Hee Choi, Y., & Yu, A. -M. (2014). ABC transporters in multidrug resistance and pharmacokinetics, and strategies for drug development. *Current Pharmaceutical Design, 20*, 793–807.

Hiramatsu, K., Cui, L., Kuroda, M., & Ito, T. (2001). The emergence and evolution of methicillin-resistant Staphylococcus aureus. *Trends in Microbiology, 9*, 486–493.

Hiramatsu, K., Katayama, Y., Matsuo, M., Sasaki, T., Morimoto, Y., Sekiguchi, A., & Baba, T. (2014). Multi-drug-resistant Staphylococcus aureus and future chemotherapy. *Journal of Infection and Chemotherapy, 20*, 593–601.

Hofmann, M., Kiecker, F., Wurm, R., Schlenger, L., Budach, V., Sterry, W., & Trefzer, U. (2006). Temozolomide with or without radiotherapy in melanoma with unresectable brain metastases. *Journal of Neuro-Oncology, 76*, 59–64.

Holmes, I., & Hilson, G. (1972). The effect of rifampicin and dapsone on experimental Mycobacterium leprae infections: Minimum inhibitory concentrations and bactericidal action. *Journal of Medical Microbiology, 5*, 251–261.

Holohan, C., van Schaeybroeck, S., Longley, D. B., & Johnston, P. G. (2013). Cancer drug resistance: an evolving paradigm. *Nature Reviews Cancer, 13*, 714–726.

Hori, S., & Hiramatsu, K. (1994). Mechanisms of antimicrobial resistance. *Nihon rinsho. Japanese Journal of Clinical Medicine, 52*, 327–331.

Hornig, J., & McGregor, A. (2014). Design and development of antivirals and intervention strategies against human herpesviruses using high-throughput approach. *Expert Opinion on Drug Discovery, 9*, 891–915.

Hussein, M. R., Haemel, A. K., & Wood, G. S. (2003a). Apoptosis and melanoma: Molecular mechanisms. *The Journal of Pathology: A Journal of the Pathological Society of Great Britain and Ireland, 199*, 275–288.

Hussein, M., Haemel, A., & Wood, G. (2003b). p53-related pathways and the molecular pathogenesis of melanoma. *European Journal of Cancer Prevention, 12*, 93–100.

Ito, T., Katayama, Y., & Hiramatsu, K. (1999). Cloning and nucleotide sequence determination of the entire mec DNA of pre-methicillin-resistant Staphylococcus aureus N315. *Antimicrobial Agents and Chemotherapy, 43*, 1449–1458.

James, S. H., & Prichard, M. N. (2014). Current and future therapies for herpes simplex virus infections: mechanism of action and drug resistance. *Current Opinion in Virology, 8*, 54–61.

Janoly-Dumenil, A., Rouvet, I., Bleyzac, N., Morfin, F., Zabot, M.-T., & Tod, M. (2012). A pharmacodynamic model of ganciclovir antiviral effect and toxicity for lymphoblastoid cells suggests a new dosing regimen to treat cytomegalovirus infection. *Antimicrobial Agents and Chemotherapy, 56*, 3732–3738.

Jenkins, F. J., Rowe, D.T., Rinaldo, C. R. Jr. (2003). Herpesvirus infections in organ transplant recipients. *Clinical and Vaccine Immunology, 10*(1), 1–7. https://doi.org/10.1128/cdli.10.1.1-7.2003. PMID: 12522031; PMCID: PMC145294.

Jevons, M. P. (1961). "Celbenin"-resistant staphylococci. *British Medical Journal, 1*, 124.

Jiang, Y.-C., Feng, H., Lin, Y.-C., & Guo, X.-R. (2016). New strategies against drug resistance to herpes simplex virus. *International Journal of Oral Science, 8*, 1–6. https://doi.org/10.1038/ijos.2016.3

Kalal, B. S., Upadhya, D., & Pai, V. R. (2017). Chemotherapy resistance mechanisms in advanced skin cancer. *Oncology reviews*, 11.

Kang, S. (2005). The mechanism of action of topical retinoids. *Cutis, 75*, 10.

Karimkhani, C., Dellavalle, R. P., Coffeng, L. E., Flohr, C., Hay, R. J., Langan, S. M., Nsoesie, E. O., Ferrari, A. J., Erskine, H. E., & Silverberg, J. I. (2017). Global skin disease morbidity and mortality: An update from the global burden of disease study 2013. *JAMA Dermatology, 153*, 406–412.

Katzung, B. G. (2017). *Basic and clinical pharmacology 14th Edition*, McGraw Hill Professional.

Kewitz, S., Stiefel, M., Kramm, C. M., & Staege, M. S. (2014). Impact of O6-methylguanine-DNA methyltransferase (MGMT) promoter methylation and MGMT expression on dacarbazine resistance of Hodgkin's lymphoma cells. *Leukemia Research, 38*, 138–143.

Kim, J., Ochoa, M.-T., Krutzik, S. R., Takeuchi, O., Uematsu, S., Legaspi, A. J., Brightbill, H. D., Holland, D., Cunliffe, W. J., & Akira, S. (2002). Activation of toll-like receptor 2 in acne triggers inflammatory cytokine responses. *The Journal of Immunology, 169*, 1535–1541.

Kim, H., Suzuki, H., Matsuoka, M., Matsuba, T., Yokoyama, K., Nakajima, C., & Suzuki, Y. (2011). Molecular mechanism of the acquisition of new-quinolone resistance in Mycobacterium leprae and M. tuberculosis and rapid differentiation methods for resistant bacilli. *Nihon Hansenbyo Gakkai zasshi= Japanese journal of leprosy: official organ of the Japanese Leprosy Association, 80*, 17–27.

Kozar, I., Margue, C., Rothengatter, S., Haan, C., & Kreis, S. (2019). Many ways to resistance: How melanoma cells evade targeted therapies. *Biochimica et Biophysica Acta (BBA)-Reviews on Cancer, 1871*, 313–322.

Krishnan, G. (1976). Comparison of two concentrations of tretinoin solution in the topical treatment of acne vulgaris. *The Practitioner, 216*, 106–109.

Kulkarni, V., & Seydel, J. (1983). Inhibitory activity and mode of action of diaminodiphenylsulfone in cell-free folate-synthesizing systems prepared from Mycobacterium lufu and Mycobacterium leprae. *Chemotherapy, 29*, 58–67.

LA, C. P. (2009). Mechanism of drug sensitivity and resistance in melanoma. *Current Cancer Drug Targets, 9*, 391–397.

Laikova, K. V., Oberemok, V. V., Krasnodubets, A. M., Gal'chinsky, N. V., Useinov, R. Z., Novikov, I. A., Temirova, Z. Z., Gorlov, M. V., Shved, N. A., & Kumeiko, V. V. (2019). Advances in the understanding of skin cancer: Ultraviolet radiation, mutations, and antisense oligonucleotides as anticancer drugs. *Molecules, 24*, 1516.

Lakhundi, S., & Zhang, K. (2018). Methicillin-resistant Staphylococcus aureus: molecular characterization, evolution, and epidemiology. *Clinical Microbiology Reviews,* 31.

Langston, A. A., Redei, I., Caliendo, A. M., Somani, J., Hutcherson, D., Lonial, S., Bucur, S., Cherry, J., Allen, A., & Waller, E. K. (2002). Development of drug-resistant herpes simplex virus infection after haploidentical hematopoietic progenitor cell transplantation. *Blood, the Journal of the American Society of Hematology, 99*, 1085–1088.

Lavania, M., Singh, I., Turankar, R. P., Ahuja, M., Pathak, V., Sengupta, U., Das, L., Kumar, A., Darlong, J., Nathan, R., & Maseey, A., (2018). Molecular detection of multidrug-resistant Mycobacterium leprae from Indian leprosy patients. *Journal of Global Antimicrobial Resistance, 12*, 214–219.

Leiter, U., & Garbe, C. (2008). Epidemiology of melanoma and nonmelanoma skin cancer—the role of sunlight. *Sunlight, vitamin D and skin cancer.* Springer.

Levy, C., Khaled, M., & Fisher, D. E. (2006). MITF: master regulator of melanocyte development and melanoma oncogene. *Trends in Molecular Medicine, 12*, 406–414.

Levy, J. M. M., Zahedi, S., Griesinger, A. M., Morin, A., Davies, K. D., Aisner, D. L., Kleinschmidt-Demasters, B., Fitzwalter, B. E., Goodall, M. L., & Thorburn, J. (2017). Autophagy inhibition overcomes multiple mechanisms of resistance to BRAF inhibition in brain tumors. *eLife, 6*, e19671.

Leyden, J. J., Shalita, A., Thiboutot, D., Washenik, K., & Webster, G. (2005). Topical retinoids in inflammatory acne: a retrospective, investigator-blinded, vehicle-controlled, photographic assessment. *Clinical Therapeutics, 27*, 216–224.

Li, N., Luo, M., Fu, Y. J., Zu, Y. G., Wang, W., Zhang, L., Yao, L. P., Zhao, C. J., & Sun, Y. (2013). Effect of corilagin on membrane permeability of Escherichia coli, Staphylococcus aureus and Candida albicans. *Phytotherapy Research, 27*, 1517–1523.

Link, W. (2019). *Principles of cancer treatment and anticancer drug development*, Springer.

Livermore, D. M. (2000). Antibiotic resistance in staphylococci. *International Journal of Antimicrobial Agents, 16*(Suppl 1), S3–10. https://doi.org/10.1016/s0924-8579(00)00299-5.

Loomba, P. S., Taneja, J., & Mishra, B. (2010). Methicillin and vancomycin resistant S. aureus in hospitalized patients. *Journal of global infectious diseases, 2*, 275.

Lowe, J. (1950). Treatment of leprosy with diamino-diphenyl sulphone by mouth. *Lancet, 258*, 145–150.

Lowrence, R. C., Subramaniapillai, S. G., Ulaganathan, V., & Nagarajan, S. (2019). Tackling drug resistance with efflux pump inhibitors: From bacteria to cancerous cells. *Critical Reviews in Microbiology, 45*, 334–353.

Lucena, S. R., Zamarrón, A., Carrasco, E., Marigil, M. A., Mascaraque, M., Fernández-Guarino, M., Gilaberte, Y., González, S., & Juarranz, A. (2019). Characterisation of resistance mechanisms developed by basal cell carcinoma cells in response to repeated cycles of Photodynamic Therapy. *Scientific Reports, 9*, 1–15.

Lynde, C. W., & Andriessen, A. (2014). Cohort study on the treatment with dapsone 5% gel of mild to moderate inflammatory acne of the face in women. *Skinmed, 12*, 15–21.

Magiorakos, A.-P., Srinivasan, A., Carey, R., Carmeli, Y., Falagas, M., Giske, C., Harbarth, S., Hindler, J., Kahlmeter, G., & Olsson-Liljequist, B. (2012). Multidrug-resistant, extensively drug-resistant and pandrug-resistant bacteria: an international expert proposal for interim standard definitions for acquired resistance. *Clinical Microbiology and Infection, 18*, 268–281.

Mandalà, M., & Romano, E. (2019). Mechanisms of Drug Resistance in Cancer Therapy. ISBN 978-3-030-10507-5. https://www.springer.com/gp/book/9783030105068.

Mannervik, B., Castro, V. M., Danielson, U. H., Tahir, M. K., Hansson, J., & Ringborg, U. (1987). Expression of class Pi glutathione transferase in human malignant melanoma cells. *Carcinogenesis, 8*, 1929–1932.

Margolis, D. J., Bowe, W. P., Hoffstad, O., & Berlin, J. A. (2005). Antibiotic treatment of acne may be associated with upper respiratory tract infections. *Archives of Dermatology, 141*, 1132–1136.

Markley, J. L., & Wencewicz, T. A. (2018). *Tetracycline-Inactivating Enzymes. Frontiers in Microbiology, 9*, 1058.

Mascaraque, M., Delgado-Wicke, P., Nuevo-Tapioles, C., Gracia-Cazaña, T., Abarca-Lachen, E., González, S., Cuezva, J. M., Gilaberte, Y., & Juarranz, Á. (2020). Metformin as an Adjuvant to Photodynamic Therapy in Resistant Basal Cell Carcinoma Cells. *Cancers, 12*, 668.

Matono, T., Nagashima, M., Mezaki, K., Motohashi, A., Kutsuna, S., Hayakawa, K., Ohmagari, N., & Kaku, M. (2018). Molecular epidemiology of β-lactamase production in penicillin-susceptible Staphylococcus aureus under high-susceptibility conditions. *Journal of Infection and Chemotherapy, 24*, 153–155.

McGuinness, W. A., Malachowa, N., & Deleo, F. R. (2017). Focus: infectious diseases: vancomycin resistance in Staphylococcus aureus. *The Yale Journal of Biology and Medicine, 90*, 269.

McLaughlin, J., Watterson, S., Layton, A. M., Bjourson, A. J., Barnard, E., & McDowell, A. (2019). Propionibacterium acnes and acne vulgaris: new insights from the integration of population genetic, multi-omic, biochemical and host-microbe studies. *Microorganisms, 7*, 128.

Meynadier, J., & Alirezai, M. (1998). Systemic antibiotics for acne. *Dermatology, 196*, 135–139.

Middleton, M. R., Lunn, J., Morris, C., Rustin, G., Wedge, S., Brampton, M., Lind, M. J., Lee, S. M., Newell, D., & Bleehen, N. (1998). O 6-methylguanine-DNA methyltransferase in pretreatment tumour biopsies as a predictor of response to temozolomide in melanoma. *British Journal of Cancer, 78*, 1199–1202.

Middleton, M. R., Grob, J., Aaronson, N., Fierlbeck, G., Tilgen, W., Seiter, S., Gore, M., Aamdal, S., Cebon, J., & Coates, A. (2000). Randomized phase III study of temozolomide versus dacarbazine in the treatment of patients with advanced metastatic malignant melanoma. *Journal of Clinical Oncology, 18*, 158–158.

Miller, W. R., Munita, J. M., & Arias, C. A. (2014). Mechanisms of antibiotic resistance in enterococci. *Expert Review of Anti-Infective Therapy, 12*, 1221–1236.

Mishra, H., Mishra, P. K., Ekielski, A., Jaggi, M., Iqbal, Z., & Talegaonkar, S. (2018). Melanoma treatment: from conventional to nanotechnology. *Journal of Cancer Research and Clinical Oncology, 144*, 2283–2302.

Mitscher, L. A. (1978). The chemistry of the tetracycline antibiotics. Medicinal research series 9, Marcel Dekker, New York.

Mohan, S. V., & Chang, A. L. S. (2014). Advanced basal cell carcinoma: Epidemiology and therapeutic innovations. *Current Dermatology Reports, 3*, 40–45.

Momper, J., Zhao, Y., Shapiro, R., Schonder, K., Gao, Y., Randhawa, P., & Venkataramanan, R. (2013). Pharmacokinetics of low-dose cidofovir in kidney transplant recipients with BK virus infection. *Transplant Infectious Disease, 15*, 34–41.

Moral, A., Palou, J., Lafuente, A., Molina, R., Piulachs, J., Castel, T., Trias, M., & Group, M. (1997). Immunohistochemical study of alpha, mu and pi class glutathione S transferase expression in malignant melanoma. *British Journal of Dermatology,* 136**,** 345–350.

Morales, E., Cots, F., Sala, M., Comas, M., Belvis, F., Riu, M., Salvadó, M., Grau, S., Horcajada, J. P., & Montero, M. M. (2012). Hospital costs of nosocomial multi-drug resistant Pseudomonas aeruginosa acquisition. *BMC Health Services Research, 12*, 122.

Morello, A. M., Downing, D. T., & Strauss, J. S. (1976). Octadecadienoic acids in the skin surface lipids of acne patients and normal subjects. *Journal of Investigative Dermatology, 66*.

Morfin, F., & Thouvenot, D. (2003). Herpes simplex virus resistance to antiviral drugs. *Journal of Clinical Virology, 26*, 29–37.

Muthusamy, V., & Piva, T. J. (2010). The UV response of the skin: A review of the MAPK, NFκB and TNFα signal transduction pathways. *Archives of Dermatological Research, 302*, 5.

Nguyen, F., Starosta, A. L., Arenz, S., Sohmen, D., Dönhöfer, A., & Wilson, D. N. (2014). Tetracycline antibiotics and resistance mechanisms. *Biological Chemistry, 395*, 559–575.

Ogungbenro, K., Matthews, I., Looby, M., Kaiser, G., Graham, G., & Aarons, L. (2009). Population pharmacokinetics and optimal design of paediatric studies for famciclovir. *British Journal of Clinical Pharmacology, 68*, 546–560.

Olszanski, A. J. (2014). Current and future roles of targeted therapy and immunotherapy in advanced melanoma. *Journal of Managed Care Pharmacy, 20*, 346–356.

Oztoprak, N., Cevik, M. A., Akinci, E., Korkmaz, M., Erbay, A., Eren, S. S., Balaban, N., & Bodur, H. (2006). Risk factors for ICU-acquired methicillin-resistant Staphylococcus aureus infections. *American Journal of Infection Control, 34*, 1–5.

Pan, D., & Coen, D. M. (2012). Quantification and analysis of thymidine kinase expression from acyclovir-resistant G-string insertion and deletion mutants in herpes simplex virus-infected cells. *Journal of Virology, 86*, 4518–4526.

Papaioannou, N. E., Beniata, O. V., Vitsos, P., Tsitsilonis, O., & Samara, P. (2016). Harnessing the immune system to improve cancer therapy. *Annals of translational medicine,* 4.

Patel, P. M., Suciu, S., Mortier, L., Kruit, W. H., Robert, C., Schadendorf, D., Trefzer, U., Punt, C. J., Dummer, R., & Davidson, N. (2011). Extended schedule, escalated dose temozolomide versus dacarbazine in stage IV melanoma: final results of a randomised phase III study (EORTC 18032). *European Journal of Cancer, 47*, 1476–1483.

Pearson, T., Rees, R., & Waters, M. (1975). Sulphone resistance in leprosy: A review of one hundred proven clinical cases. *The Lancet, 306*, 69–72.

Pebody, R., Andrews, N., Brown, D., Gopal, R., de Melker, H., François, G., Gatcheva, N., Hellenbrand, W., Jokinen, S., & Klavs, I. (2004). The seroepidemiology of herpes simplex virus type 1 and 2 in Europe. *Sexually Transmitted Infections, 80*, 185–191.

Piret, J., & Boivin, G. (2011). Resistance of herpes simplex viruses to nucleoside analogues: mechanisms, prevalence, and management. *Antimicrobial Agents and Chemotherapy, 55*, 459–472.

Piret, J., & Guy, B. (2011). Resistance of herpes simplex viruses to nucleoside analogues: Mechanisms, prevalence, and management. *Antimicrobial Agents and Chemotherapy, 55*(2), 459–472. https://doi.org/10.1128/AAC.00615-10.

Piret, J., & Boivin, G. (2014). Antiviral drug resistance in herpesviruses other than cytomegalovirus. *Reviews in Medical Virology, 24*, 186–218.

Piskin, S., & Uzunali, E. (2007). A review of the use of adapalene for the treatment of acne vulgaris. *Therapeutics and Clinical Risk Management, 3*, 621.

Polycarpou, A., Walker, S. L., & Lockwood, D. N. (2013). New findings in the pathogenesis of leprosy and implications for the management of leprosy. *Current Opinion in Infectious Diseases, 26*, 413–419.

Puhvel, S., Barfatani, M., Warnick, M., & Sternberg, T. (1964). Study of antibody levels to corynebacterium acnes: In the serum of patients with acne vulgaris, using bacterial agglutination, agar gel immunodiffusion, and immunofluorescence techniques. *Archives of Dermatology, 90*, 421–427.

Puig, S., & Berrocal, A. (2015). Management of high-risk and advanced basal cell carcinoma. *Clinical and Translational Oncology, 17*, 497–503.

Rathi, S. K. 2011. Acne vulgaris treatment: the current scenario. *Indian Journal of Dermatology, 56*, 7.

Reardon, J. E., & Spector, T. (1989). Herpes simplex virus type 1 DNA polymerase. Mechanism of inhibition by acyclovir triphosphate. *Journal of Biological Chemistry, 264*, 7405–7411.

Reibel, F., Cambau, E., & Aubry, A. (2015). Update on the epidemiology, diagnosis, and treatment of leprosy. *Medecine Et Maladies Infectieuses, 45*, 383–393.

Reygaert, W. (2009). Methicillin-resistant Staphylococcus aureus (MRSA): molecular aspects of antimicrobial resistance and virulence. *Clinical Laboratory Science, 22*, 115.

Reygaert, W. C. (2018). An overview of the antimicrobial resistance mechanisms of bacteria. *AIMS Microbiology, 4*, 482.

Rogalski, W. (1985). Handbook of experimental pharmacology, edited by JJ Hlavka & JH Boothe, Vol. 78. Springer.

Ross, J. I., Eady, E. A., Cove, J. H., & Cunliffe, W. J. (1998). 16S rRNA mutation associated with tetracycline resistance in a gram-positive bacterium. *Antimicrobial Agents and Chemotherapy, 42*, 1702–1705.

Rünger, T. M., Emmert, S., Schadendorf, D., Diem, C., Epe, B., & Hellfritsch, D. (2000). Alterations of DNA repair in melanoma cell lines resistant to cisplatin, fotemustine, or etoposide. *Journal of Investigative Dermatology, 114*, 34–39.

Sagransky, M., Yentzer, B. A., & Feldman, S. R. (2009). Benzoyl peroxide: A review of its current use in the treatment of acne vulgaris. *Expert Opinion on Pharmacotherapy, 10*, 2555–2562.

Sanada, M., Takagi, Y., Ito, R., & Sekiguchi, M. (2004). Killing and mutagenic actions of dacarbazine, a chemotherapeutic alkylating agent, on human and mouse cells: effects of Mgmt and Mlh1 mutations. *DNA Repair, 3*, 413–420.

Sánchez, D. P. (2003). MDR efflux pumps and antimicrobial resistance. Revista espanola de quimioterapia: publicacion oficial de la Sociedad Espanola de Quimioterapia, 16, 172.

Sardana, K., & Garg, V. K. (2014). Antibiotic resistance in acne: is it time to look beyond antibiotics and P ropionobacterium acnes? *International Journal of Dermatology, 53*, 917–919.

Schubert, A., Gentner, E., Bohn, K., Schwarz, M., Mertens, T., & Sauerbrei, A. (2014). Single nucleotide polymorphisms of thymidine kinase and DNA polymerase genes in clinical herpes simplex virus type 1 isolates associated with different resistance phenotypes. *Antiviral Research, 107*, 16–22.

Scollard, D. M., et al. (2006). The continuing challenges of leprosy. *Clinical Microbiology Reviews, 19*(2), 338–381. https://doi.org/10.1128/CMR.19.2.338-381.2006.

Seth, D., Cheldize, K., Brown, D., & Freeman, E. E. (2017). Global burden of skin disease: inequities and innovations. *Current Dermatology Reports, 6*, 204–210.

Shalita, A. A., Weiss, J., Chalker, D., Ellis, C., Greenspan, A., Katz, H., Kantor, I., Millikan, L., Swinehart, T., & Swinyer, L. (1996). A comparison of the efficacy and safety of adapalene gel 0.1% and tretinoin gel 0.025% in the treatment of acne vulgaris: a multicenter trial. *Journal of the American Academy of Dermatology, 34*, 482–485.

Sharma, A. (2011). Antimicrobial resistance: No action today, no cure tomorrow. *Indian Journal of Medical Microbiology, 29*, 91.

Shepard, C. C., Rees, R. J., Levy, L., Pattyn, S. R., Baohong, J., & Dela Cruz, E. (1986). Susceptibility of strain of Mycobacterium leprae isolated prior to 1977 from patients with previously untreated lepromatous leprosy. *International Journal of Leprosy and Other Mycobacterial Diseases, 54*, 11–15.

Soengas, M. S., & Lowe, S. W. (2003). Apoptosis and melanoma chemoresistance. *Oncogene, 22*, 3138–3151.

Stapleton, P. D., & Taylor, P. W. (2002). Methicillin resistance in Staphylococcus aureus: mechanisms and modulation. *Science Progress, 85*, 57–72.

Tan, A. W., & Tan, H.-H. (2005). Acne vulgaris: a review of antibiotic therapy. *Expert Opinion on Pharmacotherapy, 6*, 409–418.

Thanassi, D. G., Suh, G., & Nikaido, H. (1995). Role of outer membrane barrier in efflux-mediated tetracycline resistance of Escherichia coli. *Journal of Bacteriology, 177*, 998–1007.

Thiboutot, D., Gollnick, H., Bettoli, V., Dréno, B., Kang, S., Leyden, J. J., Shalita, A. R., Lozada, V. T., Berson, D., & Finlay, A. (2009). New insights into the management of acne: An update from the Global Alliance to Improve Outcomes in Acne group. *Journal of the American Academy of Dermatology, 60*, S1–S50.

Toriyama, K., Inoue, T., Suzuki, T., Kobayashi, T., & Ohashi, Y. (2014). Necrotizing keratitis caused by acyclovir-resistant herpes simplex virus. *Case Reports in Ophthalmology, 5*, 325–328.

Townsend, D. M., & Tew, K. D. (2003). The role of glutathione-S-transferase in anti-cancer drug resistance. *Oncogene, 22*, 7369–7375.

Vaishnavi, C., Ganguly, N., Kumar, B., Kaur, S., & Chakravarti, R. (1985). Primary dapsone resistance in leprosy. *Indian Journal of Leprosy, 57*, 507–513.

Varghese, A. (2020). Acne vulgaris and antimicrobial resistance: a review. *Reviews in Medical Microbiology,* 31, 150–154.

Velez, R., & Sloand, E. (2016). Combating antibiotic resistance, mitigating future threats and ongoing initiatives. *Journal of Clinical Nursing, 25*, 1886–1889.

Villagra, N. A., Fuentes, J. A., Jofré, M. R., Hidalgo, A. A., García, P., & Mora, G. C. (2012). The carbon source influences the efflux pump-mediated antimicrobial resistance in clinically important Gram-negative bacteria. *Journal of Antimicrobial Chemotherapy, 67*, 921–927.

Vossen, M. G., Gattringer, K.-B., Jäger, W., Kraff, S., & Thalhammer, F. (2014). Single-dose pharmacokinetics of cidofovir in continuous venovenous hemofiltration. *Antimicrobial Agents and Chemotherapy, 58*, 1952–1955.

Vowels, B. R., Yang, S., & Leyden, J. J. (1995). Induction of proinflammatory cytokines by a soluble factor of Propionibacterium acnes: implications for chronic inflammatory acne. *Infection and Immunity, 63*, 3158–3165.

Walker, S. L., Lockwood, D. N. J. (2006). The clinical and immunological features of leprosy. *British Medical Bulletin, 77–78*(1), 103–121. https://doi.org/10.1093/bmb/ldl010.

Walsh, T. R., Efthimiou, J., & Dréno, B. (2016). Systematic review of antibiotic resistance in acne: an increasing topical and oral threat. *The Lancet Infectious Diseases, 16*, e23–e33.

Watson-Jones, D., Wald, A., Celum, C., Lingappa, J., Weiss, H. A., Changalucha, J., Baisley, K., Tanton, C., Hayes, R. J., & Marshak, J. O. (2010). Use of acyclovir for suppression of human immunodeficiency virus infection is not associated with genotypic evidence of herpes simplex virus type 2 resistance to acyclovir: analysis of specimens from three phase III trials. *Journal of Clinical Microbiology, 48*, 3496–3503.

Whitby, M., McLaws, M. L., & Berry, G. (2001). Risk of death from methicillin-resistant Staphylococcus aureus bacteraemia: a meta-analysis. *Medical Journal of Australia, 175*, 264–267.

White, C., & Franco-Paredes, C. (2015). Leprosy in the 21st century. *Clinical Microbiology Reviews, 28*, 80–94.

Whitley, R. J., & Roizman, B. (2001). *Herpes Simplex Virus Infections. the Lancet, 357*, 1513–1518.

Williams, D. L., Spring, L., Harris, E., Roche, P., & Gillis, T. P. (2000). Dihydropteroate synthase of Mycobacterium leprae and dapsone resistance. *Antimicrobial Agents and Chemotherapy, 44*, 1530–1537.

Williams, D. L., & Gillis, T. P. (2004). Molecular detection of drug resistance in Mycobacterium leprae. *Leprosy review, 75*(2), 118–130.

Wolf, J. (2001). An update of recent clinical trials examining adapalene and acne. *Journal of the European Academy of Dermatology and Venereology, 15*, 23–29.

Yenikomshian, M. A., et al. (2015). The epidemiology of herpes zoster and its complications in Medicare cancer patients. *BMC infectious diseases, 15*(106). https://doi.org/10.1186/s12879-015-0810-6.

Zaenglein, A. L. (2018). Acne vulgaris. *New England Journal of Medicine, 379*, 1343–1352.

Zárate Villarroel, S. G., Morales, P., Świderek, K., Bolanos-Garcia, V. M. & Bastida, A. (2019). A molecular modeling approach to identify novel inhibitors of the major facilitator superfamily of efflux pump transporters.

Chapter 8
Drugs Resistance in Lungs Diseases

Muhammad Salman Sajid, Shafaq Saleem, Fahmida Jabeen, Batool Fatima, Adeela Saeed, Dilshad Hussain, Saadat Majeed, Muhammad Naeem Ashiq, and Muhammad Najam-ul-Haq

Abstract Asthma is projected to be the third cause of death in all communities without age, gender, or race limits, along with Chronic Obstructive Pulmonary Disease (COPD). Patients treated with inhaled corticosteroids, intramuscular triamcinolone, omalizumab, mepolizumab, and dupilumab can live normal lives. Bronchiectasis is a chronic airway condition of epithelial and cartilage damage characterized by irritation, inflammation, sputum expectoration and dilatation of airway walls. Macrolides are medicines composed of twelve or more elements with a macrocyclic lactone ring. Because of their excellent tissue penetration and efficacy against respiratory pathogens, macrolides are the first-line drugs for respiratory infections. Bronchitis causes bronchial swelling and infection of bronchium. Bronchi walls bloat with extra sticky mucus. In- and out- lung circulation is partially blocked. Oral azole antifungal treatment is used for all types of aspergillosis, including allergic bronchopulmonary aspergillosis (ABPA). After myocardial infarction and stroke, pulmonary embolism (PE) is the third most common cause of death in the cardiovascular system. Inorganic nitrite decreases the matrix metalloproteinase-9 in acute PE. During acute PE, L-arginine, and NO donor diethylenetriamine/nonoate are inactive. As an inhaled medicine, ethyl nitrite can induce methemoglobin formation. Shortness of breath at rest, degenerate tachycardia with force, relative hypoxemia, and tachypnea are all symptoms of PE. Peribronchial cuffing (cardiac asthma) may induce wheezing. The treatment for cardiogenic PE has been intravenous morphine. In patients with hypertension and PE, angiogensin translating enzyme inhibitors should lower preload and afterload. Nesiritide raises CO levels while lowering vascular resistance and PCP levels.

M. S. Sajid · D. Hussain · S. Majeed · M. N. Ashiq · M. Najam-ul-Haq (✉)
Institute of Chemical Sciences, Bahauddin Zakariya University, Multan 60800, Pakistan
e-mail: najamulhaq@bzu.edu.pk

S. Saleem · F. Jabeen · A. Saeed
Department of Chemistry, The Women University, Multan 60000, Pakistan

B. Fatima
Department of Biochemistry, Bahauddin Zakariya University, Multan 60800, Pakistan

S. Ahmed et al. (eds.), *Biochemistry of Drug Resistance*,
https://doi.org/10.1007/978-3-030-76320-6_8

8.1 Asthma

Asthma is a common chronic respiratory disease that causes airway hyper responsiveness, reversible airflow obstruction, and airway inflammation produce coughing, wheezing, and breathlessness. Inflammatory cells such as dendritic cells, mast cells, eosinophils, basophils, lymphocytes, and structural cells (fibroblasts, epithelial cells, and smooth muscle cells) interplay complexly in asthma. Within the bronchial mucosa, airway inflammation triggers inflammatory cells which are type 2 T-helper (Th2) cells that secrete interleukin (IL)-4, IL-5, and IL-13, eosinophils, and mast cells (Chung et al., 2014).

Asthma can affect anyone, nevertheless of their age and gender. By 2020, asthma, along with Chronic Obstructive Pulmonary Disease (COPD), is expected to be the third leading cause of death (Hossny et al., 2019). While asthma cannot be cured, it can be managed with proper treatment and people with asthma can live normal lives. If preventive medications are not taken, the lungs become weakened. They may be more sensitive to asthma attacks. Preventer medicines are:

Inhaled corticosteroids: Inhaled corticosteroids go straight to lungs and give fewer side effects than pills. Budesonide (Pulmicort®), fluticasone (Flovent®), ciclesonide (Alvesco®), fluticasone furoate (Arnuity®), mometasone (Asmanex®), and beclomethasone (Qvar®) are few examples.

Advair®: Made of a corticosteroid (fluticasone/Flovent®) plus a Long-Acting Beta2-Agonists (salmeterol/Serevent®).

For the treatment of asthma and chronic obstructive pulmonary disease, inhaled corticosteroids (ICS) and adrenergic agonists (LABA) are safe and long-acting b2-effective anti-inflammatory and bronchodilator agents, respectively. In asthma and COPD, the use of fixed-dose ICS/LABA combination medications recommended as maintenance therapy (Muneswarao et al., 2019). Advair Diskus (GlaxoSmithKline, Research Triangle Park, NC) is an ICS/LABA combination drug (fluticasone propionate [FP]/salmeterol [as xinafoate]; FPS) recommended for asthmatic patients that are not monitored with ICS alone, as well as COPD patients who are at high risk of exacerbations (Haughie et al., 2020).

Symbicort®: Made of a corticosteroid (budesonide / Pulmicort®) plus a long-acting Beta 2-Agonists (formoterol/Oxeze).

After the completion of SYGMA (SYmbicort®) which is given in Mild Asthma trial, a new therapeutic strategy is available that is for milder asthma patients, provides the anti-inflammatory treatment considering the high dependency on SABA. According to this trial, budesonide/formoterol 160/4.5 μg/dose is superior to SABA in asthma control and decreases asthma exacerbation rate by 64% ($p < 0.001$). SYGMA 1 and 2 trials also demonstrate that budesonide/formoterol 160/4.5 μg/dose is superior compared to budesonide in preventing asthma exacerbations while the cumulative dose of budesonide is reduced by $\geq 75\%$ (Avdeev et al., 2019). Glucocorticoids (GCs) are the efficient anti-inflammatory therapy for asthma (Palumbo et al., 2020).

Breo®: Made of corticosteroid (fluticasone) plus long-acting Beta 2-Agonists (vilanterol).

Zenhale®: Made of corticosteroid (mometasone) plus long-acting Beta 2-Agonists (formoterol) (Oba & Lone, 2014).

8.2 Severe Asthma in Children

Millions of children and adolescents with asthma are treated according to regional, national, and international guidelines (Fishwick et al., 2008). The goal is to increase preventive measures in response to worsening symptoms and treatment reduced once symptoms have been stabilized. Childhood asthma treatment is linked to less-than-ideal symptom control (Rabe et al., 2000) and asthma exacerbations are still common such as in UK, a child is admitted to hospital in every 20 min due to asthma exacerbation (Asthma, 2017).

Asthma in children is related to increased morbidity and mortality. There are two subcategories of severe asthma. The first group comprises a significant number of children with severe asthma who are experiencing symptoms as a result of factors such as inadequate or non-adherence to medication and/or poor inhaler technique. These kids and teenagers have "difficult-to-treat asthma". "Therapy-resistant asthma" is the second subcategory of severe asthma, and it describes children and adolescents whose symptoms are resistant or partly resistant to standard therapies including long-acting β agonists, inhaled and oral corticosteroids, and leukotriene modifiers. Individuals must be properly classified because the treatment approaches for these two subcategories of serious asthma are different. The "difficult to treat asthma" is treated based on education and psychology, while pharmacological therapies are used to treat "therapy-resistant asthma". The basis of asthma treatment is ICSs, leukotriene receptor antagonists, inhaled long acting β agonists, and theophylline (Suruki et al., 2017). Treatments include long term antireflux, oral corticosteroids and intramuscular omalizumab (an anti-IgE antibody).

8.2.1 Systemic Corticosteroids

In children, short bursts of oral corticosteroid (OCS) therapy (3 or 5 days) decrease the severity and duration of asthma attacks (Smith et al., 2003). This therapy has been used on children and adolescents who have been receiving OCS treatment for more than one month and have chronic poorly controlled asthma symptoms. Despite the fact that "maintenance" OCS is recommended in asthma guidelines, there is no evidence that it is effective.

8.2.2 Intramuscular Triamcinolone

Intramuscular triamcinolone can be used to diagnose steroid-responsive asthma and treat asthma where adherence to inhaled or/and oral corticosteroids is a concern (Gamble et al., 2009). There are only a few case studies in children who have received this systemic therapy with different doses of triamcinolone (Krupp et al., 2016). The largest case series of triamcinolone use in treating asthma symptoms was included in the third phase of US SARP III. Triamcinolone treatment lowers the exhaled nitric oxide and blood eosinophil count. In 111 children with severe asthma, there is improvement in symptoms and a non-significant trend of improved prebronchodilator FEV1.

8.2.3 Omalizumab

Since 2009, omalizumab, an anti-IgE antibody, has been used to treat serious asthma in children aged 6 to 11 years, as well as older children and adults. At least ten pediatric trials have looked into the effectiveness of omalizumab in the treatment of persistent asthma, severe asthma, and moderate-to-severe atopic asthma. Controlling asthma exacerbations (especially during the September peak), ICS dose, and symptoms is safe and effective with this treatment. When infliximab, an antibody specific for tumor necrosis factor, was first introduced, there were concerns about anaphylaxis and an increased risk of malignancies, but a recent review found that the risk of anaphylaxis is 0.2% per therapy and that there is no excess of malignancies (Chipps et al., 2017). Omalizumab is however not suitable for all patients (Busse et al., 2011).

8.2.4 Mepolizumab

Mepolizumab, an anti-interleukin-5 molecule, was approved for use in children aged 6 to 11 years old in Europe in August 2018, and in older children in the United States and Europe since 2015. Adults' experiences with this "mab" have been mixed, with research showing that it improves asthma outcomes while also causing serious headaches (Menzella et al., 2016). Mepolizumab was found to be effective in eliminating asthma exacerbations (Menzella et al., 2016). The licensing for children in Europe is based on safety data from a single study of 36 children aged 6 to 11. An FEV1/FVC ratio of 0.8, and blood eosinophilia (>0.15 109/l on recruitment or > 0.3 109/l in the year prior to recruitment) were all required for inclusion.

8.2.5 *Dupilumab*

Dupilumab is an IL4 and IL13 antibody that has been approved in the United States and the European Union for the treatment of serious asthma in people aged 12 and up. The LIBERTY ASTHMA QUEST study enrolled 1902 people aged 12 and up who were randomly assigned to one of two subcutaneous dupilumab doses given at two weekly intervals over a 12-month period. Inclusion criteria comprised an additional controller; medium-high dose ICS treatment; bronchodilator reversibility (>12% or 200 mls); an asthma attack, symptom control (Asthma Control Questionnaire score $\geq$ 1.5); and FEV1 < 8% (or $\leq$90% for 12-17-year old); demanding systemic glucocorticosteroid therapy in the previous year (Castro et al., 2018).

8.3 Cystic Fibrosis/Bronchiectasis

Bronchiectasis is an airway disease characterized by infection, inflammation, sputum expectoration, and dilatation of the airway walls with epithelial and cartilage damage. Chronic infection and inflammation with failure to clear secretions lead to permanent destruction of airway support cartilage causing bronchiectasis (Rubin, 2003). The common cause of bronchiectasis in children is cystic fibrosis (CF). The nature and the properties of CF airway secretions are like those of non-CF bronchiectasis (Bush et al., 2006).

In cystic fibrosis patients, aerosolized antibiotics are used to treat gram-negative infections. In people with CF and Pseudomonas aeruginosa airway infection, the aerosol tobramycin or colistin may improve pulmonary function, lower bacterial counts and reduce exacerbations of lung disease. Bacteria cause bronchiectasis, which causes airway obstruction. There are other causes of childhood bronchiectasis, but cystic fibrosis (CF) is the most common. Pulmonary complications and progression of disease in non-CF bronchiectasis is like CF bronchiectasis. Patients with bronchiectasis are treated by aerosolized tobramycin solution for inhalation (TSI) (Rubin, 2008).

Antibiotics are in oral and nebulized forms include erythromycin, roxithromycin, ceftazidime, clarithromycin, amoxicillin, oxytet-racycline, tobramycin, azithromycin and penicillin (Hernández et al., 2002). Aztreonam, colistin and Inhaled tobramycin have been shown to be beneficial in studies (Haworth et al., 2014).

Drugs with a macrocyclic lactone ring of twelve or more elements are known as macrolides. Due to tissue penetration and effectiveness against respiratory pathogens, they are the first-line therapies for respiratory infections. They bind to the 50S subunit of the ribosome in a reversible manner, preventing bacterial duplication. For staphylococci, H influenzae, and streptococci, they are bacteriostatic (Bearden & Rodvold, 1999).

8.3.1 Aminoglycosides

Tobramycin is aminoglycoside used as an aerosol nebulized antibiotic and treats CF patients (Van Westreenen & Tiddens, 2010). In 1998, inhalation solution, tobramycin was approved (TOBI® Novartis Pharmaceuticals, Basel, Switzerland) (Rose & Neale, 2010) with PARI LC Plus nebulizer (Geller, 2009), and to avoid toxicity and to improve lung function measured as FEV1- volume exhaled as forced expiration, tobramycin (300 mg) twice/day administered in alternating twenty eight day on/off cycles (Dudley et al., 2008).

8.3.2 Fluoroquinolones

To avoid drug resistance and to improve patient condition, fluoroquinolones might be administered when the tobramycin inhalation solution cycle off. They are developed for inhalation, particularly the ciprofloxacin and levofloxacin (Ballmann et al., 2011). In contrast to aminoglycosides, their antibacterial activity not reduced when diffusion is low into mucus (Kesser & Geller, 2009). Ciprofloxacin is second-generation fluoroquinolone and is widely available antibiotic against aerobic gram-positive and gram-negative bacteria (Gaspar et al., 2013). Despite its side effects, like sunlight sensitivity, (Banerjee & Stableforth, 2000) and emergence of resistance (Doring et al., 2000), it is a safe and effective alternative therapy for acute pulmonary exacerbations, have been introduced in 1985 as an oral treatment for patients of CF. Levofloxacin is a gram-positive and gram-negative antibacterial fluoroquinolone from the third generation. It's a drug that can be taken orally as well as intravenously (Gaspar et al., 2013).

8.3.3 Beta-Lactams

Beta-lactams fight pseudomonal infections in patients of CF including Anti-pseudomonal carbapenems (meropenem, doripenem, and imipenem–cilastatin), aztreonam, cephalosporins (cefepime and ceftazidime) and penicillins (ticarcillin–clavulanate and piperacillin–tazobactam) (Illamola et al., 2018). Beta-lactams antibacterial activity is not reduced in CF patient's sputum, unlike aminoglycosides and similar to fluoroquinolones (Dudley et al., 2008). Aztreonam, which is effective against aerobic gram-negative bacteria, is accessible as an aerosolized (for chronic P. aeruginosa infection in the airways) formulation and intravenous (for acute pulmonary exacerbations) formulation. Carbapenems are antibiotics that are given intravenously to treat acute pulmonary exacerbations. They are effective against aerobic, anaerobic, gram-negative and gram-positive bacteria (Illamola et al., 2018). Only cefepime and ceftazidime, among the cephalosporins active against aerobic gram-positive and gram-negative bacteria, have activity against P. aeruginosa. Exacerbations of P. aeruginosa are treated intravenously. Penicillin is used to treat anaerobic, gram-negative and gram-positive bacteria, as well as P. aeruginosa, and is given intravenously in cases of acute pulmonary exacerbations (Bhatt, 2013).

8.3.4 *Colistin*

Colistin, also known as polymyxin E, was first marketed in the 1950s but has been largely ignored due to reports of neurotoxicity and nephrotoxicity. It was brought back by multi-resistance to other antibiotics, which prompted pharmacokinetic studies and a re-evaluation of dosing schemes (Couet et al., 2012). Colistin is available as colistin sulphate for oral administration and as an inactive prodrug, colistimethate, colistin methanesulphonate, or sodium salt, administered intravenously or through inhalation, alone or in combination, to eliminate early infections and stabilize chronic infections (Oliveira et al., 2009). Because of throat irritation, cough, and severe bronchoconstriction, colistin sulphate is not used for inhaled treatment. Endobronchial P. aeruginosa infection is treated with colistimethate saline inhalation solution (Westerman et al., 2004).

8.3.5 *Other Antibiotics*

Antibiotics may be used to treat CF lung disease caused by S. aureus, P. aeruginosa, and other bacteria. Maltophilia, as well as S. aureus is a type of bacteria. Fosfomycin is active against Pseudomonas spp. and multidrug resistant Pseudomonas (gram-positive and gram-negative bacteria). Fosfomycin calcium or fosfomycin trometamol are oral formulations, and fosfomycin disodium is an intravenous formulation. Antibiotic resistance, intolerance, and side effects result from repeated use of the same antibiotics. Because fosfomycin's action is distinctive (Faruqi et al., 2008), it can be used in combination with other antibiotics in these patients, (Van Westreenen & Tiddens, 2010) avoiding cross-resistance (Faruqi et al., 2008). For pulmonary exacerbations in CF patients colonized by multi-resistant P. aeruginosa, intravenous fosfomycin in combination with other antibiotics has minimal side effects (Mirakhur et al., 2003). In CF patients with P. aeruginosa colonization, azithromycin therapy increases lung function and decreases pulmonary exacerbations. The antimicrobial properties of azithromycin and the growth of resistance in the CF airway microbiome are related to azithromycin failure. Minocycline (by oral administration) may play a "adjunct role" in the antimicrobial treatment of multidrug-resistant respiratory pathogens in CF patients, either alone or in combination with other antibiotics (Grinwis et al., 2010). Minocycline is effective against B. cepacia, S. maltophilia, and, to a lesser extent, P. aeruginosa isolates cultured from the respiratory tracts of CF patients (Kurlandsky & Fader, 2000). S. aureus is treated with clindamycin and rifampicin either orally or intravenously. MRSA infections in CF patients are treated with linezolid, an oxazolidinone antibiotic (with acute exacerbations) (Keel et al., 2011). Cross-resistance with other antibacterial agents is extremely unlikely (Dotis et al., 2010). Linezolid also has good tissue penetration (especially in the respiratory tract), which is beneficial in the treatment of CF (Di Paolo et al., 2010). When both S. aureus and S. pneumoniae are present, an oral combination of amoxicillin and clavulanic acid may be used (Döring et al., 2012).

8.4 Bronchitis

Bronchi are the air passages connecting trachea (windpipe) to tiny air sacs (alveoli) in lungs. Body absorbs oxygen to breathe in via the air sacs. Bronchitis means swelling and inflammation of bronchi. Bronchi walls are occupied with extra sticky mucus. Airflow of lungs into and out is partly blocked causing cough.

There are two kinds of bronchitis:

- **Acute bronchitis** causing little sickness and gets better in two to three weeks.
- **Chronic bronchitis** does not go away where a cough with mucus remains for three months of the year.

Aspergillus bronchitis (or aspergillary bronchitis, as it was first called) (Pepys et al., 1959) may occur in patients with underlying pulmonary or airway pathology (Maschmeyer et al., 2007), especially in the context of lung transplantation (Westney et al., 1996). Oral azole antifungal therapy is used for all forms of aspergillosis, including allergic bronchopulmonary aspergillosis (ABPA). Long-term therapy may increase the risk of resistance. Itraconazole and voriconazole resistance with reduced susceptibility to posaconazole in Aspergillus fumigatus is also observed (Howard et al., 2010).

In comparison to allopathic drugs, ethnomedicinal plants are more effective and have fewer side effects. Leguminosae (six plants), Lamiaceae (five plants), and Asteraceae are the most commonly used plant families in the Himalayan region for treating bronchitis (four plants). Due to their abundance in the study area and the presence of secondary metabolites such as flavonoids in Leguminosae, these families have a large number of medicinal plants (Chew et al., 2011); monoterpenoids, flavonoids and triterpenoids of Lamiaceae (Chiang et al., 2005); and sesquiterpene lactones in the Asteraceae (Chiang et al., 2005). These plants not only treat bronchitis but also other human diseases (Radha et al., 2013). These plant families are also used in other countries as well (Rathore et al., 2007). Medicinal plants used to treat bronchitis are herbs (58%), shrubs (24%) and trees (18%) because of their efficiency against human pathogens in outdated medicine system (Pala et al., 2010). Almost all plant components are antiviral, but the flowers, bark, leaves, and roots are the most effective. In the formation of recipes, ingredients such as water, sugar, and honey are used to decrease the bitter taste as well as the toxic effects (Mala et al., 2012).

8.5 COPD (Chronic Obstructive Pulmonary Disease)

COPD is a major cause of illness and death (Chapman et al., 2006). This causes around 2.75 million deaths annually, and the number is increasing (Lopez et al., 2006). COPD treatment decreases risk by preventing exacerbations and improving symptoms (Duffy et al., 2005). Early disease diagnosis, smoking control programs (Samet, 2005) and lung surgery of patients with emphysema (Deslee et al., 2014) can reduce mortality. In COPD, pulmonary inflammation is prominent. Anti-inflammatory drugs for example inhaled corticosteroids have slight or no effect on decline of lung function but may reduce exacerbations (Alsaeedi et al., 2002), particularly when combined with an inhaled long-acting beta-agonist (Man et al., 2003). Inhaled corticosteroids minimize the rate of mortality amongst patients with COPD (Sin et al., 2005) and that addition of long-acting beta-agonist may enhance this effect. The combination of inhaled corticosteroid fluticasone propionate and the long-acting beta-agonist salmeterol would minimize mortality among patients with COPD, as compared with typical care. COPD trial over 3-year period included placebo controlled, double-blind, randomized, and parallel-group (Soriano et al., 2002).

Acute exacerbations of chronic obstructive pulmonary disease (AECOPD) demanding hospitalization are related to 6% risk of in-hospital mortality. 35% of survived are readmitted in 3 months and are at 12% death risk (Hartl et al., 2016; Chow et al., 2015). Long treatment with 250 mg azithromycin daily inhibits AECOPD by minimizing exacerbation rate and enhancing the inter-exacerbation break (Ni et al., 2015). Validation parameters include dose (500 mg three times weekly), protection concerns related to long-term use in COPD population (e.g. the induction of antibiotic resistance (Li et al., 2014)), cardiac toxicity (Giudicessi & Ackerman, 2013), and ototoxicity (Giudicessi & Ackerman, 2013), ideal dose, treatment duration and target population. In COPD (AECOPD), azithromycin prevents acute exacerbations.

8.6 Chronic Cough

Cough is a self-protective reaction which releases extra secretions or foreign material from the respiratory tract. A chronic cough persists for eight weeks (Ojoo et al., 2013). Cough is the predominant complaint for medical consultant. Treatment-resistant cough can effect quality of life and weaken both somatically and psychologically (Kaushik et al., 2015). Herbal medicines can treat respiratory system disorders. The antibacterial Glycyrrhiza glabra L. fraction contains licorice as respiratory drug (Rahman et al., 2018). In traditional Persian medicine (TPM), licorice (Glycyrrhiza glabra L.) has been used for preparing cough mixtures with therapeutic herbs (Mosaffa-Jahromi, 2016). Avicenna and Rhazes in this herbal medicine are antitussive (Ibn Sina, 2005). Licorice (Liquiritin apioside and Liquiritin) has suppressed cough in mice model (Kuang et al., 2018).

Opioids such as morphine, diamorphine, and codeine are antitussives that activate brainstem opioid receptors as well as peripherally located opioid receptors on sensory nerve endings in the airways. Gastrointestinal symptoms, physical dependence and respiratory depression are also common side effects of these doses. Morphine and diamorphine are addictive, so they can help patients with terminal illnesses like bronchial carcinoma with a heavy cough. Low doses of weaker opioids like codeine are found in opiate-containing cough mixtures (MacRedmond & O'Connell, 1999). The dextro isomer of the opiate levomethorphan has no sedative or analgesic properties. In the United States, dextromethorphan is used as an antitussive and is as effective as codeine. Menthol as antitussive hinders citric acid induced cough slightly compared to lignocaine however it does not affect oropharynges (Morice et al., 1994; Sant'Ambrogio et al., 1991).

8.7 Blocked Lung Artery (Pulmonary Embolus)

After myocardial infarction and stroke, pulmonary embolism (PE) is the third most common cardiovascular cause of death (Wendelboe & Raskob, 2016). PE rates have risen year after year in recent decades, despite a decrease in deaths (Konstantinides et al., 2016). Although the number of deaths has decreased, there are still a number of patients with smaller emboli and PE-related complications (Jiménez et al., 2018). PE remains the leading cause of death in hospitalized patients (Jiménez et al., 2016).

Nitric oxide-releasing aspirin, inorganic nitrite, and sildenafil have advantages in animal models of acute PE (Souza-Costa et al., 2005). In acute PE, inorganic nitrite lessens the increase in matrix metalloproteinase-9 and has anti-oxidant properties in acute PE (Dias-Junior et al., 2008). In acute PE, L-arginine and NO donor diethylenetriamine/nonoate both are inactive. The effects of inhaled NO in PE have unsure hemodynamic development with no improvement in blood gases (Bhat et al., 2015). Ethyl nitrite inhalation may induce methemoglobin formation (Reynolds et al., 2018). The two organic mono nitrites, 2-hydroxy-1-methyl ethyl nitrite and 2-hydroxy propyl nitrite (henceforth referred to as PDNO) show improved vasodilator selectivity toward the pulmonary circulation as compared to nitroglycerin (Nilsson et al., 2018).

Statins, proven useful in treating arterial thrombosis, also have effect in patients with VTE. 20 mg rosuvastatin reduces the occurrence of symptomatic VTE when compared to placebo. 43% decrease in the risk of VTE (hazard ratio 0.57, 95% Ci 0.37–0.86, $P = 0.007$) without a difference between unprovoked or provoked VTE is observed in rosuvastatin group. VTE rates are 0.18 and 0.32 in rosuvastatin and placebo groups, respectively, with larger benefits for DVT (Glynn et al., 2009) than for PE. Hence, there is a link between VTE and arterial thrombosis.

8.8 Pulmonary Edema

Increased fluid in the interstitial or alveolar spaces of the lung parenchyma is associated to pulmonary edema (PE). Edema fluid may be transudative, means it doesn't contain many cells, proteins, or mediators. PE gets attention of cardiologists, pulmonologists, and physiologists (Laennec, 1962). Shortness of breath at rest is a symptom of acute PE, which can progress to force tachypnea, tachycardia, and hypoxemia. Peribranchial coughing may cause wheezing (cardiac asthma). If the edema is serious, coughing is common. Foamy or pink-tinged sputum may be vomited by patients. Hypoxemia can result from lactic acidemia, especially in patients with poor peripheral perfusion and low blood pressure.

A thin layer of alveolar lining fluid exists in the normal lung, which does not impede gas exchange. Peri broncho vascular lymphatics eliminate fluid and offer reserve volume. High pulmonary capillary pressure (PCP) with normal alveolar-capillary barrier and increased alveolar capillary permeability are the two types of acute PE. The pulmonary venous pressure has risen as a result of acute cardiogenic volume overload edema. With a high LAP (>25 mm Hg), alveolar flooding occurs. Pheochromocytoma and tocolytic medications (intravenous terbutaline and beta-adrenergic agonists) may cause more pulmonary venoconstriction than arteriolar constriction, which can raise PCP. Acute respiratory distress syndrome causes permeability edema (ARDS) (Fan et al., 2018).

8.8.1 *Vasodilators*

Nitrates are used to decrease preload and afterload in acute heart failure with cardiogenic edema. Nitrates are also valuable for pulmonary edema and hypertension. High-dose intravenous nitrates, followed by sublingual nitrates, are recommended by experts (Corstiaan & Brugts, 2015). Nitrates treat hypertensive and normotensive patients. Nitroprusside is an arterial vasodilator with venodilation. The 1970s saw the introduction of nitroprusside as a treatment for patients with high LV filling pressures, acute myocardial infarction and poor cardiac output. It necessitates the use of an arterial catheter to maintain constant arterial pressure. Vasodilators can worsen oxygenation by lowering pulmonary vascular resistance, interfering with homeostatic ventilation and perfusion mechanisms. Antagonists can decrease pulmonary capillary pressure, improve cardiac efficiency, and lower systemic vascular resistance.

8.8.2 Morphine and Opiates

Acute cardiogenic pulmonary edema is treated with intravenous morphine. Preload, afterload, and heart rate are all reduced. Anxiety, serum catecholamine levels, overactivity, afterload, and myocardial oxygen consumption are all reduced. Opiates can decrease ventilatory drive and should be used with caution in an emergency because they can increase the need for mechanical ventilation or/and ICU admission.

8.8.3 Angiogensin Converting Enzyme Inhibitors

Angiogensin translating enzyme inhibitors can help with hypertension and pulmonary edema by lowering preload and afterload. Therapy starts with low intravenous dose. If tolerated, then used orally. Long treatment recovers patients with heart failure and acute myocardial infarction.

8.8.4 Natriuretic Peptides

Nesiritide increases CO levels while reducing vascular resistance and PCP. There is less dyspnea in the patients. When nesiritide is compared to intravenous nitrates for acute congestive heart failure, nesiritide is more effective, but statistics show that nesiritide has a higher mortality rate. As a result, nesiritide is not recommended as a first-line treatment for infarction or acute myocardial ischemia.

8.8.5 Inotropic and Inodilator Drugs

Positive inotropes such as dobutamine and dopamine increase the risk of arrhythmias and myocardial oxygen demand. Milrinone is a phosphodiesterase-3 inhibitor that improves the contractility of the heart muscle.

8.9 Treatment of Arrhythmias and/or Atrioventricular Resynchronization

Tachyarrhythmias shorten the time it takes for the LV to fill, raise the LAP, and lower the CO. High LAP can be caused by sinus tachycardia, atrial tachyarrhythmias or atrial fibrillation. Digoxin, intravenous beta-blockers or acute cardioversion may be used to stop this. Patients with atrial fibrillation and a rapid ventricular response may benefit from digoxin (Gheorghiade & Braunwald, 2009).

8.10 Tuberculosis

In developing countries, tuberculosis (TB) is the common infectious disease (Nasiri et al., 2014). 8.6 million people contracted TB in 2012, with 1.3 million dying as a result (Organization, 2013). As the number of drug-resistant strains increased, tuberculosis spread (Shamaei et al., 2009). Antibiotic resistance in a specified population of patients infected with Mycobacterium tuberculosis requires knowledge of geographic variations (Bahrmand et al., 2009). Rifampin (RIF), ethambotol (EMB), streptomycin (SM), and Isoniazid (INH) are the first-line chemotherapeutic medicines used to treat TB (Mohammadi et al., 2002). Resistance to INH and RIF is a concern because it necessitates the use of second-line drugs, which are more toxic and costly than the first-line regimen (Merza et al., 2011).

8.11 Varied Lung Diseases

8.11.1 COVID-19—Coronavirus

In the starting months of 2020, a novel coronavirus appeared from Wuhan, China and started spreading rapidly around the world (Lippi et al., 2020). This virus caused an infection called COVID-19 which resulted in severe pneumonia and death (Jiang et al., 2020). This virus transmits easily and has led to worldwide pandemic (Chinazzi et al., 2020). As nations fight to get the existing wave of epidemics under control, there is a second wave in the fall or winter of 2020. It perhaps has joined with epidemics of other seasonal respiratory infections that has overcome healthcare systems of various countries (Chinazzi et al., 2020).

8.11.2 Atelectasis

Atelectasis has many complications such as dyspnea, changes in local ventilation and blood flow ratio, irritant cough, sputum, and pulmonary infection. These complications make atelectasis a refractory disease in clinic. Macrolide antibiotics treat MP pneumonia, however, resistance to macrolide antibiotics has been reported from various regions of the world (Su et al., 2020).

8.12 Lung Cancer

Lung cancer is the most diagnosed cancer worldwide in both men and women and it is also the leading cause of cancer related deaths (Miller et al., 2018). The non-small cell lung cancer (NSCLC) is the major (80%-85%) lung cancer (Nawaz & Webster, 2016). Comparing to the other lung cancer subtypes, NSCLC has worse prognosis especially when diagnosed at advanced stages. In addition to surgery, NSCLC can be treated by either radiotherapy and/or chemotherapy (Chen et al., 2014).

8.13 Sarcoidosis

- Sarcoidosis can affect any organ but is most commonly found in the intrathoracic lymph nodes and lungs.
- A compatible clinical presentation and presence of noncaseating granuloma on histopathologic examination and exclusion of other causes of granulomatous inflammation are all used to diagnose sarcoidosis.
- Because spontaneous remission is common, treatment is reserved for patients with symptoms or progressive organ dysfunction/damage.
- In refractory/recurrent cases, glucocorticoids are used first, followed by glucocorticoid-sparing agents and biological agents (Ungprasert et al., 2019).

8.14 Pleural Effusion

Dyspnea is a typical symptom of pleural effusion (PE). The size of the effusion is related to the degree of dyspnea. Large PE occupy space in the chest that would usually be occupied by pulmonary parenchyma, resulting in a reduction in lung volume. When a pleural effusion (even a massive one) is drained, the lung volumes do not alter immediately. The transition to more tense respiratory muscles, especially the diaphragm, is reflected in the clinical improvement of dyspnea after PE (Thomas et al., 2015).

If a patient has lung cancer with pleural effusion, but no malignant cells found in the tapped pleural effusion fluid, a VATS or "internal-medical" thoracoscopy should be performed before treatment like surgery and radiotherapy (Porschen et al., 2015). Thoracoscopy enables the direct inspection of pleural surface, targeted tissue sampling, and, if necessary, pleurodesis (a procedure causing the two pleural surfaces to adhere to each other). If the patient simultaneously has hemoptysis, bronchial obstruction, or an intrapulmonary mass seen on thoracic imaging study, bronchoscopy is indicated.

- **Pneumonia**. An alveoli infection, caused by viruses or bacteria, as well as the coronavirus.

- **Tuberculosis**. Caused by the Mycobacterium tuberculosis.
- **Emphysema**. When the fragile links between the alveoli are broken, this occurs. The most common cause is smoking.
- **Pulmonary edema**. Fluid leaks from the lung's small blood vessels into the air sacs and the surrounding area. Heart failure and back pressure in the lung's blood vessels cause one type. Injuries to the lungs can also result in fluid leakage.
- **Lung cancer**. It happens in lung or near the air sacs.
- **Acute respiratory distress syndrome (ARDS)**. This is a severe pulmonary injury caused by a serious illness. One example is COVID-19.
- **Pneumoconiosis**. This is a type of lung injury caused by inhaling harmful substance to the lungs. Black lung disease caused by coal dust and asbestosis caused by asbestos dust.

8.15 Conclusion

Epidemiologic data may reveal the cause, patterns of incidence, and risks of developing asthma and other lung diseases, as well as their associated morbidity. Despite the fact that multiple factors have been identified, better understanding the disease's increased prevalence and natural history will require efforts that include basic genetic, pathophysiologic and epidemiologic study. Acute pulmonary embolism can strike suddenly and without warning, making it difficult to diagnose. Different lung diseases like Cystic Fibrosis/Bronchiectasis, Asthma, Bronchitis, chronic obstructive pulmonary disease, Chronic Cough, pulmonary embolus, pulmonary edema and Tuberculosis and their epidemiology are discussed in this chapter. Treatment may lower the risk of death, and primary treatment is generally successful.

References

Alsaeedi, A., et al. (2002). The effects of inhaled corticosteroids in chronic obstructive pulmonary disease: A systematic review of randomized placebo-controlled trials. *The American Journal of Medicine, 113*(1), 59–65.

Asthma, U. (2017). Asthma facts and FAQs.

Avdeev, S., et al. (2019). Budesonide/formoterol Turbuhaler® as needed in mild asthma: results of SYGMA-1 and SYGMA-2 trials (SYmbicort® given as needed in Mild Asthma). *Russian Pulmonology, 29*(4), 419–427.

Bahrmand, A. R., et al. (2009). High-level rifampin resistance correlates with multiple mutations in the rpoB gene of pulmonary tuberculosis isolates from the Afghanistan border of Iran. *Journal of Clinical Microbiology, 47*(9), 2744–2750.

Ballmann, M., et al. (2011). Therapeutic approaches to chronic cystic fibrosis respiratory infections with available, emerging aerosolized antibiotics. *Respiratory Medicine, 105*, S2–S8.

Banerjee, D., & Stableforth, D. (2000). The treatment of respiratory Pseudomonas infection in cystic fibrosis. *Drugs, 60*(5), 1053–1064.

Bearden, D., & K. Rodvold. (1999). Penetration of macrolides into pulmonary sites of infection. *Infections in Medicine 16*(7), 480A–484A.

Bhat, T., et al. (2015). Inhaled nitric oxide in acute pulmonary embolism: a systematic review. *Reviews in Cardiovascular Medicine, 16*(1), 1–8.

Bhatt, J. M. (2013). Treatment of pulmonary exacerbations in cystic fibrosis. *European Respiratory Review, 22*(129), 205–216.

Bush, A., et al. (2006). Mucus properties in children with primary ciliary dyskinesia: comparison with cystic fibrosis. *Chest, 129*(1), 118–123.

Busse, W. W., et al. (2011). Randomized trial of omalizumab (anti-IgE) for asthma in inner-city children. *New England Journal of Medicine, 364*(11), 1005–1015.

Castro, M., et al. (2018). Dupilumab efficacy and safety in moderate-to-severe uncontrolled asthma. *New England Journal of Medicine, 378*(26), 2486–2496.

Chapman, K., et al. (2006). Epidemiology and costs of chronic obstructive pulmonary disease. *European Respiratory Journal, 27*(1), 188–207.

Chen, Z., et al. (2014). Non-small-cell lung cancers: a heterogeneous set of diseases. *Nature Reviews Cancer, 14*(8), 535–546.

Chew, Y. L., et al. (2011). Assessment of phytochemical content, polyphenolic composition, antioxidant and antibacterial activities of Leguminosae medicinal plants in Peninsular Malaysia. *BMC Complementary and Alternative Medicine, 11*(1), 12.

Chiang, L. C., et al. (2005). Antiviral activities of extracts and selected pure constituents of Ocimum basilicum. *Clinical and Experimental Pharmacology and Physiology, 32*(10), 811–816.

Chinazzi, M., et al. (2020). The effect of travel restrictions on the spread of the 2019 novel coronavirus (COVID-19) outbreak. *Science, 368*(6489), 395–400.

Chipps, B. E., et al. (2017). Omalizumab in children with uncontrolled allergic asthma: review of clinical trial and real-world experience. *Journal of Allergy and Clinical Immunology, 139*(5), 1431–1444.

Chow, L., et al. (2015). Hospital management of acute exacerbations of chronic obstructive pulmonary disease. *Journal of Hospital Medicine, 10*(5), 328–339.

Chung, K. F., et al. (2014). International ERS/ATS guidelines on definition, evaluation and treatment of severe asthma. *European Respiratory Journal, 43*(2), 343–373.

Corstiaan, A., & Brugts, J. J. (2015). Impact of intravenous nitroglycerin in the management of acute decompensated heart failure. *Current Heart Failure Reports, 12*(1), 87–93.

Couet, W., et al. (2012). Colistin pharmacokinetics: the fog is lifting. *Clinical Microbiology and Infection, 18*(1), 30–39.

Deslee, G., et al. (2014). Lung volume reduction coil treatment for patients with severe emphysema: A European multicentre trial. *Thorax, 69*(11), 980–986.

Di Paolo, A., et al. (2010). Pharmacological issues of linezolid. *Clinical Pharmacokinetics, 49*(7), 439–447.

Dias-Junior, C. A., et al. (2008). Sildenafil improves the beneficial haemodynamic effects of intravenous nitrite infusion during acute pulmonary embolism. *Basic & Clinical Pharmacology & Toxicology, 103*(4), 374–379.

Doring, G., et al. (2000). Antibiotic therapy against Pseudomonas aeruginosa in cystic fibrosis: S European consensus. *European Respiratory Journal, 16*(4), 749–767.

Döring, G., et al. (2012). Treatment of lung infection in patients with cystic fibrosis: Current and future strategies. *Journal of Cystic Fibrosis, 11*(6), 461–479.

Dotis, J., et al. (2010). Use of linezolid in pediatrics: a critical review. *International Journal of Infectious Diseases, 14*(8), e638–e648.

Dudley, M. N., et al. (2008). Aerosol antibiotics: considerations in pharmacological and clinical evaluation. *Current Opinion in Biotechnology, 19*(6), 637–643.

Duffy, N., et al. (2005). Intravenous aminophylline in patients admitted to hospital with non-acidotic exacerbations of chronic obstructive pulmonary disease: a prospective randomised controlled trial. *Thorax, 60*(9), 713–717.

Fan, E., et al. (2018). Acute respiratory distress syndrome: advances in diagnosis and treatment. *JAMA, 319*(7), 698–710.

Faruqi, S., et al. (2008). Fosfomycin for Pseudomonas-related exacerbations of cystic fibrosis. *International Journal of Antimicrobial Agents, 32*(5), 461.

Fishwick, D., et al. (2008). Standards of care for occupational asthma. *Thorax, 63*(3), 240–250.

Gamble, J., et al. (2009). The prevalence of nonadherence in difficult asthma. *American Journal of Respiratory and Critical Care Medicine, 180*(9), 817–822.

Gaspar, M., et al. (2013). Pseudomonas aeruginosa infection in cystic fibrosis lung disease and new perspectives of treatment: a review. *European Journal of Clinical Microbiology & Infectious Diseases, 32*(10), 1231–1252.

Geller, D. E. (2009). Aerosol antibiotics in cystic fibrosis. *Respiratory Care, 54*(5), 658–670.

Gheorghiade, M., & Braunwald, E. (2009). Reconsidering the role for digoxin in the management of acute heart failure syndromes. *JAMA, 302*(19), 2146–2147.

Giudicessi, J. R., & Ackerman, M. J. (2013). Azithromycin and risk of sudden cardiac death: guilty as charged or falsely accused? *Cleveland Clinic Journal of Medicine, 80*(9), 539.

Glynn, R. J., et al. (2009). A randomized trial of rosuvastatin in the prevention of venous thromboembolism. *New England Journal of Medicine, 360*(18), 1851–1861.

Grinwis, M. E., et al. (2010). Macrolide and clindamycin resistance in Streptococcus milleri group isolates from the airways of cystic fibrosis patients. *Antimicrobial Agents and Chemotherapy, 54*(7), 2823–2829.

Hartl, S., et al. (2016). Risk of death and readmission of hospital-admitted COPD exacerbations: European COPD Audit. *European Respiratory Journal, 47*(1), 113–121.

Haughie, S., et al. (2020). Equivalent systemic exposure to fluticasone propionate/salmeterol following single inhaled doses from Advair Diskus and Wixela Inhub: Results of three pharmacokinetic bioequivalence studies. *Journal of Aerosol Medicine and Pulmonary Drug Delivery, 33*(1), 34–42.

Haworth, C. S., et al. (2014). Inhaled colistin in patients with bronchiectasis and chronic Pseudomonas aeruginosa infection. *American Journal of Respiratory and Critical Care Medicine, 189*(8), 975–982.

Hernández, C., et al. (2002). Pulmonary function and quality of life in relation to bronchial colonization in adults with bronchiectasis not caused by cystic fibrosis. *Medicina Clinica, 118*(4), 130–134.

Hossny, E., et al. (2019). Challenges of managing food allergy in the developing world. *World Allergy Organization Journal, 12*(11), 100089

Howard, S., et al. (2010). Azole resistance in allergic bronchopulmonary aspergillosis and Aspergillus bronchitis. *Clinical Microbiology and Infection, 16*(6), 683–688.

Ibn Sina, A. (2005). *Al-Qanun fi'l-tibb (Canon of medicine)* (pp. 370–385). Alamy Le-Al-Matbooat Institute.

Illamola, S. M., et al. (2018). Population pharmacokinetics of amikacin in adult patients with cystic fibrosis. *Antimicrobial Agents and Chemotherapy, 62*(10), e00877–18.

Jiang, F., et al. (2020). Review of the clinical characteristics of coronavirus disease 2019 (COVID-19). *Journal of General Internal Medicine, 35*(5), 1545–1549.

Jiménez, D., et al. (2016). Trends in the management and outcomes of acute pulmonary embolism: analysis from the RIETE registry. *Journal of the American College of Cardiology, 67*(2), 162–170.

Jiménez, D., et al. (2018). Epidemiology, patterns of care and mortality for patients with hemodynamically unstable acute symptomatic pulmonary embolism. *International Journal of Cardiology, 269*, 327–333.

Kaushik, N., et al. (2015). Prevalence of chronic cough in a population based survey, Eur Respiratory Soc.

Keel, R. A., et al. (2011). Pharmacokinetics of intravenous and oral linezolid in adults with cystic fibrosis. *Antimicrobial Agents and Chemotherapy, 55*(7), 3393–3398.

Kesser, K. C., & Geller, D. E. (2009). New aerosol delivery devices for cystic fibrosis. *Respiratory Care, 54*(6), 754–768.

Konstantinides, S. V., et al. (2016). Management of pulmonary embolism: An update. *Journal of the American College of Cardiology, 67*(8), 976–990.

Krupp, N. L., et al. (2016). Monthly triamcinolone acetonide for severe, refractory, life-threatening Asthma in Children. *Journal of Allergy and Clinical Immunology, 137*(2), AB212.

Kuang, Y., et al. (2018). Antitussive and expectorant activities of licorice and its major compounds. *Bioorganic & Medicinal Chemistry, 26*(1), 278–284.

Kurlandsky, L. E., & Fader, R. C. (2000). In vitro activity of minocycline against respiratory pathogens from patients with cystic fibrosis. *Pediatric Pulmonology, 29*(3), 210–212.

Laennec, R. T. H. (1962). *A treatise on the disease [s] of the chest*, Published under the auspices of the Library of the New York Academy of ....

Li, H., et al. (2014). Meta-analysis of the adverse effects of long-term azithromycin use in patients with chronic lung diseases. *Antimicrobial Agents and Chemotherapy, 58*(1), 511–517.

Lippi, G., et al. (2020). Coronavirus disease 2019 (COVID-19): the portrait of a perfect storm. *Annals of Translational Medicine, 8*(7), 497.

Lopez, A., et al. (2006). Chronic obstructive pulmonary disease: Current burden and future projections. *European Respiratory Journal, 27*(2), 397–412.

MacRedmond, R., & O'Connell, F. (1999). Treatment of persistent dry cough: if possible, treat the cause; if not, treat the cough. *Monaldi Archives for Chest Disease= Archivio Monaldi per le Malattie del Torace, 54*(3), 269–274.

Mala, F. A., et al. (2012). Ethno-medicinal survey of Kajinaag range of Kashmir Himalaya, India. *International Journal of Pharma and Bio Sciences, 3*(2), 442–449.

Man, S. P., et al. (2003). Contemporary management of chronic obstructive pulmonary disease: clinical applications. *JAMA, 290*(17), 2313–2316.

Maschmeyer, G., et al. (2007). Invasive aspergillosis. *Drugs, 67*(11), 1567–1601.

Menzella, F., et al. (2016). Mepolizumab for severe refractory eosinophilic asthma: evidence to date and clinical potential. *Therapeutic Advances in Chronic Disease, 7*(6), 260–277.

Merza, M. A., et al. (2011). Anti-tuberculosis drug resistance and associated risk factors in a tertiary level TB center in Iran: a retrospective analysis. *The Journal of Infection in Developing Countries, 5*(07), 511–519.

Miller, K. D., et al. (2018). Cancer statistics for hispanics/latinos, 2018. *CA: A Cancer Journal for Clinicians, 68*(6), 425–445.

Mirakhur, A., et al. (2003). Fosfomycin therapy for multiresistant Pseudomonas aeruginosa in cystic fibrosis. *Journal of Cystic Fibrosis, 2*(1), 19–24.

Mohammadi, F., et al. (2002). Recovery of mycobacteria from clinical specimens and assessing drug susceptibility test of Mycobacterium Tuberculosis specimens by mycobacteria growth indicator tube (MGIT). *TANAFFOS, 1*(3), 35–44.

Morice, A., et al. (1994). Effect of inhaled menthol on citric acid induced cough in normal subjects. *Thorax, 49*(10), 1024–1026.

Mosaffa-Jahromi, M. (2016). Properties Related to Bile as Viewed in Makhzan ol-Adviya. *Iranian Journal of Medical Sciences, 41*(3 Suppl), S50.

Muneswarao, J., et al. (2019). It is time to change the way we manage mild asthma: an update in GINA 2019. *Respiratory Research, 20*(1), 183.

Nasiri, M. J., et al. (2014). Drug resistance pattern of Mycobacterium tuberculosis isolates from patients of five provinces of Iran. *Asian Pacific Journal of Tropical Medicine, 7*(3), 193–196.

Nawaz, K., & Webster, R. M. (2016). *The non-small-cell lung cancer drug market*. Nature Publishing Group.

Ni, W., et al. (2015). Prophylactic use of macrolide antibiotics for the prevention of chronic obstructive pulmonary disease exacerbation: a meta-analysis. *PLoS ONE, 10*(3), e0121257.

Nilsson, K. F., et al. (2018). Organic mononitrites of 1, 2-propanediol act as an effective NO-releasing vasodilator in pulmonary hypertension and exhibit no cross-tolerance with nitroglycerin in anesthetized pigs. *Drug Design, Development and Therapy, 12*, 685.

Oba, Y., & Lone, N. A. (2014). Comparative efficacy of inhaled corticosteroid and long-acting beta agonist combinations in preventing COPD exacerbations: A Bayesian network meta-analysis. *International Journal of Chronic Obstructive Pulmonary Disease, 9*, 469.

Ojoo, J. C., et al. (2013). Management of patients with chronic cough using a clinical protocol: A prospective observational study. *Cough, 9*(1), 1–7.

Oliveira, M. S., et al. (2009). Polymyxin B and colistimethate are comparable as to efficacy and renal toxicity. *Diagnostic Microbiology and Infectious Disease, 65*(4), 431–434.

Organization, W. H. (2013). *Global tuberculosis report 2013*. World Health Organization.

Pala, N. A., et al. (2010). Traditional uses of medicinal plants of Pauri Garhwal, Uttrakhand. *Nature and Science, 8*(6), 57–61.

Palumbo, M. L., et al. (2020). Chronic stress and glucocorticoid receptor resistance in asthma. *Clinical Therapeutics, 42*(6), 993–1006.

Pepys, J., et al. (1959). Clinical and immunologic significance of Aspergillus fumigatus in the sputum. *American Review of Respiratory Disease, 80*(2), 167–180.

Porschen, R., et al. (2015). S3-Leitlinie Diagnostik und Therapie der Plattenepithelkarzinome und Adenokarzinome des Ösophagus (Langversion 1.0–September 2015, AWMF-Registernummer: 021/023OL). *Zeitschrift Für Gastroenterologie, 53*(11), 1288–1347.

Rabe, K., et al. (2000). Clinical management of asthma in 1999: the Asthma Insights and Reality in Europe (AIRE) study. *European Respiratory Journal, 16*(5), 802–807.

Radha, B., et al. (2013). Diversity and availability status of ethno-medicinal plants in the lohba range of kedarnath forest division (KFD), Garhwal Himalaya. *Global Journal of Research on Medicinal Plants & Indigenous Medicine, 2*(4), 198.

Rahman, H., et al. (2018). Glycyrrhiza glabra HPLC fractions: identification of Aldehydo Isoophiopogonone and Liquirtigenin having activity against multidrug resistant bacteria. *BMC Complementary and Alternative Medicine, 18*(1), 140.

Rathore, B., et al. (2007). Indian herbal medicines: Possible potent therapeutic agents for rheumatoid arthritis. *Journal of Clinical Biochemistry and Nutrition, 41*(1), 12.

Reynolds, J. D., et al. (2018). Pharmacologic targeting of red blood cells to improve tissue oxygenation. *Clinical Pharmacology & Therapeutics, 104*(3), 553–563.

Rose, L. M., & Neale, R. (2010). Development of the first inhaled antibiotic for the treatment of cystic fibrosis. *Science Translational Medicine, 2*(63), 63mr4.

Rubin, B. K. (2003). Overview of cystic fibrosis and non-CF bronchiectasis. Seminars in respiratory and critical care medicine, Copyright© 2003 by Thieme Medical Publishers, Inc., 333 Seventh Avenue, New ….

Rubin, B. K. (2008). Aerosolized antibiotics for non-cystic fibrosis bronchiectasis. *Journal of Aerosol Medicine and Pulmonary Drug Delivery, 21*(1), 71–76.

Samet, J. M. (2005). Smoking kills: experimental proof from the Lung Health Study. *Annals of Internal Medicine, 142*(4), 299–301.

Sant'Ambrogio, F. B., et al. (1991). Effect of l-menthol on laryngeal receptors. *Journal of Applied Physiology, 70*(2), 788–793.

Shamaei, M., et al. (2009). First-line anti-tuberculosis drug resistance patterns and trends at the national TB referral center in Iran—eight years of surveillance. *International Journal of Infectious Diseases, 13*(5), e236–e240.

Sin, D. D., et al. (2005). Inhaled corticosteroids and mortality in chronic obstructive pulmonary disease. *Thorax, 60*(12), 992–997.

Smith, M., et al. (2003). Corticosteroids for hospitalised children with acute asthma. *Cochrane Database of Systematic Reviews*, 1, CD002886.

Soriano, J., et al. (2002). Survival in COPD patients after regular use of fluticasone propionate and salmeterol in general practice. *European Respiratory Journal, 20*(4), 819–825.

Souza-Costa, D. C., et al. (2005). l-Arginine attenuates acute pulmonary embolism-induced oxidative stress and pulmonary hypertension. *Nitric Oxide, 12*(1), 9–14.

Su, D.-Q., et al. (2020). Clinical Analysis of 122 Cases with Mycoplasma Pneumonia Complicated with Atelectasis: A Retrospective Study. *Advances in Therapy, 37*(1), 265–271.

Suruki, R. Y., et al. (2017). The frequency of asthma exacerbations and healthcare utilization in patients with asthma from the UK and USA. *BMC Pulmonary Medicine, 17*(1), 74.

Thomas, R., et al. (2015). Physiology of breathlessness associated with pleural effusions. *Current Opinion in Pulmonary Medicine, 21*(4), 338.

Ungprasert, P., et al. (2019). Clinical manifestations, diagnosis, and treatment of sarcoidosis. *Mayo Clinic Proceedings: Innovations, Quality & Outcomes, 3*(3), 358–375.

Van Westreenen, M., & Tiddens, H. A. (2010). New antimicrobial strategies in cystic fibrosis. *Pediatric Drugs, 12*(6), 343–352.

Wendelboe, A. M., & Raskob, G. E. (2016). Global burden of thrombosis: epidemiologic aspects. *Circulation Research, 118*(9), 1340–1347.

Westerman, E. M., et al. (2004). Effect of nebulized colistin sulphate and colistin sulphomethate on lung function in patients with cystic fibrosis: a pilot study. *Journal of Cystic Fibrosis, 3*(1), 23–28.

Westney, G. E., et al. (1996). Aspergillus infection in single and double lung transplant recipients. *Transplantation, 61*(6), 915–919.

Chapter 9
Neurological Disorders: Biochemistry of Drug Resistance and Future Challenges

Muhammad Sheeraz Ahmad, Salma Batool, Arshad Islam, Ammara Jabeen, Azka Noureen, Sana Shamshad, Tayyaba Zainab, Maryam Shahid, and Warda Ahmed

Abstract Neurological disorders are the major health problem worldwide and expected to grow exponentially in coming years. One of the most challenging clinical aspect of neurological disorders is the development of resistance towards drugs in clinics. Failure to achieve the required therapeutic effect with the available drugs pose severe health complications besides economic burden and there is a dire need to either combat the resistance or develop novel drugs. Drugs for Epilepsy and Parkinson's disease have faced major resistance challenges. Drug resistance in neurological disorders is met at various levels including genome or proteome level affecting cellular transporters and signaling pathways, alteration of neuronal drug targets, insufficient drug distribution inside brain and abnormalities in of drug metabolism. A combinatorial approach, biochemical understanding of the signaling pathways and cellular actions of drugs, drug metabolism and personalized therapeutic approaches have proven useful to combat drug resistance. The foremost hindrance in reversal of drug resistance is the inadequate understanding of drug metabolism and mechanism of action for existing and new drugs in different neurological disorders. Future treatment strategies must be focused towards repurposing and repositioning of drugs with modern and personalized therapeutic approaches to target the molecular basis of the disease progression.

Keywords Neurological disorders · Treatment · Mechanism of action · Drug resistance · Resistance mechanism

---

M. S. Ahmad (✉) · A. Jabeen · A. Noureen · S. Shamshad · T. Zainab · M. Shahid · W. Ahmed
University Institute of Biochemistry and Biotechnology/National Center of Industrial Biotechnology (NCIB), PMAS-Arid Agriculture University Rawalpindi, Rawalpindi, Pakistan
e-mail: dr.sheeraz@uaar.edu.pk

S. Batool
Department of Biochemistry, Faculty of Life Sciences, University of Central Punjab Lahore, Lahore, Pakistan

A. Islam
Department of Pathology, Lady Reading Hospital Medical Teaching Institution, Peshawar, KP, Pakistan

S. Ahmed et al. (eds.), *Biochemistry of Drug Resistance*,
https://doi.org/10.1007/978-3-030-76320-6_9

9.1 Introduction

Neurological disorders are ailments of central nervous system (CNS) and peripheral nervous system (PNS), the former are diseased conditions of brain and spinal cord, while latter encompasses somatic and autonomic nerve system (sympathetic and parasympathetic systems). Human brain is the most intricate organ, comprising distinct regions with specific cell types including neurons, oligodendrocytes, astrocytes and microglia. The physiological function of differentially expressed cells in various regions is essential to maintain the homeostatic balance through cross talk and electrical signal transmission. Structural alteration and dysregulation of cellular pathways within different regions and cell types results in numerous neurological disorders (Baldini, 1997; Siuly & Zhang, 2016). According to neurological literature, more than 600 brain disorders such as Epilepsy, Alzheimer's disease (AD), Schizophrenia, Parkinson's disease (PD), Amyotrophic lateral sclerosis (ALS), migraine and others are prevalent around the world (Cottler et al., 2016).

Over the past few decades, neuropharmacological interventions with the effective therapeutic agents has revolutionized the neuroscientific field and medicine. Innovative research and identification of key regulatory molecules, which can be targeted for preventing the progression of early state of disease into chronic and/or more debilitating state, has been groundbreaking. Multiple drugs have been designed and repurposed to treat the neurological diseases effectively. Each drug administered has variable metabolism and mechanism of action. Some drug crosses the blood brain barrier (BBB), such as Levodopa targets dopamine receptors and act as a precursor for dopamine production in Parkinson disease. In epilepsy, Verapamil (phenylalkylamine calcium channel blocking agent) targets the p-glycoprotein (Pgp) transporters residing at BBB. The other class of medicine is used by manipulating the particular receptors involved in triggering the drug transportation into the brain. Tetrabenazine (TBZ) act as a high affinity inhibitor of pre/post synaptic receptor depleting the uptake of dopamine by 40% in Huntington's disease (HD). Similarly, Bumetanide and Fenfluramine target the 5-recepotors to treat epilepsy and PD respectively. Additionally, medicines like Amantadine targeting n-methyl-d-aspartate (NDMA) receptors in PD, whereas Angiotensin receptor blocker are employed to target Angiotensin II (AT_1) receptor to treat Alzheimer's disease (AD) (Stricker & Arteaga, 2015). Apart from this, various drugs and other diseases also imparts drug resistance in treating the neurological disorders (Gottesman & Pastan, 1993).

For neuropharmacological agent to be effective in the brain, administered drug must cross the BBB and ferebrospinal fluid barrier (blood-CSF), the protective barriers to protect the sanctity of the brain from foreign invaders and regulate the microenvironment of the brain. All interfaces serve to modulate and shield the microenvironment of the brain, such as physical and metabolic barriers, are composed of a monolayer of brain capillary endothelial cells, interconnected by strained junctions. In BBB, the presence of these junctions between the endothelial cells induces impermeability to capillary wall that constrict the intake of circulating molecules through brain. In CSF blood barrier, stiff junctions reside between the epithelial cells of the choroid plexus in brain-CSF whereas the tight junction in Brain-inner CSF-barrier reside between the neuro-ependymal cells creasing the ventricular facet. For brain outer CSF barrier, junctions are present between the endothelial cells of the arachnoid vessels, thus limiting the transport of drug molecule into the brain (Abbott et al., 2002; Daneman & Prat, 2015). Although the barriers are essential to maintain the integratory of brain, neuropharmacological field is facing the challenges including:

(i) Complex dynamics of BBB
(ii) Drug permeability at BBB
(iii) Lack of clinical trials for systemic drug delivery
(iv) Limitation of animal models for drug trials
(v) Genetics dysfunction (Pardridge, 1999; Spector, 2000; Strazielle & Ghersi-Egea, 2016)

Most emerging challenge of this era for neuropharmacology is the resistant of drugs imparted by various membrane transporters and receptors. There are multiple efflux and influx transport proteins present in human brain regulating the translocation of drug molecules across the membrane. Two principle transporter families are ATP-binding cassette (ABC) transporter serving as efflux transporters, whereas solute carrier (SLC) functions as influx transporters. Thus, more than 49 ABC transporters divided into 7 subfamilies (encoded by ABCA/MRP1, ABCB/MRP2, ABCC/MRP3, ABCD/MRP4, ABCE/MRP5, ABCF/MRP6, ABCG/ BCRP genes) and more than 400 SLC transporters genes are subdivided into 50 subfamilies(He et al., 2009; Li & Shu, 2014; Lin et al., 2015; Ross & Nakanishi, 2010). Based on the studies, ABC transporters play role in imparting safeguard to cellular moieties against cytotoxic effects of naturally occurring toxins and xenobiotics by limiting the entry into cells and promoting the ejection of the toxic substances from cells and tissue. Apart from protection, ABC transporters family also plays role in imparting resistance against multiple drugs. Amongst all ABC transporter genes, ABCB1/MDR1, ABCC/1MRP), ABCC2/MRP, and ABCG1/BCRP encodes for multiple drug resistance product limiting the drug penetration in tissues like brain, testes and placenta (Dean et al., 2001; You & Morris, 2006).The multidrug resistant transporters are over-expressed in brain. The over-expression of these efflux transporter prevents the drug disposition in tissue, thus, imparting resistance against neurological drugs. The most distinctive efflux transporters Pgp multidrug resistant (MDR) transporters(encoded by ABCB1/MDR1), which confer drugs resistance by overexpressing itself in BBB and blood-CSF barriers preventing efficacy of the drug in brain (Gottesman et al., 2002).

In various neurological disorders, the genetic variation and polymorphism/mutation in regions encoding drug transporters can also alter the efflux/influx mechanism of different molecules across the membrane. Since expression of the drug transporters is under the control of strong transcriptional factor, therefore, slight genetic variation can lead to the malfunctioning of transporters and receptors which contributes in defective uptake of drug molecule (Drozdzik et al., 2003; Petros & Evans, 2004). In Addition, numerous membrane receptors, such as G protein-coupled receptors (GPCRs), gamma-aminobutyric acid (GABA) receptors, N-methyl-D-aspartate (NMDA) receptor among others, are responsible for the translocation of drug molecule across the cell and also impart inhibitory affect with respect to the type of drug molecule administered in various diseases (Blanke & Van Dongen, 2008; Congreve & Marshall, 2010; Olsen & DeLorey, 1999). In a nutshell, the structural alterations and/or network changes at the molecular level, biochemical molecules, growth factors, surface proteins or receptors prevents the intake of drug into the brain cell, resulting drug resistance in various neurological disorders (Löscher & Potschka, 2005).

Despite improvements, effective delivery of pharmacological agent to the target site, escaping the protective nature of brain barriers and combating drug resistance remains a challenge. Thus, in this chapter we will describe the role and mechanism of various membrane transporters and receptors imparting the resistance against various therapeutic drugs in neurological disorders like PD, Epilepsy, Schizophrenia and HD.

9.2 Huntington's Disease

Huntington's disease (HD) is an inherited disease that leads to breakdown of nerve cells progressively. It is rare hereditary disease that causes depression, indigent co-ordination, involuntary movements, irritability, working movement, trouble in learning and decisions making. It is polyglutamine neurodegenerative disorder which begins in nearly forties, mainly caused by expansion of cytosine-adenine-guanine (CAG) repeats in Huntington (HTT) gene. HTT gene is located at chromosome number 4 on short arm (4p63) and contains CAG repeats. Mutated HTT gene expression leads to abnormal protein production, with approximately 3100 amino acids (349 kDa depends upon numbers of CAG-repeats). Mutant HTT possesses consecutive glutamine residues in the NH_2-terminal and disturbs the normal pattern of protein encoding (Bates, 2003; Labbadia & Morimoto, 2013).

It is inherited as an autosomal dominant manner, transmitted from one generation to next generation by following Mendelian fashion. Numerous genetic, epigenetic and other factors attribute the onset of this genetic disorder like genotypic variation of N-methyl-d-aspartate (NMDA) receptor subunit (NR_2B, and NR_2A) accounts for genetic variation around 5–13% (Arning et al., 2005). The length of CAG-repeat is a prominent factor in the manifestation of HD, the CAG repeats >35 is associated with the disease onset (Creus-Muncunill et al., 2019; Gil & Rego, 2008; Vonsattel et al., 1987).

The HD neuropathologically is gradual deterioration of striatum and its functional activities. For the assessment of severity of HD, Vonsattel grading system (0–4) is applied, in which neuronal loss within caudate nucleus is detected. In brain structure, the degree of striatal degradation correlated with non-striatal atrophy, for instance, in grade-3 and 4 the thalamus, globusplallidus, cerebral cortex, substantianigra, cerebellium, white matter and subthalamic can be affected. Previous studies elucidate that the HD patients eventually loss the somatostatin-positive neurons and orexin-secreting neurons in the hypothalamus region. Eventually in severe cases, brain weight loss is apparent by 40% (Vonsattel et al., 1987).

9.2.1 *Treatment of Huntington Diseases and Drug Resistance*

Comprehensive mechanistic research on the HD elucidated the potential targets for the treatment of autosomal disorder. Currently, clinical trials are carried out on symptoms basis like olanzapine blocks the dopamine receptor. Numerous drugs are used to treat the HD non-motor attributes. Many curative strategies, including tetrabenazine, antisense oligonucleotide (ASO) and gene therapy through cluster regularly short palindromic repeats (CRISPR)/Cas system on stem cell level link towards HD drugs development (Bachoud-Lévi, 2019; Shannon & Fraint, 2015; Wertz et al., 2020). The list of various drugs, target sites and current state for the treatment of HD are shown in Table 9.1.

9.2.1.1 Tetrabenazine (TBZ) Resistance Developed Through Interaction of P-Glycoprotein and Hyperactivation of CYP-Enzyme

Chorea is an involuntary movement condition in the body of HDpatients. TBZ reduces harmful effects caused by the hyper-activity of neurotransmitter dopamine. TBZ perform its functional activities by couple of mechanism. Firstly, it decreases the amount of dopamine released at the synapse. Signaling pathways responsible for the activation of neurotransmitter, a special proteins vesicular monoamine transporters (VMATs), packs the dopamine into vesicles. At that time, TBZ makes association with VMATs and blocks the entrance of dopamine into vesicles that cannot release into synapse, it would reduce the activation of other nerve cells. Secondly, TBZ interact with dopamine receptors, and causes hindrance for dopamine for further signaling mechanism (Fig. 9.1). It may help in blocking the dopamine on the both side of synapse, that effectively treats choreic movement disorders(Frank et al., 2008; Kaur et al., 2016).

TBZ may cause adverse effects as dopaminergic signaling hyperily activated in the HD patients. Studies suggest that administered dose of 50 mg/day reduced the symptoms of chorea adversely. Furthermore, it would be metabolized by isoenzyme CYP2D6. TBZ and reserpine both inhibits Vesicular Monoamine transporter 2 (VMAT2) their accompanying administration may cause irreversible binding VMAT2. TBZ-drug resistance is developed by the interaction of P-glycoprotein and interaction and hyper activation of CYP-enzyme. These may metabolite TBZ after oral administration and it would eliminate through renal route (Block et al., 2010).

Table 9.1 Huntington disorder (HD) therapeutic development

S#	Drug name	Target site and function	Current phase	References
1	Tetrabenazine	Interact with VMAT2 and blocks the dopamine entrance into vesicles for signaling	Approved for chorea (symptoms of HD)	(Wang et al., 2010)
2	Novel Viral Vectors (AAV1, AAV5, AAV9, AAV-PHP.B, CREATE)	Novel Viral Vectors (AAV1, AAV5, AAV9, AAV-PHP.B, CREATE) Widespread transduction of cells	II-phase of trial Showed efficacy in primate and rodent models	(Samaranch et al., 2017)
3	Riluzole	Glutamate release inhibitor	Does not show efficacy in human trails	(Landwehrmeyer et al., 2007)
4	Memantine	Antagonist of extrasynaptic N-methyl-D-aspartate (NMDA) receptors and t diminishes striatal cell death, hinders disease progression and improves cognitive function	Trial phase	(Beister, 2004)
5	Minocycline	Tetracycline analog and can cross the blood–brain barrier (BBB) and inhibits the expression of caspase-3 and caspase-1	Rodent trial	(Thomas et al., 2004)
6	CoQ10	Mitochondrial dysfunction	Attained futility in human trial	(McGarry et al., 2017)
7	Suberoylanilide hydroxamic acid	Transcriptional deregulation histone deacetylase inhibitor	Showed efficacy in a rodent model	(Hockly, 2003)
8	Chaperonins	Decrease mHTT aggregation	Showed efficacy in a rodent model	(Thulasiraman et al., 1999)
9	ASO approach (IONIS-HTTRX, Peptide conjugated ASOs)	Reduction in HTT mRNA and protein	Showed efficacy in a rodent model	(Kordasiewicz et al., 2012)
10	CRISPR-Cas9	Excision of CAG repeat and, reduction of mutant HTT	Showed efficacy in a rodent model	(Yang et al., 2017)

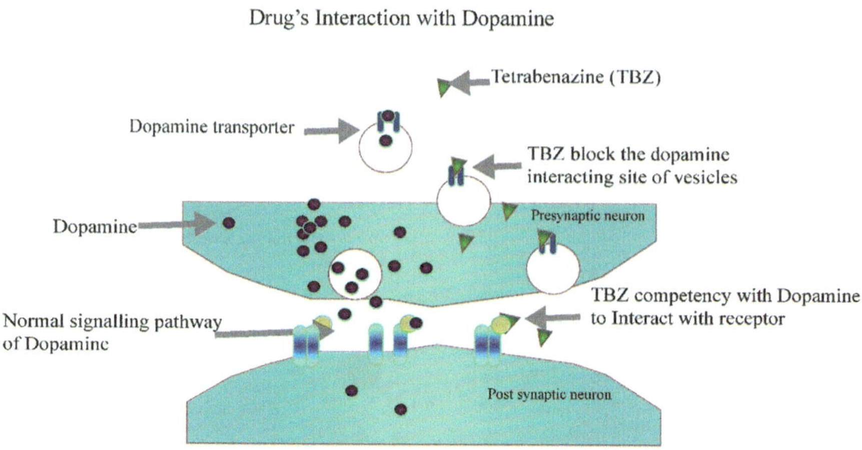

Fig. 9.1 Two mechanism of TBZ: Mechanism 1; TBZ compete with dopamine and bind with VMATs, parallel some dopamine entering into vesicles, while in mechanism 2; TBZ binds with nerve cell surface receptor and alter the signaling pathways (Modified from Kaur et al., 2016).

9.2.2 Tominersen (IONIS-HTTRx)

A therapeutic approach to combat the HD by IONIS-HTTRx. IONIS-HTTRx is an ASO, developed to arrest the mutated Huntington-messenger RNA (Fig. 9.2). In this mechanism, RNase H1-antisense mRNA degrade the HD-mRNA that demolished the synthesis of Huntington protein. Degraded toxic proteins may accumulated into the cell that leads towards the autophagy or apoptosis of neuron cell (van Roon-Mom et al., 2018). Tolerable amount of tominersen is administered into cerebrospinal fluid (CSF). Further pharmacodynamics and pharmacokinetics effects and the disease course varies from patient to patients. Precisely Tominersen suppress neurodegenerative mechanism of the patient. Scientists are continuously exploring the mechanism of drug interaction and its response with neurodegenerative cell's machinery. Abundant amount of toxic protein would be produced during the administration of drug and it may alter the signaling pathway that may block the drug interaction. Now the drug is at its 3rd level clinical trial base (Dash & Mestre, 2020; Godinho et al., 2015).

9.3 Parkinson Disease (PD)

According to an estimate, PD is quotidian neurological disorder after Alzheimer's that hits roughly 3% of the world population which is more than equal to sixty-five years of age. Neuronal loss in the substantianigra (SNpc brain) leads to striatal dopamine insufficiency, and intracellular incorporations having masses of α-synuclein are the neuropathological signs of PD. Albeit clinical finding depends on the presence of

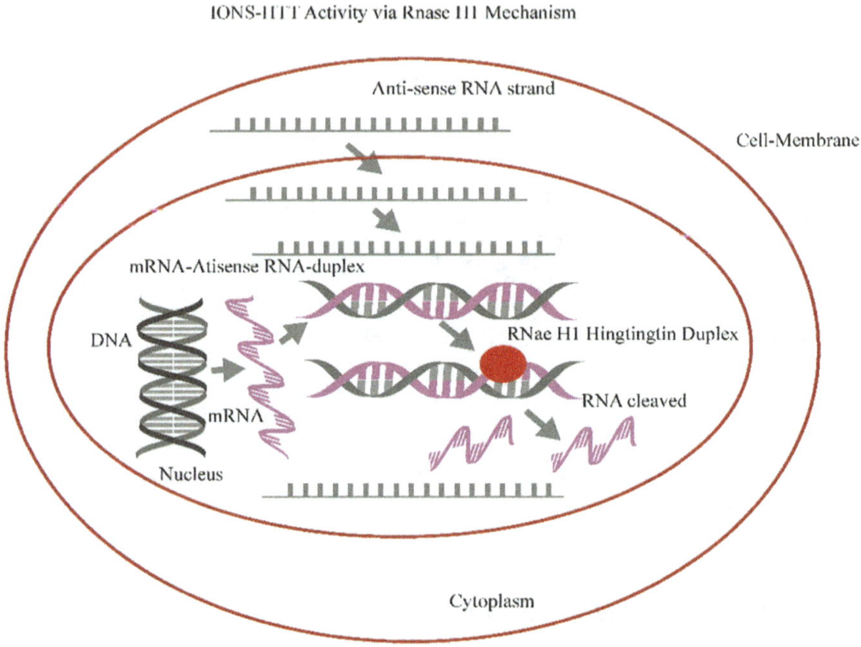

Fig. 9.2 IONS-HTT activity via RNase H1 mechanism. IONS-HTT RX degrade the huntington mRNA with the help of RNase H1 (Modified from MacLeod & Crooke, 2017)

bradykinesia and other cardinal motor symptoms include walking abnormality, cold, blinking of eyes, difficulties in speech etc. PD is related with numerous non-motor manifestations (such as daydreaming, frequent urge of urination, depression, anxiety etc.) that add to inability of speaking (Poewe et al., 2017).

9.3.1 Mechanisms of Disease and Genetics

There are several genes associated with the mechanism of progression of disease but SNCA was identified first. This gene encodes α-synuclein, upon mutation A53T was identified which confers increase in misfolding tendency of α-synuclein and its access accumulation than wild-type (Polymeropoulos et al., 1997). DJ-1 is another gene linked to PD which has an essential regulatory role in calcium flux mechanism in mitochondrion leads to protection from neuronal cells due to pace making activity of dopaminergic neurons. Lysosomal dysfunction is regulated by ATP132A gene and mutation in this gene results in early onset of PD. Additional experimental studies are needed to clarify the mechanism of PD neuronal pathways (Bonifati et al, 2003; Dehay et al., 2012). The treatment strategies used for Parkinson's disease are shown in Fig. 9.3 (Armstrong & Okun, 2020).

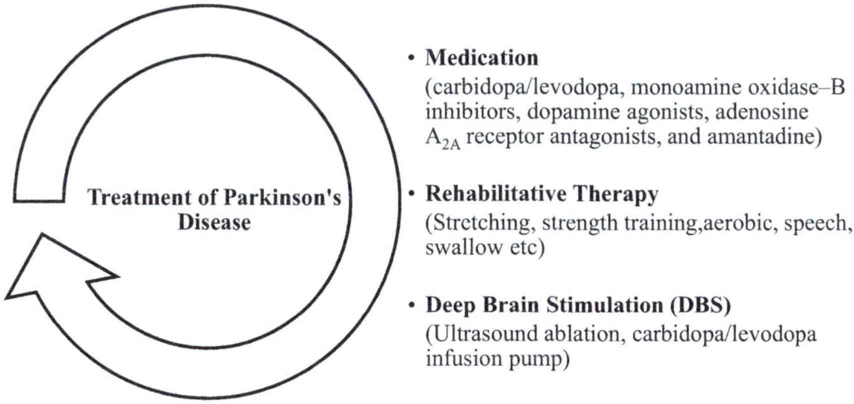

Fig. 9.3 Treatment options available for parkinson disease (Armstrong & Okun, 2020)

9.3.2 Resistance Mechanism of PD Drugs

9.3.2.1 Levodopa

Levodopa is the most commonly used drug for the treatment of motor symptoms in PD. Levodopa is orally administered and almost 95% of the drug is decarboxylated in liver and gut tissues, only 1–2% drug cross the BBB where it is converted into dopamine in remaining dopaminergic neurons of SNpc. However, peripheral activation of dopamine receptors results in nausea and hypotension. Carbidopa a decarboxylase inhibitor of dopamine has always been used in combination with levodopa to avoid these side effects. Other side effects include daydreaming, unbalancing, uncontrolled movements, sleepiness, disarray, hypersexuality, dyskinesias, psuedoresistance etc. Dopamine constant supply is not linked to these complications whether it is related to unwanted stimulation of striatal dopamine receptors in broken phasic manner (Fox et al., 2011).

9.3.2.2 Mechanism of Action of Levodopa Induced Activation of Dopamine and NMDA Receptors Leads to Retrieval of Normal Cardinal Motor Function and Dyskinesias Respectively

Dopaminergic neurons loss in PD leads to dopamine deficiency which confers to abnormal cardinal motor functions. $_{L}$-Dopa converts to dopamine by amino acid decarboxylase (AADC) enzyme and become available for dopamine receptors (DR) (LeWitt & Fahn, 2016; Olanow et al., 2006) activation of DR initiates a signaling cascade results in retrieval of normal cardinal motor functions (Fig. 9.4).

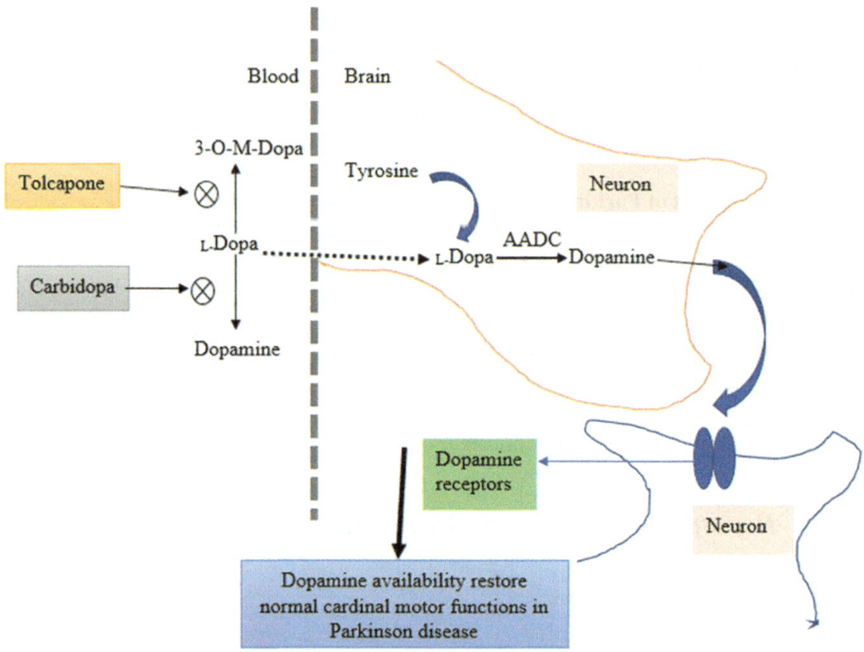

Fig. 9.4 Levodopa mechanism of action in Parkinson's Disease. Catechol-O-methyltransferase (COMT), 3-O-methyl-DOPA (3-O-M-DOPA) (Modified from Poewe et al., 2017)

Levodopa abnormal dyskinetic action was analyzed at presynaptic and postsynaptic levels of neuron. At the presynaptic level increase in concentration of dopamine was observed. Two other important mechanism of presynaptic phase is production of glutamate linked with expression of endocannabinoid CB1 and glutamate receptors. However, endocannabinoids (eCBs) decreases the glutamate quantity at presynaptic terminal. Postsynaptic level activation results in variety of cellular downstream signaling cascades triggered by NMDA receptor activation (probably with involvement of receptors with NR2B subunits), dopamine receptors D1, and metabotropic glutamate receptors. At postsynaptic phase activated NMDA receptor and glutamate receptors initiate the downstream signaling process leads to increase in concentration of secondary messenger (Ca^{2+} ion) results in activation of RAS/RAF/MEK pathway (Calabresi et al., 2010; Cenci, 2014). Activation of this pathway results in variety of physiological outcomes including dyskinesias (Fig. 9.5).

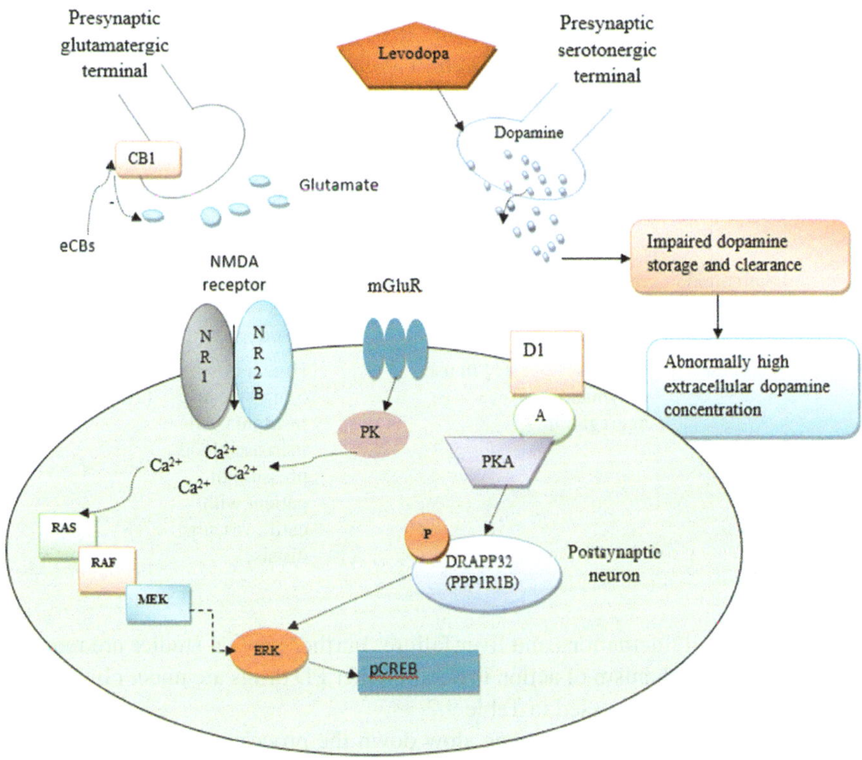

Fig. 9.5 Mechanism of Action of Levodopa in Dyskinesias at Presynaptic and Postsynaptic Level: Aadenylylcyclase (AC). Calcium ion (Ca^{2+}). Endocannabinoid receptor 1 (CB1). cAMP response element-binding protein (pCREB). Dopamine receptor (D1). Protein phosphatase 1 regulatory (inhibitor) subunit 1B (PPP1R1B). Endogenous cannabinoid (eCB). Extra cellular signal-regulator kinase (ERK) glutamate (Glu). Metabotropic glutamate receptor (mGluR). Protein Kinase A (PKA). Protein Kinase C (PKC) (Modified from Calabresi et al., 2010)

9.3.3 Monoamine Oxidase B (MAO-B) Inhibitors and Catechol-O-Methyl Transferase (COMT) Inhibitors

MAO-B is an inhibitor of decarboxylation of dopamine to increase its availability in order to initiate dopaminergic stimulation. Its early treatments in clinical trials exhibit a delay in appearance of motor symptoms but have no effect to modify the ongoing biological history of PD. Further studies needed to elaborate its significance and resistance mechanism.

Tolcapone drug is an inhibitor of COMT enzymes and administered in combination with levodopa in order to increase its half-life (Fig. 9.4). Previous studied shows that it helps to decrease in motor fluctuations caused by levodopa. But its continuous use raises the chances of liver failure (Poewe et al., 2017).Recent studied conducted by Moreira et al., (2020) observed no significant effect of tolcapone on

Table 9.2 Parkinson drugs current phase of trials and function for future use

Sr. No	Drug name	Function	Current phase	Limitations	Reference
1	Buspirone	α1-Adrenergic and 5-HT1A receptor agonist	Phase III trial in patients	Worsened motor function	Schneider et al., (2020)
2	Eltoprazine	Agonist of 5 HT1B receptor	Phase II is ongoing	Oscillatory activities not understood yet	Wang et al., (2019)
3	Safinamide	Inhibition of MAOB activity	Approval phase	Prolong motor effects of levodopa	Blair and Dhillon, (2017)
4	Caffeine	Decrease α-synuclein aggregation	In use	Higher dose 5 mg/kg cause insomnia and increase blood pressure in patient with cardiovascular disease	Negida et al., (2017)

levodopa induced fluctuations and liver failure. Further clinical studies are required to elucidate its mechanism of action in detail. Other PD drugs are under clinical trial at various phases as discussed in Table 9.2.

Currently available PD drugs can slow down the process of neurodegeneration but cannot completely stop it completely. It has been seen in patients that early administration of drugs and other pharmacological treatments have a good control in PD related symptoms at initial stages, while in later stages of the disease, it develops levodopa induced side effects (dysarthria, shivering, uncontrolled movement of limbs, dizziness, dysphagia, excessive urination, cold etc.). Advanced stages of PD are often associated with reduction in life span and mental disability. Moreover, complications arise with pharmacological treatments cannot be managed in advanced stages of this disease (Balestrino & Schapira, 2020).

9.4 Epilepsy

Epilepsy is a chronic sickness of CNS and remains one of the world's oldest recognized condition. It is characterized by abnormal brain activity causing frequent seizures. Seizures are brief episodes of involuntary body movement and sometimes loss of consciousness and sensation may also occur. Seizure happened due to extreme electrical discharges in different parts of brain cells. Some of the people having epilepsy simply stare for a few seconds during seizure, while others have the small lapses of unconsciousness or jerks to extreme-prolonged seizures. Generally, up to 10% of people worldwide have one seizure during lifetime. For an epilepsy diagnosis at least two seizures are generally required, they can differ in frequency of seizures, from less than one seizure annually to numerous per day. Surgery and some medications can control seizures for majority of people with epilepsy (Berg, 2004; Holmes, 2004).Patients with epilepsy are successfully treated with antiepileptic drugs (AEDs) and up to 70% of people with epilepsy have completely controlled their seizures with AEDs (French et al., 2004). There are about 26 AEDs used for the treatment of seizures, and different AEDs used to treat different type of seizures. However, ~10–20% of patients are pharmacoresistant to these drugs (Sillanpää et al., 1999).

9.4.1 Molecular Mechanisms of Epilepsy Drug Resistance Through Ion Channels, Transporters, Receptors and Metabolic Enzymes

Permeability of AED through the BBB may be different in medically intractable epilepsy patients (Löscher & Schmidt, 2004). Drug used for antiepileptic activity targets one or more molecules inside the brain, as shown in Fig. 9.6 (Sonneveld, 2001). These targets in the brain cells contain ion channels, transporters, receptors and different metabolic enzymes which are used in the uptake, release of neurotransmitters in the brain (Rogawski & Löscher, 2004).

Drug resistance involves the *MDR1* gene that codes for P-glycoprotein (MDR1 or PgP1), an adenosine triphosphate (ATP)-dependent transporter protein which help to takes drug out of cells and helps to shield normal and tumor cells (Volk et al., 2004). Endothelial cells in brain give expression of *MDR1* gene that translates Pgp, which has not been detected in neurons of normal human brain. Expression of MDR–associated protein (MRP) changed in epileptic tissue. Different AEDs have structures similarity to Pgp substrates which determined that these ubiquitous molecules have

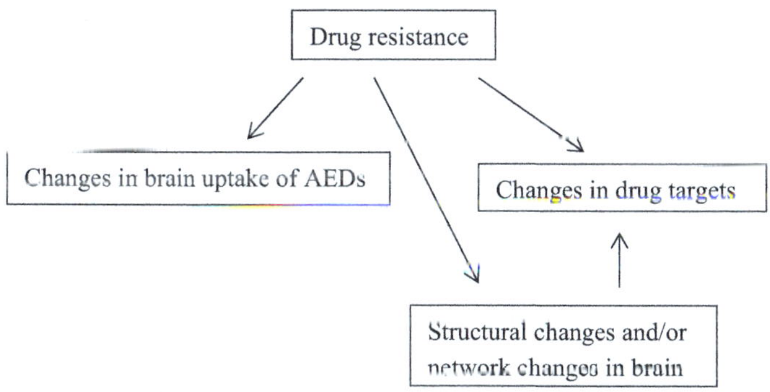

Fig. 9.6 Drug resistance mechanism for epilepsy and other neurological disorders

major role in pharmacoresistance. Differential expression of MDR gene characterized different forms of epilepsy which allow modification of drug to treat individual patients (Perucca, 1998).

Voltage-gated Na^+ ion channels are important for the initiation of seizure. AEDs such ascarbamazepine (CBZ) orlamotrigine (LTG) shows their effects on calcium and sodium channels (Löscher, 1998). They help to reduce the intensity of Na^+ current and shift the voltage by inactivation into a hyperpolarizing direction, so decreasing the number of Na^+ channels available and slow the retrieval from inactivation of ion channels. This mechanism is important for the treatment of epilepsy because it causes reduction of Na^+-channel available during high frequency seizure discharges. Loss of Na^+ channel drug sensitivity explains the development of drug-resistant epilepsy.

Inhibition by GABA receptors is an important AED target (Rogawski & Löscher, 2004). Type A receptors ($GABA_A$) facilitate the number of inhibitory neurotransmissions inside brain, and modifications in these receptors are involved in numerous neurologic diseases. $GABA_A$ receptors have seven different subunit families. Subtypes of $GABA_A$-receptor in the brain are consisting of alpha 1, beta 2, and gamma 2 subunits (Möhler et al., 2002). Studies have shown that large changes in $GABA_A$-receptors expression are associated with epilepsy. Significant decrease of potency occurs for dissolution of seizures using AEDs such as BZDs and phenobarbital (PB) which enhance $GABA_A$-mediated inhibition during pilocarpine-induced status epilepticus (Cohen et al., 2003). Thus, during epilepticus, rapid modulation of $GABA_A$ receptors result in pharmacoresistance to different AEDs that increase $GABA_A$-mediated receptor inhibition (Jones et al., 2002).

Alterations in drug target and drug transporter are the major cause of pharmacoresistance to AEDs beside this there are other mechanism of drug resistance in epileptic patients that need to identify. In the past three decades 20 new antiepileptic drugs have been developed which showed similar results to the study conducted decades earlier having similar pharmacoresistance in a study published by Lerche, (2020).

9.5 Schizophrenia

Schizophrenia is heterogeneous neuropsychiatric disorder (sometimes chronic if not treated) that affects person's thinking, feelings and behavior (Gillespie et al., 2017). Most common schizophrenic symptoms contain psychological factors including perception discernment, linguistic, social interaction, feelings, expression and power of using will, hallucinations, delusion, thought disorder which include inferential thinking, and cognitive symptoms which includes problem in attention, concentration and memory (Andreasen et al., 1995). Approximately 0.5–1% of the population is affected by schizophrenia (Vitaet al., 2019). Schizophrenia may be caused by environmental factors, which includes use of cannabis during adolescence, certain injuries and infections, parental age and lack of proper nutrients during pregnancy and genetic which include a range of common and unusual genetic variants (Altamura et al., 2005).

Table 9.3 Schizophrenia drugs and resistance

Sr. No.	Drug name	Target site	Limitations	References
01	Aripiprazole	Dopaminergic and serotonin 1^A receptor	Dopamine super sensitivity	Papakostas, (2005)
02	Asenapine	GABA receptor	Agranulocytosis, increased prolactin secretion	Frånberg et al., (2008), Kramlinger & Post, (1989)
03	Brexpiprazole	Dopaminergic D_2 and serotonin 5-HT_{1A}	Change in metabolic parameters	Fornaro et al., (2019), Newcomer et al., (2018)
04	Clozapine	Dopaminergic D_2, GABA and serotonin 5-HT_{1A} receptor	Dopamine super sensitivity	Gillespie et al., (2017)

9.5.1 *Treatment of Schizophrenia*

Antipsychotic medication is available and revolutionized for the treatment of schizophrenia. The first-generation antipsychotic medications like Aripiprazole, Asenapine, and Brexpiprazole etc. are too effective to treat maximum signs and symptoms at the lowermost possible dose. The psychiatrist or doctor might try different medications with different doses or combinations of doses over time to get the desired results but response to antipsychotic drugs are highly variable from person to person (Table 9.3) (Vita et al., 2019), due to differential dopamine (DA) and serotonin (5HT)release and metabolism (Altamura et al., 2005). Alteration in T-cell function have a major role in schizophrenia, which cause the activation of inflammatory response system. The differential inflammatory response leads to atypical antipsychotic treatment response. Approximately one third of the patients showed rare or no response to the medication and called treatment resistant schizophrenia (TRS) (Andreasen et al., 1995). Other medications like clozapine (considered as the second generation antipsychotic drug) also may help, but improvement in the symptoms of schizophrenia may be noticed after several weeks of medication. Unfortunately, drugs used for the treatment of schizophrenia can cause life threatening side effects, individuals with schizophrenia may be reluctant to take them. The neuroimaging is highly helpful in providing in vivo assessment of the expression and function of neuroreceptors (Altamura et al., 2005). Based on the differential treatment response, schizophrenia may be categorized into two distinct subtypes (Farooq et al., 2013); treatment responsive schizophrenia and treatment resistant schizophrenia (Gillespie et al., 2017).

9.5.1.1 Dopamine D2 Receptor (DRD2) Mediated Drug Resistance in Schizophrenia

Patients may be categorized based on the level of the neurotransmitter (as neurotransmitter are the core of the neuropsychiatric disorders). Most studied neurotransmitters in the schizophrenia are dopamine, serotonin and glutamine. Treatment resistant schizophrenic patients show higher level of baseline presynaptic dopamine (Krebs et al., 1998) and low level of homovanilic acid (HVA) (a metabolite of dopamine) in plasma as compared to the treatment responsive schizophrenic patients, Fig. 9.7 (Abi-Dargham et al., 2000; Garver et al., 2000).

The dopamine D2 receptor (DRD2) blockade is main known action for antipsychotic drugs (Kapur & Seeman, 2000; Seeman et al., 1976). The dopamine super sensitivity theory suggests that dopaminergic changes due to the continuous blockade of DRD2 leads to adverse symptoms which are impossible to treated by the initial medication (Chouinard et al., 2017; Suzuki et al., 2015) and thus need a continuous increase in the doses of antipsychotic drugs (Oda et al., 2015) The continuous DRD2 receptor blockade due to dopaminergic changes lead to increase in the DRD2 receptor density (Chouinard et al., 2017). In turn, increases in doses of antipsychotic drug to treat severe symptoms are supposed to cause the further increase in DRD2 density, causing amplified dopamine super sensitivity, and results in the reemergence of symptoms, Fig. 9.8 (Iyo et al., 2013).

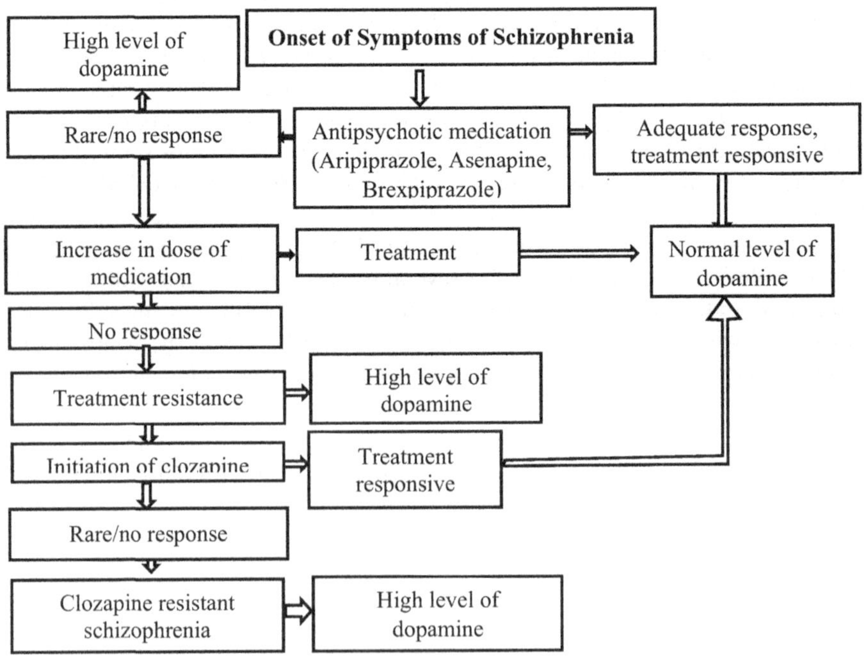

Fig. 9.7 Clinical pathway for the confirmation of treatment resistant schizophrenia (TRS)

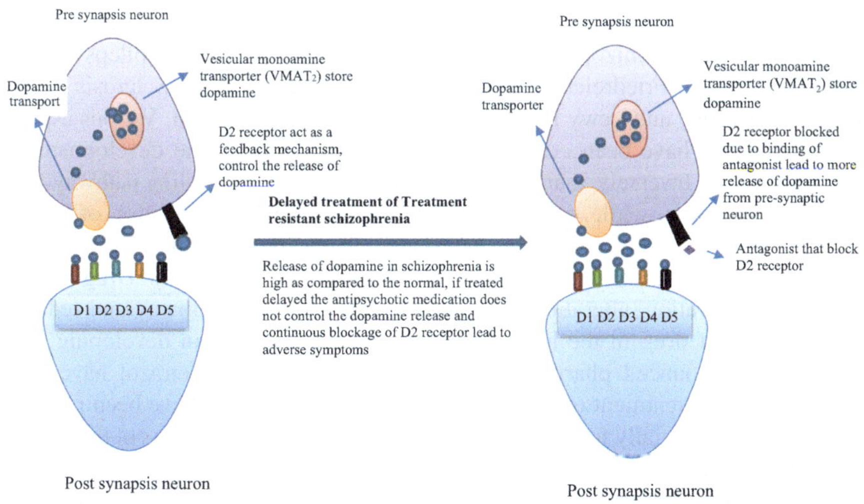

Fig. 9.8 Delayed treatment of treatment resistant schizophrenia (TRS). Abnormalities in regulation of glutamate may possibly play a major role in TRS (Demjaha et al., 2014). In the anterior cingulate cortex, level of Glutamate is higher in patients with TRS as compared with healthy controls or treatment responsive schizophrenic patients (Mouchlianitis et al., 2016). In future multidimensional approach like dose dependent treatment and neuroimaging may help in the optimization of response, reduce relapses and provide valuable treatment for treatment resistant schizophrenia

The omission of antipsychotic dopamine blockade to treat symptoms in some schizophrenic patients suggests that other neurotransmitters like glutamate plays an important role in the etiology of TRS (Schwartz et al., 2012). The hyperactivity of dopamine is downstream of GABA and glutamate dysregulation (Howes et al., 2015). N-methyl D-aspartate (NMDA) receptor dysfunction on GABA interneurons leads to the disinhibition of cortical or hippocampal glutamate neurons sticking out to the basal ganglia. This hyper activation of glutamate neurons, in turn, the activity of dopaminergic projections is stimulated to the striatum from the midbrain, leads to onset of positive symptoms of schizophrenia (Lodge, 2009; Schwartz et al., 2012). It was observed that schizophrenic patients display an aggravation of positive symptoms after administration of phencyclidine (PCP), an NMDA receptor antagonist, and that PCP mimics the positive effects of schizophrenia in healthy volunteers (Domino & Luby, 2012).

9.6 Conclusion

Neurological disorders are the major health problem worldwide, especially in the developing countries. The disease mechanisms and treatment strategies have been extensively studied over the past to combat lethal neurological disorders. Numerous drugs with boosted therapeutic values for the treatment of multiple neurological disorders, like HD, Schizophrenia, Alzheimer, Parkinson's disease, epilepsy, spinal muscular atrophy, Friedreich's Ataxia and Amyotrophic Lateral Sclerosis Spinal Muscular Atrophy and Lewy Body Dementia have been developed. Various aspects of neurogenesis have been comprehensively investigated for the development of drug expedition. Inversely, many drugs interact with several signaling pathways and impose side-effects, while some drugs are modulated by the activation of several mechanism in the body developing resistance. The drug resistance emergence is major obstacle for lowering the prevalence of neurological disorders. Drug modification or inactivation, alteration of target sites, or metabolic pathway and reduced sustainability of drug in the body are major potential factors in developing drug resistance. Pronounced pharmaco-resistance against pentylenetetrazol have been observed during treatment of epilepsy. The resistance-response have been reported in many drugs especially for levodopa and TBZ. The complete analysis of the disease mechanisms and drug metabolism are paramount to combat drug resistance and for the development of novel and effective drugs. Modern potential strategies like pharmacogenomics, personalized therapeutic approaches, pro-neurogenic treatment at stem cell level, drug targeting and modern molecular biology tools like CRISPR/Cas could be applied to overcome these challenges.

References

Abbott, N. J., Khan, E. U., Rollinson, C. M., Reichel, A., Janigro, D., Dombrowski, S. M., Dobbie, M. S. & Begley, D. J. (2002) March. Drug resistance in epilepsy: The role of the blood-brain barrier. In *Novartis foundation symposium* (pp. 38–53). John Wiley; 1999.

Abi-Dargham, A., Rodenhiser, J., Printz, D., Zea-Ponce, Y., Gil, R., Kegeles, L. S., Weiss, R., Cooper, T. B., Mann, J. J., Van Heertum, R. L., & Gorman, J. M. (2000). Increased baseline occupancy of D2 receptors by dopamine in schizophrenia. *Proceedings of the National Academy of Sciences, 97*(14), 8104–8109.

Altamura, A. C., Bassetti, R., Cattaneo, E., & Vismara, S. (2005). Some biological correlates of drug resistance in schizophrenia: a multidimensional approach. *The World Journal of Biological Psychiatry, 6*(sup2), 23–30.

Andreasen, N. C., Arndt, S., Alliger, R., Miller, D., & Flaum, M. (1995). Symptoms of schizophrenia: Methods, meanings, and mechanisms. *Archives of General Psychiatry, 52*(5), 341–351.

Armstrong, M. J., & Okun, M. S. (2020). Diagnosis and treatment of Parkinson disease: a review. *JAMA, 323*(6), 548–560.

Arning, L., Kraus, P. H., Valentin, S., Saft, C., Andrich, J., & Epplen, J. T. (2005). NR2A and NR2B receptor gene variations modify age at onset in Huntington disease. *Neurogenetics, 6*(1), 25–28.

Bachoud-Lévi, A. C., Ferreira, J., Massart, R., Youssov, K., Rosser, A., Busse, M., Craufurd, D., Reilmann, R., De Michele, G., Rae, D., & Squitieri, F. (2019). International Guidelines for the treatment of Huntington's. *Frontiers in Neurology, 10*, 710.

Baldini, N. (1997). Multidrug resistance-a multiplex phenomenon. *Nature Medicine, 3*(4), 378–380.

Balestrino, R., & Schapira, A. H. (2020). *Parkinson Disease. European Journal of Neurology, 27*(1), 27–42.

Bates, G. (2003). Huntingtin aggregation and toxicity in Huntington's disease. *The Lancet, 361*(9369), 1642–1644.

Beister, A., Kraus, P., Kuhn, W., Dose, M., Weindl, A. & Gerlach, M. (2004). The N-methyl-D-aspartate antagonist memantine retards progression of Huntington's disease. In *Focus on extrapyramidal dysfunction* (pp. 117–122). Springer.

Berg, A. T. (2004). Understanding the delay before epilepsy surgery: who develops intractable focal epilepsy and when? *CNS Spectrums, 9*(2), 136–144.

Blair, H. A., & Dhillon, S. (2017). Safinamide: A review in Parkinson's disease. *CNS Drugs, 31*(2), 169–176.

Blanke, M. L. & Van Dongen, A. M. (2008). 13 activation mechanisms of the NMDA receptor. *Biology of the NMDA Receptor*, 283.

Block, R. C., Dorsey, E. R., Beck, C. A., Brenna, J. T., & Shoulson, I. (2010). Altered cholesterol and fatty acid metabolism in Huntington disease. *Journal of Clinical Lipidology, 4*(1), 17–23.

Bonifati, V., Rizzu, P., Van Baren, M. J., Schaap, O., Breedveld, G. J., Krieger, E., Dekker, M. C., Squitieri, F., Ibanez, P., Joosse, M., & Van Dongen, J. W. (2003). Mutations in the DJ-1 gene associated with autosomal recessive early-onset parkinsonism. *Science, 299*(5604), 256–259.

Calabresi, P., Di Filippo, M., Ghiglieri, V., Tambasco, N., & Picconi, B. (2010). Levodopa-induced dyskinesias in patients with Parkinson's disease: filling the bench-to-bedside gap. *The Lancet Neurology, 9*(11), 1106–1117.

Cenci, M. A. (2014). Presynaptic mechanisms of l-DOPA-induced dyskinesia: the findings, the debate, and the therapeutic implications. *Frontiers in Neurology, 5*, 242.

Chouinard, G., Samaha, A. N., Chouinard, V. A., Peretti, C. S., Kanahara, N., Takase, M., & Iyo, M. (2017). Antipsychotic-induced dopamine supersensitivity psychosis: pharmacology, criteria, and therapy. *Psychotherapy and Psychosomatics, 86*(4), 189–219.

Cohen, A. S., Lin, D. D., Quirk, G. L., & Coulter, D. A. (2003). Dentate granule cell GABAA receptors in epileptic hippocampus: Enhanced synaptic efficacy and altered pharmacology. *European Journal of Neuroscience, 17*(8), 1607–1616.

Congreve, M., & Marshall, F. (2010). The impact of GPCR structures on pharmacology and structure-based drug design. *British Journal of Pharmacology, 159*(5), 986–996.

Cottler, L. B., Zunt, J., Weiss, B., Kamal, A. K., & Vaddiparti, K. (2016). Building global capacity for brain and nervous system disorders research. Nature 527. *S207–S213.* (http://www.nature.com/doifinder/10.1038/nature16037). Accessed, 24.

Creus-Muncunill, J., Badillos-Rodríguez, R., Garcia-Forn, M., Masana, M., Garcia-DíazBarriga, G., Guisado-Corcoll, A., Alberch, J., Malagelada, C., Delgado-García, J. M., Gruart, A., & Pérez Navarro, E. (2019). Increased translation as a novel pathogenic mechanism in Huntington's disease. *Brain, 142*(10), 3158–3175.

Daneman, R., & Prat, A. (2015). The blood–brain barrier. *Cold Spring Harbor Perspectives in Biology, 7*(1), a020412.

Dash, D., & Mestre, T. A. (2020). Therapeutic update on Huntington's disease: symptomatic treatments and emerging disease-modifying therapies. *Neurotherapeutics*, 1–15.

Dean, M., Hamon, Y., & Chimini, G. (2001). The human ATP-binding cassette (ABC) transporter superfamily. *Journal of Lipid Research, 42*(7), 1007–1017.

Dehay, B., Ramirez, A., Martinez Vicente, M., Perier, C., Canron, M. H., Doudnikoff, E., Vital, A., Vila, M., Klein, C., & Bezard, E. (2012). Loss of P-type ATPase ATP13A2/PARK9 function induces general lysosomal deficiency and leads to Parkinson disease neurodegeneration. *Proceedings of the National Academy of Sciences, 109*(24), 9611–9616.

Demjaha, A., Egerton, A., Murray, R. M., Kapur, S., Howes, O. D., Stone, J. M., & McGuire, P. K. (2014). Antipsychotic treatment resistance in schizophrenia associated with elevated glutamate levels but normal dopamine function. *Biological Psychiatry, 75*(5), e11–e13.

Domino, E. F., & Luby, E. D. (2012). Phencyclidine/schizophrenia: One view toward the past, the other to the future. *Schizophrenia Bulletin, 38*(5), 914–919.

Drozdzik, M., Bialecka, M., Mysliwiec, K., Honczarenko, K., Stankiewicz, J., & Sych, Z. (2003). Polymorphism in the P-glycoprotein drug transporter MDR1 gene: A possible link between environmental and genetic factors in Parkinson's disease. *Pharmacogenetics and Genomics, 13*(5), 259–263.

Farooq, S., Agid, O., Foussias, G., & Remington, G. (2013). Using treatment response to subtype schizophrenia: Proposal for a new paradigm in classification.

Fornaro, M., Fusco, A., Anastasia, A., Cattaneo, C. I., & De Berardis, D. (2019). Brexpiprazole for treatment-resistant major depressive disorder. *Expert Opinion on Pharmacotherapy, 20*(16), 1925–1933.

Fox, S. H., Katzenschlager, R., Lim, S. Y., Ravina, B., Seppi, K., Coelho, M., Poewe, W., Rascol, O., Goetz, C. G., & Sampaio, C. (2011). The movement disorder society evidence-based medicine review update: Treatments for the motor symptoms of Parkinson's disease. *Movement Disorders, 26*(S3), S2–S41.

Frånberg, O., Wiker, C., Marcus, M. M., Konradsson, Å., Jardemark, K., Schilström, B., Shahid, M., Wong, E. H., & Svensson, T. H. (2008). Asenapine, a novel psychopharmacologic agent: Preclinical evidence for clinical effects in schizophrenia. *Psychopharmacology (berl), 196*(3), 417–429.

Frank, S., Ondo, W., Fahn, S., Hunter, C., Oakes, D., Plumb, S., Marshall, F., Shoulson, I., Eberly, S., Walker, F., & Factor, S. (2008). A study of chorea after tetrabenazine withdrawal in patients with Huntington disease. *Clinical Neuropharmacology, 31*(3), 127–133.

French, J. A., Kanner, A. M., Bautista, J., Abou-Khalil, B., Browne, T., Harden, C. L., Theodore, W. H., Bazil, C., Stern, J., Schachter, S. C., & Bergen, D. (2004). Efficacy and tolerability of the new antiepileptic drugs II: Treatment of refractory epilepsy: Report of the therapeutics and technology assessment subcommittee and quality standards subcommittee of the American academy of neurology and the American epilepsy society. *Neurology, 62*(8), 1261–1273.

Garver, D. L., Holcomb, J. A., & Christensen, J. D. (2000). Heterogeneity of response to antipsychotics from multiple disorders in the schizophrenia spectrum. *Journal of Clinical Psychiatry, 61*(12), 964–972.

Gil, J. M., & Rego, A. C. (2008). Mechanisms of neurodegeneration in Huntington's disease. *European Journal of Neuroscience, 27*(11), 2803–2820.

Gillespie, A. L., Samanaite, R., Mill, J., Egerton, A., & MacCabe, J. H. (2017). Is treatment-resistant schizophrenia categorically distinct from treatment-responsive schizophrenia? *A Systematic Review. BMC Psychiatry, 17*(1), 12.

Godinho, B. M., Malhotra, M., O'Driscoll, C. M., & Cryan, J. F. (2015). Delivering a disease-modifying treatment for Huntington's disease. *Drug Discovery Today, 20*(1), 50–64.

Gottesman, M. M., & Pastan, I. (1993). Biochemistry of multidrug resistance mediated by the multidrug transporter. *Annual Review of Biochemistry, 62*(1), 385–427.

Gottesman, M. M., Fojo, T., & Bates, S. E. (2002). Multidrug resistance in cancer: Role of ATP–dependent transporters. *Nature Reviews Cancer, 2*(1), 48–58.

He, L., Vasiliou, K., & Nebert, D. W. (2009). Analysis and update of the human solute carrier (SLC) gene superfamily. *Human Genomics, 3*(2), 195.

Hockly, E., Richon, V. M., Woodman, B., Smith, D. L., Zhou, X., Rosa, E., Sathasivam, K., Ghazi-Noori, S., Mahal, A., Lowden, P. A., & Steffan, J. S. (2003). Suberoylanilidehydroxamic acid, a histone deacetylase inhibitor, ameliorates motor deficits in a mouse model of Huntington's disease. *Proceedings of the National Academy of Sciences, 100*(4), 2041–2046.

Holmes, G. L. (2004). Epidemiology and clinical mechanisms of drug resistance in epilepsy: new horizons in the development of antiepileptic drugs: The search for new targets: a conference review. *Epilepsy Research, 60*, 77–159.

Howes, O., McCutcheon, R., & Stone, J. (2015). Glutamate and dopamine in schizophrenia: an update for the 21st century. *Journal of Psychopharmacology, 29*(2), 97–115.

Iyo, M., Tadokoro, S., Kanahara, N., Hashimoto, T., Niitsu, T., Watanabe, H., & Hashimoto, K. (2013). Optimal extent of dopamine D2 receptor occupancy by antipsychotics for treatment of dopamine supersensitivity psychosis and late-onset psychosis. *Journal of Clinical Psychopharmacology, 33*(3), 398–404.

Jones, D. M., Esmaeil, N., Maren, S., & Macdonald, R. L. (2002). Characterization of pharmacoresistance to benzodiazepines in the rat Li-pilocarpine model of status epilepticus. *Epilepsy Research, 50*(3), 301–312.

Kapur, S., & Seeman, P. (2000). Antipsychotic agents differ in how fast they come off the dopamine D2 receptors. Implications for atypical antipsychotic action. *Journal of Psychiatry and Neuroscience, 25*(2), 161.

Kaur, N., Kumar, P., Jamwal, S., Deshmukh, R., & Gauttam, V. (2016). Tetrabenazine: Spotlight on drug review. *Annals of Neurosciences, 23*(3), 176–185.

Kordasiewicz, H. B., Stanek, L. M., Wancewicz, E. V., Mazur, C., McAlonis, M. M., Pytel, K. A., Artates, J. W., Weiss, A., Cheng, S. H., Shihabuddin, L. S., & Hung, G. (2012). Sustained therapeutic reversal of Huntington's disease by transient repression of huntingtin synthesis. *Neuron, 74*(6), 1031–1044.

Kramlinger, K. G., & Post, R. M. (1989). The addition of lithium to carbamazepine. Antidepressant efficacy in treatment-resistant depression. *Archives of general psychiatry, 46*(9), 794–800.

Krebs, M. O., Sautel, F., Bourdel, M. C., Sokoloff, P., Schwartz, J. C., Olie, J. P., Loo, H., & Poirier, M. F. (1998). Dopamine D 3 receptor gene variants and substance abuse in schizophrenia. *Molecular Psychiatry, 3*(4), 337–341.

Labbadia, J., & Morimoto, R. I. (2013). Huntington's disease: underlying molecular mechanisms and emerging concepts. *Trends in Biochemical Sciences, 38*(8), 378–385.

Landwehrmeyer, G. B., Dubois, B., de Yébenes, J. G., Kremer, B., Gaus, W., Kraus, P. H., Przuntek, H., Dib, M., Doble, A., Fischer, W., & Ludolph, A. C. (2007). Riluzole in Huntington's disease: A 3-year, randomized controlled study. *Annals of Neurology: Official Journal of the American Neurological Association and the Child Neurology Society, 62*(3), 262–272.

Lerche, H. (2020). Drug-resistant epilepsy—time to target mechanisms. *Nature Reviews Neurology*, 1–2.

LeWitt, P.A., & Fahn, S., (2016). Levodopa therapy for Parkinson disease: A look backward and forward. *Neurology, 86*(14 Supplement 1), pp. S3–S12.

Li, Q., & Shu, Y. (2014). Role of solute carriers in response to anticancer drugs. *Molecular and Cellular Therapies, 2*(1), 1–14.

Lin, L., Yee, S. W., Kim, R. B., & Giacomini, K. M. (2015). SLC transporters as therapeutic targets: emerging opportunities. *Nature Reviews Drug Discovery, 14*(8), 543–560.

Lodge, D. J., Behrens, M. M., & Grace, A. A. (2009). A loss of parvalbumin-containing interneurons is associated with diminished oscillatory activity in an animal model of schizophrenia. *Journal of Neuroscience, 29*(8), 2344–2354.

Löscher, W. (1998). New visions in the pharmacology of anticonvulsion. *European Journal of Pharmacology, 342*(1), 1–13.

Löscher, W., & Potschka, H. (2005). Role of drug efflux transporters in the brain for drug disposition and treatment of brain diseases. *Progress in Neurobiology, 76*(1), 22–76.

Löscher, W., & Schmidt, D. (2004). New horizons in the development of antiepileptic drugs: the search for new targets. *Epilepsy Research, 60*(2–3), 77–159.

MacLeod, A. R., & Crooke, S. T. (2017). RNA therapeutics in oncology: advances, challenges, and future directions. *The Journal of Clinical Pharmacology, 57*, S43–S59.

McGarry, A., McDermott, M., Kieburtz, K., de Blieck, E. A., Beal, F., Marder, K., Ross, C., Shoulson, I., Gilbert, P., Mallonee, W. M., & Guttman, M. (2017). A randomized, double-blind, placebo-controlled trial of coenzyme Q10 in Huntington disease. *Neurology, 88*(2), 152–159.

Möhler, H., Fritschy, J. M., & Rudolph, U. (2002). A new benzodiazepine pharmacology. *Journal of Pharmacology and Experimental Therapeutics, 300*(1), 2–8.

Moreira, C. G., Morawska, M. M., Baumann, A., Masneuf, S., Linnebank, M., Sommerauer, M., Landolt, H. P., Noain, D., & Baumann, C. R. (2020). Improved functional and histochemical outcomes in l-DOPA plus tolcapone treated VMAT2-deficient mice. *Neuropharmacology, 181*, 108353.

Mouchlianitis, E., Bloomfield, M. A., Law, V., Beck, K., Selvaraj, S., Rasquinha, N., Waldman, A., Turkheimer, F. E., Egerton, A., Stone, J., & Howes, O. D. (2016). Treatment-resistant schizophrenia patients show elevated anterior cingulate cortex glutamate compared to treatment-responsive. *Schizophrenia Bulletin, 42*(3), 744–752.

Negida, A., Elfil, M., Attia, A., Farahat, E., Gabr, M., Essam, A., Attia, D., & Ahmed, H. (2017). Caffeine; the forgotten potential for Parkinson's disease. *CNS & Neurological Disorders-Drug Targets (Formerly Current Drug Targets-CNS & Neurological Disorders), 16*(6), 652–657.

Newcomer, J. W., Eriksson, H., Zhang, P., Weiller, E., & Weiss, C. (2018). Changes in metabolic parameters and body weight in brexpiprazole-treated patients with acute schizophrenia: Pooled analyses of phase 3 clinical studies. *Current Medical Research and Opinion, 34*(12), 2197–2205.

Oda, Y., Kanahara, N., & Iyo, M. (2015). Alterations of dopamine D2 receptors and related receptor-interacting proteins in schizophrenia: The pivotal position of dopamine supersensitivity psychosis in treatment-resistant schizophrenia. *International Journal of Molecular Sciences, 16*(12), 30144–30163.

Olanow, C. W., Obeso, J. A., & Stocchi, F. (2006). Continuous dopamine-receptor treatment of Parkinson's disease: Scientific rationale and clinical implications. *The Lancet Neurology, 5*(8), 677–687.

Olsen, R. W., & DeLorey, T. M., (1999). GABA receptor physiology and pharmacology. *Basic Neurochemistry: Molecular, Cellular and Medical Aspects.*

Papakostas, G. I., Petersen, T. J., Kinrys, G., Burns, A. M., Worthington, J. J., Alpert, J. E., Fava, M., & Nierenberg, A. A. (2005). Aripiprazole augmentation of selective serotonin reuptake inhibitors for treatment-resistant major depressive disorder. *Journal of Clinical Psychiatry, 66*(10), 1326–1330.

Pardridge, W. M. (1999). Blood-brain barrier biology and methodology. *Journal of Neurovirology, 5*(6), 556–569.

Perucca, E. (1998). Pharmacoresistance in Epilepsy. *CNS Drugs, 10*(3), 171–179.

Petros, W. P., & Evans, W. E. (2004). Pharmacogenomics in cancer therapy: is host genome variability important? *Trends in Pharmacological Sciences, 25*(9), 457–464.

Poewe, W., Seppi, K., Tanner, C. M., Halliday, G. M., Brundin, P., Volkmann, J., Schrag, A. E., & Lang, A. E. (2017). Parkinson disease. *Nature Reviews Disease Primers, 3*(1), 1–21.

Polymeropoulos, M. H., Lavedan, C., Leroy, E., Ide, S. E., Dehejia, A., Dutra, A., Pike, B., Root, H., Rubenstein, J., Boyer, R., & Stenroos, E. S. (1997). Mutation in the α-synuclein gene identified in families with Parkinson's disease. *Science, 276*(5321), 2045–2047.

Rogawski, M. A., & Löscher, W. (2004). The neurobiology of antiepileptic drugs. *Nature Reviews Neuroscience, 5*(7), 553–564.

Ross, D. D., & Nakanishi, T. (2010). Impact of breast cancer resistance protein on cancer treatment outcomes. In *Multi-drug resistance in cancer* (pp. 251–290). Humana Press.

Samaranch, L., Blits, B., San Sebastian, W., Hadaczek, P., Bringas, J., Sudhakar, V., Macayan, M., Pivirotto, P. J., Petry, H., & Bankiewicz, K. S. (2017). MR-guided parenchymal delivery of adeno-associated viral vector serotype 5 in non-human primate brain. *Gene Therapy, 24*(4), 253–261.

Schneider, R. B., Auinger, P., Tarolli, C. G., Iourinets, J., Gil-Díaz, M. C., & Richard, I. H. (2020). A trial of buspirone for anxiety in Parkinson's disease: Safety and tolerability. *Parkinsonism & Related Disorders.*

Schwartz, T. L., Sachdeva, S., & Stahl, S. M. (2012). Glutamate neurocircuitry: Theoretical underpinnings in schizophrenia. *Frontiers in Pharmacology, 3*, 195.

Seeman, P., Lee, T., Chau-Wong, M., & Wong, K. (1976). Antipsychotic drug doses and neuroleptic/dopamine receptors. *Nature, 261*(5562), 717–719.

Shannon, K. M., & Fraint, A. (2015). Therapeutic advances in Huntington's disease. *Movement Disorders, 30*(11), 1539–1546.

Sillanpää, M., Jalava, M., & Shinnar, S. (1999). Epilepsy syndromes in patients with childhood-onset seizures in Finland. *Pediatric Neurology, 21*(2), 533–537.

Siuly, S., & Zhang, Y. (2016). Medical big data: neurological diseases diagnosis through medical data analysis. *Data Science and Engineering, 1*(2), 54–64.

Sonneveld, P. (2001). Multidrug resistance in haematological malignancies. *Journal of Internal Medicine, 249*(S741), 47–60.

Spector, R. (2000). Drug transport in the mammalian central nervous system: Multiple complex systems. *Pharmacology, 60*(2), 58–73.

Strazielle, N., & Ghersi-Egea, J. F. (2016). Potential pathways for CNS drug delivery across the blood-cerebrospinal fluid barrier. *Current Pharmaceutical Design, 22*(35), 5463–5476.

Stricker, T., & Arteaga, C. L. (2015). Drug-resistant brain metastases: a role for pharmacology, tumor evolution, and too-late therapy. *Cancer Discovery, 5*(11), 1124–1126.

Suzuki, T., Kanahara, N., Yamanaka, H., Takase, M., Kimura, H., Watanabe, H., & Iyo, M. (2015). Dopamine supersensitivity psychosis as a pivotal factor in treatment-resistant schizophrenia. *Psychiatry Research, 227*(2–3), 278–282.

Thomas, M., Ashizawa, T., & Jankovic, J. (2004). Minocycline in Huntington's disease: A pilot study. *Movement Disorders: Official Journal of the Movement Disorder Society, 19*(6), 692–695.

Thulasiraman, V., Yang, C. F., & Frydman, J. (1999). In vivo newly translated polypeptides are sequestered in a protected folding environment. *The EMBO Journal, 18*(1), 85–95.

VanRoon-Mom, W. M., Roos, R. A., & de Bot, S. T. (2018). Dose-dependent lowering of mutant Huntingtin using antisense oligonucleotides in Huntington disease patients. *Nucleic Acid Therapeutics, 28*(2), 59–62.

Vita, A., Minelli, A., Barlati, S., Deste, G., Giacopuzzi, E., Valsecchi, P., Turrina, C., & Gennarelli, M. (2019). Treatment-resistant schizophrenia: Genetic and neuroimaging correlates. *Frontiers in Pharmacology, 10*, 402.

Volk, H. A., Burkhardt, K., Potschka, H., Chen, J., Becker, A., & Löscher, W. (2004). Neuronal expression of the drug efflux transporter P-glycoprotein in the rat hippocampus after limbic seizures. *Neuroscience, 123*(3), 751–759.

Vonsattel, J. P., Myers, R. H., Stevens, T. J., Ferrante, R. J., Paskevich, P. A., Richardson, E. P., & Bird, E. D. (1987). Huntington's disease: Neuropathological grading. In *The basal ganglia II* (pp. 515–531). Springer

Wang, H., Chen, X., Li, Y., Tang, T. S., & Bezprozvanny, I. (2010). Tetrabenazine is neuroprotective in Huntington's disease mice. *Molecular Neurodegeneration, 5*(1), 18.

Wang, Q., Chen, J., Li, M., Lv, S., Xie, Z., Li, N., Wang, N., Wang, J., Luo, F., & Zhang, W. (2019). Eltoprazine prevents levodopa-induced dyskinesias by reducing causal interactions for theta oscillations in the dorsolateral striatum and substantia nigra pars reticulate. *Neuropharmacology, 148*, 1–10.

Wertz, M. H., Mitchem, M. R., Pineda, S. S., Hachigian, L. J., Lee, H., Lau, V., Powers, A., Kulicke, R., Madan, G. K., Colic, M., & Therrien, M. (2020). Genome-wide In Vivo CNS screening identifies genes that Modify CNS neuronal survival and mHTT toxicity. *Neuron.*

Yang, S., Chang, R., Yang, H., Zhao, T., Hong, Y., Kong, H. E., Sun, X., Qin, Z., Jin, P., Li, S., & Li, X. J. (2017). CRISPR/Cas9-mediated gene editing ameliorates neurotoxicity in mouse model of Huntington's disease. *The Journal of Clinical Investigation, 127*(7), 2719–2724.

You, G., & Morris, M. E. (2006). *Drug transporters: Molecular characterization and role in drug disposition.* John Wiley and Sons.

Chapter 10
Drug Resistance in Kidney Diseases

Nadia Rashid, Dilshad Hussain, Saima Ashraf, Nabila Bashir, Saadat Majeed, Muhammad Naeem Ashiq, and Muhammad Najam-ul-Haq

Abstract Chronic kidney diseases are defined as the persistent malfunction of kidneys. Drug resistance in kidney diseases is a concern in health care. Scientists are finding out therapies to treat renal infections. Several drugs and therapeutics are in the market but pathogens are getting immuned to these therapies. Some pathogens are multi-drug resistant. Selection of the best immunosuppressive therapy is a complex method. Finding the balance between their advantages and disadvantages is the key. Dosing and potency must be considered for patients with long-term use to avoid antimetabolic effects regarding diuretics including renal failure, hypomagnesemia, idiosyncratic reactions, and severe allergic reactions. This chapter highlights the drug resistance in kidney diseases which include bacterial resistance, viral resistance, kidney cancer, and kidney transplant. Some suggestions to prevent drug resistance in kidney are also included.

10.1 Introduction

In the past few years, kidney diseases are spreading around the globe. Aging is important factor related to chronic kidney disease (CKD). These diseases are predominantly present among women and African Americans (Nagata et al., 2010; Zhang & Rothenbacher, 2008). Recent studies elucidate that impaired kidney function is increasing at drastic rate (McCullough et al., 2012). The range of prevalence of iKF in China is 1.7% and 8.1% in United States. Early diagnosis can contain or reduce the progression of kidney diseases (Cusumano & Bedat, 2008). CKD must be attended to avoid complications like impaired kidney function and kidney failure and ultimately death. Renal transplant can increase life expectancy and is cost-effective as compared to

N. Rashid · S. Ashraf · N. Bashir · S. Majeed · M. N. Ashiq · M. Najam-ul-Haq (✉)
Institute of Chemical Sciences, Bahauddin Zakariya University, Multan 60800, Pakistan
e-mail: najamulhaq@bzu.edu.pk

D. Hussain
International Center for Chemical and Biological Sciences, HEJ Research Institute of Chemistry, University of Karachi, Karachi 75270, Pakistan

S. Ahmed et al. (eds.), *Biochemistry of Drug Resistance*,
https://doi.org/10.1007/978-3-030-76320-6_10

dialysis (Port et al., 1993). In United States, 547,982 patients are treated for end-stage renal diseases while 17,413 kidney transplants were performed in the year 2008 (Kim et al., 2010). Many drugs in combinations are used by medical physicians. Steroids i.e., cyclosporine (CsA) or tacrolimus (TAC) can be used in combination with calcineurin inhibitors or with signal inhibitors. The steroids can also be used in combination with antimetabolites and mycophenolate mofetil (Kasiske et al., 2010). Physicians need to consider the adverse condition of individual while deciding correct and optimal dose or strategy of immunosuppressive in treating a patient (Goldfarb, 2005). Usually patients of kidney disease are trialed by giving multiple medicines to cure kidney damage. Some drugs aggravate kidney damage if used long time. Clinical practice guideline (C.P.G) has not determined any pharma kinetic, drug-drug interaction, and drug reception,etc. A random combination of drugs can be given to the patient or sometime herbal treatment (Bailie et al., 2005).

P-glycoprotein induces multidrug resistance (Gottesman & Pastan, 1993). It is not true that the existence of MDR is always correlated with cancer. MDR is also found in normal cell lines. In the case of kidney, they are located in proximal tubules, collecting tubules and loops of Henley. These resistant genes coexist in nephron and function with Pgp for transportation of xenobiotics (Ernest et al., 1997). These proteins have range of substrates. Some of its substrates are involved in the excretion of renal metabolites (Speeg et al., 1992). Kidney transporters of drugs are affected by the Pgp and its substrates. For example, the rate of removal of digoxin through urine is blocked by using verapamil, quinidine, and daunorubicin but remains unaffected by para-aminohippurate and tetraethylammonium. It can thus be concluded that Pgp influences the transportation rate (Pritchard & Miller, 1997).

In chronic kidney disease, the efflux rate of drug to intestine is decreased due to activation of MDR proteins. Studies have revealed that bioavailability of p-glycoprotein is elevated that in turn reduces the activity of CYP3A, enzymes of the first pass of drug metabolism. Those drugs that are metabolized by other than the kidney such as liver and digestive tract are also affected metabolically in kidney disease. Similarly, some of liver enzymes of phase 2 are reduced in kidney disease that cause lowered drug metabolism in liver (Nolin et al., 2008). Diuretics are frequently prescribed for kidney diseases. They function by controlling blood pressure in kidney patients. With the sodium channels, non-osmotic transport of sodium ions is increased to the skin and muscles to control hypertension. Thiazine is a diuretic prescribed to the patients even at later stages of CKD. Sodium removal from body is increased by this drug (Agarwal & Sinha, 2012).

Patients of CKD die of heart failure more frequently than that of the last stage of kidney failure. The major factor of cardiac failure as well as aggravation of CKD is dyslipidemia. Statins are used for controlling the levels of cholesterol in preserving kidney. They are the lipid controlling medicines that function via non-cholesterol-mediated mechanisms. They control GFR lowering, decrease heart damage, and lower proteinuria in CKD patients (Agarwal, 2007). Statins work as immunomodulator in CKD patients. They slow the endothelial dysfunction, and reduce unusual change in permeability of proteins (O'Driscoll et al., 1997). They also suppress blood pressure (Golomb et al., 2008). Statins are metabolized through the liver however eliminated by the kidney. In organ transplant, statins are co-administered with cyclosporine. In that case, chances of accumulation of toxins in kidney are increased. That is why statins dose adjustment and drug-drug interactions are monitored (Launay-Vacher et al., 2005).

10.2 Genetic Basis of Kidney Disorders

Scientists have been trying to identify the genetic basis of kidney diseases for the past 25 years. As a result, medical professionals are equipped to identify the molecular basis of kidney diseases for new therapeutic drugs to control the further spread (Vivante & Hildebrandt, 2016). Advancements in the genetic renal physiology has provided information about the disease. Genes mutations are responsible for monogenic disorders of the kidney. These mutations can alter proteins structures, receptors, enzymes, transcription factors and structural components (Cañadas-Garrc et al., 2019). Moreover, kidney is a complex organ containing specialized cells. Any disorder can disturb blood pressure, balance of electrolytes, water and oxygen supply in blood, acid–base homeostasis, hormonal imbalance, metabolic clearance of secretions in blood and the central nervous system (Emma et al., 2016). Advances in medical sciences have made renal transplant possible where a patient dies rarely with kidney disorders. However, such a patient can survive but with underlying health condition (Eddy, 2014).

10.3 Antiviral Drugs and Kidney Disease

According to the Infectious Disease Society of America, there is high risk of acute renal failure in HIV infection at BII level. When CD4 count is more than 200 cells/mm^3 along with an HIV RNA level of more than 4000 copies/mL, coupled with a family history of hypertension, diabetes and viral infection like hepatitis C, the chances of renal failure are doubled as compared to that of in normal conditions (Gupta et al., 2005). These individuals must be screened for proteinuria to assess CKD. Early detection of CKD can be advantageous (Muller et al., 2015). Infection of HIV-1 can affect the kidney directly or indirectly (Fig. 10.1). The drugs metabolites used against HIV, Co-infections of HIV, and tumor formation, are associated with kidney problems (Marras et al., 2002).

HIV 1 leads to infectious cell production in glomerular epithelial cells and renal-tubules. It is the common pathological condition associated with HIV-1 (Ross & Klotman, 2004). HIV associated nephropathy (HIVAN), acute renal failure comes with HIV infections (Gupta et al., 2005). When antiretroviral drugs are taken for long-span, they lead to chronic kidney disease. Antiretroviral such as tenofovir, indinavir, and atazanavir have increased risk of CKD (Mocroft et al., 2010). The kidney function is disturbed in CKD, causing the storage of unmetabolized compounds and

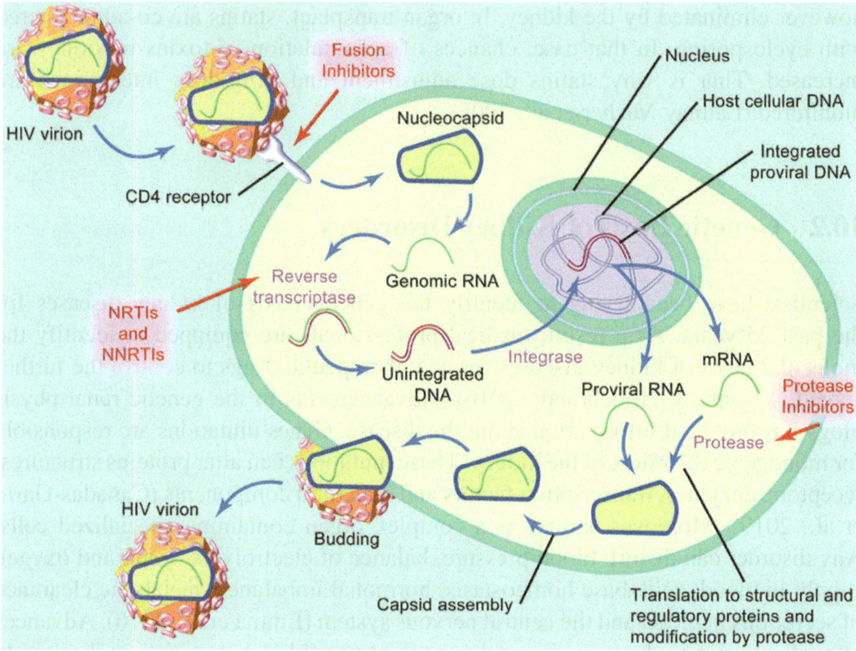

Fig. 10.1 Replication life cycle of HIV and sites of antiretroviral drug action in kidney (Berns & Kasbekar, 2006)

metabolic products. Same happens with the drugs and accumulation of byproducts of drugs producing unpredictable pharmacological effects. For example, nabbing is observed in CKD patients due to the accumulation of norepinephrine. Thus dosing amount of drug should be considered in CKD (Matzke et al., 2011). In CKD patients, the drug metabolism, functionality as well as the response is altered. Dose management is compulsory for desired results of drug in CKD patients. Dialysis can improve the pharmacokinetics of a drug (Naud et al., 2012).

Multidrug resistance proteins are distributed unevenly when compared to phosphor glycol protein in kidney epithelium of pig undergone induction with human MRP cDNA (Bréchot et al., 1998). Corticosteroids function in convoluted tubules of the kidney. When human kidney having MRP transfused on it,is subjected to corticosteroid hormones, it activates 17b-d-glucuronosyl estradiol leukotriene C4 and 17b-estradiol (Stride et al., 1997). Releasing cortisol without metabolizing, it requires functions of these hormones (Peng et al., 1999). The creatinine clearance is observed before prescribing the medicine to CKD patients. Inappropriate calibration of dose to CKD patients leads to therapeutic failure. More than 40% CKD patients of late stages get hyper dosing of drugs due to reduced renal creatinine clearance (Gabardi & Abramson, 2005). A variety of microbes is found in urinary tract. Different kinds of urinary systems have their kinds of microbes. These microbes have role in drug metabolism. CKD patient undergone kidney transplantation experiences new phyla of microbes. These microbial colonies are more resistant to antibiotic treatment (Rani et al., 2017). Patients undergone kidney transplants may exhibit resistance against antiviral drugs. One of the known antiviral, Anti-cytomegalovirus is given to the subject. Therapeutic failure of Anti-cytomegalovirus is observed in the patients having 5.25 log10 copies/mL. GFR is also lowered in the patients having this resistance (Couzi et al., 2012).

HIV associated with kidney diseases like HIVAN are controlled by host-specific factors. African Americans are 18 times more susceptible to HIV associated nephropathy as compared to other nations (Bhimma et al., 2013). This is because of the variation at MYH9 locus of chromosome 22. This gene encodes myosin which is an essential component of non-muscle myosin IIA (Kopp et al., 2008). This variation is associated with hypertensive nephrosclerosis and HIVAN in African Americans (Kao et al., 2008). Hence, most of the data about HIV-linked kidney diseases is related to African Americans. Scientists need to conduct studies on other races of populations as well. Role of MYH9 in HIV-associated kidney disease be clarified to develop effective therapy (Naicker & Fabian, 2010).

10.4 Bacterial Resistance in Kidney Diseases

Antibiotic resistance is found in pathogens of urinary tract infections. The bacterial strains are resistant to antibiotics such as streptomycin and ampicillin. Chances of MDR are increased because of improper medication such as wide use of antibiotics, prescribing antibiotics without examination, and many others (Sahm et al., 2001). UTI bacterial strains withstand most antibiotics. Some of the strains are vulnerable to antibiotics. UTI microbes are resistant to SXT and nalidixic acid (Rella & Haas, 1982). Urine sample studies of UTI patients have confirmed that bacterial strains of E. coli, Staphylococcus aureus and Proteus sppare are involved in this infection. E. coli and Pseudomonas aeruginous show MDR to antibiotics. Staphylococcus aureus and some others are affected by Gentamycin and ciprofloxacin (Manikandan et al., 2011).

UTI is common in young pregnant women because of E. coli. This bacteria is resistant to several antibiotics. Some of the antibiotics show intermediate resistance and streptomycin and kanamycin are effective against UTI (Khan & Zaman, 2006). In polycystic kidney disease, proliferation rate of kidney cells as well as fluid secretion occurs due to MRP3 down expression. Targeted MPR3 upregulation leads to enhanced functioning of PKD drugs (Chang et al., 2015).

The gene-specific for ABCB1 encodes MDR-1 or P glycoprotein. If this gene is overexpressed, it can lead to drug resistance (Holohan et al., 2013). Therefore, the levels of ABCB1 genes should be monitored. In patients with high level of this gene, secondary treatments should be employed which include the use of metformin in addition to primary treatments (Gold & Moellering Jr, 1996). The drug design to block the mechanism of bacterial resistance should be employed. For example, bacterial resistance against beta-lactam antibiotics such as penicillin can be compensated by the use of nafcillin which are not susceptible to destruction by beta-lactamases, the enzymes responsible for the breakdown of beta lactamase (Walsh, 2000). It can also be managed with beta-lactam antibiotics in addition to drugs that block beta-lactamases such as clavulanic acid so that antibiotics can work before they are targeted by bacterial components (Dean et al., 2005). The new drug molecules inhibit bacterial efflux molecules causing resistance to not only one but multiple antibiotic molecules such as beta-lactams, quinolones, chloramphenicol by exporting bacterial molecules out of cell membranes. A combination of antibiotics is employed to fight against bacteria resistant to one or more types of antibiotics (Bennett et al., 2009).

Multiple occurrences of severe conditions are regarded as nephritic syndrome like hypercholesterolemia, hypo albuminuria, strong proteinuria, or edema. Number of factors cause glomerular injury. One of these factors is idiopathic factor. Following are the anatomical mutant of idiopathic nephritic syndrome INS (Fig. 10.2);

- Membranous nephropathy
- Focal segmental glomerulosclerosis (FSGS)
- Minimal change disease (MCD) (Orloff et al., 2005).

Cyclosporine exhibits anti-proteinuria function by altering the rate of permeability through glomerulus. Factors affected by the dosing of cyclosporine include GFR, permeability, and the charge of glomerular membranes (Meyrier et al., 1994).

Epigenetic modulations affect CKD expression and its spread. By targeting kidney genes, induction of epigenetic change causing alteration of the drug targets can be made (Hayashi & Itoh, 2015). When kidney shows Glomerulus filtration rate less than 60 mL per minute for three months or more, it is regarded as kidney disease. Serum creatinine is a parameter to predict kidney damage. There are chemical formulas and equations for the estimation of Glomerular filtration rate and the kidney disease such

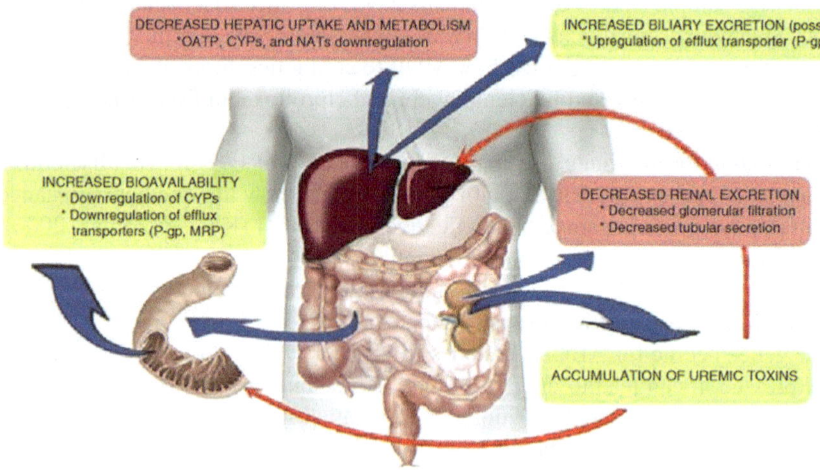

Fig. 10.2 Drug metabolism in kidney diseases (Nolin et al., 2008)

as Cockcroft Gault formulae (Levey et al., 2005). GFR is estimated by the serum albumin to creatinine ratio considering the age, and sex which helps in diagnosis. Another indicator of chronic kidney diseases is proteinuria whose levels show the extent of kidney damage. According to the severity of kidney damage, a strategy for the treatment, staging, and referring to nephrologists is decided (Levey et al., 2003). Glomerulonephritis occurs in variety of ways. Some of the types of glomerulonephritis are coupled with proteinuria, decreased renal working, and hematuria. In rare cases, proteinuria leads to nephritic syndrome. In hematuria, the urine suspended particles are observed with wrecked red blood cells. Similarly, glomerulonephritis can be estimated by the elevated concentrations of excreted proteins in urine and reduced renal function (Couser, 1999).

In kidney biopsy, immune reactions occur on the site where immunoglobin gets stacked, causing immune complex-mediated injury. These injuries are categorized into three classes:

- Mesangial pattern
- Epithelial pattern
- Endothelial pattern.

The antibiotic resistance is connected with two families of efflux pumps. They are responsible for the increasing resistance against antibiotics. The sensing of substrate is weak in few efflux pumps and they allow number of substrates to pass through which cause resistance against multiple drugs. The transfer of number of substrate can be reduced with variety of compositions. The targeting of transcription regulators reduces antimicrobial resistance. For this, the mutation of certain gene is required in bacteria through different methods. The resistance of antibiotics is generated either through plasmids or chromosomes of bacteria. The mechanism of action of bacteria and bacterial resistance is shown in Fig. 10.3.

Organisms have different methods to induce resistance. The fungi having pathogenic species have various modes and methods to create resistance in comparison to the resistance in bacteria. The cells of fungus do support the collection of drugs; they allow the passage of unrecognized substrate and initiate ATP independent mechanism of diffusion. Additionally, the efflux pump mutates targeted gene which alters the structural composition of protein target gene and lessens the affinity. In another method, the overexpressed target of drugs can generate resistance by following the passage of stress regulation.

Human diseases not caused by the infections of microorganisms, drug resistance is still experienced during treatment like in chemotherapy. The efflux pump has a role in chemotherapy. Apart from contributing to antimicrobial resistance mechanisms, the efflux pump reduces chances of piling up the chemo drug in cancer cells. This efflux pump belongs to ABC super family and has resistant proteins. The deactivation of anticancer drug can be done with enzymatic activity of glutathione s-transferase which forms thither compound after the catalysis of combination of glutathione with xenobiotic. The changes in targeted drug can cause resistance against particular drug

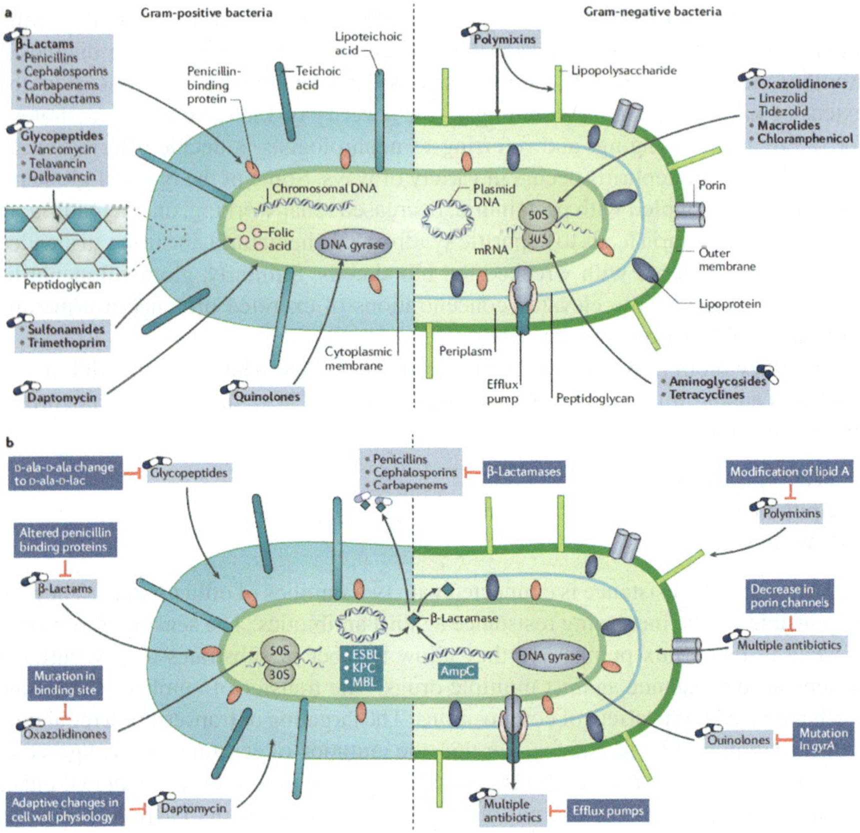

Fig. 10.3 Mechanisms of action of antibacterial agents and antibacterial resistance (Wang et al., 2019)

in cancer cells which is different in normal cells. In breast cancer, effected cancer cells are deficient in estrogen and become independent of estrogen for growth, that's why endocrine therapy is recommended. The cancer cells have reduced level of permeability towards chemo drugs due to reduced levels of cholesterol and varying combinations of lipids in comparison to normal cells.

10.5 Drugs Resistance in Kidney Cancer

Drug prescriptions to the patients with GFR less than 60 mL/min must be looked. Some classes of drugs are prohibited in patients with CKD or advanced stage. The prohibition is due to the altered function of kidney for drug metabolism. As a result, chances of piling up of toxins produced by incomplete drug metabolism are increased (Keller, 2008). Renal tumors of epithelial cells can be chromophobe, types I and II papillary, and oncocytoma. Each kind of tumor has different kind of genetic origin. Cancer of clear renal cell type is due to the inherited gene von Hippel-Lindau. Proto-oncogene c-Met is involved in genetic papillary renal carcinoma. Similarly, the gene present on chromosome 17 short arm participate in chromophobe renal carcinoma and oncocytoma (Linehan et al., 2003).

Malignancies related to kidney known as renal cell carcinoma (RCC) show resistance to the drugs. Chemo as well as radiotherapy fails to deal with RCC (Chan & Marshall, 1998). Multidrug resistance is observed via increased expression of efflux pumps. These efflux channels influence the function of chemo agents and alter the response of tumors against these drugs (O'connor, 2007). The enlarged expression of MDR-1 Pgp/P-170, MRP1 (ABCC1), and MDR-1 (ABCB1) efflux pump, the chemotherapeutic agents rapidly form tumor. This efflux rate is too fast to carry the desired chemotherapeutic effects in tumors (O'driscoll et al., 2007). Among these MDR-1, P-GP is the most resistant causing gene expression in kidney cancer (Bak et al., 1990). Its expression overlaps with phenotypically differentiated tumors of the kidney (Kanamaru et al., 1989). Studies are in progress regarding MRP-1 role in causing resistance of RCC. There is a probability of MDR-1 P-gp and MRP-1 co-function as markers of multidrug resistance. These gene products and some other translation products may affect the efflux pumps, producing resistance in kidney tumors (Walsh et al., 2009).

10.6 Anti-cytomegalovirus (CMV) Drug Resistance

Patients undergone solid organ transplant become prone to the virus known as Cytomegalovirus (CMV) of infections. Ganciclovir, acyclovir, or valacyclovir are some antivirals prescribed after transplantations. As the period goes long for the intake of these antivirals, chances of resistance also increase. To cope up with the resistance of CMV, the immunity of host should be increased (Fig. 10.4). Other than oral regimes, the bioavailable regimes can be used (Chou, 2001). Tissue-invasive diseases and resiatance against therapies are common in drug-resistant CMV patients. It has been reported that some mutations are less pathogenic than the others in non-lung transplant recipients which is a sign of heterogeneity and asymptomatic infections. Challenge in such cases is the designing of therapeutic approach which maintain the balance between reduction of immunosuppression and the severity of infection against the risks of drug toxicity. Some prognostic factors are the early and careful intervention, minimization of the risk of resistance and effective monitoring (Vaz et al., 2014).

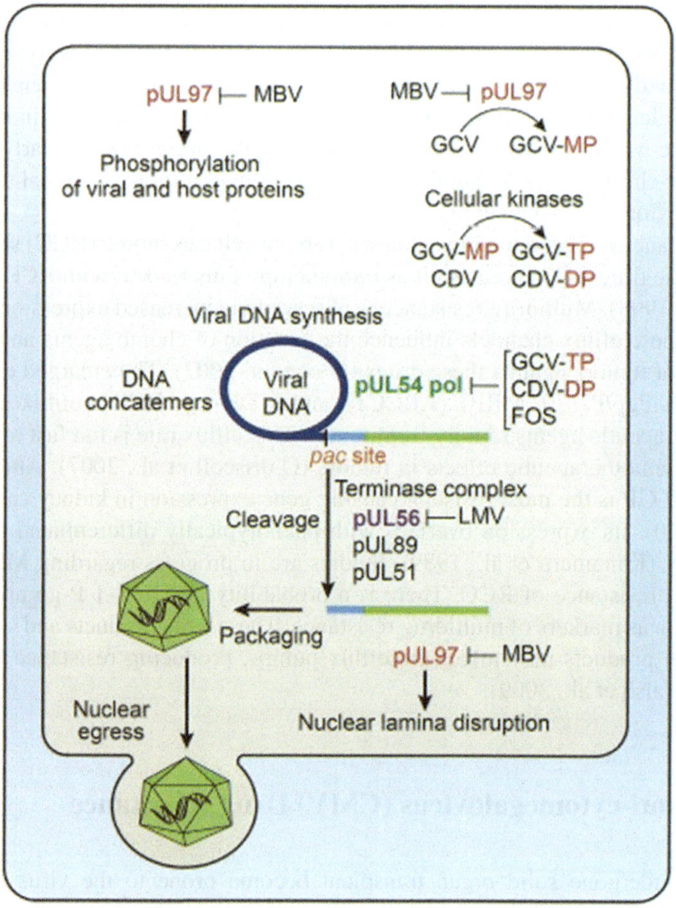

Fig. 10.4 Action mechanism of anti-cytomegalovirus drugs. MBV (maribavir) blocks pUL97 activity as well as GCV first phosphorylation. Consequently, anti cytomegalovirus drug action is inhibited due to the antagonist action of maribavir (Piret & Boivin, 2019)

10.7 Drug Resistance Prevention

The way to prevent or manage drug resistance is to take drugs as prescribed by a physician or health care provider. No doses should be exceeded or skipped during treatment. People having a specific disease like TB should narrate his doctor if they have any sort of problems in taking a specific drug for that disease (Teicher, 1994). In either case, health care providers can also help in preventing drug resistance by quick diagnosis, providing guidelines for treatment, and monitoring the response of patient to a specific treatment (Cohen, 1992). Another way is to avoid contact with patients already having resistance against specific drug or vaccine. The clinical network depends on the instruction of its prescribers, and self-guideline offered to antimicrobial stewardship which at clinics may appear as an antimicrobial stewardship program. Relying on the social setting, Government can teach people on the significance of prohibitive utilization of anti-infection agents for human clinical use, however dissimilar to opiates, there is no guideline of its utilization (Bloland & Organization, 2001). Infection prevention is the effective strategy as there are few alternatives to antibiotics in case of extensive resistance or pre-resistant infection. If the localized infection is diagnosed, it can be exempted by removal or excision. In case of systemic infection, only authenticated and generic measurements will be used, which boost up the immune system using antibodies or immunoglobin molecules (Peters, 1970). Usage of bacteriophage viruses that infect and vanish bacteria are also the alternatives and is a developing area of treatment. Moreover, it is suggested to develop new antibiotics with time as the selection of resistant bacteria cannot be eliminated. With the usage of specific antibiotic substance, the survival of bacteria is possible which will get resistant and in this case a new drug should be employed (Little et al., 2002). Consequently, the resistant gene is promoted in the living organisms. In addition to producing new anti-toxins, new strategies should be employed to keep public safe from events of resistance. New techniques are being tried, for example, UV light medicines and bacteriophage (Wolstenholme et al., 2004).

Inhibitors of PAMP metabolic pathway can result in the killing of bacteria having no toxic or immunological effects on host body, while the ones not exhibiting this pathway may result in activation of innate cellular pathway and proliferation at local sites resulting in inflammation in six to seven hours. Lymphocytes acting as active agents for killing microbes also result in the induction of apgoptosis or microptosis. The pore-forming peptide called as gyrunilysin causes permeabilization of membranes of microbes resulting in the intake of enzymes that cause an intake of granzymes which cleaves bacterial proteins responsible for the survival of bacterial components (Kage et al., 2002). As chemical inhibitors of bacterial components specific for PAMP pathway are not available commercially, it is suggested to screen or find out organisms or pathways which specifically target these enzymes. The validated components are under study and their use at clinical level is being granted. The prototype inhibitors also result in the reduction of gram-negative bacteria (D'Alessandro & Buttiëns, 2001). These inhibitions cause toxicity by interacting with bacterial respiration and bacterial cell wall components. Moreover, these effects are produced at clinical trial stages during working or culturing drug resistant bacteria (D'Alessandro & Buttiëns, 2001).

It is not possible to vanish the effect of antimicrobial compounds and resistance via the new resistant bacteria will continue to arise. For example, if a patient is treated with a primary dosage of clindamycin, ciprofloxacin will be used in combination or in the next cycle as immunotherapy (Shi et al., 2010). The production of new compounds or combinations is vital. A route is developed for finding synergistic and safety of combined antibiotics. The immunization of populaces is carried out. The conditions legitimize quickened improvement and do influenza immunizations (Fernando et al., 2017). Conditions apply to anti-microbial improvements for regular bacterial contaminations. Mix of previous anti-microbials can lead to minimal effort for effective medicines (Catry et al., 2003).

10.8 Conclusion

Kidney is a complex organ containing complex and specialized cells. Any disorder in such organ can disturb processes like blood pressure, balance of electrolyte, water and oxygen supply in blood, acid–base homeostasis, hormonal imbalance, metabolic clearance of secretions in blood, which can affect central nervous system. Dvelopment of drug resistance in kidney diseases is common phenomenon. A variety of nephrological problems are treated with antifungal, antiviral and antibacterial diseases. Some infections in kidney are associated with diseases in some other parts of the body such as HIV. As kidney performs the blood filteration in the body, all the leftovers of drugs and their bi-products pass through the kidney. They are linked with the development of resitance against treatment therapies. Majority of drug resitance pattern are linked with genetic mutation. Structure of proteins, enzymes, receptors, and their transcription factors is changed as a result of such mutations which lead to resistance. Some mutations also cause multi-drug resistance, causing problems in designing and prediction of therapies for kidney infections treatment. Understanding of the mechanism of action of drug, drug dosage, patient history, environmental conditions and connection to other diseases can be useful in designing new, effective and sustainable therapies for the treatment of kidney diseases. Keeping an eye on ever evolving mutations in pathogens can have impact on the outcome of futures vaccines and drug designs.

References

Agarwal, R., & Sinha, A. D. (2012). Thiazide diuretics in advanced chronic kidney disease. *Journal of the American Society of Hypertension, 6*, 299–308.

Agarwal, R. (2007). Effects of statins on renal function. In *Mayo Clinic Proceedings, 2007* (pp. 1381–1390). Elsevier.

Bak, M. Jr, Efferth, T., Mickisch, G., Mattern, J., & Volm, M. (1990). Detection of drug resistance and P-glycoprotein in human renal cell carcinomas. *European Urology, 17*, 72–75.

Bailie, G. R., Uhlig, K., & Levey, A. S. (2005). Clinical practice guidelines in nephrology: Evaluation, classification, and stratification of chronic kidney disease. *Pharmacotherapy: The Journal of Human Pharmacology and Drug Therapy, 25*, 491–502.

Bennett, D. E., Camacho, R. J., Otelea, D., Kuritzkes, D. R., Fleury, H., Kiuchi, M., Heneine, W., Kantor, R., Jordan, M. R., & Schapiro, J. M. (2009). Drug resistance mutations for surveillance of transmitted HIV-1 drug-resistance: 2009 update. *PLoS ONE, 4*, e4724.

Berns, J. S., & Kasbekar, N. (2006). Highly active antiretroviral therapy and the kidney: An update on antiretroviral medications for nephrologists. *Clinical Journal of the American Society of Nephrology, 1*, 117–129.

Bhimma, R., Purswani, M. U., & Kala, U. (2013). Kidney disease in children and adolescents with perinatal HIV-1 infection. *Journal of the International AIDS Society, 16*, 18596.

Bloland, P. B., & Organization, W. H. (2001). Drug resistance in malaria. World Health Organization.

Bréchot, J.-M., Hurbain, I., Fajac, A., Daty, N., & Bernaudin, J.-F. (1998). Different pattern of MRP localization in ciliated and basal cells from human bronchial epithelium. *Journal of Histochemistry & Cytochemistry, 46*, 513–517.

Cañadas-Garre, M., Anderson, K., Cappa, R., Skelly, R., Smyth, L. J., McKnight, A. J., & Maxwell, A. P. (2019). Genetic susceptibility to chronic kidney disease–some more pieces for the heritability puzzle. *Frontiers in Genetics, 10*, 453.

Catry, B., Laevens, H., Devriese, L. A., Opsomer, G., & de Kruif, A. (2003). Antimicrobial resistance in livestock. *Journal of Veterinary Pharmacology and Therapeutics, 26*, 81–93.

Chan, D. Y., & Marshall, F. F. (1998). Surgery in advanced and metastatic renal cell carcinoma. *Current Opinion in Urology, 8*, 369–373.

Chang, E., Park, E. Y., Woo, Y. M., Kang, D.-H., Hwang, Y.-H., Ahn, C., & Park, J. H. (2015). Restoring multidrug resistance-associated protein 3 attenuates cell proliferation in the polycystic kidney. *American Journal of Physiology-Renal Physiology, 308*, F1004–F1011.

Chou, S. (2001). Cytomegalovirus drug resistance and clinical implications. *Transplant Infectious Disease, 3*, 20–24.

Cohen, M. L. (1992). Epidemiology of drug resistance: Implications for a post—antimicrobial era. *Science, 257*, 1050–1055.

Couser, W. G. (1999). Glomerulonephritis. *The Lancet, 353*, 1509–1515.

Couzi, L., Helou, S., Bachelet, T., Moreau, K., Martin, S., Morel, D., Lafon, M., Boyer, B., Alain, S., & Garrigue, I. (2012). High incidence of anticytomegalovirus drug resistance among D+ R− kidney transplant recipients receiving preemptive therapy. *American Journal of Transplantation, 12*, 202–209.

Cusumano, A. M., & Bedat, M. C. G. (2008). Chronic kidney disease in Latin America: Time to improve screening and detection. *Clinical Journal of the American Society of Nephrology, 3*, 594–600.

D'Alessandro, U., & Buttiëns, H. (2001). History and importance of antimalarial drug resistance. *Tropical Medicine & International Health, 6*, 845–848.

Dean, M., Fojo, T., & Bates, S. (2005). Tumour stem cells and drug resistance. *Nature Reviews Cancer, 5*, 275–284.

Eddy, A. A. (2014). Overview of the cellular and molecular basis of kidney fibrosis. *Kidney International Supplements, 4*, 2–8.

Emma, F., Montini, G., Parikh, S. M., & Salviati, L. (2016). Mitochondrial dysfunction in inherited renal disease and acute kidney injury. *Nature Reviews Nephrology, 12*, 267–280.

Ernest, S., Rajaraman, S., Megyesi, J., & Bello-Reuss, E. N. (1997). Expression of MDR1 (multidrug resistance) gene and its protein in normal human kidney. *Nephron, 77*, 284–289.

Fernando, S. A., Gray, T. J., & Gottlieb, T. (2017). Healthcare-acquired infections: Prevention strategies. *Internal Medicine Journal, 47*, 1341–1351.

Gabardi, S., & Abramson, S. (2005). Drug dosing in chronic kidney disease. *Medical Clinics, 89*, 649–687.

Gold, H. S., & Moellering JR, R. C. (1996). Antimicrobial-drug resistance. *New England Journal of Medicine, 335*, 1445–1453.

Goldfarb, D. A. (2005). Immunosuppressive drugs for kidney transplantation. *The Journal of Urology*.

Golomb, B. A., Dimsdale, J. E., White, H. L., Ritchie, J. B., & Criqui, M. H. (2008). Reduction in blood pressure with statins: Results from the UCSD Statin Study, a randomized trial. *Archives of Internal Medicine, 168*, 721–727.

Gottesman, M. M., & Pastan, I. (1993). Biochemistry of multidrug resistance mediated by the multidrug transporter. *Annual Review of Biochemistry, 62*, 385–427.

Gupta, S. K., Eustace, J. A., Winston, J. A., Boydstun, I. I., Ahuja, T. S., Rodriguez, R. A., Tashima, K. T., Roland, M., Franceschini, N., & Palella, F. J. (2005). Guidelines for the management of chronic kidney disease in HIV-infected patients: Recommendations of the HIV Medicine Association of the Infectious Diseases Society of America. *Clinical Infectious Diseases, 40*, 1559–1585.

Hayashi, K., & Itoh, H. (2015). Transcription factors and epigenetic modulation: Its therapeutic implication in chronic kidney disease. *Archivum Immunologiae Et Therapiae Experimentalis, 63*, 193–196.

Holohan, C., van Schaeybroeck, S., Longley, D. B., & Johnston, P. G. (2013). Cancer drug resistance: An evolving paradigm. *Nature Reviews Cancer, 13*, 714–726.

Kage, K., Tsukahara, S., Sugiyama, T., Asada, S., Ishikawa, E., Tsuruo, T., & Sugimoto, Y. (2002). Dominant-negative inhibition of breast cancer resistance protein as drug efflux pump through the inhibition of S-S dependent homodimerization. *International Journal of Cancer, 97*, 626–630.

Kanamaru, H., Kakehi, Y., Yoshida, O., Nakanishi, S., Pastan, I., & Gottesman, M. M. (1989). MDR1 RNA levels in human renal cell carcinomas: Correlation with grade and prediction of reversal of doxorubicin resistance by quinidine in tumor explants. *JNCI: Journal of the National Cancer Institute, 81*, 844–849.

Kao, W. L., Klag, M. J., Meoni, L. A., Reich, D., Berthier-Schaad, Y., Li, M., Coresh, J., Patterson, N., Tandon, A., & Powe, N. R. (2008). MYH9 is associated with nondiabetic end-stage renal disease in African Americans. *Nature Genetics, 40*, 1185–1192.

Kasiske, B. L., Zeier, M. G., Chapman, J. R., Craig, J. C., Ekberg, H., Garvey, C. A., Green, M. D., Jha, V., Josephson, M. A., & Kiberd, B. A. (2010). KDIGO clinical practice guideline for the care of kidney transplant recipients: A summary. *Kidney International, 77*, 299–311.

Keller, F. (2008). *Nephrotoxische Störungen und nierenabhängige Arzneimittel*. Elsevier.

Khan, A. U., & Zaman, M. S. (2006). Multiple drug resistance pattern in urinary tract infection patients in Aligarh.

Kim, K.-H., Fonda, J. R., Lawler, E. V., Gagnon, D., & Kaufman, J. S. (2010). Change in use of gadolinium-enhanced magnetic resonance studies in kidney disease patients after US Food and Drug Administration warnings: A cross-sectional study of Veterans Affairs Health Care System data from 2005–2008. *American Journal of Kidney Diseases, 56*, 458–467.

Kopp, J. B., Smith, M. W., Nelson, G. W., Johnson, R. C., Freedman, B. I., Bowden, D. W., Oleksyk, T., McKenzie, L. M., Kajiyama, H., & Ahuja, T. S. (2008). MYH9 is a major-effect risk gene for focal segmental glomerulosclerosis. *Nature Genetics, 40*, 1175–1184.

Launay-Vacher, V., Izzedine, H., & Deray, G. (2005). Statins' dosage in patients with renal failure and cyclosporine drug–drug interactions in transplant recipient patients. *International Journal of Cardiology, 101*, 9–17.

Levey, A. S., Coresh, J., Balk, E., Kausz, A. T., Levin, A., Steffes, M. W., Hogg, R. J., Perrone, R. D., Lau, J., & Eknoyan, G. (2003). National Kidney Foundation practice guidelines for chronic kidney disease: Evaluation, classification, and stratification. *Annals of Internal Medicine, 139*, 137–147.

Levey, A. S., Eckardt, K.-U., Tsukamoto, Y., Levin, A., Coresh, J., Rossert, J., Zeeuw, D. D., Hostetter, T. H., Lameire, N., & Eknoyan, G. (2005). Definition and classification of chronic kidney disease: A position statement from Kidney Disease: Improving Global Outcomes (KDIGO). *Kidney International, 67*, 2089–2100.

Linehan, W. M., Walther, M. M., & Zbar, B. (2003). The genetic basis of cancer of the kidney. *The Journal of Urology, 170*, 2163–2172.

Little, S. J., Holte, S., Routy, J.-P., Daar, E. S., Markowitz, M., Collier, A. C., Koup, R. A., Mellors, J. W., Connick, E., & Conway, B. (2002). Antiretroviral-drug resistance among patients recently infected with HIV. *New England Journal of Medicine, 347*, 385–394.

Manikandan, S., Ganesapandian, S., Singh, M., & Kumaraguru, A. (2011). Emerging of multidrug resistance human pathogens from urinary tract infections. *Current Research in Bacteriology, 4*, 9–15.

Marras, D., Bruggeman, L. A., Gao, F., Tanji, N., Mansukhani, M. M., Cara, A., Ross, M. D., Gusella, G. L., Benson, G., & D'Agati, V. D. (2002). Replication and compartmentalization of HIV-1 in kidney epithelium of patients with HIV-associated nephropathy. *Nature Medicine, 8*, 522–526.

Matzke, G. R., Aronoff, G. R., Atkinson, A. J., Jr, Bennett, W. M., Decker, B. S., Eckardt, K.-U., Golper, T., Grabe, D. W., Kasiske, B., & Keller, F. (2011). Drug dosing consideration in patients with acute and chronic kidney disease—a clinical update from Kidney Disease: Improving global outcomes (KDIGO). *Kidney International, 80*, 1122–1137.

McCullough, K., Sharma, P., Ali, T., Khan, I., Smith, W. C., Macleod, A., & Black, C. (2012). Measuring the population burden of chronic kidney disease: A systematic literature review of the estimated prevalence of impaired kidney function. *Nephrology Dialysis Transplantation, 27*, 1812–1821.

Meyrier, A., Noël, L.-H., Auriche, P., Callard, P., & Broneer, D. (1994). Long-term renal tolerance of cyclosporin a treatment in adult idiopathic nephrotic syndrome. *Kidney International, 45*, 1446–1456.

Mocroft, A., Kirk, O., Reiss, P., de Wit, S., Sedlacek, D., Beniowski, M., Gatell, J., Phillips, A. N., Ledergerber, B., & Lundgren, J. D. (2010). Estimated glomerular filtration rate, chronic kidney disease and antiretroviral drug use in HIV-positive patients. *AIDS, 24*, 1667–1678.

Muller, E., Barday, Z., Mendelson, M., & Kahn, D. (2015). HIV-positive–to–HIV-positive kidney transplantation—results at 3 to 5 years. *New England Journal of Medicine, 372*, 613–620.

Nagata, M., Ninomiya, T., Doi, Y., Yonemoto, K., Kubo, M., Hata, J., Tsuruya, K., Iida, M., & Kiyohara, Y. (2010). Trends in the prevalence of chronic kidney disease and its risk factors in a general Japanese population: The Hisayama Study. *Nephrology Dialysis Transplantation, 25*, 2557–2564.

Naicker, S., & Fabian, J. (2010). Risk factors for the development of chronic kidney disease with HIV/AIDS. *Clinical Nephrology, 74*, S51–S56.

Naud, J., Nolin, T. D., Leblond, F. A., & Pichette, V. (2012). Current understanding of drug disposition in kidney disease. *The Journal of Clinical Pharmacology, 52*, 10S-22S.

Nolin, T., Naud, J., Leblond, F., & Pichette, V (2008). Emerging evidence of the impact of kidney disease on drug metabolism and transport. *Clinical Pharmacology & Therapeutics, 83*, 898–903.

O'connor, R. (2007). The pharmacology of cancer resistance. *Anticancer Research, 27*, 1267–1272.

O'Driscoll, G., Green, D., & Taylor, R. R. (1997). Simvastatin, an HMG–coenzyme a reductase inhibitor, improves endothelial function within 1 month. *Circulation, 95*, 1126–1131.

O'Driscoll, L., Walsh, N., Larkin, A., Ballot, J., Ooi, W., Gullo, G., O'Connor, R., Clynes, M., Crown, J., & Kennedy, S. (2007). MDR1/P-glycoprotein and MRP-1 drug efflux pumps in pancreatic carcinoma. *Anticancer Research, 27*, 2115–2120.

Orloff, M. S., Iyengar, S. K., Winkler, C. A., Goddard, K. A., Dart, R. A., Ahuja, T. S., Mokrzycki, M., Briggs, W. A., Korbet, S. M., & Kimmel, P. L. (2005). Variants in the Wilms' tumor gene are associated with focal segmental glomerulosclerosis in the African American population. *Physiological Genomics, 21*, 212–221.

Peng, K. C., Cluzeaud, F., Bens, M., Duong Van Huyen, J.-P., Wioland, M. A., Lacave, R., & Vandewalle, A. (1999). Tissue and cell distribution of the multidrug resistance-associated protein (MRP) in mouse intestine and kidney. *Journal of Histochemistry & Cytochemistry, 47*, 757–767.

Peters, W. (1970). Chemotherapy and drug resistance in malaria. *Chemotherapy and Drug Resistance in Malaria.*

Piret, J., & Boivin, G. (2019). Clinical development of letermovir and maribavir: Overview of human cytomegalovirus drug resistance. *Antiviral Research, 163*, 91–105.

Port, F. K., Wolfe, R. A., Mauger, E. A., Berling, D. P., & Jiang, K. (1993). Comparison of survival probabilities for dialysis patients versus cadaveric renal transplant recipients. *JAMA, 270*, 1339–1343.

Pritchard, J. B., & Miller, D. S. (1997). Renal secretion of organic cations: A multistep process. *Advanced Drug Delivery Reviews, 25*, 231–242.

Rani, A., Ranjan, R., McGee, H. S., Andropolis, K. E., Panchal, D. V., Hajjiri, Z., Brennan, D. C., Finn, P. W., & Perkins, D. L. (2017). Urinary microbiome of kidney transplant patients reveals dysbiosis with potential for antibiotic resistance. *Translational Research, 181*, 59–70.

Rella, M., & Haas, D. (1982). Resistance of Pseudomonas aeruginosa PAO to nalidixic acid and low levels of beta lactam antibiotics: Mapping of chromosomal genes. *Antimicrobial Agents and Chemotherapy, 22*, 242–249.

Ross, M. J., & Klotman, P. E. (2004). HIV-associated nephropathy. *AIDS, 18*, 1089–1099.

Sahm, D. F., Critchley, I. A., Kelly, L. J., Karlowsky, J. A., Mayfield, D. C., Thornsberry, C., Mauriz, Y. R., & Kahn, J. (2001). Evaluation of current activities of fluoroquinolones against gram-negative bacilli using centralized in vitro testing and electronic surveillance. *Antimicrobial Agents and Chemotherapy, 45*, 267–274.

Shi, D., Hao, Y., Zhang, A., Wulan, B., & Fan, X. (2010). Antimicrobial Resistance of Staphylococcus Aureus Isolated From Bovine Mastitis in China. *Transboundary and Emerging Diseases, 57*, 221–224.

Speeg, K. V., Maldonado, A. L., Liaci, J., & Muirhead, D. (1992). Effect of cyclosporine on colchicine secretion by the kidney multidrug transporter studied in vivo. *Journal of Pharmacology and Experimental Therapeutics, 261*, 50–55.

Stride, B. D., Grant, C. E., Loe, D. W., Hipfner, D. R., Cole, S. P., & Deeley, R. G. (1997) Pharmacological characterization of the murine and human orthologs of multidrug-resistance protein in transfected human embryonic kidney cells. *Molecular Pharmacology, 52*, 344–353.

Teicher, B. A. (1994). Hypoxia and drug resistance. *Cancer and Metastasis Reviews, 13*, 139–168.

Vaz, R., Barros, F., Tavares, I., Bustorff, M., Ferreira, I., & Pestana, M. (2014). Ganciclovir-resistant cytomegalovirus infection in renal transplantation. *Clinical Kidney Journal, 7*, 210–213.

Vivante, A., & Hildebrandt, F. (2016). Exploring the genetic basis of early-onset chronic kidney disease. *Nature Reviews Nephrology, 12*, 133.

Walsh, N., Larkin, A., Kennedy, S., Connolly, L., Ballot, J., Ooi, W., Gullo, G., Crown, J., Clynes, M., & O'Driscoll, L. (2009). Expression of multidrug resistance markers ABCB1 (MDR-1/P-gp) and ABCC1 (MRP-1) in renal cell carcinoma. *BMC Urology, 9*, 1–7.

Walsh, C. (2000). Molecular mechanisms that confer antibacterial drug resistance. *Nature*, 406, 775–781.

Wang, T. Z., Kodiyanplakkal, R. P. L., & Calfee, D. P. (2019). Antimicrobial resistance in nephrology. *Nature Reviews Nephrology, 15*, 463–481.

Wolstenholme, A. J., Fairweather, I., Prichard, R., Von Samson-Himmelstjerna, G., & Sangster, N. C. (2004). Drug resistance in veterinary helminths. *Trends in Parasitology, 20*, 469–476.

Zhang, Q.-L., & Rothenbacher, D. (2008). Prevalence of chronic kidney disease in population-based studies: Systematic review. *BMC Public Health, 8*, 117.

Chapter 11
Drugs Resistance in Heart Diseases

Adeel Mahmood, Maryam Eqan, Saher Pervez, Rimsha Javed, Rizwan Ullah, Arshad Islam, Ajmal Khan, Atif Amin Baig, Gotam Kumar, Muhammad Abaid-Ullah, and Muhammad Rafiq

Abstract Heart is a vital blood pumping organ of human body. Cholesterol buildup, smoking, obesity and high blood pressure causes heart malfunctioning leading to heart diseases. Coronary artery disease (CAD), cardiac arrhythmia, myocardial infarction, congenital heart diseases, cardiomyopathy are some among many cardiovascular diseases. Several drugs are clinically available to treat these cardiac disorders, for example, antiplatelet drugs have been useful in ischemic stroke, CAD and peripheral arterial disease. Moreover, aspirin is commonly used in myocardial infarction, while atorvastatin, a drug used to lower cholesterol, deters with cytochrome P450-mediated metabolism of clopidogrel that results in the resistance against clopidogrel (antithrombotic drug). Atropine is used in cardiac dysrhythmia and heart block

A. Mahmood (✉) · M. Eqan · S. Pervez · R. Javed
Department of Environmental Sciences, GC Women University, Sialkot, Pakistan

R. Ullah
Department of Zoology, Mirpur University of Science and Technology (MUST), Mirpur Azad Kashmir, Pakistan

A. Islam
Department of Pathology, Government Lady Reading Hospital Medical Teaching Institute, Peshawar, KP, Pakistan

A. Khan
Department of Biotechnology, Bacha Khan University, Charsadda, KP, Pakistan

A. A. Baig
Faculty of Medicine, University Sultan Zainal Abidin, Kuala Terengganu, Malaysia

G. Kumar
Nuclear Medicine Oncology & Radiotherapy Institute (NORIN), Nawabshah, Pakistan

M. Abaid-Ullah
Department of Life Sciences, Khawaja Fareed University of Engineering and Information Technology, Rahim Yar Khan, Pakistan

M. Rafiq
Department of Physiology and Biochemistry, Cholistan University of Veterinary and Animal Sciences, Bahawalpur 63100, Pakistan

S. Ahmed et al. (eds.), *Biochemistry of Drug Resistance*,
https://doi.org/10.1007/978-3-030-76320-6_11

while adenosine has depressant effects on atropine. The antibiotic, benzathine penicillin G, is an important antibiotic for the treatment of group A streptococcal infections associated with rheumatic heart disease. Furthermore, beta blockers, thiazides, ACE inhibitors, diuretics are also used for several other heart disorders. Resistance has been reported against some of the drugs, such as some antiplatelet drugs and antithrombotic agents. Resistance towards aspirin is influenced by several factors such as clinical, biological and genetic aspects. COX1-A1 mutation and glycoprotein IIIa (GPIIIa) gene polymorphism is linked to aspirin resistance and could possibly be involved in it. However, drug resistance against atorvastatin, atropine and benzathine penicillin G is still unknown.

Keywords Heart diseases · Infections · Chemotherapeutics · Drug resistance · Mechanism of resistance

11.1 Introduction

Heart, one of the most essential muscular organs of human body, is involved in pumping of blood to all the body parts. However, certain risk factors such as high levels of fats (cholesterol) and blood pressure, and obesity among others may cause heart malfunctioning also known as heart or cardiac diseases. Most common heart diseases include congenital heart diseases, dilated cardiomyopathy, coronary artery disease (CAD), cardiac arrhythmia, myocardial infarction and hypertension. Each of them has different symptoms and treatment. For some of them medication does wonders and for others patient are prescribed for particular heart surgeries. Most common medicines that are currently prescribed for the treatment of heart diseases include angiotensin converting enzyme (ACE) inhibitors, angiotensin II receptor, beta and calcium channel blockers, digoxin, thiazide, aspirin, diuretics and others. Many of these drugs are now facing resistance, however the mechanism behind the resistance is yet unknown. Although the reason behind the resistance may the over-prescription of drugs or some genetic factors.

11.2 Types of Heart Diseases

11.2.1 Congenital Heart Disease

Congenital heart disease (CHD) is an anatomic deformity of heart or vessels which occurs during intrauterine development irrespective of age at presentation (Rao, 2012). This deformity occurs at birth which leads to disturb the physiology of the newborn's heart and also results in an irregular flow of blood throughout the heart and all other body parts. These diseases cause minor (a slight hole in the heart) to major (malformed parts of heart) problems among the children (CDC, 2019). Reller et al. (2008) reported the death of approximately one individual in every 110 births in United States, out of which approximately 25% of cases were diagnosed with critical congenital heart defects (CCHDs) (Botto et al., 2001). The risk of disease observed is much greater if the mother has CHD and the risk tend to increase from 1–10% (Nora et al., 1988). Jenkins et al. (2007) identified the origin of non-inherited CHD, however they are yet generally undetermined (Rao, 2012).

Abnormal genetic/chromosomal or environmental aspects like the effect of specific drugs or organic solvents on child's mother or related infections including viral infections (rubella and influenza), maternal smoking during pregnancy period, diabetes, overweight/obesity are among the few genetic or non-inherited risk factors associated with CHD.

Among various types of CHDs, atrial septal defect, atrioventricular septal defect and ventricular septal defect are typical while coarctation of the aorta, double-outlet right ventricle, d-transposition of the great arteries, ebstein anomaly, hypoplastic left heart syndrome, interrupted aortic arch, pulmonary atresia, single ventricle, tetralogy of fallot, total anomalous pulmonary venous return, tricuspid atresia and truncus arteriosus are considered to be serious CHDs (CDC, 2019; Rao, 2012). According to the Center for Disease Control and Prevention (2019), the most common symptom in CHDs patient include blue-tinted nails/lips, rapid/unsettled breathing and fatigue when nurturing and drowsiness occurs in babies.

CHDs have caused a number of deaths both in children and adults. Gilboa et al. (2010) accounted infant mortality to be 48.1% of total mortality consequential from CHD; amongst individuals who endured the earliest year of life and 76.1% of fatalities happened in the course of maturity (≥18 years of age). Oliver et al. (2017) reported death of 59% by cardiovascular disease (194 fatalities) and 41% due to non-cardiovascular causes (137 fatalities). Respiratory disease and cancer are reported to be the main cause of non-CV mortality. Including patients' death due to CV, heart failure was the prominent manner of mortality (37%), subsequently sudden death (17%), perioperative (11%) and various cardiovascular diseases (10%) (Fig. 11.1).

11.2.2 Drugs Used in Congenital Heart Diseases

(a) *ACE Inhibitors and Angiotensin Receptor Blockers*

ACE inhibitors play a significant role in the process to overcome heart failure via blood vessels relaxation and reduction in blood pressure, which consequently stabilize the flow of blood (Cardiosmart, n.d.). Captopril, benazepril (lotensin), fosino pril, lisinopril (prinivil, zestril), quinapril (accupril), enalapril (vasotec), moexipril, perindopril, trandolapril and ramipril (altace) are few ACE inhibitors enlisted by Mayo Clinic Staff (2019).

Moreover, the ARBs (Angiotensin II receptor blockers) and ACE inhibitors have been reported to obstruct RAAS (renin–angiotensin–aldosterone system) with valuable effects on patients with cardiovascular (CV) risk factors only (hypertension,

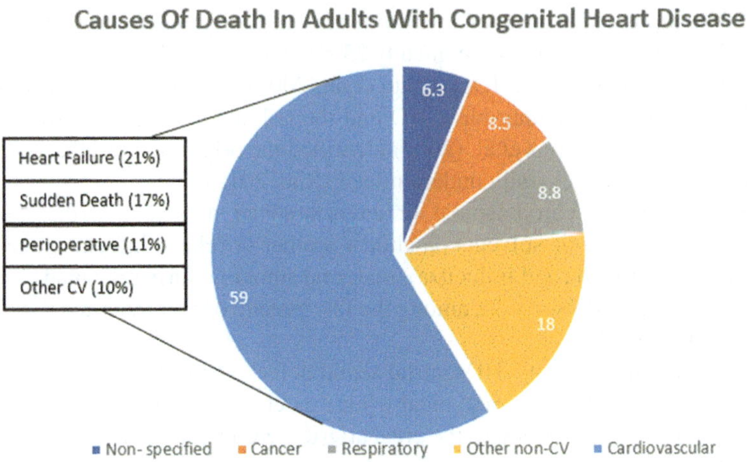

Fig. 11.1 Major causes of deaths in adult CHD patients

diabetes) and with various heart diseases (heart failure and coronary artery disease) (Asenjo et al., 2017). ACE inhibitors prevent the function of protein (enzyme) that initiates blood vessels to narrow down, which therefore regulate the opening of blood vessels. It, therefore, results in the decrease of blood pressure, which consequently leads toward regulated heart pumping. Moreover, these inhibitors also assist the body to discharge water and salt (sodium) which also help to reduce the blood pressure.

(b) ***Action Mechanism of ACE Inhibitors and ARBs***

Renin, secreted by juxtaglomerular apparatus of human kidney, transforms angiotensinogen into angiotensin-I, where ACE inhibitors further modifies it into angiotensin-II. It is then disseminated/distributed in lungs and results in the cessation of a powerful vasodilator named as bradykinin. The AT1 receptor of the kidney, clogged by ARBS, is used by the Angiotensin-II to function. In addition, only the ACE inhibitors (not the ARBs) helps enhance the bradykinin production, which in return aids in the control of main side-effects of the drugs, for example dry cough and angioedema) (Banik et al., 2014). ACE inhibitors also help in decreasing the concomitants of renin-angiotensin system. Renin, released by renal cortex (specifically the juxtaglomerular apparatus), works on angiotensinogen of plasma in order to signify the angiotensin-I. Angiotensin-I, in the lungs, is consequently transformed by the ACE inhibitor into angiotensin-II in the lungs, which directly restricts vascular cells (smooth muscle) stimulating adrenal gland resulting in the discharge of epinephrine as well as aldosterone. ACE inhibitors, with the help of angiotensin-II, also inhibit vasoconstriction (Agasti, 2011). Furthermore, ACE inhibitors also decrease the kinins deprivation rate, triggering vasodilation directly. The higher bradykinin intensity ends in a magnified production of prostaglandins via arachidonic acid alteration into it, also convincing the vasodilators (Agasti, 2011) (Fig. 11.2).

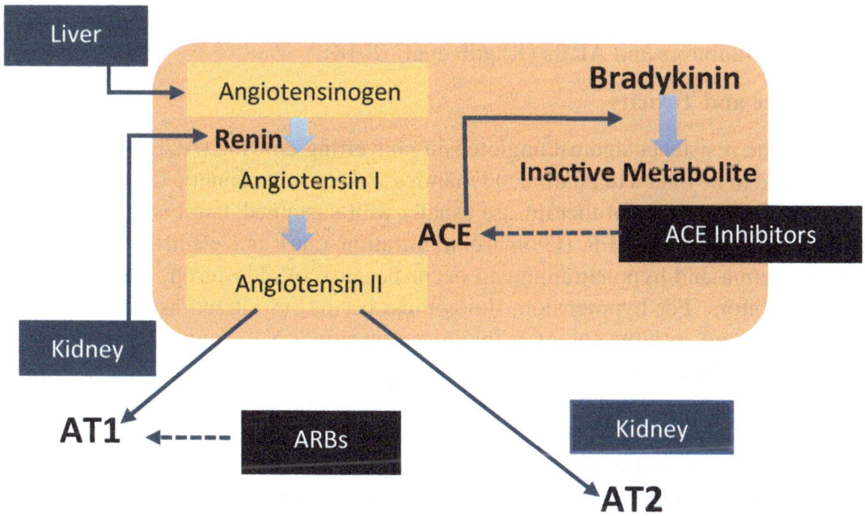

Fig. 11.2 Mechanism of action of ACE: the inhibition of angiotensin-II formation by ACE inhibitors

ACE inhibitors then bound to the ACE active site and competitively inhibit it. ACE is observed in the circulation system but its suppression in tissue is considered significantly crucial. ACE is reported to be mainly vented into the lungs, as well as in the vascular endothelium and other vital organs such as brain, kidney, adrenal gland and heart. Briefly, ACE inhibitors decrease the angiotensin-II and aldosterone concentrations. However, it does not constrain the angiotensin-II effect that is molded through ACE-independent related routes. However, with the provision of persistent medical care, angiotensin-II and aldosterone could be reversed to the 'pretreatment' times. It stipulates that irregular and life-threatening variations in blood pressure are ascribed to impediment of angiotensin-II-mediated vasoconstriction, antidiuretic hormone (ADH)-interceded sodium/water preservation and aldosterone. Though the long-standing therapeutic and curative benefits attained by using ACE inhibitors might be supplemental in better NO production as vasodilatory prostaglandins production (Hoyer et al., 1993). ACE inhibitors may have an impact endothelial role and gradual advancement of atherosclerosis through bradykinin-induced NO expansion (Lonn et al., 1994; Khatib et al., 2018).

ARBs, also called angiotensin-II receptor blockers, consist of variety of drugs such as irbesartan, candesartan, telmisartan, losartan and valsartan (antihypertensive and help prevent kidney problems and heart attack). These drugs alienate the angiotensin-II effects at the AT1 (angiotensin-II type 1) receptor. ARBs bind themselves to the AT1 receptor competitively and detach progressively (Burnier, 2001). Contrastingly, ACE inhibitors, ARBs hinder angiotensin-II effect generated from each channel; nonetheless the clinical importance of ACE-independent production/manufacture of angiotensin-II is still not certain and clear. Irrespective of the pharmacological variations, the effects observed during clinical trials, were found almost similar for both ACE inhibitors and ARBs (Khatib et al., 2018).

Resistance and Toxicity

Therapeutic resistance against angiotensin-converting enzyme (ACE) reticence has been found to be linked to pharmacodynamics and pharmacokinetic factors. Unequal dosages of ACE inhibitor therapy are usually well-sustained, but it is likely to cause hypotension, drop-in GFR (Glomerular filtration rate) and electrolyte disorders. Hyperkalemia and hyponatremia can occur because of aldosterone obstruction by ACE inhibitors. For hypotension, fluids must be the prioritized as therapy. There is absolutely no antidote for ACE inhibitor poisoning. Naloxone might be used in captopril overdoses with specific achievement and could be an alternative if intravenous fluids are not effective. Correspondingly, vasopressors are also the option for the therapy of hypotension (Lip et al., 1995).

Use of ACE inhibitors in clinics is predominantly beneficial antihypertensive in whites with an average age pf 40–45, however, in black patients relatively higher resilience has been observed even in old-aged patients, but addition of diuretics to the regimen could help to overcome resistance. High-salt diet also induces resistance which persuades hypertension in spite of a low renin state (Kester et al., 2012).

Digoxin

Digoxin, a primordial cardiovascular chemotherapeutic used now-a-days (MacLeod-Glover et al., 2016), is a clinically prescribed in order to administer atrial fibrillation and heart failure manifestations (Ziff et al., 2016). Digoxin, an FDA approved drug (1954) is recorded as a cardiac glycoside (FDA, 2019).

Digoxin is stipulated/prescribed under the following circumstances (FDA, 2019):

(1) in mild-to-moderate heart failure (myocardial infraction) in adult patients.
(2) to enhance the myocardial contraction in heart failure cases in youngers/children.
(3) to retain/control ventricular rate in adult atrial fibrillation patients.

Digoxin is often administered in combination with some diuretic or ACE blocker in adults heart failure patients upon clinical observation and feasibility to obtain optimal impacts (FDA, 2019). Digoxin, a known cardiac (or digitalis) glycoside, which is strongly related group of drugs having some similar sound effects on the myocardium. The origin of this medicine the *Digitalis lanata* leaves.

Mechanism of Action

Digoxin reduces/suppresses the activity of an enzyme, called 'sodium–potassium ATPase', which controls/regulates the quantities of both potassium and sodium ions within the cells. This enzyme suppression gives rise to a surplus intracellular sodium concentration, which through the sodium-calcium exchange stimulation result in an increase intracellular calcium concentration. The significant impacts of digoxin give rise to immediate actions on cardiac muscle, in addition to the indirect actions on the cardiovascular system intervened by effects on the autonomic nervous system.

The autonomic effects include:

(1) A vagomimetic action, playing an important role in enhancing the impacts of digoxin on the sinoatrial and atrioventricular (AV) nodes.
(2) Baroreceptor sensitization, which as a result strengthen the associated inhibition potential and decrease the supportive nervous system activity and renin-angiotensin system for a particular rise in mean arterial pressure.

Pharmacological effect of these outcomes (direct and indirect) are:

(1) accelerated myocardial systolic spasm with comparatively more strength than normal (positive inotropic action);
(2) decreased stimulation degree of the sympathetic nervous system and renin-angiotensin system (neurohormonal deactivating effect);
(3) deceleration of the heart rate and reduced transmission velocity through the AV node (vagomimetic effect).

Its positive inotropic as well as the neurohormonal disabling effect and vagomimetic measures are considered important for the use of digoxin in cases of heart failure and atrial arrhythmias, respectively. High digoxin dose may intensify its sympathetic discharge from the central nervous system (CNS). This propagation in sympathetic activity can probably be a significant aspect in digitalis toxicity.

Prolonged digoxin chemotherapy lowers the risk of intensifying of heart failure, as demonstrated by the heart failure-related hospital and emergency care admissions and the necessity for allied heart failure therapy (Fig. 11.3).

Resistance and Toxicity

No resistance has been found against digoxin; however, its toxicity has been reported cases of chronic or acute clinical setting. In younger adults, the acute toxicity is predicted to occur after an acute overdose. Common symptoms include; vomiting,

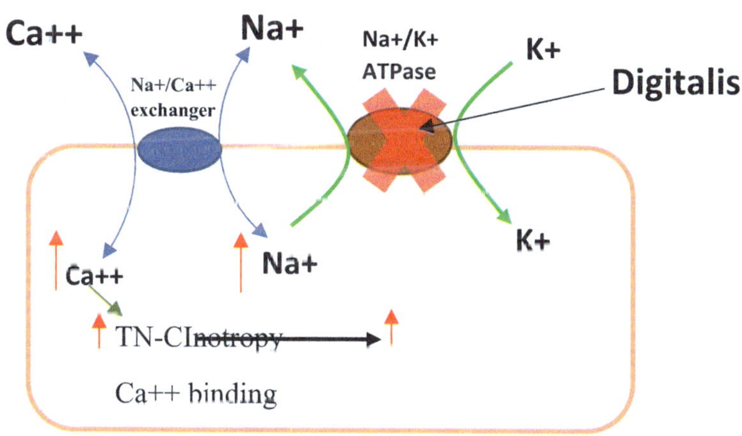

Fig. 11.3 Mechanism of action of digoxin

nausea, dysrhythmias and hyperkalemia. The toxic effects of digoxin typically develop at old age due to the insufficient renal clearance of digoxin, affecting renal function or drug to drug interactions. Nausea, malaise and weakness are typical symptoms in chronic digoxin toxicity (Levine, 2017).

11.2.3 Dilated Cardiomyopathy

Dilated cardiomyopathy (DCM) is one of the most widespread disease or condition, usually contracting by grown-ups between 20 and 60 years old. It disturbs the ventricles of heart and atria, the lower and upper chambers of the heart, respectively. The condition (CDM) is typically diagnosed primarily in the left-ventricle, the main pumping heart's chamber, resulting in the initiation of heart muscle to widen, in which condition it becomes thinner and also extends. Appropriately, the inner part of the chamber expands. The issue often goes to the right ventricle and subsequently to the atria. As the heart chambers dilate, the heart muscle does not have a contract typically and will not be able to pump blood effectively. As the heart turn out to be weaker, heart failure can occur. Prevalent indicators of heart failure involve shortness of breath, fatigue and swelling of the ankles, feet, legs, abdomen and veins in the neck.

Dilated cardiomyopathy may also cause heart valve problems, arrhythmias (irregular heartbeats) and blood clots in the heart. Some illnesses, circumstances and elements may cause several diseases, such as: coronary heart disease, heart attack, high blood pressure, diabetes, thyroid disease, viral hepatitis and HIV, infections, especially viral contagions that inflame the heart muscle, alcohol, hitches through the last month of pregnancy or within 5 months of birth, certain toxins such as cobalt, certain drugs (such as cocaine and amphetamines) and two medicines consumed to treat cancer (doxorubicin and daunorubicin).

The most common symptoms of dilated cardiomyopathy comprises of fatigue, shortness of breath (dyspnea) when the person is active or lying down, reduced ability to exercise, swelling (edema) in legs, ankles and feet, swelling of abdomen owed to fluid buildup (ascites), chest pain, extra or rare sounds heard when your heart beats (heart murmurs) (Mayo Clinic, 2018).

There are various types of cardiomyopathy other than dilated cardiomyopathy:

- Hypertrophic Cardiomyopathy
- Restrictive Cardiomyopathy
- Arrhythmogenic Right Ventricular Dysplasia.

11.2.3.1 Drugs Used in Cardiomyopathy

There are number of drugs used for the treatment and prevention of cardiomyopathy.

Table 11.1 shows the drugs and general mechanism for treating cardiomyopathy.

Table 11.1 Drugs used in cardiomyopathy

S. No.	Drugs	General mechanism
1	*Diuretics (water tablets)*	Lessen the build-up of fluid on the lungs or the ankles by urging the kidneys to get rid of water as urine
2	*ACE inhibitors (angiotensin-converting enzyme inhibitors)*	Ease the smooth muscle around the blood vessels to lessen the workload on the heart and lower the amount of the blood making it simpler for the heart to work
3	*Beta-blockers*	Lessen the rate and strength of the heart's contraction by decreasing stimulus of adrenaline (which would normally make the heart beat faster)
4	*Anti-arrhythmic medication*	Reduces irregular heart rhythms and help to control the normal rhythm
5	*Anticoagulants (blood thinners)*	Might be employed in people with arrhythmias to decrease the probability of blood clotting, which could lead to a stroke
6	*Angiotensin II Receptor Blockers (ARBs)*	Dilate (enlarge) the blood vessels which assist to lower the blood pressure and may be utilized if the person is not able to tolerate ACE inhibitors
7	*Ivabradine*	Helps to decelerate the heartbeat. It may be use by people incompetent to take beta blockers
8	*Thiazide*	Help to get rid of salt (sodium) and water from the body. It mainly affects kidneys to discharge more sodium into your urine
9	*Sacubitril and valsartan (brand name Entresto)*	It slow down the heartbeat. It may be use by people unable to take beta blockers

Thiazide

Thiazide diuretics are FDA-authorized medicines that prevent resorption of 3–5 percent of sodium present in the luminal of the distal convoluted tubules of the kidney nephron. Diuretics also referred to as water pills, get rid of salt (sodium) and water in the body. Thiazide diuretics are deemed to promote natriuresis and diuresis (Mayo Clinic, 2019). Chlorthalidone, hydrochlorothiazide (HCTZ), and indapamide are the most used thiazide as shown in Table 11.2.

Hydrochlorothiazide (HCTZ) is another utmost frequently recommended antihypertensive drug in the USA, with roughly 47.5 million prescriptions distributed in 2008. Chlorthalidone and HCTZ are FDA-approved for a medical practice to manage primary hypertension. The intermediaries are used as only the working methods of therapy or in combination with some other antihypertensive drugs to escalate the therapy response in relentlessly patients of hypertensive effectually. It has also been used as an adjunct therapy in edema correlated with hepatic cirrhosis, chronic heart

Table 11.2 Commonly used thiazides

Diuretics	
Thiazides	Chlorothiazide (Diuril)
	Chlorthalidone
	Indapamide
	Hydrochlorothiazide (Microzide)
	Metolazone

failure (CHF), estrogen, and corticosteroid therapy. The following drugs are also used to operate edema, initiated by different forms of renal dysfunction, such as acute glomerulonephritis, chronic renal failure, and nephrotic syndrome.

Mechanism of Action

Thiazide diuretics regulate hypertension by the inhibition of chloride (Cl−) and sodium (Na+) ions reabsorption from the distal convoluted tubules of the nephrons. This is done by thiazide-sensitive Na^{+}–Cl^{-} symporter hinderance (Duarte et al., 2010). Though these drugs lower arterial pressure effectively, it has prolonged confusions (Hughes et al., 2004). Thiazides lower cardiac output severely by reducing the plasma amount and extracellular fluid (ECF) and, but ECF amount reverts to approaching standards during vasodilation supervenes and chronic use (Pecker et al., 1990). At a constant state, the influence of thiazides becomes prevalent to the vasodilation than the contraction of the volume. Based on the physiological properties and sensing struggle, the volume of the ECF reduced during chronic treatment, most of the establishments propose that the chief machinery by which thiazide diuretics lessen pressure on the arteries by the direct vasodilatation, perchance arbitrated by modifications in vascular ion conveyance (Kaplan et al., 2002; Papademetriou et al., 2003; Sica et al., 2008). In others, however, accentuation of salt exhaustion is required (Pecker et al., 1990). This implies that vasodilatation is subordinate to ECF volume contraction. Thiazides use in the final phase of renal diseases is not successful in many studies. (Bennett et al., 1997).

Considerable effort has been made in organizing the mechanisms through which thiazides dilate blood vessels. The drugs could affect the vascular smooth muscle's membrane ion flux, for example. Thiazides widen Ca-activated K ion channels in vascular smooth muscle cells cause the hyperpolarization and vasorelaxation in vitro. (Pickkers & Hughes, 1995). Hydrochlorothiazide causes moderate dilation of blood vessels of the human forearm in vivo, although this effect is seen at a higher dosage than when the drug is taken orally. (Pickkers et al., 1998). The result appears to be attributed, at least to some extent, to hydrochlorothiazide's ability to suppress carbonic anhydrase, which alkalinizes the cell. (Pickkers et al., 1999). Since the carbonic anhydrase–inhibits the efficacy of thiazide diuretics with the variation of analogs, the vascular effects of these medications also vary.

Including confirmation of significant vasodilation, the key action of thiazide diuretics is to inhibit NaCl co-transporter (NCC; gene symbol SLC12A3) in the kidney's distal convoluted tubules. The following drugs are found a unique blockers of that kind of proteins since not blockage effects against the amiloride-sensitive Na channels (Ellison, 1987) or furosemide-sensitive Na–K–2Cl co-transporter (Velázquez & Wright, 1986).

The NCC is vented by the cells of distal convoluted tubules of rodents (Obermüller, 1995), rabbits (Bachmann et al., 1995), and humans (Obermüller et al., 1995). Although there is some indication that it might be expressed in bone (Dvorak et al., 2007) and intestine (Bazzini et al., 2005), it is not articulated by cardiac tissue or vascular smooth muscle (Gamba et al., 1994). The mutation in SLC12A3 is due to the Gitelman syndrome (GS) (Simon et al., 1996), hypokalemia, and alkalosis syndrome. These alterations that interrupt NCC's function (Kunchaparty et al., 1999), lessen arterial pressure by ≈8 mm Hg (Cruz et al., 2001; Sartori et al., 2007) an influence comparable to decrease the arterial pressure that comes during hypertension action of thiazide. Interestingly, vasodilation rather than ECF volume reduction mediates hypotension in GS (Calo et al., 1996, 2000, 2001, 2002), admitting the SLC12A3 is articulated to the cells of kidney, not by the cells of smooth muscles. Individuals with GS have not produced or reduced peripheral resistance, and vascular hypo-reactivity (Calo et al; 2002). Angiotensin II signaling is muffled, with decrease in downstream cellular activities, such as inositol triphosphate and intracellular Ca release and the expression of the Gq-binding protein α-subunit (Calo et al; 2002).

Though potassium deficiency has been identified as one of the contributing factors to vasodilatation (Calo et al; 2002), it remains unclear to be the primary cause, although dietary potassium storage, instead of shortage, dilates vessels and lowers blood pressure (Geleijnse et al., 2003). Therefore, it is common that the GS blood pressure is poor because of the deficiency of renal salt that induces secondary vasodilation. (Ellison & Loffing, 2009; Alon, 2018; Pourafshar et al., 2018) (Fig. 11.4).

Resistance and Toxicity

The resistance against diuretic is well-understood as the inability to accomplish the therapeutic reduction in edema after the diuretic's highest dose is utilized. Table 11.3 outlines the leading causes of diuretic resistance (Hoom et al., 2015), consisting of

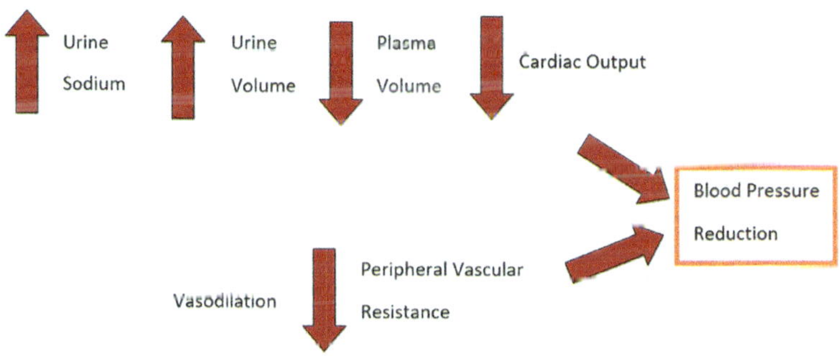

Fig. 11.4 Mechanism of action of thiazide

Table 11.3 Common causes of diuretic resistance

Common causes of diuretic resistance	Description
Nonadherence to suggested sodium and/or fluid limitation	–
Inaccurate diagnosis (eg, venous or lymphatic edema)	–
Medication not getting into the kidney	• Nonadherence • Weak absorption • Dosage too small or too sporadic
Inadequate kidney reaction to the drug	• Minimal glomerular filtration rate • Initiation of the renin-angiotensin system • Reduced effectual intravascular quantity despite the high total extracellular fluid volume • Usage of nonsteroidal anti-inflammatory drugs • Nephron adaptation
Decreased diuretic excretion	• Tubular uptake of diuretic compromised by uremic toxins • Reduced functional kidney mass • Reduced kidney blood flow

Source Hoorn et al. (2015)

inadequate attachment to drug therapy, diet, pharmacokinetics, and compensative escalations in Na re-uptake in nephron locations that are not being obstructed by the diuretic (Ellison et al., 1989). The identification of the reason behind the cause of diuretic resistance is important because it illuminates the potential for involvement. Diuretic defiance, for example, is commonly used to manage successfully by mixing a loop diuretic with a separate form of diuretic (Ellison, 1991).

The toxic effect seems to be dose-related; improved dosage leads to increased severity causing a negative impact. Loop diuretics obstruct sodium absorption in Henle's loop's dense ascending limb by constricting the Na–K-2Cl carrier. This results in diminished focusing intensity in the renal medulla and, as a result, natriuresis (resulting in hyponatremia), kaliuresis (resulting in hypokalemia), and water loss (resulting in hypovolemia) (Friedman, 1988). The subsequent hypovolemia triggers the secretion of antidiuretic hormone (ADH), inducing reabsorption of free water and worsening hyponatremia, hypokalemia, and hypocalcemia. (Sonnenblick et al., 1993).

Thiazides inhibit sodium resorption in the distal complicated tubules. This improves sodium transport to the collecting tubules, where it is substituted for K^+ (Ellison & Loffing, 2009). Adverse effects may trigger hypovolemia, hyponatremia, and hypokalemia; thiazide diuretics may cause symptomatic hypercalcemia (Friedman & Bushinsky, 1999). Potassium-sparing diuretics are discerning aldosterone antagonists that bind on a competitive basis to the exchange transporters of Na^+–K^+ in the distal tubule that verdicts preserve potassium, heightened secretion of sodium and water. This poisoning leads to hyperkalemia.

Beta Blockers

Beta-blockers (β-blockers) are a diversified class of antihypertensive medications that have competitive antagonistic action with β-adrenoreceptors (β1, β2, and β3). This prevents the "fight or flight" response triggered by noradrenaline and adrenaline. Pharmacological differences, such as β-adrenergic selectivity, dual receptor activities, and lipid solubility, distinguish each beta-blocker. Scientific research is expanding into the advanced approaches by which these characteristics translate into medical outcomes for patients (BPAC, 2017; Ladage et al., 2013). The receptor selectivity, vasodilating properties, inherent sympathomimetic activity (ISA), and metabolism of β-blockers differ. Recent cohort beta-blockers give the appearance of having an unbiased metabolic activity. Three agents have been shown to increase stamina by up to 30 percent in systolic heart failure, owing to their vasodilating effects or/and β1-blocking and/or. The function of β-blockers in the treatment of diastolic heart disease remains unknown. β-blocker in the treatment of coronary heart disease also provides a proven survival benefit, the cardioprotective function being largely attributed to a decrease in heart rate (Frauke & Hans, 2010).

β-blockers and Ca channel blockers are recommended as first line anti-anginal medications (NICE, 2011). There is no proof that one β-blocker outperforms another or that β-blockers are superior to Ca channel blockers (Shu et al., 2012). The choice of β-blocker is largely determined by the prescriber's expertise, patient's comorbidities, the patient's preference for the pace of dosing, and the patient's probability of compliance with the medication. A cardioselective β -blocker, such as metoprolol succinate or bisoprolol, would have the greatest impact and the least number of side effects. β-blockers that reduce the resting rate of the heart compared to others (due to ISA) have no tendency to be used for angina, for example, pindolol and celiprolol (BPAC, 2017). β-blockers can induce vasoconstriction because β2 adrenergic receptors may cause vascular smooth muscle constriction. As a result of blocking alpha-adrenergic receptors, third-generation β-blockers induce vasodilation. Consequently, nonselective beta-blockers are expected to have antihypertensive effects.

β-receptors are classified into three groups: beta-1 (β1), beta-2 (β2), and beta-3 (β3). The heart's β1 receptors exercise the heart's function. Through their complex position in multiple organ systems, β2 receptors manage different facets of metabolic processes in the field and promote smooth muscle relaxation. β3 receptors cause fat cell breakdown and have no therapeutic relevance at the moment. The inhibition by β-blocking medications of these receptors is used to treat a variety of diseases. (do Vale, 2019). β-blockers are important medicines that are used as first-line therapies in several chronic and acute medical problems. If they are remunerated, congestive heart patients are given β-blockers. The carvedilol, β-blockers bisoprolol, and metoprolol succinate have been chosen. (Farzam & Jan, 2020). Clinically, β-blockers are classified into two groups;

- "cardioselective" β-blockers (Relatively selective β1 blockers), or
- block both β1and β2 receptors (non-selective blockers).

Table 11.4 Classes of beta blockers

Class	Drugs
β1-selective	Nebivolol Acebutolol Metoprolol Atenolol Betaxolol Esmolol Bisoprolol
Non-selective β1/β2	Carteolol Timolol Carvedilol Nadolol Penbutolol Pindolol Sotalol Propranolol Labetalol

In addition to beta-blockade, such beta-blockers have supplemental mechanisms that lead to their narrow pharmacologic profiles. Table 11.4 lists the two groups of β-blockers in combination with compounds. The medicinal applications mentioned in the table provide both off-label and on-label advantages of β-blockers. For example, the FDA may only allow a given β-blocker to treat hypertension; however, doctors sometimes prefer to prescribe angina drugs due to the class-action benefit that β-blockers have for angina.

Mechanism of Action

Epinephrine, norepinephrine, and catecholamines bind to β1 receptors, increasing transmission velocity and cardiac automaticity. β1 receptors also stimulate the release of renin, which causes an increase in blood pressure. Attaching to β2 receptors, on the other hand, causes increased physiological effects such as glycogenolysis as well as smooth muscle relaxation.

β-blockers act differently, especially for various receptors, and hence the results provided depend on the receptor(s) types clogged as well as the involvement of organ system. Certain β-blockers often bind to α-receptors to a specific extent, causing them to create a range of medical effects as used in particular circumstances. As β-blockers bind to the β1 and β2 receptors, and block their effects. As a result, the inotropic and chronotropic effects on the heart are suppressed, and the pulse rhythm slows. β-blockers often lower blood pressure by a variety of mechanisms, including reduced cardiac activity and renin. The adverse inotropic and chronotropic effects reduce oxygen use, thus curing angina after the usage of β-blocker. These medications often produce a powerful antiarrhythmic effect and lengthen atrial refractory times.

β-1 selective and non-selective β-blockers are the two types of β-blockers. Furthermore, there are β2 and β3 specific drugs; neither has an established medicinal target. Non-selective is tightly linked to both β1 and β2 receptors and causes antagonizing special effects through receptors. Sotalol, carvedilol, labetalol, and propranolol are a few examples. β1 receptor-selective antagonists, such as bisoprolol, atenolol, esmolol, and metoprolol attach only to β1 receptors and are hence cardio-selective (Gorre et al., 2010; Rehsia and Dhalla, 2010; Machackova et al., 2011).

In certain patients, beta-blockers induce anxiety and sleep disruptions by decreasing melatonin excretion (Fares, 2011). α-1 receptors improve cardiac chronotropic and cause vasoconstriction; that causes antagonism effect at the α-1 receptors which increased heart beating rate and elevated blood pressure. Antagonism at the α-1 receptor, on the other hand, causes vasodilation and a negative chronotropic effect, lowering the pulse rate and blood pressure. β-blockers, such as bucindolol, carvedilol, and labetalol with the effect of non-selective beta receptor blockage, have additional α-1 receptor blockage action. This inhibitory effect of β-blockers to the α-1 receptor have a greater therapeutic effect on hypertension treatment (Weir, 2009) (Figs. 11.5 and 11.6).

Resistance and Toxicity

No resistance is reported related to beta-blockers. However, its toxicity exists. Glucagon is the antidote for beta-blocker overdose. It is particularly valuable in beta-blocker-stimulated cardiotoxicity. Another way of medication is cardiac pacing if glucagon collapses (Farzam & Jan, 2020).

Angiotensin II Receptor Blockers

We have already discussed it above. This drug shows therapeutic resistance against angiotensin-converting enzyme (ACE) reticence linked to pharmacodynamics and pharmacokinetics.

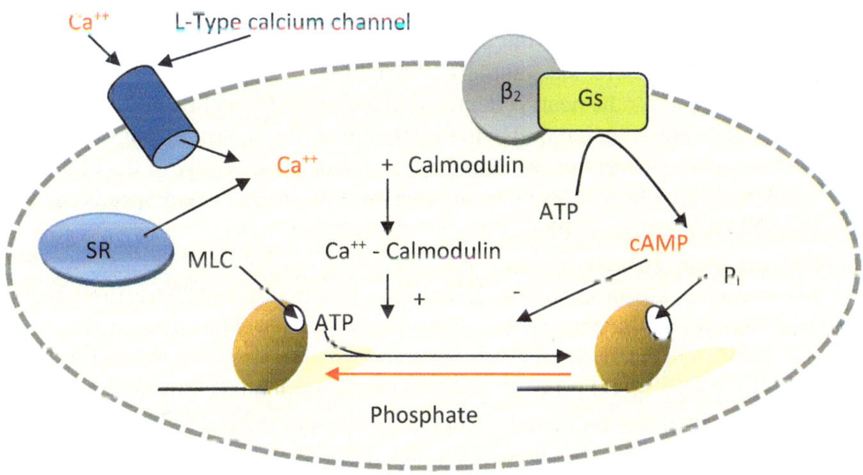

Fig. 11.5 Mechanism of action of beta-blockers

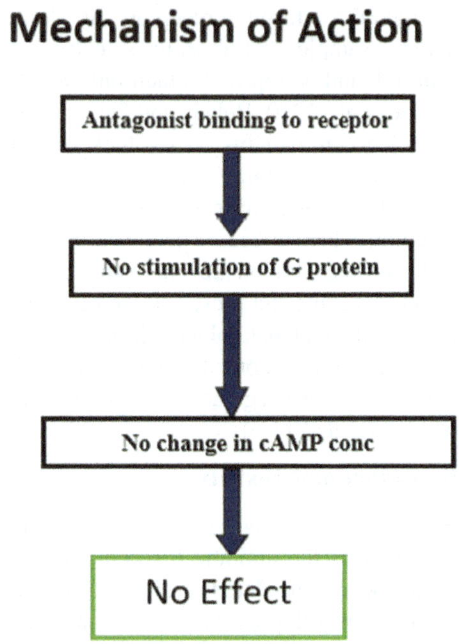

Fig. 11.6 Flowchart of mechanism of action of beta blockers

11.2.4 Coronary Artery Disease (CAD)

CAD is characterized by thinning or obstruction of the coronary arteries and is normally caused by atherosclerosis. Atherosclerosis (also known as "clogging" or "hardening" of the arteries) is the deposition of fats and cholesterol deposits (called plaques) on the inner walls of the arteries. These plaques may block the flow of blood to the heart muscle, substantially obstructing it or causing abnormal artery sound and function. As the heart is deprived of sufficient blood flow, it loses oxygen and the dynamic nutrients it requires to function properly. It induces angina or chest pain. A heart attack (heart muscle damage) can occur if the supply of the blood to a portion of the heart muscle is entirely stopped, or if the heart's energy needs become significantly greater than its blood supply (AHA, 2017).

There are many risk factors for CAD, including non-modifiable (uncontrolled) risk factors such as advanced age, male gender, a family history of heart disease, and ethnicity. The modifiable risk factors (controlled), such as exposure to tobacco and cigarette smoke, elevated blood cholesterol and triglycerides, unregulated diabetes (HbA1c >7.0), high blood pressure (140/90 mmHg or higher), physical inactivity, obese (BMI greater than 30 kg/m2) or being overweight (BMI 25–29 kg/m2), uninhibited tension or rage, and an unsanitary diet. The risk of developing CAD rises as risk factors grow.

CAD is also known as ischemic heart disease. CAD is caused by the accumulation of plaque in the blood vessels walls that supply blood to the heart (known as coronary arteries) and other body parts. Plaque is the accumulation of cholesterol (fats) and other contaminants in the blood arteries. Plaque accumulation begins within the blood arteries, which narrow over time, resulting in partial or total obstruction of the flow of the blood. This is referred to as atherosclerosis. A heart attack is believed to be the first symptom of CAD.

Symptoms of a heart attack include:

- Weakness, a cold sweat, nausea (feeling sick to the stomach), or light-headedness
- Arm or shoulder pain or stiffness
- Shortness of breath
- CAD will damage the heart muscle over time. This will result in cardiac failure, a debilitating disease in which the heart is unable to pump blood normally (CDC, 2019).

11.2.4.1 Drugs Used for Coronary Artery Disease

The following drugs can be used for the treatment of CAD, including:

- Aspirin.
- Cholesterol-modifying medications.
- Angiotensin-converting enzyme (ACE) inhibitors and angiotensin-II receptor blockers (ARBs).
- Nitroglycerin.
- Ranolazine.
- Calcium channel blockers.
- Beta blockers.

Aspirin

Aspirin, an acetylated salicylate (acetylsalicylic acid) is considered among the nonsteroidal anti-inflammatory drugs (NSAIDs). The signs and symptoms of inflammation are lessen bythese agents and reveal wide ranging pharmacologic activities, including analgesic, antipyretic and antiplatelet properties.

Mechanism of Action

Aspirin work by inhibiting platelet activation and aggregation (Vane, 1971). The chief action mechanism includes irreversible inhibition of the platelet-dependent enzyme cyclooxygenase (COX), thus preventing the synthesis of prostaglandins. Two COX isoenzymes, COX-1 and COX-2 are identified. In platelets, the COX-1 enzyme generates a powerful promoter of platelet aggregation namely thromboxane A2. Thus, aspirin irreversibly inactivate COX-1, which consequently impeding the generation of thromboxane A2, acquires a prospective antiplatelet effect (Masferrer et al., 1990).

Platelet activation and aggregation with successive clotting chute perform serious roles in the commencement of acute occlusive vascular events, such as MI and occlusive cerebrovascular accident (CVA) (Fuster et al., 1992). Since platelets do not come up with nucleus and thus cannot rejuvenate COX, they turn out to be an exceptional objective for antithrombotic therapy, while aspirin demonstrates both urgent and lasting impacts on platelets (Vane & Botting, 2003).

Other mechanisms of aspirin in CVD may also operate. Aspirin blocks the realization of COX- dependent vasoconstrictors, which contribute in endothelial dysfunction in atherosclerosis (Husain et al., 1998). Consequently, up gradation of endothelial dysfunction with aspirin may expand vasodilation, lessen thrombosis and prevent progression of atherosclerosis. Furthermore, aspirin reduces the inflammatory reaction in patients with coronary artery disease (Ridker et al., 1997) and may hamper the advancement of atherosclerosis by safeguarding low-density lipoprotein as a result of oxidation (Steer et al., 1997) (Fig. 11.7).

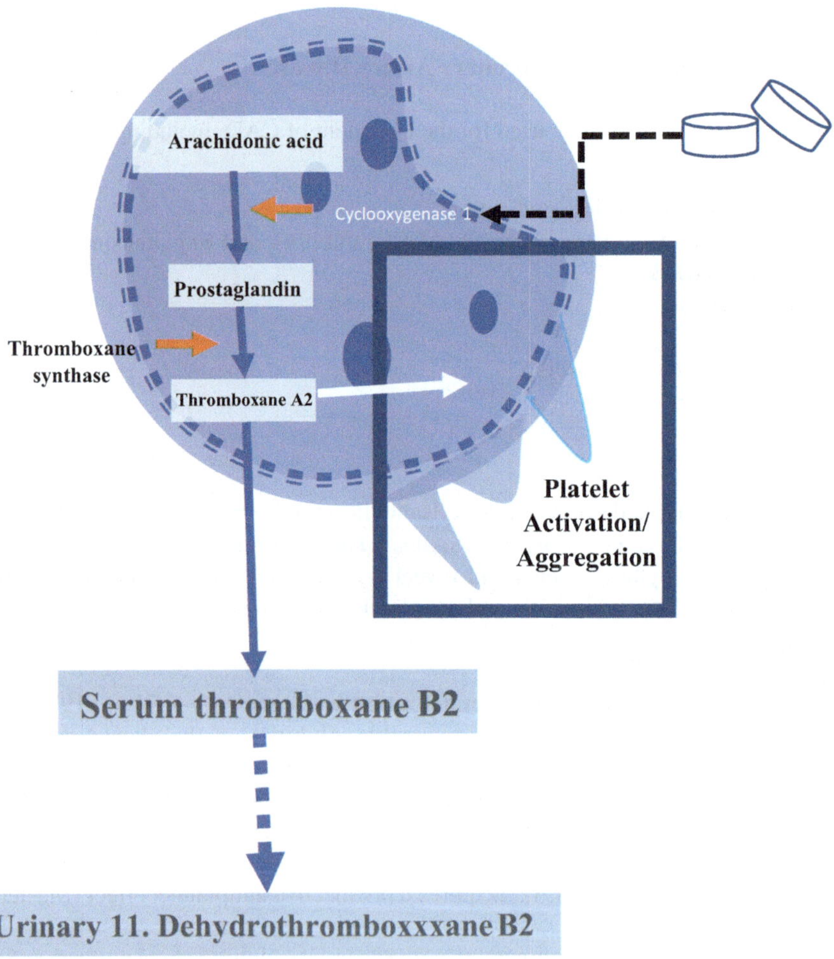

Fig. 11.7 Mechanism of action of aspirin

Resistance and Toxicity

Aspirin resistance is the incompetence of aspirin to reduce platelet fabrication of thromboxane A_2 and thus platelet stimulation and accumulation. Escalating levels of aspirin resistance might correlate autonomously with hoarding risk of cardiovascular events. Aspirin resistance can be differentiated by laboratory tests of platelet thromboxane A_2 production or platelet role that is reliant on platelet thromboxane production. Prospective sources of aspirin resistance consist of insufficient dose, drug interfaces, genetic polymorphisms of COX-1 and other genes engaged in thromboxane biosynthesis, up regulation of non-platelet causes of thromboxane biosynthesis, and amplified platelet yield. Aspirin resistance can be overwhelmed by curing the root or roots, and condensed by abating thromboxane creation and action and hindering other pathways of platelet stimulation (Hankey & Eikelboom, 2006).

The intensely lethal dose of aspirin is typically considered larger than 150 mg per kg of body mass. Mild toxicity arises at doses up to 300 mg/kg, serious toxicity ensues between 300 and 500 mg/kg, and a theoretically lethal dose is greater than 500 mg/kg. The first sign of salicylate toxicity is frequent hyperventilation and respiratory alkalosis due to medullary stimulation. Metabolic acidosis imprints and an augmented anion rift results from accrual of intracellular lactate as well as excretion of bicarbonate by the kidney to compensate for respiratory alkalosis. Preliminary ABG testing habitually reveals this miscellaneous respiratory alkalosis and metabolic acidosis. Raised body temperature may be observed. Serious hyperthermia may happen in severe instances. Vomiting and hyperpnoea as well as hyperthermia add to fluid loss and dehydration. Extreme metabolic acidosis, seizures, coma, pulmonary edema, and cardiovascular malfunction may befall with serious poisoning.

After vast aspirin utilizations, antagonistic gut cleansing is desired, involving gastric lavage, repetitive doses of activated charcoal and consideration of whole bowel irrigation. Give IVF to substitute fluid losses caused by tachypnea, vomiting and fever. For modest intoxications, IV sodium bicarbonate is given to alkalinize the urine. It also supports to salicylate excretion by trapping the salicylate in its ionized form. For serious poisoning consider emergency hemodialysis to reduce the salicylate more rapidly and restore acid–base balance and fluid status (OpenAnaesthesia, n.d.).

Nitroglycerin

Nitroglycerin, otherwise known as glyceryl trinitrate (GTN) is used for heart failure, high blood pressure, anal fissures and painful periods to cure, and inhibit chest pain caused by reduced blood flow to the heart (angina) or due to the amusing use of cocaine. Nitroglycerin is a vasodilator drug used for the treatment of chest pain and high blood pressure (Kyle et al., 2019, FDAa, n.d.). It was first sanctioned in 2000 and depending on the dosage, it is currently sold by Pfizer, and other companies (Kyle et al., 2019; FDAc, n.d.). This medicine is obtainable in a variety of ways, involving a spray form, sublingual tablet form, intravenous form, extended-release tablet form, and transdermal form. Besides treating heart diseases, it is also utilized as an ointment to cure the pain that accompanies anal fissures. FDA approved the rectal ointment form of nitroglycerin in 1955 (FDAb, n.d.).

Nitroglycerin helps in relaxation of vascular smooth muscles that results in arteriolar and venous dilatation. This reduces the cardiac preload and afterload and lessens coronary artery spasm, reducing systemic vascular endurance as well as systolic and diastolic blood pressure (FDAa, n.d.).Relief of anginal symptoms is sourced by the drop of cardiac work by nitroglycerin with certain aids from arteriolar dilatation effects (Kyle et al., 2019).

Mechanism of Action

Nitroglycerin is transformed by mitochondrial aldehyde dehydrogenase (mtALDH) to nitric oxide (NO), which accelerates the enzyme guanylate cyclase (Chen et al., 2002).The synthesis of cyclic guanosine monophosphate (cGMP) indicates the enzyme (Ferreira & Mochly-Rosen, 2012; McNiff et al., 1981) activating a cascade of protein kinase-dependent phosphorylation events in smooth muscles. This ultimately cause dephosphorylation of the myosin light chain of smooth muscles, relaxing and increasing the blood flow in veins, arteries and cardiac tissue (Ferreira & Mochly-Rosen, 2012; Kyle et al., 2019). The above procedures result in declined work of the heart, decreased blood pressure, relief of anginal symptoms, and increased blood flow to the myocardium (Ferreira & Mochly-Rosen, 2012; Kyle et al., 2019; Sanofi, n.d.). An invitro study conducted by (Madhani et al., 2003) using mouse aorta recommends that nitric oxide (an activated metabolite of nitroglycerin) aim at the natriuretic peptide receptors.

Resistance and Toxicity

NTG tolerance is nitrate-specific in patients with coronary artery disease and is associated with affirmation of impaired NTG bioconversion. Tolerance was correlated with additional O^{2-} generation, but brief preferment of O^{2-} did not disturb NTG sensitivity, signifying improved NO annihilation because O_2^- has negligible influence on tolerance (Sage et al., 2000).

A range of hemodynamic effects can occur as a result of nitroglycerin overdose. Broad-spectrum effects may include vertigo, fever, flushed skin and diaphoresis. Syncope, dyspnea, decreased heart rate, or palpitations occurs as preliminary symptoms of cardiorespiratory disorders. Paralysis, seizures, coma, and death occurs in neurologic manifestations (FDAd, n.d.). No antidotes have been identified in case of nitroglycerin overdose, neither it is recognized that its metabolites can be detached from the circulation. (FDAa, n.d.; FDAd, n.d.).The lower limbs are upraised and an intravenous infusion of normal saline or other fluids is administered if essential in case of consequent hypotension from nitroglycerin overdose, to maintain central fluid volume (FDAa, n.d.; FDAd, n.d.).

11.2.5 Cardiac Arrhythmia

Cardiac arrhythmias is electrical disproportion in heart that results in the instability of heart beat, either too slow (bradycardias) or unusually rapid (tachycardias) or create an irregular and disturbed heart beat (Kastor & Saba, 2009; AHA, 2016; Curran & Sheppard, 2011). In arrhythmic patient, in addition to disruption of heartbeat some other symptoms also appear including weakness, shortness of breath, light headedness, dizziness, fainting and rarely chest pain (Cruickshank, 2008; Curran & Sheppard, 2011). Cardiac arrhythmias is a common heart disease and is associated with the disturbance of the cardiovascular system (Delisle et al., 2004). Arrhythmia can be a consequence of any unusual state affecting the myocardium and any procedure which interfere with the normal functioning of heart valves. It leads toward many metabolic disturbances and in this association with the sodium Na^+ and potassium K^+ ions interchange across the myocardial cell membrane, Ca^{++} and Mg^{++} concentration are all of great importance (Resnekov, 1964). Arrhythmia can also occur when electrical signals produce from another part of the heart in addition to the normal signals from the special nerve cells and in turn cause disruption of normal rhythms of heart (Curran & Sheppard, 2011). There are different types of cardiac arrhythmia depending upon the heart rate speed and location of impulse generation. These types are discussed in Table 11.5. (AHA, 2016; Curran & Sheppard, 2011).

11.2.5.1 Drugs for Cardiac Arrhythmia

The drugs used for cardiac arrhythmia are known as antiarrhythmic drugs. Antiarrhythmic drugs are substances which influence cardiac ionic channels and lead towards the alternation of cardiac action potential or its generation or transmission (Dhein, n.d.). Cardiac rhythm can be normalized if unusual non homogeneity of excitation, conduction or repolarization is prohibited. To achieve regular rhythms antiarrhythmic drugs can be used to restrict cardiac ionic channels. These drugs have been categorized by Vaughan Williams (Borchard et al., 1989). Vaughan Williams categorized these drugs into four classes depending upon its similar toxicity mechanism (Abramowicz et al., 2004; Roden, 2000).

The classification of antiarrhythmic drugs is as follows:

(1) Class 1—Na^+ channel blockers, further divided into 1a (transitional), 1b (rapid), 1c (slow);
(2) Class 2—Beta-adrenoceptor antagonists;
(3) Class 3—Potassium (K^+) channel blockers;
(4) Class 4—Calcium (Ca^{2+}) channel blockers (Dan et al., 2018).

There were some limitation in Vaughan Williams approach than modern classification previously proposed by Sicilian Gambit. Sicilian Gambit has classified the antiarrhythmic drugs depending on their mechanism of action, the way they interact with arrhythmia causing substrates to suppress arrhythmias (Roden, 2000).

Table 11.5 Types of cardiac arrhythmia

Disease	Types	Description
Cardiac arrhythmia	Atrial fibrillation	Totally irregular and disturbed atrial activity
	Ventricular fibrillation	Disordered contraction of the lower chambers of the heart
	Sinus Bradycardia	Slow heart rate than normal (<60 bpm)
	Sinus Tachycardia • Supraventricular Tachycardia • Ventricular Tachycardia • Atrial Tachycardia	Rapid heart rate than normal (>100 bpm) • 150 beats per minute in upper heart chamber, common in young adults • Higher heart rate that starts in lower chamber, generally severe and results from serious heart diseases • Fast heart rate that initiate in upper chamber and is conducted into atria
	Sinus pauses (asystole)	Absence of electrical impulses in heart
	Conduction Disorders	Disruption in the normal heart beat
	Premature contraction • Premature atrial contraction • Premature ventricular contraction	Premature and early generation of heart beat • Early generation of heart beat in atria • Early generation of heart beat in ventricle
	Atrial ectopic rhythm	The generation of electrical impulse in atria comes from the wrong place
	Other rhythm disorders	

Verapamil

Verapamil, a synthetic papaverine derivative (Singh et al., 1978) is a phenylalkylaminean. It is representative of class IV antiarrhythmic agent (Dhein, n.d) and is recognized as calcium channel blocker due to its mode of action. It is reported that verapamil is most extensively used calcium blocker for the cure of cardiac arrhythmia and is potentially effective for the treatment of paroxysmal supraventricular tachycardia (Vohra, 1982).

Mode of Action

As verapamil is a calcium antagonist, it restrict the inflow of calcium ions via channels into myocardial cells and cardiovascular smooth muscle cells. The reduction in calcium inflow via channel influence the electrical activity of sinoatrial (SA) node and atrioventricular (AV) node (FDA, 2016). Verapamil retains the capability to reduce transmission beyond the AV junction and to extend the refractory period inside the

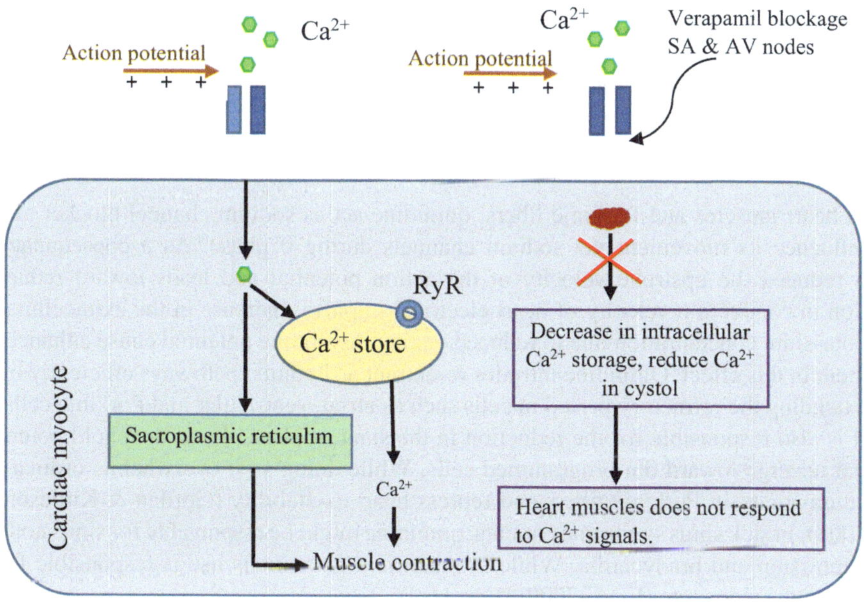

Fig. 11.8 Verapamil, mode of action as an antiarrhythmic agent

node at medically recommended measures (Gordon & Kittleson, 2008; FDA, 2016). Extension of conduction and refractoriness are traditional ways to cure arrhythmias (Gordon & Kittleson, 2008). In atrial fibrillation or atrial flutter patients it results in the decrease in ventricular rate and return at the AV node and restore ordinary sinus rhythm paroxysmal supraventricular tachycardias patient. Verapamil has no influence on the transmission across accessory bypass tracts, ordinary atrial action potential or intraventricular conduction period, but it is responsible for reduction in amplitude and velocity of depolarization conduction in depressed atrial fibers (FDA, 2016) (Fig. 11.8).

Resistance and Toxicity

No resistance has been found against verapamil however, it has toxic impacts associated with it. Verapamil has ability to reduce cardiac contractility and is responsible for marginal vasodilation. Intake of verapamil should be avoided by the myocardial failure and heart failure patients because of its ability to cause cardiac impacts. Even sick sinus syndrome patients should not use verapamil because it is responsible for depressed automaticity in the diseased or damaged tissues. Sometimes it can cause severe hypotension and cardiovascular collapse in dogs (Gordon & Kittleson, 2008).

Quinidine

Quinidine is a naturally occurring antiarrhythmic agent of class 1a (Yang et al., 2009) and is obtained from the bark of cinchona plant, a native of South America. It act as a sodium channel blocker and cause hindrance in the rapid inflow of sodium current

(Tsai-Turton, 2014). Quinidine was among the initially used drugs to recover from arrhythmias (Roden, 2000; Yang, et al., 2009) and is especially used for reduction of atrial fibrillation, and destruction of ventricular arrhythmias (Tsai-Turton, 2014; FDA, 2009).

Mode of Action

In heart muscles and Purkinje fibers, quinidine act as sodium channel blocker and influence its movement via sodium channels during 0 phase. As a consequence it reduced the upstroke velocity of the action potential and leads toward reduction in conduction velocity of heart electrical impulse. Increase in the extracellular potassium concentration due to reduced resting membrane potential cause enhancement in this effect. Quinidine intrudes re-entrant arrhythmic pathways efficiently by extending the refractory period in cells such as atrial, ventricular and Purkinje cells. It is also responsible for the reduction in the slant of phase 4 and threshold potential upsurge toward 0in programmed cells. While doing so it overwhelms ordinary automaticity in Purkinje fibers and repress heart excitability (Gordan & Kittleson, 2008). In sick sinus syndrome patients, quinidine might be responsible for sinus node depression and bradycardia. While in maximum patients its use is responsible for rise in the sinus rate (FDA, 2009) (Fig. 11.9).

Drug Resistance and Toxicity

There is no single report which indicates quinidine drug resistance in humans but it has been reported in yeast cells. A study was conducted to determine drug resistance of quinidine, ketoconazole and stereoisomer quinine by using *Saccharomyces*

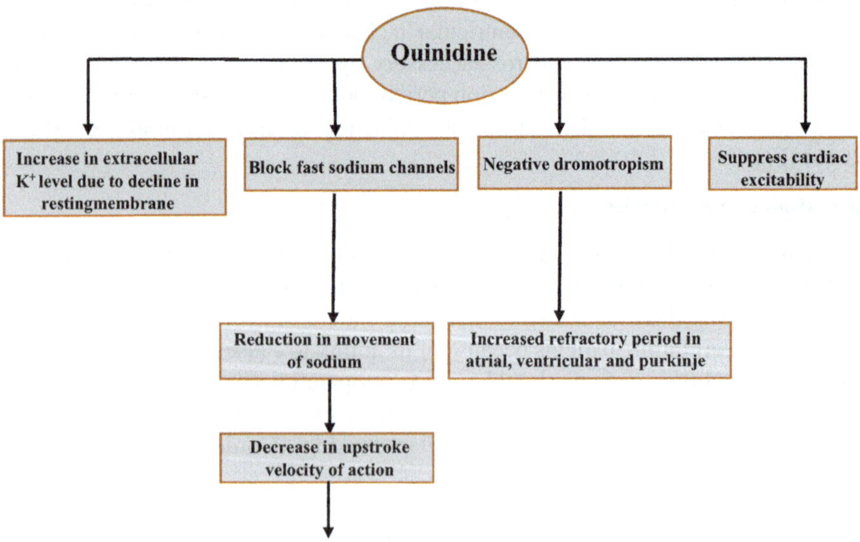

Fig. 11.9 Influence of quinidine on arrhythmias

cerevisiae open reading frame *YIL120w*, a multidrug resistance determining factor. It concludes that yeast occupied resistance to ketoconazole and quinidine, but not resistant to the stereoisomer quinine. It revealed that sudden exposure of yeast to the drug was associated with the loss of cell viability (Nunes et al., 2001).

The use of quinidine is particularly advantageous for treatment of cardiac arrhythmia. However, its use give rise to some severe side effects including diarrhea, vomiting, QT prolongation and torsades de pointes, and thrombocytopenia even at low level dose. Quinidine can cause increase in digoxin level to potentially toxic levels (Abramowicz et al., 2004).

Adenosine

Adenosine is an endogenous purine nucleoside (Camm et al., 1992) which plays a significant role in physiological and pathophysiological function of pacemaker, particular conductive tissues, operational myocardium and coronary arteries (Szentmiklósi et al., 2015). Adenosine can be produced during hypoxia and also formed in cytosol in minute quantity (Slotwiner & Lerman, 1998).

Mode of Action

Adinosine derived from adenosine triphosphate (ATP) binds with the A1 receptor leads toward activation of Potassium channel and inhibit cyclic adenosine monophosphate (cAMP) as a consequence of reduced Ca^{++} inflow. Stimulation of K^{+} channel is responsible for the hyperpolarization of atrial myocardium, reduction in action potential duration and declined diastolic depolarization of the sinus node (Paul & Pfammatter, 1997). At sinus node adenosine is responsible for the decrease in the discharge rate of sinoatrial node cell, the blockage of sinoatrial node exit, and a pacer swing from the sinoatrial node to the nearby crista terminalis cells (Belardinelli et al., 1998; West & Belardinelli, 1985). Adenosine has ability to reduce the amplitude and duration along with the cardiac upstroke velocity of action potential in N cells of AV node (Clemo & Belardinelli, 1986). As a result it is responsible for the stimulation of AV block and re-establish ordinary sinus pace in paroxysmal supraventricular tachycardia patients (Paul & Pfammatter, 1997; RxList, n.d) (Fig. 11.10).

Resistance and Toxicity

No resistance has been found against antiarrhythmic adenosine. However, adenosine give rise to severe impacts as a consequence of stimulation of adenosine receptors establish on vascular tissue, leads towards vasodilation. It is associated to cause symptoms of skin flushing, light headedness, vomiting, sweating, anxiety and unconsciousness but these symptoms are for short time due to small half-life of the adenosine. But it is also responsible for the severe cardiac impacts including premature atrial and ventricular contractions, blockage of AV, cardiac ischemia, hypotension and sustained asystole (Singh & McKintosh, 2020).

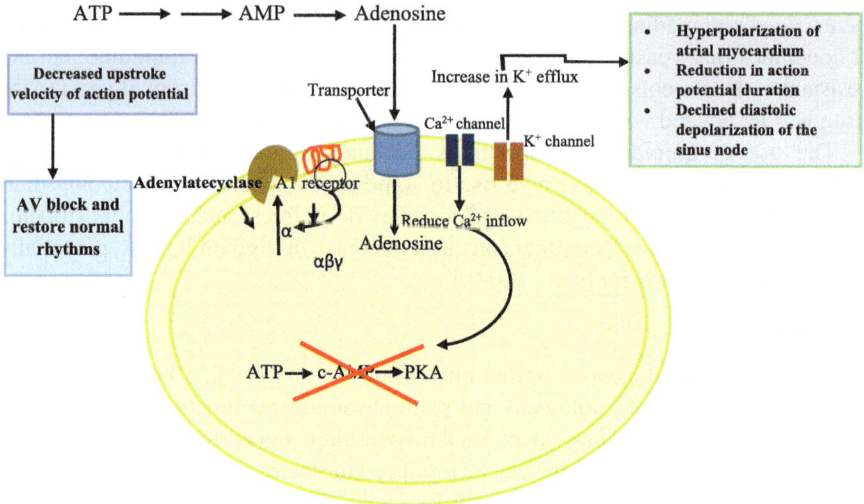

Fig. 11.10 Synthesis of adenosine by ATP pathway and itsmechanism of action to treat arrhythmia

11.2.6 Myocardial Infarction

Myocardial infraction is also known as heart attack (Saleh, & Ambrose, 2018), it is central myocardial cell death due to persistent ischemia. After 10–15 min of ischemia some ultrastructure changes have been seen such as reduced cellular glycogen, and relaxed myofibrils and sarcolemmal distraction (Jennings & Ganote, 1974). Acute myocardial infraction may originate by the reduction in coronary blood flow. The reduced oxygen supply results in cardiac ischemia. The declined coronary blood flow is due to different factors; thrombosis caused by the rupture of atherosclerotic plaques leads toward acutely decreased coronary blood flow, coronary artery embolism, cocaine-prompted ischemia, coronary segmentation and coronary vasospasm. As a consequence, myocardial infraction may lead towards irreversible damage to cardiac muscles owing to deficiency of oxygen as well as diminish functioning of diastolic and systolic and make the patient more susceptible to arrhythmias. In myocardial patients some symptoms may appear including signs of ischemia, alterations in new ST-segment or a left bundle branch block (LBBB), manifestation of Q waves on ECG, new regional wall motion anomaly and intracoronary thrombus at autopsy or angiography (Mechanic & Grossman, 2019). After acceptance of universal definition, myocardial infraction was categorized into five types in 2007 (Saaby et al., 2013).

- **Myocardial infraction type 1**—MI triggered by atherothrombotic coronary artery disease (CAD) and typically by plaque disturbance or rupture.
- **Myocardial infraction type 2**—Ischemic myocardial injury initiated by difference between oxygen supply and demand.

- **Myocardial infraction type 3**—Abrupt cardiac death that embraces indicators of myocardial ischemia, for instance ECG changes, but cause death prior to a blood sample indicates increase in serum markers.
- **Myocardial infraction type 4a**—Myocardial infraction correlated with percutaneous coronary intervention, PCI.
- **Myocardial infraction type 4b**—Myocardial infraction relevant to stent thrombosis and meet the criteria of type 1.
- **Myocardial infraction type 4c**—Myocardial infraction restenosis and meet criteria of type 1.
- **Myocardial infraction type 5**—Myocardial infraction related to Coronary artery bypass grafting, CABG.

11.2.6.1 Drugs for Myocardial Infraction

To treat myocardial infraction different types of drugs are used discussed in Table 11.6. These are (1) **Vasodilators**—dilate blood vessels which causes decrease in vascular resistance, in turn reduce blood pressure (2) **Cardiac depressant drugs**—reduce heart rate and contractility which diminishes heart output and atrial pressure. It reduces the work load and oxygen demand in heart. (3) **Antithrombotic**—stop development of thrombus (4) **Thrombolytic**—stimulate production of plasminogen to dissolve blood clots. (5) **Analgesics**—to decrease pain (Klabunde, 2007).

Table 11.6 Classification of drugs used for myocardial infraction

MI treating drugs classification	Subclasses	Drugs
Vasodilators	Nitrodilators	Isosorbide dinitrate, Isosorbide mononitrate, Nitroglycerin, Erythrityl tetranitrate, Pentaerythritol tetranitrate,
	Angiotensin converting enzyme (ACE) inhibitors	Benazepril, Captopril, Enalapril, Fosinopril, Lisinopril, Moexipril,
	Angiotensin receptor blockers (ARBs)	Atacand, Avapro, Benicar, Cozaar, Diovan, Micardis, Teveten
Cardiac depressant drugs	Β-blockers	Acebutolol, Atenolol, Bisoprolol, Lopressor, Nadolol, Nebivolol, Propranolol
Antithrombotic	Anticoagulant	Apixaban, Dabigatran, Edoxaban, Enoxaparin, Heparin, Rivaroxaban
	Antiplatelet drugs	Clopidogrel,Ticagrelor, Prasugrel, Dipyridamole, Aspirin, Ticlopidine, Eptifibatide
Thrombolytics	Plasminogen activators	Alteplase, Reteplase, Tenecteplase
Analgesics		Morphine

Source Klabunde (2007)

Lopressor

Lopressor also called Metoprolol tartrate is a lipophilic (Morris & Dunham, 2020), beta-1-selective beta-blocker (FDA, 2013). Metoprolol plays substantial role to reduce risk of abrupt cardiovascular mortality, cardiac death along with myocardial infraction (Lopez-Sendon et al., 2004). In addition to myocardial infraction Lopressor is also effective against hypertension and angina pectoris (FDA, 2007).

Mode of Action

As lopressor is a β1-selective adrenergic receptor blocker, it blocks the β1 receptors while influencing β2 receptors either at a minimum or zero level at the oral dose below 100 mg in adults (Morris & Dunham, 2020). The actual mode of action of Metoprolol for the treatment of myocardial infraction is not known (Drugs.com, 2019). Though, at higher plasma concentration more than 100 mg it also obstructs beta2-adrenoreceptors, predominately found on bronchial and vascular musculature. Clinical pharmacological studies have revealed that inhibition of β receptors activity may leads to the decrease in heart beat and heart output even at rest and after exercise, decline of systolic blood pressure after doing hard job, suppression of isoproterenol-persuaded tachycardia, and fall of reflex orthostatic tachycardia (FDA, 2013) (Fig. 11.11).

Resistance and Toxicity

The available evidences indicate that no resistance have been found against Lopressor, however, some evidences revealed toxicity related to the use of metoprolol in some special cases. The toxicity of metoprolol is interrelated with the level

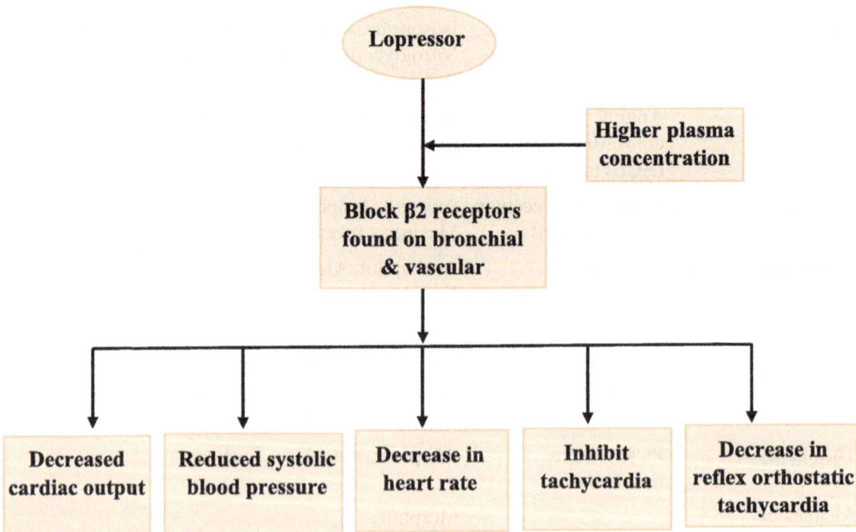

Fig. 11.11 Impact of leprossor on myocardial infraction patients (*Source* FDA, 2013)

of its dose in blood and tissues (Shore et al., 1981). The overdose of metoprolol in coronary artery disease patients is associated with the emergence of acute myocardial reinfraction and activation of myocardial ischemia and dysrhythmia (Unverir et al., 2007). Lopressor is also responsible for the sever health issues including heart block, hypotension, bronchospasm, heart failure exacerbation, bradycardia, tiredness, glucose intolerance, hypoglycemia etc. (Morris & Dunham, 2020).

Alteplase

Alteplase is a synthetic thrombolytic agent produced by using genetic engineering recombinant DNA technology (Reed et al., 2020). Alteplase is an enzyme also called activase- a tissue plasminogen activator (tPA) clinically used for the treatment of myocardia infraction for the breakdown of thrombin in coronary arteries to decrease mortality and morbidity such as cholesterol embolism. In addition to myocardial infection, it is also effective against acute ischemic stroke, pulmonary embolism and angioedema (FDA, 2015).

Mode of Action

Alteplase, a fibrinolytic agent (Reed et al., 2020) and thrombolytic agent is a serine protease which is associated with the transformation of plasminogen to plasmin, a proteolytic enzyme by hydrolyzing the arginine-valine bond in plasminogen (Hughes et al., 2020). This alternation produces limited plasmin in the absence of fibrin. When alteplase entered into the systemic circulation at clinically recommended effective concentration it binds with the fibrin in thrombus and alters the entrapped plasminogen to plasmin. It results in the initiation of fibrinolysis with restricted systemic proteolysis (FDA, 2015) and allow for the termination of clots and cause restoration of blood flow (Hughes et al., 2020) (Fig. 11.12).

Resistance and Toxicity

No evidences of alteplase resistance have been found, but serious and irreversible effects of alteplase have been reported including bleeding, angioedema, anaphylaxis and fever (Reed et al., 2020). The seriousness of issues correlates with the concentration of exposure. Alteplase associated bleeding is considered as a more serious issue and includes intracranial bleeding, retroperitoneal bleeding, gastrointestinal (GI) bleeding, respiratory bleeding, genitourinary and bleeding through venous cut downs, arterial ruptures (Hughes et al., 2020).

11.2.7 Hypertension

Hypertension is the condition of elevated blood pressure, which is directly linked with the increased risk of cardiovascular diseases (Ripley et al., 2018). The effects of hypertension increases to greater extent with a minor increase in blood pressure. The increase of every 20 mm Hg in systolic blood pressure and increase of 10 mm Hg in

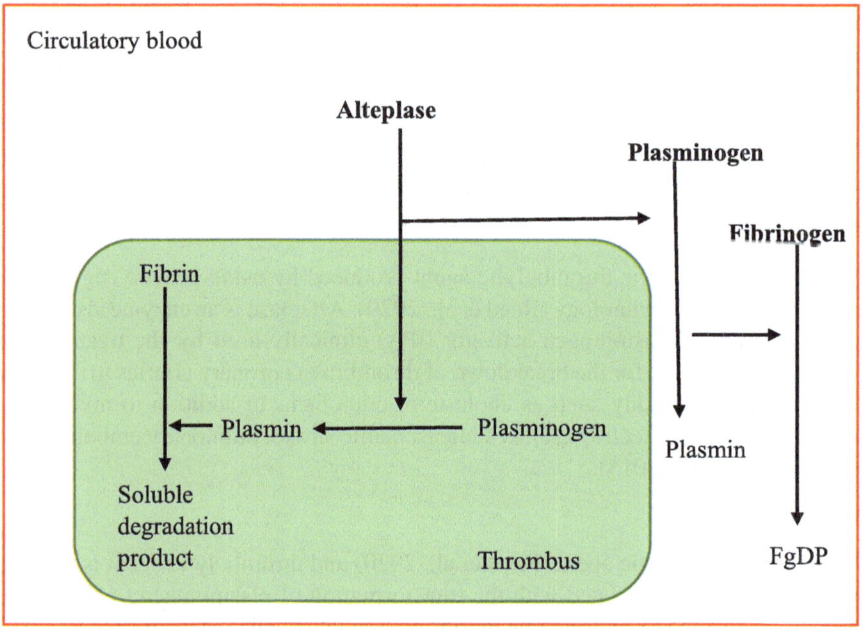

Fig. 11.12 Mechanism of action of alteplase (Kikuchi et al., 2018)

diastolic blood pressure after starting from blood pressure of 115/75 mm Hg multiplying the possibility of mortality from stroke, cardiac disease, or other vascular disease (Lewington et al., 2002). Different interrelated factors are responsible for the development of hypertension and these factors contribution towards high blood pressure may varies between individuals. These factors includes salt consumption, obesity and insulin hindrance, the renin-angiotensin system, the sympathetic nervous system, genetics, endothelial disturbance, low birth weight and intrauterine nourishment, and neurovascular abnormalities. The imbalance between the heart output and peripheral vascular resistance is also associated with elevated blood pressure (Beevers et al., 2001).

Lisinopri

Lisinopril is an angiotensin-converting enzyme (ACE) inhibitor with extensive half-life. It is active hydrophilic compound and cannot be degraded by the liver (Lopez et al., 2020). Due to these characteristics it has ability to interfere with any component of the renin–angiotensin–aldosterone system separately (Sukalo et al., 2016). FDA permitted it to be used for the cure of hypertension. In addition Lisinopril is also demonstrated to be effective against cardiac failure, along with the myocardial infarction and diabetic nephropathy prophylaxis (Sukalo et al., 2016).

Mode of Action

Lisinopril an ACE inhibitor obstructs angiotensin-converting enzyme (ACE) in both human and animals. ACE is a catalyzer which speeds up the transformation of angiotensin I to theangiotensin II(vasoconstrictor). Angiotensin II leads toward activation of aldosterone secretion by the adrenal cortex (FDA, 2008). Lisinopril act as a competitor of ACE and inhibits modification of angiotensin I to angiotensin II. As a consequence reduction in plasma angiotensin II may occur which is responsible for diminished vasopressor activity and to decline aldosterone discharge which subsequently cause decline in the Sodium reabsorption in collecting duct cause decline in the flux of potassium (Lopez et al., 2020).

Elimination of angiotensin II undesirable response on renin secretion is associated with improved plasma renin activity. ACE is similar to kininase, an enzyme that has ability to breakdown bradykinin and higher level of bradykinin, a strong vasodepressor peptide, play a significant role in the remedial effects of Lisinopril remains to be clarified. While lisinopril drops high blood pressure due to suppression of the renin–angiotensin–aldosterone system, it is also advantageous even in hypertensive patient with low-renin hypertension (FDA, 2008; Lopez et al., 2020) (Fig. 11.13).

Resistance and Toxicity

Antihypertensive lisinopril has the ability to stabilize high blood pressure in all races, but the black hypertensive patients are somehow resistant or had a minor reaction to use lisinopril alone than white patients. A study revealed that in hypertensive patients with ordinary kidney function treated with the Lisinopril for upto six months, serum potassium level was increased in them, nearly 0.1 mEq/L. Conversely, about 15% of patients had escalations more than 0.5 mEq/L and around 6% had a reduction more than 0.5 mEq/L. While patients treated with both lisinopril and hydrochlorothiazide for the same duration had a reduced potassium level of 0.1 mEq/L almost 4% of patients had intensifications larger than 0.5 mEq/L and nearly 12% had a decline more than 0.5 mEq/L (FDA, 2008).

Lisinopril as an angiotensin converting enzyme blocker is effective against high blood pressure. There are many evidences which indicates lisinopril is also effective to increase survival in cardiac failure and interruption in the development to clinical cardiac failure in patients with left ventricular impairment after myocardial infarction (Reid et al., 1993). However, use of lisinopril is also liable to cause few hostile side effects including hypotension, hyperkalemia and impaired renal function (Warner & Rush, 1988). In addition, it also cause cough and angioedema in 5%-35% and up to 0.7% respectively in treated patients (Bezalel et al., 2015). All these side effects are more common in people having renal, autoimmune, or collagen vascular diseases (Lopez et al., 2020).

Atenolol

Atenolol is a hydrophilic β1-selective adrenergic antagonist and is generally used for the treatment of hypertention as well as angina pectoris, and acute myocardial infarction (Aronson, 2016). Atenolol is potent and usually well accepted in patients

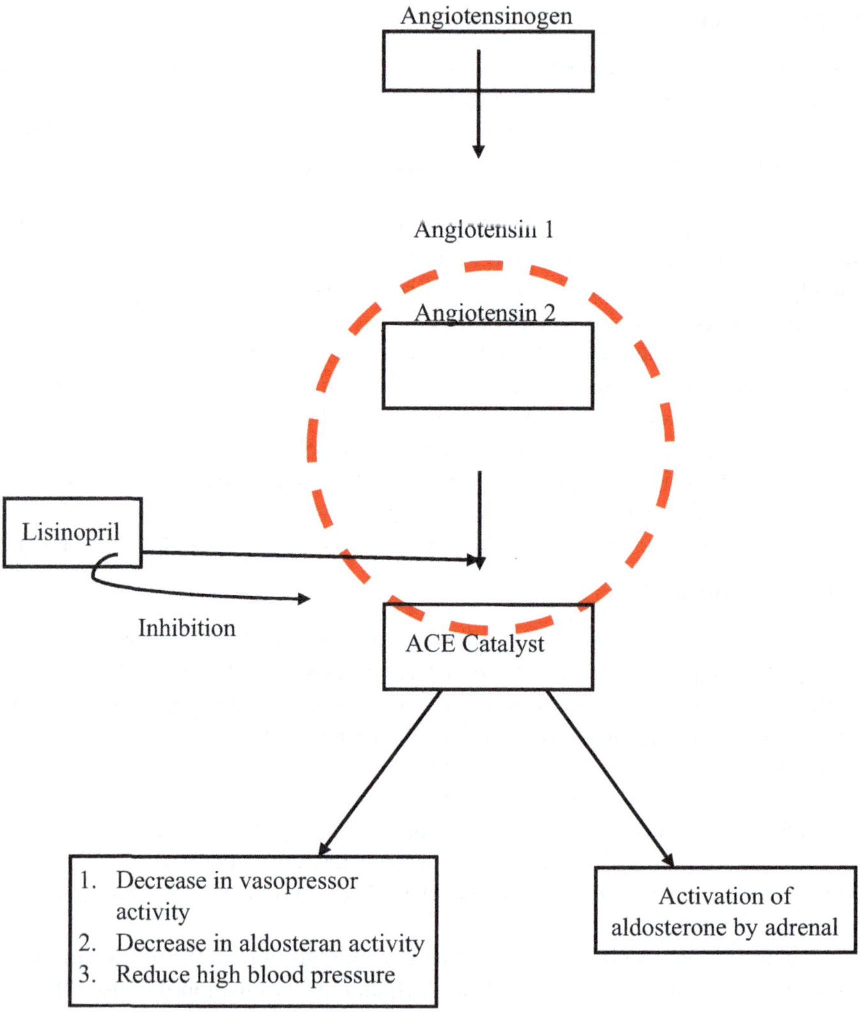

Fig. 11.13 Treatment process of high blood pressure via lisinopril

with all degrees of hypertension. (Wadworth et al., 1991) But it is not recommended to be used during emergency (Actavis Pharma Company, 2018).

Mode of Action

Atenolol as a beta inhibitor may act by selectively binding to the β1-adrenergic receptor present in vascular smooth muscle and the heart (Rehman et al., 2020). It is reported that atenolol binds with β1-adrenergic receptor 26 times more than β2-receptors (Abrahamsson et al., 1988), selective binding with β1 receptors results in cardioselectivity in heart. However, at therapeutic doses some binding also take place at β2 and β3 receptors, but the consequences intervened by antagonizing these are considerably decreased from non-selective agents. G_s joined, β1 and β2 receptors diminishes action of adenylyl cyclase and its downward signaling through cyclic adenosime monophosphate and protein kinase A (PKA) due to their antagonistic activities (DrugBank, 2020). PKA in cardiomyocytes stimulates L-type calcium channels and ryanodine receptors which results in upsurge in intracellular calcium and prompt the ryanodine receptors to discharge calcium stored in the sarcoplasmic reticulum and increased contractility. PKA is also responsible for the interruption of contraction by phosphorylating phospholamban which leads to upsurge in the affinity of sarcoplasmic reticulum Ca^{2+} ATPase to intensify reuptake of calcium into the sarcoplasmic reticulum. Both of these processes results in the decrease in contraction, it permits for the rapid cycling and subsequently greater heart beat with amplified contractility. L-type calcium channels are responsible for the cardiac depolarization and its stimulation can cause increase in regularity of action potentials and feasibly prevalence of ectopic potentials (DrugBank, 2020). The comparable inhibitory processes also take place in the smooth muscle ofbronchi to facilitate relaxation such as phosphorylation of myosin light-chain kinase, decreasing its attraction for Ca^{2+} (Billington & Penn, 2003). PKA also obstructs the excitatory G_q coupled passageway through phosphorylation of inositol trisphosphate receptor and phospholipase C leads to blockage of intracellular calcium discharge. Atenolol can therefore results in increased bronchoconstriction as an antagonism to this activity.

Resistance and Toxicity

No resistances have been found against the atenolol. But, It is considered a safest beta blocker however some adverse complications are seldomly reported such as low blood pressure even after single use of atenolol oral dose (Kholeif & Isles, 1989) organic brain syndrome (Arber, 1988) cholestasis and cutaneous vasculitis (Aronson, 2016).It is also associated with the serious metabolic effects, especially adverse variations in glucose, triglycerides and uric acid, all of which can have long-term negative cardiovascular consequences.

References

Abrahamsson, T., Ek, B., & Nerme, V. (1988). The beta 1- and beta 2-adrenoceptor affinity of atenolol and metoprolol. A receptor-binding study performed with different radioligands in tissues from the rat, the guinea pig and man. *Biochemical Pharmacology, 37*(2), 203–208.

Abramowicz, M., Zuccotti, G., Pflomm, J.-M., Rizack, M. A., Hansten, P. D., Hirsch, J., et al. (2004). Treatment Guidelines. *The Medical Letter, 2*(27), 75–82.

Actavis Pharma Company. (2018). Product monograph including patient medication information [pr]Act Atenolol, viewed 14 April 2020. Retrieved from: https://pdf.hres.ca/dpd_pm/00046253.PDF

Agasti, T. K. (2011). *Textbook of anaesthesia for postgraduates*. Jaypee Brothers, Medical Publishers. Available at: https://www.academia.edu/39321102/Textbook_of_Anaesthesia_for_Postgraduates_Textbook_of_Anaesthesia_for_Postgraduates

Alon, U. S. (2018). The effects of diuretics on mineral and bone metabolism. *Pediatric Endocrinology Reviews, 15*(4), 291–297.

American Heart Association (AHA). (2016). Arrhythmia, viewed 9 July 2020. Available at: https://www.heart.org/en/health-topics/arrhythmia/about-arrhythmia

American Heart Association. (2017). Heart disease and stroke statistics 2017 update. *Circulation.* Available at: http://circ.ahajournals.org/content/early/2017/01/25/CIR.0000000000000485

Arber, N. (1988). Delirium induced by atenolol. *British Medical Journal, 297*(6655), 1048.

Aronson, J. K. (2016). Atenolol. *Meyler's side effects of drug*, (p. 737). Elsevier publishers.

Bachmann, S., Velázquez, H., Obermüller, N., Reilly, R. F., Moser, D., & Ellison, D. H. (1995). Expression of the thiazide-sensitive NaCl cotransporter by rabbit distal convoluted tubule cells. *The Journal of Clinical Investigation, 96*, 2510–2514.

Banik, R. K., & Berger, J. S. (2014). *Peripheral Vasodilators Contents.* https://doi.org/10.1007/978-1-4614-8948-1_16

Bazzini, C., Vezzoli, V., Sironi, C., Dossena, S., Ravasio, A., Dėbiasi, S., Garavaglia, M., Rodighiero, S., Meyer, G., Fascio, U., Furst, J., Ritter, M., Botta, G., & Paulmichl, M. (2005). Thiazide-sensitive NaCl-cotransporter in the intestine: possible role of hydrochlorothiazide in the intestinal Ca2 uptake. *Journal of Biological Chemistry, 280*, 19902–19910.

Beevers, G., Lip, G. Y. H., & O'brien, E. (2001). The pathophysiology of hypertension. *British Medical Journal, 322*(7291), 912–916.

Bezalel, S., Mahlab-Guri, K., Asher, I., Werner, B., & Sthoeger, Z. M. (2015). Angiotensin-converting enzyme inhibitor-induced angioedema. *The American Journal of Medicine, 128*(2), 120–125.

Billington, C. K., & Penn, R. B. (2003). Signaling and regulation of G protein-coupled receptors in airway smooth muscle. *Respiratory Research, 4*(1), 2.

Borchard, U., Berger, F., & Hafner, D. (1989). Classification and action of antiarrhythmic drugs. *European Heart Journal, 10*(suppl E), 31–40.

Botto, L. D., Correa, A. & Erickson, J. D. (2001). Racial and temporal variations in the prevalence of heart defects. *Pediatrics, 107*(3), E32. Available at: www.pediatrics.org/cgi/content/full/107/3/e32

BPAC. (2017). Beta-blockers for cardiovascular conditions: one size does not fit all patients. Available at: https://bpac.org.nz/2017/docs/beta-blockers.pdf

Calo, L., Ceolotto, G., Milani, M., Pagnin, E., van den Heuvel, L. P., Sartori, M., Davis, P. A., Costa, R., & Semplicini, A. (2001). Abnormalities of Gq-mediated cell signaling in bartter and gitelman syndromes. *Kidney International, 60*, 882–889.

Calo, L., D'Angelo, A., Cantaro, S., Bordin, M. C., Favaro, S., Antonello, A., & Borsatti, A. (1996). Increased urinary NO_2–/NO_3– and cyclic guanosine monophosphate levels in patients with Bartter's syndrome: Relationship to vascular reactivity. *American Journal of Kidney Diseases, 27*, 784–789.

Calo, L., Davis, P. A., & Semplicini, A. (2000). Control of vascular tone in the syndromes of bartter and gitelman. *Critical Reviews in CLinical Laboratory Sciences, 37*, 503–522.

Calo, L., Davis, P. A., & Semplicini, A. (2002). Reduced content of alpha subunit of Gq protein content in monocytes of bartter and gitelman syndromes: Relationship with vascular hyporeactivity. *Kidney International, 61*, 353–354.

Camm, A. J., Malcolm, A. D., & Garratt, C. J. (1992). Adenosine and cardiac arrhythmias; The preferred treatment for supraventricular tachycardia. *BMJ, 305*, 3–4.

Cardiosmart. (n.d). Heart failure: ACE inhibitors. Available at: https://www.cardiosmart.org/~/media/Documents/Fact%20Sheets/en/zp3950.ashx

Center for Disease Control and Prevention. (2019). Coronary artery disease (CAD). Available at: https://www.cdc.gov/heartdisease/coronary_ad.htm

Chen, Z., Zhang, J., & Stamler, J. S. (2002). Identification of the enzymatic mechanism of nitroglycerin bioactivation. *Proceedings of the National Academy of Science USA, 99*(12), 8306–8311.

Clemo, H. F., & Belardinelli, L. (1986). Effect of adenosine on atrioventricular conduction: I. Site and characterization of adenosine action in the guinea pig atrioventricular node. *Circulation Research, 59*, 427–436.

Cruickshank, J. (2008). Initial management of cardiac arrhythmias. *Australian Family Physician, 37*(7), 516–520.
Cruz, D. N., Simon, D. B., Nelson-Williams, C., Farhi, A., Finberg, K., Burleson, L., Gill, J. R., & Lifton, R. P. (2001). Mutations in the NaCl cotransporter reduce blood pressure in humans. *Hypertension, 37*, 1458–1464.
Curran, T., & Sheppard G. (2011). Module 3: Cardiac arrhythmias: Mechanisms of arrhythmias—Atrial, ventricular, conduction and ST changes. *Cardiology*, 1–74.
Dan, G-A., Martínez-Rubio, A., Agewall, S., Boriani, G., Borggrefe, M., Gaita, F., Gelder, I. V., Gorenek, B., Kaski, J. C., Kjeldsen, K., Gregory Lip, Y. H., Merkely, B., Okumura, K., Piccini, J. P., Potpara, T., Poulsen, B. K., Saba, M., Savelieva, I., Tamargo, J. L., Wolpert, C., & ESC Scientific Document Group. (2018). Antiarrhythmic drugs-clinical use and clinical decision making: A consensus document from the European heart rhythm association (EHRA) and European society of cardiology (ESC) working group on cardiovascular pharmacology, endorsed by the heart rhythm society (HRS), Asia-Pacific heart rhythm society (APHRS) and international society of cardiovascular pharmacotherapy (ISCP). *Europace*, 0, 1–42.
Delisle, B. P., Anson, B. D., Rajamani, S., & January, C. T. (2004). Biology of cardiac arrhythmias: Ion channel protein trafficking. *Circular Research, 94*(11), 1418–1428.
Dhein, S. (n.d.). Antiarrhythmic drugs. *Encyclopedia of Molecular Pharmacology*, 96–102.
do Vale, G. T., Ceron, C. S., Gonzaga, N. A., Simplicio, J. A., & Padovan, J. C. (2019). Three generations of β-blockers: History, class differences and clinical applicability. *Current Hypertension Reviews, 15*(1), 22–31.
Druds.com. (2019). Metoprolol, viewed 13 July 2020. Available at: https://www.drugs.com/pro/metoprolol.html#:~:text=By%20blocking%20catecholamine%2Dinduced%20increases,term%20management%20of%20angina%20pectoris
DrugBank. (2020). Atenolol, viewed 14 July 2020. Retrieved from: https://www.drugbank.ca/drugs/DB00335
Dvorak, M. M., de Joussineau, C., Carter, D. H., Pisitkun, T., Knepper, M. A., Gamba, G., Kemp, P. J., & Riccardi, D. (2007). Thiazide diuretics directly induce osteoblast differentiation and mineralized nodule formation by interacting with a sodium chloride co-transporter in bone. *Journal of the American Society of Nephrology, 18*, 2509–2516.
Ellison, D. H., & Loffing, J. (2009). Thiazide effects and adverse effects: Insights from molecular genetics. *Hypertension, 54*(2), 196–202.
Ellison, D. H. (1991). The physiologic basis of diuretic synergism: Its role in treating diuretic resistance. *Annals of Internal Medicine, 114*(10), 886–894.
Ellison, D. H., Velázquez, H., & Wright, F. S. (1987). Thiazide sensitive sodium chloride cotransport in the early distal tubule. *American Journal of Physiology, 253*, F546–F554.
Ellison, D. H., Velázquez, H., & Wright, F. S. (1989). Adaptation of the distal convoluted tubule of the rat. Structural and functional effects of dietary salt intake and chronic diuretic infusion. *Journal Clinical Investigation, 83*(1), 113–126.
Fares, A. (2011). Night-time exogenous melatonin administration may be a beneficial treatment for sleeping disorders in beta blocker patients. *Journal Cardiovascular Diseases Research, 2*(3), 153–155.
Farzam, K., & Jan, A. (2020). *Beta blockers*. Treasure Island (FL): StatPearls Publishing; 2020. Available at: https://www.ncbi.nlm.nih.gov/books/NBK532906/
FDA. (2019). Approved drug products: Lanoxin (digoxin) oral tablets. Available at: https://www.accessdata.fda.gov/drugsatfda_docs/label/2019/020405s015lbl.pdf
FDA. (n.d.). Approved drug products: Apadaz (benzhydrocodone and acetaminophen) tablets. Retrieved 2020 from: https://www.accessdata.fda.gov/drugsatfda_docs/label/2018/208653s002s003lbl.pdf
Ferreira, J. C., & Mochly-Rosen, D. (2012). Nitroglycerin use in myocardial infarction patients. *Circulation Journal, 76*(1), 15–21.
Friedman, P. A. (1988). Biochemistry and pharmacology of diuretics. *8*(3), 198–212.

Friedman, P. A., & Bushinsky, D. A. (1999). Diuretic effects on calcium metabolism. *Seminars in Nephrology, 19*(6), 551–556.
Fuster, V., Badimon, L., Badimon, J. J., & Chesebro, J. H. (1992). The pathogenesis of coronary artery disease and the acute coronary syndromes. *New England Journal of Medicine, 326*(4), 242–250.
Gamba, G., Miyanoshita, A., Lombardi, M., Lytton, J., Lee, W. S., Hediger, M. A., & Hebert, S. C. (1994). Molecular cloning, primary structure, and characterization of two members of the mammalian electroneutral sodium–(potassium)–chloride cotransporter family expressed in kidney. *Journal of Biological Chemistry, 269*, 17713–17722.
Geleijnse, J. M., Kok, F. J., & Grobbee, D. E. (2003). Blood pressure response to changes in sodium and potassium intake: A metaregression analysis of randomised trials. *Journal of Human Hypertension, 17*, 471–480.
Gilboa, S. M., Salemi, J. L., Nembhard, W. N., Fixler, D. E., & Correa, A. (2010). Mortality resulting from congenital heart disease among children and adults in the United States, 1999 to 2006. *Circulation, 122*(22), 2254–2263.
Gordon, S. G., & Kittleson, M. D. (2008). Drugs used in the management of heart disease and cardiac arrhythmias. *Small Animal Clinical Pharmacology*, 380–457. https://doi.org/10.1016/b978-070202858-8.50019-1
Gorre, F., & Vandekerckhove, H. (2010). Beta-blockers: Focus on mechanism of action. Which beta-blocker, when and why? *Acta Cardiologica, 65*(5), 565–570.
Frauke, G., & Hans, V. (2010). Beta-blockers: Focus on mechanism of action. Which beta-blocker, when and why? *Acta Cardiologica, 65*, 565–570.
Hankey, G. J., & Eikelboom, J. W. (2006). Aspirin resistance. *The Lancet, 367*(9510), 606–617.
Hoorn, E. J., Wilcox, C. S., & Ellison, D. H. (2015). Diuretics. In K. Skorecki, G. Chertow, P. Marsden, M. Taal & A. Yu. (Eds.), *Brenner and rector's the kidney*, (Vol. 10, pp. 1702–1734). Elsevier, 2.
Hughes, A. D. (2004). How do thiazide and thiazide-like diuretics lower blood pressure. *Journal Renin Angiotensin Aldosterone System, 5*, 155–160.
Hughes, R. E., Tadi, P., & Bollu, P. C. (2020) TPA Therapy. In: StatPearls [Online]. Treasure Island (FL), StatPearls Publishing, viewed 14 July 2020. Available from: https://www.ncbi.nlm.nih.gov/books/NBK482376/
Husain, S., Andrews, N. P., Mulcahy, D., Panza, J. A., & Quyyumi, A. A. (1998). Aspirin improves endothelial dysfunction in atherosclerosis. *Circulation, 97*(8), 716–720.
Jenkins, K. J., Correa, A., & Feinstein, J. A. (2007). Noninherited risk factors and congenital cardiovascular defects: Current knowledge. A scientific statement from the American heart association council on cardiovascular disease in the young. *Circulation, 115*, 2995–3014.
Jennings, R. B., & Ganote, C. E. (1974). Structural changes in myocardium during acute ischemia. *Circulation Research, 35*, 156–172.
Kaplan, N. M. (2002). Treatment of hypertension: Drug therapy. In *Kaplan's clinical hypertension*, (pp. 237–338). Lippincott Williams and Wilkins.
Kastor, J. A., & Saba M. M. (2009). *Cardiac arrhythmias*. Willey. Available at: https://doi.org/10.1002/9780470015902.a0002112.pub2
Kester, M., Karpa, K. D., & Vrana, K. E., (2012). *Cardiovascular system*, (pp. 125–151). Elsevier's Integrated Review Pharmacology.
Khatib, R., & Wilson, F. (2018). Pharmacology of medications used in the treatment of atherosclerotic cardiovascular disease. *Encyclopedia of cardiovascular research and medicine*, (pp. 68–88).
Kholeif, M., & Isles, C. (1989). Profound hypotension after atenolol in severe hypertension. *British Medical Journal, 298*(6667), 161–162.
Kikuchi, K., Setoyama, K., Tanaka, E., Otsuka, S., Terashi, T., Nakanishi, K., Takada, S., Sakakima, H., Ampawong, S., Kawahara, K-I., Nagasato, T., Hosokawa, K., Harada, Y., Yamamoto, M., Kamikokuryo, C., Kiyama, J., Morioka, M., Ito, T., Maruyama, I., & Tancharoen, S. (2018). Uric

acid enhances alteplase-mediated thrombolysis as an antioxidant. *Scientific reports*, Available at: https://www.nature.com/articles/s41598-018-34220-1?proof=true19

Klabunde, R. E. (2007). Cardiovascular pharmacology concepts: The pharmacologic treatment of myocardial infarction, viewed 13 July 2020. Available at: https://www.cvpharmacology.com/clinical%20topics/myocardial%20infarction-3

Kunchaparty, S., Palcso, M., Berkman, J., Velázquez, H., Bernstein, P., Reilly, R. F., & Ellison, D. H. (1999). Defective processing and expression of the thiazidesensitive NaCl cotransporter as a cause gitelman's syndrome. *American Journal of Physiology Renal Physiology, 277*, F643–F649.

Kyle, H., & Derek, J. (2019). *Nitroglycerin*. Stat Pearls publishing. Available at: https://www.ncbi.nlm.nih.gov/books/NBK482382/

Ladage, D., Schwinger, R. H. G., & Brixius, K. (2013). Cardio-selective beta-blocker: Pharmacological evidence and their influence on exercise capacity. *Cardiovascular TherapEutics, 31*, 76–83.

Levine, M. (2017). Digoxin overdose, cardiac glycoside toxicity, digoxin toxicity, digoxin poisoning. Available at: https://www.cancertherapyadvisor.com/home/decision-support-in-medicine/critical-care-medicine/digoxin-overdose-cardiac-glycoside-toxicity-digoxin-toxicity-digoxin-poisoning/

Lewington, S., Clarke, R., Qizilbash, N., Peto, R., & Collins, R. (2002). Age-specific relevance of usual blood pressure to vascular mortality: A meta-analysis of individual data for one million adults in 61 prospective studies. *Lancet, 360*, 1903–1913.

Lip, G. Y., & Ferner, R. E. (1995). Poisoning with anti hypertensive drugs: Angiotensin converting enzyme inhibitors. *Journal of Human Hypertension, 9*(9), 711–715.

Lopez, O. E., Parmar, M., Pendela, V. S., & Terrell, J. M. (2020). Lisinopril. In: StatPearls [Online]. Treasure Island (FL): StatPearls Publishing, viewed 13 July 2020, Available from: https://www.ncbi.nlm.nih.gov/books/NBK482230/

Lopez-Sendon, J., Swedberg, K., Mcmurray, J., Tamargo, J., A. Maggioni, P., Dargie, H., Tendera, M., Waagstein, F., Kjekshus, J., Lechat, P., & Torp-Pedersen, C. (2004). Expert consensus document on β-adrenergic receptor blockers. The task force on beta-blockers of the european society of cardiology. *European Heart Journal, 25*(15), 1341–1362.

Machackova, J., Sanganalmath, S. K., Elimban, V., & Dhalla, N. S. (2011). β-adrenergic blockade attenuates cardiac dysfunction and myofibrillar remodelling in congestive heart failure. *Journal of Cellular and Molecular Medicine, 15*(3), 545–554.

Macleod-Glover, N., Mink, M., Yarema, M., & Chuang, R. (2016). Digoxin toxicity: Case for retiring its use in elderly patients? *Canadian Family Physician, 62*(3), 223–228.

Madhani, M., Scotland, R. S., Macallister, R. J., & Hobbs, A. J. (2003). Vascular natriuretic peptide receptor-linked particulate guanylate cyclases are modulated by nitric oxide-cyclic GMP signalling. *British Journal of Pharmacology, 139*(7), 1289–1296.

Masferrer, J. L., Zweifel, B. S., Seibert, K., & Needleman, P. (1990). Elective regulation of cellular cyclooxygenase by dexamethasone and endotoxin in mice. *Journal of Clinical Investigation, 86*(4), 1375–1379.

Mayo Clinic. (2018). Dilated cardiomyopathy. Available at: https://www.mayoclinic.org/diseases-conditions/dilated-cardiomyopathy/symptoms-causes/syc-20353149#:~:text=Dilated%20cardiomyopathy%20is%20a%20disease,as%20a%20healthy%20heart%20can

McNiff, E. F., Yacobi, A., Young-Chang, F. M., Golden, L. H., Goldfarb, A., & Fung, H. L. (1981). Nitroglycerin pharmacokinetics after intravenous infusion in normal subjects. *Journal of Pharmaceutical Sciences, 70*(9), 1054–1058.

Mechanic, O. J., & Grossman, S. A. (2019). Acute myocardial infarction. In: StatPearls [Online]. Treasure Island (FL), StatPearls Publishing, viewed 12 July 2020, Available from: https://www.ncbi.nlm.nih.gov/books/NBK459269/

Morris, J., & Dunham, A. (2020). Metoprolol. In: StatPearls [online]. Treasure Island (FL): StatPearls Publishing, viewed, 13 July 2020, Available at: https://www.ncbi.nlm.nih.gov/books/NBK532923/

National Institute for Health and Care Excellence. (2011). Stable angina: management. Available from: www.nice.org.uk/guidance/CG126 (Accessed June, 2016).

Nitroglycerin in Dextrose FDA label d. Retrieved 2020 from: https://www.accessdata.fda.gov/drugsatfda_docs/label/2014/019970s015lbl.pdf

Nitroglycerin pump spray monograph. Retrieved 2020 from: http://products.sanofi.ca/en/nitrolingual-pumpspray.pdf

Nitrostat® (Nitroglycerin) FDA Label a. Retrieved 2020 from: https://www.accessdata.fda.gov/drugsatfda_docs/label/2014/021134s007lbl.pdf

Nora, J. J., & Nora, A. H. (1988). Update on counseling the family with a first-degree relative with a congenital heart defect. *American Journal of Medical Genetics, 29*, 137.

Nunes, P. A., Tenreiro, S., Sá-correia, I. (2001). Resistance and adaptation to quinidine in *Saccharomyces cerevisiae*: Role of *QDR1* (*YIL120w*), encoding a plasma membrane transporter of the major facilitator superfamily required for multidrug resistance. *Antimicrobial Agents Chemotherapy*, *45*(5), 1528–1534.

Obermüller, N., Bernstein, P. L., Velázquez, H., Reilly, R., Moser, D., Ellison, D. H., & Bachmann, S. (1995). Expression of the thiazide-sensitive NaCl cotransporter in rat and human kidney. *American Journal of Physiology*, *269*, F900–F910

Oliver, J. M., Gallego, P., Gonzalez, A. E., Garcia-Hamilton, D., Avila, P., Alonso, A., & Fernandez-Aviles, F. (2017). Impact of age and sex on survival and causes of death in adults with congenital heart disease. *International Journal of Cardiology, 245*, 119–124.

OpenAnaesthesia. (n.d.). Aspirin toxicity: Treatment. Retrieved 2020 from: https://www.openanesthesia.org/aspirin_toxicity_treatment/

Papademetriou, V., Sica, D. A., & Izzo, J. L. (2003). Thiazide and loop diuretics. In J. L. Izzo Jr. & H. R. Black (Eds.), *Hypertension primer* (pp. 411–414). Lippincott Williams and Wilkins.

Paul, T., Pfammatter, J.-P. (1997). Adenosine: An effective and safe antiarrhythmic drug in pediatrics. *Pediatric Cardiology*, *18*(2), 118–126.

Pecker, M. S. (1990). Pathophysiologic effects and strategies for long-term diuretic treatment of hypertension. In J. H. Laragh, B. M. Brenner (Eds.), *Hypertension: Pathophysiology, diagnosis, and management*, (pp. 2143–2168). Raven Press.

Pickkers, P., Garcha, R. S., Schachter, M., Smits, P., & Hughes, A. D. (1999). Inhibition of carbonic anhydrase accounts for the direct vascular effects of hydrochlorothiazide. *Hypertension, 33*, 1043–1048.

Pickkers, P., & Hughes, A. D. (1995). Relaxation and decrease in [Ca2]i by hydrochlorothiazide in guinea-pig isolated mesenteric arteries. *British Journal of Pharmacology, 114*, 703–707.

Pickkers, P., Hughes, A. D., Russel, F. G., Thien, T., & Smits, P. (1998). Thiazide-induced vasodilation in humans is mediated by potassium channel activation. *Hypertension, 32*, 1071–1076.

Pourafshar, N., Alshahrani, S., Karimi, A., & Soleimani, M. (2018). Thiazide therapy in chronic kidney disease: Renal and extra renal targets. *Current Drug Metabolism, 19*(12), 1012–1020.

Rao, P. S. (2012). Congenital heart defects, a review. *Congenital heart disease—Selected aspects*, (pp. 3–44).

Rectiv FDA label b. Retrieved 2020 from: https://www.accessdata.fda.gov/drugsatfda_docs/label/2011/021359s000lbl.pdf

Reed, M., Kerndt, C. C., & Nicolas, D. (2020). Alteplase. In StatPearls [Online]. Treasure Island (FL): StatPearls Publishing, viewed Jan 2020, Available from: https://www.ncbi.nlm.nih.gov/books/NBK499977/

Rehman, B., Sanchez, D. P., Shah, S. (2020). Atenolol. In StatPearls [Online]. Treasure Island (FL): StatPearls Publishing, 14 July 2020, Available from: https://www.ncbi.nlm.nih.gov/books/NBK539844/

Rehsia, N. S., & Dhalla, N. S. (2010). Mechanisms of the beneficial effects of beta-adrenoceptor antagonists in congestive heart failure. *Experimental and Clinical Cardiology, 15*(4), e86-95.

Reid, J. L., Macfadyen, R. J., Squire, I. B., & Lees, K. R. (1993). Blood pressure response to the first dose of angiotensin-converting enzyme inhibitors in congestive heart failure. *The American Journal of Cardiology, 71*(17), 57E-60E.

Reller, M. D., Strickland, M. J., Riehle-Colarusso, T., Mahle, W. T., & Correa, A. (2008). Prevalence of congenital heart defects in metropolitan Atlanta, 1998–2005. *Journal of Pediatrics, 153*(6), 807–813.

Resnekov, L. (1964). Cardiac arrhythmias. *Postgraduate Medical Journal, 40*, 381–391.

Ridker, P. M., Cushman, M., Stampfer, M. J., Tracy, R. P., & Hennekens, C. H. (1997). Inflammation, aspirin, and the risk of cardiovascular disease in apparently healthy men. *New England Journal of Medicine, 336*(14), 973–979.

Ripley, T. L., Barbato, A., Bress, A., Nickerson, D. L., & Rice, K. (2018). Hypertension. *Cardiology*, 7–28.

Roden, D. M. (2000). Electrophysiology: Antiarrhythmic drugs: From mechanisms to clinical practice. *Heart, 84*, 339–346.

RxList. (n.d.). Drug description: Adenocard IV, viewed 12 July 2020, Retrieved from: https://www.rxlist.com/adenocard-drug.htm#description

Saaby, L., Poulsen, T. S., Hosbond, S., Larsen, T. B., Diederichsen, A. C. P., Hallas, J., Thygesen, K., & Mickley, H. (2013). Classification of myocardial infarction: Frequency and features of type 2 myocardial infarction. *The American Journal of Medicine, 126*(9), 789–779.

Sage, P. R., de la Lande, I. S., Stafford, I., Bennett, C. L., Phillipov, G., Stubberfield, J., & Horowitz, J. D. (2000). Nitroglycerin tolerance in human vessels: Evidence for impaired nitroglycerin bioconversion. *Circulation, 102*(23), 2810–2815.

Saleh, M., & Ambrose, J. A. (2018). Understanding myocardial infarction. *F1000Res.* 7, F1000 Faculty Rev-1378. https://doi.org/10.12688/f1000research.15096.1

Sartori, M., Parotto, E., Bonso, E., Semplicini, A., Palatini, P., Pessina, A. C., & Calo, L. A. (2007). Autonomic nervous system function in chronic hypotension associated with bartter and gitelman syndromes. *American Journal of Kidney Diseases, 49*, 330–335.

Shore, E. T., Cepin, D., & Davidson, M. J. (1981). Metoprolol overdose. *Annals of Emergency Medicine, 10*(10), 524–527.

Shu, D. F., Dong, B. R., & Lin, X. F. (2012). Long-term beta blockers for stable angina: Systematic review and meta-analysis. *European Journal of Preventive Cardiology, 19*, 330–341.

Sica, D. A. (2008). Thiazide and loop diuretics. In J. L. Izzo Jr, D. A. Sica, H. R. Black (Eds.), *Hypertension primer*, (pp. 439–442). Lippincott Williams and Wilkins.

Simon, D. B., Nelson-Williams, C., Bia, M. J., Ellison, D., Karet, F. E., Molina, A. M., Vaara, I., Iwata, F., Cushner, H. M., Koolen, M., Gainza, F. J., Gitelman, H. J., & Lifton, R. P. (1996). Gitelman's variant of bartter's syndrome, inherited hypokalemic alkalosis, is caused by mutations in the thiazide-sensitive NaCl cotransporter. *Nature Genetics, 12*, 24–30.

Singh, B. N., Ellrodt, G., & Peter, C. T. (1978). Verapamil: A review of its pharmacological properties and therapeutic use. *Drugs, 15*, 169–197.

Singh, S., & Mckintosh, R. (2020). Adenosine. In StatPearls [Online]. Treasure Island (FL): StatPearls Publishing, viewed 12 July 2020, Available from: https://www.ncbi.nlm.nih.gov/books/NBK519049/#

Slotwiner, D. J., & Lerman B. B. (1998). Adenosine and cardiac arrhythmias. In G. Burnstock, T G. Dobson Jr, B. T. Liang, J. Linden (Eds.), *Cardiovascular biology of purines*, (pp. 126–142). Developments in Cardiovascular Medicine, 209. Springer.

Sonnenblick, M., Friedlander, Y., & Rosin, A. J. (1993). Diuretic-induced severe hyponatremia. Review and analysis of 129 reported patients. *Chest, 103*(2), 601–606.

Steer, K. A., Wallace, T. M., Bolton, C. H., & Hartog, M. (1997). Aspirin protects low density lipoprotein from oxidative modification. *Heart, 77*(4), 333–337.

Sukalo, A., Deljo, D., Krupalija, A., Zjajo, N., Kos, S., Curic, A., Divkovic, G., Hubjar, S., Smailagic, M., Hodzic, E., Marjanovic, D., & Medjedovic, S. (2016). Treatment of hypertension with combination of lisinopril/hydrochlorothiazide. *Medical Archives (sarajevo, Bosnia and Herzegovina), 70*(4), 299–302.

Szentmiklósi, A. J., Galajda, Z., Cseppent, Á., Gesztelyi, R., Susán, Z., Hegyi, B., & Nánási, P. P. (2015). The Janus face of adenosine: Antiarrhythmic and proarrhythmic actions. *Current Pharmaceutical Design, 21*, 965–976.

Tsai-Turton, M. (2014). Quinidine. *Encyclopedia of Toxicology*, 16–18. https://doi.org/10.1016/b978-0-12-386454-3.00055-5
U.S. Food and Drugs Administration (FDA). (2007). Lopressor, viewed 13 July 2020. Retrieved from: https://www.accessdata.fda.gov/drugsatfda_docs/label/2008/017963s061,018704s020lbl.pdf
U.S. Food and Drugs Administration (FDA). (2008). ZESTRIL (Lisinopril), viewed 14 July 2020, Retrieved from: https://www.accessdata.fda.gov/drugsatfda_docs/label/2009/019777s054lbl.pdf
U.S. Food and Drugs Administration (FDA). (2009). Quinidine Gluconate extended-release tablets USP, viewed 11 July 2020, Retrieved from: https://www.accessdata.fda.gov/drugsatfda_docs/label/2010/089338s036lbl.pdf
U.S. Food and Drugs Administration (FDA). (2013). Lopressor metoprolol tartrate injection, USP, viewed 13 July 2020, Retrieved from: https://www.accessdata.fda.gov/drugsatfda_docs/label/2013/018704s026lbl.pdf
U.S. Food and Drugs Administration (FDA). (2016). Verapamil, viewed 11 July 2020, Retrieved from: https://www.accessdata.fda.gov/drugsatfda_docs/label/2016/018925s010lbl.pdf.
Unverir, P., Topacoglu, H., Bozkurt, S., & Kaynak, F. (2007). Cardiovascular toxicity due to metoprolol poisoning in a patient with coronary artery disease. *British Journal of Clinical Pharmacology, 64*(5), 694–697.
Vane, J. R. (1971). Inhibition of prostaglandin synthesis as a mechanism of action for aspirin-like drugs. *Nature, 231*(25), 232–235.
Vane, J. R., & Botting, R. M. (2003). The mechanism of action of aspirin. *Thrombosis Research, 10*(5–6), 255–258.
Velázquez, H., Wright, F. S. (1986). Effects of diuretic drugs on Na, Cl, and K transport by rat renal distal tubule. *American Journal of Physiology, 250,* F1013–F1023.
Vohra, J. (1982). Verapamil in cardiac arrhythmias: An overview. *Clinical Experiment Pharmacology Physiology Suppl., 6,* 129-34.
Wadworth, A. N., Murdoch, D., Brogden, R. N. (1991). Atenolol. A reappraisal of its pharmacological properties and therapeutic use in cardiovascular disorders. *Drugs, 42*(3), 468–510.
Warner, N. J., & Rush, J. E. (1988). Safety profiles of the angiotensin-converting enzyme inhibitors. *Drugs, 35*(Supplement 5), 89–97.
Weir, M. R. (2009). Beta-blockers in the treatment of hypertension: Are there clinically relevant differences? *Postgraduate Medicine, 121*(3), 90–8.
West, G. A., & Belardinelli, L. (1985). Sinus slowing and pacemaker shift caused by adenosine in the rabbit SA node. *Pflugers Archiv, 403,* 66–74.
Yang, F., Hanon, S., Lam, P., & Schweitzer, P. (2009). Quinidine revisited. *American Journal of Medicine, 122*(4), 317–321.
Ziff, O. J., & Kotecha, D. (2016). Digoxin: The good and the bad. *Trends in Cardiovascular Medicine, 26*(7), 585–595.

Chapter 12
Drug Resistance in Liver Diseases

Shehla Munir, Muhammad Ibrahim, Nor Hayati Othman, Atif Amin Baig, Mogana Das Murti, Nik Soriani, Muhammad Zeeshan Ahmed, Farhan Younus, and Sarfraz Ahmed

Abstract Currently, many liver-associated diseases are being cured by a variety of drugs. In this chapter, some of the infectious liver diseases have been addressed where the development of resistance against certain therapeutic drugs has already been reported. Some of these diseases are hepatocellular carcinoma, hepatitis B, and

S. Munir · M. Ibrahim
Department of Biochemistry, Bahauddin Zakariya University, Multan 6100, Pakistan
e-mail: ibrahimjam@bzu.edu.pk

N. H. Othman
Department of Pathology, School of Medical Sciences, UniversitiSains Malaysia, 16150 Kubang Kerian, Kelantan, Malaysia
e-mail: hayatikb@usm.my

A. A. Baig
Faculty of Medicine, University Sultan ZainalAbidin, Kuala Terengganu, Malaysia
e-mail: atifamin@unisza.edu.my

M. Das Murti
Department of Oral Pathology, School of Dental Sciences, Universiti Sains Malaysia, 16150 Kubang Kerian, Kelantan, Malaysia
e-mail: moganadasmurtey@usm.my

N. Soriani
Department of Chemical Pathology, School of Medical Sciences, Universiti Sains Malaysia, 16150 Kubang Kerian, Kelantan, Malaysia
e-mail: niksoriani@usm.my

M. Z. Ahmed
Institute of Biochemistry and Biotechnology, University of the Punjab, Quaid-i-Azam campus, Lahore 54590, Punjab, Pakistan

F. Younus
Centre for Interdisciplinary Research in Basic Sciences, International Islamic University, Islamabad 44000, Pakistan
e-mail: farhan.younas@iiu.edu.pk

S. Ahmed (✉)
Department of Basic Sciences, University of Veterinary and Animal Sciences Lahore, Narowal 51600, Pakistan
e-mail: sarfraz.ahmed@uvas.edu.pk

S. Ahmed et al. (eds.), *Biochemistry of Drug Resistance*,
https://doi.org/10.1007/978-3-030-76320-6_12

infectious liver cirrhosis. Another disease "liver abscess" with an unexplored drug resistance mechanism has also been revealed along with multidrug resistance (MDR) of human hepatocarcinoma cells. Drugs, for instance, Cholchicine, Actinomycin D, Ectoposide, Doxorubicin, and Vinblastin have been experimentally assessed for their resistance against five human hepatoma cell lines. Some of these drugs have more resistance while others exhibit less. In course of hepatitis B, the resistance against different nuclcic acids (NAs) has been depicted. This resistance may induce severe liver complications and carcinoma. In the course of liver cirrhosis, sufferers are highly receptive to bacterial infections which provoke acute complexities in cirrhosis. To address this problem mainly antibiotics are used, but multi-drug resistant organisms (MDROs) trigger resistance against some of these drugs. Antibiotic resistance such as Carbapenem has been notified in Enterobacteriaceae species when this drug was used to cure liver cirrhosis. Moreover, the drug rifaximin has been proven to be efficacious in inhibiting spontaneous bacterial peritonitis (SBP) in cirrhosis.

Keywords Liver diseases · Drug resistance · Mechanism of drug resistance · Action mechanism

12.1 Introduction

Liver is a vital organ responsible for detoxification and is actively involved in removing microbes, harmful synthetic substances, and metabolic waste from the body. Chronic infections, extreme fat deposition, excessive alcohol intake, cholestasis, and biologically transformed metabolites can lead to hepatic injuries which can induce liver degeneration and inflammation. Prolonged liver injury can cause chronic liver disease involving the multistage process of fibrosis, hepatocellular carcinoma (HCC), and cirrhosis. As the chronic stage commences, leukocytes penetrate the liver, with zonal or focal necrosis, structural disarray, and obliteration of hepatocytes (Chatterjee & Mitra, 2015).

According to a report from the Centre for Disease Control and Prevention, 2007, chronic liver infection, HCC, and cirrhosis were enlisted among the top 15 death-causing illnesses in the USA. In 2010, 31,802 individuals died of hepatic disorders, excluding 18,910 deaths caused by malignant growth in the liver. Alcoholic cirrhosis accounted for half of all the patients hospitalized for liver cirrhosis. It is currently perceived that the expanding frequency of chronic liver infections is related to HCC or cirrhosis as well as fundamental impairments, for example, non-encephalopathy-related cognitive disability, chronic fatigue, sleep disturbances, and vasomotor symptoms (Chatterjee & Mitra, 2015).

Despite advancements in screening and prevention approaches, diagnostic strategies, and utilization of new treatment innovations, morbidity due to liver ailments continues to increase. Various treatment techniques are available, although surgical resection or orthotopic liver transplantation (OLT) is considered advantageous. Distinctive treatment methodologies incorporate radiofrequency ablation

(RFA), microwave ablation (MWA), cryoablation, radiation remedy, fundamental chemotherapy, percutaneous ethanol infusion, transarterial chemoembolization, and microscopically engaged therapy plans. Adapting certain precautions may avert HCC. These include HBV immunization, utilization of secure infusion practices, standard screening of blood tests, treatment and information on alcohol addiction and intravenous medication consumers, and the inception of antiviral drugs. Different nucleotide/nucleoside derivatives are utilized to treat hepatitis B contamination while antibiotics such as methicillin, vancomycin, and quinolone are utilized for curing liver cirrhosis. Metronidazole is the most unique operator accessible against anaerobes and likely going to be of genuine treatment of veritable diseases because of these living beings (Balogh et al., Balogh, Victor, et al., 2016).

One of the serious issues of reduced efficacy of these medications towards various liver diseases is the development of resistance against these drugs. Mechanisms of drug resistance incorporate target modification, enzymatic modifications of these drugs, and avoidance of anti-microbial entrance or efflux pump. Some ABC transporters (efflux pumps) inclusive of multidrug resistance-related proteins have been reported in liver progenitor cells and hepatocytes in acute liver disorders (Borght et al., 2006; Ros et al., 2003).

12.2 Hepatocellular Carcinoma

Hepatocarcinoma (HCC) is the most prevalent cancer of the liver with chronic liver disease and cirrhosis being major risk factors. HCC is commonly diagnosed in patients at advanced stages of the disease. The patients are monitored every six months and treatment can be via surgical operation based on the size and location of the tumor size and liver function or nonsurgical approaches such as transarterial-chemo embolism and transarterial radiation. Successful surgical removal has been reported in patients with a single tumor and healthy liver function. Liver transplantation is recommended for patients having decompensated liver ailment and tumor growth of not more than 5 cm or otherwise three nodules smaller than 3 cm. However, liver transplantation is much limited by the shortage of organ donors (Balogh et al., Balogh, David Victor, et al., 2016).

The worldwide incidence of HCC and associated mortality are still increasing (ADD REFS) despite improvements in screening, diagnosis, disease management, and prevention. This is attributed to chronic Hepatitis B and C infections, which are independent risk quotients for the liver cirrhosis event (CHECK latest GLOBOCAN, IARC, American Cancer Society, etc.). Alcoholism is also a crucial risk factor. Diagnosis is affirmed without pathological evidence. Techniques such as Magnetic Resonance Imaging (MRI), Computerised Tomography (CT) Ultrasound can be used for screening besides the use of serological markers i.e. α-fetoprotein at 6-month pauses. Among various treatment options, OLT or surgical abscission is beneficial for treatment. Different treatment strategies include RFA, cryoablation, MWA, radiation remedy, percutaneous ethanol injection, systemic chemotherapy, transarterial chemoembolization, and molecular treatment plans. Tumor size, extrahepatic spread, tumor prevailing area, and critically important liver functions are the factors that affect the selection of a treatment plan. HCC is a highly aggressive type of cancer that occurs as a result of cirrhosis and typically prevails in progressive tiers. HCC may be averted by taking precautionary measures along with HBV vaccination, use of secure injection practices, regular screening of blood samples, treatment and knowledge of alcoholism and intravenous drug consumers, and initiation of antiviral medications. Endured advances in each surgical and non-surgical treatment strategy have proven considerable results in overall survival (Balogh et al., Balogh, Victor, et al., 2016).

12.2.1 Multidrug Resistance Phenotype

HCC is classified among the most commonplace cancers in females and males around the globe (Shen et al., 1991). Vinblastine, colchicine, actinomycin D, etoposide (VP-16), and doxorubicin drugs have been reported to be transported via P-glycoprotein efflux pump (Abraham et al., 1993; Endicott & Ling, 1989; Scala et al., 1995). The cytotoxicity of these drugs was tested against five different human hepatoma cell lines in comparison to the human epidermoid carcinoma KB-3-1 cells. All five HCC lines were found to be 2 to5-folds more sensitive to colchicine as compared to KB-3-1 cells. However, the BEL-7404 cell line showed 9.6 times more resistance against colchicine in comparison to KB-3-1 cells, whilst the other four cell lines revealed slightly more resistance to this drug (Table 12.1). The tested cell lines were even highly resistant against VP-16, except SMMC-7721 cells. Other drugs, i.e. methotrexate, cis-platinum, mitomycin C, 5-fluorouracil (5FU), melphalan, and 6-mercaptopurine (6MP) which are not considered to be the best substrates for P-glycoprotein were also tested for their cytotoxic effect (Table 12.2). The cell lines revealed 11–15 times more resistance against mitomycine compared to KB-3-1 cells.

Table 12.1 Relative drug resistance of human hepatocarcinoma calls to MDR drugs

Relative resistance						
Chemicals	KB-3–1	BEL-7402	QGY-7703	BEL-7404	BEL-7406	SMMC-7721
Colchicine	1.0	1.8	1.2	9.6	1.5	1.3
Vinblastine	1.0	2.7	1.7	3.1	2.2	2.7
Actinomycin D	1.0	2.2	3.6	2.4	1.9	4.8
Doxorubicin	1.0	4.5	3.0	5.0	4.0	2.4
VP-16	1.0	6.4	8.7	6.4	4.5	1.4

Relative resistance is expressed as the IC_{50} of the hepatoma cell lines divided by the IC_{50} of KB-3-1 which was arbitrarily assigned a value of unity

Table 12.2 Relative drug resistance of human hepatocarcinoma calls to NON-MDR drugs

Relative resistance						
Chemicals	KB-3-1	BEL-7402	QGY-7703	BEL-7404	BEL-7406	SMMC-7721
Cis-Platinum	1.0	2.7	10.3	4.7	1.0	5.8
Mitomycin C*	1.0	13.3	11.1	16.6	11.1	12.1
Melphalan	1.0	2.7	2.3	3.3	3.3	3.4
Methothrexate	1.0	1.3	1.3	1.8	1.2	13.0
5FU	1.0	2.6	6.0	1.1	3.2	2.8
6MP	1.0	2.5	2.8	14.5	1.8	ND

Relative resistance is expressed as the IC_{50} of the hepatoma cell lines divided by the IC_{50} of KB-3-1 which was arbitrarily assigned a value of unity
*Mytomycin C may be poorly transported by P-glycoprotein

Some of the cell lines showed broad multidrug resistance to all drugs tested. These outcomes verify the clinical impression of HCCis, being resistant against commonly administered anti-cancer drugs (Abraham et al., 1993). The wide range of resistance indicated the involvement of various transport systems specifically for hydrophobic products. The data indicates the involvement of inherent drug resistance in the teste HCC cells since the cells were obtained from patients who did not undergo earlier than surgical procedures (Shen et al., 1991; Seeger & Veen, 2009).

12.2.2 ATP Binding Cassette (ABC) Transporters and Drug Resistance

ABC transporters are the most important members of the family of transmembrane proteins. They are involved in the active transport of molecules, including sugars, metal ions, peptides, amino acids, proteins, metabolites, and hydrophobic compounds into and outside the cell (Higgins, 1992), (Dean et al., 2001). There are 49 members of the ABC transporter genes in humans, which can be categorized into seven subfamilies, primarily based on ATP-binding proteins and sequence homology (Dean et al., 2001). The ABC genes can be of any three types depending upon their domains: comprising of two nucleotide-binding folds (NBF) domains, two transmembranes (TM) domains, called full transporters, or comprised of one of each NBF and TM, called half transporters (Hyde et al., 1990). Whilst NBFs bind and hydrolyze ATP, thus developing the motional force, TM is responsible for the translocation (Seeger & Veen, 2009). ABC mainly protects the cells by avoiding the deposition of toxic compounds, cells become resistant to drugs because of the active flushing out of the drugs from the cells. A significant amount of literature is available stating the role of ABC transporters in multidrug transport at the molecular level (Seeger & Veen, 2009). Many regular stem cells and most cancer cells over-express specialized ABC transporters (Dean, 2009). Some ABC transporters, inclusive of multidrug resistance-related protein-1 (ABCC1/MRP1), multidrug resistance 1 (ABCB1/MDR1/PGY1), breast cancer resistance protein (ABCG2/BCRP/MXR), and MDR-related protein-3 (ABCC3/MRP3) have been reported in liver progenitor cells and hepatocytes in chronic hepatic disorders (Borght et al., 2006; Ros et al., 2003). The protein has vast substrates specificities and is responsible for development of resistance against a range of medications including colchicines, vinblastine and doxorubicin among others and many more (Dean et al., 2001).

12.2.3 Mechanism of Doxorubicin Antitumor Effects

Doxorubicin is an anthracycline-containing antibiotic with significant antitumor properties. It intercalates between DNA base pairs causing DNA strand breaks and as a result signaling cascades initiated, which blocks cell apoptosis. Although different mechanisms have been reported for the anticancer effect of the drug, the best understood is its interaction with the topoisomerase IIα (TOP2A) (Tewey et al., 1984). This enzyme causes DNA denaturation and creates and repairs protein-bound double-stranded DNA breaks (Chen et al., 2013). Doxorubicin settles the cleaved strand, stifling the finish of the cycle bringing about various protein-bound DSBs (Tewey et al., 1984). Cells also face several negative ramifications of DSBs, such as triggering caspase-dependent apoptosis, which activate the master controllers $FOXO_3$ and p53 and suppresses signalling cascades which result in alterations in the proportion of anti-apoptotic and pro-apoptotic Bcl-2 proteins (Roos & Kaina, 2013). The cytotoxicity of doxorubicin can be hindered by blocking this DNA damaging effect (Wilson et al., 2009). Some of the mechanisms reviewed in the cytotoxicity of doxorubicin include incorporation and arrangements of TOP2A independent DNA adducts (Fig. 12.1) (Swift et al., 2006), blockage of DNA, and RNA synthesis, and ROS generation in the mitochondria, which triggers apoptosis (Gewirtz, 1999).

12.2.4 Molecular Mechanisms of Doxorubicin Resistance

Cancer cells can develop resistance to doxorubicin when the drug fails to reach sufficient intracellular concentration, DNA damaging capacity is diminished and veiled transduction of DNA damage to apoptosis. Doxorubicin is hydrophobic and can translocate passively without any specific transporter. Accumulation of the drug can be avoided by active efflux. Overexpression of efflux pumps in a range of different tumors leads to multidrug resistance. Before their physiological functions came to light, these proteins were first recognized for their pathological functions. Presently, they have been reported to have transport functions in various tissues. Several ABC transporters are involved in the transport of different organic ions. Normally, these transporters are present in significant amounts in hepatic cells, but overexpression

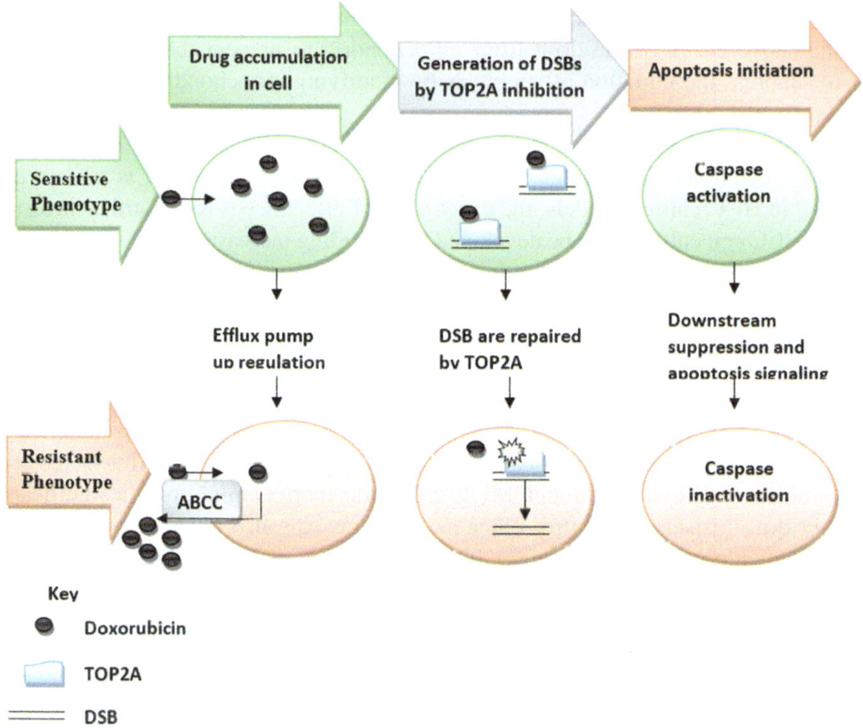

Fig. 12.1 General mechanisms of resistance to doxorubicin in HCC. Doxorubicin first enters the cell and then prevents the repairing of TOP2A-generated DSBx in DNA molecule, enhancing TOP2A-bound DSBs. Over-expression and mutations in TOP2A results in continued TOP2A performance in resistant cells. Finally, DNA damage stimulates apoptotic signalling mechanism causing release of cytochrome C from mitochondria which leads to activation of caspase and ultimately cell death. Downregulation of the apoptotic factors and upregulation of anti-apoptotic proteinaceous factors prevents the completion of apoptotic cycle in resistant cells, ABC: ATP-binding cassette; DSB: Double-stranded DNA break

in HCC leads to chemotherapy resistance. Several transcriptional factors involving members of the SP family, and NF-Y regulates the expression of ABC transporters (Scotto, 2003). p53 suppresses the expression of ABC family proteins (Johnson et al., 2001). Transcriptional factors such as AP-1(Osborn & Chambers, 1996) and NF-κB (Kuo et al., 2002) are fit for their over-expression. A COX-2 enzyme inhibitor, celecoxib down-regulates MDR1 in multidrug-resistant HCC cells (Fantappiè et al., 2007; Mazzanti et al., 2009). At HCC, three different subfamilies of ABC transporters, ABCC (MRP proteins), ABCG (BCRP proteins), and ABCB (the MDR proteins) may contribute to resistance against doxorubicin. Although they have diverse substrate specificity during normal physiological functions, they have all indicated a capacity to transport doxorubicin (Ng et al., 2000; Nies et al., 2001). MDR1, MRP1, MRP2, and MRP3 are completely transcribed in HCC. MDR1 proteins are expressed in 80–90% of HCC's (Ng et al., 2000). Proteins of the MDR family have been reported to be expressed in a functional form on mitochondrial membranes, they may be involved in the active efflux of the drug from mitochondria to prevent the DNA damage, or to suppress the initiation of apoptosis by modifying mitochondrial permeability (Solazzo et al., 2009).

Expression of MDR1 has been reported to conversely affect the response to systemic chemotherapy (Ng et al., 2000). Although the mechanism is quite well studied in HCC culture models, the exact level of effect needs to be further investigated. HCC cell cultures can be developed by providing increasing concentrations of doxorubicin. This strategy reliably initiates the expression of MDR1 and other ABC family proteins (Cheung et al., 2011; Ueda et al., 1987). Overexpression of such pumps leads to drug resistance, as efflux pump inhibitors i.e. cyclosporine A and verapamil could restore susceptibility to doxorubicin (Shiraga et al., 2001). Keeping in mind the presence of many other unrelated resistance mechanisms such as the activity of other efflux pumps, pharmacological synergy among drugs, the deficiency of significant normal physiological capacities, verapamil cannot effectively restore doxorubicin sensitivity. Although MDR efflux transporters might play a significant role in drug resistance in-vitro, other mechanisms can also be of significance in human infection (Fig. 12.1) (Kerr et al., 1986).

12.2.5 Resistance Mechanism of Etoposide

Camptothecin and etoposide-induced apoptosis is significantly blocked by tunicamycin. Camptothecin and etoposide either alter the topoisomerase levels or downregulates ATM expression, tunicamycin does not affect any of the mentioned mechanisms. Drug-trapped topoisomerase DNA complexes and DNA DSBs may lead to tunicamycin-mediated resistance. These two drugs can upregulate the protein expression of cell-cycle regulators and initiate the cleavage of proteins of the Bcl-2 family. The above-mentioned mechanisms can be nullified by tunicamycin.G1 checkpoint arrest initiated by tunicamycin can lead to increased drug resistance. Curcumin can also induce G1 arrest, which can have a reasonable impact on the resistance profile. The resistance against topoisomerase inhibitors is prompted by tunicamycin by overexpressing GRP 78 and G1 arrest. Resistance can be developed even during combinational therapy of chemo-medications and Chinese herbal prescriptions which induce endoplasmic reticulum stress and also potentially cell cycle arrest in carcinogenic cells (Tsend-Ayush et al., 2017).

12.3 Hepatitis B

Hepatitis B virus (HBV) has infected more than 350 million people globally and is one of the major causes of liver cancer and other hepatic diseases. HBV, a double-stranded

DNA virus belongs to the Hepadnaviridae family. The genome of HBV, which reproduces with the RNA transitional state can integrate into the host´s genome. Infected cells remain infected because of some of these exceptional features of HBV. Several serological and virological tests have been developed to diagnose and treat HBV-related disorders and chronic Hepatitis B virus infections. A range of liver disorders that can be caused by HBV infection includes: intense to chronic cirrhosis, hepatitis, and hepatic carcinoma. Intense HBV infection can occur both as asymptomatic and with clear signs of hepatitis. Most viral infections can be overcome, however, 5–10% can lead to persistent infections. Numerous persistently contaminated people have mild liver sickness with little mortality. Others with persistent HBV contamination create active disease, which eventually may lead to cirrhosis and hepatic carcinoma. Continuous monitoring and therapeutic intercessions are required for such patients. HBV can also cause other extrahepatic signs, such signs are not very common, but it can be tricky to analyze such signs. Inability to predict the outcome and advancement of HBV infection and the underestimated importance of understanding cellular, molecular, hereditary, and immunological parameters of the diverse disorders related to HBV are some of the areas which need to be comprehended to treat these infections (Liang, 2009).

Hepatitis B infection is a significant reason for mortality around the world. According to the findings of WHO, ~2000 million people are suffering from HBV, and out of these 350 million people have chronic HBV infections. Approximately, 65 million out of these chronically ill patients are estimated to die of liver disorders (Kane, 1995).

The drugs utilized for HBV treatment and their mechanism of resistance is given below.

12.3.1 Molecular Mechanisms of Resistance Against (Nucleic Acids) NAs

HBV exhibits higher replication and mutagenesis rates of roughly 1×10^{-5} substitution/base/cycle (King et al., 1999). The HBV genome is quite small having base pairs of 3200, all conceivable single base point mutations can be generated every day. The reverse transcriptase of HBV lacks proofreading capability, that's why mutations can emerge at an alarming rate. Before treatment, both single and double mutations are presumably present in viral infections. The likelihood of a mutation during treatment relies upon the drug capacity to reduce replication of the virus. Consequently, the drug with minimum antiviral property doesn't apply substantial force on the infectious virus, and the possibility of drug resistance is low. Then concealment of replication of virus permits a low chance for maximum resistance to arise because mutagenesis depends upon replication. NAs hinder viral replications however, don't remove existing infection or influence cDNA in any significant manner. Monotherapy applying normal antiviral action and coordinated at a single target would bring about the most elevated resistance to the drug. The main treatment routine ought to focus on various targets to lessen the danger of choosing such species that are drug-resistant. During replication, resistance arises within the sight of the drug selection. In this manner, if total concealment of replication is accomplished, resistance will not be a problem. Different issues that rise the drug resistance genetically hinder the advancement of harmful mutations, viral replication space, a system of drug resistance, and different host factors engaged with controlling viral replication (Ghany & Liang, 2007; Weber et al., 2002).

NAs can be seen as prodrugs because they should be activated for their antiviral action through a phosphorylation cycle to their nucleotide diphosphate or nucleoside triphosphates that functions as a polymerase inhibitor. Cellular kinases phosphorylated the NAs to nucleoside-monophosphates that are further phosphorylated to other derivatives like diphosphates and triphosphates (De Clercq, 2002). Three classifications of specific agents are presently accessible: non-cyclic phosphonate nucleotides, L-nucleosides, and cyclopentane deoxyguanosine analogs (Fig. 12.2).

(i) **Lamivudine**

Lamivudine is the main NA used for treating chronic hepatitis B and has a place with a derivative of L-nucleosides (Fig. 12.4a). Resistance against Lamivudine emerges 14–24% every year.rtM204V and rtM204I are two essential mutations related to antiviral resistance to lamivudine (Table 12.3). A system of resistance has been anticipated dependent on the structure of HIV-polymerase (Huang et al., 1998; Kohlstaedt et al., 1992). Lamivudine binds to the outside of the polymerase (palm area) formed to some extent by buildup rt204 (Allen et al., 1998). The initial rtM204V/I mutations are anticipated to decrease the available distance between the lamivudine and polymerase because the methyl group (–CH_3) on the β-expanded chain of isoleucine/valine/encroaches on the sulfurentity (–S) in the L-oxathiolane ring of lamivudine (Allen et al., 1998; Das et al., 2001; Lee & Chu, 2001).

These modifications have two principle impacts: (1) steric hindrance to diminish attachment of lamivudine to certain polymerases and (2) decreased catalytic action to join lamivudine triphosphate into viral DNA that is replicating (Allen et al., 1998; Das et al., 2001; Lee & Chu, 2001). These mutations don't influence the attachment of the regular substrate such as cytosine, to the viral polymerases, and, hence, the polymerase enzyme is as yet dynamic in the replication of the virus. Resistance due to the rtM204V/I mutations is a usual problem to every single L-nucleoside and is by all accounts reliant on the configuration of the NA (Lee & Chu, 2001; Yang et al., 2005). In-vitro examines with un-mutant and mutant HBV-DNA builds uphold the configurational predictions and exhibit a diminished affectability for lamivudine with more than 10,000 times increase in inhibitory focus (IC_{50}) for the two basic mutations of lamivudine as compared to the un-mutant virus (Allen et al., 1998; Ono et al., 2001). The rtM204V/I ± rtL180M mutations are sensitive to entecavir and adefovir but it is more resistant to other L-nucleosides (Brunelle et al., 2005; Ono et al., 2001).

Certain rtL180M mutations exist with the rtM204V type mutation, and these mutations re-establishes replication of HBV near normal levels and enlarge resistance

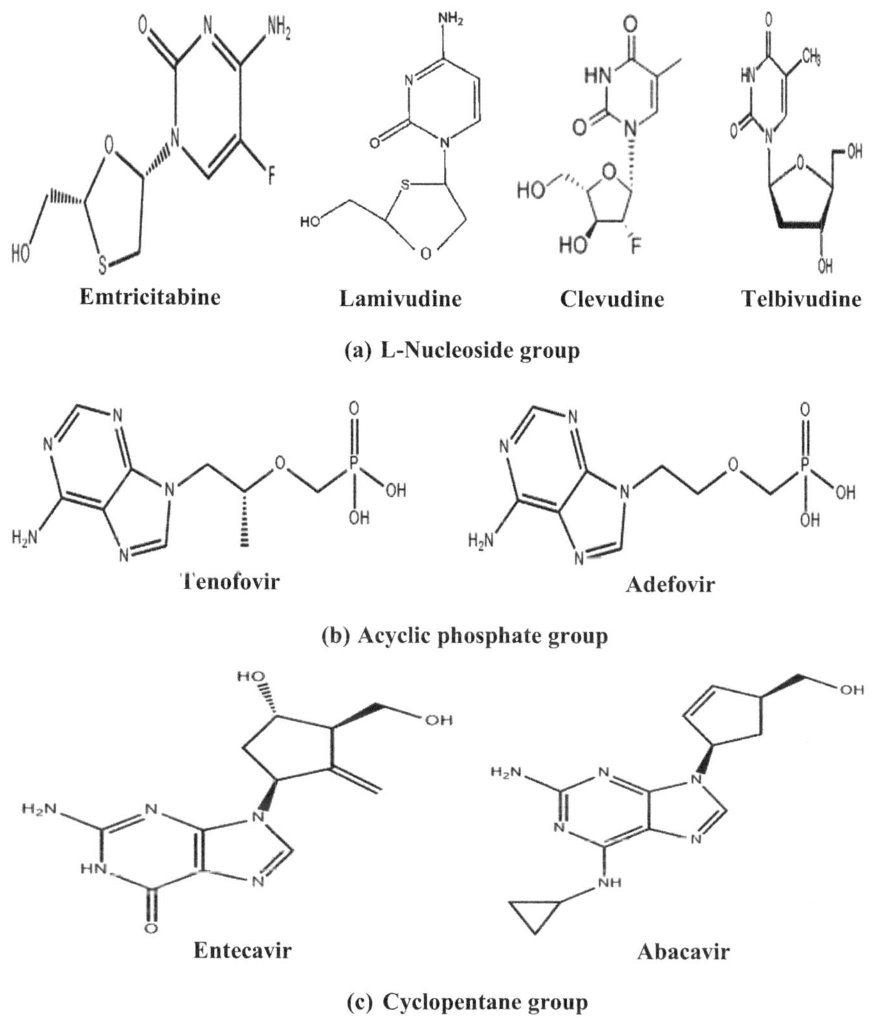

Fig. 12.2 Chemical structure of the three main categories of nucleoside derivatives having anti-HBV action. **a** Lamivudine, emtricitabine, telbivudine, and clevudine are nucleoside derivatives with β-L-configuration. **b** Tenofovirdisoproxilfumarateand Adeofovirdipivoxilare acyclic nucleotide derivatives. **c** Abacavir and Entecavir are carbocyclic derivatives of 2′-deoxyguanosine. Entecavir has anti-HBV, a little bit anti-HIV, and other DNA-containing viruses, whileAbacovir is an anti-HIV-1 and anti-HIV-2 activity. Entecavir inhibits specific HBV-polymerase through many mechanisms with the inclusion of primer, reverse transcription, and elongation of DNA

Table 12.3 Hepatitis B virus polymerase nucleoside and nucleotide-resistant mutations (Ghany & Liang, 2007)

Drug	Domain A	Domain B	Domain C	Domain D	Domain E
Lamivudine	rtL80V/I[a]	rtI169T, [a]rtV173L	rtM204V/I/S	–	–
		[a]rtL180M			
		[b]rt181T, [a]rtT184S[a]	[b]rtQ215S[a]		
Telbivudine	–	–	rtM204I[b]	–	–
Adefovir	rtV84M, [a]rtS85A[a]	rtA181T/V[b]	rtV214A, [a]rtQ215S[a]	rtI233V, [b]rtN236T, [b]rtP237H[a]	–
Tenofovir[c]	–	rtA194T[a]	–	–	–
Entecavir[d]	–	rtI169T, [b]rtS184S/A/I/L/F/G[b]	rtS202G/I[b]	–	rtM250V[b]

[a] Secondary mutations
[b] Primary resistant mutations
[c] Resistant mutations reported but not confirmed
[d] In the presence of lamivudine resistant mutations

against lamivudine. Molecular mechanism proposes that rtL180M type mutation is near to rtM204 type mutation on a nearby DNA helix characterized by 173–189 residues near to 204 and further influence the lamivudine attachment by presenting an unbending branched-chain in the oxathiolane ring of lamivudine (Das et al., 2001).

The triple mutations rtV173L, rtL180M, and rtM204V have been found in patients of liver transplantation. In-vitro data show that the rtV173L improves replication of the rtlL180M/rtM204V type mutant but doesn't influence the sensitivity of the polymerase towards lamivudine (Brunelle et al., 2005). Molecular modeling examination reveals that this type of mutation is situated underneath the template strand of HBV nucleic acid. Replacement of leucine to valine builds viability of the polymerase enzyme by improving the location of the template strand (Delaney et al., 2003).

(ii) **Telbivudine**

Telbivudine is a type of NA that is approved for curing chronic hepatitis B and has been proved to be more efficacious as compared to lamivudine in inhibiting HBV nucleic acid levels in-vivo. In clinical trials, just rtM204I type mutations arose. The purpose behind this is not completely evident but may identify with the higher strength of telbivudine (Lai et al., Lai, Gane, et al., 2005; Lai, Leung, et al., 2005).

(iii) **Adefovir**

Adefovirdipivoxil is a guanosine derivative with a non-cyclic linker. Improvement of resistance from adefovir is minimum than lamivudine, with a 2% reported rate at two years and 29% at five years (Hadziyannis et al., 2006; Yang et al., 2002). The structural similitude of adefovir to its substrate molecule (dATP), along with little linker, permits it more noteworthy availability to the polymerase enzyme and may suggest its minimum resistance rate as compared to lamivudine (Westland et al., 2003). The primary mutations related to adefovir resistance are rtI233V and rtN236T type mutations in D-domain and rtA181V type mutations in B-domain (Angus et al., 2003; Schildgen et al., 2006; Locarnini et al., 2005; Ghany & Liang, 2007). The resistance against adefovir is related to three particular sites of the HBV-polymerases: (1) mutations, including A and D-domains (rtN236T, rtI233V, rtP237H, rtS85A, rtN236T/D, and rtV84M), (2) mutations, including B-domain (rtA181V/T), and (3) The C-D interdomain (rt Q215S and rtV214A). The clinical importance of these auxiliary mutations remains unknown (Angus et al., 2003).

Molecular modeling of D and A domain mutations uncovers a potential clarification for the proper mechanism of well-known adefovir rtN236T mutations, which are external to the enzyme catalytic site, to prompts adefovir resistance (Angus et al., 2003). Domains D and A are believed to be associated with the binding of dNTPs (Ghany & Liang, 2007; Yadav & Chu, 2004). The rtN236T type mutations may disturb the specific type of hydrogen bond somewhere in the range of rtN236, rtS85, rtN236, and adefovir diphosphate, resulting in diminishing the coupling affinity for adefovir (Bartholomeusz, 2004; Yadav & Chu, 2004).

(iv) **Tenofovir**

Tenofovir is a non-cyclic nucleotide having a similar structure as a defovirdipivoxil (De Clercq, 2007). It possesses a comparable process of activity and anti-viral resistant profile like adefovir (van Bömmel et al., 2004). Tenofovir is also dynamic against the initial lamivudine-resistant virus-like adefovir. For use in chronic hepatitis B, tenofovir has not been approved and as a result, restricted data is available about its antiviral resistance profile (Delaney et al., 2006; van Bömmel et al., 2006).

(v) **Entecavir**

Entecavir is a derivative of cyclopentane compounds and possesses a particular potency against the harmful disease of HBV. It is a carbocyclic compound having 2-deoxyguanosine inclusive of the vinyl group that replaced oxygen of tiny furanose ring. It influences different functions of the enzyme polymerase (Innaimo et al., 1997). It is one of the most efficacious drugs of the clinically available drugs against HBV (Innaimo et al., 1997). The rate of resistance against entecavir in NA-naive subjects is very low (Chang et al., 2006; Colonno et al., 2006). Moreover, in the lamivudine-resistant patients, the rate of resistance is 15% at 3 years (Colonno et al., 2006). Research data demonstrates that the initial lamivudine-resistant rtM204V/I type mutations are by all accounts a pre-requisite for improvement of entecavir resistance (Tenney et al., 2004). Two patterns for developing resistance have been reported:rt184G + rtS202I + rtM204V + rtL180M and rtM250V + rtI169T + M204V + L180M (Tenney et al., 2004).

Molecular-modelling depending on the HIV-reverse transcriptase analogue proposes two potential pathways with a contrasting mechanism for entecavir resistance. (Tenney, 2003; Warner, 2003). Mutations at the sites of rt169 and rt250 influence the primer attachment site of the reverse transcriptase enzyme that joins at the 3′ end of the template DNA (Tenney, 2003; Warner, 2003). The in-vitro assay shows that the rtM250V without other modifications presents a very low resistance to entecavir, however, within the sight of the lamivudine rtM204V type mutation, prompts more than 250 times to enhance resistance (Tenney et al., 2004).

12.4 Drug Resistance in Infectious Liver Cirrhosis

12.4.1 What is Liver Cirrhosis?

Liver cirrhosis is an end-stage illness that is the result of different chronic liver infections. The infections that cause liver cirrhosis progress gradually for quite a long while. Treatment for these chronic diseases is accessible and subsequently, liver cirrhosis can be prevented. Liver cirrhosis is one of the significant reasons for hepatocellular carcinoma and thus avoidance from liver cirrhosis causes prevention from HCC (Manns et al., 2010).

Cirrhosis is a diseased condition in which liver tissues are harmed and the liver can't play its ordinary function as the normal tissues are replaced by damaged tissues influencing blood flow through the liver (Murphy et al., 2013).

12.4.2 Antibiotics Resistance in Liver Cirrhosis

Information gathered during a survey for anti-biotic resistance in sufferers of liver cirrhosis gave the outcomes as anti-microbial resistant organisms with resistance to following anti-biotics:

- Methicillin-resistant Staphylococcus aureus (MRSA),
- Vancomycin-resistant Enterococcus (VRE),
- Quinolone-resistant gram-negative rods (QRGNR) (isolated resistance to a quinolone), extended-spectrum β-lactamase–producing Enterobacteriaceae (ESBL) (resistance to ceftazidime or ceftriaxone), (Falagas et al., 2006).

The general mechanism of action and resistance to some of these drugs is given underneath.

12.4.2.1 Methicillin

Methicillin has just been found in people that have utilized or taken this drug. It belongs to penicillin, which is resistant to penicillinase, but highly susceptible to a penicillin-binding specific protein. It is administered by injection because gastric acid inactivates it (Hartman & Tomasz, 1984).

Mechanism of Action of Methicillin

Methicillin performs its function by inhibiting the bacterial cell wall synthesis by suppressing the transpeptidase enzyme activity which is engaged in cross-linking of the linear chains of peptidoglycan polymers that make up a significant segment of the Gram-positive bacterial cell wall. This enzyme activity cross-links the D-alanyl-alanine peptides which are utilized in peptidoglycan production. Methicillin hinders the transpeptidase by binding to this enzyme which becomes inaccessible for the real substrate and can't cross-link the peptide units. In this manner, the bacterial cell wall isn't appropriately synthesized and its permeability is influenced causing spillage of the cell substances and resulting causes the bacterial cell death (Hartman & Tomasz, 1984).

Mechanism of Resistance to Methicillin

Methicillin-resistant staphylococcus aureus (MRSA) created resistance against methicillin by the synthesis of change penicillin-binding proteins (PBP 2a) that are resistant to inhibition by the β-lactam antibiotics that inhibit PBP enzymes. PBP 2a can synthesize bacterial cell walls within the sight of concentration of such antibiotic agents, for example, methicillin. The synthesis of these modified penicillin-binding proteins PBP2a is completed by the mecA gene expression (Brown & Reynolds, 1980). This specific gene is situated in a special portion of DNA referred to as the staphylococcal chromosomal cassette (SCC). At an adequate amount of β-lactam-antibiotic, especially methicillin, synthesis of these changed proteins is enhanced and they render the bacterial cell resistance against methicillin as the cell wall production stays unaffected within the sight of PBP 2a (Fig. 12.3) (Hartman & Tomasz, 1984).

12.4.2.2 Vancomycin

Vancomycin is an antibiotic that is structurally belonged to glycopeptide. It predominantly acts by preventing cell wall synthesis of bacterial species. It is extraordinarily effective against specific gram-positive bacteria and has been known to be powerful against resistant strains of MRSA. Vancomycin is known as the last medication, when all other treatment options have run out. Moreover, bacteria have started to resisting vancomycin (Nagarajan, 1991).

Mechanism of Action of Vancomycin

Vancomycinshows its activity by hindering the cell wall synthesis of microbes. The peptidoglycan layer is tough due to its cross-linking structure. During the biosynthesis of the peptidoglycan layer, monomers of N-acetylglucosamine and N-acetylmuramic acid get embedded into the membrane. Reorganization of the peptide cross-links occurs by the transpeptidase enzyme activity (Hammes & Neuhaus, 1974). Vancomycin attaches to the building blocks of the peptidoglycan (i.e. NAM and NAG) and prevent the transpeptidase enzyme from acting on these newly formed building blocks and in this way prevents peptidoglycan cross-linking. Thus, vancomycin tends to make the peptidoglycan layer more permeable. This makes the bacterial cell content release out and the ultimate death of the microbes. Mutations in the transpeptidase can prompt enhanced resistance to vancomycin (Nagarajan, 1991).

Mechanism of Resistance to Vancomycin

The process of resistance of nonheritable glycopeptide i.e., Vancomycin resistance in the Van-A associate degreed Van-B enterococci was discovered as a group of genes

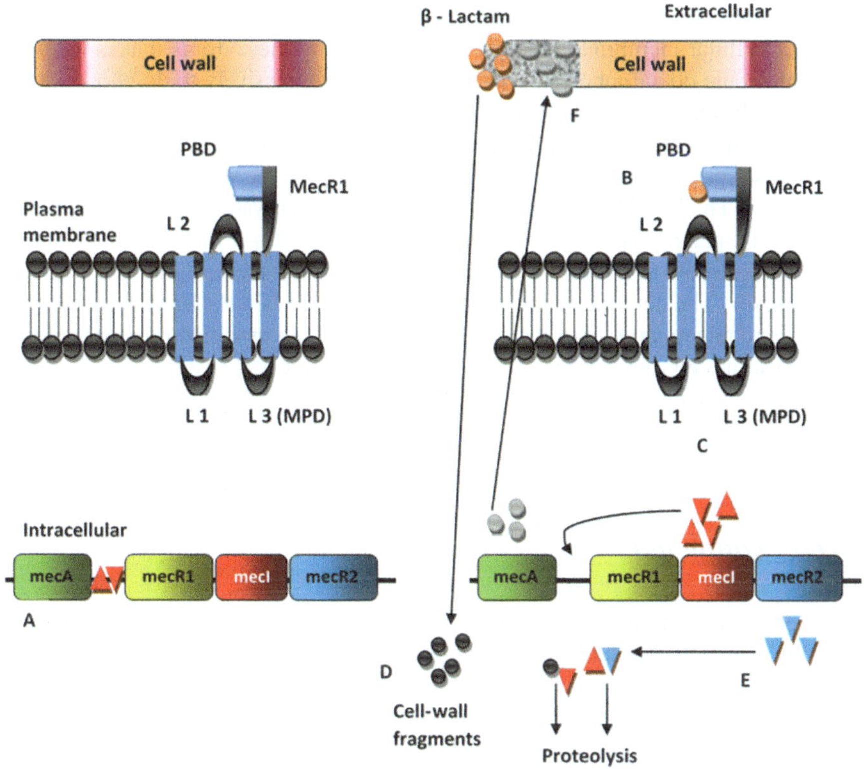

Fig. 12.3 General mechanism of methicillin resistance. **a** When β-lactams are absent, transcription from the *mec*operator is prevented by the attachment of the repressor MecI to this site. **b** β-lactams are recognized by their attachment to the MecR1 having penicillinbinding domain (PBD). **c** This association stimulates the autolytic activation of the MecR1 having intracellular metalloproteinase domain (MPD) within L3. **d** Cytoplasmic cell-wall segments, probably generated by the disturbance of cell-wall synthesis by β-lactams, performing the function of coactivator that attaches to MecI, disturb its interaction with the *mec*operator, and develops its proteolytic degradation. **e** A second anti-repressor, encoded by mecR2 in the mec complex, is transcribed by β-lactams, and the resulting protein, MecR2, also attaches to MecI, leading to its degradation. **f** The proteolysis of MecI induces mecA transcription, the synthesis of PBP2a, and the expression of methicillin resistance

encryption, a modified production mechanism for accumulation of cell membrane antecedents that poorly binds to the Vancomycin (Arthur et al., 1994, Eliopoulos and Gold, 2001). Vancomycin resistance (aside from Van-E types) ends with the depsipeptide d alanyl-d-lactate (d-Ala-d-Lac). Vancomycinattaches to modified molecules at 0.001 times as compared to its binding to local PG precursor (Bugg et al., 1991). This change of the target region for antibiotics having glycopeptides is accompanied by numerous proteins that sense the drug existence, make a drug-resistant target site. In Van-A type enterococci, e.g., the Van-Rsupermolecule, and the Van-Ssupermolecule type a two-component regulating system (Arthur et al.,

1992). The existence of Vancomycin, teicoplanin, or may probably a lot of perturbation of cellular membrane precursors stimulated by these drugs makes the Van-Ssupermolecule do autophosphorylation and then progressively to do phosphorylation of Van-R. Despite enhancing the expression of van-S and van-R, phosphorylated Van-Rsupermolecule attaches to the promoter area for van-HAX, acquiring transcription of those specific sequences, resulting in excitation of the whole system (Baptista et al., 1996). The Van-Hsupermolecule changes over pyruvate to d-lactate, which is joined with d-alanine by the Van-A ligase to produce d-Ala-d-Lac. VanY is an adjuvant supermolecule that eliminates the d-Ala residues at the terminal site from the PG precursor. The activity of the Van-Zsupermolecule is not clear, but it helps in resistance to teicoplanin. Like local PG precursors, these modified precursor blocks into the useful cell membranes. The vanB group is practically sort of vanA group, but it contrasts in its regulation (Fig. 12.4) (Baptista et al., 1996; Evers & Courvalin, 1996).

12.4.3 *Quinolones*

Quinolones are the aromatic compounds obtained from the hetero-bicyclic quinoline, the name of which derived from the oily substance got from the alkaline refining of quinine. Since quinine was isolated from Cinchona bark in 1811, numerous other quinoline derivatives have been separated from natural sources (Gerhardt, 1842).

12.4.3.1 *Mechanism of Action of Quinolones*

Quinolone shows its activity by attaching to complexes between polymer and topoisomerase-IV or gyrase. After attachment, quinolones carry a conformational modification within the enzyme. This enzyme destroys the polymer and quinolone does not permit the re-joining of broken polymer strands. This enzyme catalyst acts on the polymer, leading to the synthesis of quinolone-gyrase-DNA complex. This complex inhibits polymer replication (Khodursky & Cozzarelli, 1998). This complex inhibits polymers reversibly and cell growth and is believed to be liable for the natural process of the quinolones (Zhao et al., 1997).

It is proved that death occurs from the discharge of polymer ends from complexes. Evidence suggests that discharge of polymer ends, not merely leads to the synthesis of complexes, but also responsible for the antibiotic activity of quinolones, obtained from geological phenomena analysis of microorganism nucleoids (Chen et al., 1996). Moreover, the deadly activity of certain oxolinic acid is often inhibited by Chloromycetin, but only a partial impact is seen with an antibiotic. This means that a supermolecule issue is concerned within the free polymer ends of quinolone–gyrase–DNA type complexes which quinolones have an extra mode of action insensitive to chloramphenicol (Chen et al., 1996).

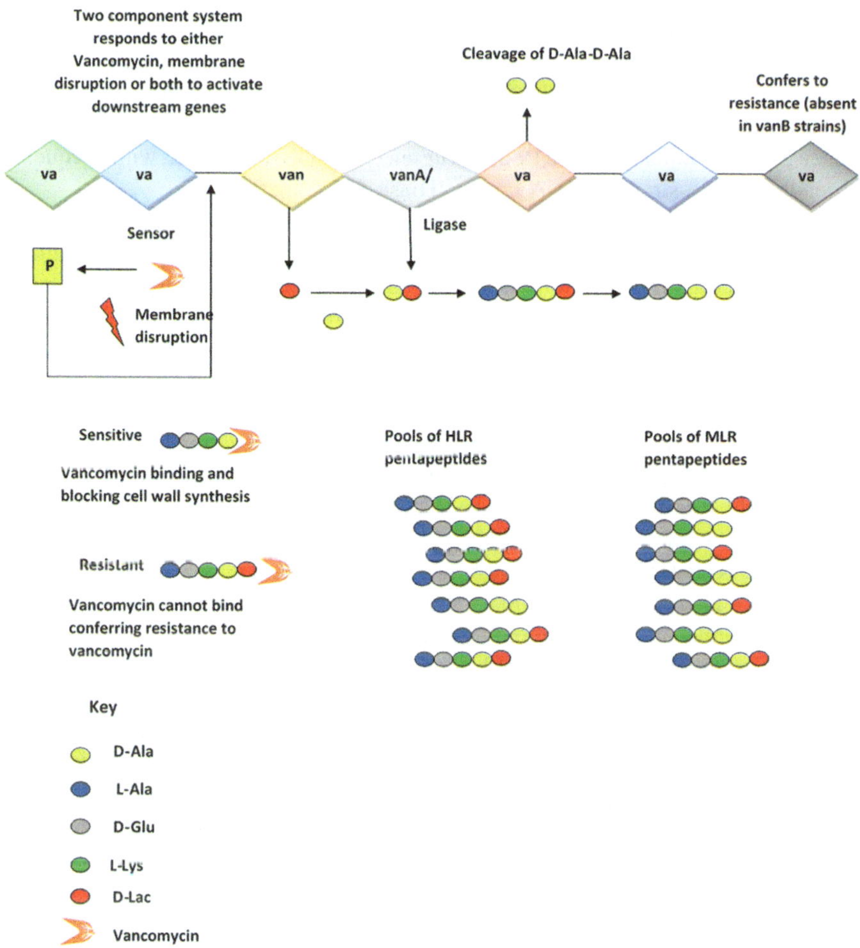

Fig. 12.4 General mechanism of resistance to vancomycin

12.4.3.2 Mechanism of Resistance to Quinolones

Three major proper mechanisms of resistance against quinolones are presently identified: mutations that modify the drug targets, mutations that lessen the accumulation of the drug, and plasmids that shield cells from the deadly impacts of quinolones (Hooper, 2003). The quinolone targets are the bacterial enzymes i.e.DNA gyrase and DNA topoisomerase IV (Drlica & Zhao, 1997). Each of the massive, complex enzymes is made up of two subunits. The DNA gyrase subunitsareGyr-A, a 97 kDasupermoleculethat is encoded by the gyr A sequence, and Gyr-B, a 90k Dasupermolecule encoded by the gyr-B sequence. DNA-gyrase adds negative supercoils into the molecule of DNA, will take away each positive and negative supercoils, and may arrange circular molecules. DNA topoisomerase-IV may take away negative and positive supercoils and is higher at decatenation than gyrases. Two enzymes work along with replication, transcription, recombination, and DNA repairing. The enzymes break double-stranded DNA, proceed a second DNAhelix through the break, which is then resealed (Kampranis et al., 1999). Quinolones block the reaction and induce gyrase or topoisomerase IV as a drug-enzyme–DNA complex, with a sequence of deadly, double-stranded DNA breaks (Hiasa & Shea, 2000).

A few bacteria are ready to operate with a single DNAgyrase; however, most of the bacteria have specific types of enzymes. For instance, the gyrase enzyme of gram-negative bacteria is mostly inhibited by the quinolones as compared to topoisomerase IV, wherein the gram-positive type of bacteria, topoisomerase IV is mostly targeted compared to gyrase. Resultantly, mutations take place initially in gyrA of gram-negative type bacteria, however, these happen initially in the parC of gram-positive type of bacteria. The resistance inclusive of organic compound substitutions in some regions of the GyrA or ParC named as "quinolone resistance determining region" (QRDR). These regions happen on activator having specific DNA binding surface (Cabral et al., 1997) and, for some bacteria like E. coli, the DNA gyrase consists of amino acids between locations 106 and 51 (Friedman et al., 2001). The QRDR in the DNA gyrase is close to amino acid 122 that makes a covalent bond with the phosphate type group of DNA within the primary strand-breaking specific reaction. Contrarily, in some S. aureus and S. pneumoniae, primary mutations take place often in parC, while, in extremely resistant bacterial strains, extra mutations present in the gyrA and parC (Eliopoulos, 2004; Ng et al., 1996). Once initial mutations have minimized the status of DNAgyrasein a gram-negative bacteria, extra mutations in the gyrA, gyrB, or parC will cause resistance, although, they might be ineffectual in bacteria with wild type of GyrA. Supporting this type of model, double and single substitutions within the QRDR in E. coli gyrase are diminished the quinolones binding to the specific complex of enzyme and DNA (Barnard & Maxwell, 2001; Willmott & Maxwell, 1993). Mutations could largely impair the target activator, thus, inhibit the enzyme and DNA complexes formation (Ince & Hooper, 2003).

12.5 Pyogenic Liver Abscess

The pyogenic liver is a typical intra-abdominal disease. Generally, *Escherichia* (*E.*) *coli* has been the transcendent causative agent (Lederman & Crum, 2005). Pyogenic liver abscess is an essential and frequently dangerous condition. It is mainly caused by basic sickness of the biliary framework, but, no likely issue can be distinguished. Its determination requires a high record of the liver (Ahronowitz et al., 2012).

12.5.1 *Bacteriology*

The affectability of societies in pyogenic liver ulcers ranges from 60 to 80%. Sterile societies are normal in patients getting an expansive range of antimicrobials before waste. The most generally distinguished causative living beings in the PLA are anaerobes and oxygen-consuming gram-negative bacteria, especially *E. coli* and *Klebsiella* (*K.*) *pneumonia* (Sharara & Rockey, 2002). Much of the time is related to gas-framing abscesses; all the more as of late, its serotype K1 was observed to be connected with liver damage and the entanglement of endophthalmitis, particularly in diabetic patients (Fung et al., 2002).

12.5.2 **K. Pneumoniae** *Genotype K1: An Emerging Pathogen that Causes Septic Ocular or Central Nervous System Complications from Pyogenic Liver Abscess*

K. pneumoniae causes urinary tract contaminations and different contaminations in hospitalized people whose unsusceptibility is anticipated by fundamental ailments, for example, diabetes mellitus (Podschun & Ullmann, 1998). Since 1986, analysts in Taiwan and a few different zones have noticed a group of disorders having pyogenic liver abscess that is confounded by metastatic endophthalmitis or central nervous system (CNS) diseases (Casanova et al., 1989; Fang et al., 2000; Liu et al., 1991).

12.5.3 *Metronidazole Medicate Utilized for Treatment of Liver Abscess Disease*

Metronidazole is the most dynamic operator accessible against anaerobes and probably going to be of real treatment of genuine diseases because of these living beings. Even though the absence of similar trials in numerous sites of utilization makes it hard to plainly express the relative treatment of metronidazole, contrasted and different drugs, for example, clindamycin, chloramphenicol, or penicillin, it is, in any case, an exceptionally compelling specialist in the treatment and counteractive action of anaerobic diseases (Reysset, 1996).

12.5.3.1 Mechanisms of Action and Resistance to Metronidazole

Mitronidazole is dynamic against an assortment of protozoa and other microscopic organisms. It steps in the cell by dispersion and is enacted in either a particular organelle or cytoplasm, though resistant cells are inadequate in tranquilize the activation. Metronidazole particle is modified to free radical. This type of drug is very cytotoxic and can connect with DNA particles. The real procedure of activity has not been completely explained rather incorporates the resistance of DNA and its damage by oxidation, resulting in DNA strands break that prompt DNA damage and ultimately cell death. Metronidazole entity nonspecifically attaches to bacterial DNA particle and inactivates it, however, cell lysis does not occur (Diniz et al., 2000; Land & Johnson, 1999). Cells need electron-transport proteins having adequate oxidation/reduction potential; in this way, the drug is active against microaerophils, for example, H. pylori. Moreover, re-oxidation may happen within the sight of atomic oxygen and may modify the compound to its unique inert form (Reysset, 1996). Electron benefactors associated with decrease procedure fluctuate, depends upon the micro-organisms. In anaerobes, ferredoxin and flavodoxin, which get the electrons from the pyruvate-ferredoxin oxidoreductase complex, assume critical parts, whereas different proteins and electron exchange segments may likewise be associated with the procedure. These electron acceptors have lower reduction potential than metronidazole particles and in this way give their electrons to the drug (Reysset, 1996). In *H. pylori*, a different component is by all accounts engaged with metronidazole inactivation. Metronidazole-resistance clones are regularly transformed with the rdxA gene (Land & Johnson, 1999; Moore & Salama, 2005).

A few factors of resistance of metronidazole in anaerobes have been evaluated. These systems contrast in bacteria, but the main reason for resistance is diminished take-up of the drug or modified decrease proficiency. These two mechanisms work together; diminished action of enzyme nitroreductase prompts diminished take-up of the drug. Different systems incorporate active efflux, drug inactivation, and repairment of DNA damage (Land & Johnson, 1999). Particularnitroimidazolesresistantgenes (nim) presenting resistance to nitroimidazoles have been disengaged in various genera of gram-positive and gram-negative anaerobic microscopic organisms, including Bacteroides species (Löfmark et al., 2005; Theron et al., 2004). The exchange of these genes gives resistance against metronidazole (Breuil et al., 1989). The nim genes encode an elective reductase enzyme which can change over nitroimidazole to a specific nontoxic subsidiary, subsequently evading the harmful impact that causes breakage of the DNA (Leiros et al., 2004; Reysset, 1996). Up to this point, seven individual genes from nim-A to nim-G have been identified, even though the identification of new variations shows the presence of a higher assortment of these genes in the anaerobic group than was at first anticipated. Concentrates on the predominance of the nim genes recorded an overexpression of nim-A in anaerobes (Leiros et al., 2004). The nim genes typically present on specific plasmids, but have additionally located on the chromosome of bacteria. The inclusion components are portable and thought to be associated with the versatility of prokaryotic genomes. These have appointed a part in a few resistance causing gene expression in some species of Bacteroides (Whittle et al., 2002).

12.5.3.2 Proposed Mechanisms of Resistance to Metronidazole in Anaerobic Bacteria

The relatedness of the nim genes are not generally connected with the resistance, and their real effect on clinically important metronidazole resistance is not yet clearly identified. nim-Negative strains, communicating abnormal state resistance, are sporadically separated, demonstrating the significance of extra components of resistance. In any case, the nim genes essentially build the danger of lessened vulnerability to metronidazole (Löfmark et al., 2005). Among >1500 examined clinical strains in an investigation by Löfmark et al. (2005), 2 particular locations regarding vulnerability could be perceived (Löfmark et al., 2005). Along these lines, a critical relative danger of metronidazole resistance (MIC, ~32 mg/L; as characterized by the Clinical and Laboratory Standards Institute) among nim-positive strains was uncovered (chances proportion, 26; 95% certainty, 4.6–147), while the hazard for diminished weakness (MIC, ~8 mg/L) was considerably higher among nim-positive strains (chances proportion, 53; 95% certainty, 19–147). This last breakpoint partitions the ordinary location of vulnerable disengages from those communicating resistance determinants and is likewise acknowledged as the European breakpoint for metronidazole resistance (Cole et al., 2011).

Different mechanisms that may add to the resistance in Bacteroides species incorporate efflux pumps. Some or no information may exist on any efflux structure in Bacteroidesspecies, yet overexpression of the efflux pumps is regularly associated with multidrug resistance in different species and for different anti-biotics. Pumbwe et al. (2006) recommend that efflux overexpression assumes a part in metronidazole protection and can make half of the resistance against fluoroquinolones and high resistance against β-lactams in clinical Bacteroides strains. This mechanism could assume a vital part in the expanded number of secluded clinical multidrug-resistant strains that do not have any of the known nim genes. Extra clinical tests are required. The modification of DNA repair structure assuming a part in metronidazole resistance in H. pylori is another not entirely contemplated component in Bacteroides strains. Overexpression of the compounds associated with the procedure corresponds with diminished antibiotic weakness. A solid applicant is the recA protein; mutants that are delicate to oxygen and the activity of metronidazole. Different genes in the rec family have likewise been recommended for a part in DNA repair, although they have not been studied (Sander et al., 2001).

The procedure of inducing resistance to metronidazole is another element of this drug that might possess clinical complications. Reversible type and irreversible type of resistance have been incited in strains of Bacteroides with the utilization of inhibitory core of metronidazole (Gal & Brazier, 2004; Löfmark et al., 2005). Subinhibitorycore of metronidazole can meddle with the cell properties of individuals from the B. fragilis gathering, with suggestions for their relatedness with the host defensive system (Diniz et al., 2003). In an in vitro test, the resistance against metronidazole was appeared to be related to the prevalent activity of H. pylori, giving assurance against anti-microbial treatment (Lai et al., 2006).

12.5.3.3 Reactions

Reactions have sometimes been accounted for in patients treated with metronidazole for the anaerobic disease. The utilization of a higher than normal amount has brought about periodic occasions of parasthesiae of the feet and hands, fringe neuropathy, and epileptiform seizures. Additionally examines are expected to decide the recurrence and nature of harmful impacts liable to be experienced with the higher doses of metronidazole (Brogden et al., 1978).

12.6 Conclusions

The liver is the largest detoxifying gland and is viably associated with clearing harmful microorganisms, toxic synthetic substances, metabolic waste products from the body, and developing homeostasis. Hepatic injury caused by constant exposure of the liver to infections and alcohols, fat deposition, cholestasis, and biologically transformed metabolites can prompt liver diseases like HCC, cirrhosis, hepatitis B, etc. A variety of liver diseases have been treated through antibiotics, antiviral, and nucleotide/nucleoside derivatives, etc. Regardless of advancement in using different potential drugs as treatment for these liver diseases, pervasiveness and morbidity are continuously arising due to the development of resistance to these drugs. The majority of drug resistance pattern in liver diseases is connected to ABC transporters-efflux pumps and gene mutation which is also the main cause of multi-drug resistance. In the future, new effective and sustainable therapies for the treatment of liver diseases can be introduced by just understanding the mode of action of the drug, drug dosage, patient history, and adverse reactions of drugs.

References

Abraham, E. H., Prat, A. G., Gerweck, L., Seneveratne, T., Arceci, R. J., Kramer, R., Guidotti, G., & Cantiello, H. F. (1993). The multidrug resistance (mdr1) gene product functions as an ATP channel. *Proceedings of the National Academy of Sciences of the United States of America, 90*, 312–316.

Ahronowitz, I., Harp, J., & Shinkai, K. (2012). Etiology and management of pyoderma gangrenosum. *American Journal of Clinical Dermatology, 13*, 191–211.

Allen, M. I., Deslauriers, M., Andrews, C. W., Tipples, G. A., Walters, K. A., Tyrrell, D. L., Brown, N., & Condreay, L. D. (1998). Identification and characterization of mutations in hepatitis B virus resistant to lamivudine. *Hepatology, 27*, 1670–1677.

Angus, P., Vaughan, R., Xiong, S., Yang, H., Delaney, W., Gibbs, C., Brosgart, C., Colledge, D., Edwards, R., & Ayres, A. (2003). Resistance to adefovir dipivoxil therapy associated with the selection of a novel mutation in the HBV polymerase. *Gastroenterology, 125*, 292–297.

Arthur, M., Depardieu, F., Snaith, H. A., Reynolds, P. E., & Courvalin, P. (1994). Contribution of VanY D, D-carboxypeptidase to glycopeptide resistance in enterococcus faecalis by hydrolysis of peptidoglycan precursors. *Antimicrobial Agents and Chemotherapy, 38*, 1899–1903.

Arthur, M., Molinas, C., & Courvalin, P. (1992). The VanS-VanR two-component regulatory system controls synthesis of depsipeptide peptidoglycan precursors in enterococcus faecium BM4147. *Journal of Bacteriology, 174*, 2582–2591.

Balogh, J., David Victor III, E. H. A., Burroughs, S. G., Boktour, M., Saharia, A., Li, X., Ghobrial, R. M. & Monsour Jr, H. P. (2016a). Hepatocellular carcinoma: A review. *Journal of Hepatocellular Carcinoma, 3*, 41.

Balogh, J., Victor, D., Asham, E. H., Burroughs, S. G., Boktour, M., Saharia, A., Li, X., Ghobrial, R. M., & Monsour, H. P. (2016). Hepatocellular carcinoma: A review. *Journal of Hepatocellular Carcinoma, 3*, 41–53.

Baptista, M., Depardieu, F., Courvalin, P., & Arthur, M. (1996). Specificity of induction of glycopeptide resistance genes in enterococcus faecalis. *Antimicrobial Agents and Chemotherapy, 40*, 2291–2295.

Barnard, F. M., & Maxwell, A. (2001). Interaction between DNA gyrase and quinolones: Effects of alanine mutations at GyrA subunit residues ser83and Asp87. *Antimicrobial Agents and Chemotherapy, 45*, 1994–2000.

Bartholomeusz, A. (2004). Molecular modelling of hepatitis B virus polymerase and adefovir resistance identifies three clusters of mutations. *Hepatology, 40*, 246A.

Borght, S. V., Libbrecht, L., Katoonizadeh, A., Pelt, J. V., Cassiman, D., Nevens, F., Lommel, A. V., Petersen, B. E., Fevery, J., & Jansen, P. L. (2006). Breast cancer resistance protein (BCRP/ABCG2) is expressed by progenitor cells/reactive ductules and hepatocytes and its expression pattern is influenced by disease etiology and species type: Possible functional consequences. *Journal of Histochemistry and Cytochemistry, 54*, 1051–1059.

Breuil, J., Dublanchet, A., Truffaut, N., & Sebald, M. (1989). Transferable 5-nitroimidazole resistance in the Bacteroides fragilis group. *Plasmid, 21*, 151–154.

Brogden, R., Heel, R., Speight, T., & Avery, G. (1978). Metronidazole in anaerobic infections: A review of its activity, pharmacokinetics and therapeutic use. *Drugs, 16*, 387–417.

Brown, D. F., & Reynolds, P. E. (1980). Intrinsic resistance to β-lactam antibiotics in staphylococcus aureus. *FEBS Letters, 122*, 275–278.

Brunelle, M. N., Jacquard, A. C., Pichoud, C., Durantel, D., Carrouée-Durantel, S., Villeneuve, J. P., Trépo, C., & Zoulim, F. (2005). Susceptibility to antivirals of a human HBV strain with mutations conferring resistance to both lamivudine and adefovir. *Hepatology, 41*, 1391–1398.

Bugg, T. D., Wright, G. D., Dutka Malen, S., Arthur, M., Courvalin, P., & Walsh, C. T. (1991). Molecular basis for vancomycin resistance in enterococcus faecium BM4147: Biosynthesis of a depsipeptide peptidoglycan precursor by vancomycin resistance proteins VanH and VanA. *Biochemistry, 30*, 10408–10415.

Cabral, J. H. M., Jackson, A. P., Smith, C. V., Shikotra, N., Maxwell, A., & Liddington, R. C. (1997). Crystal structure of the breakage–reunion domain of DNA gyrase. *Nature, 388*, 903.

Casanova, C., Lorente, J. A., Carrillo, F., Pérez-Rodríguez, E., & Núñez, N. (1989). Klebsiella pneumoniae liver abscess associated with septic endophthalmitis. *Archives of Internal Medicine, 149*, 1467–1467.

Chang, T.-T., Gish, R. G., de Man, R., Gadano, A., Sollano, J., Chao, Y.-C., Lok, A. S., Han, K.-H., Goodman, Z., & Zhu, J. (2006). A comparison of entecavir and lamivudine for HBeAg-positive chronic hepatitis B. *New England Journal of Medicine, 354*, 1001–1010.

Chatterjee, R., & Mitra, A. (2015). An overview of effective therapies and recent advances in biomarkers for chronic liver diseases and associated liver cancer. *International Immunopharmacology, 24*, 335–345.

Chen, C.-R., Malik, M., Snyder, M., & Drlica, K. (1996). DNA gyrase and topoisomerase IV on the bacterial chromosome: Quinolone-induced DNA cleavage. *Journal of Molecular Biology, 258*, 627–637.

Chen, S. H., Chan, N.-L., & Hsieh, T. S. (2013). New mechanistic and functional insights into DNA topoisomerases. *Annual Review of Biochemistry, 82*, 139–170.

Cheung, S. T., Cheung, P. F., Cheng, C. K., Wong, N. C., & Fan, S. T. (2011). Granulin-epithelin precursor and ATP-dependent binding cassette (ABC) B5 regulate liver cancer cell chemoresistance. *Gastroenterology, 140*(344–355), e2.

Cole, M., Unemo, M., Hoffmann, S., Chisholm, S., Ison, C., & van de Laar, M. (2011). The European gonococcal antimicrobial surveillance programme, 2009. *Eurosurveillance, 16*, 19995.

Colonno, R. J., Rose, R. E., Pokornowski, K., Baldick, C. J., Klesczewski, K., & Tenney, D. (2006). Assessment at three years shows high barrier to resistance is maintained in entecavirtreated nucleoside naive patients while resistance emergence increases over time in lamivudine refractory patients. *Hepatology, 44*, 229A-230A.

Das, K., Xiong, X., Yang, H., Westland, C. E., Gibbs, C. S., Sarafianos, S. G., & Arnold, E. (2001). Molecular modeling and biochemical characterization reveal the mechanism of hepatitis B virus polymerase resistance to lamivudine (3TC) and emtricitabine (FTC). *Journal of Virology, 75*, 4771–4779.

de Clercq, E. (2002). Strategies in the design of antiviral drugs. *Nature Reviews Drug Discovery, 1*, 13.

de Clercq, E. (2007). The acyclic nucleoside phosphonates from inception to clinical use: historical perspective. *Antiviral Research, 75*, 1–13.

Dean, M. (2009). ABC transporters, drug resistance, and cancer stem cells. *Journal of Mammary Gland Biology and Neoplasia, 14*, 3–9.

Dean, M., Hamon, Y., & Chimini, G. (2001). The human ATP-binding cassette (ABC) transporter superfamily. *Journal of Lipid Research, 42*, 1007–1017.

Delaney, W. E., Ray, A. S., Yang, H., Qi, X., Xiong, S., Zhu, Y., & Miller, M. D. (2006). Intracellular metabolism and in vitro activity of tenofovir against hepatitis B virus. *Antimicrobial Agents and Chemotherapy, 50*, 2471–2477.

Delaney, W. E., Yang, H., Westland, C. E., Das, K., Arnold, E., Gibbs, C. S., Miller, M. D., & Xiong, S. (2003). The hepatitis B virus polymerase mutation rtV173L is selected during lamivudine therapy and enhances viral replication in vitro. *Journal of Virology, 77*, 11833–11841.

Diniz, C., Arantes, R., Cara, D., Lima, F., Nicoli, J., Carvalho, M., & Farias, L. (2003). Enhanced pathogenicity of susceptible strains of the Bacteroides fragilis group subjected to low doses of metronidazole. *Microbes and Infection, 5*, 19–26.

Diniz, C. G., Santos, S. G., Pestana, A. C. N., Farias, L. M., & Carvalho, M. A. R. (2000). Chromosomal breakage in the B. fragilis group induced by metronidazole treatment. *Anaerobe, 6*, 149–153.

Drlica, K., & Zhao, X. (1997). DNA gyrase, topoisomerase IV, and the 4-quinolones. *Microbiology and Molecular Biology Reviews, 61*, 377–392.

Eliopoulos, G. M. (2004). Quinolone resistance mechanisms in pneumococci. *Clinical Infectious Diseases, 38*, S350–S356.

Eliopoulos, G. M., & Gold, H. (2001). Vancomycin-resistant enterococci: mechanisms and clinical observations. *Clinical Infectious Diseases, 33*, 210–219.

Endicott, J. A., & Ling, V. (1989). The biochemistry of P-glycoprotein-mediated multidrug resistance. *Annual Review of Biochemistry, 58*, 137–171.

Evers, S., & Courvalin, P. (1996). Regulation of VanB-type vancomycin resistance gene expression by the VanS (B)-VanR (B) two-component regulatory system in Enterococcus faecalis V583. *Journal of Bacteriology, 178*, 1302–1309.

Falagas, M. E., Koletsi, P. K., & Bliziotis, I. A. (2006). The diversity of definitions of multidrug-resistant (MDR) and pandrug-resistant (PDR) Acinetobacter baumannii and Pseudomonas aeruginosa. *Journal of Medical Microbiology, 55*, 1619–1629.

Fang, C.-T., Chen, Y.-C., Chang, S.-C., Sau, W.-Y., & Luh, K.-T. (2000). Klebsiella pneumoniae meningitis: Timing of antimicrobial therapy and prognosis. *QJM, 93*, 45–53.

Fantappiè, O., Solazzo, M., Lasagna, N., Platini, F., Tessitore, L., & Mazzanti, R. (2007). P-glycoprotein mediates celecoxib-induced apoptosis in multiple drug-resistant cell lines. *Cancer Research, 67*, 4915–4923.

Friedman, S. M., Lu, T., & Drlica, K. (2001). Mutation in the DNA gyrase A gene ofEscherichia coli that expands the quinolone resistance-determining region. *Antimicrobial Agents and Chemotherapy, 45*, 2378–2380.

Fung, C., Chang, F., Lee, S., Hu, B., Kuo, B. I., Liu, C., Ho, M., & Siu, L. (2002). A global emerging disease of Klebsiella pneumoniae liver abscess: Is serotype K1 an important factor for complicated endophthalmitis? *Gut, 50*, 420–424.

Gal, M., & Brazier, J. (2004). Metronidazole resistance in Bacteroides spp. carrying nim genes and the selection of slow-growing metronidazole-resistant mutants. *Journal of Antimicrobial Chemotherapy, 54*, 109–116.

Gerhardt, C. (1842). Untersuchungen über die organischen Basen. *European Journal of Organic Chemistry, 42*, 310–313.

Gewirtz, D. (1999). A critical evaluation of the mechanisms of action proposed for the antitumor effects of the anthracycline antibiotics adriamycin and daunorubicin. *Biochemical Pharmacology, 57*, 727–741.

Ghany, M., & Liang, T. J. (2007). Drug targets and molecular mechanisms of drug resistance in chronic hepatitis B. *Gastroenterology, 132*, 1574–1585.

Hadziyannis, S. J., Tassopoulos, N. C., Heathcote, E. J., Chang, T. T., Kitis, G., Rizzetto, M., Marcellin, P., Lim, S. G., Goodman, Z., & Ma, J. (2006). Long-term therapy with adefovir dipivoxil for HBeAg-negative chronic hepatitis B for up to 5 years. *Gastroenterology, 131*, 1743–1751.

Hammes, W. P., & Neuhaus, F. C. (1974). On the mechanism of action of vancomycin: inhibition of peptidoglycan synthesis in Gaffkya homari. *Antimicrobial Agents and Chemotherapy, 6*, 722–728.

Hartman, B. J., & Tomasz, A. (1984). Low-affinity penicillin-binding protein associated with beta-lactam resistance in Staphylococcus aureus. *Journal of Bacteriology, 158*, 513–516.

Hiasa, H., & Shea, M. E. (2000). DNA gyrase-mediated wrapping of the DNA strand is required for the replication fork arrest by the DNA gyrase-quinolone-DNA ternary complex. *Journal of Biological Chemistry, 275*, 34780–34786.

Higgins, C. F. (1992). ABC transporters: From microorganisms to man. *Annual Review of Cell Biology, 8*, 67–113.

Hooper, D. C. (2003). Mechanisms of quinolone resistance. *Quinolone antimicrobial agents*, (3rd edn). American Society of Microbiology

Huang, H., Chopra, R., Verdine, G. L., & Harrison, S. C. (1998). Structure of a covalently trapped catalytic complex of HIV-1 reverse transcriptase: Implications for drug resistance. *Science, 282*, 1669–1675.

Hyde, S. C., Emsley, P., Hartshorn, M. J., Mimmack, M. M., Gileadi, U., Pearce, S. R., Gallagher, M. P., Gill, D. R., Hubbard, R. E., & Higgins, C. F. (1990). Structural model of ATP-binding proteing associated with cystic fibrosis, multidrug resistance and bacterial transport. *Nature, 346*, 362.

Ince, D., & Hooper, D. C. (2003). Quinolone resistance due to reduced target enzyme expression. *Journal of Bacteriology, 185*, 6883–6892.

Innaimo, S. F., Seifer, M., Bisacchi, G. S., Standring, D. N., Zahler, R., & Colonno, R. J. (1997). Identification of BMS-200475 as a potent and selective inhibitor of hepatitis B virus. *Antimicrobial Agents and Chemotherapy, 41*, 1444–1448.

Johnson, R. A., Ince, T. A., & Scotto, K. W. (2001). Transcriptional repression by p53 through direct binding to a novel DNA element. *Journal of Biological Chemistry, 276*, 27716–27720.

Kampranis, S. C., Bates, A. D., & Maxwell, A. (1999). A model for the mechanism of strand passage by DNA gyrase. *Proceedings of the National Academy of Sciences, 96*, 8414–8419.

Kane, M. (1995). Global programme for control of hepatitis B infection. *Vaccine, 13*(Suppl 1), S47–S49.

Kerr, D., Graham, J., Cummings, J., Morrison, J., Thompson, G., Brodie, M., & Kaye, S. (1986). The effect of verapamil on the pharmacokinetics of adriamycin. *Cancer Chemotherapy and Pharmacology, 18*, 239–242.

Khodursky, A. B., & Cozzarelli, N. R. (1998). The mechanism of inhibition of topoisomerase IV by quinolone antibacterials. *Journal of Biological Chemistry, 273*, 27668–27677.
King, R. W., Ladner, S. K., Miller, T. J., Zaifert, K., Perni, R. B., Conway, S. C., & Otto, M. J. (1999). Inhibition of Human Hepatitis B Virus Replication by AT-61, a Phenylpropenamide derivative, alone and in combination with (−) β-l-2′, 3′-Dideoxy-3′-Thiacytidine. *Antimicrobial Agents and Chemotherapy, 43*, 726.
Kohlstaedt, L., Wang, J., Friedman, J., Rice, P., & Steitz, T. (1992). Crystal structure at 3.5A resolution of HIV-1 reverse transcriptase complexed with an inhibitor. *Science, 256*, 1783–1790.
Kuo, M. T., Liu, Z., Wei, Y., Lin-Lee, Y.-C., Tatebe, S., Mills, G. B., & Unate, H. (2002). Induction of human MDR1 gene expression by 2-acetylaminofluorene is mediated by effectors of the phosphoinositide 3-kinase pathway that activate NF-κB signaling. *Oncogene, 21*, 1945.
Lai, C., Gane, E., Liaw, Y., Thongsawat, S., Wang, Y., Chen, Y., Heathcote, J., Rasenack, J., Bzowej, N. & Naoumov, N. (2005b). Telbivudine (LdT) versus Lmivudine for chronic hepatitis B: First-year results from the international phase III GLOBE trial. *Hepatology*.
Lai, C.-H., Kuo, C.-H., Chen, P.-Y., Poon, S.-K., Chang, C.-S., & Wang, W.-C. (2006). Association of antibiotic resistance and higher internalization activity in resistant helicobacter pylori isolates. *Journal of Antimicrobial Chemotherapy, 57*, 466–471.
Lai, C.-L., Leung, N., Teo, E.-K., Tong, M., Wong, F., Hann, H.-W., Han, S., Poynard, T., Myers, M., & Chao, G. (2005). A 1-year trial of telbivudine, lamivudine, and the combination in patients with hepatitis B e Antigen—Positive chronic hepatitis B. *Gastroenterology, 129*, 528–536.
Land, K. M., & Johnson, P. J. (1999). Molecular basis of metronidazole resistance in pathogenic bacteria and protozoa. *Drug Resistance Updates, 2*, 289–294.
Lederman, E. R., & Crum, N. F. (2005). Pyogenic liver abscess with a focus on Klebsiella pneumoniae as a primary pathogen: An emerging disease with unique clinical characteristics. *The American Journal of Gastroenterology, 100*, 322.
Lee, K., & Chu, C. K. (2001). Molecular modeling approach to understanding the mode of action of L-nucleosides as antiviral agents. *Antimicrobial Agents and Chemotherapy, 45*, 138–144.
Leiros, H.-K.S., Kozielski-Stuhrmann, S., Kapp, U., Terradot, L., Leonard, G. A., & Mcsweeney, S. M. (2004). Structural Basis of 5-nitroimidazole antibiotic resistance the crystal structure of NimA from deinococcus radiodurans. *Journal of Biological Chemistry, 279*, 55840–55849.
Liang, T. J. (2009). Hepatitis B: The virus and disease. *Hepatology (Baltimore, Md.), 49*, S13-S21.
Liu, Y.-C., Yen, M.-Y., & Wang, R.-S. (1991). Septic metastatic lesions of pyogenic liver abscess: Their association with Klebsiella pneumoniae bacteremia in diabetic patients. *Archives of Internal Medicine, 151*, 1557–1559.
Locarnini, S., Qi, X., Arterburn, S., Snow, A., Brosgart, C., Currie, G., Wulfsohn, M., Miller, M., & Xiong, S. (2005). Incidence and predictors of emergence of HBV mutations associated with ADV resistance during 4 Years of ADV therapy for patients with chronic hepatitis B. *Journal of Hepatology, 42*, 17.
Löfmark, S., Fang, H., Hedberg, M., & Edlund, C. (2005). Inducible metronidazole resistance and nim genes in clinical Bacteroides fragilis group isolates. *Antimicrobial Agents and Chemotherapy, 49*, 1253–1256.
Manns, M. P., Czaja, A. J., Gorham, J. D., Krawitt, E. L., Mieli-Vergani, G., Vergani, D., & Vierling, J. M. (2010). Diagnosis and management of autoimmune hepatitis. *Hepatology, 51*, 2193–2213.
Mazzanti, R., Platini, F., Bottini, C., Fantappiè, O., Solazzo, M., & Tessitore, L. (2009). Down-regulation of the HGF/MET autocrine loop induced by celecoxib and mediated by P-gp in MDR-positive human hepatocellular carcinoma cell line. *Biochemical Pharmacology, 78*, 21–32.
Moore, J. M., & Salama, N. R. (2005). Mutational analysis of metronidazole resistance in Helicobacter pylori. *Antimicrobial Agents and Chemotherapy, 49*, 1236–1237.
Murphy, S. L., Xu, J. & Kochanek, K. D. (2013). Deaths: Final data for 2010.
Nagarajan, R. (1991). Antibacterial activities and modes of action of vancomycin and related glycopeptides. *Antimicrobial Agents and Chemotherapy, 35*, 605.
Ng, E. Y., Trucksis, M., & Hooper, D. C. (1996). Quinolone resistance mutations in topoisomerase IV: Relationship to the flqA locus and genetic evidence that topoisomerase IV is the primary

target and DNA gyrase is the secondary target of fluoroquinolones in Staphylococcus aureus. *Antimicrobial Agents and Chemotherapy, 40*, 1881–1888.

Ng, I. O., Liu, C. L., Fan, S. T., & Ng, M. (2000). Expression of P-glycoprotein in hepatocellular carcinoma: A determinant of chemotherapy response. *American Journal of Clinical Pathology, 113*, 355–363.

Nies, A. T., König, J., Pfannschmidt, M., Klar, E., Hofmann, W. J., & Keppler, D. (2001). Expression of the multidrug resistance proteins MRP2 and MRP3 in human hepatocellular carcinoma. *International Journal of Cancer, 94*, 492–499.

Ono, S. K., Kato, N., Shiratori, Y., Kato, J., Goto, T., Schinazi, R. F., Carrilho, F. J., & Omata, M. (2001). The polymerase L528M mutation cooperates with nucleotide binding-site mutations, increasing hepatitis B virus replication and drug resistance. *The Journal of Clinical Investigation, 107*, 449–455.

Osborn, M. T., & Chambers, T. C. (1996). Role of the stress-activated/c-Jun NH2-terminal protein kinase pathway in the cellular response to adriamycin and other chemotherapeutic drugs. *Journal of Biological Chemistry, 271*, 30950–30955.

Podschun, R., & Ullmann, U. (1998). Klebsiella spp. as nosocomial pathogens: Epidemiology, taxonomy, typing methods, and pathogenicity factors. *Clinical Microbiology Reviews, 11*, 589–603.

Pumbwe, L., Chang, A., Smith, R. L., & Wexler, H. M. (2007). BmeRABC5 is a multidrug efflux system that can confer metronidazole resistance in Bacteroides fragilis. *Microbial Drug Resistance, 13*, 96–101.

Pumbwe, L., Glass, D., & Wexler, H. M. (2006). Efflux pump overexpression in multiple-antibiotic-resistant mutants of Bacteroides fragilis. *Antimicrobial Agents and Chemotherapy, 50*, 3150–3153.

Reysset, G. (1996). Genetics of 5-nitroimidazole resistance in Bacteroides species. *Anaerobe, 2*, 59–69.

Roos, W. P., & Kaina, B. (2013). DNA damage-induced cell death: From specific DNA lesions to the DNA damage response and apoptosis. *Cancer Letters, 332*, 237–248.

Ros, J. E., Libbrecht, L., Geuken, M., Jansen, P. L., & Roskams, T. A. (2003). High expression of MDR1, MRP1, and MRP3 in the hepatic progenitor cell compartment and hepatocytes in severe human liver disease. *The Journal of Pathology, 200*, 553–560.

Sander, P., Papavinasasundaram, K., Dick, T., Stavropoulos, E., Elliott, K., Springer, B., Colston, M. J., & Böttger, E. C. (2001). Mycobacterium bovis BCG recADeletion mutant shows increased susceptibility to DNA-damaging agents but wild-type survival in a mouse infection model. *Infection and Immunity, 69*, 3562–3568.

Scala, S., Budillon, A., Zhan, Z., Cho-Chung, Y. S., Jefferson, J., Tsokos, M., & Bates, S. E. (1995). Downregulation of mdr-1 expression by 8-Cl-cAMP in multidrug resistant MCF-7 human breast cancer cells. *Journal of Clinical Investigation, 96*, 1026–1034.

Schildgen, O., Sirma, H., Funk, A., Olotu, C., Wend, U. C., Hartmann, H., Helm, M., Rockstroh, J. K., Willems, W. R., & Will, H. (2006). Variant of hepatitis B virus with primary resistance to adefovir. *New England Journal of Medicine, 354*, 1807–1812.

Scotto, K. W. (2003). Transcriptional regulation of ABC drug transporters. *Oncogene, 22*, 7496.

Seeger, M. A. & van Veen, H. W. (2009). Molecular basis of multidrug transport by ABC transporters. *Biochimica et Biophysica Acta (BBA)-Proteins and Proteomics, 1794*, 725–737.

Sharara, A. I., & Rockey, D. C. (2002). Pyogenic liver abscess. *Current Treatment Options in Gastroenterology, 5*, 437–442.

Shen, D., Lu, Y., Chin, K., Pastan, I., & Gottesman, M. (1991). Human hepatocellular carcinoma cell lines exhibit multidrug resistance unrelated to MRD1 gene expression. *Journal of Cell Science, 98*, 317–322.

Shiraga, K., Sakaguchi, K., Senoh, T., Ohta, T., Ogawa, S., Sawayama, T., Mouri, H., Fujiwara, A., & Tsuji, T. (2001). Modulation of doxorubicin sensitivity by cyclosporine A in hepatocellular carcinoma cells and their doxorubicin-resistant sublines. *Journal of Gastroenterology and Hepatology, 16*, 460–466.

Solazzo, M., Fantappiè, O., D'amico, M., Sassoli, C., Tani, A., Cipriani, G., Bogani, C., Formigli, L. & Mazzanti, R. (2009). Mitochondrial expression and functional activity of breast cancer resistance protein in different multiple drug-resistant cell lines. *Cancer Research, 69*, 7235–7242

Swift, L. P., Rephaeli, A., Nudelman, A., Phillips, D. R., & Cutts, S. M. (2006). Doxorubicin-DNA adducts induce a non-topoisomerase II–mediated form of cell death. *Cancer Research, 66*, 4863–4871.

Tenney, D. (2003). Hepatitis B virus resistance to entecavir in volves novel changes in the viral polymerase. *Hepatology, 40*, 245A.

Tenney, D., Levine, S., Rose, R., Walsh, A., Weinheimer, S., Discotto, L., Plym, M., Pokornowski, K., Yu, C., & Angus, P. (2004). Clinical emergence of entecavir-resistant hepatitis B virus requires additional substitutions in virus already resistant to Lamivudine. *Antimicrobial Agents and Chemotherapy, 48*, 3498–3507.

Tewey, K., Rowe, T., Yang, L., Halligan, B., & Liu, L. (1984). Adriamycin-induced DNA damage mediated by mammalian DNA topoisomerase II. *Science, 226*, 466–468.

Theron, M. M., Janse Van Rensburg, M. N., & Chalkley, L. J. (2004). Nitroimidazole resistance genes (nimB) in anaerobic gram-positive cocci (previously Peptostreptococcus spp.). *Journal of Antimicrobial Chemotherapy, 54*, 240–242.

Tsend-Ayush, A., Zhu, X., Ding, Y., Yao, J., Yin, L., Zhou, J., & Yao, J. (2017). Lactobionic acid-conjugated TPGS nanoparticles for enhancing therapeutic efficacy of etoposide against hepatocellular carcinoma. *Nanotechnology, 28*, 195602.

Ueda, K., Cardarelli, C., Gottesman, M. M., & Pastan, I. (1987). Expression of a full-length cDNA for the human "MDR1" gene confers resistance to colchicine, doxorubicin, and vinblastine. *Proceedings of the National Academy of Sciences, 84*, 3004–3008.

van Bömmel, F., Wünsche, T., Mauss, S., Reinke, P., Bergk, A., Schürmann, D., Wiedenmann, B., & Berg, T. (2004). Comparison of adefovir and tenofovir in the treatment of lamivudine-resistant hepatitis B virus infection. *Hepatology, 40*, 1421–1425.

van Bömmel, F., Zöllner, B., Sarrazin, C., Spengler, U., Hüppe, D., Möller, B., Feucht, H. H., Wiedenmann, B., & Berg, T. (2006). Tenofovir for patients with lamivudine-resistant hepatitis B virus (HBV) infection and high HBV DNA level during adefovir therapy. *Hepatology, 44*, 318–325.

Warner, N. (2003). Molecular modeling of entecavir resistant mutations in the hepatitis B virus polymerase selected during therapy. *Hepatology, 40*, 245A.

Weber, O., Schlemmer, K.-H., Hartmann, E., Hagelschuer, I., Paessens, A., Graef, E., Deres, K., Goldmann, S., Niewoehner, U., & Stoltefuss, J. (2002). Inhibition of human hepatitis B virus (HBV) by a novel non-nucleosidic compound in a transgenic mouse model. *Antiviral Research, 54*, 69–78.

Westland, C. E., Yang, H., Delaney, W. E., Gibbs, C. S., Miller, M. D., Wulfsohn, M., Fry, J., Brosgart, C. L., & Xiong, S. (2003). Week 48 resistance surveillance in two phase 3 clinical studies of adefovir dipivoxil for chronic hepatitis B. *Hepatology, 38*, 96–103.

Whittle, G., Shoemaker, N., & Salyers, A. (2002). The role of Bacteroides conjugative transposons in the dissemination of antibiotic resistance genes. *Cellular and Molecular Life Sciences CMLS, 59*, 2044–2054.

Willmott, C., & Maxwell, A. (1993). A single point mutation in the DNA gyrase A protein greatly reduces binding of fluoroquinolones to the gyrase-DNA complex. *Antimicrobial Agents and Chemotherapy, 37*, 126–127.

Wilson, T., Johnston, P., & Longley, D. (2009). Anti-apoptotic mechanisms of drug resistance in cancer. *Current Cancer Drug Targets, 9*, 307–319.

Yadav, V., & Chu, C. K. (2004). Molecular mechanisms of adefovir sensitivity and resistance in HBV polymerase mutants: A molecular dynamics study. *Bioorganic and Medicinal Chemistry Letters, 14*, 4313–4317.

Yang, H., Qi, X., Sabogal, A., Miller, M., Xiong, S., & Delaney, W. E. (2005). Cross-resistance testing of next-generation nucleoside and nucleotide analogues against lamivudine-resistant HBV. *Antiviral Therapy, 10*, 625.

Yang, H., Westland, C. E., Delaney, W. E., Heathcote, E. J., Ho, V., Fry, J., Brosgart, C., Gibbs, C. S., Miller, M. D., & Xiong, S. (2002). Resistance surveillance in chronic hepatitis B patients treated with adefovir dipivoxil for up to 60 weeks. *Hepatology, 36*, 464–473.

Zhao, X., Xu, C., Domagala, J., & Drlica, K. (1997). DNA topoisomerase targets of the fluoroquinolones: A strategy for avoiding bacterial resistance. *Proceedings of the National Academy of Sciences, 94*, 13991–13996.

Chapter 13
Drug Resistance in Cancer

Batool Fatima, Maryam Jamil, Dilshad Hussain, Adeela Saeed, Fahmida Jabeen, Muhammad Salman Sajid, and Muhammad Najam-ul-Haq

Abstract Cancer drug resistance is one of the obstacles in cancer treatment thus leading cause of death in cancer. Drug resistance in cancer means the evolution of tumor cells against therapeutic strategies. In each kind of resistance, every time there is a different pattern of resistivity. Two basic modes of resistance are found in the neoplastic cells that can be inherent or acquired. A large number of factors are influencing these two modes including mutations, efflux and influx variations, tumor cell heterogeneity, microenvironment of the cancer stem cells, microRNAs, long non-coding regions of RNAs, apoptotic failure, and many others. Evolution in tumor cells may be a result of increased ability of DNA repair, alteration in target molecules, involvement of kinase inhibitor, or topoisomerase inhibitors inactivity. It is the need of the hour to utilize other fields of science to avoid cancer drug resistance. Advanced knowledge of bioinformatics, genomics, proteomics, and nanotechnology is the way forward. Drugs with increased efficacy, decreased efflux pumps,high specificity, and targeted effects are required to be developed. Nanocarriers can play a role to avoid cancer drug resistance. Nano dosage of cancer drugs can prove more specific, lesstoxic, and can increase therapeutic effects.

B. Fatima (✉) · M. Jamil
Department of Biochemistry, Bahauddin Zakariya University, Multan 60800, Pakistan
e-mail: batoolfatima@bzu.edu.pk

D. Hussain
International Center for Chemical and Biological Sciences, HEJ Research Institute of Chemistry, University of Karachi, Karachi 75270, Pakistan

A. Saeed · F. Jabeen
Department of Chemistry, The Women University Multan, Multan, Pakistan

M. S. Sajid · M. Najam-ul-Haq
Institute of Chemical Sciences, Bahauddin Zakariya University, Multan 60800, Pakistan

S. Ahmed et al. (eds.), *Biochemistry of Drug Resistance*,
https://doi.org/10.1007/978-3-030-76320-6_13

13.1 Introduction

13.1.1 Drug Resistance in Cancer

Drugs that previously respond to a certain type of malignancy become unable to function properly in some cases. This is known as drug resistance. The same thing happens during the chemotherapies of cancer that fail cancer treatment (Pastan & Gottesman, 1987). Two general factors are involved in the failure of drug therapy in cancer. One of them is the host response and the other is the specific epigenetic alterations of the target. Decreased absorption of elevated metabolism (efflux) resulting in low serum accumulation, poor bearing capacity especially in aged hosts, doses below optimal levels, transportation failure of drugs are the common factors influencing drug resistance. Every patient has its genetic makeup that determines the metabolism of the drug depending on the origin of cancer. The specific phenotype of patients also determines the level of activation of oncogenes and tumor suppressor genes. As a result, every patient responds differently to a drug. To overcome the metabolic failure of a single drug, a combination of structurally unrelated drugs is given to the patient. This has solved the problem of drug failure to some extent. In some cases, the tumors get resistant to drugs causing multiple drug resistance (MDR). MDR occurs due to changes that reduce the serum level of drugs within cells. They do so by limiting uptake, affecting the permeability of membrane lipids such as ceramide. Ultimately the programmed cell death that was to be done by the drugs does not occur and the common detoxification of drug goes on leading to cancer cell survival (Gottesman, 2002).There are two main types of therapeutic failure in cancer.

- Intrinsic resistance
- Acquired resistance.

Intrinsic factors in some types of tumors are produced naturally by large quantities of chemicals that work against chemotherapeutic agents. The drug pathway involved in intrinsic drug resistance is difficult to describe and requires a lot of work to investigate. An experiment was conducted by taking a sample from treated tumor cell lines exposed to the anti-cancer drug, Doxorubicin. Feedback of the target cells towards treatment agents, the concentration of drugs within body fluid, increased metabolism rate of drug, rapid removal from the body, and elevated levels of protein synthesis that control apoptosis processes were observed (Gariboldi et al., 2003).

In acquired resistance tumors are targeted by the therapeutic agents which undergo mutations. These mutations lead to resistance in tumor cells against that drug. Acquired immunity is caused by factors such as activation of alternative mechanisms that transmit compensatory signals for developing adaptive features in tumors (Holohan et al., 2013). Drug resistance leading to treatment failure can occur due to 8 factors.

13.1.2 High Level of Drug Efflux

Chemotherapeutic agents may lose their functionality due to a major transport system known as ATP binding protein (Dean). They work by increasing the efflux rate of compounds that were introduced inside the cell to kill tumor cells. As a result, the drug can not perform its desired cytotoxicity inside stem cells. Work is being done to use ABC transporters inhibitors coupled with chemotherapeutic agents (Kathawala et al., 2015). For an organ, it is compulsory to maintain the solidity of stem cells. ATP binding transporters are the transporters that help stem cells for its interactivity. The same transporters are involved in tumor stem cell protection thus producing drug resistance. These transports work by the ATP-binding cassette (ABC) gene family. Various works over tumor cells' innate drug resistance have shown that ABC transporters are the causative agents for it. Therapy failure or even tumor cells repair may occur due to the expression of these transporters (Dean, 2009) (Fig. 13.1).

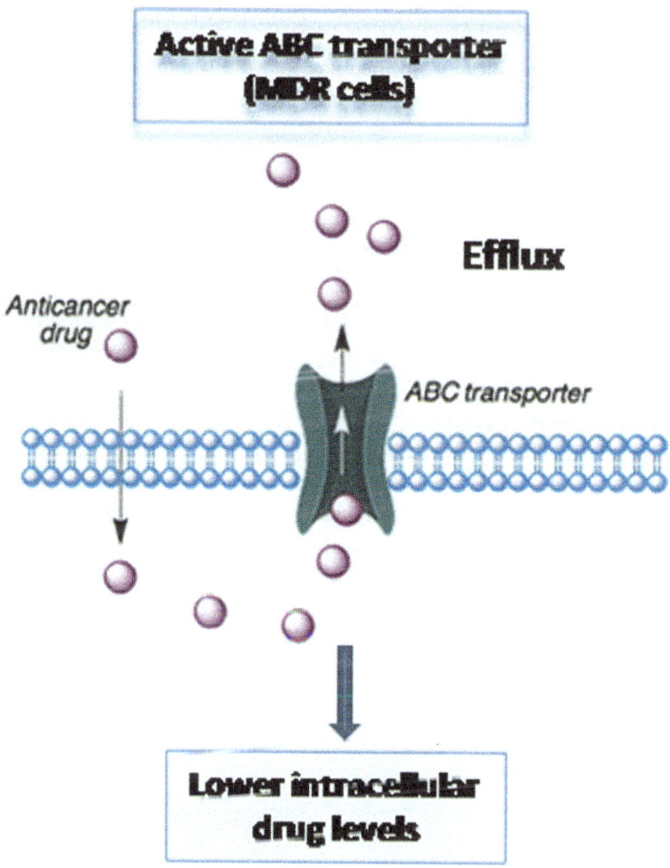

Fig. 13.1 The higher level of drug efflux

13.1.3 Lower Level of Drug Efflux

SLC transporters are the secondary carriers that influence the transportation of therapeutic agents against cancer and other mediators inside the cell across the outer membrane. ABC transporters are the major transporters as mentioned previously. Anticancer drugs such as antifolate, methotrexate (MTX) move inside the cell via hypo folate carrier (SLC transporters). However several mutations and downregulation of these transporters cause antifolate resistance (Joyce et al., 2015). Active transport is the main mode of absorption of the drug. Normal channeling of drugs and other such compounds is necessary for the maintenance of the internal system. For this purpose, various transporters such as active, passive, and vesicular transport (exosomes) are helpful. When the disease conditions such as cancer are encountered, abnormal trafficking of extracellular compounds occurs. It leads to the misexpression of a variety of proteins and microRNAs. One of these hurdles in the transportation of drug across the membrane is exosome-mediated expulsion. It results in therapeutic failure by not letting them into the cell or performing them their proper action. Studies have shown that cancer chemotherapies may remain ineffective due to desmoplastic reaction by increased fibroblastic growth that occurs due to exosome-secreted proteins. This is known as exosome-mediated drug resistance (Azmi et al., 2013) (Fig. 13.2).

13.1.4 Mutations in Tumor Cells (Epigenetic Alteration)

Intratumor heterogeneity (ITH) is another factor producing therapeutic failure. It gives evolutionary genetic makeup to the cells. ITH also helps in revealing the molecular mechanisms and cancer cells' genomic machinery (McGranahan & Swanton, 2017). As are the result of acquired resistance, the formation of resistant clones may occur in some cases. These clones may be expressed in the previous model of cancer with low intensity but at the next-generation, they appear with tremendous frequency as a response to targeted therapies. Thus intratumor heterogeneity may be a leading cause of drug resistance in cancer (Burrell & Swanton, 2014). Deep sequencing of tumor cells' genes provide san approach to the evolutionary dynamics. It reveals drug resistance and tumor relapse. Spatial genetic heterogeneity involving the branching pattern of subclones produces therapeutic implications. Studies have shown that targeted therapy is done for mutations that are dominant in the tumor within the founder clone. Whenever new mutations occur, they evolve with resistance to the therapy (Easwaran et al., 2014).

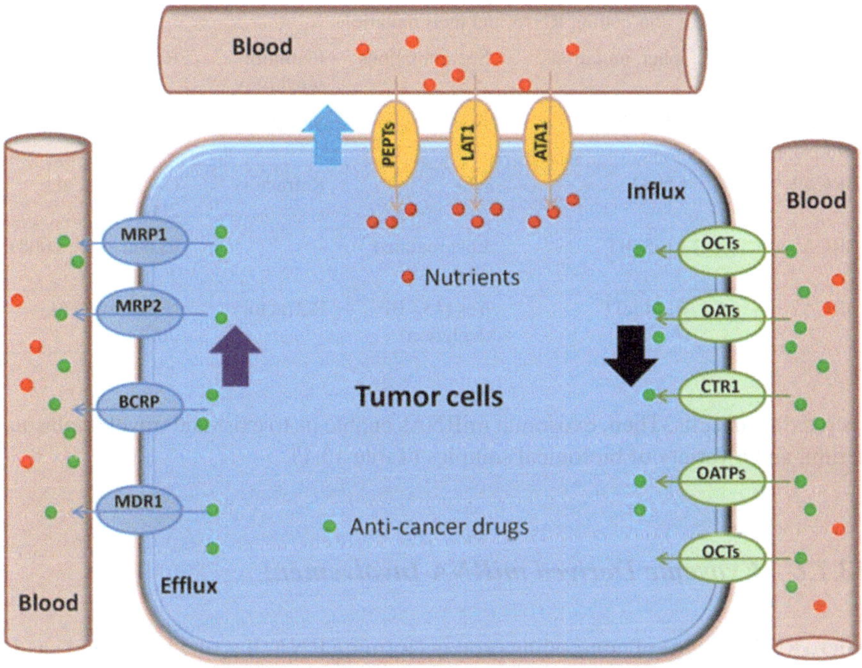

Fig. 13.2 Illustration of a lesser number of SLC transporters expressed and a greater number of ABC transporters influencing influx and efflux rate respectively (Li & Shu, 2014).

13.1.5 Non-coding RNA Involvement

The central dogma of protein synthesis is followed by the stepwise reading of DNA into RNA and RNA into proteins is carried by non-coding regions of RNA known as non-coding RNA. They resemble mRNA structurally. They are involved in different metabolic reactions. These RNAs control gene expression according to the response of encountered stimuli. When non-coding RNAs are deregulated, disorders such as responsiveness to the drugs, loss of sensitivity, and malignancies may occur (Majidinia & Yousefi, 2016). Genetic heterogeneity within tumor cells influences the results of chemotherapy for curing the same type of cancer. As gene expression is governed by non-coding RNAs, it also influences cell division. Thus non-coding RNAs can be regarded as a marker for the indication of malignancies, even can predict the stage of cancer (Malek et al., 2014). There are two main pathways in which circulating miRNAs can be released from donor cells into body fluids. The first, most recently discovered pathway, occurs via the conjugation of miRNAs with RNA-binding proteins, such as Ago2. The complex can be released from the cell and circulates in body fluids. The more investigated pathway occurs through the budding of membranes and the creation of small membrane vesicles called exosomes that are

Table 13.1 Detail of non-coding RNAs and their mutations

Mutated gene	Inher. mutations	Sec. mutations	Platinum sensitivity	References
BRCA1	E1250X	No	Sensitive	(Swisher et al., 2008)
BRCA1	L598X	NO	Refractory	(Swisher et al., 2008)
BRCA2	8765dlIT	Lost mutated allele	–	(Sakai et al., 2008)
BRCA2	6174delT	Yes (137 BP depletion)	Refractory	(Edwards et al., 2008)

secreted from cells. Then, exosomal miRNAs enable us to exist and travel in plasma, serum, and a variety of biological samples (Table 13.1).

13.1.6 Exosome Derived miRNA Involvement

Micro RNA is a type of non-coding proteins forming RNA. It is tiny in size and has 19 to 25nucleotides. It is synthesized by a congenital way or transcription such as protein by capping, polymerization, and splicing. The expression of miRNA is involved in the development of resistance, particularly in cancer. Such type of miRNA induced resistance is found in breast cancer, colon cancer, ovarian cancer, lung cancer, neuroblastoma, Squamous cell carcinoma, and gastric cancer (Ma et al., 2010). miRNA governs many biological processes of the cell such as cell cycle, division, and proliferation. It can access tumor cells inside the body as well as shows the therapeutic response in tumor cells. It is also involved in cell survival during the therapies used for cancer treatment. It thus develops resistance in the target cells. More or less expression of miRNA produces resistance in the tumor cells. It is revealed that six types of miRNA are either down-regulated or up-regulated in the cells showing resistance to therapies. i.e. one is down regulated in one type of cancer and up-regulated in another. For example, MCSF celllines show reduced expression of 130a miRNA which produces chemo resistance in these cells (Ma et al., 2010) (Table 13.2).

Table 13.2 Drug resistance due to miRNA involved in cancer

miRNA	Expression level	Cancer type	Resistance against	References
miRNA 451	Altered	Breast cancer	Doxorubicin	(Kovalchuk et al., 2008)
miRNA 200c	Downregulated	Breast cancer	Epirubicin	(Chen et al., 2012)
miRNA 130a	Up-regulated	Ovarian cancer	Cisplatin	(Yang et al., 2012)

13.1.7 Apoptotic Arrest

Chemotherapeutic agents work either by interrupting the normal functioning of a tumor cell or by the physiological death of tumor cells. Programmed cells known as apoptosis works by the proenzymes known as zymogens. These proenzymes get activated by the induction of specific cell surface receptors "death receptors" due to mitochondrial alterations and various such types of metabolic reactions. Some inhibitors often present in cancer cells seize the apoptotic machinery that is activated by the anticancer agents. As a result of apoptotic inhibition caused by IAP (Inhibitor of apoptotic proteins), cells continue to survive even under the treatment with chemotherapies. This leads to cancer drug resistance (Kaufmann & Vaux, 2003). Galectins area family of 14 participants of B-galactoside binding protein involved in cell survival and metabolic activities. It is involved in anti-apoptotic functions causing tumor drug resistance (Fukumori et al., 2007).

For the activation of the apoptotic pathway, there is a threshold determined by the relative ratio of pro- to antiapoptotic members of Bc1-2 and Bcl-xL family. When a cell is Null, Bax-mitochondrial regulated apoptosis does not occur. Apoptosis is reduced due to overexpression of Bcl-xL or Bcl-2 and its low expression leads to apoptotic damage to DNA (Pommier et al., 2004) (Fig. 13.3).

The mantle cell lymphoma model is used for the description of autophagy and related phenomenon. This study reveals that two interlinked systems of TGM2-NFKB (transglutaminase) signaling pathways and interleukin 6 are particular participants in MCL. In this type of melanoma, drug resistance is developed in the cells

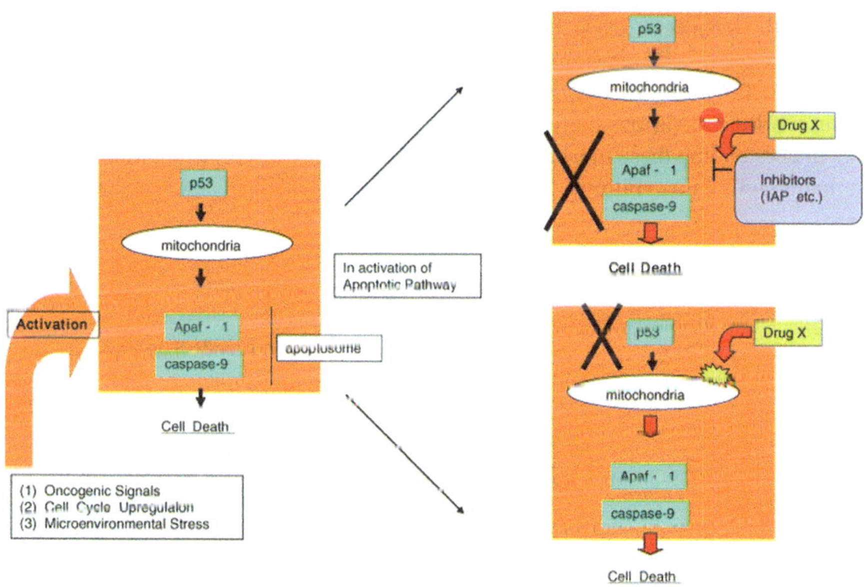

Fig. 13.3 Failure of apoptotic machinery (Mashima & Tsuruo, 2005)

by signal induction to TGM2-NFKB facilitated by the cytokine IL6. This causes an increase in Interleukin 6 secretion and imparts protection to the cancer cell from autophagy. Consequently, drug resistance is developed in MCL cells (Zhang & McCarty, 2017).

13.1.8 Drug Inactivation

CYP1B1 is a human protein present in tumor cells. It is a kind of cytochrome p450 that can metabolize xenobiotics and carcinogenic medicines. It alters the biology of chemotherapeutic agents. Flutamide is one of the well known anticancer agents which are biologically transformed by CYP1B1 in cancer cells. Thus CYP1B1 may resist the activity of anticancer causing therapeutic failure (Kim et al., 2006).

13.1.9 Multiple Drug Resistance

Tumor stem cells have self-repair abilities as compared to the normal tissues. Several environmental, as well as intrinsic properties of the cell, impart resistance in tumor cells against genotoxins. Certain kinds of proteins such as glycoprotein are expressed that resist the induced apoptosis, maintenance of cell integrity, and decreasing the rate of diffusion of cytotoxic agents. Blood stream is the intrinsic resistive features of the tumor cells. Many genetic changes also activate immune responses against therapeutic agents (Baguley, 2010).

When cancer cells resist the effects of anti-cancer drugs, varying in structures as well as the functions are known as multidrug resistance. It functions by the simultaneous working of various inhibitory pathways against the drugs. Studies have shown that understanding these mechanisms can be helpful to overcome the drug resistance in malignant tumors (Filipits, 2004) (Table 13.3).

Table 13.3 ABC transporters modulators leading to multidrug resistance (Kathawala et al., 2015)

Natural compounds	Curcumin, Gingerol, Epigallocatecingallate
Vascular growth Factor Receptor TKIs	Vandetanib, Sorafenib, Motesanib
Phosphodiesterase-5 inhibitors	Vardenafil, Sildenafil
Marine compounds	Ergosterol A, Bryostatin 1, Sipholanetriterpenoids
Miscellaneous TKIs	Crizotinib, Vemurafenib, Masitinib

When the number of confined regions of chromosome number increases, its called gene amplification. It is also found in tumors. Gene amplification is involved in acquired drug resistance (Albertson, 2006). A spontaneous rate of the application occurs in all eukaryotic cells. However, this amplification is enhanced manifolds by the action of cytotoxicity agents and inhibitors. In tumor cells, several extra chromosomal by products of gene amplifications provide conditions that enable tumor cells to survive in stressful conditions. Tumor cells get more malignant due to increased clonal populations of the cell. It ultimately leads to the proliferation of tumor cells which are more evolved to harmful conditions. Drug resistance in these tumor cells occurs due to the increased ability of genes to amplify in treated patients (Alitalo and Schwab, 1986).

In brain malignancies such as glioblastoma, DNA repair is a leading cause of therapeutic failure. Temozolomide and some other alkylating agents are given to the patient for treatment purposes. Studies have shown that in the case of glioblastomas, cells restore their activities with even an increased tendency of DNA damage recovery. Cell goes to rehabilitation by different mechanisms of DNA restoration. One of these mechanisms is restoring the structural abilities of alkylated base pairs of DNA. It leads to the development of resistance in tumor cells (Johannessen et al., 2008). A similar kind of DNA damage occurs in chronic lymphocytic leukemia (CLL). In this case, DNA recovery takes place due to an increased level of DNA-PK (DNA protein kinase) as a reaction to chlorambucil treatment. It is also revealed that this DNA repair fastens due to the activation of foci. These foci are correlated with rapid DNA repair and produce drug resistance against chlorambucil in CLL (Panasci et al., 2001).

13.2 Mechanisms of Cancer Drug Resistance

- Resistance due to anti-metabolites
- Resistance due to spindle poisons
- Resistance due to topoisomerase inhibitors
- Resistance due to drugs that target tyrosine kinase receptors and some other kinases (Rebucci & Michiels, 2013).

Targeted destroying of folates has been a mode of curing the disease for a long time. Antifaoltes are used for the purpose. Pemetrexed is a novel antifolate. It blocks thymidine synthase. Another antifolate that targets dihydrofolate reductase is pralatrexate. However antifolate resistance is a hurdle to chemotherapy for cancer. Transportation variation including changes in efflux and influx of antifolate is involved in antifolate resistance. By targeting antifolate, resistance can be helpful to deal with anticancer drug resistance (Gonen & Assaraf, 2012).

For regular mitotic cell division, synthesis of the mitotic spindle is compulsory. The formation of these spindle fibers takes place by the tubulin proteins. Microtubules formation occurs by the polymerization of tubulin. Microtubules are increasing in

Table 13.4 Drug Inhibition due to different tubulins (Wood et al., 2001)

Treating type of cancer	Spindle poisons	Affect	References
Acutelymphoblastic leukemia and Hodgkin's disease	Vincristine 16 and vinblastine 17	Inhibit the assembly of tubulin	(Hadfield et al., 2003)
Leukemia, breast cancer, and colorectal cancer, and ovarian cancer	Withaferin A (WA	Block spindle assembly checkpoint	(Das et al., 2014)
HeLa (cervical carcinoma), U2OS (osteosarcoma)	Nocodazole and taxol	Kinetochore attachment checkpoint (KAC), inhibit spindle microtubule assembly	(Brito & Rieder, 2009)

length during mitotic division. Spindle poisons are the drugs that target this tubulin to destroy the spindles. When these spindles are targeted by chemotherapeutic agents, the mitotic cell divisions are stopped resulting in tumor cell deaths. Before the complete decay of tumor cells, resistance may develop in the target tissues. This resistance is the result of increased efflux rate or alteration in tubulin expression. It is one of the leading hurdles in chemotherapeutic failure (Matson & Stukenberg, 2011) (Table 13.4).

Nuclear enzymes topoisomerase I and II are being used for the chemotherapy. These enzymes are crucial for cell functioning. So the drugs containing inhibitors of these enzymes (Topoisomerase II inhibitors) are developed for treating many kinds of tumors. These drugs have different modes of actions, metabolic rates, as well as a variety of mechanisms for resistance development. The elevated rate of detoxification and lowered activation rate of these drugs is one of the leading causes of cancer drug resistance. However, topoisomerase inhibitors have reliability as it can cure a wide range of tumors and not shown multidrug resistance (Sinha, 1995). Camptothecin (topoisomerase I inhibitor) or Amsacrine (topoisomerase II inhibitor) are the two drugs that are seen with resistance in leukemia. Studies have shown that the levels of these inhibitors are reduced in the targeted cells. Hypermethylation is also observed on the gene of inhibitors causing gene mutation and ultimately drug resistance in targeted cells (Tan et al., 1989).

TAM is a family of tyrosine kinase receptors consisting of three members Tyro3, Axl, and Mer TK. This family of proteins is involved in tumorigenesis. It also shows tumor drug resistance by acting as a target against therapeutic agents. By targeted blockage of signaling through these receptors, one can achieve inhibition of biological activities in tumor cells resulting in reduced chances of drug resistance (Vouri & Hafizi, 2017). Epidermal growth factor receptor (EGFR) tyrosine kinase inhibitor (TKIs) in non-small-cell lung cancer (NSCLC) has antagonist activity in advance stages when given to patients with chemotherapy. Preclinical analysis has shown that EGFR TKIs produce cell cycle inhibition in that cell having wild type EGFR leading to decreased activity of chemotherapeutic drugs. In NSLCS, tumor cells mostly consist of wild-type EGFR. When a particular signal of EGFR TKI comes in contact with a chemotherapeutic agent, it reduces the effectiveness of the drug in tumor cells (Davies et al., 2006).

Two most common cancer types found to be frequent resistance to chemotherapy are.

13.2.1 Ovarian Cancer Drug Resistance

Ovarian cancer is the cancer of adult females. Various chemotherapeutic strategies have been developed for the treatment of this cancer. Alkylating agents, natural products, and platinum complexes are widely used as a drug for chemotherapy. However, with the gradual exposure of tumors with these chemotherapeutic agents, resistance is developed. Following levels of resistance are found during chemotherapy:

- The development of membrane P-170 glycoprotein causes resistance. This resistance makes natural biochemical agents useless in chemotherapy.
- Higher concentrations of glutathione (Wong et al.) and glutathione S-transferase are involved in the removal of alkylation agents and platinum complexes from the body rapidly.
- Increasing the dosage of platinum complexes is also a major cause of drug resistance.
- Tumor cells DNA may develop the ability of repair leading to chemotherapy failure in the case of alkylation agent and platinum complexes (Johnson et al., 1993).

13.2.2 Prostate Cancer Drug Resistance

Prostate cancer has tumors that are androgen-dependent at the early stages. It is one of those cancers which gain drug resistance earlier. Specific forms of P450, epoxide hydrolase and different forms of GST are involved in prostate cancer. GSTS overexpression is one of the major causes of producing therapeutic failure. GST has three types which are GST-a, GST-p, and GST-n found in normal prostate cells. However, GST-n is indicated as a marker for prostate cancer as it is not expressed in prostate cells with malignancy, even the cell lines of cancer are having GST-n. Moreover, the two important forms of P450 i.e. CYP 1A and CYP3A make anti-cancer drugs useful for prostate cancer "anti androgen drug flutamide" more effective. CYPIA has a particular anti-androgen effect in prostate cancer. Various forms of P450 are involved in the metabolic activation of this drug leading to drug resistance in prostate cancer (Murray et al., 1995).

13.2.3 Breast Cancer Drug Resistance

In breast cancer, ABC transporters strongly affect the drug resistance proteins by changing the drug metabolism including efflux, influx, and removal from the body (Ross & Nakanishi, 2010). Breast cancer resistance protein (BCRP and ABCG2) are ABC transports involved in breast cancer drug resistance. BCRP play a crucial role in coping with all kind of invaders such as toxic agents at hut brain and placenta also. BCRP maintains the level of folates and heame in the cell as well as helps in the survival of normal stem cells to withstand the oxygen-deficient environment making a cell more resistant to encounter severe conditions. These characteristics of BCRP enable the cancer cells to withstand the hypoxic conditions. BCRP treats chemotherapeutic agents as undesired xenobiotics, causing resistance in tumor cells of breast cancer (Natarajan et al., 2012). Growth inhibition of breast cancer MCF-7 cells by the application of doxorubicin is studied widely. Doxorubicin functionality is found to be effective for Bcl-2 and Raf-1 mainly. An elevated level of Bcl-2 expression produces resistances against doxorubicin. Similarly Raf-1 increases level decreases the sensitivity of breast cancer cells for doxorubicin. However, the Raf-1 induced breast cancer drug resistances against doxorubicin is less than that of Bcl-2 over expression (Davis et al., 2003).

13.2.4 Colon Cancer Drug Resistance

Like other types of cancer drug resistance, colon cancer drug resistance also involves the multiple steps of the ABC transporters pathway. Various types of ABC proteins are found to be over-activated that increases the efflux rate of the drugs. Certain repressors and modulators are given to enhance the absorbance of therapeutic drugs against colon cancer. Another factor of colon cancer drug resistance is a delay in apoptosis. To deal with drug resistance due to blocked apoptosis, BH3 is given to induce cell death via autophagy in tumor cells (Hu et al., 2016). CD133 (glycoprotein) is a marker of colon cancer that is involved in the growth of the tumor cells.CD133 and IL-4 cytokine association study reveal that IL-4 cytokine helps CD133 in making cancer step cells to withstand against apoptotic cell death caused by chemotherapeutic agents. Thus IL-4 inhibitors are also necessarily should be prescribed to avoid colon cancer drug resistance (Todaro et al., 2008).

13.2.5 Lung Cancer Drug Resistance

Drug resistance during lung cancer treatment is a result of acquired and inherent resistance. Genomic study of lung cancer stem cells reveals that genetic alterations leading to biochemical changes in the structure of the tumor cells make them more

Table 13.5 Some chemotherapeutic drugs and their respective resistant tumors

Drug	Resistant tumors	References
Doxorubicin	Ovarian and colon cancer	(Goldsmith et al., 1995)
Enzalutamide	Prostate cancer	(Salami et al., 2018)
Adriamycin and Vincristine	Lung cancer	(Twentyman et al., 1987)
Cisplatin	Testicular cancer	(Koster et al., 2010)

resistant to chemotherapy (Nishio et al., 1999). A few types of lung cancer have exhibited innate drug resistance, even at the time of first exposure to the chemotherapeutic agents produces resistance in tumor cells (Kim, 2016). Non-small cell lung cancer (N-SCLC) is naturally resistant to all kinds of drugs even without exposure. The participation of rapid DNA damage repair in Non-small cell lung cancer (N-SCLC) is keenly studied and rapid recovery of DNA is strongly associated with drug resistance in lung cancer (Zeng-Rong et al., 1995) (Table 13.5).

13.3 How to Combat Cancer Drug Resistance?

13.3.1 Targeted Autophagy

Autophagy is self-eating which means the destruction of tumor cells on its own. Hand by hand it is an acquired response of tumor cells for survival when there is a condition of stress such as drug therapy used in cancer. When a cell develops autophagy with the help of drugs, it becomes able to eat away its tumor cell. Similarly, the hindrance in autophagy may promote the growth of tumor cells. Thus a well-guided autophagy system can present a novel therapeutic way of treating cancer (Yang et al., 2011). Studies have shown that autophagy can influence both tumors spreading as well as tumor killing. Here two conditions can help cure cancer. One is to block autophagy in normal cells and the other is an elevated rate of autophagy in tumor cells. So the available extracellular conditions faced by the cells and stage of cancer, the strategy will be selected either to inhibit or promote autophagy. For example in the early stages of cancer, autophagy should be increased while at the spreading stage of cancer, autophagy has to be ceased. Some tumor-promoting factors and tumor-suppressing factors are involved in autophagy. These factors indicate the functioning of autophagy. miRNA and autophagy suppressers can be used to establish sensitivity in the desensitize tumor cells. Combined with cellular toxic drugs, autophagy can overcome drug resistance (Kumar et al., 2015).

13.3.2 Co-delivery of siRNA and Chemotherapeutic Drugs

MDR produces drug resistance due to two basic methods. One is elevated drug efflux through cell surface or by increased apoptotic arrest. When nano amounts of drugs using nanotechnology are delivered to tumor cells, these problems can be controlled. It happens by increasing the solubility of drugs or decreasing the number of toxins produced by them. When drugs are coated with the nanoparticles, its efflux rate is reduced. As a result, the level of the drug inside the target cell is increased. The same nano-encapsulations are used for the delivery of siRNA (silencing/short interfering RNA) that can reduce antiapoptotic actions in tumor cells. When siRNA is co-delivered with other drugs, it prevents MDR. This nano-coating protects the siRNA from degradation by the action of RNAase. It also prevents them from renal clearance (Tsouris et al., 2014). However studies are needed to understand the nanotechnology-based drug delivery, their mode of action and effective dosage cope up with multidrug resistance (Gao et al., 2012).

13.3.3 Combinational Drug Therapy

Combinational drug treatment is one of the reliable methods used for cancer treatment. To combat drug resistance and lowered drug toxicity on MDR, nanotechnology-based drugs are given simultaneously. The targeted and particular dosage of chemotherapeutic drugs in this combination produces beneficial therapeutic anticancer effects with reduced drug-associated toxicity. It is one of the advanced ways to overcome drug resistance (Parhi et al., 2012).

13.3.4 Therapeutic Drugs Usage

Quinoline derivatives consisting of 24 newly synthesized members are under study in contrast to the multidrug resistance. This class of compounds has rings having aryl groups in hydrophobic regions that deviate from the normal plane is a completely active state. Thus allowing bonding with p-170 glycoprotein and reversing the drug resistance. A similar type of resistance reversal is found in piperazine. These compounds can inverse the resistance response in tumors. These drugs do so by the elevated levels [3H] Vincristine in K562/ADM cells (Suzuki et al., 1997).

Anticancer drugs such as 7-ethyl-10-hydroxycamptothecin, mitoxantrone, and topotecan are failed in curing cancer due to the presence of two proteins, BCRP and ABCG2 that resist the curing capability when given against breast cancer. Acrylonitrile derivatives such as YHO-13177 and its prodrug YHO-13351 have shown the capability of reversing drug resistance induced by these proteins. These derivatives work by increasing the efficiency of SN-38, mitoxantrone, and topotecan given to the patient in the lung (A549/SN4),colon(HCT116/BCRP), and breast cancer. YHO-13177 increases the concentration of Hoechst 33342 within the cell. Hoechst 33342 is a substrate of BCRP.YHO-13177 is formed from YHO-13351 immediately after administration when given to mice. YHO-13351 combination with irinotecan lowers the spreading of the tumor as well as the growth of tumors in mice that are induced with BCRP-transduced murine leukemia. Not only the growth of the tumor is suppressed by this combination but also increases the lifetime of affected mice. So it is suggested that YHO-13177 and its prodrug YHO-13351 can prove the drug against cancer drug resistance induced by BCRP/ABCG2 (Yamazaki et al., 2011).

13.3.5 Use of Natural Compounds

Natural compounds are focused on minimizing the toxic effects of chemotherapy as well as to compete with cancer cells' drug resistance. Studies have shown that these compounds can slow down the gene expression of resistance proteins. They enhance the sensitivity of tumor cells. The natural compounds affect multiple cell lines. Many natural compounds including herbs compounds can overcome cancer cell drug resistance. Detailed studies are needed as various herbal extracts have proved effective against drug restart tumor cells. These combined natural and synthetic compounds can help reverse drug resistance (Kars et al., 2008). Nitrogen derivatives and natural alkaloids CBT-01® (tetrandrine) are also involved in cancer drug resistance. They do so by inhibition of P-GPefflux pump and some other gates of flow blockage (Joshi et al., 2017).

13.4 Proteomic and Genomic Analysis of Cancer Drug Resistance

Eternal cancer cells have been used for the genomic analysis of tumor cells. Even the effects of drugs as well as drug resistance can be estimated by these cell lines. Heterogeneity and rapid mutation effects of the drug on tumor cells can be predicted by genomic profiling such as NCI-cell lines in gene mapping of the genome. So, an abundant number of gene proofing can help study drug resistance (Hu & Zhang, 2016).

Drug resistance in breast cancer is one of the major causes of chemo failure. How it functions particularly is still a mystery. MCF-7 and its adriamycin-resistant counterpart MCF-7/ADR are the two factors of cancer cell lines. They are studied using nanoflow liquid chromatography and tandem mass tags (TMT) quantitative mass spectrometry. Bioinformatics tools are used to identify functionally active proteins and networks. All these techniques help to separate the normal and abnormal cells based on resistivity. 80 DEPs are the clusters of proteins that are different in their expression. Six of them show a different pathway of signaling. These DEPs are associated with chemo resistance, tumor spreading, metastasis, and apoptosis. These differentially expressed proteins are resistant to the drug and chosen for further analysis (Wang et al., 2015). Proteomic approaches can be used as biomarkers for chemotherapy resistance. Antibody microarray is a well discussed proteomic strategy (Hodgkinson et al., 2010).

13.5 Future Recommendations

Combinational therapy having a blocking capacity of energy supply must be brought into use. Energy supply is mandatory for the tumor cell that cannot be evaded, unlike any other factor that can be replaced. For the sake of energy accumulation, glucose is a more efficient source. Thus glucose inhibitors can prove more frequent tumor cells death as cancer cells begin to die rapidly in glucose deprivation. Tumor microenvironment (TME) can also reverse or avoid drug resistance in cancer. One of the major TME is pH gradient. As tumor cells develop acidic pH gradient in extracellular via reverse pH gradient, inhibitors of proton pumps can also prove beneficial for the tumor cell destruction by making them more sensitive to chemotherapeutic drugs (Wang et al., 2019). Another field that needs to be work upon is enhancing the cytotoxicity of chemotherapeutic drugs. By increasing the cytotoxic effects of the drug, reversing the drug resistance can be achieved. solid lipid nanoparticles (SLN) by solvent diffusion method can also give results (Miao et al., 2013). Combinational therapy using BRAF inhibitors co-administered with some inhibitors from BRAF/RTKs/MEK mixture of inhibitor families give dependable results. Similarly, dose continuity and discontinuity cycles for cancer cells can be helpful for the inhibition of growth of resistant tumor cells and control the dynamic spreading of cancer (Workman et al., 2013).

References

Albertson, D. G. (2006). Gene amplification in cancer. *TRENDS in Genetics, 22*, 447–455.

Alitalo, K. & Schwab, M. 1986. Oncogene amplification in tumor cells. *Advances in cancer research*. Elsevier.

Azmi, A. S., Bao, B., & Sarkar, F. H. (2013). Exosomes in cancer development, metastasis, and drug resistance: a comprehensive review. *Cancer and Metastasis Reviews, 32*, 623–642.

Brito, D. A., & Rieder, C. L. (2009). The ability to survive mitosis in the presence of microtubule poisons differs significantly between human nontransformed (RPE-1) and cancer (U2OS, HeLa) cells. *Cell Motility and the Cytoskeleton, 66*, 437–447.

Burrell, R. A., & Swanton, C. (2014). Tumour heterogeneity and the evolution of polyclonal drug resistance. *Molecular Oncology, 8*, 1095–1111.

Chen, J., Tian, W., Cai, H., He, H., & Deng, Y. (2012). Down-regulation of microRNA-200c is associated with drug resistance in human breast cancer. *Medical Oncology, 29*, 2527–2534.

Das, T., Roy, K. S., Chakrabarti, T., Mukhopadhyay, S., & Roychoudhury, S. (2014). Withaferin A modulates the Spindle Assembly Checkpoint by degradation of Mad2–Cdc20 complex in colorectal cancer cell lines. *Biochemical Pharmacology, 91*, 31–39.

Davies, A. M., Ho, C., Lara, P. N., Jr., Mack, P., Gumerlock, P. H., & Gandara, D. R. (2006). Pharmacodynamic separation of epidermal growth factor receptor tyrosine kinase inhibitors and chemotherapy in non–small-cell lung cancer. *Clinical Lung Cancer, 7*, 385–388.

Davis, J. M., Navolanic, P. M., Weinstein-Oppenheimer, C. R., Steelman, L. S., Hu, W., Konopleva, M., Blagosklonny, M. V., & Mccubrey, J. A. (2003). Raf-1 and Bcl-2 induce distinct and common pathways that contribute to breast cancer drug resistance. *Clinical Cancer Research, 9*, 1161–1170.

Dean, M. (2009). ABC transporters, drug resistance, and cancer stem cells. *Journal of Mammary Gland Biology and Neoplasia, 14*, 3–9.

Easwaran, H., Tsai, H.-C., & Baylin, S. B. (2014). Cancer epigenetics: tumor heterogeneity, plasticity of stem-like states, and drug resistance. *Molecular Cell, 54*, 716–727.

Edwards, S. L., Brough, R., Lord, C. J., Natrajan, R., Vatcheva, R., Levine, D. A., Boyd, J., Reis-filho, J. S., & Ashworth, A. (2008). Resistance to therapy caused by intragenic deletion in BRCA2. *Nature, 451*, 1111–1115.

Fukumori, T., Kanayama, H.-O., & Raz, A. (2007). The role of galectin-3 in cancer drug resistance. *Drug Resistance Updates, 10*, 101–108.

Gao, Z., Zhang, L., & Sun, Y. (2012). Nanotechnology applied to overcome tumor drug resistance. *Journal of Controlled Release, 162*, 45–55.

Gariboldi, M. B., Ravizza, R., Riganti, L., Meschini, S., Calcabrini, A., Marra, M., Arancia, G., Dolfini, E., & Monti, E. (2003). Molecular determinants of intrinsic resistance to doxorubicin in human cancer cell lines. *International Journal of Oncology, 22*, 1057–1064.

Goldsmith, M. E., Gudas, J. M., Schneider, E., & Cowan, K. H. (1995). Wild type p53 stimulates expression from the human multidrug resistance promoter in a p53-negative cell line. *Journal of Biological Chemistry, 270*, 1894–1898.

Gonen, N., & Assaraf, Y. G. (2012). Antifolates in cancer therapy: structure, activity and mechanisms of drug resistance. *Drug Resistance Updates, 15*, 183–210.

Gottesman, M. M. (2002). Mechanisms of cancer drug resistance. *Annual Review of Medicine, 53*, 615–627.

Hadfield, J. A., Ducki, S., Hirst, N., & Mcgown, A. T. (2003). Tubulin and microtubules as targets for anticancer drugs. *Progress in Cell Cycle Research., 5*, 309–326.

Hodgkinson, V. C., Eagle, G. L., Drew, P. J., Lind, M. J., & Cawkwell, L. (2010). Biomarkers of chemotherapy resistance in breast cancer identified by proteomics: current status. *Cancer Letters, 294*, 13–24.

Holohan, C., van Schaeybroeck, S., Longley, D. B., & Johnston, P. G. (2013). Cancer drug resistance: an evolving paradigm. *Nature Reviews Cancer, 13*, 714–726.

Hu, T., Li, Z., Gao, C.-Y., & Cho, C. H. (2016). Mechanisms of drug resistance in colon cancer and its therapeutic strategies. *World Journal of Gastroenterology, 22*, 6876.

Hu, X., & Zhang, Z. (2016). Understanding the genetic mechanisms of cancer drug resistance using genomic approaches. *Trends in Genetics, 32*, 127–137.

Johannessen, T.-C.A., Bjerkvig, R., & Tysnes, B. B. (2008). DNA repair and cancer stem-like cells–potential partners in glioma drug resistance? *Cancer Treatment Reviews, 34*, 558–567.

Johnson, S. W., Ozols, R. F., & Hamilton, T. C. (1993). Mechanisms of drug resistance in ovarian cancer. *Cancer, 71*, 644–649.

Joshi, P., Vishwakarma, R. A., & Bharate, S. B. (2017). Natural alkaloids as P-gp inhibitors for multidrug resistance reversal in cancer. *European Journal of Medicinal Chemistry, 138*, 273–292.

Joyce, H., Mccann, A., Clynes, M., & Larkin, A. (2015). Influence of multidrug resistance and drug transport proteins on chemotherapy drug metabolism. *Expert Opinion on Drug Metabolism & Toxicology, 11*, 795–809.

Kars, M. D., İşeri, O. D., Gunduz, U., & Molnar, J. (2008). Reversal of multidrug resistance by synthetic and natural compounds in drug-resistant MCF-7 cell lines. *Chemotherapy, 54*, 194–200.

Kathawala, R. J., Gupta, P., Ashby, C. R., Jr., & Chen, Z.-S. (2015). The modulation of ABC transporter-mediated multidrug resistance in cancer: a review of the past decade. *Drug Resistance Updates, 18*, 1–17.

Kaufmann, S. H., & Vaux, D. L. (2003). Alterations in the apoptotic machinery and their potential role in anticancer drug resistance. *Oncogene, 22*, 7414.

Kim, E. S. (2016). *Chemotherapy resistance in lung cancer*. Springer.

Kim, D.-S., Park, S.-S., Nam, B.-H., Kim, I.-H., & Kim, S.-Y. (2006). Reversal of drug resistance in breast cancer cells by transglutaminase 2 inhibition and nuclear factor-κB inactivation. *Cancer Research, 66*, 10936–10943.

Koster, R., di Pietro, A., Timmer-Bosscha, H., Gibcus, J. H., van den Berg, A., Suurmeijer, A. J., Bischoff, R., Gietema, J. A., & de Jong, S. (2010). Cytoplasmic p21 expression levels determine cisplatin resistance in human testicular cancer. *The Journal of Clinical Investigation, 120*, 3594–3605.

Kovalchuk, O., Filkowski, J., Meservy, J., Ilnytskyy, Y., Tryndyak, V. P., Vasyl'f, C., & Pogribny, I. P. (2008). Involvement of microRNA-451 in resistance of the MCF-7 breast cancer cells to chemotherapeutic drug doxorubicin. *Molecular Cancer Therapeutics, 7*, 2152–2159.

Kumar, A., Singh, U. K., & Chaudhary, A. (2015). Targeting autophagy to overcome drug resistance in cancer therapy. *Future Medicinal Chemistry, 7*, 1535–1542.

Li, Q., & Shu, Y. (2014). Role of solute carriers in response to anticancer drugs. *Molecular and Cellular Therapies, 2*, 15.

Ma, J., Dong, C., & Ji, C. (2010). MicroRNA and drug resistance. *Cancer Gene Therapy, 17*, 523–531.

Majidinia, M., & Yousefi, B. (2016). Long non-coding RNAs in cancer drug resistance development. *DNA Repair, 45*, 25–33.

Malek, E., Jagannathan, S., & Driscoll, J. J. (2014). Correlation of long non-coding RNA expression with metastasis, drug resistance and clinical outcome in cancer. *Oncotarget, 5*, 8027.

Mashima, T., & Tsuruo, T. (2005). Defects of the apoptotic pathway as therapeutic target against cancer. *Drug Resistance Updates, 8*, 339–343.

Matson, D. R., & Stukenberg, P. T. (2011). Spindle poisons and cell fate: a tale of two pathways. *Molecular Interventions, 11*, 141.

Mcgranahan, N., & Swanton, C. (2017). Clonal heterogeneity and tumor evolution: past, present, and the future. *Cell, 168*, 613–628.

Miao, J., Du, Y.-Z., Yuan, H., Zhang, X.-G., & Hu, F.-Q. (2013). Drug resistance reversal activity of anticancer drug loaded solid lipid nanoparticles in multi-drug resistant cancer cells. *Colloids and Surfaces b: Biointerfaces, 110*, 74–80.

Murray, G. I., Taylor, V. E., Mckay, J. A., Weaver, R. J., Ewen, S. W., Melvin, W. T., & Burke, M. D. (1995). The immunohistochemical localization of drug-metabolizing enzymes in prostate cancer. *The Journal of Pathology, 177*, 147–152.

Natarajan, K., Xie, Y., Baer, M. R., & Ross, D. D. (2012). Role of breast cancer resistance protein (BCRP/ABCG2) in cancer drug resistance. *Biochemical Pharmacology, 83*, 1084–1103.

Nishio, K., Nakamura, T., Koh, Y., Suzuki, T., Fukumoto, H., & Saijo, N. (1999). Drug resistance in lung cancer. *Current Opinion in Oncology, 11*, 109.

Panasci, L., Paiement, J.-P., Christodoulopoulos, G., Belenkov, A., Malapetsa, A., & Aloyz, R. (2001). Chlorambucil drug resistance in chronic lymphocytic leukemia: the emerging role of DNA repair. *Clinical Cancer Research, 7*, 454–461.

Parhi, P., Mohanty, C., & Sahoo, S. K. (2012). Nanotechnology-based combinational drug delivery: an emerging approach for cancer therapy. *Drug Discovery Today, 17*, 1044–1052.

Pastan, I., & Gottesman, M. (1987). Multiple-drug resistance in human cancer. *New England Journal of Medicine, 316*, 1388–1393.

Pommier, Y., Sordet, O., Antony, S., Hayward, R. L., & Kohn, K. W. (2004). Apoptosis defects and chemotherapy resistance: molecular interaction maps and networks. *Oncogene, 23*, 2934–2949.

Rebucci, M., & Michiels, C. (2013). Molecular aspects of cancer cell resistance to chemotherapy. *Biochemical Pharmacology, 85*, 1219–1226.

Ross, D. D. & Nakanishi, T. 2010. Impact of breast cancer resistance protein on cancer treatment outcomes. *Multi-Drug Resistance in Cancer.* Springer.

Sakai, W., Swisher, E. M., Karlan, B. Y., Agarwal, M. K., Higgins, J., Friedman, C., Villegas, E., Jacquemont, C., Farrugia, D. J., & Couch, F. J. (2008). Secondary mutations as a mechanism of cisplatin resistance in BRCA2-mutated cancers. *Nature, 451*, 1116–1120.

Salami, J., Alabi, S., Willard, R. R., Vitale, N. J., Wang, J., Dong, H., Jin, M., Mcdonnell, D. P., Crew, A. P., & Neklesa, T. K. (2018). Androgen receptor degradation by the proteolysis-targeting chimera ARCC-4 outperforms enzalutamide in cellular models of prostate cancer drug resistance. *Communications Biology, 1*, 1–9.

Sinha, B. K. (1995). Topoisomerase inhibitors. *Drugs, 49*, 11–19.

Suzuki, T., Fukazawa, N., San-Nohe, K., Sato, W., Yano, O., & Tsuruo, T. (1997). Structure– activity relationship of newly synthesized quinoline derivatives for reversal of multidrug resistance in cancer. *Journal of Medicinal Chemistry, 40*, 2047–2052.

Swisher, E. M., Sakai, W., Karlan, B. Y., Wurz, K., Urban, N., & Taniguchi, T. (2008). Secondary BRCA1 mutations in BRCA1-mutated ovarian carcinomas with platinum resistance. *Cancer Research, 68*, 2581–2586.

Tan, K., Mattern, M. R., Eng, W.-K., Mccabe, F. L. & Johnson, R. K. 1989. Nonproductive rearrangement of DNA topoisomerase I and II genes: correlation with resistance to topoisomerase inhibitors. *JNCI: Journal of the National Cancer Institute,* 81, 1732–1735.

Todaro, M., Perez Alea, M., Scopelliti, A., Medema, J. P., & Stassi, G. (2008). IL-4-mediated drug resistance in colon cancer stem cells. *Cell Cycle, 7*, 309–313.

Tsouris, V., Joo, M. K., Kim, S. H., Kwon, I. C., & Won, Y.-Y. (2014). Nano carriers that enable co-delivery of chemotherapy and RNAi agents for treatment of drug-resistant cancers. *Biotechnology Advances, 32*, 1037–1050.

Twentyman, P., Fox, N., & White, D. (1987). Cyclosporin A and its analogues as modifiers of adriamycin and vincristine resistance in a multi-drug resistant human lung cancer cell line. *British Journal of Cancer, 56*, 55.

Vouri, M., & Hafizi, S. (2017). TAM receptor tyrosine kinases in cancer drug resistance. *Cancer Research, 77*, 2775–2778.

Wang, X., Zhang, H., & Chen, X. (2019). Drug resistance and combating drug resistance in cancer. *Cancer Drug Resistance, 2*, 141–160.

Wang, Z., Liang, S., Lian, X., Liu, L., Zhao, S., Xuan, Q., Guo, L., Liu, H., Yang, Y., & Dong, T. (2015). Identification of proteins responsible for adriamycin resistance in breast cancer cells using proteomics analysis. *Scientific Reports, 5*, 1–11.

Wood, K. W., Cornwell, W. D., & Jackson, J. R. (2001). Past and future of the mitotic spindle as an oncology target. *Current Opinion in Pharmacology, 1*, 370–377.

Wong, H. L., Rauth, A. M., Bendayan, R., Manias, J. L., Ramaswamy, M., Liu, Z., Erhan, S. Z., & Wu, X. Y. (2006). A new polymer–lipid hybrid nanoparticle system increases cytotoxicity of doxorubicin against multidrug-resistant human breast cancer cells. *Pharmaceutical Research, 23*, 1574–1585.

Workman, P., Al-Lazikani, B., & Clarke, P. A. (2013). Genome-based cancer therapeutics: targets, kinase drug resistance and future strategies for precision oncology. *Current Opinion in Pharmacology, 13*, 486–496.

Yamazaki, R., Nishiyama, Y., Furuta, T., Hatano, H., Igarashi, Y., Asakawa, N., Kodaira, H., Takahashi, H., Aiyama, R., & Matsuzaki, T. (2011). Novel acrylonitrile derivatives, YHO-13177 and YHO-13351, reverse BCRP/ABCG2-mediated drug resistance in vitro and in vivo. *Molecular Cancer Therapeutics, 10*, 1252–1263.

Yang, Z. J., Chee, C. E., Huang, S., & Sinicrope, F. A. (2011). The role of autophagy in cancer: therapeutic implications. *Molecular Cancer Therapeutics, 10*, 1533–1541.

Yang, L., Li, N., Wang, H., Jia, X., Wang, X., & Luo, J. (2012). Altered microRNA expression in cisplatin-resistant ovarian cancer cells and upregulation of miR-130a associated with MDR1/P-glycoprotein-mediated drug resistance. *Oncology Reports, 28*, 592–600.

Zeng-Rong, N., Paterson, J., Alpert, L., Tsao, M.-S., Viallet, J., & Alaoui-Jamali, M. A. (1995). Elevated DNA repair capacity is associated with intrinsic resistance of lung cancer to chemotherapy. *Cancer Research, 55*, 4760–4764.
Zhang, H., & Mccarty, N. (2017). Tampering with cancer chemoresistance by targeting the TGM2-IL6-autophagy regulatory network. *Autophagy, 13*, 627–628.

Chapter 14
Drugs Resistance and Treatment Failure in HIV and/or AIDS

Fahmida Jabeen, Muhammad Salman Sajid, Adeela Saeed, Batool Fatima, Dilshad Hussain, Saadat Majeed, Muhammad Naeem Ashiq, and Muhammad Najam-ul-Haq

Abstract Human immunodeficiency virus (HIV) is an epidemic that has compelled the clinical sciences to profusely develop and modify this rapidly evolving pathogen. AIDS is fatal and incurable disease which severely affects immune system. The treatment for AIDS is a blend of two or more drugs known as antiretroviral drugs (ARV) and the therapy is cumulatively called Highly Active Anti-Retroviral Therapy (HAART). Although ARV drugs are successful and treatment is adapted with known efficacy, the major impediment resulting in treatment failure is drug resistance in HIV infected patients. Specific mutations in HIV genome result in the development of drug resistance against recommended therapy. These DNA mutations predict drug resistance against stated drug however there are number of factors contributing to the development of drug resistance. The main HIV targeted regions include developing countries due to deprived resources, poor clinical follow up and disease control management. Phenotype testing in link to computational modulation can be helpful in selecting the best treatment regimens however subsequent viral mutations and range of drug combinations can result in multifaceted scenario. Competing HIV along with evolving drug resistance challenges the efficiency of treatment and provokes more clinical trials towards threshold of failure.

F. Jabeen (✉) · A. Saeed
Department of Chemistry, The Women University Multan, Multan, Pakistan

M. S. Sajid · S. Majeed · M. Naeem Ashiq · M. Najam-ul-Haq
Institute of Chemical Sciences, Bahauddin Zakariya University, Multan 60800, Pakistan

B. Fatima
Department of Biochemistry, Bahauddin Zakariya University, Multan 60800, Pakistan

D. Hussain
International Centre for Chemical and Biological Sciences, HEJ Research Institute of Chemistry, University of Karachi, Karachi 75270, Pakistan

S. Ahmed et al. (eds.), *Biochemistry of Drug Resistance*,
https://doi.org/10.1007/978-3-030-76320-6_14

14.1 Introduction

One of the major targets of Joint United Nations Programme on HIV/AIDS (UNAIDS) and World Health Organization (WHO) is the implementation of antiretroviral therapy to control the fatal health threat of AIDS by 2030. Prevention of HIV-1 transmission, increasing survival and improving the health of infected individuals can be achieved by sustained viral suppression. A recent report published by WHO in 2019 stated an increase in the frequency and recurrence of acquired, transmitted and drug-resistant HIV-1 in ART-naive individuals. This generates a worldwide panic in health administrators considering drug resistance as a major impediment in fight against HIV-1 (Amangaldiyevaet al., 2019).

HIV Drug resistance (HIVDR) is the ability of HIV to undergo genetic mutation and self-replication in the presence of antiretroviral drugs nullifying drug activity and causing failure of treatment (Armenia et al., 2020). This results in increased infection and spread of drug resistant HIV. Therefore, the efficiency of adopted therapy is compromised leaving limited therapeutic options for viral suppression, HIV occurrence, fatality, and morbidity (Hamers, 2018).

It is accepted that no antiretroviral drug is resistance proof at present. It is a natural phenomenon where HIV drug resistance develops due to the selective strain caused by adopted drug therapy or from the immune system. The major aspect which promotes viral mutation and develops drug resistance is inadequate suppression of HIV replication by the drug. This scenario frequently builds up concurrent to non-conformity of ARV therapy. After mutation, resistant virus strain spreads rapidly leading to poor efficacy of ARV therapy. Currently transmission of HIV resistance strains is one of the major concerns in developing countries with adopted ARV therapy (Davy-Mendez et al., 2018).

Control on resistance can be achieved by careful selection of treatment, probability of predictable drug resistance and appropriate monitoring. The drug resistance monitoring can help to stand out the drug resistance and timely alterations of ARV therapy in HIV-infected patient facing resistance, as the individual is more prone to HIV spread if similar medications are continued. The new mutated form in favourable environment will flourish at rapid rate and favour further HIV replication resulting in dominant HIV type in the body. The prescribed medication and/or therapy is thus ineffective due to complete drug resistance.

14.2 Current Medication and Related Mutations

There are two step targeted goals of ARV drugs (i) promote virus suppression and (ii) decrease HIV transmission. HIV medication reduces the viral load on patients' body thus giving a boost to immune system for fast recovery. It is not logical to assume complete removal of virus from the body, but the immune system is strengthened to fight infections and regain control of the body. Once the viral suppression is achieved

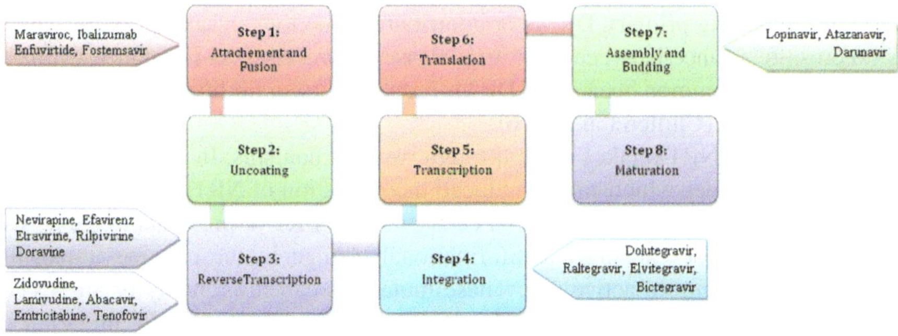

Fig. 14.1 Schematic representation of FDA-approved anti-HIV drugs inhibiting the stages of HIV-1 life cycle

by ARVs, the chances of virus transmission are minimal to other individuals, thus slowing down the breakout in a community (Günthardet al., 2019).

There are several types of antiretroviral drugs which are distinguished based on their mode of actions, features, and functions. The key role is to block or alter enzymatic processes complimentary to HIV replication. Nucleoside reverse transcriptase inhibitors (NRTIs) block the reverse transcriptase (RT) enzyme whereas another class of ARVs binds to reverse transcriptase causing changes in structural conformation known as non-nucleoside reverse transcriptase inhibitors (NNRTIs) (Margolis, 2014). Integrase inhibitors (IN) disable a protein called integrase necessary for insertion of HIV genetics into CD4 T-cells while protease inhibitors (PIs) block protease enzyme required during viral replication (Babikeret al., 2011).

HIV infection spreads to healthy CD4 cells by binding at sites on cell membrane. The second set of ARVs interferes with the ability of HIV to infect CD4 immune system cells. Fusion inhibitors (FIs) block HIV from entering cells whereas CCR5 antagonists and post-attachment inhibitors block binding sites on CD4 cells. Figure 14.1 depicts eight stages of HIV infection in a cell. Each step can be controlled or altered or blocked by ARVs to achieve viral suppression leading to successful treatment.

14.3 Mechanism of HIV Drug Resistance

The emergence of HIV drug resistant variants depends on factors like extent of viral replication because of incomplete suppressive therapy, the acquisition of specific drug resistance mutation (DRM) and its effect on drug receptiveness and virus replication. Exposure to ARVs can add to the development of resistance by HIV variants and patients may even have resistant viral strains to drug therapy. Drug resistance reduces susceptibility to single or multiple classes of antiretroviral drugs complicating HIV therapy. Primary DRMs decrease drug susceptibility and accessory DRMs strengthen variants containing primary DRMs or contribute reductions in susceptibility. There is no cross-resistance between HIV drug classes however cross-resistance within a class of drug is common because of DRMs reducing susceptibility to multiple ARVs of the same class (Clutteret al., 2016).

Resistance to NRTIs takes place through two mechanisms. In first case, there is mutation of residues which causes reduced incorporation of NRTI into the growing DNA chain. Mutations occur at actual catalytic site of RT or proximal to the active catalytic site of RT causing a conformational change in the enzyme impairing binding of HIV drug to active site. These mutations can resist many or all NRTIs (Melikianet al., 2012) (Table 14.1). Increased resistance with M184V mutation is observed in cytosine analogue, lamivudine. For guanosine analogue, abacavir, two to three concomitant mutations (e.g. M184V, L74V) are necessary for resistance. The second NRTI resistance mechanism is associated with reduced binding of drug to the end of DNA chain resulting in drug removal. These RT mutations cause ATP or pyrophosphate present in high concentration within the cell to bind at the active site next to bound drug. ATP liberates drug by attacking bond between drug and DNA, thus terminating its effect (Thigpenet al., 2012).

NNRTIs increase reverse transcriptase rigidity and prevent its normal polymerization function. NNRTIs have few side effects as compared to NRTIs however they are more susceptible to resistance and limit its use as monotherapy of HIV infection. Resistance to NNRTIs occurs through mutation of hydrophobic RT residues present in their binding pockets. The mode of binding by NNRTIs is essentially to the similar region of RT, thus mutation in this area affect the efficiency of NNRTIs analogues (Melikian et al., 2014).

Resistances to protease inhibitors (PIs) is the result of amino acid mutations arising within or proximal of catalytic binding site to the drug. Replacement of major amino acids in protease enzyme changes the affinity of reverse transcriptase for binding protease inhibitors (Boenderet al., 2016). The conformational changes involve alteration in the geometry of catalytic site. Mutations at these catalytic sites result in enlargement of this region reducing the binding of PIs as compared to other endogenous peptides (Wensinget al., 2017).

Resistance to CCR5 or CXCR4 is offered by a shift in co-receptor usage or by alterations in the regions of viral envelope resulting in interaction between gp120 and the co-receptor. An enhanced affinity of gp120 for CCR5 increases competition of HIV-1 with CCR5 inhibitors for bind to CCR5 co-receptor. Most of CCR5 antagonist-resistant strains preserve CCR5 co-receptor where mutations at regions of gp120 likeV3, V2, C2 and C4 are responsible for drug-resistant phenotype. Enfuvirtide binds to HR1 region blocking the formation of six-helix bundle structure essential for the fusion process. The resistance to enfuvirtide involves changes in three amino acids (positions 36–38) of HR1 region of gp41 (Su et al., 2015).

Table 14.1 FDA approved antiretroviral drugs including their roles, features, modes of action, resistance, and considerations

Type of drug	Key role	Approved FDA drugs	Features	Modes of action	Resistance	Considerations
Nucleoside/nucleotide reverse transcriptase inhibitors (NRTIs)	Compel HIV virus to exploit damaged versions of building blocks to decrease replication and infection spread to healthy cells	Approved nucleoside analogues: Abacavir Didanosine, Emtricitabine, Lamivudine, Stavudine, Zidovudine, Zalcitabine Approved Aytosine Analogues: Tenofoviralafenamide, Tenofovir disoproxil fumarate	Structural nucleoside analogues of DNA nucleotides Prevent reverse transcription of HIV genome Inhibiting the action of HIV-1 RT Inhibit viral replication	Functions as chain terminators during extension of DNA in reverse transcription process by HIV RT* Allow correct base-pairing and incorporation into DNA chain Replace hydroxyl group necessary for addition of next nucleotide by non-reactive chemical group	Changes in primary structure of RT increasing selectivity of RT to incorporate native nucleotide over NRTI-triphosphate (e.g. K65R, K70E, L74V, M184V and Q151M) Resistance is caused by alteration in primary structure of RT that facilitates the deletion of chain-terminators (e.g. D67N, T215F/Y, M41L, K70R, K219Q and L210W)	Mitochondrial toxicity Different metabolism with each drug

(continued)

Table 14.1 (continued)

Type of drug	Key role	Approved FDA drugs	Features	Modes of action	Resistance	Considerations
Non-nucleoside Reverse Transcriptase Inhibitors (NNRTIs) (Himmel and Arnold, 2020)	Reduction of polymerase activity by altering the spatial configuration of substrate binding site by creating hydrophobic pocket proximal at active site	First generation: Delavirdine, Efavirenz, Nevirapine Second generation: Doravirine, Etravirine (ETR), Rilpivirine(RPV)	No requirement of phosphorylation for activation No overlapping sites of resistance, mutations and carry out inhibition mechanism in combination with NRTIs	Target the transcription of HIV life cycle Bind HIV RT and block complementary HIV DNA elongation Inhibit DNA polymerization Decrease RT and nucleoside binding affinity Prevents viral replication	First-generation NNRTI mutations: Offer low genetic barrier to resistance Single mutation in the binding site, decrease the drug activity Associated with mutations at multiple codons K103N or Y181C mutation causing clinical failure	Prolonged dual and triple therapies using ETR or RPV with INSTIs to improve adherence and safety
Protease Inhibitors (PIs)	Block formation of proteins for HIV replication	First generation: Indinavir, Nelfinavir, Saquinavir, Ritonavir, Amprenavi Second generation: Atazanavir, Darunavir, Fosamprenavir Lopinavir + Ritonavir, Tipranavir	Boost PI monotherapy (Ritonavir-boosted PIs) in practice Suppressors of HIV replication Genetic barrier to drug resistance	Systematic cleaving of individual proteins (Gag and Gag –pol) into functional subunits vital for viral capsid formation during replication and infection spread	Multiple protease inhibitor resistance mutations Alteration in replication resulting in poor drug activity	Efficient resistance testing for PI Investigation of alternative resistance mechanisms in PI drug resistance

(continued)

Table 14.1 (continued)

Type of drug	Key role	Approved FDA drugs	Features	Modes of action	Resistance	Considerations
Integrase strand transfer inhibitors (INSTIs) (Trivedi et al., 2020)	Block protein necessary for HIV DNA attachment to healthy cell DNA	Bictegravir, BIC (combined with other drugs as Biktarvy)	Latest approval More efficient Used in clinic as a single tablet fixed-dose combination	Inhibits IN strand transfer activity	R263K and M184V resistance mutations	Clinical investigation in combination with ARV drugs is needed
		Dolutegravir(DTG)	Single, long term anti-HIV medication	Binds to the active site of HIV-1 bringing β4-α2 loop of enzyme in contact with extension in linker region which leads to conformational changes disengaging IN** from dApresent at the 3′ end of virus	TG resistance mutations	Evaluation of virological failure and drug resistance mutation Screening of IN gene to detect genetic alterations andto indicate therapeutic failure of INIs

(continued)

Table 14.1 (continued)

Type of drug	Key role	Approved FDA drugs	Features	Modes of action	Resistance	Considerations
		Elvitegravir (EVG)	Superior therapeutic efficacy Blockagainst multidrug resistant strains of HIV	Recognize IN-DNA binding to interface between 3′ end of viral DNA and enzyme Inhibition of strand transfer reaction Integration of viral dsDNA into host genome	Mutation E92Q nreduces EVG susceptibility by 33-fold Other mutations resulting in lower susceptibility (T66I/A/K, T124A, P145S, and Q148K/R) Overlaps mutation Q148K Low barrier to resistance compared to other INIs Limited therapeutic efficiency	Available in fixed dose combination Clinical studies for more reliable fixed dose combination are needed
		Raltegravir (RAL)	Approved for salvage therapy*** Active against multidrug resistant isolates of HIV	Interact with catalytic residues (D64, D116, E152, T66, E92, Y143, Q148, and N155) of HIV-1 IN	N155H and Q148K/R/H mutations resulting in complete resistance	Less effective therapeutic option in salvage therapy Use is limited to first-line cART for treatment

(continued)

Table 14.1 (continued)

Type of drug	Key role	Approved FDA drugs	Features	Modes of action	Resistance	Considerations
Fusion Inhibitors (FIs)	Prevent HIV penetration to healthy cells	Enfuvirtide	T20 peptide Homologous part of gp41 encoded fusion Used in combination for infected patients having HIV-1 replication	Engage in coil to coil interaction (HR1 region) in viral glycoprotein gp41 resulting in blockade of virus particle with outer cell membrane	Changes in amino acid triad (GIV) at positions 36–38 in HR1 region of gp41	Development of T20 like peptides with improved efficacy FIs to target alternative gp41 domain
CCR5 Antagonist Inhibitors	Inhibit CCR5 binding suppressing viral entry into the cell	Maraviroc	Maraviroc therapy for HIV patients under treatment Not active against HIV strains with CXCR4, dual****, or mixed tropism	Selective binding to CCR5 at the membrane of CD4 cells (T-cells) and prevent interaction of HIV-1 gp120 and CCR5 to inhibit viral entry into cell	Mutations in V3 loop of gp120	Poorly understood resistance Mechanism study is underway

(continued)

Table 14.1 (continued)

Type of drug	Key role	Approved FDA drugs	Features	Modes of action	Resistance	Considerations
Post-Attachment Inhibitor or Monoclonal Antibody	Inhibit spreading of body's HIV infected cells from spreading the virus	Ibalizumab-uiyk (Trogarzo)	For HIV-1 infection in treated patients with multidrug resistance and failed ARTs	CD4 directed post attachment of HIV-1 Bind to domain 2 of CD4 Interfere with post-attachment steps Preventing cell to cell viral transmission by blocking HIV-1 from infecting CD4 + T cells	Reduce expression or loss of N-linked glycosylation sites in V5 loop of envelope glycoprotein 120	Prolonged clinical efficacy, safety, and resistance data is under investigation

*_RT_ reverse transcriptase, **_IN_ Integrase, ***Salvage therapy: Treatment given when HIV or disease is non-responsive to *standard* treatments/therapy, ****Selective binding of HIV to a particular co-receptor on the surface of a host CD4 cell as tropic via CCR5 co-receptor (R5-tropic) or the CXCR4 co-receptor (X4-tropic) or both (dual-tropic)

14.4 Drug Resistance Testing

Genotypic and phenotypic drug resistance tests monitor ARV drugs in naïve or matured HIV patients. Several other drug resistance tests are available, but the sensitivity, specificity, and reproducibility are not well established. TrueGene™ is an FDA approved ARVs drug resistance test.

In genotypic resistance testing, the genetic structure/genotype of an infected HIV patient is examined. Blood sample is analysed to highlight the genetic mutations causing drug resistance (Llibreet al., 2010). For example, an individual with HIV strain having mutation at M184V in reverse transcriptase gene will suffer drug resistance for lamivudine (Epivir®) and emtricitabine (Emtriva®). They are thus ineffective against such HIV forms. A positive test for mutation helps therapistsin choosing ARV combination for disease control without facing drug resistance (Rheeet al., 2015).

Phenotypic testing is more accurate as it measures the phenotype of acquired HIV strain in response to specific antiretrovirals drugs. The test results determine drug dosage and combinations required to contain specific HIV strain and prevent its replication. ARVs are routine medicines for infected as well as recovered patients however monitoring is necessary as HIV can mutate at any stage of therapy (Paredeset al., 2010).The factors responsible for drug resistance are still elusive. It is difficult to list factors that promote mutations and form resistant HIV strain. An interruption in treatments like pregnancy or breastfeeding can initiate and propel HIV mutations and drug resistance (Günthardet al., 2019).

Drug resistance against first generation NNRTIs is immense in developing countries where they are prescribed as first line regimen in HIV treatment. Clinical developments have provided second generation NNRTIs to overcome HIV resistant strains. There are rare cases of occurrence for multidrug-resistant HIV-1 variants. The surfacing of triple class drug-resistant virus is however common. The resistance to at least one NRTIs, one NNRTIs and one PIs has imposed challenge at clinical level for HIV-1 management (Jadeet al., 2008). According to reports, the occurrence of triple class drug resistant HIV strain ranges from 3 to 10% in ARV treated patients. The frequency is higher in patients with initial treatment by ART regimens, particularly in prenatal infection in infants and young ones. These resistant HIV variants are linked to failure of second line PI-based ARVs. This increases fatality of HIV and its outburst compromising the clinical outcomes and emphasizing on development of new ARV drugs (Walter and Kredo, 2017).

In 2007, clinical trials led to approval of integrase strand transfer inhibitors (INSTIs) and CCR5 antagonists for HIV patients with developed resistance against second-line treatment regimens. Recent addition of Dolutegravir (INSTI) offers genetic barrier to resistance with improved tolerance and is recommended as first line treatment regimen (Boffitoet al., 2020). The new classes have reduced the cumu lative drug resistance among HIV patients owing to the antiretroviral therapy. Still the development of multidrug resistance cannot be avoided in patients with poor virological response to treatment.

The mortality rate due to HIV has decreased after the highly active antiretroviral therapy (HAART) (Quinn, 2008). The therapy decreases viral load and improves CD4 count in individuals undergoing treatment (Hull and Montaner, 2013).Although clinical outcomes have been achieved by HAART, virological and immunological treatment failures have resulted in disease outburst and increased risk of HIV transmission. In terms of testing, virological failure is the viral load above 1000 copies per mL of plasma using two consecutive viral load measurements after 3 months. The immunological failure decreases in CD4 T cell count below or near to baseline, constant low levels of CD4 below 100 cells/mm^3 for adult or below 200 cells/mm^3 in children under the age of 5 years. Occurrence or recurrence of advanced WHO clinical stage after 6 months of therapy needs clinical therapy (Sidibéet al., 2016).

Globally, UNAIDS has targeted HAART with viral suppression of 90% in HIV infected individuals by 2030. The systematic analysis of HIV treatment can help in monitoring the viral suppression and HIV prevention (Levi et al., 2016). According to WHO, improved HAART strategies, regularity in prescribed medication, prevention of drug to drug interaction, enhanced knowledge of infected patients towards HAART, timely testing of drug resistance, and implementation of effective food and nutrition policy (Endalamaw et al., 2020) will help against HIV epidemic.

14.5 Next Generation Medicine in HIV

Cabotegravir (CAB, GSK744) has carbamoyl pyridone backbone. It is structurally and mechanistically like DTG (Zhouet al., 2018). It has shown results in Phase-II clinical trials. CAB is monthly or quarterly administered (Rusconiet al., 2017), orally or injected as nano-formulation. The clinical trials have resulted in improved tolerance with limited side effects. The monthly dosage of CAB + RPV formulation offers better virological control. Mutations in integrase gene resist CAB (Whitfield et al., 2016), limiting the choice including INSTIs in follow-up therapy. *In vitro* study has however supported the improved genetic barrier of CAB for clinical efficacy in prolonged follow up therapy.

MK-2048 had been developed by Merck & Co. in early 2009 and has completed the phase-II trial in 2014. MK-2048 inhibits viral isolates with R263K-resistant mutation in cell-free integrase strand transfer assay. Currently, daily MK-2048 dosage (30 mg) is being evaluated in combination with vicriviroc of Phase-I trial (Hoesley et al., 2019). There is no cross-resistance to RAL- and EVG-resistant mutations, supporting its potency as second generation INSTIs (Hareet al., 2010). Low half-life (1–1.5 h) and reduced efficacy due to G118R and E138K mutations has limited (Van Wesenbeecket al., 2011) its inclusion as second generation INSTIs.

The nano-formulations with targeted drug release in AIDS offer improved specificity and prolonged therapeutic effects (Mitragotri et al., 2014). ARV drugs are prepared in gold and diamond-based nano-formulations (Kevadiyaet al., 2018) with high surface area for better absorption and drug binding. The nano-formulations have sustained drug release overcoming the uneven drug distribution and toxicity to the host. Nano-formulated DTG and CAB given once in 2-monthsare under investigation (Sillmanet al., 2018). Nano-formulations may reduce pressure on immune system which results in increased drug resistant mutations in HIV strains.

14.6 Broadly Neutralizing Antibodies

Vaccine Research Centre (VRC) at National Institute of Allergy and Infectious Disease (NIAID) has initiated antibodies development and their testing against HIV. Antibodies treat with fewer side effects and multiple modification modes for prolonged life in host. Antibodies have eliminated the HIV resistant strains from infected human cells at laboratory level and given the name of broadly neutralizing antibodies (bNAbs). These antibodies excel HIV treatment by: (i) binding to the virus particle, preventing its entry into CD4 cells and promote elimination; (ii) adhering to HIV infected cell and engaging immune system in fighting off the infection and (iii) forming a complex with HIV particle, stimulating the immune cells similar to a vaccine and leaving an imprint for future safety. Immune system is either prepared prior to viral spread or eliminated prior to replication (Kumaret al., 2018). Clinical trials show viral suppression by infusing bNAbs to HIV patients. Failure is faced when individuals develop resistant HIV strain against one bNAb. In vivo studies prove viral suppression by combination of either multiple bNAbs or bNAbs with ARVs like ARTs. Modifications to known bNAbs treat and prevent number of drug resistant HIV strains. Studies demonstrate the life of antibody in host, blocking efficacy, efficiency in triggering immune system and number of HIV strains for which the antibody responds (Yubinet al., 2020).

14.7 Therapeutic HIV Vaccines

A therapeutic vaccine is designed for HIV infections of infected individuals. Such vaccine stimulates the immune system for controlling recurrence of disease and thereby ending the antiretroviral therapies. The approach provides viral remission in case of undetectable levels of HIV without antiretroviral therapy (Vieillardet al., 2019).The therapeutic vaccine has controlled HIV infection at the early stage or after halting ART. It can equip the immune system for prolonged HIV infection control. NIAID has funded for better understanding of immune responses for sustainable HIV suppression (Dimiet al., 2019).

14.8 Future Directions for Developing Daily HIV Drugs

New drugs have been designed for antiretroviral therapy being effective against HIV resistant strains. NIAID research has developed inhibitors to target the same stage of HIV lifecycle as PIs via alternate mechanism. The other stages of HIV lifecycle are also targeted where the inhibitor fostemsavir attaches to gp120 protein on virus and blocks its entry into immune cells. Capsid assembly inhibitors may halt the viral capsid during replication.

14.8.1 Future Perspective

In developing countries, HIV related mortality has decreased due to potent combinations of antiretroviral (ARV) therapy. Emergence of resistance to ARV drugs has challenges. With advancement of therapy and new drug additions, the increased life span for infected individuals is expected. Natural evolution in HIV genetic makeup has rendered limitations, particularly the HIV resistant mutations for ARV drugs whereby limiting the permanent cure. Continuous upgradation of HIV therapies is thus required. Aprovirus incorporated into the host genome is impossible to remove from host cells and IN inhibitors are the clinical options. The binding to active site increases resistance and must be considered in upcoming clinical administrations.

14.9 Conclusion

Recent studies have demonstrated the prevalence and risk factors in failure of HIV therapeutics. Countries with poor health system are more prone to drug resistance and are high risk of virus transmission to other communities. Antiretroviral therapy must be applied after consideration of complementary treatment failure. This will economize the treatment using first-line ART regimen rather than moving to more expensive second-line ART regimen. Factors need to be tackled that promote drug resistance resulting in treatment failure. Underprivileged HIV health services, late recognition of treatment failure, delayed introduction of HAART, malnourishment, adverse reactions linked to ART and emerging psychological issues are the challenges. In the absence of frequent monitoring of therapeutic drugs or routine resistance testing, the HIV infected individuals are at risk of failed treatment.

References

Amangaldiyeva, A., Davlidova, S., & Baiserkin, B. (2019). Implementation of antiretroviral therapy (ART) in former Soviet Union countries. *AIDS Resistant Therapy, 16*, 35–43.

Armenia, D., Di Carlo, D., Flandre, P., Bouba, Y., Borghi, V., Forbici, F., Bertoli, A., Gori, C., Fabeni, L., Gennari, W., Pinnetti, C., Mondi, A., Cicalini, S., Gagliardini, R., Vergori, A., Bellagamba, R., Malagnino, V., Montella, F., Colafigli, M. ... Santoro, M. M. (2020). HIV MDR is still a relevant issue despite its dramatic drop over the years. *Journal of Antimicrobial Chemotherapy, 75*, 1301–1310.

Babiker, A., Castro Nee Green, H., & Compagnucci, A. (2011). First-line antiretroviral therapy with a protease inhibitor versus non-nucleoside reverse transcriptase inhibitor and switch at higher versus low viral load in HIV-infected children: An open-label, randomised phase 2/3 trial. *Lancet Infectious Disease, 11*, 273–283.

Boender, T.S., Hamers, R. L., Ondoa, P., Wellington, M., Chimbetete, C., Siwale, M., Maksimos, E. E. F. L., Balinda, S. N., Kityo, C. N., Adeyemo, T. A., Akanmu, A. S., Mandaliya, K., Botes, M. E., Stevens, W., Rinke De Wit, T. F., & Sigaloff, K. C. E. (2016). Protease inhibitor resistance in the first 3 years of second-line antiretroviral therapy for HIV-1 in Sub-Saharan Africa. *The Journal of Infectious Diseases, 214*, 873–883.

Boffito, M., Waters, L., Cahn, P., Paredes, R., Koteff, J., Wyk, J. V., Vincent, T., Demarest, J., Adkison, K., & Quercia, R. (2020). Perspectives on the barrier to resistance for Dolutegravir + Lamivudine, a two-drug antiretroviral therapy for HIV-1 infection. *AIDS Research and Human Retroviruses, 36*, 13–18.

Clutter, D. S., Jordan, M. R., Bertagnolio, S., & Shafer, R. W. (2016). HIV-1 drug resistance and resistance testing infection. *Genetics and Evolution, 46*, 292–307.

Davy-Mendez, T., Eron, J. J., Brunet, L., Zakharova, O., Dennis, A. M., & Napravnik, S. (2018). New antiretroviral agent use affects prevalence of HIV drug resistance in clinical care populations. *AIDS, 32*, 2593–2603.

Dimi, S., Zucman, D., Chassany, O., Lalanne, C., Prazuck, T., Mortier, E., Majerholc, C., Aubin-Auger, I., Verger, P., & Duracinsky, M. (2019). Patients' high acceptability of a future therapeutic HIV vaccine in France: a French paradox? *BMC Infectious Disease, 19*, 401.

Endalamaw, A., Mekonnen, M., Geremew, D., Fikadu, A., Yehualashet, F. A., Tesera, H., & Habtewold, T. D. (2020). HIV/AIDS treatment failure and associated factors in Ethiopia: Meta-analysis. *BMC Public Health, 20*, 82.

Günthard, H. F., Calvez, V., Paredes, R., Pillay, D., Shafer, R. W., Wensing, A. M., Jacobsen, D. M., & Richman, D. D. (2019). Human immunodeficiency virus drug resistance: 2018 recommendations of the international antiviral society-USA panel. *Clinical Infectious Disease, 68*, 177–187.

Hamers, R. L., Rinke De Wit, T. F., & Holmes, C. B. (2018). HIV drug resistance in low-income and middle-income countries. *Lancet HIV, 5*, e588–e596.

Hare, S., Vos, A. M., Clayton, R. F., Thuring, J. W., Cummings, M. D., & Cherepanov, P. (2010). Molecular mechanisms of retroviral integrase inhibition and the evolution of viral resistance. *Proceedings of the National Academy of Sciences of the United States of America, 107*, 20057–20062.

Himmel, D. M., & Arnold, E. (2020). Non-nucleoside reverse transcriptase inhibitors join forces with integrase inhibitors to combat HIV. *Pharmaceuticals, 13*, 122.

Hoesley, C. J., Chen, B. A., Anderson, P. L., Dezzutti, C. S., Strizki, J., Sprinkle, C., Heard, F., Bauermeister, I., Hall, W., Jacobson, C., Berthiaume, J., Mayo, A., Gundacker, H., Richardson-Harman, N., & Piper, J. (2019). Phase 1 safety and pharmacokinetics study of MK-2048/Vicriviroc (MK-4176)/MK-2048A intravaginal rings. *Clinical Infectious Disease, 68*, 1136–1143.

Hull, M. W., & Montaner, J. S. (2013). HIV treatment as prevention: The key to an AIDS-free generation. *Journal of Food and Drug Analysis, 21*, S95–S101.

Jade, G., Marie-Laure, C., & Constance, D. (2008). HIV-1 resistance to first- and second-generation non-nucleoside reverse transcriptase inhibitors. *AIDS Reviews, 11*, 165–173.

Kevadiya, B. D., Woldstad, C., Ottemann, B. M., Dash, P., Sajja, B. R., Lamberty, B., Morsey, B., Kocher, T., Dutta, R., Bade, A. N., Liu, Y., Callen, S. E., Fox, H. S., Byrareddy, S. N., McMillan, J. M., Bronich, T. K., Edagwa, B. J., Boska, M. D., & Gendelman, H. E. (2018). Multimodal theranostic nanoformulations permit magnetic resonance bioimaging of antiretroviral drug particle tissue-cell biodistribution. *Theranostics, 8*, 256–276.

Kumar, R., Qureshi, H., Deshpande, S., & Bhattacharya, J. (2018). Broadly neutralizing antibodies in HIV-1 treatment and prevention. *Therapeutic Advance in Vaccines and Immunotherapy, 6*, 61–68.

Levi, J., Raymond, A., Pozniak, A., Vernazza, P., Kohler, P., & Hill, A. (2016). Can the UNAIDS 90–90-90 target be achieved? A systematic analysis of national HIV treatment cascades. *BMJ Global Health, 1*, e000010.

Llibre, J. M., Schapiro, J. M., & Clotet, B. (2010). Clinical implications of genotypic resistance to the newer antiretroviral drugs in HIV-1-infected patients with virological failure. *Clinical Infectious Disease, 50*, 872–881.

Margolis, A. M., Heverling, H., Pham, P. A., & Stolbach, A. (2014). A review of the toxicity of HIV medications. *Journal of Medical Toxicology, 10*, 26–39.

Melikian, G. L., Rhee, S. Y., Taylor, J., Fessel, W. J., Kaufman, D., Towner, W., Troia-Cancio, P. V., Zolopa, A., Robbins, G. K., Kagan, R., Israelski, D., & Shafer, R. W. (2012). Standardized comparison of the relative impacts of HIV-1 reverse transcriptase (RT) mutations onnucleoside RT inhibitor susceptibility. *Antimicrobial Agents Chemotherapy, 56*, 2305–2313.

Melikian, G. L., Rhee, S. Y., Varghese, V., Porter, D., White, K., Taylor, J., Towner, W., Troia, P., Burack, J., Dejesus, E., Robbins, G. K., Razzeca, K., Kagan, R., Liu, T. F., Fessel, W. J., Israelski, D., & Shafer, R. W. (2014). Non-nucleoside reverse transcriptase inhibitor(NNRTI) cross-resistance: Implications for preclinical evaluation of novel NNRTIs and clinical genotypic resistance testing. *Journal of Antimicrobial Chemotherapy, 69*, 12–20.

Mitragotri, S., Burke, P. A., & Langer, R. (2014). Overcoming the challenges in administering biopharmaceuticals: Formulation and delivery strategies. *Nature Reviews of Drug Discovery, 13*, 655–672.

Paredes, R., Lalama, C. M., Ribaudo, H., Schackman, B. R., Shikuma, C., Giguel, F., Meyer, W. A., 3RD, Johnson, V. A., Fiscus, S. A., D'aquila, R. T., Gulick, R. M., & Kuritzkes, D. R. (2010). Pre-existing minority drug-resistant HIV-1 variants, adherence, and risk of antiretroviral treatment failure. *The Journal of Infectious Diseases, 201*, 662–671.

Quinn, T. C. (2008). HIV epidemiology and the effects of antiviral therapy on long-term consequences. *AIDS, 22*, S7–S12.

Rhee, S. Y., Jordan, M. R., Raizes, E., Chua, A., Parkin, N., Kantor, R., Van Zyl, G. U., Mukui, I., Hosseinipour, M. C., Frenkel, L. M., Ndembi, N., Hamers, R. L., Rinke De Wit, T. F., Wallis, C. L., Gupta, R. K., Fokam, J., Zeh, C., Schapiro, J. M., Carmona, S. … Shafer, R. W. (2015). HIV-1 drug resistance mutations: potential applications for point-of-care genotypic resistance testing. *PloS One, 10*, e0145772.

Rusconi, S., Marcotullio, S., & Cingolani, A. (2017). Long-acting agents for HIV infection: Biological aspects, role in treatment and prevention, and patient's perspective. *NewMicrobiology, 40*, 75–79.

Sidibé, M., Loures, L., & Samb, B. (2016). The UNAIDS 90–90–90 target: A clear choice for ending AIDS and for sustainable health and development. *Journal of International AIDS Society, 19*, 21133.

Sillman, B., Bade, A. N., Dash, P. K., Bhargavan, B., Kocher, T., Mathews, S., Su, H., Kanmogne, G., Poluektova, L., Gorantla, S., Mcmillan, J., Gautam, N., Alnouti, Y., Benson, E., & Gendelman, H. (2018). Creation of a long-acting nanoformulated dolutegravir. *Natural Communication, 9*, 443.

Su, Y., Chong, H., Qiu, Z., Xiong, S., & He, Y. (2015). Mechanism of HIV-1 resistance to short-peptide fusion inhibitors targeting the Gp41 pocket. *Journal of Virology, 89*, 5801–5811.

Thigpen, M. C., Kebaabetswe, P. M., Paxton, L. A., Smith, D. K., Rose, C. E., Segolodi, T. M., Henderson, F. L., Pathak, S. R., Soud, F. A., Chillag, K. L., Mutanhaurwa, R., Chirwa, L. I.,

Kasonde, M., Abebe, D., Buliva, E., Gvetadze, R. J., Johnson, S., Sukalac, T., Thomas, V. T. … Group, T. D. F. S. (2012). Antiretroviral preexposure prophylaxis for heterosexual HIV transmission in Botswana. *The New England Journal of Medicine, 367*, 423–434.

Trivedi, J., Mahajan, D., Jaffe, R. J., Acharya, A., Mitra, D., & Byraredy, S. N. (2020). Recent advances in the development of integrase inhibitors for HIV treatment. *Current HIV/AIDS Reports, 17*, 63–75.

Van Wesenbeeck, L., Rondelez, E., Feyaerts, M., Verheyen, A., Van der Borght, K., Smits, V, Cleybergh, C., De Wolf, H., Van baelen, K., & Stuyver, L. J. (2011). Cross-resistance profile determination of two second-generation HIV-1 integrase inhibitors using a panel of recombinant viruses derived from raltegravir-treated clinical isolates. *Antimicrobial Agents and Chemotherapy,* 55, 321–325.

Vieillard, V., Combadière, B., Tubiana, R., Launay, O., Pialoux, G., Cotte, L., Girard, P. M., Simon, A., Dudoit, Y., Reynes, J., Rockstroh, J., Garcia, F., Gatell, J., Devidas, A., Yazdanpanah, Y., Weiss, L., Fätkenheuer, G., Autran, B., Joyeux, D. … Katlama, C. (2019). HIV therapeutic vaccine enhances non-exhausted $CD4^+$ T cells in a randomised phase 2 trial. *npj Vaccines, 4*, 1–25.

Walter, J., & Kredo, T. (2017). Boosted protease inhibitor (PI) monotherapy for treating HIV/AIDS. *The Cochrane Database of Systematic Reviews, 8*, CD008171.

Wensing, A. M. J., Fun, A., & Nijhuis, M. (2017). HIV protease inhibitor resistance. In M. Gotte, A. Berghuis, G., Matlashewski, M. Wainberg, & D. Sheppard (Eds.), *Handbook of antimicrobial resistance*. https://doi.org/10.1007/978 1 4939-0694-9_28

Whitfield, T., Torkington, A., & van Halsema, C. (2016). Profile of cabotegravir and its potential in the treatment and prevention of HIV-1 infection: Evidence to date. *HIV AIDS, 8*, 157–164.

Yubin, L., Liu, Y., Cao, W., Sun, M., & Li, T. (2020). Broadly neutralizing antibodies for HIV-1: efficacies, challenges and opportunities. *Emerging Microbes & Infections, 9*, 194–206.

Zhou, T., Su, H., Dash, P., Lin, Z., Dyavar Shetty, B. L., Kocher, T., Szlachetka, A., Lamberty, B., Fox, H. S., Poluektova, L., Gorantla, S., Mcmillan, J., Gautam, N., Mosley, R. L., Alnouti, Y., Edagwa, B., & Gendelman, H. E. (2018). Creation of a nanoformulated cabotegravir prodrug with improved antiretroviral profiles. *Biomaterials, 151*, 53–65.

Chapter 15
Drug Resistance in Reproductive Diseases

Dilshad Hussain, Adeela Saeed, Hafiza Nadia Rashid, Batool Fatima, Muhammad Salman Sajid, Fahmida Jabeen, and Muhammad Najam-ul-Haq

Abstract The ineffectiveness of drugs against certain disease is termed as drug resistance. Drug resistance is a common phenomenon and a primary concern for health care professionals. Resistance to any particular drug(s) leads to the development of new therapies and vaccines. Drug resistance also exists in all types of reproductive system infections including bacterial infections, viral infections, and fungal infections. Bacterial resistance is found in Gonorrhea, Chlamydia, Syphilis, Chancroid, and Prostatitis. Viral resistance is associated with Genital herpes, Genital warts, and AIDS. Two main types of fungal infections (Candidiasis and Trichomoniasis) are also prone to drug resistance. The drug resistance ability of pathogens is different in regions, environments, and populations. Some pathogens are multi-drug resistant which cause problems in treatment. Each pathogen has its mechanism of action but most of these resistances are associated with genetic mutations caused during the interaction of drugs with targets. In this chapter, the effect of drug resistance on reproductive system diseases is discussed. Various aspects of resistance, possible alternatives, and future directions related to drug resistance are evaluated.

15.1 Introduction

Disorders, infections, congenital abnormalities affecting the human reproductive system or reproductive tract are known as reproductive diseases. Primarily there are two types of infections; Sexually transmitted diseases (STDs) and non-STDs

D. Hussain (✉)
International Center for Chemical and Biological Sciences, HEJ Research Institute of Chemistry, University of Karachi, Karachi 75270, Pakistan

A. Saeed · F. Jabeen
Department of Chemistry, The Women University Multan, Multan, Pakistan

H. N. Rashid · M. S. Sajid · M. Najam-ul-Haq
Institute of Chemical Sciences, Bahauddin Zakariya University, Multan 60800, Pakistan

B. Fatima
Department of Biochemistry, Bahauddin Zakariya University, Multan 60800, Pakistan

S. Ahmed et al. (eds.), *Biochemistry of Drug Resistance*,
https://doi.org/10.1007/978-3-030-76320-6_15

(Shaskolskiy et al., 2016). STDs are the infections of human reproductive system spread primarily through person-to-person sexual contact while non-STDs are endogenous infections of the genital organs. There may be factors causing reproductive diseases, such as abnormal hormone production (by ovaries, testes, or by other endocrine glands, for instance; thyroid, pituitary, or adrenals). Genetic or congenital abnormalities, tumors, infections, or disorders of unknown cause may also cause such diseases. There is health concern about non-STDs and STDs and approximately 1 billion new cases of STDs are reported each year according to WHO (Taylor et al., 2017). Drug resistance of pathogens results in immunity against drugs and difficulty in treatment. Reduced susceptibility to third-generation cephalosporins and association of other diseases such as staphylococcal infection, malaria, and tuberculosis, along with *N. gonorrhoeae* strains are the concerns during treatment (Josey & Schwebke, 2008). Thus, identification of pathogens and the etiology of infection is necessary. In this chapter, the drug resistance patterns of reproductive system diseases along with mechanisms and molecular determinants of drug resistance against pathogens are encompassed.

15.2 Infections in Reproductive System

15.2.1 Bacterial Infections

Bacteria can access reproductive organs and cause diseases of reproductive system. Following are some reproductive diseases caused by bacteria. Pathogenic infections in male reproductive systems are shown in Fig. 15.1.

Gonorrhea is asexually transmitted disease caused by Gram-negative diplococcus Neisseria gonorrhea often called gonococcus or GC. Gonorrhea is also colloquially known as the clap. The bacteria cause urethritis by attaching itself to the epithelial cells of male and female urethra. More extreme symptoms are shown in males than do females, with a penile discharge and burning with urination. Antibiotics such as tetracycline, penicillin, ceftriaxone, and azithromycin, and others are the treatments. Gonorrhea's complications may result in sterility, pelvic inflammatory disease (PID) or infections, etc. Newborns may have to suffer eye infection called gonococcal ophthalmia who are subjected to N. gonorrhea during track through the birth canal (Workowski & Bolan, 2015). Infection can be prevented by treating with silver nitrate and/or erythromycin shortly after birth. Untreated gonorrhea may spread throughout the body affecting the joints and heart valves (Edwards & Butler, 2011).Reduced access of drugs to the target site and alteration of target site itself by either chromosomal or extra chromosomal elements (plasmids) are the mechanisms of antibiotic resistance. The changes in porin proteins reduce permeability of cell envelope and antibiotic destruction before its interaction with the

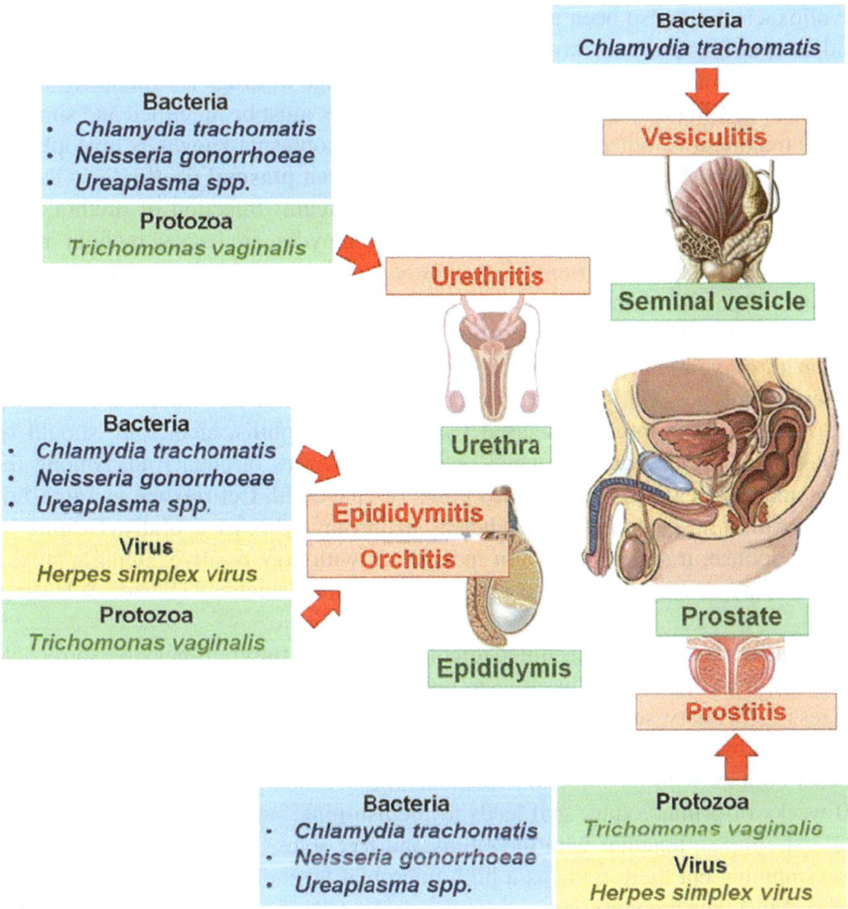

Fig. 15.1 Various microbial pathologies of the male reproductive tract (Sengupta et al., 2020)

target causing reduced access of antibiotic to active site. Multiple resistance mechanisms can coexist in an organism and a single strain can be resistant to many antibiotics. Plasmid-mediated resistance spreads more rapidly in case of N. gonorrhea than chromosomally mediated resistance (Unemo & Shafer, 2011).

Chlamydia is an infection caused by a member of chlamydia group of bacteria i.e. Chlamydia trachomatis. To differentiate it from gonorrhea, this disease is referred to as nongonococcal urethritis. Many a times, chlamydia is asymptomatic. Several million people suffer annually from this disease having frequent desire to urinate, pain during urination, itching, tenderness, abdominal pain, and a watery discharge. To treat this, Tetracycline is used. The condition may be complicated by PID resulting in sterility. Newborns may suffer chlamydial ophthalmia (Vasilevsky et al., 2014). Macrolides and tetracyclines are the drugs for Chlamydiosis. Ofloxacin and levofloxacin have also been used as alternates. The causative agent of urealyticum and ureaplasma appears as co-infection with Chlamydia. Due to this co-infection, its treatment becomes challenging, although wide range of drugs is available. For the treatment of Chlamydia, additional microorganisms must be identified and simultaneous treatment of other infections be made. Mycoplasma known as mycoplasma hominis causes **Mycoplasmal urethritis** while **urea plasmal urethritis** is caused by a mycoplasma known as Ureaplasma urealyticum. Infection of urethra, with similar symptoms as those of gonorrhea and chlamydia, are caused by both organisms. To treat these conditions, Tetracycline is used, and PID may cause complications. The microorganisms are its major source and it also appears as co-infection with Chlamydia. Tetracyclines (doxycycline) and macrolides (josamycin and azithromycin) are the drugs for this infection. Both ofloxacin and levofloxacin (fluoroquinolones) can be used as alternative remedy for this infection (Morrone, 2020).Frequent co-infection caused by herpes or syphilis, chancroid, should be treated byantibiotic with presumed or confirmed cases of chancroid. Among the recommended regimes of confirmed cases of chancroid, Ceftriaxone is a first-line drug as intramuscular injection whereas azithromycin is used orally. Despite its good response, it suffers failure in individuals with HIV-positive. Being a single-dose therapy, Azithromycin and ceftriaxone are more beneficial (Larsen & Hwang, 2010).

Syphilis is a sexually transmitted infection caused by the gram-negative spirochete *Treponema pallidum*. This disease was once known as "great pox". Syphilis has four stages i.e. (primary, secondary, latent, or tertiary) and the symptoms depend on these stages. Primary syphilis causes skin ulceration called a chancre. It is acquired by sexual contact with an infectious person. The secondary stage develops after 4–10 weeks of primary stage and leads to reddish-pink, non-itchy rash on the trunk and extremities, together with the palms and feet soles. The latent phase presents no symptoms but there remains a high microbial level. This phase can persist for years. Most severe symptoms are shown at the tertiary stage which can be fatal. Deadly lesions called gummas are developed at many locations and may cause tissue damage. Syphilis of cardiovascular system as well as the central nervous system is highly dangerous. Syphilis can be transmitted to the fetus from an infected mother (Bai et al., 2008). Among available drugs, the main drug used for Syphilis treatment is benzathine penicillin G. Others include tetracycline, doxycycline, macrolides, and ceftriaxone as third-generation drugs. Sometimes patients are allergic to penicillin where its use is avoided and replaced by other antibiotics mentioned above. For the treatment of early syphilis, azithromycin is more effective than penicillin G.

Chancroid is a sexually transmitted disease (STD) of reproductive tract caused by a small, gram-negative rod-shaped bacterium, *Haemophilus ducreyi* called as chancroid. A genital organs' swollen and painful ulcer with infection of lymph nodes called buboes are the disease characteristics. Chancroid is also known as soft chancre and tetracycline can treat this disease. A strain of Chlamydia trachomatis causes this sexually transmitted disease **Lymphogranuloma venereum (LGV)** which presents lesions at the infection site followed by swollen lymph nodes called buboes. Such changes may progress to abscesses, necrosis, and fistulas. The infectious agent breaks in the skin or passes through the epithelial layer of mucous membranes to enter the body. Antibiotics, generally doxycycline, tetracycline, or erythromycin may be used for the treatment of LGV (Arnold et al., 2013).

Acute or chronic **prostatitis** can be caused by gram-negative bacteria such as *Escherichia coli*, *Klebsiella*, *Pseudomonas*, *Enterobacter*, *Proteus* as well as gram-positive bacteria such as *Staphylococcus aureus*. Symptoms like chills, groin, fever, and lower back pain, urination issues, and general body aches arise in case of acute prostatitis, on the other hand, chronic prostatitis is a rare condition which may cause urinary tract infections (UTIs) leading to cystitis. It may sometimes be asymptomatic. Acute prostatitis requires immediate treatment otherwise, it may cause complications (Ramakrishnan & Salinas, 2010). **Pelvic Inflammatory Disease** (PID) is due to inflammation of female reproductive organs including the ovaries, uterus, and fallopian tubes. Fever, unusual discharge, irregular menstrual bleeding, lower abdominal pain, painful intercourses are the symptoms of PID. Infectious agent such as virus, bacteria, and fungi may cause pelvic inflammatory disease but C. trachomatis, and N. gonorrhoea are the infectious agents for PID (Soper, 2010).

15.2.2 Viral Infections

Genital herpes is caused by a DNA icosahedral virion, the herpes simplex virus. Mostly type II strain of the virus is present in genital herpes. Pain with urination and vesicles filled with fluid on genital organs are the symptoms of genital herpes. Newborns can be affected at the time of birth or before birth too, as the virus is capable of crossing placenta and affecting the fetus. **Genital warts** is highly contagious STD caused due to many types of human papillomaviruses, such virus contain DNA. Warts can be smooth, flat, or enlargement of the dermal papillae (finger like projections). Few methods are available to treat visible warts otherwise, there is no cure for this viral disease (Tamer et al., 2019).

Acquired immune deficiency syndrome **(AIDS)** is caused by a human immunodeficiency virus (HIV) and is a disease of human immune system. At the start of disease, patient experiences illness as in influenza for a short period, and then this disease may prolong asymptomatically. AIDS is mostly transmitted sexually and is also transmitted by hypodermic needles, contaminated blood transfusions, and can be transferred to infants from mother during pregnancy, delivery, or through breastfeeding. HIV is not transmitted through saliva or tears. AIDS does not have any cure or vaccine, but the use of antiretroviral may reduce disease complications and death risk (Park et al., 2009).

15.2.3 Fungal Infections

A disease of the reproductive tract caused by yeast Candida Albicans is known as **Candidiasis**. Vulvovaginal candidiasis infections accompany the vagina's mucous membranes. Overuse of antibiotics may lead to this disease. Symptoms include dysuria, irritation and vulval soreness, itching, edema, lesions similar to those in thrush, yellowish, cheesy discharge, as well as severe pruritis. Anti-fungal drugs like nystatin, clotrimazole, and miconazole can treat vulvovaginal candidiasis (Pappas et al., 2018). **Trichomoniasis** sometimes referred to as "trich", is the only protozoal disease of reproductive tract caused by single-celled Flagellate Trichomonas vaginalis. This protozoan grows along the mucosa of reproductive tract and results in profuse and internal discomfort, inflammation of cervix, urethra, and vagina, green-yellow discharge with a foul odor, and rarely lower abdominal pain may occur. This disease is sexually transmitted. Metronidazole is used for the therapy of Trichomoniasis (Schwebke & Burgess, 2004).

Antifungal drug resistance can be microbiological or clinical. Such drug resistance developmental mechanisms are complex. Antifungal drug resistance also depends on hosts and microbial factors. The host immune system is an important factor. Up-regulation of pharmaceutical transporters, alteration of drug target, and cellular changes caused induced by stress response are drug resistance mechanisms. Inhibition of drug binding by altering the target through the substitution of amino acids is an azole drug resistance mechanism in candida. More than 140 substitutions have been reported in resistant strains. Mostly antifungal resistance is acquired through multiple mechanisms in different sequential steps. Production of matrix is a key factor in drug resistance by *C. Albicans* and other candida species. Loss of heterozygosity (LOH) generates phenotypic diversity in *C. Albicans*. Enhanced mRNA stability and chromosomal rearrangements also contribute to the azole resistance mechanism in *C. Albicans*. All resistance mechanisms are not followed by all fungal strains (Zhang et al., 2020).

15.3 General Mechanism of Drug Resistance in Reproductive Diseases

> The defence strategies acquired by microorganisms to survive against antibiotics are termed as resistance mechanisms

There is diversity in the mechanisms of resistance in bacteria. Due to resistance, the modification of existing drugs and evolution of new drugs is of high concern. Adequate disease therapy and in-time detection are the only way to treat and control disease spread. The selection of treatment is challenging due to increased resistance to pathogens. Bacteria DNA helps in developing resistance against antibiotics. Mostly, plasmids contain resistance genes. Bacteria can share their DNA with other microorganisms to become resistant. Following resistance mechanisms are adopted by microorganisms:

- Microorganisms can change or limit entryways for the drug. For example, gram-negative bacteria can use the outer membrane to selectively stop drug entry.
- By using the pumps inside their cell walls, they can push out the entered drug. In this way, they get rid of antibiotics.
- By using the enzymes and proteins, microorganisms can break down the drugs and can change or destroy their structures.
- As most drugs target specific part of bacterium, few microbes can change the drug targets, so, drugs do not fit and cannot perform their activities.
- By developing new cell processes, the microbes can bypass the drug effect.

15.4 Drug Resistance in Reproductive System

15.4.1 Bacterial Drug Resistance

Increase in pathogenic drug resistance to antimicrobial agents is a global concern. Drugs and strains lose their effectiveness as pathogens immune to their activity. Such drugs are replaced after certain period and new strains effectively respond. This cycle continues and with every passing day, new drugs are developed to overcome the problem. With this rise in the number of drugs, pathogens also develop resistance to multiple drugs, termed as multidrug resistance (MDR) (Unemo & Nicholas, 2012). This type of MDR has been found in both gram-negative and gram-positive bacteria. Drug-resistance is associated with socio-economic growth of population, population ecology of infectious disease, and antibiotic consumption by individuals in the population (Fig. 15.2).

It has been more than 70 years that syphilis is treated by penicillin's and till now not a single penicillin-resistant is found and no clinically failed attempt has been reported for the treatment of syphilis using penicillin-related drugs. Risk of resistance however cannot be ignored. Recently, a study on resistance to antibiotics by *T. pallidum* is reported. Results show no resistance to b-lactamantibiotics, and no mutation in *tp47* and *tromp* genesis found (Kubanova et al., 2013). Similarly, *T. pallidum* does not show any resistance to tetra cyclines. Some resistance of *T. pallidum* is reported to macrolides which are the alternative to penicillin. In 1977, erythromycin and azithromycin resistant strains (Street strain 14) were identified. The resistance was due to the absence of plasmids and transposons in that strain responsible for resistance. Resistance is originated because of chromosomal mutation in 23S rRNA gene (Stamm, 2012).

The common pathogen for developing resistance against antibiotics in 10 20 years is *N. gonorrhoeae. N. gonorrhoeae* can mutate and develop multiple drug resistance. Several strains of this pathogen are resistant to almost every clinical drug used to treat

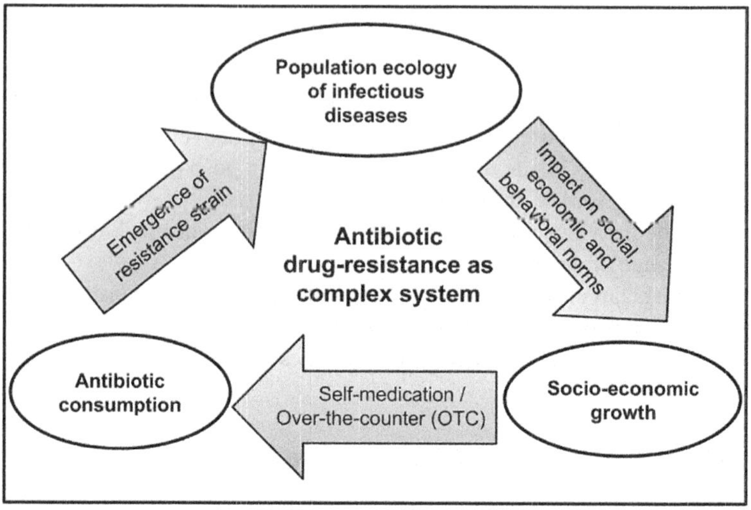

Fig. 15.2 Antibiotic drug-resistance as an integrated system in the population (Malik & Bhattacharyya, 2019)

gonorrhea, including azithromycin, spectinomycin, macrolides, tetracyclines, fluoroquinolones, cephalosporins, and b-lactams. These resistant strains are the global concern as there have been cases of *N. gonorrhoeae* which are incurable and untreatable till 2021. There is no progress in the development of new drugs or vaccines for such strains (Unemo, 2015). On the other hand, mechanism of drug resistance in Gonococci are well-known; the reduction of membrane permeability, alteration of drug binding sites, increased drug efflux, and the inactivation of antibiotic (Fig. 15.3). Such new characteristics in bacteria are imparted via chromosomal mutations, especially horizontal gene transfer and plasmids carrying the determinants of resistance. Gene transfer occurs from other species of *Neisseria.* The effective type of mutation increases the activity of flux pumps and alter membrane permeability. Both of these mechanisms develop resistance against variety of drugs. Penicillin treatment against Gonococci was started in 1940 and was effective for 20 years. Later, the activity of penicillin decreased due to chromosomal mutations. Over the years, other drugs including azithromycin, tetracycline, and erythromycin are also applied (Unemo & Shafer, 2014).

Ciprofloxacin, spectinomycin, ofloxacin, and β-lactam antibiotics have been used for Gonococcal infection. Azithromycin and ceftriaxone are currently used. The use of both generations of β-lactams has decreased due to the widespread penicillin-resistant gonorrhea. Ceftriaxone remained effective due to resistance against lactamases. For the patients allergic to β-lactams, Spectinomycin is useful but not used as first-line drug. Dual therapy using two antimicrobial agents (cephalosporin plus azithromycin) with different action mechanisms improves efficiency. Azithromycin

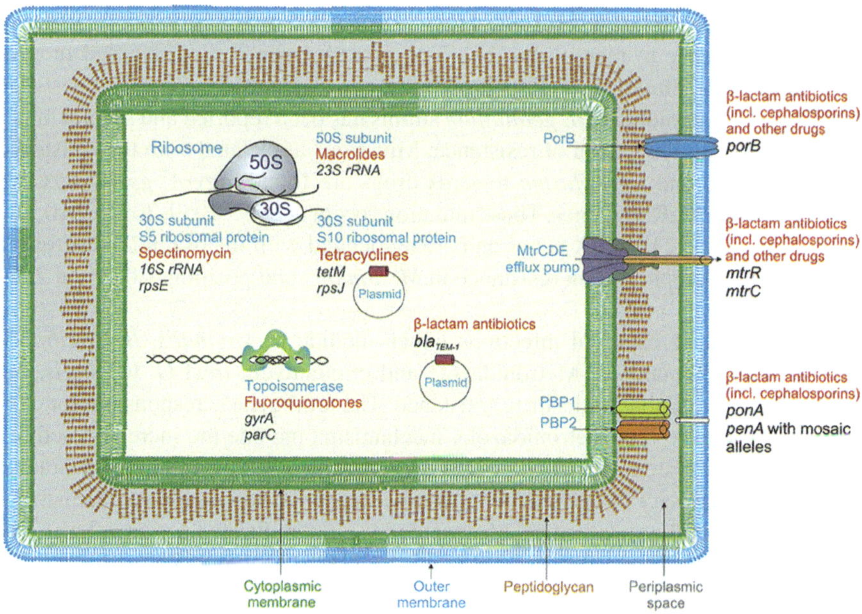

Fig. 15.3 A general scheme for antimicrobial resistance determinants in *N. gonorrhoeae* (Shaskolskiy et al., 2016)

is preferred over doxycycline as a second antimicrobial drug because of its single-dose therapy and ubiquity of gonococcal resistance to tetracycline (Control and Prevention, 2012).

Drug resistance and its resistance mechanism for *C. trachomatis* is contradictory. There have been cases of multidrug resistance in *C. trachomatis* (Kohlhoff & Hammerschlag, 2015). In patients with chlamydial infections, treatment with doxycycline and macrolides was unsuccessful due to multi-drug resistant strains resistant to azithromycin, doxycycline, josamycinand, ofloxacin, and spiramycin. Several clinical biomarkers have been linked to macrolides and fluoroquinolones, but the relationship of these markers with antibiotic therapy for *Chlamydia* is yet to be understood. Although, no mutations in *gyrA*, *gyrB*, *parC*, and *pare* genes are found, these pathogens are resistant to fluoroquinolones and macrolides. Chlamydia usually develops phenotypic resistance, known as heterotypic resistance, in which a small proportion of pathogen develops resistance. One of the characteristics of this type of resistance is that the isolated strains from patients are sensitive to the same drugs (Kohlhoff & Hammerschlag, 2015).

There is no cell wall in the members of *Mycoplasma* and *Ureaplasma* family. They have natural susceptibility and resistance towards all types of b-lactam antibiotics. Sulfonamides are also ineffective in these classes of pathogens at physiological concentrations as there is no synthesis of folic acid due to the absence of metabolic pathways. About 73–97% resistance rate is reported for erythromycin and tetracycline. Resistance to macrolides including moxifloxacin is also reported in clinical and microbial treatments (Redelinghuys et al., 2014). More recently, resistance against fluoroquinolone in *M. genitalium* strains has been reported and several mutations in DNA are the reason of resistance. Mutations associated with the resistance of *Mycoplasma* and *Ureaplasma* towards drugs are found in*gyrA, gyrB, parC,* or *parE*gene and 23SrRNA genes. These mutations occur at 2057, 2059, 2062, 2610, and 2611 positions of 23SrRNA genes and are associated with lincosamide, macrolide, orketolide and streptogramin resistance in *M. hominis* and positions 2058 and 2059 in *M. genitalium.*

There are other bacterial infections which include *G. vaginalis, B. fragilis, A. vaginae, and M. mulieris.* Metronidazole and clindamycin treat *G. vaginalis,* but limited data is available for its drug resistance. The tetM gene is responsible for clindamycin resistance. For metronidazole, mechanisms include the increased activity of DNA repair system, enhanced performance of active flux for drug clearance, and enhanced activity of enzymes to consume oxygen. The reliable mechanism for metronidazole is the inactivation of genes with nitro-reductase activity (Dhand & Snydman, 2009). *B. fragilis* possesses multi-drug resistance to clindamycin, amoxicillin, metronidazole, and imipenem and the common mechanism is the production of b-lactamases. *A. vaginae* exhibits high-level resistance to metronidazole by the presence and activation of *nim*-genes encoding that causes DNA breakage. *M. mulier* shows resistance to metronidazole from 50 to 81% associated totetQ gene.

15.4.2 Viral Drug Resistance

External genital warts (EGWs), also termed as condylomata acuminate (CA), are STDs spread across the World. In USA, more than 1 million cases are reported each year. EGWs are associated with human papillomavirus (HPV) which are more than 120 subtypes of HPV. HPVs with more than 40 subtypes cause infections associated with genital warts and cervical cancer (Yanofsky et al., 2012). HPVs are divided into three categories depending on their capability to develop intraepithelial dysplasias; high risk, medium risk, and low risk. Low-risk HPV types are 6 and 11 as they are not associated with cervical cancers. 90% of these sub-types form genital wart. Genotypes 16 and 18 of HPVs are high-risk or called oncogenic types as they develop cervical dysplasia. One of the evidence is 70% formation of squamous cell carcinomas (SCC) in cervix. Some of the types of HPVs are medium risk which include 59, 56, 45, and 31. These genotypes are associated to squamous neoplasms but not linked with cervical cancers (Izadian et al., 2004). The number of HPV-infected patients has increased in the last decades and in USA, more than 20 million people are infected. The reason is the increase in number of sexual partners. Most infected persons are below the age of 25 years.

Instead of virus eradication, warty tissue is removed. Along with the topical therapeutics, it can be physically removed by laser, cryotherapy, or electrocautery. Photodynamic therapy can also be made. 5-fluorouracil/epinephrine with high concentration of 5-fluorouracil is an effective therapy. Other choices include activation of immune system or improvement in the delivery of therapeutic compounds at infection site. Imiquimod (imidazoquinolinamines) 5% cream as a modifier for immune response induce endogenous cytokines and interferon. 0.05% solution of podophyllotoxin and 0.15% cream treat EGWs which inhibit the mitotic division by binding with cellular microtubules. Method is somewhat safe and effective. 80–90% solution of trichloroacetic acid (TCA) is also recommended for warts treatment. Vaccination can prevent genital warts instead of being treated (Singh, 2020).

Main cause of genital ulcersis genital herpes. Genital herpes are associated with herpes simplex virus (HSV). Both HSV2 and HSV1 can cause genital herpes, however HSV2 is more linked to genital herpes as compared to HSV1. People with genital herpes infections are prone to HIV and AIDS. Clinical diagnosis of these viruses is difficult with non-specific and non-sensitive tests. Serological tests identify the virus. Acyclic guanosine, pyrophosphate analogues, and acyclic nucleoside analogue drugs treat HSV by targeting the viral DNA replication (Gupta et al., 2007). Cidofovir, valacyclovir (VCV), and foscarnet drugs belong to these categories. Acyclovir (ACV) treats HSV however they become resistant strains after long use. A gold standard for prophylaxis is ACV (9-(2-hydroxy ethoxy methyl) guanine) in treating HSV. Other nucleoside drugs include VCV, ganciclovir, and famciclovir as the first-line drug against HSV infections (Jiang et al., 2016).

Severe acute respiratory syndrome Corona Virus 2019 (SARS-COV-2) is also associated with reproductive diseases. COVID-19 virus can target male reproductive systems because of co-expression of ACE2 receptor and the protease (TPPRSS) (Fig. 15.4). Virus expression can lead to fertility issues in males. Due to the absence of ACE proteins in female reproductive system, SARS-COV-2 is not attached, hence, no infection in females (Zupin et al., 2020).

HIV in humans causing AIDS was incurable for some time. Several antiviral drugs treat HIV including retroviral treatments (ARTs). HIV virions have also become tolerant to ARTs due to microevolution and genetic mutations. There is no certain cure to this disease. The polymerase has high frequency of mutation also coupled to the high rate of HIV replication which makes the treatment of HIV infection challenging due to drug resistance. To fight HIV, the drugs should act on viral life cycle rather than nucleoside analogues. Targeting viral protease can also treat. Acyclovir is a drug with high viral selectivity and inhibits viral polymerase more efficiently than cellular polymerase. Enfuvirtide is similar to HIV surface protein. Due to identical nature with surface protein, it competes or interferes with the fusion process of surface protein during its release. This interference prevents the new cells from infection. Enfuvirtide is especially useful than other viral drugs (Jerome, 2005).

The drug resistance in HIV infection due to several strains was reported in 1989. Primary resistance and induced resistance are the resistances associated with HIV. Primary resistance is not developed due to ART therapy but some strains are resistant to viral drugs. Primary resistance occurs when a person is infected with previous

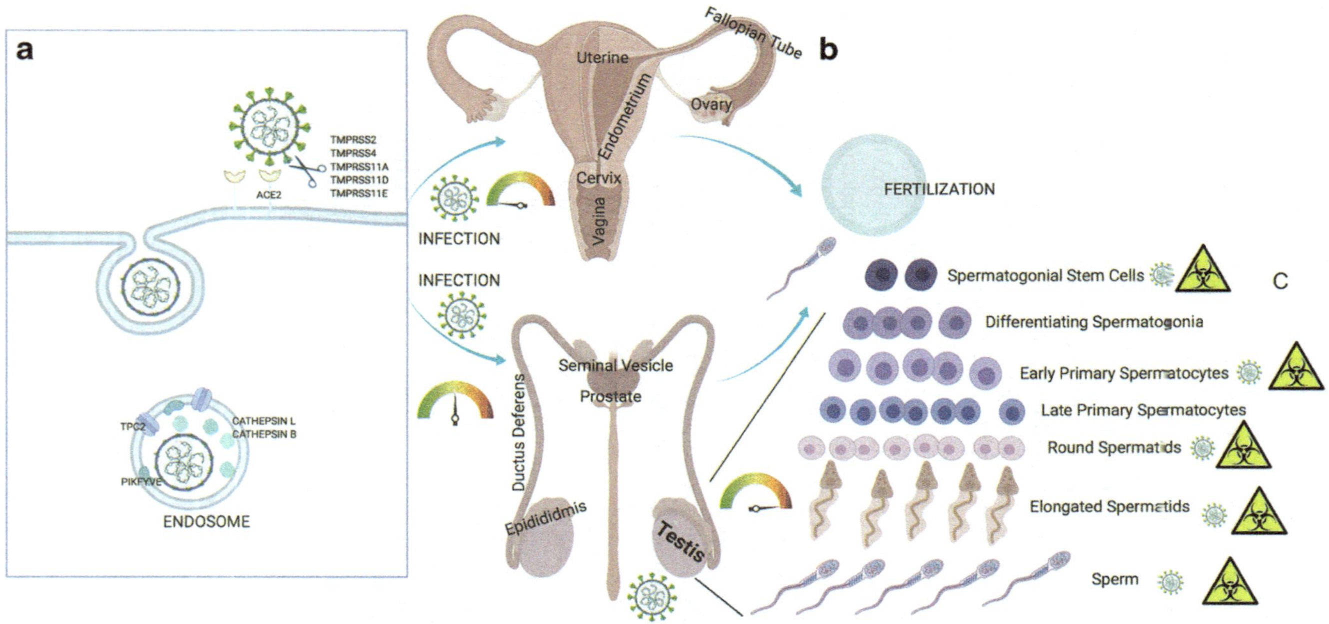

Fig. 15.4 Schematic diagram showing effect of SARS-CoV-2 on reproductive system (Zupin et al., 2020)

resistant strain of HIV. With limited options in ARTs, this resistance can cause problems in later stages of disease. Induced resistance is caused by retroviral treatments (ARTs) (Fig. 15.5).Phenotype and genotype testings are used for drug resistance testing in HIV strains. Genotype testing tests the genetic structure (genotype) of HIV taken from infected persons while phenotype testing measures the mutations, behavior, and HIV phenotype. HIV is a retrovirus with fast replication and mutates causing drug resistance. Fast mutation rates using reverse transcriptase lead to higher survival for the virus by developing new types of strains (Larder, 2001). These mutations vary in populations, hence, variety of strains may exist in different regions of world, leading to complexity in developing drugs or vaccines for the treatment. Mechanisms of drug resistance include blocking of nucleosides by mutations and adherence to the prescribed drug regiment. Lack of health facilities, missing doses of drugs during treatment, low availability of drugs in some countries, and stigmatization of HIV also replicate the depleted virus (Nikolenko et al., 2011). The way to treat HIV-infected patients is to minimize viral load. Most of the drugs have same mechanism of action, resistance is also developed against all drugs. Drug resistance in countries is different ranging from 0.2 to 76.3% for NNRTIs and NRTIs (Kuritzkes, 2011).Number of HIV infected patients with resistant strains have increased issues for World Health Organization and health care officials. There is a need to search new drugs and treatment therapies (Nachega et al., 2011).

15.4.3 *Fungal Drug Resistance*

Candidiasis and Trichomoniasis are the two fungal infections associated with reproductive system. Candidiasis is a yeast disease and resistance is common in these infections. Reasons for drug resistance include point mutations and chromosomal rearrangements (Arendrup & Patterson, 2017). These mutations change genes by blocking the binding of drugs to their targets. Resultantly, biofilms are formed in fungal strains and acquire resistance. Fluconazole treats candidiasis but 7% of blood cells are drug resistant. Resistance is associated to all species of Cadidiasis including *Candida albicans, Candida parapsilosis, Candida Auris, and Candida glabrata* (Toda et al., 2019). Echinocandins is the drug resistant for *Candida glabrata* which also resist to antifungal fluconazole. The resistance is prevailing for about 20 years and thus have limited the treatment of Candidiasis. Amphotericin B treats Candidiasis but it is toxic to serious patients. The survival rate in immunocompromised patients or elderly patients who are candidemia drug-resistant is slightly less than the other patients with candidemia. Thus, candidemia can be treated with antifungal drugs (Alexander et al., 2013).

The mutation mechanism of resistance is divided into two classes; (i) mutations changing the binding of drug with target and (ii) mutations for low intake by intracellular accumulation of drug (Carreté et al., 2018). Among antifungal drugs, azoles work by inhibiting yeast proliferation but are unable to destroy yeast cells. Localized

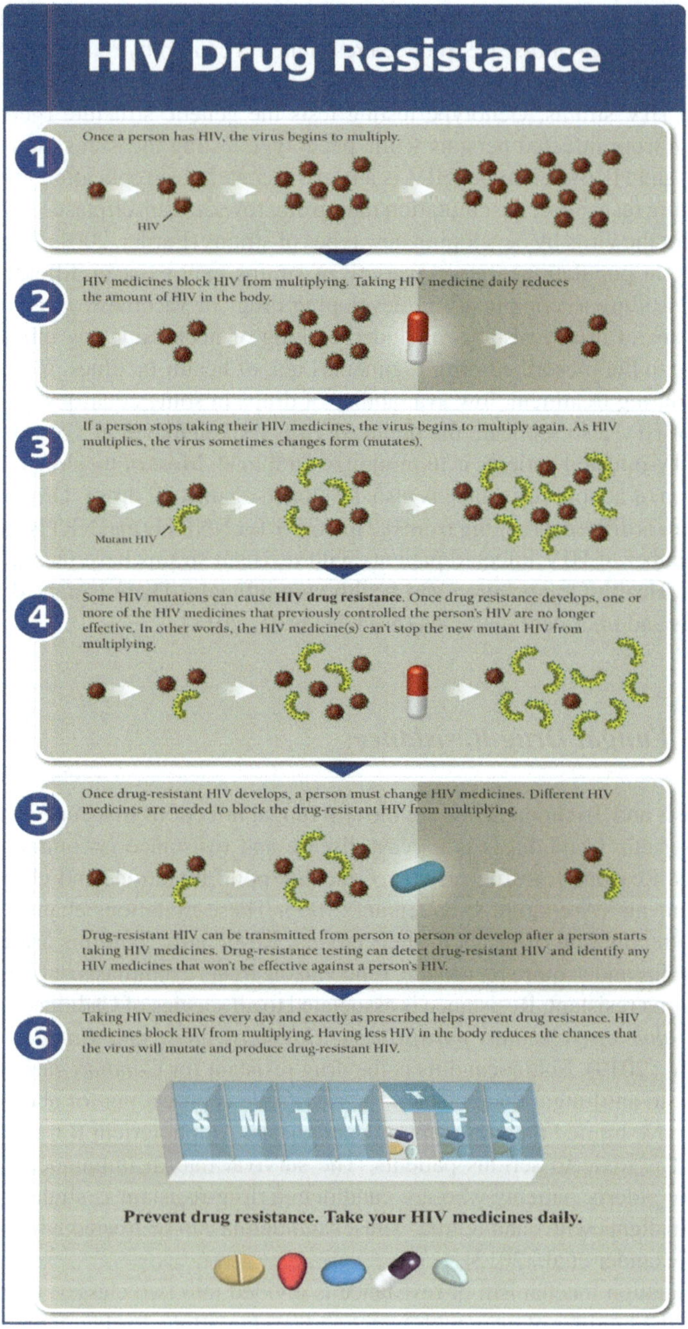

Fig. 15.5 General mechanism of drug resistance in HIV (https://hivinfo.nih.gov)

candida infections are cured by these drugs. Azoles include butoconazole, clotrimazole, miconazole, tioconazole, and Isoconazole. Weaker azoles in treating yeast infection may lead to resistant Candida species. Action mechanism in Cadniasis is shown in Fig. 15.6 (Ksiezopolska & Gabaldón, 2018). Candida also shows drugs susceptibility and epidemiology studies reveal steady increase in cases. Moreover, secondary mutations also cause resistance in Candida species. These species adapt to environmental conditions due to genome plasticity. During drug treatment, the fungal populations lead to increased allele frequencies of mutants. In prolong treatments and hospitalization, the species develop drug resistance and do not respond to treatment. Drug sequestration in biofilm matrix, the use of sub-therapeutic concentrations, and poor control of infections also induce resistance (Ben-Ami, 2018).

The prevalence of STDs is found in Trichomoniasis which leads to HIV and cervical cancer. In trichomoniasis, metronidazole is effective but upto 10% of infections are resistant to this treatment. Metronidazole is a prodrug and enters the cell via passive diffusion and competes with hydrogenase enzyme. The drug is activated after electron transfer. Free radical is formed from the nitro groups followed by impairing of molecular hydrogen resulting in hydrogen peroxide production. Finally, the free radical is converted to prodrug in aerobic conditions (Dunne et al., 2003). Tinidazole, chemically related to metronidazole is also used for trichomonas treatment. Metronidazole treats Vaginitis/Trichomoniasis caused by *T. vaginalis*. Resistance against this

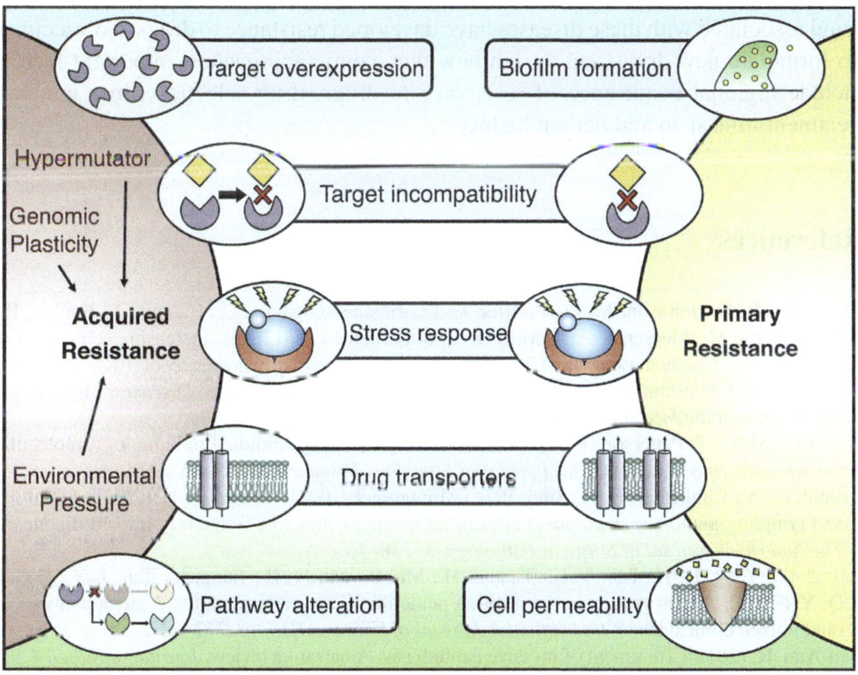

Fig. 15.6 Antifungal drug actions and resistance mechanisms (Revie et al., 2018)

drug has been reported without known mechanism. It is however proposed that drug resistance by parasite can be aerobic or anaerobic. It is unclear whether the events are primary or secondary advents of resistance. Aerobic resistance is proposed in naturally infected patients while anaerobic resistance can be developed in laboratory (Upcroft & Upcroft, 2001).

15.5 Conclusion

Development of drug resistance in humans is a natural response of evolutionary process due to the excessive use of antibiotics, antifungal agents, antiviral therapies and chemotherapies. Pathogens strains become resistant after a specific period of time via mutations or genetic changes. Some mutations can result in multi-drug resistance leading to ineffective therapeutic response. There is a continuous war between the disease and its treatment. To develop new drugs, vaccines and treatment therapies, it is necessary to understand drug mechanisms and the emergence of resistance. Modifications of existing drugs and to develop new drug targets are necessary for ever-increasing reproductive system diseases. Gonorrhea, Chlamydia, Mycoplasmal urethritis, Syphilis, Chancroid, HIV, Prostatitis, Genital herpes, Candidiasis and Trichomoniasis are some reproductive diseases. Number of therapies are being used for the treatment of these diseases with varying efficiency. The bacteria, viruses and fungi associated with these diseases have developed resistance to drugs and vaccines. To formulate new drugs and design new therapeutic approaches, important factors include regional occurrence of resistance to drugs, drug selection, drug amount, treatment time span and patient history.

References

Alexander, B. D., Johnson, M. D., Pfeiffer, C. D., Jiménez-Ortigosa, C., Catania, J., Booker, R., Castanheira, M., Messer, S. A., Perlin, D. S., & Pfaller, M. A. (2013). Increasing echinocandin resistance in Candida glabrata: Clinical failure correlates with presence of FKS mutations and elevated minimum inhibitory concentrations. *Clinical Infectious Diseases: An Official Publication of the Infectious Diseases Society of America, 56*, 1724–1732.

Arendrup, M. C., & Patterson, T. F. (2017). Multidrug-resistant candida: Epidemiology, molecular mechanisms, and treatment. *The Journal of Infectious Diseases, 216*, S445–S451.

Arnold, C. A., Limketkai, B. N., Illei, P. B., Montgomery, E., & Voltaggio, L. (2013). Syphilitic and lymphogranuloma venereum (LGV) proctocolitis: Clues to a frequently missed diagnosis. *The American Journal of Surgical Pathology, 37*, 38–46.

Bai, Z.-G., Yang, K.-H., Liu, Y.-L., Tian, J.-H., Ma, B., Mi, D.-H., Jiang, L., Tan, J.-Y., & Gai, Q.-Y. (2008). Azithromycin vs. benzathine penicillin G for early syphilis: A meta-analysis of randomized clinical trials. *International Journal of STD & AIDS, 19*, 217–221.

Ben-Ami, R. (2018). Treatment of invasive candidiasis: A narrative review. *Journal of Fungi, 4*, 97.

Carreté, L., Ksiezopolska, E., Pegueroles, C., Gómez-Molero, E., Saus, E., Iraola-Guzmán, S., Loska, D., Bader, O., Fairhead, C., & Gabaldón, T. (2018). Patterns of genomic variation in

the opportunistic pathogen Candida glabrata suggest the existence of mating and a secondary association with humans. *Current Biology, 28*(15–27), e7.
Control, C. F. D., & Prevention. (2012). Update to CDC's sexually transmitted diseases treatment guidelines, 2010: Oral cephalosporins no longer a recommended treatment for gonococcal infections MMWR. *Morbidity and Mortality Weekly Report, 61*, 590.
Dhand, A., & Snydman, D. R. (2009). *Mechanism of resistance in metronidazole*. Springer.
Dunne, R. L., Dunn, L. A., Upcroft, P., O'Donoghue, P. J., & Upcroft, J. A. (2003). Drug resistance in the sexually transmitted protozoan Trichomonas vaginalis. *Cell Research, 13*, 239–249.
Edwards, J. L., & Butler, E. K. (2011). The pathobiology of Neisseria gonorrhoeae lower female genital tract infection. *Frontiers in Microbiology, 2*, 102.
Gupta, R., Warren, T., & Wald, A. (2007). Genital herpes. *The Lancet, 370*, 2127–2137.
Izadian, Z., Niakan, M., Eftekhar, Z., Jamali, Z. M., Golalipour, F., Faghihzadeh, S., & Jalali, M. R. 2004. Human papillomairus genotype as a major determinant of the course of cervical cancer.
Jerome, K. R. (2005). The road to new antiviral therapies. *Clinical and Applied Immunology Reviews, 5*, 65–76.
Jiang, Y.-C., Feng, H., Lin, Y.-C., & Guo, X.-R. (2016). New strategies against drug resistance to herpes simplex virus. *International Journal of Oral Science, 8*, 1–6.
Josey, W., & Schwebke, J. (2008). The polymicrobial hypothesis of bacterial vaginosis causation: A reassessment. *International Journal of STD & AIDS, 19*, 152–154.
Kohlhoff, S. A., & Hammerschlag, M. R. (2015). Treatment of chlamydial infections: 2014 update. *Expert Opinion on Pharmacotherapy, 16*, 205–212.
Ksiezopolska, E., & Gabaldón, T. (2018). Evolutionary emergence of drug resistance in candida opportunistic pathogens. *Genes, 9*, 461.
Kubanova, A., Kubanov, A., Frigo, N., Volkov, I., Rotanov, S., & Suvorova, A. (2013). First experience of molecular typing and determining the antibiotic resistance of syphilis pathogen Treponema pallidum in the Russian Federation. *Vestn. Dermatol. Venerol, 3*, 34–36.
Kuritzkes, D. R. (2011). Drug resistance in HIV-1. *Current Opinion in Virology, 1*, 582–589.
Larder, B. (2001). Mechanisms of HIV-1 drug resistance. *AIDS, 15*, S27–S34.
Larsen, B., & Hwang, J. (2010). Mycoplasma, Ureaplasma, and adverse pregnancy outcomes: a fresh look. In *Infectious diseases in obstetrics and gynecology.*
Malik, B., & Bhattacharyya, S. (2019). Antibiotic drug-resistance as a complex system driven by socio-economic growth and antibiotic misuse. *Scientific Reports, 9*, 1–12.
Morrone, A. (2020). Donovanosis, chancroid, and endemic treponematoses: Clinical features and control. In Cristaudo, A. & Giuliani, M. (Eds.), *Sexually transmitted infections: Advances in understanding and management.* Springer International Publishing.
Nachega, J. B., Marconi, V. C., van Zyl, G. U., Gardner, E. M., Preiser, W., Hong, S. Y., Mills, E. J., & Gross, R. (2011). HIV treatment adherence, drug resistance, virologic failure: Evolving concepts. *Infectious Disorders Drug Targets, 11*, 167–174.
Nikolenko, G., Kotelkin, A., Oreshkova, S., & Ilyichev, A. (2011). Mechanisms of HIV-1 drug resistance to nucleoside and nonnucleoside reverse transcriptase inhibitors. *Molecular Biology, 45*, 93–109.
Pappas, P. G., Lionakis, M. S., Arendrup, M. C., Ostrosky-Zeichner, L., & Kullberg, B. J. (2018). Invasive candidiasis. *Nature Reviews Disease Primers, 4*, 1–20.
Park, B. J., Wannemuehler, K. A., Marston, B. J., Govender, N., Pappas, P. G., & Chiller, T. M. (2009). Estimation of the current global burden of cryptococcal meningitis among persons living with HIV/AIDS. *AIDS, 23*, 525–530.
Ramakrishnan, K., & Salinas, R. C. (2010). Prostatitis: Acute and chronic. *Primary Care: Clinics in Office Practice, 37*, 547–563.
Redelinghuys, M. J., Ehlers, M. M., Dreyer, A. W., Lombaard, H. A., & Kock, M. M. (2014). Antimicrobial susceptibility patterns of Ureaplasma species and Mycoplasma hominis in pregnant women. *BMC Infectious Diseases, 14*, 171.
Revie, N. M., Iyer, K. R., Robbins, N., & Cowen, L. E. (2018). Antifungal drug resistance: Evolution, mechanisms and impact. *Current Opinion in Microbiology, 45*, 70–76.

Schwebke, J. R., & Burgess, D. (2004). Trichomoniasis. *Clinical Microbiology Reviews, 17*, 794–803.

Sengupta, P., Dutta, S., Alahmar, A. T., & D'Souza, U. J. A. (2020). Reproductive tract infection, inflammation and male infertility. *Chemical Biology Letters, 7*, 75–84.

Shaskolskiy, B., Dementieva, E., Leinsoo, A., Runina, A., Vorobyev, D., Plakhova, X., Kubanov, A., Deryabin, D., & Gryadunov, D. (2016). Drug resistance mechanisms in bacteria causing sexually transmitted diseases and associated with vaginosis. *Frontiers in Microbiology, 7*.

Singh, R. (2020). *Single-cell sequencing in human genital infections*. Springer.

Soper, D. E. (2010). Pelvic inflammatory disease. *Obstetrics & Gynecology, 116*, 419–428.

Stamm, L. V. (2012). Antibiotic resistance in Treponema pallidum subsp. pallidum, the syphilis agent. In *The Pathogenic Spirochetes: Strategies for evasion of host immunity and persistence*. Springer.

Tamer, F., Yuksel, M. E., & AVCı, E. (2019). Should patients with anogenital warts be tested for genital herpes? Initial results of a pilot study. *Our Dermatology Online/Nasza Dermatologia Online, 10*.

Taylor, M. M., Korenromp, E., & Wi, T. (2017). Pathways and progress to enhanced global sexually transmitted infection surveillance. *PLoS Medicine, 14*, e1002328.

Toda, M., Williams, S. R., Berkow, E. L., Farley, M. M., Harrison, L. H., Bonner, L., Marceaux, K. M., Hollick, R., Zhang, A. Y., & Schaffner, W. (2019). Population-based active surveillance for culture-confirmed candidemia—four sites, United States, 2012–2016. *MMWR Surveillance Summaries, 68*, 1.

Unemo, M. (2015). Current and future antimicrobial treatment of gonorrhoea–the rapidly evolving Neisseria gonorrhoeae continues to challenge. *BMC Infectious Diseases, 15*, 364.

Unemo, M., & Nicholas, R. A. (2012). Emergence of multidrug-resistant, extensively drug-resistant and untreatable gonorrhea. *Future Microbiology, 7*, 1401–1422.

Unemo, M., & Shafer, W. M. (2011). Antibiotic resistance in Neisseria gonorrhoeae: Origin, evolution, and lessons learned for the future. *Annals of the New York Academy of Sciences, 1230*, E19.

Unemo, M., & Shafer, W. M. (2014). Antimicrobial resistance in Neisseria gonorrhoeae in the 21st century: Past, evolution, and future. *Clinical Microbiology Reviews, 27*, 587–613.

Upcroft, P., & Upcroft, J. A. (2001). Drug targets and mechanisms of resistance in the anaerobic protozoa. *Clinical Microbiology Reviews, 14*, 150–164.

Vasilevsky, S., Greub, G., Nardelli-Haefliger, D., & Baud, D. (2014). Genital Chlamydia trachomatis: Understanding the roles of innate and adaptive immunity in vaccine research. *Clinical Microbiology Reviews, 27*, 346–370.

Workowski, K. A., & Bolan, G. A. (2015). Sexually transmitted diseases treatment guidelines, 2015.*Morbidity and Mortality Weekly Report. Recommendations and Reports, 64*, 1.

Yanofsky, V. R., Patel, R. V., & Goldenberg, G. (2012). Genital warts: A comprehensive review. *The Journal of Clinical and Aesthetic Dermatology, 5*, 25.

Zhang, M.-R., Zhao, F., Wang, S., Lv, S., Mou, Y., Yao, C.-L., Zhou, Y., & Li, F.-Q. (2020). Molecular mechanism of azoles resistant Candida Albicans in a patient with chronic mucocutaneous candidiasis. *BMC Infectious Diseases, 20*, 1–6.

Zupin, L., Pascolo, L., Zito, G., Ricci, G., & Crovella, S. (2020). SARS-CoV-2 and the next generations: Which impact on reproductive tissues? *Journal of Assisted Reproduction and Genetics, 37*, 2399–2403.

Chapter 16
Drug Resistance in Diabetes

Adnan Khaliq, Muhammad Farhan Jahangir Chughtai, Javed Iqbal, Haq Nawaz, Samreen Ahsan, Tariq Mehmood, Atif Liaqat, Muhammad Nadeem, Nimra Sameed, Kanza Saeed, Syed Junaid-ur-Rahman, Ayesha Ali, and Shoaib Aziz

Abstract The rise in the number of diabetic patients with each passing day give rise to a disturbing situation as serious complications including renal failure and heart attack are associated with disease. Type-2diabetes is more prevalent in elders and constitute about 80–90% of diabetic patients while Type-1 is most common in children and younger with very rare incidence worldwide. It is crucial to regulate the blood glucose level in diabetic patients either through use insulin or by employing the oral antidiabetic medication or adopting healthy life style. TheType-1 diabetes individuals who required insulin therapy as mandatory medication, experienced some life-threatening consequences such as reduced insulin secretion, elevation of plasma glucose and ketoacidosis. However, Type-2diabetes can be regulated primarily through monotherapy but may require combined therapy subsequently. Therefore, a gradual incorporation of other medication with primary therapy is exceedingly recommended to achieve the desired glycemic index. Antidiabetic medication exhibits limited endurance despite of their ability to regulate or normalize plasma glucose level. Antihypertensive agents (glucocorticoid, β-blockers and nicotinic acid) that have ability to reduce blood pressure may result in the impaired glucose regulation. Glucocorticoids reduce the inflammation and regulate the blood glucose level. This chapter gives the knowledge about diabetes mechanism of drug resistance, classification, pathophysiology and its management and further research is needed to be done in order to prevent the world from this widespread disease.

A. Khaliq (✉) · M. F. J. Chughtai · S. Ahsan · T. Mehmood · A. Liaqat · N. Sameed · K. Saeed · S. Junaid-ur-Rahman · A. Ali · S. Aziz
Department of Food Science and Technology, Khwaja Freed University of Engineering and Technology, Rahim Yar Khan 64200, Pakistan
e-mail: adnan.khaliq@kfueit.edu.pk

J. Iqbal
Department of Chemical Engineering, Khwaja Freed University of Engineering and Technology, Rahim Yar Khan 64200, Pakistan

H. Nawaz
Department of Biochemistry, BahauddinZakariya University, Multan 60800, Pakistan

M. Nadeem
Department of Environmental Sciences, COMSATS University Islamabad, Vehari 61100, Pakistan

S. Ahmed et al. (eds.), *Biochemistry of Drug Resistance*,
https://doi.org/10.1007/978-3-030-76320-6_16

Keywords Diabetes mellitus · Autoimmune disorder · Oxidative stress · Krebs cycle · Hyperinsulinemia · Drug resistance

16.1 Introduction

Diabetes mellitus (DM) is the most prevailing disease known simply as 'diabetes'. It is characterized when pancreas doesn't produce enough insulin to control the amount of glucoseor sugar in blood. However, diabetes insipidus (DI) is a rare condition, in which an individual have normal blood glucose levels, but kidneys are unable to balance fluid in the body. Metabolic disorders, caused by insufficient insulin secretion or excessive glucagon secretion. It is linked with hyperglycemia (increased blood glucose level) that can further lead to the cardiovascular, renal and retinal dysfunction (American Diabetes Association, 2014).The global incidence of diabetes for adults aged 20–69 was 7.7% in 2011, which was forecasted to rise to 9% by 2030 (Wou et al., 2019; Hu & Jia, 2018).Currently, there are 422 million peoples suffering from diabetes mellitus worldwide and statistics predicts increment in this numbers upto 2030. The expeditious rise in the number of diabetic patients is associated with the increasing trends of unhealthy and sedentary life style. Hence, adaptation of healthy lifestyle and adequate medication can prevent this condition from worsening (Prabhakar & Doble, 2011).

There seems an increase in diabetic patients during the last decade especially due to the increased prevalence of Type-2 diabetes. Along with the health constrains, manysocio-economic predicaments are also linked with the proliferation of diabetes as antidiabetic medications are expansive and their cost keep on increasing annually. A vast range of antidiabetic drugs are available in market that can be administrated alone or in combination with other medication. The mode of action of these drugs varies with the type and dose of medication. Most of the antidiabetic drugs are administered to reduce plasma glucose level by regulating glucose metabolism. The factors responsible for the development of diabetes such as reduced insulin sensitivity and action can also be controlled by these medications, but incapable to compensate certain detrimental effects. So, this field required extensive and continuous research along with the development of novel drugs (Feingold, 2020).

It is arduous to achieve the glycemic control in some of diabetic patients. Therefore, the scientific community is suggested to apply upmost efforts in order to improve the utilization of existing drugs and emphasize on most appropriate medication. There are numerous antidiabetic medications available for the treatment with varied mechanism of action. Obesity is an associated factor for the development of diabetes. The development of some anti-obesity drugs has been found to be effective against the development of diabetes (Marin-Penalver et al., 2016).

In case of diabetes, the body is unable to regulate the blood glucose level. The liver metabolites all the digestible carbohydrates taken in the food into glucose that is released into blood stream. Plasma glucose level is regulated by insulin, a hormone produced by the beta cells of Islets of Langerhans of pancreas, by increasing the cellular uptake of glucose to fulfill the cellular requirement of energy. In case of diabetes, there is insufficient production or inefficient utilization of insulin resulting in an elevation in blood sugar level, a physiological condition known as Hyperglycemia finally results in energy starvation and cell damage (Siddiqui et al., 2013).

Antihypertensive medication impairs glucose regulation due to the presence of β-blockers, glucocorticoid, and nicotinic acid. The β-blockers create a hurdle in insulin secretion from pancreases by interacting with β-adrenoceptors nerve signals. The glucocorticoid negatively regulates the glucose hemostasis resulting in elevated blood glucose levels that may lead to the Type-2diabetes. Prolonged antihypertensive medication may result in impaired incretin function auxiliary to the progression of diabetes. The loss of incretin effect is associated with the impaired b-cell functioning, insulin resistance or glucose intolerance. (Meier & Nauck, 2010a, b). Most of these drugs alter the cellular metabolism leading to adverse or beneficial effects that should be monitored vigilantly while prescribing any medication to diabetic patient. Moreover, utilization of multiple medications (polypharmacy regime) by some diabetic patients may expose them to many side effects (Shomali, 2012).

However, a large capital is required for the development of effectual and reliable antidiabetic medication. In order to control glycemic index in diabetic patients a lot of progress has been done and novel medications are formulated every year. Although, a lot of noticeable safety constrains are also associated with antidiabetic medication along with drug resistance, the most highlighted issue occurred due to utilization of antidiabetic medicines is increased risk of hypoglycemia, that can further lead to severe complications like dementias or myocardial infarction (Kalra et al., 2013).

16.1.1 Types of Diabetes

There are three main types of diabetes mellitus as explained below.

16.1.2 Type-1 Diabetes

Type-1 diabetes, also identified as insulin-dependent diabetes mellitus, is an immune-mediated disorder produced due to destruction and dysfunction of beta cells of pancreas to produce enough insulin. Insulin is needed by the body to enhance the cellular uptake of glucose to generate energy. The incapability of pancreas to yield sufficient insulin results in hyperglycemia and depletion of cellular energy. Prolonged hyperglycemia may lead to the serious complication of ketoacidosis due to increased production of ketone bodies from lipid metabolism in the absence of cellular glucose. As a result of ketoacidosis, the blood becomes too acidic that it affects the normal functioning of organs like liver and kidney leading to death. In case of excessive insulin, the blood glucose level drops and body progress to the state of hypoglycemia. Hypoglycemia may lead to hallucination, loss of consciousness leading to death if the condition persists for longer time (Atkinson & Eisenbarth, 2001; Blair, 2016; Devendra et al., 2004; Gillespie, 2006; Narendran et al., 2005).

16.1.3 Type-2 Diabetes

Type-2 diabetes is the most common and prevalent type of diabetes as majority of the diabetic patients (about 90%) is diagnosed with Type-2 diabetes. It is also known as non-insulin dependent diabetes mellitus (Prabhakar & Doble, 2011). It is heterogeneous metabolic disease related to the insulin resistance and impaired insulin production due to dysfunction of beta cells of pancreas resulting in the impaired glucose tolerance in the body (Blair, 2016; Nitzan et al., 2015). Type-2 diabetic patients have greater chance to develop urinary tract infection. It is most common in the over-weight and obese individual due to variation in the levels of some hormones which promote insulin resistance, increase leptin, cytokines and glucagon level and decrease adiponectin. Most of the individuals with Type-2 diabetes have been found to have family history of diabetes, hypertension, dyslipidemia, increased weight, older age, and unhealthy life style (Baynes, 2015).

16.1.4 Gestational Diabetes Mellitus

Gestational diabetes mellitus (GDM) is a type of diabetes mellitus identified in women during pregnancy or gestation period. Mostly, this type of diabetes is screened during the 3rd trimester around 28th week of gestation. Women with family history of diabetes are more likely to develop gestation diabetes. In this condition, women are unable to produce enough insulin or encounter insulin resistance due to obesity and overweight. The complications associated with GDM are premature births, congenital malformations, neonatal hypoglycaemia and intrauterine death (Baynes, 2015; Siddiqui et al., 2013).

16.2 Pathophysiology of Type-1 Diabetes

Type-1 diabetes is autoimmune disorder caused due to damage to insulin secreting islet cells of pancreas by immune system of the body. Some genetic and environmental factors also contribute to the disease. In Type-1 diabetes, the damage to pancreatic cells is caused by the $CD4^+$ (helper T cells), $CD8^+$ (cytotoxic T cells) and islet infiltrating macrophages. Type-1 diabetes is closely associated with human leukocyte antigen (HLA) found on chromosome 6. It is the first site to be linked with diabetes. In 90% of the children, suffering from Type-1 diabetes, two main combinations of HLA (DR4-DQ8/DR3-DQ20) are identified. These two HLA combinations are most commonly found in children identified with Type-1 diabetes at very young age. The family members of these children have more possibility to develop Type-1 than those who develop diabetes in older age. The presence of insulin gene on the 11th chromosome has been also found to increase the 10% more risk of developing Type-1 diabetes. Cyclic repetition of shorter forms of insulin promoter is linked with greater vulnerability to the disease. While expression of insulin in the long run is associated with the protection from the disease as they promote the elimination of insulin specific T cells during the immune tolerance induction (Atkinson et al., 2014; Daneman, 2006; Ozougwu et al., 2013).

Two variant forms of genes including CTLA-4 and PTPN22are the important susceptibility locus for Type-1 diabetes due to their action as T-cell activation suppressor. It is also observed that all these genes like HLA, CTLA-4 and PTPN22are found to be closely linked to all autoimmune diseases including diabetes and thyroid diseases. Anti-insulin and anti-islet cell antibodies, present in almost 85% of diabetic patients, reduce insulin secretion by destroying the beta cells of pancreatic after reacting with glutamic acid decarboxylase (GAD). The abnormalities in the functioning of α-cells of pancreas also contribute to Type-1 diabetes by excessive production of glycogen. Usually, the glycogen production is decreased in hyperglycemia but in patients with Type-1 diabetes the glycogen level is not even reduced by hyperglycemia. The elevated level to glycogen along with insulin deficiency worsen the metabolic disorders in Type-1 patients (Atkinson & Eisenbarth, 2001; Atkinson et al., 2014; Baynes, 2015; Gillespie, 2006).

16.3 Pathophysiology of Type-2 Diabetes

The major pathological disorders associated with Type-2 diabetes include impaired insulin secretion and insulin resistance caused due to dysfunction of beta pancreatic cells and reduced insulin action respectively. In case of the insulin resistance, the blood glucose level cannot be regulated even at normal insulin level due to inefficient utilization of insulin. Despite of increased plasma insulin concentration, the body is still unable regulate the blood glucose level as a consequence of insulin resistance leading to impaired glucose tolerance (IGT). Mode of inheritance of Type-2 diabetes is still vague except the maturity onset diabetes for young (MODY). MODY is characterized by hyperglycemia before the age of 25 due to mutation of glucosidase gene found on chromosome 7p. It can be treated without insulin for five year in the individuals having negative islet cell antibodies and is commonly found in Africans and Indian (Baynes, 2015; Ozougwu et al., 2013).

i. **Management of Type-1 diabetes**

Insulin therapy is essential for the treatment of Type-1 diabetes as decreased insulin secretion (insulinopenic) results in metabolic disorders like tissue catabolism, elevated triglyceride level (hypertriglyceridemia) and ketoacidosis that can be life intimidating. The better and long-lasting recovery is associated with reduced A1C level (glycated protein carries oxygen) that required multiple time daily injection of insulin or insulin pump therapy recommended by Diabetes Control and Complications Trial (DCCT, 2016). Long acting and fast acting insulin analogs are employed from last 25 years and have dissimilar pharmacokinetics and pharmacodynamics profiles in comparison to recombinant human insulin. Recently available long acting insulin analogues like basal insulin analogs exhibit longer duration of action and more consistent or smoother plasma concentration along with peak less activity as compared to intermediate acting neutral protamine hagedorn (NPH) insulin. While rapid acting insulin analogs are faster, quicker with short duration of action with peak action occur within first hour. Treatment of Type-1 diabetic patients with insulin analogs in comparison with human insulins is linked with fewer chances of hypoglycemia and weight gain and reduced A1C (Tricco et al., 2014). Currently two novel insulin preparations have been formulated with improved rapid action like inhaled insulin that act immediately, have faster peak, result in fast reduction in prandial excursions as compared to rapid acting analogs (RAA),may result in weight gain and fewer episodes of hypoglycemia (Russell-Jones et al., 2017). For the management of Type-1 diabetes several approaches exist with the incorporation of several forms of insulin in scheduled regime of patient in order to evade hypoglycemia, prevent the formation of acidic substance known as ketones (ketoacidosis) and helps to meet the patient glycemic index. Usually, the multiple dosage regimens are also used for the treatment of Type-1 diabetes that involves the incorporation of short acting insulins before meals along with the long acting preparations at night. The basal dose of long acting insulin is used to control fasting glucose level during night. The control of postprandial glucose excursions is done through the timely injection of prandial insulin and the best time for the administration of prandial insulin differ depending upon the carbohydrate consumption, pharmacokinetics of insulin formulation used and level of glucose in blood before meal. Therefore, the suggestions for the administration of prandial insulin dosage must be modified for each individual. The normal secretion of insulin from the body fluctuates with meal size, blood glucose level and glucose required by the tissues. Therefore, depending upon the expected need, the prandial insulin dose should be adjusted and relevant education about prandial dose adjustment must be offered to patients (Vaz et al., 2018).

Efficacy of oral and injectable drugs in addition to insulin therapy is also studied for the treatment of Type-1 diabetes. Pramlintide is injectable amylin drug permitted for the treatment of Type-1 diabetes in adults. Many studies confirm that the incorporation of pramlintide with insulin results in the 1–2 kg loss in body weight along with reduction in the hemoglobin AIC level (Edelman et al., 2006). Likewise, the drugs only used for the treatment of Type-2diabetes showed the similar results like incorporation of metformin for the treatment of Type-1 diabetic adult produce no change in glycated hemoglobin level but the minute loss in lipid profile and body weight (Meng et al., 2018). The incorporation of GLP-1 receptor with insulin therapy exhibit the minor decrease in A1C level and body weight about 3 kg as compared to the use of insulin alone (Wang et al., 2017). The risk and advantages associated with the additional medication needed to be assessed thoroughly as only pramlintide is permitted to be used for the treatment of Type-1 diabetes.

ii. **Management of Type-2 diabetes**

Type-2diabetes is the most common and prevalent in the aged patient (age more than 65 years). The factors that majorly promote the Type-2diabetes include fatness, sedentary lifestyle and age (Siddiqui et al., 2013). Oral administration of metformin must be started, as a preliminary treatment, at the diagnostic stage of diabetes with lifestyle interventions that can be discontinued in case of any adverse reaction. The advantages of metformin administration are associated with its low cost and lower risk of cardiovascular implications while the side effects include gastrointestinal complications that can be lessened by steadily varying the drug dose (Holman et al., 2008, Aroda et al., 2016). Insulin has been found to be effective alone or in combination regimen especially in case of hypertriglyceridemia, loss in weight and ketosis. Similarly, sulfonylurea has been also used effectively for the treatment of Type-2diabetic patients with uncontrolled hyperglycemia (Basu et al., 2009).

The glycemic index in case of Type-2diabetes can be maintained through monotherapy during few initial years but afterwards required combination therapy. Slow addition of other medication to metformin is usually suggested to attain the glycemic target. The good and bad effects of the suggested drug must be carefully examined before its incorporation to ensure the patient safety and expenses (Cahn and Cefalu, 2016). The use of combination therapy such as metformin and dipeptidyl peptidase 4 (DPP-4) inhibitor vildagliptin at initial stage is also evident for the quick attainment of glycemic target in comparison to the use of metformin alone (Abdul-Ghani et al., 2015). The selection of medication to be used with metformin in combined therapy depends on patient's clinical characteristics that include particular drug safety, side effects, tolerability, and cost and the associated risk of atherosclerotic cardiovascular disease (ASCVD). In case of the target glycated hemoglobin level is not achieved within three months of metformin administration, then any of these six-drug option (sulfonylurea, DPP-4 inhibitor, GLP-1 RA, SGLT2 inhibitor, thiazolidinedione, basal insulin) can be used along with metformin.

The GLP-1 RA or SGLT-2 inhibitors are recommended to be employed in glucose-lowering regime for the patients with the established high risk of ASCVD, heart attack or renal disorder (Vijan et al., 2014). Oral medications are more preferred by patients as compared to injectable drugs as the ultimate requirement of high potency injectable medicines is normal in patients having diabetes from longer period of time. The use of basal insulin, either human insulin or long-acting analogs, along with oral medication has been also found effectual for many diabetic patients (Pratley et al., 2019).

16.4 Anti-diabetic Medication-Oral and Insulin

Type-1 diabetes is also known as insulin-dependent diabetes and most common in children. The patients with Type-1 diabetes required insulin therapy in order to maintain the blood glucose level. Type-2diabetes or non-insulin dependent diabetes is most common and widespread with around 190 million of patients worldwide. Most of the Type-2diabetic patients are obese and required the oral medication in order to maintain their normal glycemic index. However, a large number of Type-2patients ultimately move towards insulin when they are unable to maintain long term glycemic index with oral medication or lifestyle interventions (Krentz & Bailey, 2005).

i. **Insulin therapy**

Type-1 diabetic patients have specifically required the insulin for the treatment, as they cannot be treated with oral antidiabetic medicines. Insulin is recognized as the most powerful agent that reduces plasma glucose level in both Type-1 and Type-2 patients. Depending upon the period of action the short acting, intermediate acting, and long acting formulations of insulin are used. Insulin association with increased risk of hypoglycemia and weight limits its dosage. Recently, many improvements have been done in formulation and treatment routine to reduce the risk of hypoglycemia (Bastaki, 2005).Insulin usually binds to the receptors in plasma membrane and organizes the target cells for combine anabolic reaction and availability of nutrients. The expression of insulin receptor is shown by somatic cells and significant role of insulin in homeostasis of glucose is characterized by its effect on liver, skeletal muscles and adipose tissues. These tissues show a distinctive effect in homeostasis metabolism and transduction pathways of tissue specific insulin signaling. In skeleton muscles, insulin increases the cellular uptake and storage of glucose in the form of glycogen. Moreover, the liver also enhances the synthesis of glycogen and lipogenesis while decreases gluconeogenesis. However, in adipose tissues the insulin decreases lipolysis and enhances the occurrences of lipogenesis and transportation of glucose (Vargas et al., 2020).

ii. **Oral anti-diabetic medication**

The antidiabetic oral medicines are only recommended for the Type-2diabetic patients whose body is still able to produce some of the insulin. The oral antidiabetic medicines are used alone or in combination with insulin in order to regulate the blood glucose level. Adoption of healthy lifestyle and consumption of balanced nutritious diet are also prescribed with such medication (Spiller & Sawyer, 2006). The oral antidiabetic medication include (1) Secretagogues agent like sulphonylureas, meglitinides and incertins that stimulate the insulin production, (2) Insulin sensitizers like metformin, (3) Thiazolidinediones that reduce the insulin resistance and (4) α-glucosidase glucose inhibitors (AGIs) that slowdowns the carbohydrate digestion thus lowering glucose level in body (Krentz & Bailey, 2005).

iii. Secretagogues agent

a. Sulphonylureas

Sulphonylureas are the insulin secretagogues agents that increase the insulin secretion by stimulating the beta pancreatic cells. They inhibits the ATP sensitive potassium channel of beta cells of pancreas resulting in the polarization of cell membrane and opening of calcium channel, thus increasing the insulin secretion (Stein et al., 2013). They are the well-known, cheap, and readily available antidiabetic medicine in most of the countries of the world. They are the best option for 60–70% patients to attain the required glycemic level except for the overweight diabetic patients having high fasting blood sugar level. The main adverse effect of this drug on elderly patients is hypoglycemia, which may lead to the dysfunction of liver and kidney. Such adverse effects are mostly visible in the hyperglycemic patients, who are given a very high initial dosage or their dose is increased too rapidly to control the hyperglycemia. In diabetic patients having mild to intermediate hyperglycemia, the preliminary dose of Sulphonylureas must be lowered to half. In order to control the postprandial hyperglycemia the medication suggested to be given 30 min before meal (Krentz & Bailey, 2005).

b. Meglitinides

Meglitinides are the fast-acting insulin regulators that stimulate the insulin secretion (insulin secretagogues) by closing the ATP sensitive potassium channels (K_{ATP}). It has short duration of action against the postprandial hyperglycemia. The repaglinide and nateglinide are its derivatives having the same mode of action but different sites for molecular binding. As compared to repaglinide, the nateglinide act more rapidly in inhibiting the K_{ATP} channel, thus have short duration of action (Kawamori et al., 2012). These medicines can be used in substitute to sulphonylureas, as there is less possibility of occurrence of hypoglycemia. Meglitinides are also the best for the diabetic patients more sensitive sulphonylureas or metformin (Omori et al., 2019). It is usually suggested to incorporate 4 mg of repaglinide and 120 mg dosage of nateglinide15 min before every meal (Hu et al., 2000; Scheen & Lefèbvre, 1998).

c. Incertins

The incertins include glucose-dependent insulinotropic polypeptide (GIP) and glucagon-like peptide-1 (GLP-1) are the gut hormones with short life span and their concentration is enhanced by dipeptidyl peptidase-4 (DPP-4) inhibitors. They are categorized as new insulin secretagogues as they enhance the insulin release and lower the glycogen secretion from beta cells of pancreas, and regulate the plasma glucose level. They are recommended for elderly patients due to low associated risk of developing hypoglycemia (Gilbert & Pratley, 2009; Mahajan & Gupta, 2010).

iv. **Insulin Sensitizers**

One of the most important features of Type-2diabetes is insulin resistance, in which the cells of the body are not able to respond to the signals of insulin and cannot utilize the glucose to get energy. The insulin resistance can lead to the serious cardiovascular complications. Therefore, to overcome or reduce insulin resistance, the insulin sensitizers like metformin and thiazolidinediones are employed. Insulin sensitizers has no influence on insulin secretion mechanism (Zangeneh et al., 2003).

a. **Metformin**

Metformin is a least expensive and widely available biguanide that act as insulin sensitizer. It is commonly used by the elderly patients as it helps to lower the blood glucose level along with reduced risk of hypoglycemia. Metformin is also suggested to the overweight or obese diabetic patients as unlike other medication (Sulphonylureas, thiazolidinediones and meglitinides), it do not promote weight gain. The mode of action of metformin involves the inhibition of AMP-activated protein kinase (AMPK)-dependent or AMPK-independent respiratory chain in mitochondria or inhibition of mitochondrial glycerophosphate dehydrogenase (Rena et al., 2017).

It can also be used in combination with insulin or other oral anti-hyperglycemic medication such as Sulphonylureas. Its utilization with balanced diet results in the reduction in fasting glucose and hemoglobin A1C level. It is advised with an initial dose is 500 mg that should be slowly increased to prevent gastrointestinal complications. It also reduces the vitamin B_{12} absorption in the body (Krentz & Bailey, 2005; Stein et al., 2013; Zangeneh et al., 2003).

b. **Thiazolidinediones**

Thiazolidinediones (TZDs) includes the compounds like pioglitazone and rosiglitazone that act as insulin sensitizers, reduce the insulin resistance and enhance its absorption in peripheral tissues. They activate the nuclear receptors known as peroxisome-proliferator-activated receptors (PPARs) that enhance the haptic glycolysis. The antidiabetic effect of TZDs achieved through the mechanism that implicates the activation of PPAR-gamma whose function is to regulate glucose metabolism. TZD-initiated enactment of PPAR-gamma modifies the transcription of genes associated with the lipids and glucose metabolism including those coding for lipoprotein lipase, fatty acyl-CoA synthase, adipocyte fatty acid binding protein, GLUT-4 glucose carrier, malic compound, fatty acid carrier protein and glucokinase (Hauner, 2002).

Both obese and non-obese patient with glucose intolerance can utilize TZDs to reduce insulin resistance without causing the weight gain and hypoglycemia. TZDs also improve the abnormalities caused by the insulin resistance like lipid anomalies and hypertension. They have slow mode of action as compared to other anti-hyperglycemic medication as it takes almost 7–28 days to initiate a response. TZDs can be used as monotherapy with dosage of 15–45 mg or can also be combined with sulphonylureas, metformin and insulin. The adverse effects of medication include hepatotoxicity, weight gain and polycystic ovarian syndrome in women (Petersen et al., 2000; Scheen & Lefèbvre, 1998; Stein et al., 2013; Tonelli et al., 2004; Zangeneh et al., 2003).

v. **α-Glucosidase inhibitors (AGIs)**

The α-Glucosidase inhibitors got the approval to be used for the treatment Type-2diabetesin 1990s. They inhibit α-Glucosidase, an upper GIT enzyme that converts the polysaccharides into monosaccharides, and thus reduce the carbohydrate absorption and lowers the blood glucose level (Joshi et al., 2015; Scheen, 2003). Acarbose, voglibose and miglitoal are the AGIs approved in Europe. Their utilization lowers the hyperinsulinemia and triglyceride concentration in the blood along with no weight gain. Acarbose can be utilized alone or in combination with other oral antidiabetic medication. However, their high cost and associated gastrointestinal complications limit their utilization. In order to reduce the adverse effect of medication, the initial dose must be 25–50 mg per day and slowly increased to 50–1000 mg thrice a day (Krentz & Bailey, 2005; Stein et al., 2013).

vi. **Other oral antidiabetic agents**

a. **Bile acid sequestrants**

Primarily, bile acid sequestrants (BAS) are approved for curing hyperlipidemia but unintentionally their anti-hyperglycemic effect unveils. Presently, colresvelam is only approved BAS that is used for the treatment of Type-2diabetes due to its plasma glucose lowering effect. It shows higher efficacy when used in combination with insulin, sulfonylureas, metformin resulting in 0.5% reduction in hemoglobin A1C. Its adverse effects include constipation that limits its recommendation to the patients having gastroparesis. The mode of action of BAS as antidiabetic drug involves their impacts on the bile acid receptors including farnesoid X receptor and G protein–coupled receptor for bile acids inside the digestive tract and farnesoid X receptor inside the liver diminishing the production of endogenous glucose (Handelsman, 2011; Staels, 2009; Staels & Kuipers, 2007).

b. **Bromocriptine**

Bromocriptine is novel medicine that has recently approved by Food and Drug Administration (FDA) in 2009. Previously, it was used for the treatment of hypothalamic malfunctioning but later HbA1C test shows its ability to lower blood glucose level from 0.4 to 0.8%. It is used alone with a dose of 4.8 mg per day or as combined therapy with insulin and other antidiabetic oral medications. It has satisfactory safety profile as it lowers the possibility of cardiovascular complications and also reduce blood glucose level, triglyceride and free fatty acid value. Its side effects include hypotension, nausea and dizziness (Pijl et al., 2000; Shivaprasad & Kalra, 2011). Bromocriptine is the novel drug in that it does not have a particular receptor that intervene its activity on glucose and lipid digestion or its effects are intervened by means of resetting of dopaminergic and approving tone inside the central nervous system (Cincotta et al., 2002).

16.5 Factors Affecting the Body's Response to Drugs

The human body is complex in nature and influenced by the absorption, distribution, and biological effects of a particular drug. This heterogeneity is helpful for an accurate assessment as well as to observe impact of suggested drugs on the patient's health. For this reason, certain drugs are administered alone or in combinations to exert their effect. This part focuses on understanding the patient's response towards drugs and to figure out the possible solution for unexpected behavior of the patient. Various oral antidiabetic drugs like GLP-1 agonists, meglitinides, dopamine-2 agonists, DPP-4 inhibitors, and amylin analogs perform their function by reducing the hepatic glucose production, stimulating the insulin secretion, delaying the intestinal digestion and absorption of carbohydrates, and assisting the insulin (Sanchez-Ibarra et al., 2018). There is diversity in the patient's responses towards these drugs that dependent on the following factors.

i. **Demographic characteristics of individuals**

The people with different body actions experience different effects for same dose of a drug. The reason could be difference in Pharmacodynamics (number of drug receptors at action site) and pharmacokinetics amounts of the drug reach the receptor sites) of drug. The difference might be inherited or acquired from suffered community. In individual difference, body size and composition, age, gender, educational status and hormonal differences exhibit significant influence to drug response (Becker et al., 2013; Emami-Riedmaier et al., 2015). The persons with larger body size possess larger vascular system, greater blood volume compared to those with a small body size. Therefore, particular dose of drug tend to be more diluted and comparatively less effective at the target site in such people. Therefore, drug doses are decided as per body dimensions and composition. The water soluble drugs respond more strongly in obese people than leaner people with same body size. The effectiveness of drug is also dependent on age of the patients. The recommended dose exhibits its effects in prolonged period in elders than younger people. The change in proportion of fats to lean tissue is prominent factor for drug absorption, metabolism as well as for excretion. In aging, efficiency of body organs is supposed to be reduced with increased likelihood of diseases (chronic) as well as receptors sensitivity is also observed (Jemal et al., 2017).

In spite of these, gender difference also affects the distribution of drug in the body. The women and men differ in their mode of drug pharmacokinetics and alcohol metabolism is good example of gender difference. The enzyme alcohol dehydrogenase is required for its metabolism to acetaldehyde. The women are deficient in alcohol dehydrogenase, so greater amount of alcohol is noted in blood stream of women as compared to men on ingestion of same quantity of alcohol (Frezza et al., 1990). In menstrual cycle the drug response also varies possibly due to secretion as well as interaction of female sex hormones with dopamine-based neurotransmitter systems (Justice & de Wit, 1999).

ii. **Genetic makeup of the patient**

The drug metabolism is also affected by genetic makeup of the patient. The genes encoded for enzyme production can exist in different forms in individual with different genetic makeup i.e. ethnic and racial groups. This phenomenon is known as polymorphism and affects the functioning of enzyme and can also make it inactive at site of action (Evans & Johnson, 2001). In Asian groups of people, alcohol dehydrogenase with reduced activity is produced to metabolize alcohol into acetaldehyde (Luczak et al., 2009). In this condition, level of acetaldehyde is increased in the body due to improper metabolism of alcohol and can cause flushing of the skin and nausea condition known as antabuse. These ethnic and genetic differences noticed in other liver function enzymes tend to metabolize and block receptor targets of drugs (Bertilsson et al., 2002). For example, variations in adenosine receptor are linked with caffeine response and dopamine transporter affect response of amphetamine (Hamidovic et al., 2010).

iii. **Personality traits of the patient**

The personality traits of the patients significantly affect the acute responses of drugs. This parameter demonstrates the influence of social characteristics on drug responses *i.e.* how much people are smokers, alcoholics, and chat-chewers in the selected population. The people with optimistic disposition and metabolism of drugs can respond greater to relief pain as compared to people with a pessimistic mode as noted in placebo study (Geers et al., 2010). The studies on the disorders linked to high rates of alcohol reported that healthy young individuals showed more intoxication in alcohol intake than men suffered with bipolar disorder. It is reported that less response to alcohol may be due to misuse and increased consumption of alcohol which interfere with serotonin genes-responsible for alcoholic intoxication (Yip et al., 2012).

iv. **Clinical history of the patient**

The factors including family history of the patients, drug taking history, no of other drugs, and duration of diabetes since diagnosis are studied in clinical profiling of the patient. The history of drug taking can affect responses due to tolerance and sensitization processes. All these characteristics also statistically influence the host respond to drugs (Horii et al., 2019).

Table 16.1 Adherence question among diabetic patients

S. No	Questions
1	Do you think, you are not careless to take your medicine?
2	Do you sometimes, you stop your anti-diabetic medicines if you feel better?
3	Do you sometime forget, to take dose of your anti-diabetic medicines?
4	Do you stop to take dose of your anti-diabetic medicines if feel worse?

v. **Adherence status of the patients**

This status is checked by Morisky's four item method and comprised of four questions (Table 16.1). A study by Jemal et al. (2017) concluded that a significant interaction ($p < 0.05$) was observed between glycemic outcomes and adherences status of patient. It was found that 70.4% of patients were adherent to their medications while remaining were not adherent.

vi. **Stressful circumstances**

The positive and direct link is reported between drug and stress response. The reinforcing effects of drugs can be prominently observed in stressful condition and vice versa (Stairs & Bardo, 2009). The drug responses in animal studies were also checked by manipulation of these stressful environments. It was concluded that respond to drugs is varied from person to person (Miczek et al., 2008).

vii. **Environmental factors**

The relationships among the drug, diet of patient, drug taking history and drug taking conditions are studied under umbrella of environmental factors. These also significantly affect drug responses in spite of genes which are itself being considered as one of an important factor. The influence of drug by environmental factors is reported as interactions and can be defined as responses of all types of drugs with possible and prevailing factors at that particular time.

16.6 Interactions Between Drug-Food and Drug-Drug

This is particularly for patients who are taking more than one type of medicines. The people taking antidepressant drugs are also suffered with and anxiety. That is why they are also prescribed for antianxiety drugs. The cumulative effect of these medications could be synergistic or antagonistic (Morgan et al., 2010). In this situation, a person's response to a drug and their influence with patient's food constituents is presented in interactions form that may occur due to lack of knowledge about ingested substances or accidentally. This kind of interactions sometime may either be beneficial or harmful for patient i.e. use of alcohol prior to benzodiazepine drugs (recommended for anxiety) is dangerous which tend to impair motor coordination and result in likelihood of an accident (Maxwell, 2012).

16.7 Interactions Between Drug, Expectations, and Personal Behavior

The expectations are indeed a person's behavior and can become dependent on the settings in which a drug is taken *i.e.* different drug effects can be experienced by same person. Sometime, expectations about dose of drugs can be altered by viewing the drug taking history of patients as well as beliefs. These beliefs can be effective or negative about expected drugs. The studies are called placebo and nocebo effect based on effectivity and negativity of recommended dose of drugs (Tracey, 2010).

16.8 Oxidative Stress: The Pathogenic Mechanism Underlying Insulin Resistance

The molecular biology studies support the hypothesis that oxidative stress alters the intracellular signaling pathways linked with insulin resistance (Dimova et al., 2019). This resistance reduces efficiency of intracellular antioxidant defensive mechanism in human body. The adipose tissues and muscles are involved in pathogenic mechanism underlying insulin resistance. The substrates continue to add in Kreb's cycle to generate reactive oxygen species (ROS) and mitochondrial NADH (mNADH) beyond the demand of the cell. This condition prevails, if energy expenditure rate is less than energy production rate. The body cell exhibits the self-defensive mechanism against harmful effects of ROS. Prevention of ROS formation is accomplished by preventing the build-up of mNADH by inhibiting insulin stimulated nutrient uptake and preventing the entrance of energetic substrates (pyruvate, fatty acids) into the mitochondria. Controversy exists whether the free fatty acid (FFA) or glucose is the primary fuel source in the over-nourished muscle and adipose tissue. In either case, an influx of substrates into the citric acid cycle generates mitochondrial acetyl-CoA and NADH. The Acetyl-CoA, either derived from glucose through pyruvate or from beta-oxidation of FFA, combines with oxaloacetate to form citrate, which enters the citric acid cycle and is converted to isocitrate. The ROS formation is inhibited, if mNADH production and entry of fatty acid/pyruvate in mitochondria is stopped. The substrate addition tends to produce NADH and mitochondrial acetyl-CoA during citric acid cycle. This acetyl-CoA is product of either beta-oxidation of FFA or glycoltic break down of glucose which, in turn, increases the production of electron donors in form of NADH (Ceriello & Motz, 2004). This production increases the membrane potential due to pumping of protons across the mitochondrial inner membrane via ETC

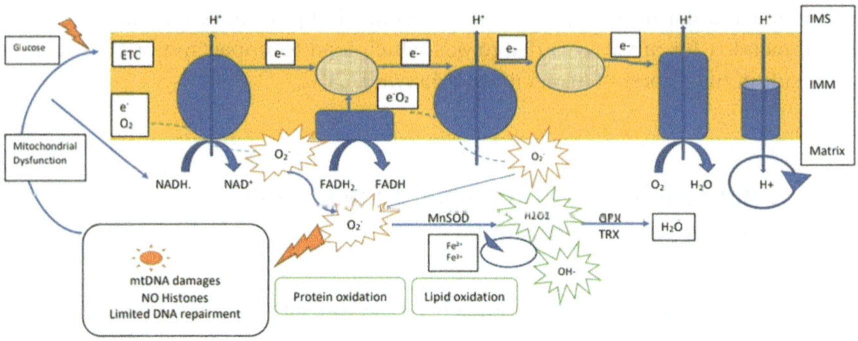

Fig. 16.1 Mechanism for formation of free radicals on influx of glucose and FFA (Cohen et al., 2012)

(electron transport chain). The increase in membrane potential inhibits the transport of electrons at complex III and lead to reduction of oxygen (O_2) in the form of superoxide radicals (Fig. 16.1).

In Kreb cycle, NADH is generated by enzymatic (isocitrate dehydrogenase) action and oxidative phosphorylation of iso-citrate. On the excessive production of NADH, proton gradient in mitochondria increases from required level, formation of superoxide free radicals starts due to transfer of one/spare electron to oxygen. This excessive production can be prevented by inhibiting the oxidation of FFA that is ultimately linked with resistance to insulin-stimulated glucose uptake in the muscle. This condition on temporary basis is called as compensatory mechanism that protects the cells from oxidative damage. For dysfunction of endothelial and beta cells, oxidative stress is considered as a pathogenic factor (Fig. 16.2). It markedly leads to blunt secretion of insulin as well as damage to mitochondria due to low free-radicals scavenging enzymes (Zhang et al., 2020).

The in-vivo studies confirmed that increased mitochondrial ROS exhibit potential to suppress glucose-induced insulin secretion in first phase of glycolysis. This secretion is directly linked with the activity of glyceraldehyde 3-phosphate dehydrogenase (GAPDH). It is reported that vitamins (E, C) and antioxidant enzymes (catalase, superoxide dismutase, and glutathione per oxidase) assist in improvement of insulin sensitivity. Conclusively, oxidative stress is assumed as mediator to induce hyperglycemia and lipotoxicity based hyperlipidemia. The modification in lifestyle and antioxidants based defensive activities can control risks of impaired glucose tolerance, insulin resistance, and other cardiovascular diseases (Molehin et al., 2020).

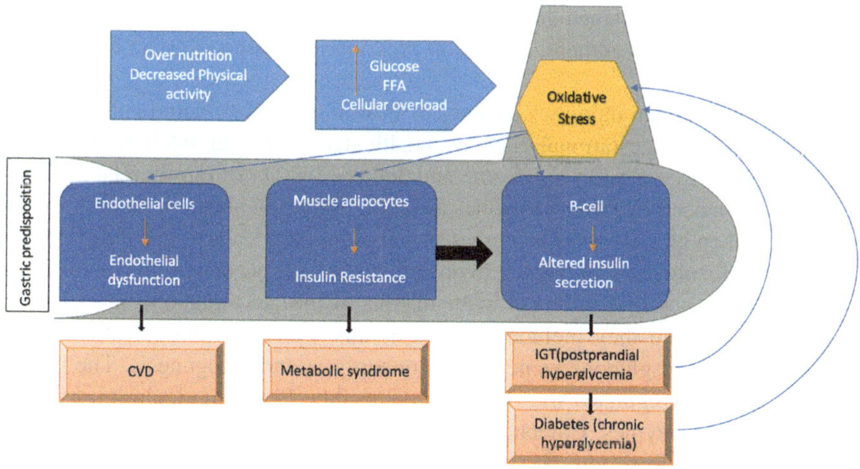

Fig. 16.2 Oxidative stress as pathogenic factor for insulin resistance and dysfunction of endothelium (Ceriello & Motz, 2004)

16.9 Drug Resistance in Diabetes

i. **Insulin Resistance**

Insulin is a peptide hormone released from the beta cells of pancreatic islets of Langerhans for the metabolism and regulation of carbohydrates, lipids, and protein. Insulin acts on glucose regulation by promoting the cellular uptake of glucose from blood liver and muscles. Insulin is also effective to promote cell division and growth by its mutagenic effects. Insulin resistance in diabetes can be explained as a condition where body cells shows a weakened biological response to the released insulin (Cefalu, 2001) resulting in reduced sensitivity to insulin mediated glucose disposal (Reaven, 2004). However, resistance to insulinis the leading factor that increases the chances of prediabetes, Type-2diabetes and gestational diabetes. In insulin resistant individual's insulin secretion is increased drastically leading to *hyperinsulinemia.*

The physiopathology of vascular disease in patients with Type-2diabetes mellitus is associated with the effects of higher glucose level on vascular tissues and clotting of blood, irregularities in blood pressure, endothelial function, lipid metabolism, fibrinolysis and inflammation (Beckman et al., 2002). Insulin resistance is also associated with other ailments including atherosclerosis and this metabolic disorder increases glucose level in the body (Kernan, 2002).

Insulin resistance is a recently discovered risk factor that can be considered as a target in the treatment of stroke. This treatment approach is of more significance particularly when there is a secondary stroke as there are about 50% of non-diabetic people suffering from ischemic stroke. This type of stroke initiates with the blockage of blood vessel with a clot in the brain. Consequently, blood supply to the brain is restricted and brain cells start dying due to insulin resistance (Kernan et al., 2002).

To prevent insulin resistance and stroke prevention the utmost care is required for diet, body weight management, and blood pressure regulation (Kernan & Inzucchi, 2004).

The factors that can increases chances of insulin resistance include obesity, high carbohydrates food, gestational diabetes, health status e.g. non-alcoholic fatty liver disease and polycystic ovary syndrome and diabetic history of family. The action of insulin is influenced by other hormones as it is dominating hormone in fed stage. It is released when glucose level is increased and ultimately leads to release of growth hormone and insulin growth factor 1 to prevent insulin-induced hypoglycemia. There are other hormones that interplay with insulin to initiate metabolic processes in the fasting state and they include glucocorticoids, glucagon and catecholamines. Glucagon promotes gluconeogenesis, glycogenolysis and ketogenesis. The insulin to glucagon ratio regulates the degree of phosphorylation or dephosphorylation of the pertinent enzymes. Catecholamines stimulate lipolysis and glycogenolysis; glucocorticoids endorse muscle catabolism, gluconeogenesis and lipolysis. Higher secretion of these hormones may show insulin resistance but only in certain stages. Insulin resistance in most cases is apparent at the cellular level by post-receptor defects in insulin signaling. The possible mechanisms include down-regulation, genetic polymorphisms of tyrosine phosphorylation of the insulin receptor, insulin receptor substrate proteins or phosphoinositol-3-phosphate (PIP-3) kinase or may be abnormalities of glucose transport protein-4 (GLUT 4) function (Wilcox, 2005).

The insulin resistance progresses to Type-2Diabetes in such a way that the pancreas of a person having insulin resistance releases extra insulin to fulfil its requirement. For some time, it will stay with normal blood glucose level but with the passage of time the pancreas becomes unable to release more insulin. If the person does not make any changes in lifestyle and eating habits, he/she would be shifted towards prediabetes that leads toType-2diabetes if not managed properly (Wilcox, 2005).

ii. Oral antidiabetic medications resistance

Oral antidiabetic medicines and insulin therapy effectively control the diabetes while the oral anti-hyperglycemic medications efficiently reduce the glycated hemoglobin A1C from 0.5–2%. Regardless the glucose regulating potential, the persistence or endurance of the antidiabetic medication is insubstantial. The oral antidiabetic medicines show 36–85% adherence except in the patients using insulin therapy (Soule Odegard & Capoccia, 2007). Glucose regulation can be impaired by the use of various antihypertensive medications including β-blockers, glucocorticoid, and nicotinic acid (Ananth et al., 2004). The β-blockers act as obstacle in reduced insulin secretion from pancreases by interacting with nerve signals on β-adrenoceptors. The glucocorticoid is an anti-inflammatory drug and regulates the glucose hemostasis but its consequences include the rise in blood glucose level and may leads to the Type-2diabetes. Thiazide is a diuretic that works by blocking the sodium absorption in the kidney leading to hypokalemia (decreased potassium concentration in the blood). The potassium is essential to regulate glucose uptake in the cells resulting in the increased plasma glucose level. Niacin or nicotinic acid is used to lower low-density lipoprotein and increase the level of high-density lipoprotein but exerts adverse effect on plasma glucose level (Chen et al., 2002). HIV protease inhibitors that are used to treat HIV but concerns are associated with its usage as it induce variations in mitochondrial functioning along with insulin resistance (Hruz, 2008).

Many oral antidiabetic agents are also used alone or in combination with insulin to control the insulin action. The sulphonylureas stimulate the insulin secretion though there seems the decrease in insulin secretory ability with the passage of time. This decrease has been named 'secondary failure' and has been credited to an unfavorable impact of sulphonynylureas on remaining pancreatic β-cell limit. The insight into the 'secondary failure' shows the possible beta cell exhaustion caused by sulphonyl urea treatment (Taverna et al., 2001).

Moreover, glucotoxicity occur as a result of increased glucose level causing loss of insulin secretory function. Glucotoxicity is attributed to several potential causes including aggregation of glycogen and other metabolites affecting interleukin II (Kaiser et al., 2003). If hyperglycemia diminished to normal levels, the condition of glucotoxicity is alterable but the dynamic loss of insulin secretory limit, that happens autonomous of glucose levels, is not reversible (Maedler et al., 2002).

Hypoglycemia linked to meglitinide, in case of chronic kidney disorder (CKD), could be ascribed to drug and non-drug factors. In case of CKD, the meglitinides and its metabolic products are discharged through kidney, aggregate in the body, and exert uremic effects on various systems with an increased half-life. These uremic toxins disturb various function of body including insulin metabolism and glucose homeostasis (Nogueira et al., 2013). However various factors resulting in hypoglycemic condition of the patient includes marginal insulin resistance and impaired beta-cell response to glucose, increased hepatic gluconeogenesis, eating disorder resulting in body weight reduction and reduction in glycogen stores, and lack of instantaneous counter-regulatory catecholamine (Chapagain et al., 2014; Leyking & Fliser, 2014; Moen et al., 2009).

Impaired incretin function is auxiliary to the progression of diabetes and it is a matter of attention that reduction in the effectiveness of incretin is linked to impaired β-cell functioning, insulin resistance or glucose intolerance. The minor variation in the response of insulin to intravenous and oral glucose was noticeably the reason of loss in certain effect with a significant difference in an increased response of insulin to intravenous glucose. In concurrence with this, the disposition index during intravenous glucose was unaltered by the intercession. Moreover, a healthy individual, tested for intravenous glucose, was found completely fit for balancing enhanced insulin secretion with reduced insulin sensitivity but still unable to up regulate his capacity to deal with orally administrated glucose. Currently, it has been recommended that the impaired incretin function in Type-2diabetes only signifies to a decreased beta cell secretory limit that would greatly influence the powerful stimulus produced by oral glucose as compared to flimsier stimulus solicited by intravenous glucose (Meier & Nauck, 2010a, b).

There also exists a situation where oral administration of glucose would not be capable to produce prominent glucose response while the administration of intravenous glucose produce response comparing to the maximum secretory limit of the beta cells. Many studies have also suggested that the blends of intravenous glucose, arginine, and glucagon were independent of glucose tolerance and significantly enhanced the insulin secretion more in combination than alone (Mann et al., 2020). Final conclusion obtained from the above discussion is that the loss of incretin effect occurred undoubtedly because of improper islet cells response to incretin. It is also an early indication of weakened glucose metabolism, possibly linked with insulin resistance, and that further irregularity is viewed as glucose intolerance. In case of Type-2diabetes, the outcomes of impaired incretin effect result in the loss of patient's ability to proficiently discard orally rather than intravenously directed glucose (Cipolla-Neto & Amaral, 2018).

16.10 Prevention and Treatment of Drug Resistance

Drugs used in the treatment of diabetes are very specific in nature and any modification in their structure may result in destructive effects leading to drug resistance. There are different mechanisms involved in resistance process like enzymatic deactivation of drugs, interchanging in the binding site of insulin, modification in metabolic pathways, and more pumping out of diabetic drugs crossways the surface of cell (Murali et al., 2014). There are hundreds of infections that occur due to diabetes (mellitus and juvenile). These infections include foot infections, gangrenous cholecystitis, and soft tissue and urinary tract infections. *Escherichia coli* are one of the main reasons of infections in diabetic patients because they have weak immunity due to poor response of T cells and neutrophils. Other organisms including *Klebsiella, Proteus, Enterobacter* species, *Staphylococcus aureus* and *Streptococcus, Enterococcus* species are also considered as causative agents in drug resistance in diabetes (Chakraborty et al., 2017).

Persistent levels of hyperglycemia or other causative factors associated with diabetes worsen the condition of patients due to attack of many infections at a time. Glycemic control of diabetic patients is deteriorated leading to immunocompromised state. Consequently, the immunocompromised condition of the diabetic patient and an excessive use of antibiotics results in drug resistance (Magee et al., 2013).

Before prevention and treatment, lets focus on the main causes of drug resistant in diabetes so that this worsen condition can be easily eradicated. Overuse of antidiabetic drugs in order to lower the glycemic level of body is considered as the main reason of drug resistance. Other reasons include incomplete course of antibacterial drugs and poor setting of infection control not only at hospital but also at home. Hygienic condition of patients and health care setting is also very vital. An excessive use of antibiotics in livestock also results in drug resistant (Chianget al., 2015).

16.11 Precautionary Measures and Management of Drug Resistant

i. **At individual level**

- Avoid the use of over the counter (OCT) drugs and use only prescribed drugs to control high level of glucose in mellitus, juvenile and gestational diabetes.
- Do not insist for antibiotics or antimicrobials if you are in no need of drugs.
- Preparation of food should be free from pathogens and safe from any kind of germs.
- Try to avoid infections by washing hands regularly and maintaining a social distancing from sick people and try to keep vaccinations as diabetic patient is very prone to infections.
- Ask questions related to your antibiotic course and learn about all the possible side effects of your prescribed drugs.

ii. **For policy makers**

- Policy makers can take the responsibility to design a plan to handle antimicrobial resistance at different health setups.
- Progress in the investigation of antibiotic-resistant infections is required by the policy makers.
- Stakeholders should be involved in strengthening the policies and programs to prevent drug resistance.
- The safe disposal and usage of quality medicines should be regulated.
- Policy makers must provide enough information related to impact of resistance.

iii. **Role of health professionals**
Health professionals must:

- Ensure the cleaning of not only environment but also all the instruments.
- Should only prescribed drugs when they are needed
- Should be report the resistant agents to the scrutiny teams.
- Must discuss with their patients about all the outcomes of drug abuse.
- Talk with patients about the causes of resistance and how to avoid them.

iv. **Healthcare industry**

- Investment in research should be done and industry should focus on the development of new antibiotics and vaccines with minimum resistance (Chakrabortyet al., 2017).

v. **Role of World Health Organization (WHO)**

The WHO takes drug resistance as a foremost priority for being the biggest stake holder in health sector. A plan was introduced in 2015 sanctioned at the World Health Assembly to tackle this problem. The aim of this plan was to prevent infections with most effective medicines.

16.12 Future Perspectives of Diabetes Treatment and Drug Resistance

In spite of being aware of beneficial consequences of healthy lifestyle, numerous individuals are unable to maintain such a pleasant and health boosting lifestyle in that existing world of modernity (Piette et al., 2001). As a severe consequence, the prevalence of diabetes mellitus is expected to expand dramatically in present era and upcoming decades. Diabetes mellitus is an ultimate cause of brain stroke, renal failure, heart attack and blindness. However, diabetes, in spite of having expanded prevalence all over the world, is still incurable (Yadav et al., 2016).

As discussed earlier, diabetes mellitus is caused by impaired secretion of insulin from pancreatic cells due to environmental and genetic (Bennett et al., 2009). The multiple lines of research are being designed in the pharmacological, biological and technological fields for the development of the durable treatment of diabetes mellitus. The technological diabetic plan has been focused on the development and delivery of glucagon and insulin via artificial pancreas. But, the development of that technique may take 10 years for being applicable and practicable. Secondly, the biological technique is struggling to make feasible transplantation of pancreas. However, the application of such type of biological techniques has some limitations regarding structural resemblances of islet-cells and graft survival after transplantation. Hence, working on these techniques has made a significant progress. The pharmacological technique appears to be most appropriate and effective in reducing global burden of diabetes mellitus (Fowler, 2007). Besides this, complementary research is being conducted for protecting insulin producing cells from damaging effect of high blood glucose level with emphasis on proactive control of energy intake and increased energy expenditures (Gold & Moellering, 2012).

16.13 Future Treatment Solutions for Diabetes

i. **Technological Solutions**

Among susceptible of diabetes, prevalence of diabetes mellitus Type-2 is common. In diabetes mellitus Type-1 lack of endogenous insulin is reported and research is under focus for delivery of insulin by artificial methodologies. In this technique artificial pancreas is used for efficient and proper delivery of insulin to blood (Gonzalez et al., 2008). The instrumentation is already in use for sensing blood glucose level, calculating insulin requirements and an automatic pump for optimum deliverance of insulin. Multiple technological challenges are need to be addressed for producing efficient and clinically useful artificial pancreas (Group, 2016).

ii. **Biological Solutions**

The scientists are continuously focusing on replacing of damaged cells with functional cells or replacement of whole pancreas. A strict donor criterion is needed to be adopted for attaining promising and excellent outcomes among diabetic patients. Usually, pancreas transplant is performed in patients suffering from end stage renal disease (Group, 2005). A less invasive choice is to replace beta cells only instead of whole pancreas because of limited availability of pancreas's donors. The isolation of these beta cells from donor pancreas and its infusion in patient portal vein is under practiced at research level. At present, this technique is still being regarded as experimental entity because scientists are hoping for the remarkable outcomes (Holohan et al., 2013).

iii. **Pharmacological Solutions**

In patients with diabetes a wide range of pharmacological treatments has been invented. The pharmacological formulas used for addressing diabetes are supphonylureas and biguanides. However, the invention of these drugs cannot be posed to be a miracle cure because of its highly associated risk of hypoglycemia. Both classes of these drugs reported to have antihyperglycemic characteristics (Group, 2001). According to the recent studies, brain plays important role in provoking eating behavior, feeling of hunger. Hence, gut-brain inter-communication can badly be affected by administration of these drugs. Therefore, a wise and monitored use of pharmacological treatment is recommended in diabetic individuals (Holohan et al., 2013).

iv. **Drug resistance**

Since inception of discovery of drugs, it was assumed that consumption of drugs is only interlinked with remedial response of different ailments. However, with the passage of time and exploration of innovative research, it become evident that the prolonged use of drugs unconditionally results in progressive decrease in their remedial affect against treatment of diseases (Nagaraju & Divakar, 2012). Hence, after long time use of drugs targeted organisms become resistance to them. The continuous use of drugs provides resistance to targeted organisms and their tendency of lethality is predominately decreased. Initially, the problem of drug resistance was tried to solve by discovering new classes of drugs and then replacing them with older ones (Rother, 2007). Unfortunately, there was no evidence that development of new drugs can maintain pace with continual increasing resistance of microorganisms.

After comprehension about mechanism of resistance of microbes it was revealed that microbes remarkably have array of tools for disposal of administered drugs. Ultimately, as new researches came out genetic mutations are being introduced in drug producing entities (White, 2004).

16.14 Association Between Diabetes Mellitus, Other Diseases and Their Drug Therapies

i. Overview of Resistance to other drugs in diabetes

The worldwide occurrence of diabetes mellitus has been proliferated promptly during the period of last 10 years due to increased trend of population inclination towards urban areas, related sluggish lifestyle and older age. According to the data presented by International Diabetes Federation (IDF), during the year 2015, 415 million peoples were suffering from Type-2diabetes globally (IDF). A considerable rise of 4.7% in the number of diabetic patients was perceptible in Mexico during the year 1998–2012 (Sistema Nacional de Vigilancia Epidemiologicaet al., 2013). Adverse effects induced by Type-2DM include failure of treatment of tuberculosis, drug resistance, hospitalization, multi drug resistance and postponement of sputum conversion rate (Thanh et al., 2015; Restrepo et al., 2008). However, probably due to the divergences in the study populaces these results are not uniform in all the studies involved (Balde et al., 2006; Magee et al., 2013).

ii. Interaction among drugs and resistance

Sulphonylurea derivatives are antidiabetic drugs while rifampin accelerates the hepatic metabolism of sulphonylurea drugs. Resultantly patient may develop hypoglycemia or hyperglycemia due to difficult dose adjustment. Metformin is an antidiabetic drug used for Type-2diabetes and its administration does not result in hypoglycemia. P450 enzymes do not metabolize metformin thus its concentration is not reduced (Sun and Scott 2011). However, metformin is a substrate for organic cation transporter and its administration to healthy individuals leads to glucose lowering affects (Cho et al., 2011). Disadvantage of administrating metformin include lactic acidosis and gastrointestinal implications (Salpeter et al., 2010).

Rifampicin efficacy is dependent on exposure to drug or maximum concentration achieved by drug (Jayaram et al., 2003). The study conducted by (Nijland et al., 2006) indicated that rifampicin exposure was two times lower in patients having diabetes along tuberculosis in comparison to patients without diabetes. Similarly, maximum plasma concentration of rifampicin achieved in diabetic patients was less as compared to that in non-diabetic patients (Peloquin, 2002). These differences due to drug pharmacokinetics can explain the drug resistance and decreased bacteriological response in diabetic patients having tuberculosis (Alisjahbanaet al., 2007).

In diabetes, drug absorption, distribution, metabolism and excretion are changed. Delayed gastric emptying may result in decreased drug absorption. Decreased exposure of rifampicin is explained by high pH dependent absorption and solubility of rifampicin in diabetic-tuberculosis patients (Kenny & Strates, 1981). Moreover, diabetic patients having tuberculosis are subjected to tuberculosis drugs resistance due to bacterial genome mutations as well.

iii. Antipsychotic Drugs resistance in Diabetes

Due to various psychological issues a large percent of current population is on antipsychotic drugs among which schizophrenia is the most prevailing issue, antipsychotic drugs are also effective against mental problem like Alzheimer and bipolar disorder. Evidence of association between neuroleptic medication and diabetes was first reported in mid 1960s (Dixon et al., 2000) by the time research also established the fact that people using atypical antipsychotics are at higher risk of developing diabetes. Studies proved that people with severe mental illness (SMI) are more prone to diabetes mellitus but reason behind this link remained a controversial mystery for a long time (Holt et al., 2005). Numerous scientists have been working to define association of severe mental illness (SMI) and diabetes. Although, the exact reason was still inconsistent, yet the studies provided a direction towards the multiple mechanisms involved. It was observed that diabetes is more common in the people using novel second-generation antipsychotic drugs (Holt and Peveler, 2006).

The patients in later age groups receiving an antipsychotic drug with a family history of obesity are more likely to develop diabetes. Traditional neuroleptics and atypical antipsychotics drugs elevate the patient's weight providing the grounds of development of Type-2diabetes (Czobor et al., 2002: Meyer 2001; UKPDS, 1998). Massive research data support the theory that antipsychotic-induced diabetes is directly linked to weight gain. Drugs like clozapine and olanzapine have outrageous tendency to cause diabetes with weight gain. Both of these factors affect the cardiac health profile of patient usually resulting in problems like coronary heart disease reported to be the prime reason of premature death in patients of schizophrenia (Bick et al., 2007). Cardiac condition is further worsened by stagnant lifestyle and smoking habits which are common in schizophrenia patients. To avoid such problems usually antipsychotics drugs like risperidone are prescribed, which does not affect patient's weight or pancreatic health, and have least chances of disease relapse in comparison to traditional neuroleptic drugs like haloperidol (Csernansky et al., 2002).

Patients of severe mental illness like schizophrenia on atypical antipsychotic medication are regarded as high-risk group for diabetes and vascular ailments. The reason is drug induced elevated weight but there is clear evidence on the impact of antipsychotic drugs on the metabolic functions of the patient. Other known causes of antipsychotics drug-induced diabetes are discord 5-HT2C or histamine H1 receptors, escalated serum leptin above normal levels, and antagonism at central dopamine D2 receptors (Seeman, 2002). In most cases antipsychotic induced diabetes is reversible upon discontinuation of medication. According to a study, antipsychotics disrupt the glucose transport mechanism into PC12 cells, resulting in the cellular escalation of glucose transporter isoforms i.e. GLUT1 and GLUT3. This condition will induce hyperglycemia, followed by homeostatic rise in insulin release. A prolonged hyper-insulinemia will eventually develop insulin resistance due to destruction of insulin receptors (Dwyer et al., 1999).

Diabetes mellitus is perilous medical condition that requires an immediate mediation along with life-long management therapies involving high dose of antidiabetic drugs. However, the severe mental illnesses cannot be controlled by antipsychotic medication. To control both conditions side by side is only possible through collaborative treatment designed by psychiatrist and diabetologist. High-fat diet is usually tapered in this condition as it will boost weight gains in patients with antipsychotic drug-induced lethargic lifestyle (Micheal et al., 2003).

iv. **Antibiotic resistance in Diabetes**

The diabetic patients with compromised immune system, disturbed glycemic index, and bacterial attacks are suspected to frequent and recurrent infections and frequently administered with antibiotic. The rate of antibiotic resistance directly depends on factors like patient history of antibiotic usage and national consumption levels of antibiotics. As the diabetics are more prone to infectious diseases, their antibiotic exposure is high that leads to elevated antibiotic resistance rate. Numerous studies have proved that regions with higher diabetes prevalence have higher antibiotic consumption rate (Nitzan et al., 2010). Diseases like diabetes mellitus also impair antibiotic pharmacokinetics. Patients of peripheral vascular disease, with disease free intervals have shown the evidence of poor antibiotic penetration in disease affected tissues, thus boasting the development of antibiotic resistance in pathogenic bacteria (Lipsky et al., 2012).

Various researches have proved clear evidence of elevated antibiotic resistance rates in diabetes mellitus patient in comparison to non-diabatic patients. In India about 70% antibiotic resistant strains are isolated from the wounds of patients with DM (Basu et al., 2009). According to a data of diabetes patient with asymptomatic bacteriuria collected in Cameroon, the causative agents of infection *Escherichi-acoli, Pseudomonas aeruginosa, Streptococcus pyogenes, Enterococcus faecalis, Klebsiella pneumoniae, Proteus mirabilis and Staphylococcus aureus*exhibits strong antibiotic resistance rates: 96.2% against amoxicillin, 70.1% against ceftriaxone and 61.3% against ciprofloxacin (Njunda et al., 2012). The studies have proved that lacking adherence to National Antibiotic Policy (NAP) is another reason of higher antibiotic drug resistance in patients of DM. In diabetes patient infections should be treated along with strict glycemic control (Koh et al., 2012). For diabetic patients after surgical procedures usually a closed-loop system involving artificial pancreases for controlling blood glucose and glycemic index is implemented (Kawahito et al., 2009). Destructive reactions in neutrophils can also be averted by improving glycemic control (UKPDS, 1998).

To avoid the problem of drug resistance in the strains like *Propionibacterium*, *corynebacterium*, and coagulase-negative *staphylococci* are only treated with antibiotics if regularly observed in patient specimens (Richard et al., 2011). Despite of the fact, there is no clearly defined link between DM and multidrug-resistant tuberculosis, the studies have proved that diabetics are at greater risk to be affected by this strain. To deal with the problem, the scientists use a combination of antidiabetic anti-tuberculosis antibiotics drug (such as rifampicin). To attain optimal regimens in diabetic patients with tuberculosis researchers are focusing on antibiotic pharmacokinetics. In diabetic patients, mixed infections caused by multidrug-resistant bacteria recently emerged resistance mechanisms in the Gram-negative Bacteria. The New Delhi metallo-β-lactamase (NDM-1) is an enzyme that makes bacteria resistant to a broad range of beta-lactam antibiotics, thus posing another therapeutic challenge. In case of these infections' novel (tigecycline) and old antibiotics (polymyxin and fosfomycin) have been found to be equally effective (Lyudmila & Ivan, 2013).

In diabetic patients, keeping a constant track of antibiotic susceptibility patterns of disease-causing bacteria facilitate optimal treatment selection. For better prophylaxis and cure of infections in diabetes, scientists are focusing on nonantibiotic supplements, probiotics and bacterial oral extracts with immune-stimulant properties. In conclusion, in diabetics to control emanation and escalation of antibiotic-resistant bacteria combined efforts government organizations, pharmaceutical companies along with researchers like microbiologists, clinicians and epidemiologists is required.

v. **Anti-tuberculosis drugs resistance in Diabetes**

Diabetes is known as a predisposing factor for developing tuberculosis and in countries having higher number of tuberculosis patients 15% of them develop diabetes (Lönnrothet al., 2010). Diabetic patients pose risk of developing tuberculosis three times more than the normal people. In coming twenty years, number of drug resistant tuberculosis and tuberculosis patients is expected to rise in middle- and low-income countries where the incidence of diabetes is also likely to be increased (Dooley and Chaisson 2009). Due to growing cases of drug resistant tuberculosis it is important to understand the relationship between drug resistant tuberculosis and diabetes in order to optimize the affected patients in future. Diabetes was identified as a major risk factor for developing drug resistance in tuberculosis patients and may result in total drug resistance and poly drug resistance against isoniazid, ethambutol, streptomycin, and Rifampin among tuberculosis patients (Song et al., 2019). Thus, diabetes increases the treatment failure and morbidity rate among tuberculosis patients. The results of several studies indicated that the tuberculosis patients with diabetes remain sputum smear positive at two to three months even after undergoing tuberculosis treatment (Baker et al., 2011).Diabetes can impact the pharmacokinetics of tuberculosis drugs. The antimicrobial drug susceptibility is determined in the following categories:

- Pan susceptible (Susceptible to all drugs)
- Resistant (Resistance towards all drugs)

- Multidrug resistant (resistance against combined drug isoniazid plus rifampin)
- Extensive drug resistance (Multidrug resistance along resistance against quinolones and injectable drugs).

vi. **Aspirin resistance in diabetes**

Aspirin or Acetylsalicylic acid (ASA) resistance is a term used to define various phenomenon that explain the aspirin thrombotic complications in patients, the aspirin inability to shorten the bleeding time or to prevent biosynthesis of thromboxane, and foreseen effects on in vitro testing of platelet function(Patrono, 2003). Aspirin resistance has been proved a serious issue in controlling stroke, cardiovascular implications and in diabetes (National Cholesterol Education Program, 2002). Aspirin dosage is recommended as primary prevention in patients with high risk of cardiovascular diseases while as secondary preventive measure in cardiovascular patients (Fox et al., 2015).

Aspirin resistance can be defined in another way as inhibition of platelets function incompletely by aspirin despite the patient is on clinically affective therapeutic dose of aspirin in that specific disease (Watala et al., 2005). Factors responsible for aspirin resistance include age, gender, aspirin dose, any metabolic disease, glycemic control (fasting blood sugar and HbAIC), inflammatory markers and endogenous level of cholesterol (Tantry et al., 2005). In diabetic patients, glycosylation of protein and increased platelet turnover is suggested as possible cause of aspirin resistance (Winocour, 1994). Hereditary factors and environment as well play important role in determining the resistance. Aspirin is antiplatelet drug and the diabetes reduces platelet sensitivity towards aspirin and affects glycemic control (Watala et al., 2005).

Some studies showed that aspirin can acetylate all hydroxyl groups and amino acids that may be susceptible into plasma proteins and platelets (Cherian & Abraham, 1993). In diabetes non enzymatic glycosylation compete with nonspecific acetylation. In diabetic patients reduced platelets sensitivity towards aspirin is related to glycemic control in patients. This association can provide a link between increased non-enzymatic glycosylation due to chronically hyperglycemia and reduced non-enzymatic acetylation on aspirin intake in diabetic patients (Watala et al., 2004). In diabetes patients increased plasma level, activity of clotting factors and variations in structure of prothrombotic clot are important factors. Activity of platelets and coagulation factors can be reduced in order to improve the clinical efficiency of aspirin by the use of angiotensin converting enzyme inhibitors (ACEI), modulation of insulin resistance and use of statins (Ajjanet al., 2008). Control of glycemic level and increased therapeutic uses of aspirin can improve the acetylation of proteins. The mechanism of aspirin resistance in diabetes has been shown in Fig. 16.3.

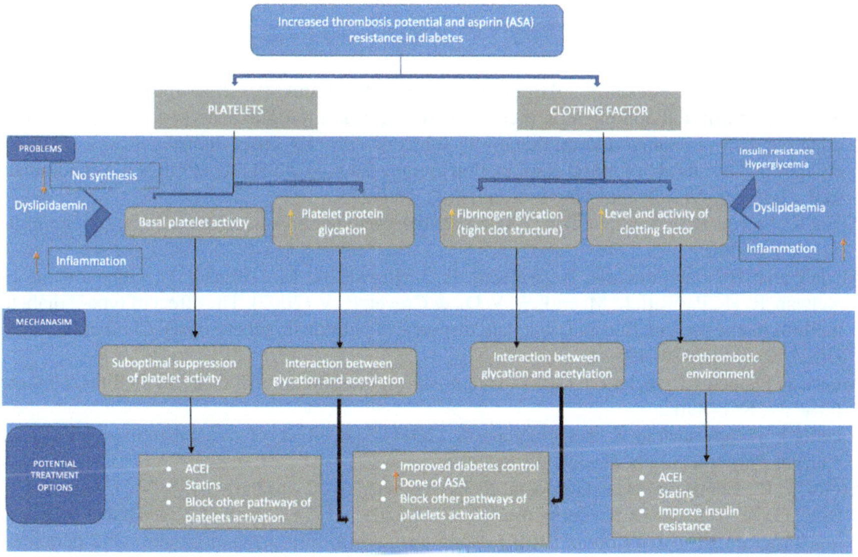

Fig. 16.3 Mechanism of aspirin resistance in diabetics and thrombotic risk (Ajjan et al., 2008)

16.15 Conclusion

Diabetes is the widespread disease associated with serious metabolic disorders. In order to prevent diabetes mellitus insulin therapy and many oral antidiabetic medications are used. Insulin therapy with different formulations is used to reduce the associated risk of hypoglycemia. At times during prolonged treatment body cells shows a weakened biological response by do not responding appropriately to the released insulin and depict a resistance for insulin. Various factors including obesity, high carbohydrates food, gestational diabetes, health status e.g. non-alcoholic fatty liver disease and polycystic ovary syndrome elevates the chances of insulin resistance. Oral antidiabetic agents are also used alone or in combination with insulin to augment insulin secretion. Prolong treatment causes β cell exhaustion resulting in elevated glucose levels causing glucotoxicity. The impaired beta-cell response to glucose results in insulin resistance. To overcome such issues research is focused on development of methodologies for artificial insulin delivery. Another novel technique involves replacement of beta cells only instead of whole pancreas. After comprehension about mechanism of resistance in microbial subjects it has been revealed that genetic mutations result in new drug producing entities to avert resistance issues in diabetes.

References

Abdul-Ghani, M. A., Puckett, C., Triplitt, C., Maggs, D., Adams, J., Cersosimo, E., & DeFronzo, R. A. (2015). Initial combination therapy with metformin, pioglitazone and exenatide is more effective than sequential add-on therapy in subjects with new-onset diabetes. Results from the efficacy and durability of initial combination therapy for type 2 diabetes (EDICT): A randomized trial. *Diabetes, Obesity and Metabolism, 17*(3), 268–275.

Ajjan, R., Storey, R. F., & Grant, P. J. (2008). Aspirin resistance and diabetes mellitus. *Diabetologia, 51*(3), 385–390.

Alisjahbana, B., Sahiratmadja, E., Nelwan, E. J., Purwa, A. M., Ahmad, Y., Ottenhoff, T. H., Nelwan, R. H., Parwati, I., Meer, J. W. V. D., & Crevel, R. V. (2007). The effect of type-2 diabetes mellitus on the presentation and treatment response of pulmonary tuberculosis. *Clinical Infectious Diseases, 45*(4), 428–435.

American Diabetes Association. (2014). Diagnosis and classification of diabetes mellitus. *Diabetes Care, 37*, S81–S90.

Ananth, J., Parameswaran, S., & Gunatilake, S. (2004). Side effects of atypical antipsychotic drugs. *Current Pharmaceutical Design*, 2219–2229.

Aroda, V. R., Edelstein, S. L., Goldberg, R. B., Knowler, W. C., Marcovina, S. M., Orchard, T. J., Bray, G. A., Schade, D. S., Temprosa, M. G., White, N.H., Crandall, J.P., & Diabetes Prevention Program Research Group. (2016). Long-term metformin use and vitamin B12 deficiency in the diabetes prevention program outcomes study. *The Journal of Clinical Endocrinology & Metabolism, 101*(4), 1754–1761.

Ashcroft, F., & Rorsman, P. (2004). Type 2 diabetes mellitus: Not quite exciting enough? *Human Molecular Genetics, 13*(suppl_1), R21–R31.

Atkinson, M. A., & Eisenbarth, G. S. (2001). Type-1 diabetes: New perspectives on disease pathogenesis and treatment. *The Lancet, 358*, 221–229.

Atkinson, M. A., Eisenbarth, G. S., & Michels, A. W. (2014). type-1 diabetes. *The Lancet, 383*, 69–82.

Baker, M. A., Harries, A. D., Jeon, C. Y., Hart, J. E., Kapur, A., Lönnroth, K., Ottmani, S. E., Goonesekera, S. D., & Murray, M. B. (2011). The impact of diabetes on tuberculosis treatment outcomes: A systematic review. *BMC Medicine, 9*(1), 81.

Bastaki, A. (2005). Diabetes mellitus and its treatment. *International Journal of Diabetes and Metabolism, 13*, 111.

Basu, S., RamchuranPanray, T., Bali Singh, T., Gulati, A. K., & Shukla, V. K. (2009). A prospective, descriptive study to identify the microbiological profile of chronic wounds in outpatients. *Ostomy/wound Management, 55*(1), 14–20.

Baynes, H. W. (2015). Classification, pathophysiology, diagnosis and management of diabetes mellitus. *Journal of Diabetes Metabolism, 6*, 1–9.

Becker, M. L., Pearson, E. R., & Tkáč, I. (2013). Pharmacogenetics of oral antidiabetic drugs. *International journal of endocrinology.*

Beckman, J. A., Creager, M. A., & Libby, P. (2002). Diabetes and atherosclerosis. *JAMA, 287*, 2570–2581.

Bennett, D. E., Camacho, R. J., Otelea, D., Kuritzkes, D. R., Fleury, H., & Kiuchi, M. (2009). Drug resistance mutations for surveillance of transmitted HIV-1 drug-resistance: 2009 update. *Plos One, 4*, 16–24.

Bertilsson, L., Dahl, M. L., Dalén, P., & Al-Shurbaji, A. (2002). Molecular genetics of CYP2D6: Clinical relevance with focus on psychotropic drugs. *British Journal of Clinical Pharmacology, 53*(2), 111–122.

Bick, P., Knoesen, N., & Castle, D. (2007). Clinical implications of the CATIE schizophrenia trials: Day-to-day management lessons for Australasian psychiatrists. *Australasian Psychiatry., 15*(6), 465–469.

Blair, M. (2016). Diabetes mellitus review. *Urologic Nursing, 36*.

Cahn, A., & Cefalu, W. T. (2016). Clinical considerations for use of initial combination therapy in type 2 diabetes. *Diabetes Care, 39*(Supplement 2), S137–S145.

Cefalu, W. T. (2001). Insulin resistance: Cellular and clinical concepts. *Experimental Biology and Medicine (maywood, N.J.), 226*, 13–26.

Ceriello, A., & Motz, E. (2004). Is oxidative stress the pathogenic mechanism underlying insulin resistance, diabetes, and cardiovascular disease? The common soil hypothesis revisited. *Arteriosclerosis, Thrombosis, and Vascular Biology, 24*(5), 816–823.

Chakraborty, A., Shenoy, S., Adhikari, P., Saralaya, V., & Rao, S. (2017). Is diabetes mellitus an important risk factor for the antibiotic resistance in extraintestinal pathogenic Escherichia coli? *Indian Journal of Pathology and Microbiology., 60*, 546–549.

Chapagain, A., Caton, P. W., Kieswich, J., Andrikopoulos, P., Nayuni, N., Long, J. H., Harwood, S. M., Webster, S. P., Raftery, M. J., Thiemermann, C., Walker, B. R., Seckl, J. R., Corder, R., & Yaqoob, M. M. (2014). Elevated hepatic 11β-hydroxysteroid dehydrogenase Type-1 induces insulin resistance in uremia. *ProcNatlAcadSci USA, 111*, 3817–3822.

Chen, D., Misra, A., & Garg, A. (2002). Lipodystrophy in human immunodeficiency virus-infected patients. *Journal of Clinical Endocrinology and Metabolism, 87*, 4845–4856.

Cherian, M., & Abraham, E. C. (1993). Glycation of human lens crystallins: Effect of age and aspirin treatment. *Ophthalmic Research, 25*, 349–354.

Chiang, C. Y., Bai, K. J., Lin, H. H., Chien, S. T., Lee, J. J., Enarson, D. A., Lee, T., & Yu, M. C. (2015). The influence of diabetes, glycemic control, and diabetes-related comorbidities on pulmonary tuberculosis. *PLoS ONE, 10*, 33–39.

Cho, S. K., Yoon, J. S., Lee, M. G., Lee, D. H., Lim, L. A., Park, K., Park, M. S., & Chung, J. Y. (2011). Rifampin enhances the glucose-lowering effect of metformin and increases OCT1 mRNA levels in healthy participants. *Clinical Pharmacology & Therapeutics, 89*(3), 416–421.

Cincotta, A. H., Hansen, B., & Shaffrir, E. (2002). Hypothalamic role in the insulin resistance syndrome. In *Insulin resistance syndrome* (pp. 271–312). Taylor and Francis.

Cipolla-Neto, J., & Amaral, F. G. D. (2018). Melatonin as a hormone: New physiological and clinical insights. *Endocrine Reviews, 39*(6), 990–1028.

Cohen, G., Sasson, S., & Riahi, Y. (2012). Free radicals and metabolic disorders. In *Encyclopedia of radicals in chemistry, biology and materials*, Online 2012. Wiley.

Csernansky, J. G., Mahmoud, R., & Brenner, R. (2002). for the Risperidone-USA 79 Study Group: A comparison of risperidone and haloperidol for the prevention of relapse in patients with schizophrenia. *New England Journal of Medicine, 346*, 16–22.

Czobor, P., Volavka, J., Sheitman, B., Lindenmeyer, J.-P., Citrome, L., McEvoy, J., Cooper, T. B., Chakos, M., & Lieberman, J. A. (2002). Antipsychotic- induced weight gain and therapeutic response: A differential association. *Journal of Clinical Psychopharmacology, 22*, 244–251.

Daneman, D. (2006). Type-1 diabetes. *The Lancet., 367*, 847–858.

Devendra, D., Liu, E., & Eisenbarth, G. S. (2004). Type-1 diabetes: Recent developments. *BMJ, 328*, 750–754.

Dimova, R., Chakarova, N., Grozeva, G., Kirilov, G., & Tankova, T. (2019). The relationship between glucose variability and insulin sensitivity and oxidative stress in subjects with prediabetes. *Diabetes Research and Clinical Practice, 158*, 107911.

Dixon, L., Weiden, P., Delahanty, J., Goldberg, R., Postrado, L., Lucksted, A., & Lehman, A. (2000). Prevalence and correlates of diabetes in national schizophrenia samples. *Schizophrenia Bulletin, 26*, 903–912.

Dooley, K. E., & Chaisson, R. E. (2009). Tuberculosis and diabetes mellitus: Convergence of two epidemics. *The Lancet Infectious Diseases, 9*(12), 737–746.

Dwyer, D. S., Pinkofsky, H. B., Liu, Y., & Bradley, R. J. (1999). Antipsychotic drugs affect glucose uptake and the expression of glucose transporters in PC12 cells. *Prog Neuro- Psychopharmacol Biol Psychiatry, 23*, 69–80.

Edelman, S., Garg, S., Frias, J., Maggs, D., Wang, Y., Zhang, B., Strobel, S., Lutz, K., & Kolterman, O. (2006). A double-blind, placebo-controlled trial assessing pramlintide treatment in the setting of intensive insulin therapy in type 1 diabetes. *Diabetes care, 29*(10), 2189–2195.

Emami-Riedmaier, A., Schaeffeler, E., Nies, A. T., Mörike, K., & Schwab, M. (2015). Stratified medicine for the use of antidiabetic medication in treatment of Type II diabetes and cancer: Where do we go from here? *Journal of Internal Medicine., 277*(2), 235–247.

Evans, W. E., & Johnson, J. A. (2001). Pharmacogenomics: The inherited basis for interindividual differences in drug response. *Annual Review of Genomics and Human Genetics., 2*(1), 9–39.

Feingold, K. R. (2020). Oral and injectable (non-insulin) pharmacological agents for Type-2 diabetes. Endotext [Internet].

Fowler, M. J. (2007). Diabetes treatment, part 2: Oral agents for glycemic management. *Diabetes Foundation, 25*, 131–134.

Fox, C. S., Golden, S. H., Anderson, C., Bray, G. A., Burke, L. E., De Boer, I. H., Deedwania, P., Eckel, R. H., Ershow, A. G., Fradkin, J., & Inzucchi, S. E. (2015). Update on prevention of cardiovascular disease in adults with Type-2 diabetes mellitus in light of recent evidence: A scientific statement from the American Heart Association and the American Diabetes Association. *Circulation, 132*(8), 691–718.

Frezza, M., di Padova, C., Pozzato, G., Terpin, M., Baraona, E., & Lieber, C. S. (1990). High blood alcohol levels in women: The role of decreased gastric alcohol dehydrogenase activity and first-pass metabolism. *New England Journal of Medicine, 322*(2), 95–99.

Geers, A. L., Wellman, J. A., Fowler, S. L., Helfer, S. G., & France, C. R. (2010). Dispositional optimism predicts placebo analgesia. *The Journal of Pain, 11*(11), 1165–1171.

Gilbert, M. P., & Pratley, R. E. (2009). Efficacy and safety of incretin-based therapies in patients with Type-2 diabetes mellitus. *European Journal of Internal Medicine, 20*, S309–S318.

Gillespie, K. M. (2006). Type-1 diabetes: Pathogenesis and prevention. *175*, 165–170.

Gold, H. S., & Moellering, R. C. (2012). Antimicrobial-Drug Resistance. *Drug Therapy, 335.*

Gonzalez, J. S., Peyrot, M., Mccarl, L. A., & Collins, E. M. (2008). Depression and diabetes treatment nonadherence: A meta-analysis. *Diabetes Care, 31*, 2398–2403.

Group, S. R. (2016). Intensive diabetes treatment and cardiovascular outcomes in type 1 diabetes: The DCCT/EDIC study 30-year follow-up. *Diabetes Care, 39*, 686–693.

Group, T. D. (2001). Influence of intensive diabetes treatment on body weight and composition of adults With type 1 diabetes in the diabetes control and complications trial. *Clinical Care, 24*, 1711–1721.

Group, T. D. (2005). Intensive diabetes treatment and cardiovascular disease in patients with type 1 diabetes. *The New England Journal of Medicine, 353*, 2643–2653.

Hamidovic, A., Dlugos, A., Palmer, A. A., & de Wit, H. (2010). Catechol-O-methyltransferase val158met genotype modulates sustained attention in both the drug-free state and in response to amphetamine. *Psychiatric Genetics, 20*(3), 85.

Handelsman, Y. (2011). Role of bile acid sequestrants in the treatment of Type-2 diabetes. *Diabetes Care, 34*, S244–S250.

Hauner, H. (2002). The mode of action of thiazolidinediones. *Diabetes/Metabolism Research and Reviews, 18*(S2), S10–S15.

Holman, R. R., Paul, S. K., Bethel, M. A., Matthews, D. R., & Neil, H. A. W. (2008). 10-year follow-up of intensive glucose control in type 2 diabetes. *New England Journal of Medicine, 359*(15), 1577–1589.

Holohan, C., Schaeybroeck, S. V., Longley, D. B., & Johnston, P. G. (2013). Cancer drug resistance: An evolving paradigm, pp. 714–726.

Holt, R. I. G., & Peveler, R. C. (2006). Antipsychotic drugs and diabetes—an application of the Austin Bradford Hill criteria. *Diabetologia, 49*, 1467–1476.

Holt, R. I. G., Bushe, C., & Citrome, L. (2005). Diabetes and schizophrenia 2005: Are we any closer to understanding the link. *Journal of Psychopharmacology, 19*, 56–650.

Horii, T., Momo, K., Yasu, T., Kabeya, Y., & Atsuda, K. (2019). Determination of factors affecting medication adherence in Type-2 diabetes mellitus patients using a nationwide claim-based database in Japan. *PloS One, 14*(10).

Horii, T., Momo, K., Yasu, T., Kabeya, Y., & Atsuda, K. (2019). Determination of factors affecting medication adherence in type 2 diabetes mellitus patients using a nationwide claim-based database in Japan. *PloS one, 14*(10), e0223431

Hruz, P. W. (2008). HIV protease inhibitors and insulin resistance: lessons from in vitro, rodent and healthy human volunteer models. *Current Opinion in HIV and AIDS, 3*, 660.

Hu, C., & Jia, W. (2018). Diabetes in China: Epidemiology and genetic risk factors and their clinical utility in personalized medication. *Diabetes, 67*(1), 3–11.

Hu, S., Wang, S., Fanelli, B., Bell, P. A., Dunning, B. E., Geisse, S., Schmitz, R., & Boettcher, B. R. (2000). Pancreatic β-cell K ATP channel activity and membrane-binding studies with nateglinide: A comparison with sulfonylureas and repaglinide. *Journal of Pharmacology and Experimental Therapeutics, 293*, 444–452.

Jayaram, R., Gaonkar, S., Kaur, P., Suresh, B. L., Mahesh, B. N., Jayashree, R., Nandi, V., Bharat, S., Shandil, R. K., Kantharaj, E., & Balasubramanian, V. (2003). Pharmacokinetics-pharmacodynamics of rifampin in an aerosol infection model of tuberculosis. *Antimicrobial Agents and Chemotherapy, 47*(7), 2118–2124.

Jemal, A., Abdela, J., & Sisay, M. (2017). Adherence to oral antidiabetic medications among Type-2 diabetic (T2DM) patients in chronic ambulatory wards of HiwotFana Specialized University Hospital, Harar, Eastern Ethiopia: A cross sectional study. *Journal of Diabetes Metabolism, 8*, 721. https://doi.org/10.4172/2155-6156. An open access journal ISSN: 2155-6156. Socio-demographic characteristics frequency% Age. 18–40.

Joshi, S. R., Standl, E., Tong, N., Shah, P., Kalra, S., & Rathod, R. (2015). Therapeutic potential of α-glucosidase inhibitors in Type-2 diabetes mellitus: An evidence-based review. *Expert Opinion on Pharmacotherapy, 16*, 1959–1981.

Justice, A. J., & de Wit, H. (1999). Acute effects of d-amphetamine during the follicular and luteal phases of the menstrual cycle in women. *Psychopharmacology (berl), 145*(1), 67–75.

Kaiser, N., Leibowitz, G., & Nesher, R. (2003). Glucotoxicity and beta-cell failure in Type-2 diabetes mellitus. *Journal of Pediatric Endocrinology and Metabolism, 16*, 5–22.

Kalra, S., Mukherjee, J. J., Venkataraman, S., Bantwal, G., Shaikh, S., Saboo, B., Das, A. K., & Ramachandran, A. (2013). Hypoglycemia: The neglected complication. *Indian Journal of Endocrinology and Metabolism, 17*(5), 819.

Kawahito, S., Kitahata, H., & Oshita, S. (2009). Problems associated with glucose toxicity: Role of hyperglycemia-induced oxidative stress. *World Journal of Gastroenterology, 15*(33), 4137–4142.

Kawamori, R., Kaku, K., Hanafusa, T., Kashiwabara, D., Kageyama, S., & Hotta, N. (2012). Efficacy and safety of repaglinide vs nateglinide for treatment of Japanese patients with Type-2 diabetes mellitus. *Journal of Diabetes Investigation, 3*, 302–308.

Kenny, M. T., & Strates, B. (1981). Metabolism and pharmacokinetics of the antibiotic rifampin. *Drug Metabolism Reviews, 12*(1), 159–218.

Kernan, W. N., & Inzucchi, S. E. (2004). Type-2 diabetes mellitus and insulin resistance: Stroke prevention and management. *Current Treatment Options in Neurology, 6*(6), 443–450.

Kernan, W. W., Inzucchi, S. E., & Viscoli, C. M. (2002). Insulin resistance and risk for stroke. *Neurology, 59*, 809–815.

Kernan, W. N., Viscoli, C. M., & Inzucchi, S. E. (2002). Insulin resistance among patients with ischemic stroke and TIA. *Stroke, 33*, 389.

Koh, G. C., Peacock, S. J., van der Poll, T., & Wiersinga, W. J. (2012). The impact of diabetes on the pathogenesis of sepsis. *European Journal of Clinical Microbiology and Infectious Diseases, 31*(4), 379–388.

Krentz, A. J., & Bailey, C. J. (2005). Oral antidiabetic agents. *Drugs, 65*, 385–411.

Leyking, S., & Fliser, D. (2014). Insulin resistance in CKD. *Clinical Journal of the American Society of Nephrology, 9*, 638–640.

Lipsky, B. A., Peters, E. J., Senneville, E., et al. (2012). Expert opinion on the management of infections in the diabetic foot. *Diabetes/metabolism Research and Reviews, 28*(Suppl. 1), 163–178.

Lönnroth, K., Castro, K. G., Chakaya, J. M., Chauhan, L. S., Floyd, K., Glaziou, P., & Raviglione, M. C. (2010). Tuberculosis control and elimination 2010–50: Cure, care, and social development. *The Lancet, 375*(9728), 1814–1829.

Luczak, S. E., Glatt, S. J., & Wall, T. L.. (2009). Meta-analyses of ALDH2 and ADH1B with alcohol dependence in Asians.

Lyudmila, B., & Ivan, M. (2013). Antibiotic resistance rates in causative agents of infections in diabetic patients: Rising concerns. *Expert Review of Antiinfective Therapy 11*(4), 411–420.

Maedler, K., Sergeev, P., & Ris, F. (2002). Glucose-induced beta cell production of IL-1beta contributes to glucotoxicity in human pancreatic islets. *The Journal of Clinical Investigation, 110*, 851–860.

Magee, M. J., Bloss, E., Shin, S. S., Contreras, C., Arbanil, H., Calderon, J., Bayona, J., Bonia, C., Yagui, M., Jave, O., & Cegielski, J. P. (2013). Clinical characteristics, drug resistance, and treatment outcomes among tuberculosis patients with diabetes in Peru. *International Journal of Infectious Diseases, 17*, 404–412.

Mahajan, R., & Gupta, K. (2010). Incretin mimetics and enhancers: New insulin secretagogues for Type-2 diabetes.

Mann, E., Sunni, M., & Bellin, M. D. (2020). Secretion of insulin in response to diet and hormones. *Pancreapedia: The Exocrine Pancreas Knowledge Base.*

Marin-Penalver, J. J., Martin-Timon, I., Sevillano-Collantes, C., & del Canizo-Gomez, F. J. (2016). Update on the treatment of Type-2 diabetes mellitus. *World Journal of Diabetes, 7*(17), 354.

Maxwell, J. C. (2012). Drunk versus drugged: How different are the drivers? *Drug and Alcohol Dependence, 121*(1–2), 68–72.

Meier, J. J., & Nauck, M. A. (2010a). Is the diminished incretin effect in Type-2 diabetes just an epi-phenomenon of impaired β-cell function? *Diabetes, 59*(5), 1117–1125.

Meier, J. J., & Nauck, M. A. (2010b). Is the diminished incretin effect in Type-2 diabetes just an epi-phenomenon of impaired beta cell function? *Diabetes, 259*, 1117–1125.

Meng, H., Zhang, A., Liang, Y., Hao, J., Zhang, X., & Lu, J. (2018). Effect of metformin on glycaemic control in patients with type 1 diabetes: A meta-analysis of randomized controlled trials. *Diabetes/Metabolism Research and Reviews, 34*(4), e2983.

Meyer, J. M. (2001). Effects of atypical antipsychotics on weight gain and serum lipid levels. *Journal of Clinical Psychiatry, 62*, 27–34.

Miczek, K. A., Yap, J. J., & Covington, H. E., III. (2008). Social stress, therapeutics and drug abuse: Preclinical models of escalated and depressed intake. *Pharmacology & Therapeutics, 120*(2), 102–128.

Moen, M. F., Zhan, M., Hsu, V. D., Walker, L. D., Einhorn, L. M., Seliger, S. L., & Fink, J. C. (2009). Frequency of hypoglycemia and its significance in chronic kidney disease. *Clinical Journal of the American Society of Nephrology, 4*, 1121–1127.

Molehin, O. R., Adefegha, S. A., & Adeyanju, A. A. (2020). Role of oxidative stress in the pathophysiology of Type 2 diabetes and cardiovascular diseases. In *Role of oxidative stress in pathophysiology of diseases* (pp. 277–297), Springer.

Morgan, C. J., Schafer, G., Freeman, T. P., & Curran, H. V. (2010). Impact of cannabidiol on the acute memory and psychotomimetic effects of smoked cannabis: Naturalistic study. Correction.

Murali, T. S., Kavitha, S., Spoorthi, J., Bhat, D. V., Bharath, A. S., Upton, Z., Ramachandra, L., Acharya, R. V., & Satyamoorthy, K. (2014). Characteristics of microbial drug resistance and its correlates in chronic diabetic foot ulcer infections. *Journal of Medical Microbiology, 63*, 1377–1385.

Nagaraju, E. V., & Divakar, G. (2012). Antibiotic susceptibility of bacterial strains. *International Journal of Advances in Pharmacy, 1*, 546–550.

Narendran, P., Estella, E., & Fourlanos, S. (2005). Immunology of Type-1 diabetes. *QJM: An International Journal of Medicine, 98*, 547–556.

National Cholesterol Education Program (US). (2002). Expert panel on detection and treatment of high blood cholesterol in adults. *Third report of the National Cholesterol Education Program (NCEP) expert panel on detection, evaluation, and treatment of high blood cholesterol in adults (adult treatment panel III)* (No. 2). National Cholesterol Education Program, National Heart, Lung, and Blood Institute, National Institutes of Health.

Nijland, H. M., Ruslami, R., Stalenhoef, J. E., Nelwan, E. J., Alisjahbana, B., Nelwan, R. H., Van Der Ven, A. J., Danusantoso, H., Aarnoutse, R. E., & Van Crevel, R. (2006). Exposure to rifampicin is strongly reduced in patients with tuberculosis and Type-2 diabetes. *Clinical Infectious Diseases, 43*(7), 848–854.

Nitzan, O., Low, M., Lavi, I., Hammerman, A., Klang, S., & Raz, R. (2010). Variability in outpatient antimicrobial consumption in Israel. *Infection, 38*(1), 12–18.

Nitzan, O., Elias, M., Chazan, B., & Saliba, W. (2015). Urinary tract infections in patients with Type-2 diabetes mellitus: review of prevalence, diagnosis, and management. *Diabetes, Metabolic Syndrome and Obesity: Targets and Therapy., 8*, 129.

Njunda, A. L., Assob, N. J. C., Nsagha, S. D., et al. (2012). Uropathogens from diabetic patients with asymptomatic bacteriuria and urinary tract infections. *Scientific Journal of Microbiology, 1*(6), 141–146.

Nogueira, C., Souto, S. B., Vinha, E., Braga, D. C., & Carvalho, D. (2013). Oral glucose lowering drugs in Type-2 diabetic patients with chronic kidney disease. *Hormones (athens, Greece), 12*, 483–494.

Omori, K., Nomoto, H., Nakamura, A., Takase, T., Cho, K. Y., Ono, K., Manda, N., Kurihara, Y., Aoki, S., & Atsumi, T. (2019). Reduction in glucose fluctuations in elderly patients with Type-2 diabetes using repaglinide: A randomized controlled trial of repaglinide vs sulfonylurea. *Journal of Diabetes Investigation, 10*, 367–374.

Ozougwu, J., Obimba, K., Belonwu, C., & Unakalamba, C. (2013). The pathogenesis and pathophysiology of Type-1 and Type-2 diabetes mellitus. *Journal of Physiology and Pathophysiology, 4*, 46–57.

Patrono, C. (2003). Aspirin resistance: Definition, mechanisms and clinical read-outs. *Journal of Thrombosis and Haemostasis, 1*(8), 1710–1713.

Peloquin, C. A. (2002). Therapeutic drug monitoring in the treatment of tuberculosis. *Drugs, 62*(15), 2169–2183.

Petersen, K. F., Krssak, M., Inzucchi, S., Cline, G. W., Dufour, S., & Shulman, G. I. (2000). Mechanism of troglitazone action in Type-2 diabetes. *Diabetes, 49*, 827–831.

Piette , J. D., Weinberger, M., & Mcphee, J. S. (2001). Impact of automated calls with nurse follow-up on diabetes treatment outcomes in a department of veterans affairs health care syste. *Clinical Care, 24*, 202–208.

Pijl, H., Ohashi, S., Matsuda, M., Miyazaki, Y., Mahankali, A., Kumar, V., Pipek, R., Iozzo, P., Lancaster, J. L., & Cincotta, A. H. (2000). Bromocriptine: A novel approach to the treatment of Type-2 diabetes. *Diabetes Care, 23*, 1154–1161.

Prabhakar, P. K., & Doble, M. (2011). Mechanism of action of natural products used in the treatment of diabetes mellitus. *Chinese Journal of Integrative Medicine, 17*, 563.

Pratley, R., Amod, A., Hoff, S. T., Kadowaki, T., Lingvay, I., Nauck, M., Pedersen, K. B., Saugstrup, T. & Meier, J. J. (2019). Oral semaglutide versus subcutaneous liraglutide and placebo in type 2 diabetes (PIONEER 4): a randomised, double-blind, phase 3a trial. *The Lancet, 394*(10192), 39–50.

Reaven, G. (2004). The metabolic syndrome or the insulin resistance syndrome? Different names, different concepts, and different goals. *Endocrinology and Metabolism Clinics of North America, 33*, 283–303.

Rena, G., Hardie, D. G., & Pearson, E. R. (2017). The mechanisms of action ofmetformin. *Diabetologia, 60*(9), 1577–1585.

Richard, J. L., Sotto, A., & Lavigne, J. P. (2011). New insights in diabetic foot infection. *World Journal of Diabetes, 2*(2), 24–32.

Russell-Jones, D., Bode, B. W., De Block, C., Franek, E., Heller, S. R., Mathieu, C., Philis-Tsimikas, A., Rose, L., Woo, V. C., Østerskov, A. B. Graungaard, T., & Bergenstal, R. M. (2017). Fast-acting insulin aspart improves glycemic control in basal-bolus treatment for type 1 diabetes: Results of a 26-week multicenter, active-controlled, treat-to-target, randomized, parallel-group trial (onset 1). *Diabetes Care, 40*(7), 943–950.

Salpeter, S. R., Greyber, E., Pasternak, G. A., & Salpeter, E. E. (2010). Risk of fatal and nonfatal lactic acidosis with metformin use in Type-2 diabetes mellitus. *Cochrane Database of Systematic Reviews, 4*.

Sanchez-Ibarra, H. E., Reyes-Cortes, L. M., Jiang, X. L., Luna-Aguirre, C. M., Aguirre-Trevino, D., Morales-Alvarado, I. A., Leon-Cachon, R. B., Lavalle-Gonzalez, F., Morcos, F., & Barrera-Saldaña, H. A. (2018). Genotypic and phenotypic factors influencing drug response in mexican patients with Type-2 diabetes mellitus. *Frontiers in Pharmacology, 9*, 320.

Scheen, A. J. (2003). Is there a role for α-glucosidase inhibitors in the prevention of Type-2 diabetes mellitus? *Drugs, 63*, 933–951.

Scheen, A. J., & Lefèbvre, P. J. (1998). Oral antidiabetic agents. *Drugs, 55*, 225–236.

Seeman, P. (2002). Atypical antipsychotics: Mechanism of action. *Canadian Journal of Psychiatry, 47*, 27–38.

Shi, L., & Tu, B. P. (2015). Acetyl-CoA and the regulation of metabolism: Mechanisms and consequences. *Current Opinion in Cell Biology, 33*, 125–131.

Shivaprasad, C., & Kalra, S. (2011). Bromocriptine in Type-2 diabetes mellitus. *Indian Journal of Endocrinology and Metabolism, 15*, 17.

Shomali, M. (2012). Diabetes treatment in 2025: Can scientific advances keep pace with prevalence? *Therapeutic Advances in Endocrinology and Metabolism, 3*(5), 163–173.

Siddiqui, A. A., Siddiqui, S. A., Ahmad, S., Siddiqui, S., Ahsan, I., & Sahu, K. (2013). Diabetes: Mechanism, pathophysiology and management-A review. *Int J Drug Dev Res, 5*, 1–23.

Sistema Nacional De Vigilancia Epidemiologica. (2013). Diabetes mellitus Type 2, primer trimester. Secretaria de SAlud, 1–5.

Song, W. M., Shao, Y., Liu, J. Y., Tao, N. N., Liu, Y., Zhang, Q. Y., Xu, T. T., Li, S. J., Yu, C. B., Gao, L., & Cui, L. L. (2019). Primary drug resistance among tuberculosis patients with diabetes mellitus: a retrospective study among 7223 cases in China. *Infection and Drug Resistance, 12*, 2397.

Soule Odegard, P., & Capoccia, K. (2007). Medication taking and diabetes: A systematic review of the literature. *The Diabetes Educator, 33*(6), 1014–1029.

Spiller, H. A., & Sawyer, T. S. (2006). Toxicology of oral antidiabetic medications. *American Journal of Health-System Pharmacy, 63*, 929–938.

Staels, B. (2009). A review of bile acid sequestrants: Potential mechanism (s) for glucose-lowering effects in Type-2 diabetes mellitus. *Postgraduate Medicine, 121*, 25–30.

Staels, B., & Kuipers, F. (2007). Bile acid sequestrants and the treatment of Type-2 diabetes mellitus. *Drugs, 67*, 1383–1392.

Stairs, D. J., & Bardo, M. T. (2009). Neurobehavioral effects of environmental enrichment and drug abuse vulnerability. *Pharmacology Biochemistry and Behavior, 92*(3), 377–382.

Stein, S. A., Lamos, E. M., & Davis, S. N. (2013). A review of the efficacy and safety of oral antidiabetic drugs. *Expert Opinion on Drug Safety, 12*, 153–175.

Sun, H., & Scott, D. O. (2011). Impact of genetic polymorphisms of cytochrome P450 2 C (CYP2C) enzymes on the drug metabolism and design of antidiabetics. *Chemico-Biological Interactions, 194*(2–3), 159–167.

Tantry, U. S., Bliden, K. P., & Gurbel, P. A. (2005). Overestimation of platelet aspirin resistance detection by thrombelastograph platelet mapping and validation by conventional aggregometry using arachidonic acid stimulation. *Journal of the American College of Cardiology., 46*(9), 1705–1709.

Taverna, M. J., Pacher, N., & Bruzzo, F. (2001). Beta-cell function evaluated by HOMA as a predictor of secondary sulphonylurea failure in Type-2 diabetes. *Diabetic Medicine, 18*, 584–588.

Tonelli, J., Li, W., Kishore, P., Pajvani, U. B., Kwon, E., Weaver, C., Scherer, P. E., & Hawkins, M. (2004). Mechanisms of early insulin-sensitizing effects of thiazolidinediones in Type-2 diabetes. *Diabetes, 53*, 1621–1629.

Tracey, I. (2010). Getting the pain, you expect: Mechanisms of placebo, nocebo and reappraisal effects in humans. *Nature Medicine, 16*(11), 1277.

Tricco, A. C., Ashoor, H. M., Antony, J., Beyene, J., Veroniki, A. A., Isaranuwatchai, W., Harrington, A., Wilson, C., Tsouros, S., Soobiah, C. Catherine, H.Y., & Straus, S. E. (2014). Safety, effectiveness, and cost effectiveness of long acting versus intermediate acting insulin for patients with type 1 diabetes: Systematic review and network meta-analysis. *BMJ, 349*.

UKPDS Study Group. (1998). Characteristics of newly presenting Type-2 diabetic patients: Preponderance and obesity at different ages. *Diabetes Medicine, 5*, 154–159.

Vargas, E., Joy, N. V., & Sepulveda, M. A. C. (2020). Biochemistry, insulin metabolic effects. StatPearls.

Vaz, E. C., Porfírio, G. J. M., Nunes, H. R. D. C., & Nunes-Nogueira, V. D. S. (2018). Effectiveness and safety of carbohydrate counting in the management of adult patients with type 1 diabetes mellitus: A systematic review and meta-analysis. *Archives of Endocrinology and Metabolism, 62*(3), 337–345.

Vijan, S., Sussman, J. B., Yudkin, J. S., & Hayward, R. A. (2014). Effect of patients' risks and preferences on health gains with plasma glucose level lowering in type 2 diabetes mellitus. *JAMA Internal Medicine, 174*(8), 1227–1234.

Wang, W., Liu, H., Xiao, S., Liu, S., Li, X., & Yu, P. (2017). Effects of insulin plus glucagon-like peptide-1 receptor agonists (GLP-1RAs) in treating type 1 diabetes mellitus: A systematic review and meta-analysis. *Diabetes Therapy, 8*(4), 727–738.

Watala, C., Golanski, J., Pluta, J., Boncler, M., Rozalski, M., Wieclawska, B., Kropiwnicka, A., & Drzewoski, J. (2004). Reduced sensitivity of platelets from Type-2 diabetic patients to acetylsalicylic acid (aspirin)—its relation to metabolic control. *Thrombosis Research, 113*, 101–113.

Watala, C., Boncler, M., & Gresner, P. (2005). Blood platelet abnormalities and pharmacological modulation of platelet reactivity in patients with diabetes mellitus. *Pharmacological Reports, 57*, 42.

Wilcox, G. (2005). Insulin and insulin resistance. *Clinical Biochemist Reviews, 26*(2), 19.

Winocour, P. D. (1994). Platelet turnover in advanced diabetes. *European Journal of Clinical Investigation., 24*(S1), 34–37.

Wou, C., Unwin, N., Huang, Y., & Roglic, G. (2019). Implications of the growing burden of diabetes for premature cardiovascular disease mortality and the attainment of the Sustainable Development Goal target. *Cardiovascular Diagnosis and Therapy., 9*(2), 140.

Yadav, K. S., Singh, B., Singh, G. V., & Rani, B. (2016). Study the pattern of drug resistance in patients of pulmonary tuberculosis having diabetes mellitus. *European Journal of Pharmaceutical and Medical Research, 5*, 290–294.

Yip, S. W., Doherty, J., Wakeley, J., Saunders, K., Tzagarakis, C., De Wit, H., Goodwin, G. M., & Rogers, R. D. (2012). Reduced subjective response to acute ethanol administration among young men with a broad bipolar phenotype. *Neuropsychopharmacology, 37*(8), 1808–1815.

Zangeneh, F., Kudva, Y. C., & Basu, A. (2003). Insulin sensitizers. *Mayo Clinic Proceedings, 78*, 471–479.

Zhang, P., Li, T., Wu, X., Nice, E.C., Huang, C., & Zhang, Y. (2020). Oxidative stress and diabetes: Antioxidative strategies. *Frontiers of Medicine*, 1–18.

Chapter 17
Drug Resistance in Rheumatological Diseases

Hassan Waqar, Ammar Arshad, Muhammad Asad Raza, Mohammad Qamar Nasir, Ahmed Shams Nasir, Arshad Islam, Ifra Idrees, Sarfraz Ahmed, Khalid Muhammad, and Atif Amin Baig

Abstract Rheumatological diseases are chronic inflammatory diseases that involve joints, skin, vessels, and other organ systems. Steroids, DMARDS, and various biological agents are used to control the disease activity by interfering at various levels of the inflammatory cascade. Long term use of the above-mentioned agents is required for treatment, which leads to resistance against these agents. Drug resistance involves various mechanisms that are not completely understood. Proposed mechanisms involve drug uptake and drug efflux transporters, e.g., P-glycoprotein, multidrug resistance protein-1, and organ anion transporters-3, Antidrug antibodies against various anti-TNF drugs, down regulation of Receptors, reprogramming of stimulatory signals, and genetic polymorphisms. Drug resistance is also complicated by non-compliance and the emergence of drug-related adverse effects. Different

H. Waqar
P.I.M.H, Lahore, Pakistan

A. Arshad · M. A. Raza
Internal Medicine, KEMU/Mayo Hospital, Lahore, Pakistan

M. Q. Nasir
Otorhinolaryngology SIMS Hospital Lahore, Lahore, Pakistan

A. S. Nasir
Orthopedics SIMS Hospital Lahore, Lahore, Pakistan

A. Islam
Department of Pathology, Lady Reading Hospital Medical Teaching Institution, Peshawar, Pakistan

I. Idrees
Shaikh Khalifa Bin Zayed Al-Nahyan Medical and Dental College Lahore, Lahore, Pakistan

S. Ahmed
Department of Biochemistry, The University of Veterinary and Animal Sciences, Lahore, Pakistan

K. Muhammad
Department of Biology, College of Science, United Arab Emirates University, Al Ain, UAE

A. A. Baig (✉)
Faculty of Medicine, University Sultan Zainal Abidin, Kuala Terengganu, Malaysia
e-mail: atifamin@unisza.edu.my

S. Ahmed et al. (eds.), *Biochemistry of Drug Resistance*,
https://doi.org/10.1007/978-3-030-76320-6_17

strategies are proposed to overcome drug resistance that includes the use of a combination of drugs that increase the intracellular concentration of drugs as well as bypass or suppress the multidrug resistance. Alternative agents with slightly different mechanisms of action but similar end targets can also be prescribed. Modification of intracellular signal transduction is also one way to overcome resistance. However, further studies are required to get more insight into the mechanisms of resistance. This will pave new ways to overcome drug resistance, helping in more effective control of disease progression. In this chapter, we will be giving an overview of major rheumatologic diseases, followed by the major drugs used to treat them, the mechanisms of action and mechanisms of resistance to these drugs, and the proposed evidence-based steps to overcome that resistance.

17.1 Introduction

Rheumatologic diseases include a spectrum of disorders that involve various inflammatory arthritides as well as other connective tissue diseases. Rheumatoid arthritis and seronegative spondyloarthritis are among the leading conditions of the inflammatory arthritides. Connective tissue disorders, including systemic lupus erythematosus (SLE), mixed connective tissue disease (MCTD), systemic sclerosis, and Sjogren's syndrome are also commonly encountered rheumatologic diseases.

Rheumatoid arthritis (RA), one of the most common rheumatologic diseases, is an autoimmune disorder with multisystem involvement linked to different antibodies (Rheumatoid Factor, anti-cyclic citrullinated peptide). It primarily affects the joints (most commonly the hands and wrists) with symmetrical joint involvement causing inflammation and leading to deformities of the affected joints with the passage of time. Moreover, it can also affect the lungs (nodules, pleuritis, pleural effusion, fibrosis, and pulmonary hypertension), heart (pericarditis, endocarditis, increased risk of atherosclerosis and myocarditis), kidneys (amyloidosis), eyes (scleritis and episcleritis), nervous system (peripheral neuropathy, mononeuritis multiplex and in extreme cases spinal cord injury due to atlantoaxial subluxation) and blood dyscrasias (anemia, thrombocytopenia, and neutropenia).

Major drugs available clinically for the RA treatment include corticosteroids, methotrexate, sulfasalazine, hydroxyl chloroquine, leflunomide, and some biological agents, e.g., infliximab, etanercept. Besides, Systemic Lupus Erythematosus (SLE) is a chronic inflammatory disease which, similar to RA, is due to autoantibody response to several self-antigens. A few of the antibodies involved in SLE manifestation include Antinuclear Antibodies (ANA), Anti double stranded-DNA (anti ds-DNA), anti–Sjögren's-syndrome-related antigen A autoantibodies (Anti-SSA) anti–Sjögren's-syndrome-related antigen B autoantibodies (Anti-SSB). Although it rarely causes joint deformities, it does cause arthralgias and myalgias. It also commonly causes skin involvement (malar rash and discoid lupus), kidney disease (lupus nephritis and acute/chronic kidney disease), pulmonary disease (pleural effusion, pleuritis, interstitial lung disease, and pulmonary hypertension), heart disease (pericarditis, myocarditis, and Libman Sacks endocarditis), neuropsychiatric (seizure and psychosis) and hematological disorders (anemia, leucopenia, and thrombocytopenia).

Acute flares are usually treated with corticosteroids and non-steroidal anti-inflammatory drugs (NSAIDs). Hydroxychloroquine (HCQ), azathioprine, cyclophosphamide, and anti-TNF agents.

Corticosteroids and disease-modifying antirheumatic drugs (DMARDs) are a class of pharmacological agents that target specific areas of the inflammatory cascade to slow down the disease process in the above-mentioned various diseases. Steroids and DMARDs are also effective in inflammatory myositis and vasculitides. In clinics, no single and effective pharmacological agent is yet available that can completely reverse or cure the disease process in all of the above-mentioned conditions.

In this chapter, the exact mechanisms of action and resistance to steroids and commonly used DMARDs with respect to their use in rheumatologic diseases are discussed.

17.2 Mechanism of Action of Antirheumatic Drugs

The basic factor behind the pathogenesis of all the rheumatologic conditions involves *autoimmunity*that triggers the inflammatory cascade. The immune system abnormally reacts to various infectious, environmental antigens(environmental what? antigens? factors?) and self-antigens, triggering a T-cell mediated inflammatory response against various organs of the body and affecting predominantly the joints, skin, muscles, and blood vessels. Other organ systems like the lungs, gastrointestinal tract, kidneys, heart, liver, and eyes are also variably affected.

Corticosteroids are used in managing flares of different rheumatologic diseases and also produce long-term immunosuppressant effects when used along with DMARDs. However, their long-term use is usually discouraged due to multiple adverse effects. Corticosteroids (Table 17.1) bind to their receptor (growth receptor α/ GR- α) in the cytoplasm forming a complex. This complex translocates to the nucleus to activate or suppress the transcription of different genes, thus decreasing the production of proteins that promote inflammation and increasing the translation

Table 17.1 Mechanism of action and mechanism of resistance to corticosteroids

Drug	Mechanism of action	Mechanism of resistance
Corticosteroids (prednisolone, prednisone, methyl prednisolone, dexamethasone)	Formation of corticosteroid-receptor complex trans-repression of pro-inflammatory and trans-activation of anti-inflammatory genes	↑ GR-β ↑ MIF ↓ HDAC2

of proteins that are anti-inflammatory (lipocortin I and mitogen-activated protein kinase phosphatase).

DMARDs are immunosuppressive and immune modulatory agents. They are broadly classified into conventional (Table 17.2) and biological agents (Table 17.3).

Table 17.2 Mechanism of action and mechanism of resistance to conventional DMARDs

Drugs	Mechanism of action	Mechanism of resistance
Methotrexate	Inhibits dihydro-folate reductase decreasing THF ↓ DNA synthesis and WBC proliferation	P-gp BCRP MRP1 OAT3
Sulfasalazine	Prevents oxidation, nitration and nitrosation	P-gp BCRP
Hydroxychloroquine	Inhibits TLR-9	No significant data found
Leflunomide	Inhibits dihydroorotate dehydrogenase ↓ pyrimidine synthesis	BCRP
Cyclophosphamide	Alkylating agent, cross-links DNA and RNA strands ↓ pro-inflammatory protein synthesis	↑ ADH

Table 17.3 Mechanism of action and mechanism of resistance to biological DMARDs

Drugs	Mechanism of action	Mechanism of resistance
Abatacept	Binds to CD-80 and CD-86 blocking the "second signal"	No significant data found
Rituximab	Anti CD-20 receptor antibody	↑ BCL-2 anti-apoptotic activity Down regulation of CD-20
Tocilizumab	IL-6 receptor blocker	No significant data found
Anakinra	IL-1 receptor blocker	No significant data found
Etanercept	Bindsto TNF-α	Polymorphism of TNF-α gene (-308 A/A or A/G genotypes)
Adalimumab	Anti-TNF α antibodies	Antidrug antibodies
Infliximab		Antidrug antibodies Polymorphism of TNF-α gene (-308 A/A or A/G genotypes) Over activity of LT- α
Certolizumab pegol		No significant data found
Golimumab		No significant data found

17.3 Conventional DMARDs

17.3.1 Methotrexate

Methotrexate (MTX) is still the first-line agent used in rheumatoid arthritis treatment. It is an anti-folate agent, which increases adenosine release from fibroblasts and inhibits leukotriene B4 synthesis by neutrophils, and limits their adhesions. It also inhibits local interleukin-1 (IL-1) production, reduces levels of IL-6 and IL-8, suppresses cell-mediated immunity, and inhibits synovial collagenase gene expression, thus suppressing the inflammatory process in the body.

17.3.2 Leflunomide

Leflunomide inhibits dihydroorotate dehydrogenase, halting the pyrimidine synthesis and ultimately blocking lymphocyte proliferation.

17.3.3 Sulfasalazine

Sulfasalazine prevents oxidation, nitration, and nitrosation, thus producing its anti-inflammatory effects.

17.3.4 Hydroxychloroquine

It inhibits toll-like receptor-9 (TLR-9), producing mild immune modulatory effects. Moreover, it also affects the activity of P-glycoprotein as discussed below (efflux pumps) (give the section number or title where it is discussed, in brackets).

17.3.5 Cyclophosphamide

It is an alkylating agent of the nitrogen mustard type 2. It exerts its cytotoxic effects mainly by cross-linking DNA as well as RNA strands, causing the inhibition of pro-inflammatory proteins.

17.4 Biological DMARDs

The emergence of various adverse effects and resistance against various conventional DMARDs with failure to control ongoing disease activity necessitated the development of biological agents in the 1990s.

Biological DMARDs (Table 17.3) are very selective in their mechanism of action. Major mechanisms of biologics include (i) Interfering with cytokine production or function, (ii) inhibiting the "second signal" required for T cell activation, and (iii) depleting B cells or inhibiting factors that activate B cells.

17.4.1 Anti-TNF Alpha Monoclonal Antibodies

Most of the anti-TNF agents are monoclonal, chimeric humanized fusion antibodies that bind to TNF-α, thus inhibiting its action, e.g., infliximab, adalimumab, certolizumab, and golimumab.

17.4.2 Etanercept

Etanercept is a protein derived from recombinant DNA by fusing uses the TNF receptor to the constant end of the IgG1 antibody. It is a fusion of TNF receptor to the constant portion of IgG-1 antibodies, which binds to TNF- α, thus inhibiting its action.

17.4.3 Abatacept

This drug, like etanercept, is a fusion protein that consists of the Fc region of IgG-1 combined with the extracellular component of CTLA-4. This binds to CD-80 and CD-86, blocking the "second signal" for T-cell activation.

17.4.4 Rituximab

It is an anti-CD-20 receptor antibody that inhibits the activation of B cells. Since CD-20 plays a pivotal role in the activation of B cells, this drug has proven to be beneficial in suppressing the inflammatory process.

17.4.5 Tocilizumab

This is a monoclonal antibody which prevents signal transduction by Interleukin-6 (serves as an important mediator in leukocyte activation) by binding to its receptor.

17.4.6 Anakinra

It is a protein that competitively binds the interleukin-1 type I receptor. IL-1, like IL-6, is an important component of the inflammatory cascade.

17.4.7 Janus Kinase (JAK) Inhibitors

Previously being used in treating hematological malignancies, these are the latest addition to the family of DMARDs. These act by inhibiting the activity of the Janus kinase family of enzymes (JAK1, JAK2, JAK3, and TYK2), thereby disrupting the JAK/STAT pathway of signal transduction within the inflammatory cells (Table 17.4).

17.5 Mechanism of Resistance Against Anti-rheumatic Drugs

Chronic nature of all rheumatologic conditions requires long-term and sometimes lifelong use of steroids and DMARDs; this has resulted in poor compliance, drug-related adverse effects, and loss of efficacy, which are all responsible for both; the discontinuation of treatment and progression of the disease. Loss of efficacy of steroids and DMARDs has been proposed to share the same mechanism of resistance as was reported for anticancer drugs (Van Der Heijden et al., 2007). Since

Table 17.4 Mechanism of action and mechanism of resistance to Novel DMARDs (JAK inhibitors)

Drugs	Mechanism of action	Mechanism of resistance
Baricitinib	Inhibit one or more of the JAK family of signal transducers (JAK1, JAK2, JAK3, TYK2)	P-gp BCRP OAT3
Tofacitinib		No significant data found
Upadacitinib		No significant data found

Abbreviations: *BCRP* breast cancer resistance protein, *MRP1* multidrug resistance protein-1, *P-gp* P-glycoprotein, *OAT3* organic anion transporter3, *TLR-9* toll-like receptor-9, *IL* Interleukin, *JAK* janus Kinase, *TYK* tyrosine kinase

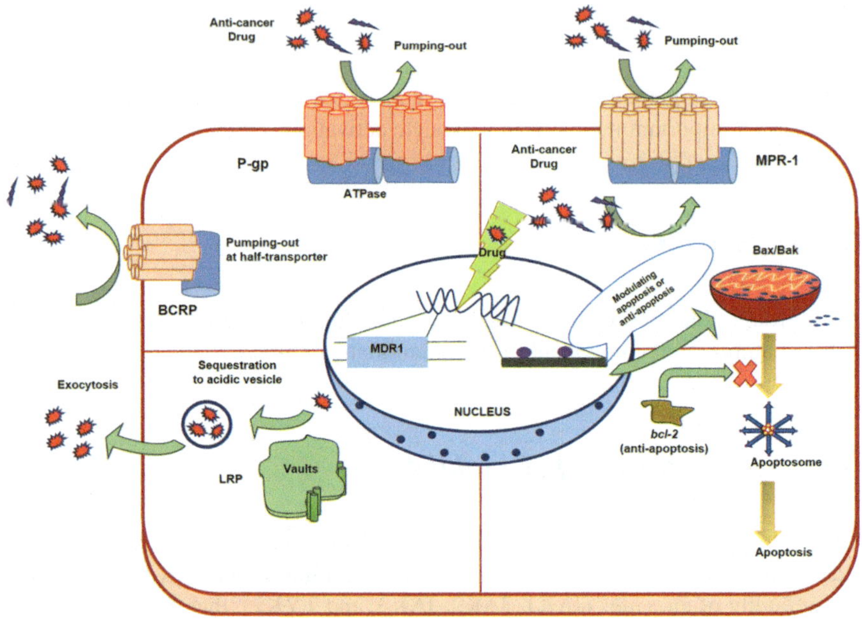

Fig. 17.1 Mechanisms of drug-resistance (such as P-gp, MRP, BCRP, LRP, and *bcl*-2), in tumor cells. Adapted from (Oh et al., 2009)

anti-TNF drugs are novel agents, their mechanisms of resistance are still not fully understood and are needed to be further investigated.

Understanding the mechanism of resistance against various (Fig. 17.1) conventional and biological DMARDs can help in devising better treatment plans and more efficient use of DMARDs.

17.5.1 Efflux Pumps

Multiple studies suggest the role of specific drug uptake and drug efflux transporters, which are a part of the superfamily of multidrug resistance (MDR) transporters (Jansen et al., 2003). There are also changes in drug metabolism within the cells and genetic polymorphism of the above-mentioned transporters and metabolic enzymes involved in resistance to steroids and conventional DMARDs.

17.5.2 P-glycoprotein (P-gp)

P-glycoprotein (also called multidrug resistance protein-1, MDR-1, or CD-243) is an ATP-dependent transmembrane efflux pump transporting a wide variety of foreign substances, including corticosteroids and DMARDs *(MTX, sulfasalazine, and baricitinib)* from lymphocytes resulting in resistance to these drugs (Diaz-Borjon et al., 2000). This is actually a part of the defence mechanism of the cells against harmful substances. It is encoded by the ABCB1 gene. Studies have also found surface expression of P-gp on peripheral lymphocytes in SLE patients and a substantial correlation between the level of expression of P-gp and disease activity (García-Carrasco et al., 2017). Moreover, other studies show that P-gp overactivity is linked with increased dose requirement for corticosteroids (Tsujimura & Tanaka, 2012) and DMARDs.

P-glycoprotein (P-gp) is also excessively expressed in tumor cells, thus leading to their resistance in chemotherapeutic agents such as topoisomerase inhibitors (etoposide and doxorubicin), microtubules targeted drugs (vinblastine and vincristine), and tyrosine kinase inhibitors (gefitinib and sunitinib). Other drugs transported by P-gp include colchicine, digoxin, and antiretroviral drugs (protease inhibitors and NNRTIs).

Some drugs, although being used for other purposes, were incidentally found to reduce P-gp levels and thus could be used as potential augmenters of the DMARDs whose function is affected by P-gp. In addition, the anti-TNF drugs and HCQ (Wang et al., 2019) have also been shown to decrease p-gp levels and are, therefore, used in combination with the older DMARDs e.g., infliximab. When given in combination with MTX, it increases the efficacy of MTX by decreasing P-gp.

17.5.3 Breast Cancer Resistance Protein (BCRP)

MDR mechanism involves breast cancer resistance protein (BCRP) overexpression in inflammatory cells on the synovial tissues in rheumatoid arthritis. This transporter is not only responsible for the efflux of *Leflunomide* (Kis et al., 2009), *MTX* (Van der Heijden et al., 2009), *sulfasalazine* (Van der Heijden et al., 2004) as well as *Baricitinib*.

17.5.4 Multidrug Resistance Associated Protein 1 (MRP1)

This is a protein which like P-gp is an ATP dependent transporter encoded by *ABCC1* gene. Apart from transporting multiple organic anions, this pump is also involved in the efflux of anticancer drugs and anti-rheumatic drugs. One study even suggests that MRP1, along with BCRP, is the major transporter of *MTX* rather than P-gp (Kremer, 2004).

17.5.5 Organic Anion Transporter 3 (OAT3)

OAT3, also known as solute carrier family 22 member 8, is a protein encoded by *SLC*22A8 gene. Unlike P-gp, it is a sodium-dependent transport protein. It is involved in efflux and excretion (in renal tubules) of organic anions, which includes *MTX, Baricitinib*, Ciprofloxacin, Penicillin G and Indomethacin, etc.

17.5.6 Antidrug Antibodies

With timecertain anti-TNF monoclonal antibodies have been found to induce antibody production in the patients requiring increased dosage. **Infliximab** (Finckh et al., 2006; Sidiropoulos & Boumpas, 2006) **and adalimumab** have been found to become less effective with time, requiring increasing dosage in a significant proportion of patients after 6 months. A proposed mechanism is the development of anti-infliximab (Finckh et al., 2010) and anti-adalimumab antibodies (Li et al., 2018) in the patients.

17.5.7 Alternative Inflammatory Mediators

Another cause of resistance to anti-TNF drugs is said to be lymphotoxin alpha (LT-α), also called TNF beta (TNF-β). This is a cytokine that is encoded by the LTA gene. It has an important role in the prevention of tumor growth and induction of apoptosis by activation of the NF-kB pathway, thus promoting immune regulation. On the other hand, LT-α signaling also promotes inflammatory responses leading to cellular and tissue damage. As pointed out earlier, biologic agents are highly selective in their action and only block TNF-α, sparing LT-α. Over activity of LT-α in some patients can cause resistance to anti-TNF agents, especially *Infliximab* (Buch et al., 2004).

17.5.8 Genetic Polymorphisms

It has been found that patients having different genotypes tend to respond differently to certain antirheumatic drugs, examples of which are *Etanercept and Infliximab*. In one study, it was seen that the gene encoding TNF-a, **−308** has 3 different genotypes, and patients with the genotype -308 G/G were found to have a better response to Infliximab or Etanercept than the patients with A/G or A/A genotypes (Guis et al., 2007).

17.5.9 Down-Regulation of Receptors/reprogramming of Stimulatory Signals

The fate of a cell hangs in a delicate balance between expressions of different counter-regulatory elements within it. It also involves multiple receptors on the cell's surface of B cells, usually which control signal transduction. Resistance to Rituximab has been reported to be linked to increased activity of BCL-2, an anti-apoptotic protein, and also to the reduced level of BAX/BAK, a complex involved in promoting apoptosis (Olejniczak et al., 2008). Down-regulation of CD20 receptors is also linked with decreasing efficacy of Rituximab. All these changes decrease the effects of Rituximab and promote lymphocyte activity, accentuating the inflammatory process.

Similarly, reduced activity of Glucocorticoid Receptor Alpha (GR-α) or its replacement by Glucocorticoid Receptor Beta (GR-β) might decrease the translocation of *corticosteroids* to the nucleus. Furthermore, increased macrophage migration inhibitory factor (MIF), competition with transcription factor activator protein1, and decreased production of histone D-acetylase 2 (HDAC2, important in the mediation of action of steroids for switching off the pro-inflammatory genes) is also a proposed mechanism of resistance to corticosteroids (Barnes, 2013). Cyclophosphamide resistance has been linked to increased activity of aldehyde dehydrogenase (ADH), which decreases the interstrand cross-links (ISCs) of DNA (Andersson et al., 1995).

17.6 Strategies Adopted to Overcome Resistance and Future Prospects

17.6.1 Reduction/Suppression/Bypass of Transmembrane Carriers

As discussed previously, the DMARDs that act as P-gp antagonist (that decrease P-gp levels) such as cyclosporine and tacrolimus, when used in combination with the DMARDs that are transported by P-gp, can augment their effects by increasing their intracellular concentration (Picchianti-Diamanti et al., 2014). For example, *etanercept* decreases P-gp levels and thus can help in increasing the efficacy of methotrexate or sulfasalazine.

Another suggestion for patients resistant to older DMARDs is to check their P-gp levels so that appropriate drugs e.g., infliximab and tacrolimus, can be prescribed for treatment of refractory RA (Tsujimura et al., 2008, 2010). A new and interesting use of nano-technology in the field of pharmacology is bypassing P-gp altogether via receptor-mediated endocytosis of drugs (Oh et al., 2009).

17.6.2 Modifying Signal Transduction

Another proposed solution includes targeting the specific signal transducers that cause the resistance in the first place, e.g., suppressing B cell-activating factor (BAFF) with anti-BAFF agents in patients with diminished response to Rituximab might help to overcome resistance (Quartuccio et al., 2008). Such a novel approach can also be applied to reverse steroid resistance by increasing HDAC2 expression by phosphoinositide 3-kinase δ inhibitors (Barnes, 2013).

17.6.3 Combination with Other Agents

Conjugation of methotrexate to poly (L-lysine) increases drug transport into the cell, thus increasing intracellular drug concentration and helping to overcome drug resistance (Ryser & Shen, 1978).

17.6.4 Alternative Drugs with Similar End-Targets

Certain drugs with slightly different mechanisms of action as the existing DMARDs but the same end results are also being studied, e.g., genetically engineered anti-CD20-hIFn-α fusion protein, which was successfully used in mice carrying B-cell lymphoma xenografts, overcoming resistance to Rituximab (Vega et al., 2015).

17.7 Conclusion

From this discussion, we can infer that there have been several advancements in the treatment of rheumatologic diseases. However, there has been reported resistance to the newer drugs as well as, the older ones. Mechanisms of resistance to these drugs are still under study. A better understanding of these mechanisms is necessary for devising new methods to overcome drug resistance. Furthermore, methods of overcoming drug resistance of anticancer chemotherapeutic drugs can also be of help in treating rheumatologic diseases.

References

Andersson, B. S., Collins, V., Kurzrock, R., Larkin, D., Childs, C., Ost, A., Cork, A., Trujillo, J., Freireich, E., & Siciliano, M. (1995). Kbm-7, a human myeloid leukemia cell line with double philadelphia chromosomes lacking normal C-Abl and Bcr transcripts. *Leukemia, 9*, 2100–2108.

Barnes, P. J. (2013). Corticosteroid resistance in patients with asthma and chronic obstructive pulmonary disease. *Journal of Allergy and Clinical Immunology, 131*, 636–645.

Buch, M. H., Conaghan, P. G., Quinn, M. A., Bingham, S. J., Veale, D., & Emery, P. (2004). True infliximab resistance in rheumatoid arthritis: A role for lymphotoxin α? *Annals of the Rheumatic Diseases, 63*, 1344–1346.

Diaz-Borjon, A., Richaud-Patin, Y., Jakez-Ocampo, J., Ruiz-Argüelles, A., & Llorente, L. (2000). Multidrug resistance-1 (Mdr-1) in rheumatic autoimmune disorders. Part II: Increased p-glycoprotein activity in lymphocytes from systemic lupus erythematosus patients might affect steroid requirements for disease control. *Joint, Bone, Spine, 67*, 40–48.

Finckh, A., Dudler, J., Wermelinger, F., Ciurea, A., Kyburz, D., Gabay, C., & Bas, S. (2010). Influence of anti-infliximab antibodies and residual infliximab concentrations on the occurrence of acquired drug resistance to infliximab in rheumatoid arthritis patients. *Joint, Bone, Spine, 77*, 313–318.

Finckh, A., Simard, J. F., Gabay, C., & Guerne, P.-A. (2006). Evidence for differential acquired drug resistance to anti-tumour necrosis factor agents in rheumatoid arthritis. *Annals of the Rheumatic Diseases, 65*, 746–752.

García-Carrasco, M., Mendoza-Pinto, C., Macías-Díaz, S., Etchegaray-Morales, I., Méndez-Martínez, S., Soto-Santillán, P., Pérez-Romano, B., Jiménez-Herrera, E. A., Guzmán-Ruiz, O., & Ruiz-Argüelles, A. (2017). Clinical relevance of p-glycoprotein activity on peripheral blood mononuclear cells and polymorphonuclear neutrophils to methotrexate in systemic lupus erythematosus patients. *Clinical Rheumatology, 36*, 2267–2272.

Guis, S., Balandraud, N., Bouvenot, J., Auger, I., Toussirot, E., Wendling, D., Mattei, J. P., Nogueira, L., Mugnier, B., & Legeron, P. (2007). Influence of −308 A/G polymorphism in the tumor necrosis factor α gene on etanercept treatment in rheumatoid arthritis. *Arthritis Care & Research, 57*, 1426–1430.

Jansen, G., Scheper, R., & Dijkmans, B. (2003). Multidrug resistance proteins in rheumatoid arthritis, role in disease-modifying antirheumatic drug efficacy and inflammatory processes: An overview. *Scandinavian Journal of Rheumatology, 32*, 325–336.

Kis, E., Nagy, T., Jani, M., Molnár, É., Jánossy, J., Ujhellyi, O., Német, K., Herédi-Szabó, K., & Krajcsi, P. (2009). Leflunomide and its metabolite A771726 are high affinity substrates of BCRP: Implications for drug resistance. *Annals of the Rheumatic Diseases, 68*, 1201–1207.

Kremer, J. M. (2004). Toward a better understanding of methotrexate. *Arthritis & Rheumatism, 50*, 1370–1382.

Li, M., Li, H., Gao, K., Wang, M., An, W., Zhu, Y., Ding, L., Wang, L., Gu, J., & Zuo, C. (2018). A simple and cost-effective assay for measuring anti drug antibody in human patients treated with adalimumab. *Journal of Immunological Methods, 452*, 6–11.

Oh, K. T., Baik, H. J., Lee, A. H., Oh, Y. T., Youn, Y. S., & Lee, E. S. (2009). The reversal of drug-resistance in tumors using a drug-carrying nanoparticular system. *International Journal of Molecular Sciences, 10*, 3776–3792.

Olejniczak, S. H., Hernandez-Ilizaliturri, F. J., Clements, J. L., & Czuczman, M. S. (2008). Acquired resistance to rituximab is associated with chemotherapy resistance resulting from decreased Bax and Bak expression. *Clinical Cancer Research, 14*, 1550–1560.

Picchianti-Diamanti, A., Rosado, M. M., Scarsella, M., Laganà, B., & D'amelio, R. (2014). P-glycoprotein and drug resistance in systemic autoimmune diseases. *International Journal of Molecular Sciences, 15*, 4965–4976.

Quartuccio, L., Fabris, M., Moretti, M., Barone, F., Bombardieri, M., Rupolo, M., Lombardi, S., Pitzalis, C., Beltrami, C. A., & Curcio, F. (2008). Resistance to rituximab therapy and local Baff overexpression in Sjögren's syndrome-related myoepithelial sialadenitis and low-grade parotid B-cell lymphoma. *The Open Rheumatology Journal, 2*, 38.

Ryser, H.J.-P., & Shen, W.-C. (1978). Conjugation of methotrexate to poly (l-lysine) increases drug transport and overcomes drug resistance in cultured cells. *Proceedings of the National Academy of Sciences, 75*, 3867–3870.

Sidiropoulos, P., & Boumpas, D. (2006). Differential drug resistance to anti-tumour necrosis factor agents in rheumatoid arthritis. Bmj Publishing Group Ltd.

Tsujimura, S., Saito, K., Nakayamada, S., & Tanaka, Y. (2010). Etanercept overcomes p-glycoprotein-induced drug resistance in lymphocytes of patients with intractable rheumatoid arthritis. *Modern Rheumatology, 20*, 139–146.

Tsujimura, S., Saito, K., Nawata, M., Nakayamada, S., & Tanaka, Y. (2008). Overcoming drug resistance induced by p-glycoprotein on lymphocytes in patients with refractory rheumatoid arthritis. *Annals of the Rheumatic Diseases, 67*, 380–388.

Tsujimura, S., & Tanaka, Y. (2012). Treatment strategy based on targeting p-glycoprotein on peripheral lymphocytes in patients with systemic autoimmune disease. *Clinical and Experimental Nephrology, 16*, 102–108.

Van Der Heijden, J., De Jong, M. C., Dijkmans, B. A., Lems, W. F., Oerlemans, R., Kathmann, I., Scheffer, G. L., Scheper, R. J., Assaraf, Y. G., & Jansen, G. (2004). Acquired resistance of human t cells to sulfasalazine: Stability of the resistant phenotype and sensitivity to non-related dmards. *Annals of the Rheumatic Diseases, 63*, 131–137.

Van Der Heijden, J. W., Dijkmans, B. A., Scheper, R. J., & Jansen, G. (2007). Drug insight: Resistance to methotrexate and other disease-modifying antirheumatic drugs—from bench to bedside. *Nature Clinical Practice Rheumatology, 3*, 26–34.

Van Der Heijden, J. W., Oerlemans, R., Tak, P. P., Assaraf, Y. G., Kraan, M. C., Scheffer, G. L., Van Der Laken, C. J., Lems, W. F., Scheper, R. J., & Dijkmans, B. A. (2009). Involvement of breast cancer resistance protein expression on rheumatoid arthritis synovial tissue macrophages in resistance to methotrexate and leflunomide. *Arthritis & Rheumatism: Official Journal of the American College of Rheumatology, 60*, 669–677.

Vega, G. G., Franco-Cea, L. A., Huerta-Yepez, S., Mayani, H., Morrison, S. L., Bonavida, B., & Vega, M. I. (2015). Overcoming rituximab drug-resistance by the genetically engineered anti-Cd20-Hifn-a fusion protein: Direct cytotoxicity and synergy with chemotherapy. *International Journal of Oncology, 47*, 1735–1748.

Wang, F., Zhang, Z., Leung, W. T., Chen, J., Yi, J., Ying, C., Yuan, M., Wang, M., Zhang, N., & Qiu, X. (2019). Hydroxychloroquine reverses the drug resistance of leukemic K562/Adm cells by inhibiting autophagy. *Molecular Medicine Reports, 20*, 3883–3892.

Chapter 18
Pherotypes in *Streptococcus Pneumoniae* and Role of CSP-1 and CSP-2 in Antibiotic Susceptibility and Resistance; Towards Development of Live Attenuated Vaccine Candidates in Inducing Netosis Based Acquired Immune Response

Saima Iqbal, Muhammad Hassan, Hassan Waqar, Sarfraz Ahmed, Arshad Islam, Ahmed Zubaidi Abd Latiff, Nordin Simbak, Shehla Munir, and Atif Amin Baig

Abstract Infections caused by *Streptococcus (S.) pneumoniae* have become one of the leading causes of mortality and morbidity worldwide. Through colonization, it causes local or systemic infection among the human population belonging to all age groups. Its spreads from the nasopharynx become the source of local infections such as sinusitis or acute otitis media. In the case of aspiration *S. pneumonia* directly spreads in the lungs and leads to pneumonia acquired from the community. More than 96 serotypes have been reported so far, so determining all serotypes' susceptibility and resistance is a cumbersome process that needs hefty effort, time, and resources. Thus, it becomes evident and necessitates that there should be some appropriate approach or technique to effectively and timely identify strains of bacteria, especially widespread ones like *S. pneumoniae*. Instead of identifying all these reported strains, identifying only two pherotypes can effectively serve the purpose of accurate identification and treatment. As a result of effective identification, the susceptibility

S. Iqbal · M. Hassan · A. Z. A. Latiff · N. Simbak · A. A. Baig (✉)
Faculty of Medicine, Universiti Sultan Zainal Abidin (UniSZA), Kuala Terengganu, Malaysia
e-mail: atifamin@unisza.edu.my

S. Iqbal
Shalamar Medical & Dental College (SMDC), Lahore, Pakistan

H. Waqar
Mayo Hospital, Lahore, Pakistan

S. Ahmed · A. Islam
Department of Pathology, Lady Reading Hospital Medical Teaching Institution, Peshawar, Khyber Pakhtunkhwa, Pakistan

S. Munir
Department of Biochemistry, Bahauddin Zakariya University, Multan 60800, Pakistan

S. Ahmed et al. (eds.), *Biochemistry of Drug Resistance*,
https://doi.org/10.1007/978-3-030-76320-6_18

and resistance of prevalent strains of *S. pneumoniae* can be easily done promptly with a minimum level of complex technology or complex protocols. It will pave the way for effective treatment of *S. pneumoniae* strains with rapid and higher recovery rates and minimal mortality rates. This chapter aims to evaluate antibiotics susceptibility and resistance to *S. pneumoniae* pherotypes called CSP-1 and CSP-2. Different methods of pherotypes determination have also been objectively evaluated.

Keywords Gram-positive bacteria · Serotypes · Pherotypes · Upper respiratory tract

18.1 Introduction

Streptococcus (S.) pneumonia or pneumococcus is a common commensal of the upper respiratory tract. Taxonomically, *S. pneumoniae* exists in a group of the pneumoniae-mitis pseudo-pneumoniae cluster which belongs to Streptococcus genus. *S. pneumoniae* is a gram-positive, lancet-shaped, usually diplococcus, however, it has also been reported to grow in chains of varying lengths (Brugger et al., 2009). Besides, it is a beta-hemolytic (aerotolerant anaerobic) as well as alpha-hemolytic (under aerobic conditions) pathogenic bacterium (Bogaert et al., 2004). *S. pneumoniae* has been reported to be the prime cause of a wide range of infections and one of the significant causes of mortality and morbidity across the globe. It causes local or systemic infections through colonization, while its spread from the nasopharynx may give rise to local infections, including acute otitis and sinusitis. In addition, aspiration may cause *S. pneumoniae* to directly spread into the lungs, and ultimately giving rise to community-acquired pneumonia (CAP). In addition, it develops complications of pleural empyema in less than 5% of CAP cases (Zhu & Lau, 2013). It is worth mentioning that invasive disease has been reported in 10–15% of CAP patients. The pneumococcus may also spread to the brain via blood which causes pneumococcal meningitis. Furthermore, arthritis, peritonitis, and osteomyelitis have also been recognized as rare manifestations of pneumococcal diseases (Camilli et al., 2017).

Pakistan is one of the developing countries threatened by the highest number of pneumococcal disease cases. Still, it is quite fearsome that research studies in this demanding aspect of research continue to remain limited in scope and validity (Bibi et al., 2018). It was further recorded that isolated strains remained susceptible to linezolid as well as chloramphenicol, while on the other hand, over 80% were found to be gentamycin resistant. Moreover, it was inferred that 19F and 18B (non-vaccine type) serotypes were the most prevalent among Pakistani pneumococcal isolates (Håvarstein et al., 1997).

More than 96 pneumococcal serotypes have been reported so far, hence determining susceptibility and resistance of all serotypes is a cumbersome process that needs hefty effort, time, and resources. Therefore, there must be an appropriate approach or technique to tackle the identification of bacterial strains, such as the widespread strains of *S. pneumoniae*, in an effective and timely manner. As a result of this rapid and precise identification, the information related to the susceptibility and resistance of *S. pneumoniae* prevalent strains can be easily obtained in minimum time with a minimum level of complex technology or complex protocols. This may further help treat the *Pneumonia* infections with comparatively maximum recovery while minimum mortality rates (Bibi et al., 2018; Claverys & Havarstein, 2002).

The pneumococci ability to undergo genetic transformation was reported earlier and it was, later on, found that its spread was varying among different age groups (Bergé et al., 2002). The competence of any bacteria to propagate itself is done through a diffusible signal. This process is called quorum-sensing and found its use for gram-positive as well as gram-negative bacteria for synchronizing the switching of genetic testing programs at a broader population level. This approach helps to achieve the desired goals that could have been otherwise unattainable by using single cells. Several biomolecules (proteins, peptides, etc.) get utilized by bacteria for regulating the mechanism of their quorum-sensing. For quorum-sensing in *S. pneumoniae*, competence-stimulating peptide (CSP) which is an unmodified 17-aminoacid peptide pheromone gets secreted. It is *comC* gene's product that gets secreted and subsequently subjected to processing by an ABC transporter (ComAB). This results in CSP accumulation in the lungs of the infected population. It has been reported that two components (ComD and its corresponding response regulator named ComE) are involved in sensing CSP accumulation and trigger the competence response (Håvarstein et al., 1997).

In pneumococci strains, numerous distinct and mature CSPs have been recognized, however, the majority of strains tend to produce only one of the variants among CSP-1 (CSP-α) or CSP-2 (CSP-β). Every CSP variant identifies a specific pherotype, while strains genetically entailing one variant are not able to respond to the signaling peptide presence owing to the *ComD* receptors specified nature for cognate CSP alleles (Pozzi et al., 1996). The competence induction, which is responsible for the genetic transformation of *S. pneumoniae* gets constituted by pherotypes, which have been classified as CSP1 and CSP2 (Johnsborg et al., 2006). These two pherotypes are quite similar as far as primary structures are concerned, but they hold high specificity to their receptors, ComD1 and ComD2. Therefore, it has been inferred that pherotype specificity directly depends on pneumococcal strain (Carrolo et al., 2009). Moreover, quorum sensing is highly related to antibiotic susceptibility, and there is a need to study the genes involved.

18.2 Detection Mechanisms of the Pherotype and Endonuclease Restriction Phenotype by PCR Method

The PCR method is one of highly recognized methods for detection of *S. pneumoniae* pherotypes. A study reported the detailed procedure regarding CSP-1 and CSP-2 gene fragments which were amplified using multiplex PCR with primers CSP_up (5'-TGA AAA ACA CAG TTA AAT TGG AAC-3'), CSP1_dn (5'-TCA AGA AAG GAT AAA GGT AGT CCT C-3') and CSP2_dn (5'-TAA AAA TCT TTC AAT CCC TAT TT-3'), that allowed the amplification of fragments of 620 bp for the CSP-1 allele and 340 bp for the CSP-2 allele. The preparation of template DNA was done by diluting the culture (9 μl) in water (441 μl), which was subsequently boiled for 2 min. Later on, those were subjected to PCR reactions using solution of 50 μl that contained template solution and reaction buffer (Biotools, Madrid, Spain), 10 mM dNTPs (Fermentas, Vilnius, Lithuania), primers, and polymerase (Invitrogen, Carlsbad, California). The process of the same PCR (30 cycles)was run for a minute at 95 °C which was followed by annealing performed at 55 °C for a half minute, and finally, incubation was done at 72 °C to complete the process of extension (Brugger et al., 2009).

The pneumococci bacteria tend to bind with fragments of naked DNA (double-stranded) which are abundantly present in the cellular environment and resultantly translocate those fragments into the single-stranded form in the cytoplasm. Multiple genes facilitate the process of recombination of bacterial chromosomes with the foreign DNA by favoring the integration that causes a permanent change in the cell genotype along with its phenotype. Furthermore, it has been recognized that competent cells had the capacity for killing the non-competent pneumococci using a mechanism that has been referred to as fratricide. This is a key property to trigger transformation in-vivo through the provision of free DNA (Johnsborg et al., 2006).

Numerous epidemiological studies on drug sensitivity of clinical and carriage isolates of *S. Pneumoniae* from developing countries such as Pakistan and Vietnam have shown their resistance towards clinically available antibiotics, including penicillin. The multidrug-resistant globally circulating clones of *S. pneumoniae* (Spain23F-1 and Taiwan19F-14) and some other related strains specific to South Asia are primarily responsible for the prevailing antimicrobial resistance (Bibi et al., 2018). Serotypes 19, 23, 14, and 6 are predominant in Vietnam and neighboring countries. However, there is a lack of concrete evidence regarding the risk of carrying antimicrobial-resistant *S. pneumoniae* to healthy children or increasing resistance to beta-lactam antimicrobials, including carbapenems. It has been reported that transfer modes of external DNA from the environment into the genome of *S. pneumoniae* is usually induced by competence-stimulating peptide (CSP). This natural transformation has been referred to as the *S. pneumoniae* quorum sensing system, which is responsible for the respective bacterial communication to regulate the expression of gene (Zhu & Lau, 2013). Moreover, CSP acts as a triggering agent for DNA transformation and ultimately induces the competency of bacteria (Bergé et al., 2002).

The emergence and evolution of pneumococcal strains that are multidrug-resistant have become a vital global concern of serious pertinence. Since 1980s, an extremely high upsurge in antibiotic-resistant pneumococci has been observed and reported worldwide and, therefore, it has been declared as globally spread pathogens. Penicillin-resistant pneumococci have also been identified in South Africa and Spain, and therefore the need for conducting studies to sort out the most effective antibiotics for pneumococci has been stressed (Kilian et al., 2008). In USA, penicillin resistance was very meager (5%) in 1990s, but it was increased manifold in 1992 to 6.6% in 1991–1992. As there is a great probability of the spread of antibiotic-resistant pneumococci across the borders and, therefore, extensive surveillance is required to control its spread to the regions where relatively infrequent has been reported so far (Camilli et al., 2017).

Antibiotics have found their use in treating ailments, including pneumococcal infections. Until 1970, pneumococcal isolates were susceptible to the prevalent antibiotics, including vancomycin, penicillin, and clindamycin. Recently, the emergence of resistance against penicillin and various other antibiotics has important implications for prevalent treatment options. The most effective treatment options for pneumococcal infections can be devised in the presence of historical findings and by conducting studies involving patients infected by susceptible strains. No verified findings are currently available of pneumococcal resistance in most of the Asian countries, especially Pakistan and China; however, such information is of great importance because resistance degree or extent alterations may impart significant influence on antimicrobial therapy. For obtaining information regarding pneumococci resistance to various antimicrobial agents, a microbiological survey was executed in China which involved 79 pneumococcal strains obtained from 4 specialized hospitals and nasopharyngeal strains (244) were isolated from various daycare centers. This study took into account the 12 antimicrobial agents *in-vitro* reactions(Zhu & Lau, 2013). The solo and co-applied ceftriaxone and cefotaxime with vancomycin remained effective in treating acute systemic infections (including meningitis) caused by numerous pneumococcal strains (Pozzi et al., 1996). However so far, considerable research and knowledge gap exists regarding the evolving resistance of pneumococcal infections against antibiotics, and recent research findings are needed to sort the most effective antibiotics against pneumococcal infections.

A study was to conducted explore any sort of association between *S. pneumoniae* pherotypes and antibiotic resistance revealed that foreign DNA acquisition by pneumococcus was the prime reason for developing resistance to a variety of antibiotics (Ramirez et al., 1997). Besides, Pneumococci resistance to antibiotics and its widespread dissemination in the local population was estimated since 1940s. It was indicated that there existed an association between most of the antibiotics (except erythromycin) with a particular pherotype (Tables 18.2). The resistance exhibited by isolates to penicillin and various other β lactams was owing to their association with CSP-1. Furthermore, it was inferred that resistance against β-lactams was evolved owing to interaction with many closely related species and genetic variations within the pneumococcal population were transmitted by genetic recombination (Guiral et al., 2005). Furthermore, penicillin-resistant isolates contained CSP-1 suggesting that gene flow tends to occur among isolates having the same pherotype.

In order to determine the susceptibility of antibiotics against pneumoniae, testing of 277 penicillin-susceptible, as well as S. Pneumoniae nasopharyngeal isolates, was done (Tables 18.1 and 18.2). The results revealed that pherotypes were completely susceptible to all the antibiotics under study. Among nasopharyngeal isolates, resistance exhibited by tetracycline, ofloxacin, erythromycin, trimethoprim-sulfamethoxazole, chloramphenicol, and resistance were 88, 74, 48, 4, and 63%, respectively. No resistance against vancomycin and rifampin was found. These findings elaborated that pneumococci-resistant penicillin was relatively meager in China in comparison to other Asian countries, while resistance against non-beta-lactams was recorded to be on the higher side compared to non-beta lactams. Moreover, E-test along with the agar dilution method proved their efficacy and reliability to identify the resistant strains (Claverys & Håvarstein, 2007). Moreover, it has been reported that pneumococcal clinical isolates started to become resistant to most of the antibiotics during the 1990s. The underlying reason behind this resistance of *S. Pneumoniae* was owing to obtaining more genetic diversity achieved through alteration of penicillin-binding protein receptors on the cell wall (Punatar et al., 2017).

The prevalence of penicillin-resistant *S.pneumoniae* in Europe (Spain, France, Greece) varied from 25 to 50%, while in Israel, it remained to be 10–25%. Over half of *S. pneumoniae* strains (56%) remained susceptible to the antibiotics under investigation. However, pneumococcal non-susceptibility of 1, 2, 17, and 27% were observed for antibiotics such as trimethoprim-sulphamethoxazole, penicillin, clindamycin, and erythromycin, respectively (Croucher et al., 2014). Interestingly, none of the isolates showed resistance to norfloxacin. Besides, pneumococcal resistance variation concerning geography within different cities (Cornejo et al., 2010). It was also inferred that various variables including age and sex of the patients along with the location of the daycare center, and frequency of antimicrobials treatments remained non-significant for non-susceptible of various *S. pneumoniae* strains. Among non-susceptible *S. pneumoniae* serotypes, the majority (67–82%) were recorded to be

Table 18.1 Antibiotics susceptibility and resistance to CSP-1 pherotype of *S. pneumoniae* (Carrolo et al., 2009)

Antibiotic	Susceptible	Resistant
Penicillin	168	28
Erythromycin	189	32
Chloramphenicol	156	59
Cefuroxime	134	68
Tetracycline	112	88
Co-trimoxazole	109	71
Clindamycin	138	102

Table 18.2 Antibiotics susceptibility and resistance to CSP-2 pherotype of *S. pneumoniae* (Carrolo et al., 2009)

Antibiotic	Susceptible	Resistant
Penicillin	168	92
Erythromycin	171	32
Chloramphenicol	185	5
Cefuroxime	111	89
Tetracycline	182	18
Co-trimoxazole	197	66
Clindamycin	180	22

present in pneumococcal conjugate vaccines. The nasopharyngeal *S. pneumoniae* susceptibility rates to antibiotics such as penicillin along with macrolides remained higher especially among the pre-school children of Lithuania. It was suggested to follow a strict policy regarding antibiotic prescription, along with strengthening vaccination campaigns, could put halt to increasing pneumococci resistance to antibiotics.

18.3 Different Methods to Identify Susceptibility and Resistance

18.3.1 Kirby Bauer Method

For the identification of Antibiotic susceptibility Kirby Bauer method is also known as disk-diffusion method in which Mueller–Hinton agar is used for the growth (Drew et al., 1972) (Fig. 18.1).

A collection of 483 invasive pneumococcal isolates were collected in Portugal for serotypes antibiotic susceptibility of these strains. All the strains of *S. pneumoniae* were cultured at 37 °C under anaerobic conditions using a semi-synthetic medium and a tryptic soy agar (TSA) (Oxoid, Hampshire, England). Those were supplemented with 5% (v/v) sterile blood of sheep, which was subsequently incubated using 5% CO_2 at 37 °C. For all isolates, serotyping, antimicrobial susceptibility and PFGE analyses were performed.

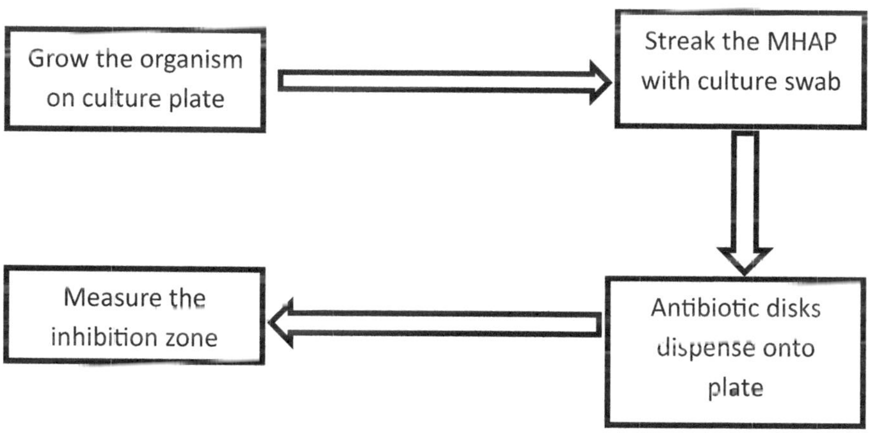

Fig. 18.1 Flowchart of Kirby Bauer method for identification of bacterial anti-biotic susceptability

18.3.2 E Test

E test was performed as per guidance provided by the manufacturer. Mueller–Hinton medium which contained the blood of sheep (5%) was utilized for testing susceptibility except for TMP-SMZ, which was performed using the blood agar plates of lysed horse.

18.3.3 Agar Dilution Test

Agar dilution was practiced as per recommendations of the National Committee for Clinical Laboratory Standards (13a). E test produced the MICs of the antimicrobial agents which were within the ideal ranges of two quality control strains (Carrolo et al., 2014; Gray, 1980).

18.4 Quorum Sensing and Immunity

Various individuals of the Mitis group stimulate specialized Competence Stimulating Peptide (CSP), which carry out genetic transformation to fulfill objectives beyond the access of individual cells (Håvarstein et al., 1996). Competence is considered a rapid response of bacteria coordinated via dispersed competence peptide, named quorum-sensing (QS), that appeared in both gram-negative as well as gram-positive bacteria (Waters & Bassler, 2005). Genetic competence, acidic tolerance, and biofilm production are organized by the CSP-based QS system, which plays a pivotal role in bacterial virulence (Allan et al., 2007). The reserved 17-amino acid-based peptide pheromone, a product of the *comC* gene, is responsible for QS in *S. pneumonia* (Claverys & Havarstein, 2002). The regulatory system of the *comC* is based on the histidine kinase receptor (ComD) and its particular response regulator (ComE) that recognizes the CSP concentration of media and triggers the specific response (Pozzi et al., 1996). Pherotypes are distinguished on CSP variant, and six well-defined CSPs have been found owing to sequence variation, but the majority of *S. pneumoniae* secrete either CSP1 or CSP2 (Allan et al., 2007). Due to the specificity of a receptor, comC1 (CSP1) carrying strain is impotent to respond in the presence of the extracellular CSP2 (Ramirez et al., 1997). Consequently, genetic transformation and antibiotic resistance are commonly found in similar pherotype than distinct ones, e.g., penicillin-resistance was observed commonly in CSP1 (Carrolo et al., 2009). In pneumococcal population, pherotype vary independently of serotype and have clonal property. More divergence in genetic material can show distinct pherotype (Carrolo et al., 2009). Competent cells can lyse the non-competent pneumococci, through the fratricide mechanism (Guiral et al., 2005). Fratricide is considered a pivotal factor in the genetic transformation of *S. pneumoniae* by uptaking naked environmental DNA of incompetent bacterial cells, which is lysed by competent bacterial cells (Baig et al., 2021; Håvarstein et al., 2006). In the pneumococcol population, the exchange of genetic material between Inter-pherotype diminishes than Intra-pherotype (Carrolo et al., 2009).

18.5 Current *S. pneumoniae* Vaccine

At present, two distinct types of pneumococcal vaccines are available in the global market against *S. pneumoniae*, i.e., the pneumococcal conjugate vaccine (PCV) and the pneumococcal polysaccharide vaccine (PPV) (Tan, 2012; McIntosh & Reinert, 2011). A 23-valent pneumococcal polysaccharide vaccine (PPV-23) comes under PPV type of vaccine, which has been approved since 1983 and is significantly carried out for prevention of 23 serotypes including 1, 2, 3, 4, 5, 6B, 7F, 8, 9N, 9V, 10A, 11A, 12F, 14, 15B, 17F, 18C, 19F, 19A, 20, 22F, 23F, and 33F (Ofori-Anyinam et al., 2017). However, PPV23 is only endorsed for children older than 2 years of age, while the PCVs are acceptable for children younger than the age of 2 years and immune-compromised elders (Nurse-Lucas et al., 2016). Currently, three PCVs (PCV-7, PCV-10, and PCV-13) are available and widely used to prevent invasive pneumococcal diseases (IPDs). The United Kingdom had implemented the 7-valent pneumococcal conjugate vaccine (PCV-7) in 2006, and it was able to target 75% of IPDs in children. PCV-7 is effective against seven pneumococcal serotypes: 4, 19F, 9V, 23F, 18C, 6B, and 14 (Nurse-Lucas et al., 2016). The rate of IPDs is reduced by 65.7% after one year, approximately 90% in 2009 and 2010 (Tocheva et al., 2013). In 2011, 10-valent pneumococcal conjugate vaccine (PCV-10) and 13-valent pneumococcal conjugate vaccine (PCV-13) was recommended for IPDs prevention from PCV-7 serotypes plus 1, 5, and 7F and PCV-10 serotypes plus 3, 6A, and 19A, respectively (Nurse-Lucas et al., 2016). Currently, no vaccine is available globally for the prevention of all serotypes. The implementation of pneumococcal vaccines is reducing IPDs burden, on the other hand, these are also increasing the non-vaccine types (NVTs). The imposition of PCV-10 and PCV-13 has reduced the serotypes 19A and 7F but abruptly increased NVTs like 22F and 33F (van der Linden et al., 2015). Consequently, advanced vaccine stratigcy should be adopted that esclate innate immune response independent of CSP and serotype. The broad antibiotic spectrum and resistance in *S. pneumoniae* demands a vaccine candidates which is independent of antibiotic resistance issues.

18.6 NETosis and Immunity

Neutrophil Extracellular Traps (NETs) are originated by neutrophils in the innate immune response to trap pathogens by their chromatin fibers and kill the pathogen with its antimicrobial protein granules. The mechanism of releasing traps to eliminate the pathogen from the host is known as NETosis (Brinkmann et al., 2004). NETs formation enhances the efficiency of microbial elimination by trapping the large size of various bacteria that cannot be eliminated by phagocytosis (Brinkmann et al., 2004). Besides its contribution to innate immune response, NETs also correlate with immune disorders such as autoimmune disease, chronic lung inflammation, and vascular disease (Zawrotniak & Rapala-Kozik, 2013).

Earlier studies revealed that the NETosis could be induced by multiple stimuli, including co-culture of neutrophils with pathogens, activate platelets, and pathogen components (Brinkmann et al., 2013). In *S. pneumoniae*, α-enolase was examined as a pivotal pathogen component in stimulating NETosis (Mori et al., 2012). In the innate immune response, α-enolase binds on neutrophils to form NETs, which trigger a mass of antimicrobial proteins and kill microbes extracellularly by trapping them in mobilized chromatin fiber (Hassan et al., 2021; Mori et al., 2012). DNAse is a specialized weapon of gram-positive bacteria to retaliate in the response of NETs. Cell surface-exposed endonuclease, produced by *endA*, enables *S. pneumoniae* to escape from NETs by degrading DNA scaffold and invade over the host (Beiter et al., 2006).

18.7 Conclusion

Different isolates analyzed led to the conclusion that it is highly unlikely that ecological fragmentation and geographic spreads could explore the serotype patterns, prevalence, and spread across the population. The model simulations also include the possibility that multiple isolates of *S. pneumoniae* having identical genotypes as well as common ancestry. It may also be inferred that CSP-2 population might be recently evolved owing to the novel pherotype's introduction into the pneumococci. However, contradictory findings reported that if it had been evolved recently, then e lower haplotype diversity for the CSP-2 population should have been observed. Furthermore, multiple serotypes and clones contained CSP-2 pherotype indicating its evolution in the distant past. Other observations also support that CSP-2 pherotype had an ancient origin that has allowed more than sufficient time for two pherotypes coalescence. Globally, pneumococci resistance against antibiotics has become widespread majorly owing to macrolides and β-lactams. Many European countries (France, Spain, etc.) reported that pneumococci (up to 40%) displayed multidrug resistance. Increasing understanding of various molecular determinants which impart resistance, cause transformation, and promote competence may assist to address and put a halt to this trend of higher genetic recombination of the pneumococci strains. Genetic exchanges occurring through natural transformations of the genetic materials enable the streptococci to adapt well in the oral as well as upper respiratory tract despite constant environmental changes. Previously, several research studies elucidated the mechanisms and identified the genes responsible for imparting competence against antibiotics and improved our understanding of natural transformations occurring in *S. pneumoniae*. Contrastingly, *S. mitis*, which has commensal resemblance underwent transformations that could be detected only by using advanced genetic tools that revealed their biology (Salvadori et al., 2016). This research aspect still has not gained the appropriate attention of the researchers and resultantly wide research and knowledge gaps are present regarding commensals. Yet, *S. mitis* genetic sequence and close relatives could explore the sources of variability which produced various serotypes of *S. pneumoniae* (Kilian et al., 2008). It is expected that future studies considering the wide diversity of *S. mitis* strains can reveal vital information regarding plasticity of pneumococcal genome as well as its pathogenicity. In addition, enhanced Netosis's ability to evoke the acquired immune response also needs thorough investigations. This integrated approach (by combining α-enolase and endonuclease) may prevent different strains of *S. pneumoniae* globally.

Acknowledgements We acknowledge all support by FRGS/1/2017/SKK06/UNISZA/02/1 under Kementerian Pendidikan Malaysia (KPM) and Universiti Sultan Zainal Abidin for our research on *S. pneumoniae*.

References

Allan, E., Hussain, H. A., Crawford, K. R., Miah, S., Ascott, Z. K., Khwaja, M. H., & Hosie, A. H. (2007). Genetic variation in comC, the gene encoding competence-stimulating peptide (CSP) in Streptococcus mutans. *FEMS Microbiology Letters, 268*, 47–51.

Baig, A. A., Zulkiflee, N. A. S. B., Hassan, M., Rohin, M. A. K. B., Johari, M. K. B. Z., Latif, A. Z. B. A., Khan, M. U., & Simbak, N. B. (2021). Narrative review: Use of competent stimulating peptide in gene transfer via suicide plasmid in streptococcus pneumoniae. *Advancements in Life Sciences, 8*, 211–216.

Beiter, K., Wartha, F., Albiger, B., Normark, S., Zychlinsky, A., & Henriques-Normark, B. (2006). An endonuclease allows streptococcus pneumoniae to escape from neutrophil extracellular traps. *Current Biology, 16*, 401–407.

Bergé, M., Langen, H., Claverys, J.-P., & Martin, B. (2002). Identification of a protein that inactivates the competence-stimulating peptide of Streptococcus pneumoniae. *Journal of Bacteriology, 184*, 610–613.

Bibi, S., Siddiqui, T. R., Hasan, S. F., & Ahmed, W. (2018). Colonizing pneumococcal serotypes in Pakistani children (<3 years) before the introduction of pneumococcal conjugate vaccine-implications for vaccine formulation. *Pakistan Journal of Medical Research, 57*, 138–143.

Bogaert, D., de Groot, R., & Hermans, P. (2004). Streptococcus pneumoniae colonisation: The key to pneumococcal disease. *The Lancet Infectious Diseases, 4*, 144–154.

Brinkmann, V., Reichard, U., Goosmann, C., Fauler, B., Uhlemann, Y., Weiss, D. S., Weinrauch, Y., & Zychlinsky, A. (2004). Neutrophil extracellular traps kill bacteria. *Science, 303*, 1532–1535.

Brinkmann, V., Goosmann, C., Kühn, L. I., & Zychlinsky, A. (2013). Automatic quantification of in vitro NET formation. *Frontiers in Immunology, 3*, 413.

Brugger, S. D., Hathaway, L. J., & Mühlemann, K. (2009). Detection of Streptococcus pneumoniae strain cocolonization in the nasopharynx. *Journal of Clinical Microbiology, 47*, 1750–1756.

Camilli, R., D'ambrosio, F., Del Grosso, M., Pimentel de Araujo, F., Caporali, M. G., Del Manso, M., Gherardi, G., D'ancona, F., & Pantosti, A. (2017). Impact of pneumococcal conjugate vaccine (PCV7 and PCV13) on pneumococcal invasive diseases in Italian children and insight into evolution of pneumococcal population structure. *Vaccine, 35*, 4587–4593.

Carrolo, M., Pinto, F. R., Melo-Cristino, J., & Ramirez, M. (2009). Pherotypes are driving genetic differentiation within Streptococcus pneumoniae. *BMC Microbiology, 9*, 191.

Carrolo, M., Pinto, F. R., Melo-Cristino, J., & Ramirez, M. (2014). Pherotype influences biofilm growth and recombination in Streptococcus pneumoniae. *PLoS ONE, 9*, e92138.

Claverys, J.-P., & Havarstein, L. S. (2002). Extracellular-peptide control of competence for genetic transformation in Streptococcus pneumoniae. *Frontiers Bioscience, 7*.

Claverys, J.-P., & Håvarstein, L. S. (2007). Cannibalism and fratricide: Mechanisms and raisons d'etre. *Nature Reviews Microbiology, 5*, 219–229.

Cornejo, O. E., McGee, L., & Rozen, D. E. (2010). Polymorphic competence peptides do not restrict recombination in Streptococcus pneumoniae. *Molecular Biology and Evolution, 27*, 694–702.

Croucher, N. J., Chewapreecha, C., Hanage, W. P., Harris, S. R., McGee, L., van der Linden, M., Song, J.-H., Ko, K. S., de Lencastre, H., & Turner, C. (2014). Evidence for soft selective sweeps in the evolution of pneumococcal multidrug resistance and vaccine escape. *Genome Biology and Evolution, 6*, 1589–1602.

Drew, W. L., Barry, A., O'Toole, R., & Sherris, J. C. (1972). Reliability of the Kirby-Bauer disc diffusion method for detecting methicillin-resistant strains of Staphylococcus aureus. *Applied Microbiology, 24*, 240–247.

Gray, B. M., Converse III, G. M., & Dillon, H. C. (1980). Epidemiologic studies of Streptococcus pneumoniae in infants: Acquisition, carriage, and infection during the first 24 months of life. *Journal of Infectious Diseases, 142*, 923–933.

Guiral, S., Mitchell, T. J., Martin, B., & Claverys, J.-P. (2005). Competence-programmed predation of noncompetent cells in the human pathogen Streptococcus pneumoniae: Genetic requirements. *Proceedings of the National Academy of Sciences, 102*, 8710–8715.

Hassan, M., Baig, A. A., Attique, S. A., Abbas, S., Khan, F., Zahid, S., Ain, Q. U., Usman, M., Simbak, N. B., & Kamal, M. A. (2021). Molecular docking of alpha-enolase to elucidate the promising candidates against Streptococcus pneumoniae infection. *DARU Journal of Pharmaceutical Sciences*, 1–12.

Håvarstein, L. S., Gaustad, P., Nes, I. F., & Morrison, D. A. (1996). Identification of the streptococcal competence-pheromone receptor. *Molecular Microbiology, 21*, 863–869.

Håvarstein, L. S., Hakenbeck, R., & Gaustad, P. (1997). Natural competence in the genus Streptococcus: Evidence that streptococci can change pherotype by interspecies recombinational exchanges. *Journal of Bacteriology, 179*, 6589–6594.

Håvarstein, L. S., Martin, B., Johnsborg, O., Granadel, C., & Claverys, J. P. (2006). New insights into the pneumococcal fratricide: Relationship to clumping and identification of a novel immunity factor. *Molecular Microbiology, 59*, 1297–1037.

Johnsborg, O., Kristiansen, P. E., Blomqvist, T., & Håvarstein, L. S. (2006). A hydrophobic patch in the competence-stimulating peptide, a pneumococcal competence pheromone, is essential for specificity and biological activity. *Journal of Bacteriology, 188*, 1744–1749.

Kilian, M., Poulsen, K., Blomqvist, T., Håvarstein, L. S., Bek-Thomsen, M., Tettelin, H., & Sørensen, U. B. (2008). Evolution of Streptococcus pneumoniae and its close commensal relatives. *PLoS ONE, 3*, e2683.

McIntosh, E. D. G., & Reinert, R. R. (2011). Global prevailing and emerging pediatric pneumococcal serotypes. *Expert Review of Vaccines, 10*, 109–129.

Mori, Y., Yamaguchi, M., Terao, Y., Hamada, S., Ooshima, T., & Kawabata, S. (2012). α-Enolase of Streptococcus pneumoniae induces formation of neutrophil extracellular traps. *Journal of Biological Chemistry, 287*, 10472–10481.

Nurse-Lucas, M., McGee, L., Hawkins, P. A., Swanston, W. H., & Akpaka, P. E. (2016). Serotypes and genotypes of Streptococcus pneumoniae isolates from Trinidad and Tobago. *International Journal of Infectious Diseases, 46*, 100–106.

Ofori-Anyinam, O., Leroux-Roels, G., Drame, M., Aerssens, A., Maes, C., Amanullah, A., Schuind, A., Li, P., Jain, V. K., & Innis, B. L. (2017). Immunogenicity and safety of an inactivated quadrivalent influenza vaccine co-administered with a 23-valent pneumococcal polysaccharide vaccine versus separate administration, in adults ≥50years of age: Results from a phase III, randomized, non-inferiority trial. *Vaccine, 35*, 6321–6328.

Pozzi, G., Masala, L., Iannelli, F., Manganelli, R., Havarstein, L., Piccoli, L., Simon, D., & Morrison, D. (1996). Competence for genetic transformation in encapsulated strains of Streptococcus pneumoniae: Two allelic variants of the peptide pheromone. *Journal of Bacteriology, 178*, 6087–6090.

Punatar, R. S., Martin, M. J., Wyatt, H. D., Chan, Y. W., & West, S. C. (2017). Resolution of single and double Holliday junction recombination intermediates by GEN1. *Proceedings of the National Academy of Sciences, 114*, 443–450.

Ramirez, M., Morrison, D. A., & Tomasz, A. (1997). Ubiquitous distribution of the competence related genes comA and comC among isolates of Streptococcus pneumoniae. *Microbial Drug Resistance, 3*, 39–52.

Salvadori, G., Durante, F., de Michele, C., Bernardi, M., & Petrella, L. (2016). A multivariate copula-based framework for dealing with hazard scenarios and failure probabilities. *Water Resources Research, 52*, 3701–3721.

Tan, T. Q. (2012). Pediatric invasive pneumococcal disease in the United States in the era of pneumococcal conjugate vaccines. *Clinical Microbiology Reviews, 25*, 409–419.

Tocheva, A. S., Jefferies, J. M., Christodoulides, M., Faust, S. N., & Clarke, S. C. (2013). Distribution of carried pneumococcal clones in UK children following the introduction of the 7-valent pneumococcal conjugate vaccine: A 3-year cross-sectional population based analysis. *Vaccine, 31*, 3187–3190.

van der Linden, M., Perniciaro, S., & Imöhl, M. (2015). Increase of serotypes 15A and 23B in IPD in Germany in the PCV13 vaccination era. *BMC Infectious Diseases, 15*, 207.

Waters, C. M., & Bassler, B. L. (2005). Quorum sensing: Cell-to-cell communication in bacteria. *Annual Review of Cell and Developmental Biology, 21*, 319–346.

Zawrotniak, M., & Rapala-Kozik, M. (2013). Neutrophil extracellular traps (NETs)-formation and implications. *Acta Biochimica Polonica, 60*.

Zhu, L., & Lau, G. W. (2013). Therapeutic potential of the Streptococcus pneumoniae competence regulon. *Expert Review of Anti-Infective Therapy, 11*, 227–229.

Chapter 19
Hypoxia, Obesity and Drug Resistance; Towards Sustainable Development Goals and Globalization

Farwa Munir, Fizza Maqbool, Shahzaib Naeem, Shaharbano, Mavra Riaz Atif, Asad Ur Rehman, Sarfraz Ahmed, Faheem Mustafa, Muhammad Hassan, Shehla Munir, and Atif Amin Baig

Abstract Globosity is one of the rising threats towards sustainable development goals and the globalization of low and medium-income countries besides rising trends in high-income countries. Obesity is the excessive fat accumulation in a body, which hinders many of the metabolic processes as well as causes secondary morbidities. In obesity, when adipose tissues are increased, there is an enhanced need for oxygen; if they don't get the appropriate amount of oxygen, they are turned into a hypoxic state, leading to activation of inflammatory markers and causes inflammation and swelling in the body. This excessive fat and hypoxia in obese patients lead to drug resistance. Hypoxia works as an obstacle in the normal influx of ions in the body cells because of which the pharmacodynamics and pharmacokinetics of the drugs are affected, and the drug cannot play an efficient role in the patient's body. Moreover, the mode of action of the drug is also hindered by fats, and when there is high-fat concentration, lipophilic drugs are absorbed more easily; contrary to it, the hydrophilic drugs are not absorbed properly.

F. Munir · F. Maqbool · S. Naeem · Shaharbano · F. Mustafa
School of Health Sciences, University of Management and Technology, Lahore, Pakistan

M. R. Atif · A. U. Rehman
Faculty of Business and Management, Universiti Sultan Zainal Abidin, Kuala Terengganu, Malaysia

S. Ahmed
Department of Biochemistry, University of Veterinary and Animal Sciences, Lahore, Pakistan

F. Munir · F. Mustafa · M. Hassan · A. A. Baig (✉)
Faculty of Medicine, Universiti Sultan Zainal Abidin, Kuala Terengganu, Malaysia
e-mail: atifamin@unisza.edu.my

S. Munir
Department of Biochemistry, Bahauddin Zakariya University, Multan 60800, Pakistan

S. Ahmed et al. (eds.), *Biochemistry of Drug Resistance*,
https://doi.org/10.1007/978-3-030-76320-6_19

19.1 Introduction

As a treatment of such resistance, the dosage of the medicine is increased or decreased accordingly. Combination drug therapies are also used to treat drug resistance. Physical inactivity, high caloric diet, and stress are the other factors that encourage weight gain and hypoxia. To cure secondary morbidities, we first have to target excessive weight and inflammation in the body that can be done by changing the lifestyle of an individual, and for that, one has to take a moderate amount of fats, calories and get involved in physical activities and focus group settings on curing the underlying issues that cause obesity and hypoxia. These measures also help improve drug resistance.

Obesity and globalization are interlinked with each other; with rising globalization trends, the food consumption patterns are changing globally, causing more and more obesity among humans. Globalization has increased the living standards of people across the globe but on the other hand, it is also responsible for economic development and increased urbanization, causing the change in food consumption patterns and more wealth being spent on obesity-causing food (Fox et al., 2019). Worldwide, obesity has doubled since the 1980s, and recent researches have predicted that obesity affects low income and middle-income countries more than the developed nations (Fox et al., 2019).

The World Bank, in its latest report, has identified trade liberalization, fiscal policies, and country wealth as one of the indirect factors that cause obesity among masses. There is no standardized formula to calculate the obesity cost because it includes not only the cost of food-related to obesity but also the health expenditures-resulting from poor health, decreased productivity of employees (again due to poor health), and other social and cultural costs attached to it. Governments worldwide are encouraged to increase taxation on unhealthy food items and subsidize healthy food in order for people to rearrange their food intake choices (Shekar and Popkin, 2020).

Hypoxia is a lack of normal level (60 mmHg) of oxygen to the body tissue and cells beside the other vital parameter of oxygen saturation at the cellular level if less than normal (90%) (Beasley et al., 2017).Breathing and cellular respiration, if depleted of oxygen (Lunt and Vander Heiden, 2011)can induce major changes in the cell. For instance, under hypoxia, regarding the sodium–potassium pump, the potassium diffused out into the extracellular fluid cause influx of sodium. This results in decreased cellular ATP and an effective decrease in the semi permeability of cell membranes via multi ions influx. Extra sodium in the cell leads to water retention as well as calcium channel dysfunction (Hoiland et al., 2016).

This entire imbalance of ion and energy deprivation as a result of low concentration of cellular oxygen results in drug resistance in the body, as it is the main site of action of the drug is in hypoxic state mode of action of the drug is altered (Wu et al., 2018).

Body as a defensive response to hypoxia leads to the induction and activation of several genes which further induce transcription factors and cause angiogenesis-metabolism of stored iron in the body to activate hemoglobin and red blood cells to carry more oxygen to the cells and increase glycolysis for high availability of energy (Zimna & Kurpisz, 2015). The primary mediator in this whole process is Hypoxia-inducible factor-alpha (HIF-α), which activates in response to hypoxia and activates the co-activator cAMP response element-binding protein, binding protein/p300 and regulates the expression of target genes. The overexpression of HIF-α also induces the inflammatory markers such as interleukins, NF-Kβ, and others, which hinder the normal functioning of the cells by causing inflammation (Arjamaa et al., 2017).

This relationship between inflammation caused by hypoxia or hypoxia caused by inflammation is becoming the cause of many major morbidities and mortalities (Ramakrishnan et al., 2014). There is an increasing trend in studies that links hypoxia and inflammation to obesity. Obesity is among the most prominent health issues of the twenty-first century. It not only adds mass into the human body but also causes many secondary comorbidities such as heart diseases, diabetes, and cancer. For a long time, obesity was considered as an accumulation of additional or excessive fat into the body, and the only major cause studied and taught for decades was the intake of a high-fat diet as the main risk factor (Engin, 2017a).

However, recent researches and studies have shown that number of factors are involved in the mechanism of the body gaining extra mass and phenotypically becoming a cause of body shaming. Some of the major factors of obesity are thought to be the lack of physical activity, high caloric diet, genetic prevalence as well as some underlying hormonal issues, especially in females (Joslyn & Haider-Markel, 2019). Also, the current focus of experts has been shifting from the basic concept of obesity to underneath inflammatory mechanism of obesity, in which the adipose tissues are inflamed and are characterized toward hypoxic shift resulting in whole-body swellings and obesity (Batatinha, 2019) (Fig. 19.1).

19.2 Effective Mechanism of Role of Hypoxia

Hypoxia is a state in which sufficient oxygen is not reaching the site. When there is deficient oxygen in a cell, it enhances the work rate to achieve the required amount of oxygen, resulting in increased heart rate and swelling in the body region that is in the hypoxic state (Serocki et al., 2018).

The mechanism that connects hypoxia and obesity is because of the release of inflammatory markers in the body, such as interleukin-6, as a result of increased adipocyte volume (Bahceci et al., 2007).

Hypoxia is induced by transcriptional factor HIF-alpha (hypoxia-inducible factor) that helps our body to adapt in an oxygen deficient environment by activating mRNA expression of different genes that stimulate glycolysis or angiogenesis in the hypoxic tissues (Zhang et al., 2009). Obesity with hyperplasia of adipose tissues comprises of majorly two forms of adipose tissues. The first one is brown adipose tissues, and the second is white adipose tissues based on their colors, as brown adipocytes have more iron-containing mitochondria on the other side the white adipocytes are the cells that have the stored form of fat, which if not burnt or used in the body

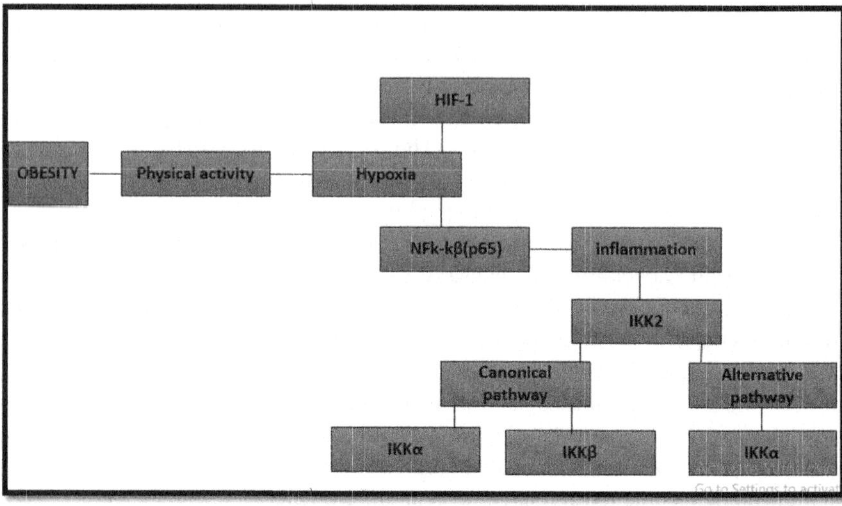

Fig. 19.1 A pathway linking obesity, hypoxia, and inflammation in the body (D'Ignazio et al., 2017)

is stored in the form of visceral fat concentrating around the vital organs causing fatty liver or others (Ikeda et al., 2018). In obesity we majorly deal with the white adipocytes, which secrete the adipokines in the blood; those adipokines have some inflammatory factors too, like IL-6, IL-B and Nf-kβ.These adipokines play a role in the inflammation of the adipocytes. The bigger the cell, the more amount of oxygen it requires for its function. This state of oxygen insufficiency leads to the activation of HIF-alpha as a molecular oxygen sensor (Jaakkola et al., 2001).

There are very few reports linking obesity to drug metabolism. Studies showed that drug metabolism in obese adults is slower than in non-obese adults (Brill et al., 2012). If seen specifically in cytochrome p450 3A4 substrate clearance is lower in obese individuals; in contrast to it the drugs majorly metabolized via UGT (uridine diphosphate glucuronosyl transferase) are rapidly cleared out from the body of obese adults (Brill et al., 2012). As obesity targets the vital systems and organs of the human body, so the pharmacokinetics and pharmacodynamics of the drug molecules are also altered in obese patients. In obese individuals, the gastric mobility and emptying are slower than the normal individuals that result in the delay of the drug action (Gérard, 2016). Fatty liver has also altered the supply of blood, and prolonged obesity affects the kidneys, so detoxification of drugs are also slowed down, varying according to the obesity type and type of drugs. Researches show that the volume of distribution of lipophilic drugs is more in obese adults but is lesser in hydrophilic drugs; the following graph shows an overview of such drugs (Jain et al., 2011) (Fig. 19.2).

When a person is obese and has a greater number of white adipocytes in his body, as well as increased volume of the adipocytes, it results in activation of inflammatory markers and HIF-α, that lead his tissues to be into a hypoxic state; showed a drug

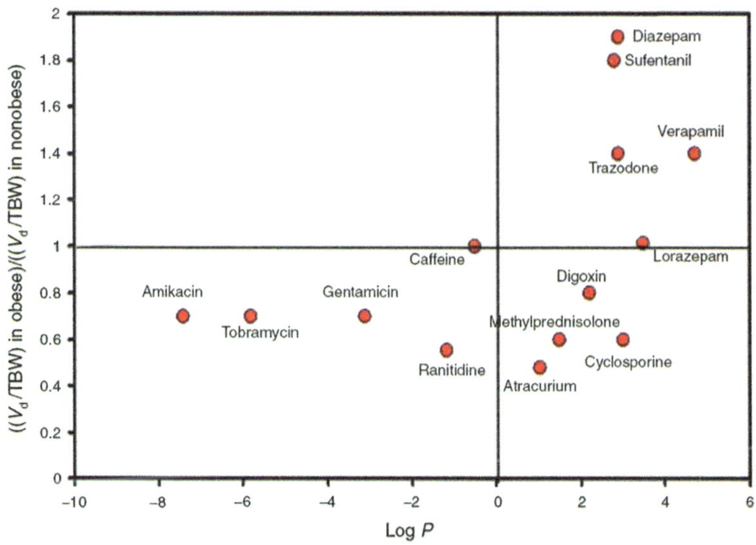

Fig. 19.2 The ratios of the volumes of distribution (Vd) of various drugs (after normalization with total body weight (TBW)) in persons with obesity relative to those without obesity, with respect to various drugs with different lipophilic and hydrophilic properties (Jain et al., 2011)

resistance and the dosage of the drug might be needed to be reframed (Smit et al., 2018).

Obesity is multifactorialin origin, and one of the factors is a high-fat diet (HFD). By definition, HFD is a diet with high-fat contents (45% of calories are sourced from fats) (Hector & Phillips, 2018). High-fat foods are energy-dense as well as palatable, so it encourages overconsumption. There is a general statement that consumption of a high-fat diet promotes obesity in the human body, which is actually true to some extent because generalization doesn't work in ruling out the cause of obesity (Hohos & Skaznik-Wikiel, 2017).

Fat and energy imbalance act as a favorable environment for obesity; in the obesity model, rat high-fat diets created an imbalance of energy intake and energy expenditure (Hall et al., 2011), and as a result of this imbalance, the metabolic environment favors the enzyme profile for lipid storage in the body.

According to a survey in US, adults who consume low fiber, high-fat diets are more likely to become overweight compared to high-fiber, low-fat diets (O'Keefe et al., 2015). Fat is a component of the diet that not only promotes adulthood obesity, but a high-fat postnatal diet also develops childhood obesity and obesity-related metabolic complications (Hou et al., 2014).

When it comes to saturated and unsaturated fats and their effects on weight; particularly diet high in saturated fats promote obesity; US population showed a significant relation of high BMI as well as obesity Genetic Risk Score (GRS) when total fat and saturated fatty acid intake were focused as a gene-diet interaction (Casas-Agustench et al., 2014).

Obese people usually chose high palatable food over normal-weight people. Human and Animal trials showed that high-fat foods in the diet could induce hyperphagia (passive overconsumption), which leads to fat storage of 50–100 g/day in humans (Klingbeil & de La Serre, 2018). Fat is the least selected option for the body to use as fuel. First, the body utilizes carbohydrates and then proteins efficiently, and by combining these two factors, overconsumption of a high-fat diet and the body's ability to use less fat as an energy substrate plays a significant role in obesity.

Advancement in studies now linked gut dysbiosis to obesity, as a high-fat diet alters the gut microbiome, so it could be one of the purposed mechanism for high-fat diet-induced obesity (Abenavoli et al., 2019).

In anin-vivo study to investigate the efficacy of the anti-adipogenic drug on five groups of male mice for a total period of 8-weeks; feeding with a control diet, a high-fat diet, or a high-fat diet with 50, 100, or 200 mg/kg body weight/day were administrated respectively. The group of mice that was subjected to a high-fat diet showed a notable increase in body weight in contrast with the group on the control diet (Green et al., 2016).

This suggested that energy imbalance, type of fat, low fiber-high-fat diet, food palatability, and gut dysbiosis, as well as in-vivo animal studies, strongly support the narrative high-fat diets induce obesity though a lot is still needed to be studied in the inflammatory aspect of this phase.

The human body stores fat in the adipose tissues as a result of lipogenesis that takes place in the presence of acetyl-CoA. The availability of acetyl-CoA is necessary to initiate the lipogenesis (Green et al., 2016). As a result of excess fat storage, the cytosol of adipocytes expands or swells up, which causes inflation of adipose cells known as adipocyte hypertrophy.

Expanded adipocytes give rise to compromised blood flow as obese people have 58% less vascular endothelial growth factor, so as a result, they have less capillary beds within the adipocytes that creates an imbalance of oxygen concentration leading to cellular hypoxia (Al-Anazi et al., 2018).

Obesity-induced inflammation puts the body's metabolic health at stress; it also contributes to insulin resistance. An increased level of inflammatory adipokines IL-1 beta, IL-6, tumor necrosis factor-alpha, and MCP-1, which are the key factor for the infiltration of macrophages, collectively invade the immune system so as a response of macrophage infiltration, it creates a generalized or local inflammation (Engin, 2017b).

Highly raised inflammatory markers produce reactive oxygen species (ROS) in the obese person causing oxidative stress (Newsholme et al., 2016); here, this fact is needed to be highlighted, adipose cells are not mere storage site for fat, but it also acts as an endocrine organ, it releases adipokines, and these in return make ROS. The potential involvement of oxidative stress and macrophages recruitment can be linked to obesity-induced adipose cell inflammation.

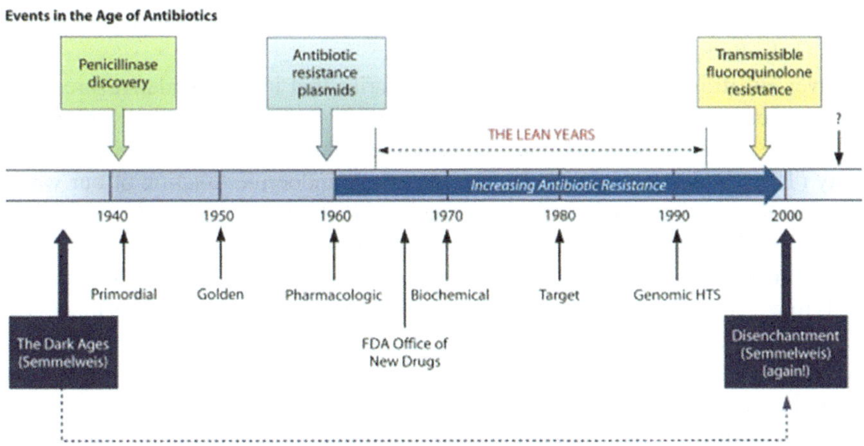

Fig. 19.3 Events in the age of antibiotics (Davies & Davies, 2010)

According to WHO, antibiotic resistance is a global threat to health which originated back in 1960 (Fig. 19.3). Misuse of antibiotic develops its resistance, as well as prolonged use of oral antibiotics cause's deleterious effects on gut microbes, with an enhanced inflammatory response. The link between reduced colonization of gut bacteria and enhanced inflammatory response is unknown.

19.3 Obesity and Nutrigenetics

Another factor that causes obesity is food, which is not directly high in fat content, but the calories consumed are more than required (Small et al., 2018). The unit of energy is calorie through which we can monitor the required energy. This unit of energy that is calories or Kcal, can measure the amount of energy that we take from food or food items. Mainly calorie intake depends upon age, height, weight, and physical activities. Mostly the recommendation of calories for women and men is 2000 and 2500, respectively (Dietary Guidelines for Americans 2015–2020). In the international system of the unit, kilocalories are equivalent to kilojoules. Approximately 1 kcal is equal to 4.2 kJ. To keep up a healthy weight, you have to adjust calories you burn through food and drink versus calories you consume via physical action (Saklayen, 2018).

Calorie admittance and metabolic energy consumption are misleading towards weight gain when an HF diet is burned-through and that the high caloric viscosity of high-fat ingestion routines is considered a major part of weight gain. People might be progressively prone to put on weight when an HF diet is consumed (Johnson et al., 2013). As per recent researches, High caloric diet consumption causes weight gain, due to which our adipocytes get inflamed. It is now obvious that adipocytes not only stays as a storage reservoir of fat but are active endocrine organs that play numerous parts in the body.

Over-nutrition gives an extra supply of calories that induce oxidative stress and redox potential in the body leading to the activation of Kupfer cells. Kupfer cells secrete the paracrine hormone that affects lipid metabolism and hepatic insulin sensitivity (Thomas & Apovian, 2017). The important endocrine function of our whole body is regulated mainly by adipose tissues in which obesity causes dysfunction of our metabolic system and causes insulin resistance (Longo et al., 2019).

A computational model of recent researches concluded that a high caloric diet leads to weight gain and enlarge adipocytes (Lombardo et al., 2016). This model has been supplemented with some anthropometrics of the individuals, such as age, mass, stature, and gender, and act out on the basis of extra calorie intake and the buildup of lipid analysis, the body mass as a role of the physique and of the capacity of the adipocytes which in turn swells and rises in number as a result of the fat buildup owing to the addition of calorie consumption (Kazak et al., 2017).

Defective adipose tissue is also correlated with metabolic syndrome in results; it's swelling of adipocytes accumulation of macrophages into adipose tissue. Type 2 diabetes mellitus and insulin resistance are caused because of obesity and ectopic fat accumulation (Sezer, 2017). Defective adipocytes are due to the destruction of adipocyte diffractions that encourages chronic inflammation of adipocytes, due to which emission of adiponectin is reduced and increases the emission of proinflammatory cytokines, i.e., interleukin (IL)-6, IL-8 and monocytes chemo-attractant protein 1 (Pietiläinen et al., 2011).

White adipose tissue and brown adipose tissue are the two main types (Rosen et al., 2006). Excessive amounts of triglycerides are stored by brown adipose tissue by draining of energy through the making of heat by brown adipose tissue. An increasing amount of brown adipose tissue is specifically found in lean individuals, while aged individuals are found with low levels of brown adipose tissue and high body mass, as adipose tissue and blood circulation have a crucial part in individual well-being. As we discussed above, the increase of adipocytes is due to the consumption of a high caloric diet that also encourages hypoxia and inflammatory signs (Fig. 19.4).

High occurrence of extreme overweight leads to accumulation or inflammation of white adipose tissue. The tissue is a substantial endocrine and weakening organ, with adipocytes, the trademark cell type, emitting a collection of protein factors, the adipokines. Increases in the emission of innumerable adipocytes happen in load, associate aggravation in white fat tissue, and the improvement of stoutness related sicknesses (Carobbio et al., 2017). This is evident, especially in humans, that hypoxia produces in fat tissue as the tissue mass grows, and the decrease in Po2 is considered to underlie the provocative reaction.

The discharge of various irritants related to adipocytes is up-regulated by hypoxia, and there is a change from oxidative breakdown to anaerobic glycolysis. Glucose usage is expanded in hypoxic adipocytes with relating increments in lactate formation (Takubo et al., 2013). Significantly, hypoxia incites insulin opposition in fat cells and prompts the improvement of fat tissue fibrosis. A large number of the reactions of adipocytes to hypoxia are started at the Partial pressure of oxygen

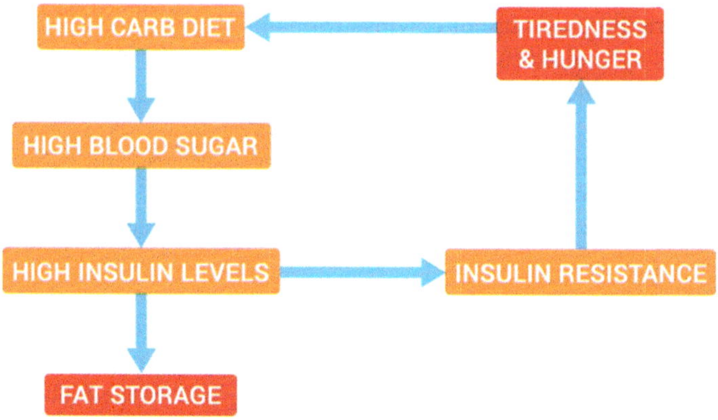

Fig. 19.4 The pathway of high-calorie food intake to fat storage in the body causing obesity

levels over the ordinary physiological reach for fat tissue (Engin, 2017b). Other cell types inside the tissue additionally react to hypoxia, with the separation of preadipocytes to adipocytes being repressed and preadipocytes being changed into leptin-emitting cells. In general, hypoxia affects the capacity of adipocytes and seems, by all explanations, to be a key factor in fat tissue inflammation.

As a course of treatment of Insulin resistance and inflammation in the body, plant extracts such as polyphenolmixes, including non-flavonoids and flavonoids, are used. These natural extracts target many inflammatory components and are becoming the focus of nutraceuticals. Polyphenols are found to be effective against chronic inflammatory diseases like obesity and diabetes (Yahfoufi et al., 2018). Targeting gene expression and proinflammatory cytokines, they inactivate NF-κβ and modulatemitogen-initiated protein kinases (MAPK) pathways while enacting the 5' adenosine monophosphate-actuated protein kinase (AMPK) pathway in fat tissue (Di Pierro et al., 2015). A survey conducted by Hardie (2015), examined the major sub-atomic instruments of dietary polyphenols and unsaturated fats, alone or in a blend, which are answerable for fat tissue-related calming impacts. The effectiveness and mode of action of the polyphenols for hypoxia and inflammation are needed to be studied in further detail, as there are many questions yet to be answered about their activities and results of their administration (Di Pierro et al., 2015).

In this modern era of hectic and very calculated routine, with current lifestyles, we need to know and take good care of our health. We have to manage the time for a healthy life and health. For this, we have to be physically active and have to do healthy exercises like cardio muscular, aerobics etc. (Heianza & Qi, 2017). Our slow and sedentary lifestyles take us towards the high risk of diabetes, inflamed adipocytes and weight gain, and many other diseases as well. If we are not active, we gain more and more fats in our body towards obesity and metabolic syndrome (Rodulfo, 2019).

Overweight people are more prone to non-communicable diseases because they eat more and more high-calorie diets. We have to cut out the high-calorie intake from our food (Barnosky et al., 2014). Especially pregnant ladies take more drinks and take high-calorie food because of cravings (Solianik et al., 2018); in this condition, a woman needs to be more and more careful about her dietary preferences as well as should maintain a pattern of physical activity.

Our population is greatly affected by this fancy food or high caloric diet due to which our population severely affected morbid diseases (Gulland, 2014), i.e., cardiovascular disease and t2dm, etc. CVD and diabetes type disease are perceived as an intense increase in emergent nations, and it's mainly due to empty calories or a high caloric diet.

So far better life we have important the people towards a healthy lifestyle for that people should be avoided from such of habits like consumption of junk food (Mesejo, 2011), for a better quality of life. We have to courage people to avoid such type of sedentary lifestyle, and they should change their eating patterns and should give them nutritional or healthy awareness to avoid such type of routine (Borel et al., 2014). As we discussed it starting from high calories diet caused inflamed adipocytes inflamed due to which metabolism can be distracted because, in metabolism, adipose tissue plays an important role; it plays an important role in lifespan in developing countries and coming within enormous proliferation in excessive weight, and correlated issues. We are not telling you to totally restrict, in case you are so fond of this high calories diet you can eat it but in a small portion or moderate amount or on the special occasion but not on a daily basis and no doubt we can't win this high calories food war against to healthy eating, but we should try it because of high calories diet due to which weight gain is gain, due to which our lifespan badly affected (Leite et al., 2009).

As we know, during pregnancy, almost all body organs are affected, but after post-partum, it will have to come back to its original position. But many of us think that we have to eat for two people during pregnancy, one for the pregnant lady other for the unborn fetus (Morton et al., 2014). But many of us just adapt this type of habit due to which our gain excess then the recommended, people start to eat junk food or empty calories instead of healthy food and their physical activity are also affected due to gestational diabetes occurred so high-calorie intake is correlated with a faster decline in insulin sensitivity and b cell consumption in women who are at higher risk of t2dm and independence adipose tissue (Longo et al., 2019). We observe the molecular alterations inherited from a history of obesity. The history of obesity is very important and plays a vital role in our health.

Obesity occurs when we eat high sugar-containing food and drink more beverages in our daily routine; obesity affects our liver and adipose tissue. Obesity leaves inflammatory effects in adipose tissues. In male adipose tissues are less than female bodies. Within seven weeks after a diet change, most obesity-associated phenotypes such as body mass, glucose intolerance, and metabolic levels get reserved, although hepatic inflammation, hepatic steatosis as well as hypertrophy and inflammation of peritoneal are not in adipocytes persisted in present obese female (Polyzos et al., 2019). So we show that obesity leaves a worsen form of inflammation in the liver and peritoneal fat, which indicates the high inflammation signs in the liver and adipose tissue, so we investigate the molecular alteration in their history of obesity we have to more work to cut down a high calorie, we have to make changes in our diet and exercise to make our healthy lifestyle.

Physical activity also has a role in overall human health (Jakicic & Davis, 2011).When a person is doing mindful physical activity with a perception to burn calories, his body and bodily systems are involved in helping him to gain his goal. The routine work is not found efficient in calorie burning as the mindful and planned physical activity shows its results (Swift et al., 2018).

A person who is eating either a high caloric diet or a high-fat diet also has a sedentary lifestyle is one step more prone to obesity (Sander et al., 2018). The physical activity helps not only to burn calories but also speed up the gastric mobility, resulting in the better absorption of the food and drugs taken as well as physical activity is done in the fresh air early morning helps to breathe in more oxygen the people who live in close dark spaces and are not facilitated with green open lawns not only get stressed but also face obesity (Westerterp, 2018). It is studied that people with a friendly environment and good social interactions involve themselves in more physical activities resulting in the burning of extra loads of energy.

Multiple cohort studies done on the human population show that the ones who spend their leisure time in physical activities are less exposed to gaining weight at later ages (Gretebeck et al., 2017). Another side of the picture is that people with loads of fats in their bodies and hypoxic bodily environment are not active enough to involve themselves in activities. This sedentary and lazy lifestyle is another challenge for the drug administration as people in such states are more exposed to stress and hopelessness (Milaneschi et al., 2019).

Additionally, the anxiety and stress in human life also play a role in adding adversity to the case of hypoxia, obesity, and drug resistance (Luppino et al., 2010). When a person is stressed, ghrelin hormone is released, which target the hypothalamus, hippocampus, and mesolimbic regions of the brain and generate the feeling of hunger and craving for carbohydrates that increase the intake of calories above the required amount, cycling into the process of extra energy storage in the body as in white adipocytes, causing their inflammation, increased need of oxygen leading to hypoxia and interrupt the drug mechanism of action. Felix-Sebastian in 2008 also proved that the psychological stress in the family also played a role in childhood obesity and explained counseling is also required with lifestyle modification in such families.

Liver and gut metabolism has a clear impact on drug bioavailability. In obesity, CYP-mediated hepatic metabolism might also be affected. Bowman et al. reported no significant difference in bioavailability of drugs (cyclosporine, dexfenfluramine, midazolam, moxifloxacin, propranolol, and trazodone) in obese and non-obese individuals; however, high absorption was observed in the case of propranolol (Bowman et al., 1986). In another study, decreased absorption rate was observed in morbid (BMI > 45) obese people upon oral administration of levothyroxine and acetaminophen (Brill et al., 2014; Lee et al., 1981). A study in China injected a fixed dose of 10,000 IU of human chorionic gonadotropin (HCG) both subcutaneously and intramuscularly in 12 moderately obese women and12 non-obese women. Low AUC was reported in obese women in both routes (Chan et al., 2003). However, evidence on the effect of obesity in drug absorption/resistance is limited (Smit et al., 2018).

19.4 Conclusion

To combat the increasing challenge of obesity and drug resistance, a change in the general lifestyle of an obese adult where he is asked for eating moderately and as per the requirements of his body. Not following the cravings of his body but only eating for the purpose of living. Minimizing fat to less than or equal to 30% of the energy need in a sedentary to a less active lifestyle shouldn't exceed calorie limits more than 2000. They should involve in aerobics such a walking, jogging, swimming where the hypoxic areas of the body can get a sufficient amount of oxygen. The ionic imbalance in those cells can be settled as the cell comes to its normal form; activity will also help to remove that water that is accumulated inside one's body. Social groups and sittings should be encouraged to be ears of people's stress and to give them relaxation of their anxieties so they can work in a better way and have an active lifestyle. These moderations can help a person to defeat overweight, stored fats, hypoxia, and drug resistance; if a person positively endures to all such practices, his body will respond to the medication more efficiently.

Obesity is spreading its roots all around the world, and the link between hypoxia and its generated inflammatory pathway is yet to be studied in more depths. Studies should be done in more depth about the genes and the mediators those are involved in the hypoxic state of tissues, and those activate inflammatory markers such as the role of HIf-α and Nf-kβ is needed to be studied in comparison with their moderation in obese and non-obese individuals to categorize the possible adaptations in obese individuals and mutation. For this purpose, gene extraction, PCR, and DNA sequencing can be applied to figure out better and deeper results. These studies will lead us to a more vivid understanding of the inflammatory pathway hypothesized in obese individuals; along with that, it will help us to devise new methods of holistic and individualized treatments.

References

Abenavoli, L., Scarpellini, E., Colica, C., Boccuto, L., Salehi, B., Sharifi-Rad, J., Aiello, V., Romano, B., de Lorenzo, A., & Izzo, A. A. (2019). Gut microbiota and obesity: A role for probiotics. *Nutrients, 11*, 2690.

Al-Anazi, A., Parhar, R., Saleh, S., Al-Hijailan, R., Inglis, A., Al-Jufan, M., Bazzi, M., Hashmi, S., Conca, W., & Collison, K. (2018). Intracellular calcium and NF-kB regulate hypoxia-induced leptin, VEGF, IL-6 and adiponectin secretion in human adipocytes. *Life Sciences, 212*, 275–284.

Arjamaa, O., Aaltonen, V., Piippo, N., Csont, T., Petrovski, G., Kaarniranta, K., & Kauppinen, A. (2017). Hypoxia and inflammation in the release of VEGF and interleukins from human retinal pigment epithelial cells. *Graefe's Archive for Clinical and Experimental Ophthalmology, 255*, 1757–1762.

Bahceci, M., Gokalp, D., Bahceci, S., Tuzcu, A., Atmaca, S., & Arikan, S. (2007). The correlation between adiposity and adiponectin, tumor necrosis factor α, interleukin-6 and high sensitivity C-reactive protein levels. Is adipocyte size associated with inflammation in adults? *Journal of Endocrinological Investigation, 30*, 210–214.

Barnosky, A., Hoddy, K., Unterman, T., & Varady, K. (2014). AP Барноский1, КК Ходди2, ТГ Унтерман1, КА Варади2. *Translational Research, 164*, 302–311.

Batatinha, H. A. P., Rosa Neto, J. C., & Krüger, K. (2019). Inflammatory features of obesity and smoke exposure and the immunologic effects of exercise. *Exercise Immunology Review, 25*, 96–111.

Beasley, R., Chien, J., Douglas, J., Eastlake, L., Farah, C., King, G., Moore, R., Pilcher, J., Richards, M., & Smith, S. (2017). Target oxygen saturation range: 92–96% versus 94–98%. *Respirology, 22*, 200–202.

Borel, A.-L., Schwebel, C., Planquette, B., Vésin, A., Garrouste-Orgeas, M., Adrie, C., Clec'h, C., Azoulay, E., Souweine, B., & Allaouchiche, B. (2014). Initiation of nutritional support is delayed in critically ill obese patients: A multicenter cohort study. *The American Journal of Clinical Nutrition, 100*, 859–866.

Bowman, S. L., Hudson, S. A., Simpson, G., Munro, J. F., & Clements, J. A. (1986). A comparison of the pharmacokinetics of propranolol in obese and normal volunteers. *British Journal of Clinical Pharmacology, 21*(5), 529–532.

Brill, M. J., van Rongen, A., Houwink, A. P., Burggraaf, J., van Ramshorst, B., Wiezer, R. J., van Dongen, E. P., & Knibbe, C. A. (2014). Midazolam pharmacokinetics in morbidly obese patients following semi-simultaneous oral and intravenous administration: A comparison with healthy volunteers. *Clinical Pharmacokinetics, 53*(10), 931–941.

Carobbio, S., Pellegrinelli, V., & Vidal-Puig, A. (2017). *Adipose tissue function and expandability as determinants of lipotoxicity and the metabolic syndrome*. Springer.

Casas-Agustench, P., Arnett, D. K., Smith, C. E., Lai, C.-Q., Parnell, L. D., Borecki, I. B., Frazier-Wood, A. C., Allison, M., Chen, Y.-D.I., & Taylor, K. D. (2014). Saturated fat intake modulates the association between an obesity genetic risk score and body mass index in two US populations. *Journal of the Academy of Nutrition and Dietetics, 114*, 1954–1966.

Chan, C. C., Ng, E. H., Chan, M. M., Tang, O. S., Lau, E. Y., Yeung, W. S., & Ho, P. C. (2003). Bioavailability of hCG after intramuscular or subcutaneous injection in obese and non-obese women. *Human Reproduction, 18*(11), 2294–2297.

D'Ignazio, L., Batie, M., & Rocha, S. (2017). Hypoxia and inflammation in cancer, focus on HIF and NF-κB. *Biomedicines, 5*, 21.

Davies, J., & Davies, D. (2010). Origins and evolution of antibiotic resistance. *Microbiology and Molecular Biology Reviews, 74*, 417–433.

Engin, A. . (2017). *The definition and prevalence of obesity and metabolic syndrome*. Springer.

Engin, A. B. . (2017). *Adipocyte-macrophage cross-talk in obesity*. Springer.

Fox, A., Feng, W., & Asal, V. (2019). What is driving global obesity trends? Globalization or "modernization"? *Globalization and Health, 15*, 32.

Green, C. R., Wallace, M., Divakaruni, A. S., Phillips, S. A., Murphy, A. N., Ciaraldi, T. P., & Metallo, C. M. (2016). Branched-chain amino acid catabolism fuels adipocyte differentiation and lipogenesis. *Nature Chemical Biology, 12*, 15–21.

Gretebeck, K. A., Sabatini, L. M., Black, D. R., & Gretebeck, R. J. (2017). Physical activity, functional ability, and obesity in older adults: A gender difference. *Journal of Gerontological Nursing, 43*, 38–46.

Gulland, A. (2014). Curbs on fast food chains could halt rise in obesity, says WHO. *BMJ: British Medical Journal (Online)*, 348.

Hall, K. D., Sacks, G., Chandramohan, D., Chow, C. C., Wang, Y. C., Gortmaker, S. L., & Swinburn, B. A. (2011). Quantification of the effect of energy imbalance on bodyweight. *The Lancet, 378*, 826–837.

Hector, A. J., & Phillips, S. M. (2018). Protein recommendations for weight loss in elite athletes: A focus on body composition and performance. *International Journal of Sport Nutrition and Exercise Metabolism, 28*, 170–177.

Heianza, Y., & Qi, L. (2017). Gene-diet interaction and precision nutrition in obesity. *International Journal of Molecular Sciences, 18*, 787.

Hohos, N. M., & Skaznik-Wikiel, M. E. (2017). High-fat diet and female fertility. *Endocrinology, 158*, 2407–2419.

Hoiland, R. L., Bain, A. R., Rieger, M. G., Bailey, D. M., & Ainslie, P. N. (2016). Hypoxemia, oxygen content, and the regulation of cerebral blood flow. *American Journal of Physiology-Regulatory, Integrative and Comparative Physiology, 310*, R398–R413.

Hou, L., Hellgren, L., Kongsted, A. H., Vaag, A., & Nielsen, M. O. (2014). Pre-natal undernutrition and post-natal overnutrition are associated with permanent changes in hepatic metabolism markers and fatty acid composition in sheep. *Acta Physiologica, 210*, 317–329.

Ikeda, K., Maretich, P., & Kajimura, S. (2018). The common and distinct features of brown and beige adipocytes. *Trends in Endocrinology & Metabolism, 29*, 191–200.

Jaakkola, P., Mole, D. R., Tian, Y.-M., Wilson, M. I., Gielbert, J., Gaskell, S. J., von Kriegsheim, A., Hebestreit, H. F., Mukherji, M., & Schofield, C. J. (2001). Targeting of HIF-α to the von Hippel-Lindau ubiquitylation complex by O2-regulated prolyl hydroxylation. *Science, 292*, 468–472.

Jakicic, J. M., & Davis, K. K. (2011). Obesity and physical activity. *Psychiatric Clinics of North America, 34*, 829–840.

Johnson, R. J., Nakagawa, T., Sanchez-Lozada, L. G., Shafiu, M., Sundaram, S., Le, M., Ishimoto, T., Sautin, Y. Y., & Lanaspa, M. A. (2013). Sugar, uric acid, and the etiology of diabetes and obesity. *Diabetes, 62*, 3307–3315.

Joslyn, M. R., & Haider-Markel, D. P. (2019). Perceived causes of obesity, emotions, and attitudes about discrimination policy. *Social Science & Medicine, 223*, 97–103.

Kazak, L., Chouchani, E. T., Lu, G. Z., Jedrychowski, M. P., Bare, C. J., Mina, A. I., Kumari, M., Zhang, S., Vuckovic, I., & Laznik-Bogoslavski, D. (2017). Genetic depletion of adipocyte creatine metabolism inhibits diet-induced thermogenesis and drives obesity. *Cell Metabolism, 26*(660–671), e3.

Klingbeil, E., & de la Serre, C. B. (2018). Microbiota modulation by eating patterns and diet composition: Impact on food intake. *American Journal of Physiology-Regulatory, Integrative and Comparative Physiology, 315*, R1254–R1260.

Lee, W. H., Kramer, W. G., & Granville, G. E. (1981). The effect of obesity on acetaminophen pharmacokinetics in man. *The Journal of Clinical Pharmacology*, *21*(7), 284–287.

Leite, N., Milano, G., Cieslak, F., Lopes, W., Rodacki, A., & Radominski, R. (2009). Effects of physical exercise and nutritional guidance on metabolic syndrome in obese adolescents. *Brazilian Journal of Physical Therapy, 13*, 73–81.

Lombardo, G. E., Arcidiacono, B., de Rose, R. F., Lepore, S. M., Costa, N., Montalcini, T., Brunetti, A., Russo, D., de Sarro, G., & Celano, M. (2016). Normocaloric diet restores weight gain and insulin sensitivity in obese mice. *Frontiers in Endocrinology, 7*, 49.

Longo, M., Zatterale, F., Naderi, J., Parrillo, L., Formisano, P., Raciti, G. A., Beguinot, F., & Miele, C. (2019). Adipose tissue dysfunction as determinant of obesity-associated metabolic complications. *International Journal of Molecular Sciences, 20*, 2358.

Lunt, S. Y., & Vander Heiden, M. G. (2011). Aerobic glycolysis: meeting the metabolic requirements of cell proliferation. *Annual Review of Cell and Developmental Biology, 27*, 441–464.

Luppino, F. S., de Wit, L. M., Bouvy, P. F., Stijnen, T., Cuijpers, P., Penninx, B. W., & Zitman, F. G. (2010). Overweight, obesity, and depression: A systematic review and meta-analysis of longitudinal studies. *Archives of General Psychiatry, 67*, 220–229.

Mesejo, A., Sánchez Álvarez, C., & Arboleda Sánchez, J. A. (2011). Guidelines for specialized nutritional and metabolic support in the critically-ill patient: Update. Consensus SEMICYUC-SENPE: Obese patient. *Nutrition Hospitalty, 26*(Suppl 2), 54–58.

Milaneschi, Y., Simmons, W. K., van Rossum, E. F., & Penninx, B. W. (2019). Depression and obesity: Evidence of shared biological mechanisms. *Molecular Psychiatry, 24*, 18–33.

Morton, S. M., Grant, C. C., Wall, C. R., Carr, P. E. A., Bandara, D. K., Schmidt, J. M., Ivory, V., Inskip, H. M., & Camargo, C. A. (2014). Adherence to nutritional guidelines in pregnancy: Evidence from the growing up in New Zealand birth cohort study. *Public Health Nutrition, 17*, 1919–1929.

Newsholme, P., Cruzat, V. F., Keane, K. N., Carlessi, R., & De Bittencourt, P. I. H. (2016). Molecular mechanisms of ROS production and oxidative stress in diabetes. *Biochemical Journal, 473*, 4527–4550.

O'Keefe, S. J., Li, J. V., Lahti, L., Ou, J., Carbonero, F., Mohammed, K., Posma, J. M., Kinross, J., Wahl, E., & Ruder, E. (2015). Fat, fibre and cancer risk in African Americans and rural Africans. *Nature Communications, 6*, 1–14.

Pietiläinen, K. H., Rog, T., Seppänen-Laakso, T., Virtue, S., Gopalacharyulu, P., Tang, J., Rodriguez-Cuenca, S., Maciejewski, A., Naukkarinen, J., & Ruskeepää, A.-L. (2011). Association of lipidome remodeling in the adipocyte membrane with acquired obesity in humans. *PLoS Biology, 9*, e1000623.

di Pierro, F., Bressan, A., Ranaldi, D., Rapacioli, G., Giacomelli, L., & Bertuccioli, A. (2015). Potential role of bioavailable curcumin in weight loss and omental adipose tissue decrease: Preliminary data of a randomized, controlled trial in overweight people with metabolic syndrome. Preliminary study. *European Review for Medical and Pharmacological Sciences, 19*, 4195–4202.

Polyzos, S. A., Kountouras, J., & Mantzoros, C. S. (2019). Obesity and nonalcoholic fatty liver disease: From pathophysiology to therapeutics. *Metabolism, 92*, 82–97.

Ramakrishnan, S., Anand, V., & Roy, S. (2014). Vascular endothelial growth factor signaling in hypoxia and inflammation. *Journal of Neuroimmune Pharmacology, 9*, 142–160.

Rodulfo, J. I. A. . (2019). Sedentarismo, la enfermedad del siglo xxi. *Clínica e Investigación En Arteriosclerosis, 31*, 233–240.

Rosen, E., Walkey, C., & Puigserver, P. (2006). Spiegelman Bm. Adipocytes as regulators of energy balance and glucose homeostasis. *Nature, 444*, 847–853.

Saklayen, M. G. . (2018). The global epidemic of the metabolic syndrome. *Current Hypertension Reports, 20*, 12.

Sander, C., Ueck, P., Mergl, R., Gordon, G., Hegerl, U., & Himmerich, H. (2018). Physical activity in depressed and non-depressed patients with obesity. *Eating and Weight Disorders-Studies on Anorexia, Bulimia and Obesity, 23*, 195–203.

Serocki, M., Bartoszewska, S., Janaszak-Jasiecka, A., Ochocka, R. J., Collawn, J. F., & Bartoszewski, R. (2018). miRNAs regulate the HIF switch during hypoxia: A novel therapeutic target. *Angiogenesis, 21*, 183–202.

Sezer, H. . (2017). *Insulin resistance, obesity and lipotoxicity*. Springer.

Shekar, M., & Popkin, B. (2020). *Obesity: health and economic consequences of an impending global challenge*, World Bank Publications.

Small, L., Brandon, A. E., Turner, N., & Cooney, G. J. (2018). Modeling insulin resistance in rodents by alterations in diet: What have high-fat and high-calorie diets revealed? *American Journal of Physiology-Endocrinology and Metabolism, 314*, E251–E265.

Smit, C., De Hoogd, S., Brüggemann, R. J., & Knibbe, C. A. (2018). Obesity and drug pharmacology: A review of the influence of obesity on pharmacokinetic and pharmacodynamic parameters. *Expert Opinion on Drug Metabolism & Toxicology, 14*(3), 275–285.

Solianik, R., Sujeta, A., & Čekanauskaitė, A. (2018). Effects of 2-day calorie restriction on cardiovascular autonomic response, mood, and cognitive and motor functions in obese young adult women. *Experimental Brain Research, 236*, 2299–2308.

Swift, D. L., McGee, J. E., Earnest, C. P., Carlisle, E., Nygard, M., & Johannsen, N. M. (2018). The effects of exercise and physical activity on weight loss and maintenance. *Progress in Cardiovascular Diseases, 61*, 206–213.

Takubo, K., Nagamatsu, G., Kobayashi, C. I., Nakamura-Ishizu, A., Kobayashi, H., Ikeda, E., Goda, N., Rahimi, Y., Johnson, R. S., & Soga, T. (2013). Regulation of glycolysis by Pdk functions as a metabolic checkpoint for cell cycle quiescence in hematopoietic stem cells. *Cell Stem Cell, 12*, 49–61.

Thomas, D., & Apovian, C. (2017). Macrophage functions in lean and obese adipose tissue. *Metabolism, 72*, 120–143.

Westerterp, K. . (2018). Changes in physical activity over the lifespan: impact on body composition and sarcopenic obesity. *Obesity Reviews, 19*, 8–13.

Wu, C., Yang, T., Liu, Y., Lu, Y., Yang, Y., Liu, X., Liu, X., Ye, L., Sun, Y., & Wang, X. (2018). ARNT/HIF-1β links high-risk 1q21 gain and microenvironmental hypoxia to drug resistance and poor prognosis in multiple myeloma. *Cancer Medicine, 7*, 3899–3911.

Yahfoufi, N., Alsadi, N., Jambi, M., & Matar, C. (2018). The immunomodulatory and anti-inflammatory role of polyphenols. *Nutrients, 10*, 1618.

Zhang, R., Wu, Y., Zhao, M., Liu, C., Zhou, L., Shen, S., Liao, S., Yang, K., Li, Q., & Wan, H. (2009). Role of HIF-1α in the regulation ACE and ACE2 expression in hypoxic human pulmonary artery smooth muscle cells. *American Journal of Physiology-Lung Cellular and Molecular Physiology, 297*, L631–L640.

Zimna, A., & Kurpisz, M. (2015). Hypoxia-inducible factor-1 in physiological and pathophysiological angiogenesis: Applications and therapies. *BioMed Research International.*

Chapter 20
Genetic and Molecular Mechanisms of Multidrug-Resistance in Uropathogens and Novel Therapeutic Combat

Muhammad Mubashar Idrees and Ali Saeed

Abstract Urinary tract infections (UTIs) affect the kidney, ureters, bladder, and urethra. UTIs are divided into pyelonephritis (inflammation of kidney), urethritis (urethral inflammation) and cystitis (inflammation or infection in bladder). UTIs are caused by gram-negative bacteria like *E.coli*, *Proteus spp.*, *Klebsiella spp.*, *Pseudomonas aeruginosa* and *Acinetobacter spp.* Gram-negative bacteria more commonly cause UTIs as compared to gram-positive bacteria. Some gram-positive bacteria including *Staphylococcus aureus,* coagulase-negative *Staphylococci* and *Enterococcus spp.* can also cause UTIs. UTIs are treated by using different groups of broad-spectrum antibiotics. Gradually, genetic mutations in essential proteins involved in the antibacterial activity such as mutations in antibiotics bind proteins (Penicillin-binding proteins, PBPs), mutations in proteins involved in antibiotic uptake (such as Porin channels), efflux system (active transport of antibiotics) or degradation (such as β-lactamases including extended-spectrum β-lactamases, penicillinases, carbapenemases and cephalosporinases). β-lactamases are produced by *Enterobacteriaceae* due to genetic mutations (such as *CTX-M, OXA, KPC* genes), which lead to the production of β-lactamases and subsequent degradation of antibiotics containing β-lactam ring. Recently, multidrug resistance uropathogens have been increased remarkably due to excessive use of antibiotics, self-medication, prolonged stay in the hospital, antibitics use without clinical diagnosis and genetic variation in these bacteria. This chapter will be focused on the pathogenesis of multidrug resistant uropathogens (such as MDR, MRSA, ESBLs, VRE) in UTIs and their virulence factor and genetic variation, which will be helpful for the development of novel anti-virulence therapies and vaccine. The molecular mechanisms involved in drug resistance and novel therapeutic targeting of uropathogens will be discussed in detail.

M. M. Idrees · A. Saeed (✉)
Institute of Molecular Biology & Biotechnology, Bahauddin Zakariya University, Multan 66000, Pakistan
e-mail: alisaeed@bzu.edu.pk

M. M. Idrees
Multan Institute of Kidney Diseases (MIKD), Multan, Pakistan

S. Ahmed et al. (eds.), *Biochemistry of Drug Resistance*,
https://doi.org/10.1007/978-3-030-76320-6_20

20.1 Urinary Tract Infections

Urinary tract infection (UTI) is one of the most common infection exist in developing countries that lead to mortality and morbidity (Noor et al., 2004). It affects almost 150 million people in one year throughout the world (Stamm & Norrby, 2001). The urinary tract contains the ureters, kidneys, urethra and bladders. Infection or inflammation in any part of the urinary tract is known as UTIs (Peach et al., 2016). Infection usually occurs due to the invasion of the human body's normal flora into the urinary tract (bladder) from urethra or blood through an ascending or hematogenous route, respectively. UTIs are responsible to increase morbidity rates among infants, females and older men of any age. Bacteria are normally present around the urethra and enter the bladder through the urethra where they multiply and move into the kidney through the ureter or wash out during urination (Foxman, 2010). Urinary tract infections have more common in females as compared to males due to the anatomical position of the urethra and rectum. Bacteria colonize the bladder and do not move into other parts of the urinary tract, then it can be painful and frustrating. On the other hand, if it spread beyond the bladder to other parts of the urinary tract, then it will have serious consequences (Foxman, 2010).

UTIs are classified into complicated and uncomplicated forms based on clinical aspects. Infection in a healthy person with no structural or neurological dysfunction in the urinary tract is categorized as uncomplicated UTIs (Nielubowicz & Mobley, 2010; Hooton, 2012). Complicated UTIs are associated with different pathological complications along with infections which include the obstruction of the urinary tract, the neurological form of UTIs lead to urine retention, compromised immune response, transplantation, renal dysfunction, pregnancy and some other like calculi, catheters for drainage that affects the urinary tract or human immune response. Indwelling catheters are involved in 70–80% of complicated UTIs as a major risk factor. For example, according to an estimate UTIs are caused by indwelling catheters approximately 1 million patients per year in USA (Lo et al., 2014). Mortality and morbidity have been increased due to catheter-associated UTIs (CAUTIs) that also lead to a blood infection. Prolong use of catheters is responsible for CAUTIs most commonly in female patients (Chenoweth et al., 2014).

20.2 Types

Urinary tract is divided into the lower and upper urinary tract. Ureters, bladder and urethra are part of the lower urinary tract and kidneys are part of the upper urinary tract (Fig. 20.3). Infection is more common in the lower part of the urinary tract as compared to the upper urinary tract. There are different types of urinary tract infection as given below.

20.2.1 Pyelonephritis

The entrance of uropathogens into the kidney and cause infection or inflammation in the upper urinary tract is called pyelonephritis (Lane & Takhar, 2011). It is a common bacterial infection that usually occurs in adult females. *E.coli* is the major causative agent of pyelonephritis followed by *Klebsiella spp., Staphylococcus saprophyticus, Candida spp., Enterococcus spp., Pseudomonas spp.,* and other *Enterobacteriaceae* including *Proteus spp.,* and *Enterobacter* (Czaja et al., 2007). Recurrent UTIs, diabetes, sexual activity and pathological condition of the urinary tract may increase the risk for pyelonephritis. Infection usually occurs through the urinary tract (Colgan et al., 2011) and very often through the bloodstream (Colgan et al., 2011).

Sign and symptoms of acute pyelonephritis may appear within a day after infection. Pyelonephritis leads to fever, abdominal pain along with vomiting and discomfort urination (Ramakrishnan & Scheid, 2005). Chronic kidney inflammation can cause weight loss, abdominal pain, fever, blood in urine and lower UTIs symptoms (Korkes et al., 2008).

Community-acquired pyelonephritis is caused by the ascension of bowel pathogens like *E. coli* and *Enterococcus faecalus* into the urinary tract. Hospital-acquired UTIs are caused by *Enterococci* and coliform bacteria along with *Klebsiella spp* and *Pseudomonas spp.* Pyelonephritis is followed by prostatitis and cystitis infection (Ramakrishnan & Scheid, 2005). Invasion of *E. coli* into bladder cells leads to the formation of intracellular bacterial communities (IBCs) resulting in biofilm formation. Biofilm protects the bacterial cells from immune response and antibiotics. Bacteria colonize into the bladder and multiply, which cause opportunistic UTIs like pyelonephritis (Hultgren, 2011).

Risk factors for pyelonephritis may be increased due to structural abnormalities in the urinary tract, diabetes, immunocompromised patients, kidney stones, use of catheterization, pregnancy and genetically running of disease in the family.

20.2.2 Cystitis

Ascension and multiplication of bacterial normal flora into the bladder through the urethra that lead to inflammation is called cystitis (Lane & Takhar, 2011). Diabetes, obesity, female gender, prior UTIs and genetic susceptibility are the major risk factors associated with cystitis (Fig. 20.1) (Hannan et al., 2012; Foxman, 2013).

20.2.3 Urethritis

Urethritis is an inflammation of the urethra (lower urinary tract). Urethra is composed of the fibromuscular tube from where urine excrete as a waste product in both males and females and semen in males.

Urethritis is most prone to infection spread through sexual activity. *Chlamydia trachomatis* and *Neisseria gonorrhoeae* are major causative agents for sexually transmitted infection (STIs) (Fig. 20.1) (Burstein & Zenilman, 1999). There are two major groups of urethritis as gonococcal and non-gonococcal urethritis (Moi et al., 2015). *Neisseria gonorrhoeae* cause gonococcal urethritis and *Chlamydia trachomatis* cause non-gonococcal urethritis (NGU). Other causative agents for infectious urethritis are *Herpes simplex virus*, Methicillin-resistant *Staphylococcus aureus* (Sharma & Singal, 1999), *Adenovirus* (Bradshaw et al., 2002), *Candida spp.* (Bedük & Manalp, 1986), *Trichomonas vaginalis* (Caini et al., 2014; Workowski & Bolan, 2015), group B *Streptococci* (Chowdhury & Pareek, 1984) and *Mycoplasma genitalium* (Control and Prevention, 2002). Non-infectious urethritis may occur due to trauma or irritation in the genital area.

The major symptom of urethritis is difficulty in urination through the urethra. Symptoms may vary from patient to patient due to causative agents. Symptoms of urethritis may appear after a few weeks in infectious disease. Non-infectious urethritis shows symptoms after a few days of infection.

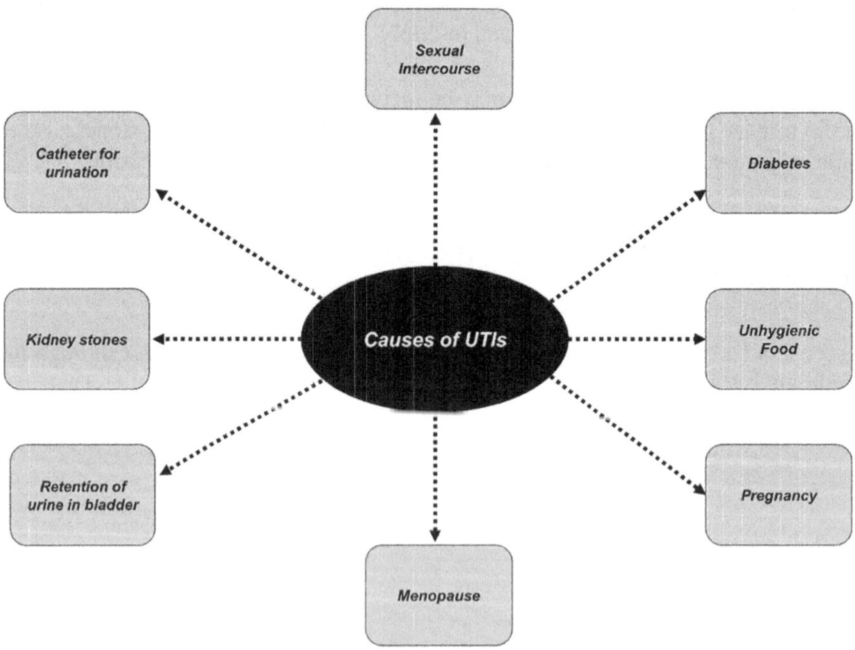

Fig. 20.1 Causes of urinary tract infection (UTIs)

20.3 Pathogenesis

The pathogenesis of each uropathogens to cause UTIs differs due to their portal of entry and their virulence factor.

20.3.1 Portal Of Uropathogens Entry

There are two types of pathogenesis by which uropathogens enter into the urinary tract and cause UTIs including ascending and haematogenous routes.

i. **Ascending route**
   Ascension of uropathogens into kidney following by urethra, bladder and then ureters is called ascending route. Ascending route is very common for uropathogens to cause UTIs as compared to others. In this route, uropathogens are present around the urethra and enter and bind with urethra due to presence of appendage like structure as pili or flagella. During sexual intercourse, these uropathogens move into the bladder from where, they enter into other parts of the urinary tract after overcoming the host immune response and biofilm formation (Fig. 20.2) (Foxman, 2010).
ii. **Hematogenous route**
   The hematogenous route is not important as compared to the ascending route for the administration of bacteria into the urinary tract and cause UTIs. It only occurs due to inflammation of renal cells, bacteria present in the blood enter into the urinary tract and cause infection (blood-borne infection) (Fig. 20.2) (Davis et al., 2011).

20.3.2 Mechanism of Urinary Tract Infections (UTIs)

Adherence of uropathogens present around the urethra is an important and initiative step for UTIs. Uropathogens move towards the bladder with the help of pili and flagella after adherence with the urethra. Host-uropathogens interaction determines either pathogen colonize and multiply or eliminate.

Bacterial adhesion with bladder occurs by recognition of receptors present on the bladder epithelium layer (uroepithelium) and induces colonization and multiplication. Uropathogenic *E. coli* (UPEC) produces toxins and proteases after invasion into the bladder epithelium layer and induces host cells to release nutrients (Fig. 20.3). Uropathogens overcome the host immune response by colonization and multiplication within the bladder. After this, uropathogens move towards the kidney through the appendage-like structure as pili or flagella and release toxins after invasion into renal epithelium resulting in tissue damage (Fig. 20.3 and Table 20.1). Finally, bacteria become able to get blood access by crossing the renal tubular membrane and cause bacteraemia.

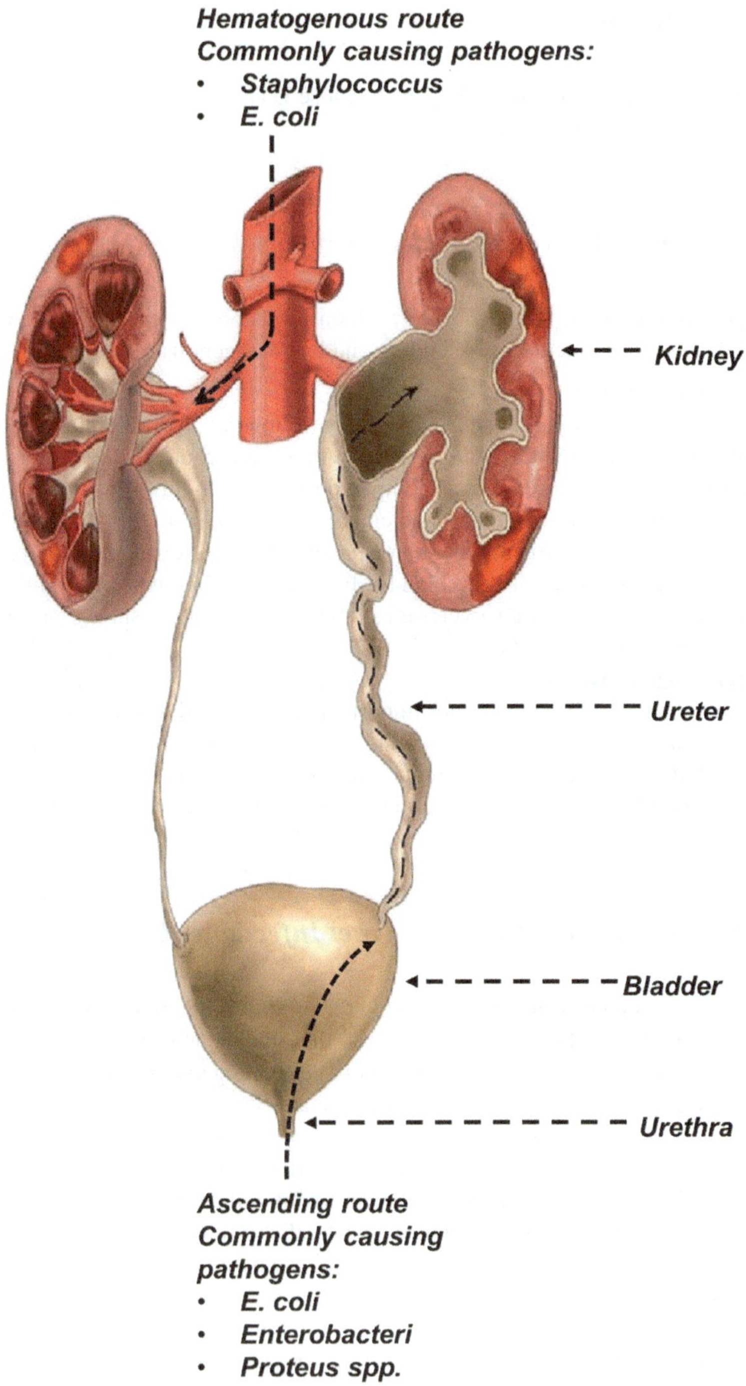

Fig. 20.2 Portal of uropathogens entry

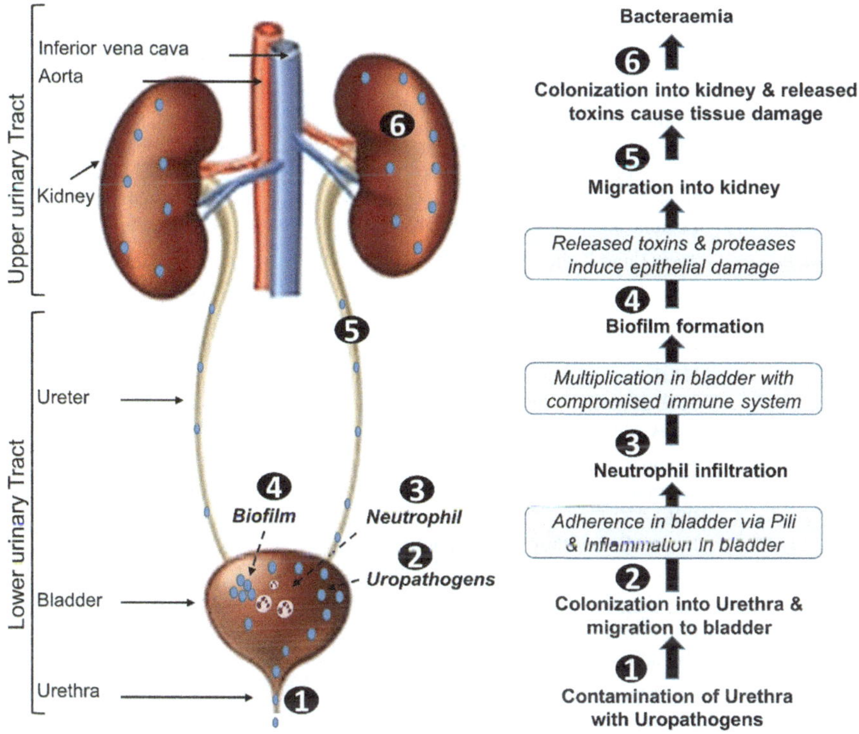

Fig. 20.3 Mechanism of UTIs caused by uropathogens in lower and upper urinary tracts

Uncomplicated UTIs causing pathogens, especially *E. coli., Staphylococcus saprophyticus* and *Klebsiella spp.* have an affinity to bind with the epithelium layer of the bladder (umbrella cells or facet cells), basal cells and intermediate cells (Table 20.1) (Khandelwal et al., 2009). Umbrella cells protect the mammalian bladder by forming the crystalline array from inflammation by urine gradients (Lee, 2011). *Klebsiella spp.* and UPEC bind with umbrella cells via uroplakin protein (major protein part of umbrella cell membrane) (Khandelwal et al., 2009). $\alpha_3\beta_1$ integrins are present on the surface of uroepithelial cells by which UPEC binds despite uroplakin receptors (Eto et al., 2007).

Initiation of complicated UTIs is occurred due to adhesion of bacteria with catheter, kidney stone or prolong stay of urine in the bladder due to obstruction in the urinary tract. Few pathogens especially UPEC can cause both uncomplicated and complicated UTIs. Major causative agents of complicated UTIs are *P. mirabilis, Enterococcus spp.* and *P. aeruginosa*. Mostly, uropathogens form a biofilm that helps in settlement and multiplication of these bacteria (Niveditha et al., 2012).

Table 20.1 Virulence factors of uropathogens and their toxins release during UTIs

Uropathogens	Adhesion factors	Immune escaping	Toxins	Uptake of iron
Escherichia coli	P pili F1C pili Type 1 pili S pili Dr adhesins	HlyA Capsular antigens CNF1 Yersiniabactin	HlyA CNF1	Aerobactin Enterobactin Salmochelin Yersiniabactin
Klebsiella pneumoniae	Type 1 and 3 pili	Capsule	ND	Aerobactin Enterobactin
Proteus mirabilis	MR/P pili PMFs NAF TaaP adhesin AipA adhesin	Capsule ZapA	Haemolysins (HpmA and HlyA) Pta	Proteobactin Yersiniabactin-related
Pseudomonas aeruginosa	Exopolysaccharides (alginate, PEL and PSL) Extracellular DNA	Capsule Elastase ExoS Phospholipase Rhamnolipids	ND	Pyochelin Pyoverdin
Enterococcus faecalis	Ace adhesin Ebp pili Esp adhesin	Epa	ND	ND
Enterococcus faecium	Esp adhesin Ebp pili	ND	ND	ND
Staphylococcus saprophyticus	SdrI adhesin Aas adhesin Uaf adhesin	ND	Aas	ND

i. **Chaperon usher pathway (CUP) pili**

Mostly uropathogens cause UTIs by adhesion and invasion into the host cell through pili and make biofilm by interbacterial interactions. Many uropathogens (especially gram-negative bacteria) as *E. coli, Proteus spp., Klebsiella spp., Pseudomonas spp.,* and *Salmonella spp.* produce large and highly conserved adhesive fibers known as CUP pili (Thanassi et al., 1998; Vallet et al., 2001).

CUP pili are composed of pilin subunits through chaperon-usher molecular pathway (Wurpel et al., 2013). Pilin subunits have no carboxyl-terminal of β-strand at the seventh position and lead to partial folding of immunoglobulins (Chorell et al., 2014; Piatek et al., 2013). Chaperons are present in periplasmic space and donate a β-strand to each pilin subunit by forming a complex to complete their immunoglobulin folding leading to an appropriate folding through donor-strand complementation. Usher assembly proteins are present over the outer membrane that is targeted by the chaperon-subunit complex where chaperon-subunit complex is differentiated into pili on outer membrane through donor-strand conversation (Geibel et al., 2013). Donated β-strand from chaperon to pilin subunit is replaced by an amino-terminal extension with next subunit.

20.4 Uropathogens

Gram-positive and negative bacteria are major causative agents for UTIs. Uropathogenic *Escherichia coli* (UPEC) is more common in complicated and uncomplicated UTIs. Other pathogens like *Klebsiella spp., Proteus spp., Pseudomonas spp., Staphylococcus aureus, Staphylococcus saprophyticus, Enterococcus spp.* and *Candida spp.* are involved in uncomplicated UTIs (Ronald, 2002; Nielubowicz & Mobley, 2010; Kline et al., 2011; Foxman, 2013). Order of prevalence for pathogens causing complicated UTIs is followed by UPEC as *P. aeruginosa K. pneumoniae, P. mirabilis, Enterococcus spp., Candida spp.*, and *S. aureus* (Fisher et al., 2011; Jacobsen et al., 2008; Levison & Kaye, 2013).

20.4.1 Gram-Negative Bacteria

Gram-negative bacteria are the major causative agent for both community and hospital-acquired UTIs. *E. coli* is the most common UTIs causing bacteria following by *Klebsiella spp., Proteus spp.,* and *Pseudomonas spp.* among the population.

i. ***E. coli***

E. coli is a gram-negative rod, facultative, aerobic and normal flora of the gastrointestinal tract. *E. coli* genomes have 38 unique CUP pilus operon and only UPEC has a genome that code 12 different types of CUP pilus (Wurpel et al., 2013). Nevertheless, the prevalence of CUP operons among different UPEC are not similar, few are present universally and others in few strains. The majority of CUP pili are encoded by UPEC that are involved in adhesion, few of them induce tropism due to recognition of receptor with the specificity of stereochemical in the upper (kidney epithelium) and lower (bladder epithelium) urinary tract (Wright & Hultgren, 2006) (Tables 20.1).

There are two types of pili (pyelonephritis pili (P pili) and type 1 pili) that are considered for the characterization of CUP pili. Type 1 pili are important for UPEC to colonize, invasion and retain in mouse bladder (Hadjifrangiskou et al., 2012). Type 1 pili with FimH7 recognize the $\alpha_1\beta_3$ and uroplakin integrins (stereochemical specificity) that initiate colonization and invasion within umbrella cells (Guiton et al., 2012; Hannan et al., 2012).

Type 1 pili with umbrella cell surface induce a signal for activation of RHO GTPases that lead to actin rearrangement and engulfing of UPEC. The Zippering mechanism is used for engulfing of UPEC by umbrella cells as present in the plasma membrane for bacteraemia (Martinez & Hultgren, 2002). The entrance of UPEC into umbrella cells lead to overcome the host immune response and become resistant against antibiotics used as a choice of drug for treatment. Nevertheless, an innate immune response plays a vital role to protect the uroepithelium from an invasion of UPEC through a removal mechanism that depends on Toll-like receptor 4 (TLR4) present over the surface of the uroepithelium. Lipopolysaccharide (LPS) induces a signal for the activation of TLR4 that further sends a signal to adenylyl cyclase 3 (AC3) for the production of cyclic AMP. CAMP initiates the exocytosis process to eliminate the vesicular UPEC to the plasma membrane (umbrella cells) (Song et al., 2009).

Contrarily, eradication of UPEC does not occur due to any reason or UPEC overcomes the eradication pathway, then bacteria can grow and multiply rapidly to form a biofilm (intracellular bacterial communities (IBC)) (Anderson et al., 2003; Hannan et al., 2010). After the maturation of biofilm communities, bacterial colonies spread into other body parts where they invade into tissue or cells and make another biofilm by IBC (Kostakioti et al., 2013). Biofilm formation is a very common pathway for clinically important UPEC and even has been identified in different mouse studies. Biofilm formation has been observed in exfoliated cells of uroepithelial taken from a patient's urine sample suffering from acute UTIs but not in the urine of healthy persons (Robino et al., 2014; Rosen et al., 2007). Invasion and biofilm formation facilitate the UPEC to grow and multiply under the different condition as blockage of the urinary tract, urination, inflammation, eradication by activation of TLR4 and removal of affected umbrella cells (Hannan et al., 2012; Schwartz et al., 2011).

UPEC can make a quiescent-intracellular reservoir (QIR) within a transitional cell (membrane-bound sections) trapped by actin. QIRs are composed of 4–10 bacterial isolates that are non-replicating and remain viable over months. These bacteria may cause recurrent UTIs after re-activation (Hannan et al., 2012). Different studies have revealed that during the turnover of uroepithelial cells (differentiation of immature cells into the umbrella cells), rearrangement of F-actin and other related signals may force the UPEC to restore from QIRs (release bacteria into the bladder) (Blango et al., 2014).

In P-pili adhesion, PapG directly binds with globosides present in kidney epithelium instead of adhesion FimH in type-1 pili (Wright & Hultgren, 2006). Additionally, PapG with TLR4 reduces the production of the antibodies by an immune response by decreasing the expression of polymeric immunoglobulin receptor (PIGR). In the end, the transportation of immunoglobulin A reduces from lamina propria into the renal epithelium (Rice et al., 2005). UPEC becomes protective due to lack of immunoglobulin A transportation in the renal epithelium and leads to ascension of bacteria into other parts of the urinary tract resulting UTIs (Ashkar et al., 2008; Rice et al., 2005).

The innate immune response against colonization and invasion of UPEC not only identifies the current infection but is also important for recognizing the host susceptibility against the following infection. Susceptibility for recurrent UTIs may increase due to a lack of effective host immune response against UPEC. Uncontrolled Lymphocyte-dependent immune response against acute infection may cause acute injury in uroepithelium cells that lead to the following infections (Hannan et al., 2010).

ii. ***Klebsiella spp.***

Klebsiella spp. is also used the type-1 pili for adhesion and biofilm formation as in UPEC (Gerlach et al., 1989). Remarkably, *K. pneumoniae* and UPEC have highly conserved and homologous adhesin FimH with dissimilar binding characteristics (Stahlhut et al., 2009). Both *Klebsiella spp.* and UPEC have FimH induced formation of biofilm that is repressed by heptyl mannose and methyl mannose, respectively (Table 20.1). Furthermore, adhesion of UPEC FimH with the bladder is stronger as compared to *Klebsiella spp.* resulting in higher UPEC titer and biofilm formation as compared to *Klebsiella spp.* in the bladder of the mouse.

Despite a weaker adhesion of FimH in the urinary tract by *Klebsiella spp.*, it has an important virulence factor during biofilm formation and colonization in both CAUTIs and UTIs (Rosen et al., 2008a; b). Additionally, *Klebsiella pneumoniae* genome encodes type-3 pili instead of type-1 and P-pili. Type-3 pili in *Klebsiella pneumoniae* are important for adhesion, colonization and IBCs formation in both UTI and CAUTIs (Guiton et al., 2012; Murphy et al., 2013; Struve et al., 2008).

iii. ***Proteus spp.***

After attachment, mannose-resistant proteus-like pili (MRP) as similar to CUP pili are produced by *Proteus mirabilis* that facilitate the colonization and formation of biofilm in kidney and bladder and is critical for biofilm formation in CAUTIs (Armbruster & Mobley, 2012; Jacobsen & Shirtliff, 2011). *P. mirabilis*-like fimbriae (PMFs) is important for colonization in kidney and bladder and non-agglutinating fimbriae (NAF) are for adhesion with uroepithelial cell (Armbruster & Mobley, 2012).

Nevertheless, the role of NAF, PMFs and their receptors are not characterized in vivo. Despite CUP pili, two other autotransporters as AipA (adhesion and invasion mediated autotransporter) and TaaP (trimeric auto-agglutinin autotransporter) are encoded by *Proteus* for kidney and bladder infection, respectively (Armbruster & Mobley, 2012).

AipA can bind with kidney and bladder cell line in humans (in vitro) but is required for infection only in the kidney (not for bladder) and Taap is for infection in the bladder of mice by *Proteus mirabilis*. Interestingly, both AipA and TaaP can bind with extracellular matrix proteins as collagen I and laminin, respectively in tropisms (Table 20.1).

20.4.2 *Gram-Positive Bacteria*

Gram-positive bacteria are not common in UTIs patients as compared to gram-negative bacteria. *Enterococcus spp.* and group B streptococci (GBS) are more common to cause UTIs among gram-positive bacteria. *Staphylococcus aureus* is common in those patients having a secondary infection as bacteraemia.

i. ***Staphylococcus***

Staphylococcus aureus (S. aureus) is not a common causative agent among the local population in UTIs (Barrett et al., 1999; Demuth et al., 1979). However, the presence of *S. aureus* in urine indicates secondary infection (bacteraemia) usually in endocarditis (Lee et al., 1978) and in rare cases, *S. aureus* causes infection through ascending route. The use of a catheter for urination may increase the risk of UTIs caused by *S. aureus* (Von Baum et al., 2002). Mostly, UTIs caused by *S. aureus* are asymptomatic. Identification of *S. aureus* has no significance among prolong use of the catheter with bacteriuria patients. It is difficult to differentiate the UTIs and asymptomatic bacteriuria among the aged population (Breitenbucher, 1984; Nicolle et al., 1988; Simor, 2001). So, UTIs causing *S. aureus* can cause bacteraemia (Muder et al., 1992; Mylotte et al., 2002; Al Mohajer, 2013).

ii. ***Streptococcus***

UTIs caused by Group B *Streptococci* (GBS) especially *Streptococcus agalactiae*are more common in children, pregnant women's and aged peoples with severe consequences (High et al., 2005). GBS are causative agents for different types of UTIs as cystitis, urethritis and pyelonephritis. These infections are more common in patients with a compromised immune system, children and pregnant women. Bacteriuria caused by GBS is more critical in perinatal infection which transmits into the infant vertically (Furfaro et al., 2018). Pyelonephritis is the most adverse type of UTIs caused by GBS and leads to drug resistance patterns (Ulett et al., 2009). The mechanism by which GBS enters the kidney through the urethra, bladder and ureters and causes of pyelonephritis are mostly unknown.

- ***Enterococcus spp.***

Enterococci also encodes for different adhesion factor as Enterococcal surface protein (Esp), collagen adhesin Ace, Enterococcal polysaccharide antigen (Epa) and endocarditis and biofilm-associated pili (Ebp). Ebp is essential for persistency in infection caused by CAUTIs (Arias & Murray, 2012; Guiton et al., 2010; Nielsen et al., 2012). Different studies have revealed that mechanical trauma during catheterization leads to immunological and histological damage of the urinary bladder resulting in strong inflammation, mucosal lesions and oedema of kidney and epithelial cells of the urinary tract (Glahn et al., 1988; Goble et al., 1989). Significantly, a model organism study (mouse) on CAUTIs has revealed that catheter mediated inflammation, damage of bladder epithelial cells, oedema in bladder and change in immune response become worst with increasing the time of catheterization (Guiton et al., 2013). Catheter facilitates the attachment and IBCs formation of *E. faecalis* in the bladder that cause persistence and distribution of *E. faecalis*into the kidney.

However, *E. faecalis* can not bind with catheter in vitro and grow in urine (Flores-Mireles et al., 2014). It was found that the catheter encourages the release and accumulation of fibrinogen in the bladder for inflammatory response and then binds with the catheter (Flores-Mireles et al., 2014). After deposition of fibrinogen over the surface of the catheter during CAUTIs, adhesion of *E. faecalis* by Ebp pili (EbpA) having N-terminal binding domain for fibrinogen, lead to colonization and IBCs formation in the bladder (Nielsen et al., 2013) (Fig. 20.4). Additionally, fibrinogen can be used by *E. faecalis* resulting in their growth and increased biofilm formation over the surface of the catheter in the bladder (Flores-Mireles et al., 2014). It has been verified by in vitro study that *E. faecalis* binds with fibrinogen deposited catheter and grow in urine having fibrinogen (Flores-Mireles et al., 2014) (Fig. 20.4).

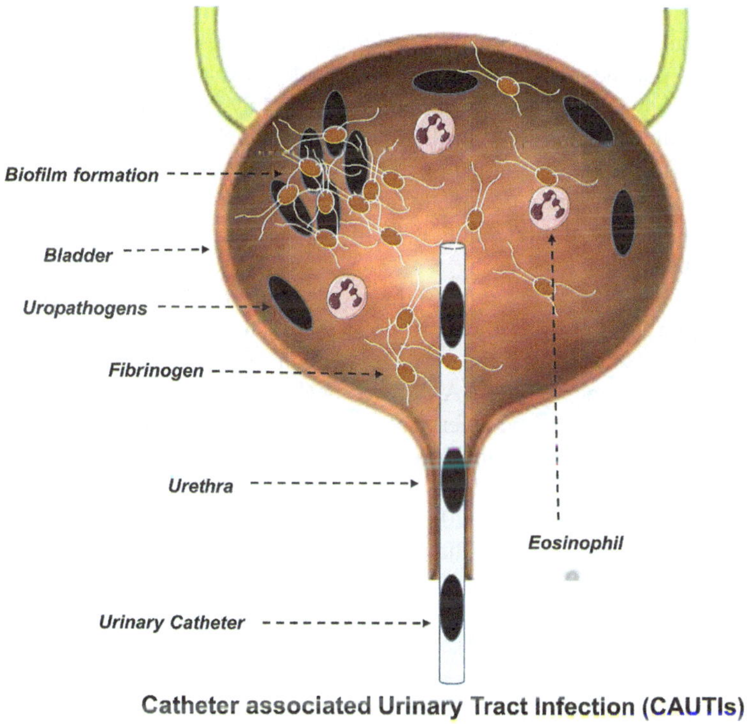

Fig. 20.4 Catheter-associated UTIs caused by uropathogens

20.5 Pathogenicity Causing Element

Nutrients for bacterial growth are not enough in the bladder environment. To obtain nutrients, uropathogens release different types of toxins and proteases that cause damage to the epithelial cells in the bladder and release nutrients that use by uropathogens for survival and colonization (Table 20.1). It also provides support to uropathogens for invasion and spreading in the urinary tract.

20.5.1 Toxins and Proteases Enzymes

α-haemolysin (HlyA) are secreted by UPEC in high concentration that oligomerizes and invades into microdomains (cholesterol-rich) present in the cell membrane through Ca^+-dependent pathway (Table 20.1). Finally, pore formation occurs in umbrella cells by UPEC and facilitates them for lysis by getting iron and other nutrients. α-haemolysin also causes exfoliation, colonization in the deeper uroepithelium layer and facilitates the dissemination of bacteria into other parts of the urinary tract. Moreover, biofilm formation uropathogens produce a high concentration of HlyA that is significant to cause infection by them (Hannan et al., 2008, 2010; Nagamatsu et al., 2015).

Cytotoxic necrotizing factor 1 (CNF1) is also secreted by UPEC. CNF1 shows their effects on actin rearrangement present in the cell membrane by using three RHO GTPases like RHOA, RAC1 and cell division control (CDC42). CNF1 binds with basal-cell adhesion molecule (BCAM) receptor (Piteau et al., 2014), present over the surface of umbrella cells, and enters the host cell. It induces the activation of RHO GTPases by deamination of glutamine that directly affects the actin reorganizations and disrupting membrane resulting in maximum internalization of uropathogens (Doye et al., 2002; Garcia et al., 2013).

Furthermore, RAC1-GTP mediated anti-apoptotic mechanism accomplishes by AKT (as PKB), phosphoinositide 3-kinase (PI3K) and nuclear factor-κB (NF-κB) that protect the colonized UPEC and niche in uroepithelium from apoptosis (Garcia et al., 2013; Miraglia et al., 2007).

Proteus toxic agglutinin (Pta) and haemolysin (HpmA) toxins are secreted by *Proteus mirabilis* that causes inflammation in tissue and ascension of uropathogens into the kidney resulting in acute infection (pyelonephritis) (Cestari et al., 2013; Jacobsen et al., 2008). HpmA (Ca^+ based cytolysin) invades into the host cell membrane by itself and forms a pore that causes Na^+ efflux (Jacobsen et al., 2008). Besides HpmA, Pta is activated only in an alkaline environment that is accompanied by urease produced by *Proteus mirabilis* (Alamuri & Mobley, 2008). Pta forms the pore in the cell membrane lead to a decrease in the osmotic stress, outflow of cytosol and break down of actin arrangement resulting in the integrity of the cell become compromised and cause kidney and bladder damage. Pta is also responsible for pathogen-pathogen interaction through autoaggregation (Alamuri & Mobley, 2008; Armbruster & Mobley, 2012).

ND, not determined; Ebp, endocarditis- and biofilm-associated; P pili, pyelonephritis-associated pili; F1C pili, type 1-like immunological group C pili; MR/P, mannose-resistant Proteus-like; Msr, methionine sulfoxide reductase; PMF, P. mirabilis-like fimbria; HpmA, haemolysin; NAF, non-agglutinating fimbria; AipA, adhesion and invasion mediated by the Proteus autotransporter; Pta, Proteus toxic agglutinin; TaaP, trimeric autoagglutinin autotransporter of Proteus; CNF1, cytotoxic necrotizing factor 1; Epa, enterococcal polysaccharide antigen; HlyA, α-haemolysin; Esp, enterococcal surface protein; UPEC, uropathogenic *Escherichia coli.* and ExoS, exoenzyme S.

Haemolytic phospholipase C and exoenzyme S (ExoS) are secreted by *P. aeruginosa* that are involved in UTI instigation and distribution of pathogens into the kidney resulting in pyelonephritis (Table 20.1). ExoS (showing GTPase activity) decreases the RAC1 activity and works with the formation of lamelliopodium that causes damage to the cell membrane. ExoS (showing ADP—ribosyltransferase activity) affects the adhesion and morphology by targeting the protein of RHO family (RAS and RalA protein) (Rocha et al., 2003).

Protease activity of elastases cause tissue damage and release nutrients (including iron) that are used for bacterial growth and colonization. Phospholipase C (α-toxin) hydrolyses the phosphatidylcholine present in the cell membrane and organ damage occurs due to compromising the cell integrity (Berka & Vasil, 1982; Meyers et al., 1992).

Quorum sensing system activated by autoinducer is involved to regulate the expression and function of all these virulence factors (Senturk et al., 2012). Small molecules (autoinducer) are accumulated to increase cell density. After achieving the threshold, autoinducer bind with transcriptional activator and initiate the virulence factor expression (Mittal et al., 2009; Senturk et al., 2012) (Table 20.1).

20.5.2 Ureases Enzymes

P. mirabilis (Armbruster & Mobley, 2012; Li et al., 2002), *K. pneumonia* (Podschun & Ullmann, 1998), *S. saprophyticus* (Gatermann et al., 1989) and *P. aureginosa* (Visca et al., 1992) genomes are encoded for the production of urease. It is important for persistence and colonization when UTI is caused by *S. saprophyticus* and *P. mirabilis* (Gatermann et al., 1989; Li et al., 2002). Break down of urea into ammonia and CO_2 by urease rise the urine pH with calcium and magnesium ammonium phosphate crystals production in urine and on the catheter (Armbruster & Mobley, 2012). Furthermore, ammonia accumulates in uroepithelial cells resulting in tissue damage directly (Coker et al., 2000). Well-known urease (Ni^{2+}-based metalloenzyme) causing UTI is produced by *P. mirabilis* that is important for bladder and kidney colonization and facilitates stone formation (Griffith et al., 1976). Urea induces the production of urease by *P. mirabilis* during their growth in urine (Kosikowska & Berlicki, 2011) (Table 20.1).

The urease produced by *P. mirabilis* is highly active and capable of urea hydrolysis and *P. mirabilis* more rapidly produced ureas in the presence of urea as compared to others microbial strains such as *Proteus vulgaris, Providencia spp.* and *Morganella spp.* (Jones & Mobley, 1987). Crystalline biofilms formation occurs due to obstruction of polysaccharide (formed by neighbour bacterial cell) by crystals produced by highly active *Proteus* enzymes (Kosikowska & Berlicki, 2011; Stickler, 2014). Antibiotics and host immune response do not show their effect on *P. mirabilis* due to crystalline biofilm production (Coker et al., 2000). It also blocks the ureter and causes ascension of bacteria into the kidney resulting in septicaemia, pyelonephritis and shock (Armbruster & Mobley, 2012).

20.5.3 Uptake of Iron

Iron in urine is not present in enough amount as required for bacterial growth and survival. Siderophore system is used by the uropathogens to get the iron from urine for their survival. Siderophore binds with iron and receptor to take up the iron from urine (Caza & Kronstad, 2013).

Yersiniabactin and aerobactin siderophores are produced by UPEC (Table 20.1). Aerobactin siderophores have a high binding affinity with iron and maintain themselves at low pH as compared to yersiniabactin (Valdebenito et al., 2006; Watts et al., 2012). Yersiniabactin protects the uropathogens from killing due to copper stress and important for IBCs formation within the urine (Chaturvedi et al., 2012).

Aerobactin and enterobactin are produced by *K. pneumonia* (Podschun & Ullmann, 1998) and *P. mirabilis* is produced proteobactin and yersiniabactin (Himpsl et al., 2010). *P. aeruginosa* produces pyoverdin and pyochelin (Visca et al., 1992). Siderphore systems are very effective for vaccine development by disturbing their function in UTIs.

20.6 Antibiotics and Their Mechanism

UTIs have a large impact and threat to economics and public health. It also affects the life value of the population having UTIs (Kostakioti et al., 2012). Recently, ciprofloxacin, co-trimoxazole and ampicillin are used as a choice of drug to treat UTIs (Foxman, 2010). However, drug resistance against uropathogens and relapse UTIs have been increased remarkably. We should establish an alternative treatment against drug-resistant uropathogens.

Many successful approaches can be developed by targeting the virulence factor of uropathogens by which they cause UTIs. Anti-virulence drug should support us theoretically to combat UTIs by neutralizing the ability of uropathogens to cause infection without affecting the normal flora of human beings because it only affects the pathway by which pathogenesis occur. Below, we will discuss the genetic and molecular mechanism of uropathogens by which they become resistant against antibiotics resulting in treatment failure and alternative drug discovery by using virulence factors. We also consider the mechanism of drug resistance in uropathogens spread among other strains and we should modify the current therapeutics measure accordingly.

20.6.1 Multidrug Resistance Uropathogens

UTIs are going to be a big challenge for scientists and physicians due to the increase of antibiotics resistance patterns among uropathogens. Acquired plasmids-encoding extended-spectrum β-lactamases (ESBLs) are present in *Klebsiella spp.* and *E. coli* that are important members of *Enterobacteriaceae.* This plasmid facilitates the pathogens to disseminate the resistance among other pathogens against third-generation cephalosporin along with other antibiotics (Bradford, 2001; Pendleton et al., 2013; Gupta & Bhadelia, 2014).

Other family members of *Enterobacteriaceae* are also produced β-lactamases (AmpC enzymes) by which they become resistant against 3rd generation cephalosporins and cephamycin along with β-lactamase inhibitors (Paterson, 2006; Pendleton et al., 2013; Gupta & Bhadelia, 2014). *Klebsiella pneumoniae* are also resistant to carbapenem due to the production of AmpC enzymes.

Multi-drug resistance patterns of uropathogens have been increased remarkably. It became a big challenge for scientists and physicians in the modern era to treat the infection caused by MDR pathogens. Family members of *Enterobacteriaceae* like *E. coli, Klebsiella spp., P. aeruginosa,* and *Proteus spp.* produce extended-spectrum β-lactamases (ESBLs) like oxacillinases (OXAs), AmpC-type β-lactamases, carbapenemases and cefotaximases (CTX-Ms) to cause multi-drug resistance in these pathogens.

i. **β-lactamases**

β-lactamases are enzymes produced by gram-negative and positive bacteria that break the β-lactam ring present in β-lactam antibiotics resulting in β-lactam therapy become ineffective. Different types of β-lactamases are produced by uropathogens as penicillinases, carbapenemases, ESBLS, AmpC and inhibitor resistance β-lactamases (IR).

- **Penicillinases**

It was the first β-lactamase discovered by Chain and Abraham from *E. coli* strains in 1940, even before introducing penicillin into the market for a therapeutic purpose (Fig. 20.5). Penicillinase has a great potential to target β-lactam ring present in penicillin specifically (Fig. 20.6).

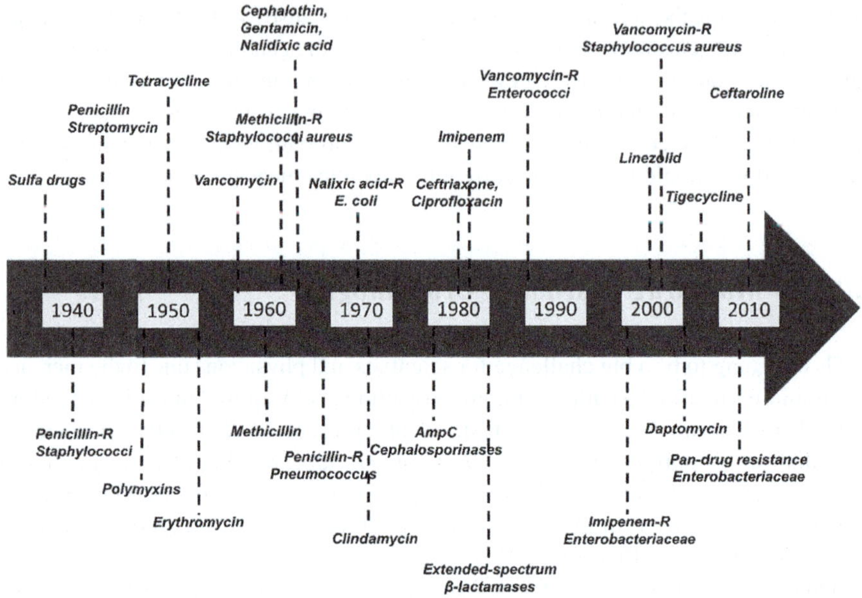

Fig. 20.5 Epidemiology of drug resistance among pathogens

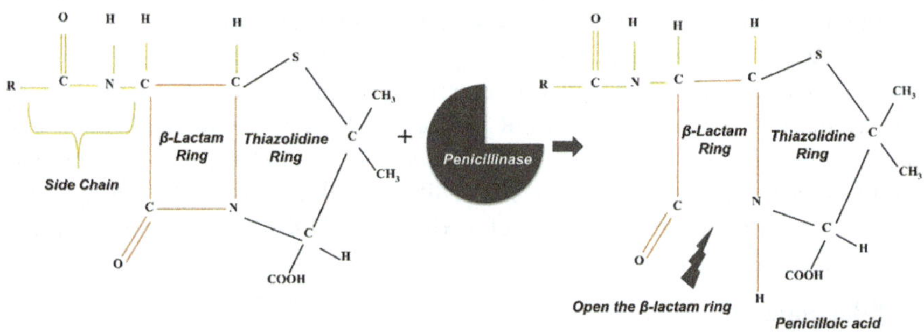

Fig. 20.6 Mechanism of penicillinase to inactivation of antibacterial drugs

- **Extended spectrum β-lactamases (ESBLs)**

ESBLs were discovered initially in *Escherichia coli* and *Klebsiella pneumoniae* but now, they have become more common among *Enterobacteriaceae* due to the common use of cephalosporin in the hospital environment and transferable ability of ESBLs coding genes together. It creates a perfect environment for bacteria to gain antibiotics resistance through gene transformation (Paterson, 2006). ESBLs are encoded by bacterial chromosomes or plasmid that have great potential to break the β-lactam ring present in cephalosporin or penicillin group antibiotics make them ineffective against uropathogens. Remarkably, the plasmid is also encoded to different antibiotic resistance genes besides ESBLs that provide antimicrobial resistance against sulphonamides, aminoglycosides and quinolones. Bacteria carrying these genes become multidrug-resistant (Gupta & Bhadelia, 2014; Pendleton et al., 2013).

- **Cefotaximases (CTX-Ms)**

CTX-Ms are encoded by a distinct plasmid which is different phylogenetically from other β-lactamases encoded plasmid and have hydrolytic activity against penicillin, cephalosporins and monobactams. Remarkably, they are also highly active enzymes against cefotaxime. CTX-Ms are more common enzymes among the community and hospital-acquired uropathogens, encoded by plasmid along with other antibiotic resistance genes (Bradford, 2001; Gupta & Bhadelia, 2014). CTX-Ms target the carbonyl carbon ring present in β-lactam antibiotics through serine residues which are conserved in β-lactamases resulting in antibiotics become ineffective due to the opening of the ring (Walther-Rasmussen & Høiby, 2004).

- **Oxacillinases (OXAs)**

OXAs are plasmid-encoded β-lactamases and are highly active enzyme against oxacillin, ampicillin and cloxacillin by attacking and opening the β-lactam ring lead to inactivation. It is also reported that OXAs are potentially active against clavulanate (β-lactam-inhibitor drug). Recently, expression of OXAs has been identified only in *Pseudomonas aeruginosa*, not seen in other pathogens (Bradford, 2001).

- **AmpC enzymes**

AmpC β-lactamases are encoded by bacterial chromosomes, not by the plasmid. AmpC is also hydrolytically active against cephalosporins, penicillins and cephamycins as well as clavulanate (β-lactamase inhibitor). It was identified early in 1980 (Fig. 20.5). Expression of AmpC is usually occurred due to exposure of β-lactam antibiotics (cephalosporins and cephamycins) (Paterson, 2006).

- **Carbapenemases**

Carbapenemases are also encoded by bacterial plasmid and are resistant to carbapenem particularly as well as broad-spectrum cephalosporins and penicillins. Carbapenemases were discovered originally in *Klebsiella pneumoniae* and spread widely among *Enterobacteriaceae* due to exchangeable plasmid. New Delhi metallo-β-lactamase (NDM-1) and *K. pneumoniae* (serine) carbapenemase (KPC) have great importance clinically, spread widely via plasmid transformation into other. Carbapenemases create the carbapenem-resistant *Enterobacteriaceae* (CRE) (Paterson, 2006). Uropathogens producing carbapenemases are highly protective by broad and extended-spectrum antibiotics, especially carbapenem (Chen et al., 2013).

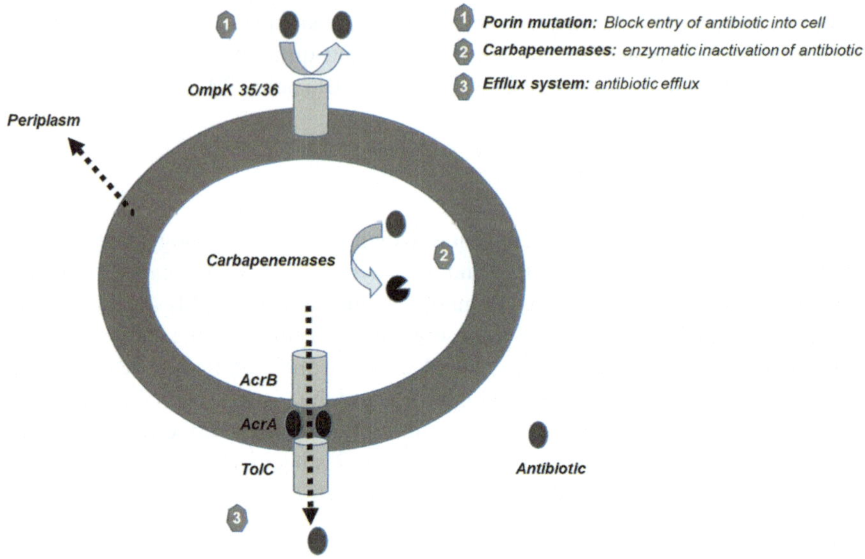

Fig. 20.7 Mechanism of carbapenem resistance in *Enterobacteriacae*

Three channels in *Enterobacteriacae* are responsible for resistance against carbapenem: breakdown of antibiotic by carbapenemases, efflux pump and mutation in porin lead to reduce the permeability of outer membrane (Fig. 20.7).

ii. **Vancomycin Resistant *Enterococci***

Multi-drug resistance (MDR) patterns among *Enterococci* are also very common as they are naturally resistant against clindamycin, penicillin, cephalosporins and trimethoprim (Gupta & Bhadelia, 2014; Pendleton et al., 2013). In current studies, it is revealed that *Enterococci spp.* becomes resistant against glycopeptides especially vancomycin which is the last choice of drug to combat with MDR uropathogens. Unambiguously, *Enterococci* developed the resistance against glycopeptides (vancomycin and teicoplanin) by using the *Van* genes that are responsible to produce penicillin-binding proteins (PBPs) like VanA, VanB, VanD, VanG and VanL (Gupta & Bhadelia, 2014; Pendleton et al., 2013). VanA (PBPs) is produced by *Enterococci* mostly by which vancomycin binds. Due to a change in cell wall configuration (d-alanine–d-lactose instead of d-alanine–d-alanine), VanA becomes unable to bind with vancomycin (Courvalin, 2006). The hike of MDR uropathogens has encouraged scientists to develop an alternative way to combat UTIs.

iii. **Methicillin Resistant *Staphylococcus aureus***

S. aureus is a gram-positive, cocci bacterium which is normally present (colonize) in our nose, skin and armpit (normal flora). *S. aureus* is also produced plasmid or chromosomal mediate β-lactamases that confer resistance to methicillin called

MRSA. First MRSA was reported in 1961. Normally, staphylococcal cassette chromosome (SCC) encodes to a penicillin-binding protein 2 (PBP2) receptor present in cell membrane of *S. aureus*, having more binding affinity and susceptibility to penicillin. Alteration of PBP to PBP2a is occurred due to mutation in the genetic, mobile part of SCC leading to *mecA* gene, having low binding affinity and resistance to penicillin or methicillin. There are five important types of SCCmec genes: *mecA, A1, A2, B* and *C* which encodes to PBP2 receptors. Prevalence of MRSA is gradually increasing due to the transformation of SCCmec gene among other species through conjugation or transduction (Foster, 2004).

20.6.2 Combination Therapies

β-lactamase antibiotics in combination with β-lactamase inhibitors are under consideration that is susceptible to β-lactamase producing organisms and *K. pneumoniae* carbapenemases (KPCs) (Mushtaq et al., 2013).This combination of antibiotics is active against carbapenem-resistant *Enterobacteriaceae* (CRE). Additionally, it has been reported that ceftazidime (third-generation cephalosporin β-lactam antibiotic) is effective against both gram-positive and gram-negative bacteria in complicated UTIs. It is also effective against β-lactamase and carbapenemases producing uropathogens, if we use it in combination with avibactam (β-lactamase inhibitor) (Zhanel et al., 2013). However, further study is required to check the activity of ceftazidime-avibactam against *Enterobacteriaceae* producing ESBL-, AmpC- and KPC-enzymes during infection because this combination is potentially susceptible to cephalosporin-resistant uropathogens. Although resistance against inhibitor antibiotics is not identified (Zhanel et al., 2013). Furthermore, the efficacy of combination drugs (antibiotic-inhibitor) depends upon the MDR pattern by each uropathogens because carbapenemases and ESBLs may induce resistance to these combinations (Mushtaq et al., 2013). β-lactam antibiotics (BAL29880 and BAL30072) with β-lactamase inhibitor (clavulanate) are active against CRE but not against KPCs, sulphydryl variables (SHVs) and AmpC enzymes producing *K. pneumonia* (Livermore & Mushtaq, 2013). However, it is difficult to say which antibiotic is effective or not against uropathogens to overcome the UTIs.

20.6.3 Anti-virulence Therapy by Targeting Bacterial Adhesion

Bacterial adhesion with the receptor is an important step to cause infection. Anti-virulence therapy, even vaccines, can be developed by targeting the bacterial adhesion factors like CUP pili. Generally, whole pili-based vaccines or anti-virulence therapy are not effective to provoke the immune system to produce antibodies that may protect from UTIs. Anyhow, it has been reported in different studies that adhesion-based therapy is effective to protect from UTIs by hindering the pathogen-host relationship. Model organism study (monkey and mouse) has revealed that chaperon adhesion complexes (FimC–FimH or PapD–PapG) based therapy are effective against the development of UTIs. FimC–FimH based vaccines were effective owing to provoke the immune system to produce antibodies that check the FimH role in the colonization of uropathogens in the bladder (Langermann et al., 1997). Additionally, anti-FimH based antibodies were not observed any effect on normal flora (*E. coli* in the gut). Alternations of these vaccines are being under investigation that enhance the immune response to a large extent (Karam et al., 2013; Savar et al., 2014).

For example, an acute inflammatory response has been induced by fusing the FimH with flagellin FliC through TLR4 signaling via MYD88 pathway. A clinical trial has been started in January 2014 to evaluate the effectiveness of FimC–FimH based vaccines in combination with adjuvants (monophosphoryl lipid A). Although UPEC, anti-virulence therapy or vaccines are being under development by using the adhesion factor of *E. faecalis* and *P. mirabilis* (Li et al., 2004). *P. mirabilis* pilus adhesin (MR/P or MrpH) based vaccines were administrated into the mouse before infections. After exposure with uropthaogens, bacteria count was lesser in vaccinated mouse organism as compared to unvaccinated animals, similar findings were also reported for UPEC (FimH-based vaccine) (Langermann et al., 1997; Li et al., 2004).

Furthermore, Ebp pilus adhesin (EbpA) based vaccines are under trials for CAUTIs causing by *E. faecalis.* In conclusion, adhesion-based vaccines is a novel technique to overcome the UTIs and MDR uropathogens throughout the world. Consequently, the molecular mechanism to cause UTIs is key to develop a vaccine against the host-pathogens relationship.

20.6.4 Anti-virulence Therapy by Targeting Bacterial Toxins and Proteases

α-haemolysin (HlyA) produced by UPEC are used for the development of anti-virulence therapy and vaccines to protect the UTIs. Effectiveness of HlyA based vaccines was observed in the mouse model having pyelonephritis that reduced the risk for renal distortion as compared to the control organism but did not control the UPEC colonization in the kidney (O'Hanley et al., 1991; Sivick & Mobley, 2010).

Furthermore, *P. mirabilis* haemolysin (HpmA)-based vaccines were ineffective in mice having UTIs to protect from colonization of bacteria. Anyhow, alkaline proteases (Pta)-based vaccine in affected mouse (UTIs) have given protection from upper UTIs but not reduced bacterial burden in the bladder (Alamuri et al., 2009). Although, toxins and proteases are useful for the development of vaccines against upper UTIs. Further studies and trials are required to evaluate the effectiveness of these toxins and enzymes-based vaccine development.

20.6.5 Anti-virulence Therapy by Targeting Bacterial Siderophores

Anti-virulence therapy or vaccine can be developed by targeting the iron taking receptors because uropathogens cannot survive without iron. Bacteria does not grow and colonize in the urinary tract without nutrients especially iron. Haem-acquisition systems and siderophores are activated to large extent during UTIs.

It is a landmark to develop a vaccine or anti-virulence therapy by targeting the iron uptaking receptor-like iron uptake transport aerobactin (IutA) receptor, ferric yersiniabactin uptake (FyuA) receptor, haem acquisition protein (Hma) and siderophore receptor as iron-responsive element A (IreA) (Brumbaugh et al., 2013). Model organism study revealed that Hma and FyuA-based vaccines are effective to protect mice from pyelonephritis and IreA and IutA-based vaccines are active to minimize the bacterial burdens in mice bladder (Alteri et al., 2009). These receptors have different expressions in different places (kidney or bladder) owing to tissue-specific protection. Other siderophores (FitA or ChuA) receptor-based vaccines were not observed to be effective in UTIs affected mouse. In conclusion, FyuA, IutA, Hma, and Ire-based vaccines are effective to block the iron supply to uropathogens resulting in protection from persistence and colonization (Alteri et al., 2009).

20.6.6 Small Molecules Targeting Urease

Urease has a great potential to target for anti-virulence therapy and vaccine development. Urease producing uropathogens (*S. saprophyticus* and *P. mirabilis*) is treated by different inhibitors that inhibit the activity of urease (Kosikowska & Berlicki, 2011). Different urease inhibitors have been characterized, many of them have low inhibitory and binding affinity but acetohydroxamic acid (AHA) inhibitor has some potential to treat UTIs. AHA inhibitor inhibits the activity of urease by changing the alkaline nature of urine and was accepted in 1983 by FDA (Kosikowska & Berlicki, 2011). Mostly, inhibitors have a drawback to causes toxicity. Teratogenicity along with muscular-integumentary and psychoneurological effects were observed after administration of AHA inhibitors. Different studies have revealed that the AHA derivative had the potential to inhibit the urease activity, but their mutagenic characteristics made them unfavorable (Kosikowska & Berlicki, 2011).

Another inhibitor (phosphoramidites) can block the urease activity producing by *P. mirabilis* and was active in mouse against infection but was not stable in a low pH environment (gastric juice) resulting ineffective treatment (Kosikowska & Berlicki, 2011).

In the end, the heterocyclic compound (benzimidazole) has great attention to treat-peptic ulcer and gastroesophageal diseases. Benzimidazole (inhibitor) deactivates the ATPase system permanently by inhibiting with proton pump system (Nagaya et al., 1989). Remarkably, this compound can also block the urease activity by occupying its active site (Kosikowska & Berlicki, 2011). Bactericidal characteristics of benzimidazole are not accompanied by inhibitor activity against urease producing by *H. pylori* (Sjøstrøm et al., 1997). Further studies are needed to verify and characterize the new urease inhibitor to treat the infection efficiently.

20.6.7 Small Molecules Targeting Bacterial Adhesion

A detailed study of the pilus adhesions factor has opened a new door for the development of novel anti-virulence therapy and vaccine to overcome the infection. Pilis-receptor and pilus assembly have a great consideration to target the pili for the development of two different novel synthetic compounds: pillicides and mannosides block the pilus assembly and pilus function, respectively.

UPEC adhesion, colonization and IBCs formation should be controlled by targeting the functionality and assembly of CUP pilus resulting in an infection controlled (Chorell et al., 2014; Cusumano et al., 2011; Piatek et al., 2013; Pinkner et al., 2006). Pellicides having 2-pyridone scaffold were used originally to block the UPEC type-1 pili assembly by hindering the chaperon-usher interaction. Further studies are being under consideration to see their effectiveness against other CUP pili (Niveditha et al., 2012). A recent study has revealed that 38 different CUP pili have been identified based on the usher phylogeny among genomes of 35 *Escherichia spp.* and 132 plasmids having total 458 operons.

A compound that targets the CUP pili must have broad-spectrum characteristics because 16 different CUP operons may present in a single *Escherichia spp.* For example, pilicide ec240-based molecules have a great affinity to bind with type-1, p- and S-pili to abolish the bacterial adhesion and colonization during infection (Greene et al., 2014). Transcriptomics and proteomics study of UPEC has revealed that ec240-based molecules decreased the expression of type-1 pili genes. Expression of type-1 pili is mediated by the oscillation of type 1 fimbriae promoter element (fimS) between ON and OFF phase. ec240-based molecules mediated the fimS OFF phase by increasing the S-fimbrial expression leading to regulatory proteins (SfaB and P-pilus protein) resulting fimS phase OFF orientation (Greene et al., 2014). However, pilicide ec240 molecules have great effectiveness for inversion of phase OFF in type-1 pili without affecting the chaperon-usher pathway. Further study has revealed that the production of other pili (Dr pili: UPEC CUP pili important for pyelonephritis) is also inhibited by pellicides (Piatek et al., 2013; Wright & Hultgren, 2006). Pellicides have a great potential to target the broad-spectrum CUP pili in uropathogens failing to attach and cause infection. Further research is needed to verify the efficacy of this molecule against CUP pili in uropathogens by using the organism (mouse having UTIs or CAUTIs).

Mannosides have a great affinity to bind with FimH receptor and inhibit the attachment of FimH with mannosylated receptor present over the bladder epithelium surface resulting in no infection due to blocking the pathogen-host relationship. It can be used as a treatment and prevention from UTIs by disturbing the Host-Pathogens interaction (Abgottspon et al., 2010; Totsika et al., 2013). Mouse model organism study has revealed that mannosides have different characteristics as a novel approach to treat UTI, orally bioavailable, strong and fast-acting to prevent and treat UTIs, inhibit colonization in the bladder and effective against MDR UPEC (Totsika et al., 2013).

20.6.8 Nanoparticle-Based Therapy

Novel therapy is required to overcome the antibiotic resistance in uropathogens either by increasing the effectiveness of current antibiotics or by developing the new antibiotics. Recently, it has been reported that different approaches like antibiotics modification, combinational therapy, antimicrobial peptide, photothermal agents and nanoparticles are effective to combat MDR uropathogens (Gebreyohannes et al., 2019). In the last decade, nanomedicine became an emerging and novel therapy to treat infectious diseases. Nanoparticles (NPs) can work in two ways either as bacteriostatic or bactericidal agents or involve in the delivery of drug agents to the target cell as a carrier (Fig. 20.8) (Baptista et al., 2018). Nano-based drugs are developed by using different constituents like chitosan, metals and surfactants (Taylor, 2011). Compounds having a net positive charge may bind with bacterial cell membrane

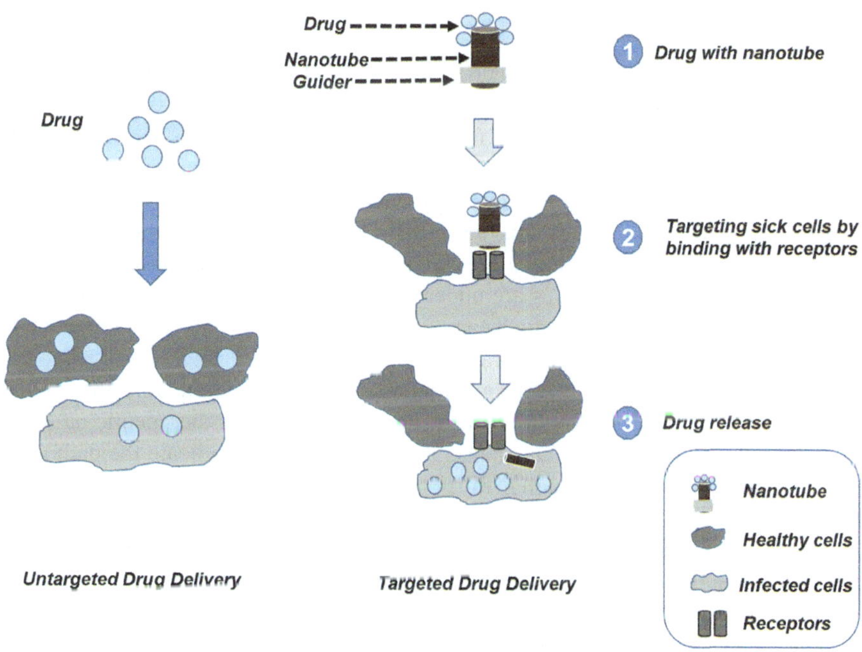

Fig. 20.8 Nanoparticle-based drug delivery to the target cells

having a negative charge, amphiphilic characteristics of nanoparticles facilitate the penetration of NPs into the cell membrane. Encapsulation of antibiotic ingredients with nanotube is a strategy to increase the elimination of bacteria and bioavailability (Chen et al., 2012).

Drug delivery to the target cell by using a nano-system has more effectiveness and fewer side effects as compared to the old technique. Nanoparticles may also defend the enzymatic attack against antibiotics and increase its availability. Nano-based drugs are helpful to improve the delivery of drugs to kill the bacteria, reduces the possibility of side effects and drug resistance. Appropriately designed nanoparticles may give a response against internal changes including pH, specific enzyme concentration or any other chemical variation which demonstrates the pathological condition (Lee et al., 2018). It is difficult to treat the infection-causing by biofilm-forming uropathogens. Biofilm is composed of extracellular polymer substances (EPS) secreted by the pathogens which are rigid and protect the bacteria from antibiotics (Mohammed et al., 2018). Some nanoparticles have the potential to penetrate the biofilm and deteriorate it resulting in bacterial elimination. Bacterial eradication is occurred due to the large acquiring of nanoparticles. More than fifty nanoparticles have been already approved for the treatment of infectious diseases caused by bacteria (Bobo et al.).

20.7 Conclusion

UTI is the second most common infectious disease leading to morbidity and mortality throughout the world. Every country spends a significant amount of money every year to combat UTIs and CAUTIs. It is a big challenge for scientists and clinicians to develop effective therapies and to combat with multi-drug resistance uropathogens. The spread of genetic factors associated with multi-drug resistant among the drug-sensitive uropathogens makes this more difficult to treat. Furthermore, recurrent UTIs demonstrate that antibiotics are not effective therapeutic options to combat all kinds of UTIs. Many studies are undergoing to optimize the treatment by using the translational mechanisms which can confer to different virulence factors without affecting the normal body microflora.

Different virulence factors of uropathogens including bacterial adhesion, toxins, ureases, proteases, and siderophores have a great potential for therapeutic targeting by anti-virulence or nanoparticles-based therapies either to eradicate the pathogens or to minimize the disease burden in the population. Molecules like pillicide, mannoside or specific nanoparticles can also be used to target the bacterial adhesion receptor or siderophores, or microbial virulence factors, which can result in impaired colonization in the urinary tract and protection from UTIs or CAUTIs.

Further studies are still required to investigate the impact of vaccination, anti-virulence and nanoparticles-based therapy on the specific pathogen targeting and possible (side-) effects on the healthy gut microbiome. The FimH-based vaccine is under phase 1 clinical trials against UTIs and the only available preventive option against UTIs, which do not affect the normal body microflora. The healthy gut microbiota is affected by most of the antimicrobial drugs or antibiotics. The chemical antimicrobial inhibitors and (inorganic-) nanoparticles showed great therapeutic potential against multi-drug resistant bacteria involved in UTIs and CAUTIs, although the literature also highlights a moderate to sever impact on healthy gut microbiome by these therapies.

In conclusion, there is an urgent need to develop novel antimicrobial therapies against multi-drug resistant bacteria involved in UTIs and CAUTIs including vaccinations, drugs targeting bacterial virulence or drug resistance pathways, nanoparticles-based antimicrobial actions, which may not alter the healthy gut microbiome in the human body. Future studies should be focused to develop novel therapeutic drugs that can also combat the different genetic factors or mechanisms of multi-drug resistance in uropathogen.

References

Abgottspon, D., Rölli, G., & Hosch, L. (2010). Development of an aggregation assay to screen FimH antagonists. *Journal of Microbiological Methods, 82*, 249–255.

Alamuri, P., Eaton, K. A., & Himpsl, S. D. (2009). Vaccination with proteus toxic agglutinin, a hemolysin-independent cytotoxin in vivo, protects against Proteus mirabilis urinary tract infection. *Infection and Immunity, 77*, 632–641.

Alamuri, P., & Mobley, H. L. (2008). A novel autotransporter of uropathogenic Proteus mirabilis is both a cytotoxin and an agglutinin. *Molecular Microbiology, 68*, 997–1017.

Alteri, C. J., Hagan, E. C., & Sivick, K. E. (2009). Mucosal immunization with iron receptor antigens protects against urinary tract infection. *PLoS Pathogens, 5*, e1000586.

Anderson, G. G., Palermo, J. J., & Schilling, J. D. (2003). Intracellular bacterial biofilm like pods in urinary tract infections. *Science, 301*, 105–107.

Arias, C. A., & Murray, B. E. (2012). The rise of the Enterococcus: beyond vancomycin resistance. *Nature Reviews Microbiology, 10*, 266–278.

Armbruster, C. E., & Mobley, H. L. (2012). Merging mythology and morphology: the multifaceted lifestyle of Proteus mirabilis. *Nature Reviews Microbiology, 10*, 743–754.

Ashkar, A. A., Mossman, K. L., & Coombes, B. K. (2008). FimH adhesin of type 1 fimbriae is a potent inducer of innate antimicrobial responses which requires TLR4 and type 1 interferon signalling. *PLoS Pathogens, 4*, e1000233.

Baptista, P. V., McCusker, M. P., & Carvalho, A. (2018). Nano-strategies to fight multidrug resistant bacteria—"A Battle of the Titans." *Frontiers in Microbiology, 9*, 1441.

Barrett, S., Savage, M., & Rebec, M. (1999). Antibiotic sensitivity of bacteria associated with community-acquired urinary tract infection in Britain. *Journal of Antimicrobial Chemotherapy, 44*, 359–365.

Bedük, Y., & Manalp, M. (1986). Detection of candidiasis in non-gonococcal urethritis resistant to therapy. *Mikrobiyoloji Bulteni, 20*, 190–195.

Berka, R. M., & Vasil, M. L. (1982). Phospholipase C (heat-labile hemolysin) of Pseudomonas aeruginosa: purification and preliminary characterization. *Journal of Bacteriology, 152*, 239–245.

Blango, M. G., Ott, E. M., & Erman, A. (2014). Forced resurgence and targeting of intracellular uropathogenic Escherichia coli reservoirs. *PLoS ONE, 9*, e93327.

Bobo, D., Robinson, K., & Islam, J. Nanoparticle-based medicines: a review of FDA-approved materials and clinical trials to.

Bradford, P. A. . (2001). Extended-spectrum β-lactamases in the 21st century: characterization, epidemiology, and detection of this important resistance threat. *Clinical Microbiology Reviews, 14*, 933–951.

Bradshaw, C., Denham, I., & Fairley, C. K. (2002). Characteristics of adenovirus associated urethritis. *Sexually Transmitted Infections, 78*, 445–447.

Breitenbucher, R. B. . (1984). Bacterial changes in the urine samples of patients with long-term indwelling catheters. *Archives of Internal Medicine, 144*, 1585–1588.

Brumbaugh, A. R., Smith, S. N., & Mobley, H. L. (2013). Immunization with the yersiniabactin receptor, FyuA, protects against pyelonephritis in a murine model of urinary tract infection. *Infection and Immunity, 81*, 3309–3316.

Burstein, G. R., & Zenilman, J. M. (1999). Nongonococcal urethritis—A new paradigm. *Clinical Infectious Diseases, 28*, S66–S73.

Caini, S., Gandini, S., & Dudas, M. (2014). Sexually transmitted infections and prostate cancer risk: A systematic review and meta-analysis. *Cancer Epidemiology, 38*, 329–338.

Caza, M., & Kronstad, J. (2013). Shared and distinct mechanisms of iron acquisition by bacterial and fungal pathogens of humans. *Frontiers in Cellular and Infection Microbiology, 3*, 80.

Cestari, S. E., Ludovico, M. S., & Martins, F. H. (2013). Molecular detection of HpmA and HlyA hemolysin of uropathogenic Proteus mirabilis. *Current Microbiology, 67*, 703–707.

Chaturvedi, K. S., Hung, C. S., & Crowley, J. R. (2012). The siderophore yersiniabactin binds copper to protect pathogens during infection. *Nature Chemical Biology, 8*, 731–736.

Chen, C., Hu, J., & Zhang, S. (2012). Molecular mechanisms of antibacterial and antitumor actions of designed surfactant-like peptides. *Biomaterials, 33*, 592–603.

Chen, Y.-H., Ko, W.-C., & Hsueh, P.-R. (2013). Emerging resistance problems and future perspectives in pharmacotherapy for complicated urinary tract infections. *Expert Opinion on Pharmacotherapy, 14*, 587–596.

Chenoweth, C. E., Gould, C. V., & Saint, S. (2014). Diagnosis, management, and prevention of catheter-associated urinary tract infections. *Infectious Disease Clinics, 28*, 105–119.

Chorell, E., Uvell, H., & Pinkner, J. S. (2014). Syntheses and biological evaluation of 2-amino-3-acyl-tetrahydrobenzothiophene derivatives; antibacterial agents with antivirulence activity. *Organic & Biomolecular Chemistry, 12*, 1942–1956.

Chowdhury, M., & Pareek, S. (1984). Urethritis caused by group B streptococci: a case report. *Sexually Transmitted Infections, 60*, 56–57.

Coker, C., Poore, C. A., & Li, X. (2000). Pathogenesis of Proteus mirabilisurinary tract infection. *Microbes and Infection, 2*, 1497–1505.

Colgan, R., Williams, M., & Johnson, J. R. (2011). Diagnosis and treatment of acute pyelonephritis in women. *American Family Physician, 84*, 519–526.

Control, C. F. D. & Prevention. (2002). Diseases characterized by urethritis and cervicitis. *Morb. Mortal. Wkly. Rep, 51*, 36–42.

Courvalin, P. . (2006). Vancomycin resistance in gram-positive cocci. *Clinical Infectious Diseases, 42*, S25–S34.

Cusumano, C. K., Pinkner, J. S., & Han, Z. (2011). Treatment and prevention of urinary tract infection with orally active FimH inhibitors. *Science Translational Medicine, 3*, 109ra115–109ra115.

Czaja, C. A., Scholes, D., & Hooton, T. M. (2007). Population-based epidemiologic analysis of acute pyelonephritis. *Clinical Infectious Diseases, 45*, 273–280.

Davis, N. F., Callanan, A., & Mcguire, B. B. (2011). Evaluation of viability and proliferative activity of human urothelial cells cultured onto xenogenic tissue-engineered extracellular matrices. *Urology, 77*, 1007. e1001–1007. e1007.

Demuth, P. J., Gerding, D. N., & Crossley, K. (1979). Staphylococcus aureus bacteriuria. *Archives of Internal Medicine, 139*, 78–80.

Doye, A., Mettouchi, A., & Bossis, G. (2002). CNF1 exploits the ubiquitin-proteasome machinery to restrict Rho GTPase activation for bacterial host cell invasion. *Cell, 111*, 553–564.

Eto, D. S., Jones, T. A., & Sundsbak, J. L. (2007). Integrin-mediated host cell invasion by type 1–piliated uropathogenic Escherichia coli. *PLoS Pathogens, 3*, e100.

Fisher, J. F., Kavanagh, K., & Sobel, J. D. (2011). Candida urinary tract infection: pathogenesis. *Clinical Infectious Diseases, 52*, S437–S451.

Flores-Mireles, A. L., Pinkner, J. S., & Caparon, M. G. (2014). EbpA vaccine antibodies block binding of Enterococcus faecalis to fibrinogen to prevent catheter-associated bladder infection in mice. *Science Translational Medicine, 6*, 254ra127–254ra127.

Foster, T. J. . (2004). The Staphylococcus aureus "superbug." *The Journal of Clinical Investigation, 114*, 1693–1696.

Foxman, B. . (2010). The epidemiology of urinary tract infection. *Nature Reviews Urology, 7*, 653–660.

Foxman, B. . (2013). Urinary tract infection syndromes: occurrence, recurrence, bacteriology, risk factors, and disease burden. *Infectious Disease Clinics of North America, 28*, 1–13.

Furfaro, L. L., Chang, B. J., & Payne, M. S. (2018). Perinatal Streptococcus agalactiae epidemiology and surveillance targets. *Clinical Microbiology Reviews, 31*, e00049-e118.

Garcia, T. A., Ventura, C. L., & Smith, M. A. (2013). Cytotoxic necrotizing factor 1 and hemolysin from uropathogenic Escherichia coli elicit different host responses in the murine bladder. *Infection and Immunity, 81*, 99–109.

Gatermann, S., John, J., & Marre, R. (1989). Staphylococcus saprophyticus urease: characterization and contribution to uropathogenicity in unobstructed urinary tract infection of rats. *Infection and Immunity, 57*, 110–116.

Gebreyohannes, G., Nyerere, A., & Bii, C. (2019). Challenges of intervention, treatment, and antibiotic resistance of biofilm-forming microorganisms. *Heliyon, 5*, e02192.

Geibel, S., Procko, E., & Hultgren, S. J. (2013). Structural and energetic basis of folded-protein transport by the FimD usher. *Nature, 496*, 243–246.

Gerlach, G.-F., Clegg, S., & Allen, B. L. (1989). Identification and characterization of the genes encoding the type 3 and type 1 fimbrial adhesins of Klebsiella pneumoniae. *Journal of Bacteriology, 171*, 1262–1270.

Glahn, B. E., Braendstrup, O., & Olesen, H. (1988). Influence of drainage conditions on mucosal bladder damage by indwelling catheters: II. Histological study. *Scandinavian Journal of Urology and Nephrology, 22*, 93–99.

Goble, N., Clarke, T., & Hammonds, J. (1989). Histological changes in the urinary bladder secondary to urethral catheterisation. *British Journal of Urology, 63*, 354–357.

Greene, S. E., Pinkner, J. S., & Chorell, E. (2014). Pilicide ec240 disrupts virulence circuits in uropathogenic Escherichia coli. *MBio, 5*.

Griffith, D. P., Musher, D. Á. & Itin, C. (1976). Urease. The primary cause of infection-induced urinary stones. *Investigative Urology, 13*, 346.

Guiton, P. S., Cusumano, C. K., & Kline, K. A. (2012). Combinatorial small-molecule therapy prevents uropathogenic Escherichia coli catheter-associated urinary tract infections in mice. *Antimicrobial Agents and Chemotherapy, 56*, 4738–4745.

Guiton, P. S., Hannan, T. J., & Ford, B. (2013). Enterococcus faecalis overcomes foreign body-mediated inflammation to establish urinary tract infections. *Infection and Immunity, 81*, 329–339.

Guiton, P. S., Hung, C. S., & Hancock, L. E. (2010). Enterococcal biofilm formation and virulence in an optimized murine model of foreign body-associated urinary tract infections. *Infection and Immunity, 78*, 4166–4175.

Gupta, K., & Bhadelia, N. (2014). Management of urinary tract infections from multidrug-resistant organisms. *Infectious Disease Clinics of North America, 28*, 49.

Hadjifrangiskou, M., Gu, A. P., & Pinkner, J. S. (2012). Transposon mutagenesis identifies uropathogenic Escherichia coli biofilm factors. *Journal of Bacteriology, 194*, 6195–6205.

Hannan, T. J., Mysorekar, I. U., & Chen, S. L. (2008). LeuX tRNA-dependent and-independent mechanisms of Escherichia coli pathogenesis in acute cystitis. *Molecular Microbiology, 67*, 116–128.

Hannan, T. J., Mysorekar, I. U., & Hung, C. S. (2010). Early severe inflammatory responses to uropathogenic E. coli predispose to chronic and recurrent urinary tract infection. *PLoS Pathog, 6*, e1001042.

Hannan, T. J., Totsika, M., & Mansfield, K. J. (2012). Host–pathogen checkpoints and population bottlenecks in persistent and intracellular uropathogenic Escherichia coli bladder infection. *FEMS Microbiology Reviews, 36*, 616–648.
High, K. P., Edwards, M. S., & Baker, C. J. (2005). Group B streptococcal infections in elderly adults. *Clinical Infectious Diseases, 41*, 839–847.
Himpsl, S. D., Pearson, M. M., & Arewång, C. J. (2010). Proteobactin and a yersiniabactin-related siderophore mediate iron acquisition in Proteus mirabilis. *Molecular Microbiology, 78*, 138–157.
Hooton, T. M. . (2012). Uncomplicated urinary tract infection. *New England Journal of Medicine, 366*, 1028–1037.
Hultgren, S. 2011. Pathogenic cascade of E. coli UTI. *UTI pathogenesis. St. Louis, Missouri: Molecular microbiology and microbial pathogenesis program*, Washington University.
Jacobsen, S., Stickler, D., & Mobley, H. (2008). Complicated catheter-associated urinary tract infections due to Escherichia coli and Proteus mirabilis. *Clinical Microbiology Reviews, 21*, 26–59.
Jacobsen, S. M., & Shirtliff, M. E. (2011). Proteus mirabilis biofilms and catheter-associated urinary tract infections. *Virulence, 2*, 460–465.
Jones, B. D., & Mobley, H. (1987). Genetic and biochemical diversity of ureases of Proteus, Providencia, and Morganella species isolated from urinary tract infection. *Infection and Immunity, 55*, 2198–2203.
Karam, M. R. A., Oloomi, M., & Mahdavi, M. (2013). Vaccination with recombinant FimH fused with flagellin enhances cellular and humoral immunity against urinary tract infection in mice. *Vaccine, 31*, 1210–1216.
Khandelwal, P., Abraham, S. N., & Apodaca, G. (2009). Cell biology and physiology of the uroepithelium. *American Journal of Physiology-Renal Physiology, 297*, F1477–F1501.
Kline, K. A., Schwartz, D. J., & Lewis, W. G. (2011). Immune activation and suppression by group B streptococcus in a murine model of urinary tract infection. *Infection and Immunity, 79*, 3588–3595.
Korkes, F., Favoretto, R. L., & Bróglio, M. (2008). Xanthogranulomatous pyelonephritis: clinical experience with 41 cases. *Urology, 71*, 178–180.
Kosikowska, P., & Berlicki, Ł. (2011). Urease inhibitors as potential drugs for gastric and urinary tract infections: A patent review. *Expert Opinion on Therapeutic Patents, 21*, 945–957.
Kostakioti, M., Hadjifrangiskou, M., & Hultgren, S. J. (2013). Bacterial biofilms: development, dispersal, and therapeutic strategies in the dawn of the postantibiotic era. *Cold Spring Harbor Perspectives in Medicine, 3*, a010306.
Kostakioti, M., Hultgren, S. J., & Hadjifrangiskou, M. (2012). Molecular blueprint of uropathogenic Escherichia coli virulence provides clues toward the development of anti-virulence therapeutics. *Virulence, 3*, 592–593.
Lane, D. R., & Takhar, S. S. (2011). Diagnosis and management of urinary tract infection and pyelonephritis. *Emergency Medicine Clinics, 29*, 539–552.
Langermann, S., Palaszynski, S., & Barnhart, M. (1997). Prevention of mucosal Escherichia coli infection by FimH-adhesin-based systemic vaccination. *Science, 276*, 607–611.
Lee, B. K., Crossley, K., & Gerding, D. N. (1978). The association between Staphylococcus aureus bacteremia and bacteriuria. *The American Journal of Medicine, 65*, 303–306.
Lee, G. . (2011). Uroplakins in the lower urinary tract. *International Neurourology Journal, 15*, 4.
Lee, S., Stubelius, A., & Hamelmann, N. (2018). Inflammation-responsive drug-conjugated dextran nanoparticles enhance anti-inflammatory drug efficacy. *ACS Applied Materials & Interfaces, 10*, 40378–40387.
Levison, M. E., & Kaye, D. (2013). Treatment of complicated urinary tract infections with an emphasis on drug-resistant gram-negative uropathogens. *Current Infectious Disease Reports, 15*, 109–115.
Li, X., Erbe, J. L., & Lockatell, C. V. (2004). Use of translational fusion of the MrpH fimbrial adhesin-binding domain with the cholera toxin A2 domain, coexpressed with the cholera toxin

B subunit, as an intranasal vaccine to prevent experimental urinary tract infection by Proteus mirabilis. *Infection and Immunity, 72*, 7306–7310.

Li, X., Zhao, H., & Lockatell, C. V. (2002). Visualization of Proteus mirabilis within the matrix of urease-induced bladder stones during experimental urinary tract infection. *Infection and Immunity, 70*, 389–394.

Livermore, D. M., & Mushtaq, S. (2013). Activity of biapenem (RPX2003) combined with the boronate β-lactamase inhibitor RPX7009 against carbapenem-resistant Enterobacteriaceae. *Journal of Antimicrobial Chemotherapy, 68*, 1825–1831.

Lo, E., Nicolle, L. E., & Coffin, S. E. (2014). Strategies to prevent catheter-associated urinary tract infections in acute care hospitals: 2014 update. *Infection Control & Hospital Epidemiology, 35*, S32–S47.

Martinez, J. J., & Hultgren, S. J. (2002). Requirement of Rho-family GTPases in the invasion of Type 1-piliated uropathogenic Escherichia coli. *Cellular Microbiology, 4*, 19–28.

Meyers, D. J., Palmer, K. C., & Bale, L. A. (1992). In vivo and in vitro toxicity of phospholipase C from Pseudomonas aeruginosa. *Toxicon, 30*, 161–169.

Miraglia, A. G., Travaglione, S., & Meschini, S. (2007). Cytotoxic necrotizing factor 1 prevents apoptosis via the Akt/IκB kinase pathway: Role of nuclear factor-κB and Bcl-2. *Molecular Biology of the Cell, 18*, 2735–2744.

Mittal, R., Aggarwal, S., & Sharma, S. (2009). Urinary tract infections caused by Pseudomonas aeruginosa: A minireview. *Journal of Infection and Public Health, 2*, 101–111.

Mohajer, M., Musher, D. M., Minard, C. G. (2013). Clinical significance of Staphylococcus aureus bacteriuria at a tertiary care hospital. *Scandinavian Journal of Infectious Diseases, 45*, 688–695.

Mohammed, Y. H. E., Manukumar, H., & Rakesh, K. (2018). Vision for medicine: Staphylococcus aureus biofilm war and unlocking key's for anti-biofilm drug development. *Microbial Pathogenesis, 123*, 339–347.

Moi, H., Blee, K., & Horner, P. J. (2015). Management of non-gonococcal urethritis. *BMC Infectious Diseases, 15*, 1–7.

Muder, R. R., Brennen, C., & Wagener, M. M. (1992). Bacteremia in a long-term-care facility: a five-year prospective study of 163 consecutive episodes. *Clinical Infectious Diseases, 14*, 647–654.

Murphy, C. N., Mortensen, M. S., & Krogfelt, K. A. (2013). Role of Klebsiella pneumoniae type 1 and type 3 fimbriae in colonizing silicone tubes implanted into the bladders of mice as a model of catheter-associated urinary tract infections. *Infection and Immunity, 81*, 3009–3017.

Mushtaq, S., Woodford, N., & Hope, R. (2013). Activity of BAL30072 alone or combined with β-lactamase inhibitors or with meropenem against carbapenem-resistant Enterobacteriaceae and non-fermenters. *Journal of Antimicrobial Chemotherapy, 68*, 1601–1608.

Mylotte, J. M., Tayara, A., & Goodnough, S. (2002). Epidemiology of bloodstream infection in nursing home residents: evaluation in a large cohort from multiple homes. *Clinical Infectious Diseases, 35*, 1484–1490.

Nagamatsu, K., Hannan, T. J., & Guest, R. L. (2015). Dysregulation of Escherichia coli α-hemolysin expression alters the course of acute and persistent urinary tract infection. *Proceedings of the National Academy of Sciences, 112*, E871–E880.

Nagaya, H., Satoh, H., & Kubo, K. (1989). Possible mechanism for the inhibition of gastric (H++ K+)-adenosine triphosphatase by the proton pump inhibitor AG-1749. *Journal of Pharmacology and Experimental Therapeutics, 248*, 799–805.

Nicolle, L., Muir, P., & Harding, G. (1988). Localization of urinary tract infection in elderly, institutionalized women with asymptomatic bacteriuria. *Journal of Infectious Diseases, 157*, 65–70.

Nielsen, H. V., Flores Mireles, A. L., & Kau, A. L. (2013). Pilin and sortase residues critical for endocarditis-and biofilm-associated pilus biogenesis in Enterococcus faecalis. *Journal of Bacteriology, 195*, 4484–4495.

Nielsen, H. V., Guiton, P. S., & Kline, K. A. (2012). The metal ion-dependent adhesion site motif of the Enterococcus faecalis EbpA pilin mediates pilus function in catheter-associated urinary tract infection. *MBio, 3*.
Nielubowicz, G. R., & Mobley, H. L. (2010). Host–pathogen interactions in urinary tract infection. *Nature Reviews Urology, 7*, 430–441.
Niveditha, S., Pramodhini, S., & Umadevi, S. (2012). The isolation and the biofilm formation of uropathogens in the patients with catheter associated urinary tract infections (UTIs). *Journal of Clinical and Diagnostic Research: JCDR, 6*, 1478.
Noor, N., Ajaz, M., & Rasool, S. A. (2004). Urinary tract infections associated with multidrug resistant enteric bacilli: Characterization and genetical studies. *Pakistan Journal of Pharmaceutical Sciences, 17*, 115–123.
O'Hanley, Á., Lalonde, G., & Ji, G. (1991). Alpha-hemolysin contributes to the pathogenicity of piliated digalactoside-binding Escherichia coli in the kidney: efficacy of an alpha-hemolysin vaccine in preventing renal injury in the BALB/c mouse model of pyelonephritis. *Infection and Immunity, 59*, 1153–1161.
Paterson, D. L. . (2006). Resistance in gram-negative bacteria: Enterobacteriaceae. *American Journal of Infection Control, 34*, S20–S28.
Peach, B. C., Garvan, G. J., & Garvan, C. S. (2016). Risk factors for urosepsis in older adults: a systematic review. *Gerontology and Geriatric Medicine, 2*, 2333721416638980.
Pendleton, J. N., Gorman, S. P., & Gilmore, B. F. (2013). Clinical relevance of the ESKAPE pathogens. *Expert Review of Anti-Infective Therapy, 11*, 297–308.
Piatek, R., Zalewska-Piatek, B., & Dzierzbicka, K. (2013). Pilicides inhibit the FGL chaperone/usher assisted biogenesis of the Dr fimbrial polyadhesin from uropathogenic Escherichia coli. *BMC Microbiology, 13*, 131.
Pinkner, J. S., Remaut, H., & Buelens, F. (2006). Rationally designed small compounds inhibit pilus biogenesis in uropathogenic bacteria. *Proceedings of the National Academy of Sciences, 103*, 17897–17902.
Piteau, M., Papatheodorou, P., & Schwan, C. (2014). Lu/BCAM adhesion glycoprotein is a receptor for Escherichia coli cytotoxic necrotizing factor 1 (CNF1). *PLoS Pathogens, 10*, e1003884.
Podschun, R., & Ullmann, U. (1998). Klebsiella spp. as nosocomial pathogens: epidemiology, taxonomy, typing methods, and pathogenicity factors. *Clinical Microbiology Reviews, 11*, 589–603.
Ramakrishnan, K., & Scheid, D. C. (2005). Diagnosis and management of acute pyelonephritis in adults. *American Family Physician, 71*, 933–942.
Rice, J. C., Peng, T., & Spence, J. S. (2005). Pyelonephritic Escherichia coli expressing P fimbriae decrease immune response of the mouse kidney. *Journal of the American Society of Nephrology, 16*, 3583–3591.
Robino, L., Scavone, P., & Araujo, L. (2014). Intracellular bacteria in the pathogenesis of Escherichia coli urinary tract infection in children. *Clinical Infectious Diseases, 59*, e158–e164.
Rocha, C. L., Coburn, J., & Rucks, E. A. (2003). Characterization of Pseudomonas aeruginosa exoenzyme S as a bifunctional enzyme in J774A. 1 macrophages. *Infection and Immunity, 71*, 5296–5305.
Ronald, A. . (2002). The etiology of urinary tract infection: traditional and emerging pathogens. *The American Journal of Medicine, 113*, 14–19.
Rosen, D. A., Hooton, T. M., & Stamm, W. E. (2007). Detection of intracellular bacterial communities in human urinary tract infection. *PLoS Medicine, 4*, e329.
Rosen, D. A., Pinkner, J. S., & Jones, J. M. (2008). Utilization of an intracellular bacterial community pathway in Klebsiella pneumoniae urinary tract infection and the effects of FimK on type 1 pilus expression. *Infection and Immunity, 76*, 3337–3345.
Rosen, D. A., Pinkner, J. S., & Walker, J. N. (2008). Molecular variations in Klebsiella pneumoniae and Escherichia coli FimH affect function and pathogenesis in the urinary tract. *Infection and Immunity, 76*, 3346–3356.

Savar, N. S., Jahanian-Najafabadi, A., & Mahdavi, M. (2014). In silico and in vivo studies of truncated forms of flagellin (FliC) of enteroaggregative Escherichia coli fused to FimH from uropathogenic Escherichia coli as a vaccine candidate against urinary tract infections. *Journal of Biotechnology, 175*, 31–37.

Schwartz, D. J., Chen, S. L., & Hultgren, S. J. (2011). Population dynamics and niche distribution of uropathogenic Escherichia coli during acute and chronic urinary tract infection. *Infection and Immunity, 79*, 4250–4259.

Senturk, S., Ulusoy, S., & Bosgelmez-Tinaz, G. (2012). Quorum sensing and virulence of Pseudomonas aeruginosa during urinary tract infections. *The Journal of Infection in Developing Countries, 6*, 501–507.

Sharma, P., & Singal, A. (1999). Methicillin-resistant Staphylococcus aureus non-gonococcal urethritis. *Acta Dermato-Venereologica, 79*, 415.

Simor, A. E. . (2001). The role of the laboratory in infection prevention and control programs in long-term–care facilities for the elderly. *Infection Control & Hospital Epidemiology, 22*, 459–463.

Sivick, K. E., & Mobley, H. L. (2010). Waging war against uropathogenic Escherichia coli: winning back the urinary tract. *Infection and Immunity, 78*, 568–585.

Sjøstrøm, J., Kühler, T., & Larsson, H. (1997). Basis for the selective antibacterial activity in vitro of proton pump inhibitors against Helicobacter spp. *Antimicrobial Agents and Chemotherapy, 41*, 1797–1801.

Song, J., Bishop, B. L., & Li, G. (2009). TLR4 mediated expulsion of bacteria from infected bladder epithelial cells. *Proceedings of the National Academy of Sciences, 106*, 14966–14971.

Stahlhut, S. G., Tchesnokova, V., & Struve, C. (2009). Comparative structure-function analysis of mannose-specific FimH adhesins from Klebsiella pneumoniae and Escherichia coli. *Journal of Bacteriology, 191*, 6592–6601.

Stamm, W. E., & Norrby, S. R. (2001). Urinary tract infections: disease panorama and challenges. *The Journal of Infectious Diseases, 183*, S1–S4.

Stickler, D. . (2014). Clinical complications of urinary catheters caused by crystalline biofilms: something needs to be done. *Journal of Internal Medicine, 276*, 120–129.

Struve, C., Bojer, M., & Krogfelt, K. A. (2008). Characterization of Klebsiella pneumoniae type 1 fimbriae by detection of phase variation during colonization and infection and impact on virulence. *Infection and Immunity, 76*, 4055–4065.

Taylor, E. (2011). TJ webster reducing infections through nanotechnology and nanoparticles. *International Journal of Nanomedicine, 6*, 1463.

Thanassi, D. G., Saulino, E. T., & Hultgren, S. J. (1998). The chaperone/usher pathway: a major terminal branch of the general secretory pathway. *Current Opinion in Microbiology, 1*, 223–231.

Totsika, M., Kostakioti, M., & Hannan, T. J. (2013). A FimH inhibitor prevents acute bladder infection and treats chronic cystitis caused by multidrug-resistant uropathogenic Escherichia coli ST131. *The Journal of Infectious Diseases, 208*, 921–928.

Ulett, K. B., Benjamin, W. H., & Zhuo, F. (2009). Diversity of group B streptococcus serotypes causing urinary tract infection in adults. *Journal of Clinical Microbiology, 47*, 2055–2060.

Valdebenito, M., Crumbliss, A. L., & Winkelmann, G. (2006). Environmental factors influence the production of enterobactin, salmochelin, aerobactin, and yersiniabactin in Escherichia coli strain Nissle 1917. *International Journal of Medical Microbiology, 296*, 513–520.

Vallet, I., Olson, J. W., & Lory, S. (2001). The chaperone/usher pathways of Pseudomonas aeruginosa: identification of fimbrial gene clusters (cup) and their involvement in biofilm formation. *Proceedings of the National Academy of Sciences, 98*, 6911–6916.

Visca, Á., Chiarini, F., & Mansi, A. (1992). Virulence determinants in Pseudomonas aeruginosa strains from urinary tract infections. *Epidemiology & Infection, 108*, 323–336.

von Baum, H., Schmidt, C., & Svoboda, D. (2002). Risk factors for methicillin-resistant Staphylococcus aureus carriage in residents of German nursing homes. *Infection Control & Hospital Epidemiology, 23*, 511–515.

Walther-Rasmussen, J., & Høiby, N. (2004). Cefotaximases (CTX-M-ases), an expanding family of extended-spectrum β-lactamases. *Canadian Journal of Microbiology, 50*, 137–165.

Watts, R. E., Totsika, M., & Challinor, V. L. (2012). Contribution of siderophore systems to growth and urinary tract colonization of asymptomatic bacteriuria Escherichia coli. *Infection and Immunity, 80*, 333–344.

Workowski, K. A. & Bolan, G. A. (2015). Sexually transmitted diseases treatment guidelines. In *MMWR. Recommendations and reports: Morbidity and mortality weekly report. Recommendations and reports* (Vol. 64, p. 1).

Wright, K. J. & Hultgren, S. J. (2006). Sticky fibers and uropathogenesis: Bacterial adhesins in the urinary tract.

Wurpel, D. J., Beatson, S. A., & Totsika, M. (2013). Chaperone usher fimbriae of Escherichia coli. *PLoS ONE, 8*, e52835.

Zhanel, G. G., Lawson, C. D., & Adam, H. (2013). Ceftazidime-avibactam: A novel cephalosporin/β-lactamase inhibitor combination. *Drugs, 73*, 159–177.

Chapter 21
Drugs Resistance Management

Nabila Bashir, Saima Ashraf, Nadia Rashid, Syed Tayyab Raza Naqvi, Saadat Majeed, Ghulam Murtaza, Muhammad Naeem Ashiq, and Muhammad Najam-ul-Haq

Abstract Drug resistance is affecting people worldwide. Drug resistance reduces the efficiency of medicine incurring a disease. An antibiotic resistance results from overuse of antibiotic in bacterial infections. In designing the drug, understanding of underlying antibiotic resistance factors is important which may include inactivation by specific enzymes, alternation in target site or metabolic pathways and reduction in the accumulation of drug molecules. It usually develops when microbes multiply and find route of fighting out drug molecules. Other reasons may include attitude towards the prescription, carelessness in precautions employed for infections, and lack of attention of both government and patients. Resistance can be combated using specific enzymes in activating the resistant genes. Surgery or medicinal intake controls antibiotic resistance in HIV. Vaccines also manage sensitivity of microbes to the drug. Generations of antibiotics are needed which recognize pathogen-associated molecular pathways (PAMPS). Alternations in drugs dosage or utilizing drugs combinations increase the efficacy of treatment. Drug resistance mechanisms must therefore be controlled for preventing the spread of many acute diseases.

21.1 What is Drug Resistance?

Drug resistance is the decrease in effectiveness of a medicine to cure any disease or disorder. This resistance can affect the people of various ages and countries. Nowadays, antibiotic-resistance is increasing and threatening our ability to cure even common infectious diseases. So, understanding the varied molecular mechanisms behind this resistance to antibiotics and other tonic drugs will benefit in the development of new drugs to fight increasing drug resistance. The development of

N. Bashir · S. Ashraf · N. Rashid · S. T. R. Naqvi · S. Majeed · M. N. Ashiq · M. Najam-ul-Haq (✉)
Institute of Chemical Sciences, Bahauddin Zakariya University, Multan 60800, Pakistan
e-mail: najamulhaq@bzu.edu.pk

G. Murtaza
Department of Pharmacy, COMSATS University Islamabad (CUI), Lahore Campus, Lahore, Pakistan

S. Ahmed et al. (eds.), *Biochemistry of Drug Resistance*,
https://doi.org/10.1007/978-3-030-76320-6_21

drug resistance is not unnatural but overdose can eventually speed it up. The large number of infections are getting harder to cure because these antibiotics are losing their impact due to regular repetition of their usage. These resistances are increasing the number of patients in hospitals and also causing wastage of money with increased rate of mortality. The term is used mainly for the cancers and microbes which have evolved with the passage of time. Antimicrobial resistance has great potential in research. In the case an organism is resistant to one or more drug, it would be referred as multi-drug resistant (Longley & Johnston, 2005). The development of antibiotic resistance comes from drugs or medication targeting only a specific bacteria which in most cases is protein. As the drug molecule is specific, any change in composition or its effect can lead to drug resistance. Examples of targeting altering microbes include staphylococcus aurous and enterococci which are vancomycin resistant and streptococcus resistant to macrolides (Gottesman, 2002).

Drug or chemical resistance is the consequence of evolution and results in pressure on living organisms. Individual organisms differ in their ability to respond to a specific drug or antigenic. Drug resistance mechanisms or traits can be passed from one generation to other resulting in a population which is more drug resistant. Unless a drug molecule employs reproduction and cell division, it is impossible to transfer the drug resistant gene (White, 2004).

The cancerous cells have attained some level of resistance during chemotherapy. Chemotherapy results in the formation of fibroblasts near tumor cells to produce large amount of proteins WNT16B. In other case, RNA also gets affected due to changes from chemotherapy. Rapid process of sharing resistance between two single celled organisms is termed as horizontal gene transfer where there is exchange of genes among two organisms, in particular with biofilm state. Similar mechanism is used by fungi using asexual reproduction termed as parasexual reproduction. Examples of drug resistant mechanism is found in number of organisms which include bacteria, viruses, arthropods, reptiles and mammals etc. (Clavel & Hance, 2004).

In domestic environments, the drug resistance arise as a result of activities like bleach, tooth brushing, mouth washing, antibiotics, antivirals, sprays and shampoos etc. On the other hand, in agricultural field, it is due to insecticides, pesticides and dips etc. Chemicals in these preparations in addition to having harmful effects may interact with target organism to produce drug resistance. There are four mechanisms by which organisms develop resistance to specific antimicrobials (Croft et al., 2006).

- Drug inactivation or modification by enzyme like beta lactamase in penicillin resistant bacteria.
- Alteration in target site i.e. change in PBP which is the binding site for penicillin in MRSA and other resistant bacteria.
- Metabolic pathway alterations, for example sulfonamide resistant bacteria do not require para amino benzoic acid as precursor or substrate for synthesis of folic acid, nucleic acids as inhibited by sulfonamide substances.
- Reducing accumulation of drug from reduced permeability, increasing active flux of drug molecules across the cell surface (Pastan & Gottesman, 1987).

21.2 How Drug Resistance Develops?

When microorganisms find a way to fight against the drug and continue to multiply their population, they become resistant towards that particular drug. Handling of diseases caused by resistant microorganisms is difficult but not impossible. Such diseases need more time and attention by doctors to get cured and the alternate drugs would be comparatively expensive but effective against germs.

This drug resistance does not mean that body has become irresponsive towards them but the microorganism have become strong enough to deal with that combination of chemicals. The level of resistance towards drugs is increasing globally which is creating panic by limiting the cure of even common infections. The increasing number of infectious diseases is becoming worse day by day as their treatment is getting difficult. The consumption of drugs for diseases without the recommendation of expert physician has become common, making difficult to kill bacteria in time by developing resistance because of excessive use of drugs. Same is the case in countries where drugs are recommended in large amounts and largely consumed by people.

21.3 Types of Antibiotics Resistances

- Antimicrobial drug resistance
- Antifungal drug resistance

21.4 Drug Resistance Sources

21.4.1 Drug Use in Humans

The introduction of resistance has variety of reasons such as,

1. the usage of antibiotics is much higher
2. there is no check and balance over prescriptions
3. non-serious and insufficient sensibility towards health issues
4. consumption of higher dosage
5. carelessness towards precautions against infections
6. the transfer speed of infection is much higher than before and multiply rapidly
7. Lack of encouragement for patients, doctors and also government to pay attention towards rising levels of resistance.

The reasons behind spreads of resistance and also its consequences must be taken into account. The basis of resistance gives information about the initiation of resistance. In case if the single base pair of DNA has undergone mutation and generated resistance, then this resistance would spread widely. Such kinds of mutations cause a strain of resistance in patients. The transfer of such strain resistances from patient to healthy people is known as the drug resistance and procedure is recognized as primary resistance method. The bacteria which are independent and participate in resistance transfer finally cause multidrug resistances. Both the transfers of resistance and its production contribute towards humans worldwide. This mechanism is also found in case of malaria where antimalarial drug is injected into the body to kill malarial parasite however results are opposite. The drug not even fails to kill the parasite but also inhibit the growth of population in the body. This is because the malarial drugs are not commonly mutagenic.

In case of tuberculosis, the mutation of genes is rapid and creates resistance in the body. Multiple numbers of drugs do not allow the resistance and kill the germs. The resistance is preventable which occurs due to improper cure, inappropriate prescription of drugs and negligible responsibility towards drug consumption. In case of malaria, the chloroquine is blindly prescribed and consumed which eventually create its resistance.

The pneumonia also has a widespread resistance problem. This resistance is not against penicillin which is transmitted by multiple mutation of DNA or by importing foreign genes. It is resistant towards sulfamethoxazole-trimethoprim which can be found in the population of affected people and it spreads due to the change in single base pair of DNAs. The resistances were studied in children affected by pneumonia. It was concluded that 42% of child population had consumed the drug weekly after being exposed to pneumonia infection while 38% of children consumed the drug after being infected by malaria. The consumption of drugs and negligence behavior of physicians towards the prescriptions of drugs has increased the drugs resistance up to 60% worldwide.

Biological research has revealed that the prescription and consumption of drugs has influenced the spread of resistance in human body. The developed countries have database of antimicrobial consumption and drug resistance. Less developed countries are not able to follow as they have lesser numbers in the stored data about use of antibiotics. This level of resistance is not rising in poorly developed countries as they are not socioeconomically strong to consume the costly antibiotic at large level. India is a large country and it has controlled the resistance by managing the prescription of antibiotics. They have less cases of pneumonia at least in villages. This lesser number is in spite of the availability of antibiotics but poverty does not allow them to consume much antibiotics and severe cases are cured in short period of time. Improving their economic status would increase the resistance rate in their country. Same would be the case of quinolone which is available in affordable price in city areas but still it has low level of resistance. A country's economic status also influences the availability of antibiotic and its consumption. Large numbers of resistances for multiple drugs are studied and recorded in developed countries where drugs are available in affordable range and prescribed in large amounts and blindly consumed by population resulting in alarming health issues.

21.5 Environment

The increase in antimicrobial resistance is unveiling the health crisis of public. The resistance of commonly used antimicrobials like quinolone, carbapenems, and cephalosporins of third and fourth generation is a matter of concern and is the multidrug resistance. This resistance is not only restricted to human drugs, but a reasonable amount is also added in the production of livestock and organic farms. This spread of resistance is followed by a specific route from food and animals to human beings. The workers of agriculture can be affected by the direct contact with livestock while normal people have other ways to get this resistance such as food products derived from livestock and use of waste as fertilizer in fields can pollute the supplies of water. Still large amounts of antimicrobials are consumed by humans and genes carrying bacterial resistance are spreading in offspring or contaminating water, food and environment through sewage.

A pandemic was spread in late nineteenth century where bacteria were multidrug resistant and considered as zoonotic. These bacteria were spreading through domestic cattle via close contact and food produced by them. It gave a blurred picture of animal derived route. This idea was tested by researchers in Scotland. They collected and isolated 200 genomes in same geographical region from animals and humans over a period of 20 years to study their sequencing in laboratory. The sequences were analyzed to clarify the histories of animals and human hosts. The research revealed that pandemics of animals and humans are independent to each other.

21.6 Fungal Cell Wall and Drug Resistance

The fungal homeostasis is dependent on cell wall. If the human cells are lacking their protective covering, they become target site for the antifungal development. The stress caused by antifungals and environment of host combines to modify the cell wall for drug resistance. The antifungals such as beta D glucan synthase inhibitor, GPI (glycosyl phosphatidyl inositol) and pathway inhibitors target the cell wall and create weapon against antifungal resistance. The infected cell wall changes the composition of infectious beta D glucan and chitin through carbon sources and creates echinocandin resistance. The populations of candida show resistance against echinocandin and chitin level increases in cell wall while other population known as aspergillums flatus shows resistance against amphotericin. The study of fungal cell dynamics is important not only to understand its interactions with host cells but also to study the treatment of infection. The developed antifungals target the host's cell wall. Many of these developed drugs act by inhaling β-D-glucan synthase. Chitin and GPI pathway inhibitors are under process.

21.7 Echinocandin

Echinocandin were first introduced as the polypeptide antibiotics derived from aspergillus nidulans. These hexapeptide molecules have chains of N-linked acyl fatty acids that interact with the lipid layers of cell membrane. This class of antifungals inhibits β-D-glucan synthase and decreases its amount in cell wall by binding to the enzymes. The complex of cell wall with beta D glucan synthase has two subunits.

1. Catalytic subunit produces glycosidic bonds.
2. GTP binding protein responsible for regulating the activity of β-D-glucan synthase.

In case of candida specie, the cell death occurs if β-D-glucan synthase is inhibited. The modified morphogenesis of hyphae is done by echinocandin to generate fungi against aspergillus species. Recently FDA has approved three echinocandin i.e. micafungin, anidulafungin and caspofungins for the treatment of infections. In comparison to other classes of antifungals, echinocandin have,

1. low level of toxicity for kidney and liver
2. drug-drug interactions are rare
3. the elimination of liver is predominant
4. Adjustment of dosage is not required in dialysis or renal failure.

The echinocandin have few limitations such as,

1. They have poor bioavailability while administering orally.
2. High rate of protein binding.
3. Lower level of CNS (central nervous system) penetration.

The β-D-glucan with improved profile of pharmacokinetics, better oral administration and bioavailability are under study. The fungal cell wall is modified due to the stress of fungal exposure and microenvironment of host. The in vitro study has proved that change in microenvironment of host around infectious site will cause yeast cell to modify cell wall and create antifungal resistance. The cells growing in blood serum have shown changes in structure of cell wall and content of chitin and β-D-glucan.

The development of tolerance is an intermediate step for antifungal resistance. A selective response is given by the cells influenced by the drugs to create resistance. The exposure of echinocandin generates stress over cell walls to inhibit the synthesis of β-D-glucan and accelerate cellular factors for chitin production. KC, HOG and Ca^{2+} calcineurin pathways are accelerated in response to damage of cell wall and synthesis of chitin. The antifungal stress causes another adaptive mechanism called as parodoxal effect. This effect is due to increase in fungal growth after getting exposed to excess amounts of antifungals beyond threshold limit. Similarly, paradoxical effect is also connected to intracellular signaling pathways for rebuilding of cell walls as the concentration of chitin and β-D-glucan increases.

In recent times, the scientific advancements of fungal cell wall have put forward new modalities of therapy in case of fungal infections and helped to study the antifungal resistance mechanisms. In comparison to other molecules, new antifungals are safely targeting the cell walls. β-D-glucan has strong interactions with echinocandin, and its resistant infections can be treated. The pathway inhibitors have good profiles of pharmacokinetics and are active against pathogenic species including multi resistant species. It makes them unique antifungals and will be launched in near future. Microenvironment around the site of infection causes modifications in fungal cell wall to produce antifungal resistance. Due to the exposure of echinocandin, a stress is induced on cell wall which forms tolerant cells with high level of chitin. For the phenomenon of antifungal tolerance, PKC, HOG and calcineurin pathways are difficult components and they must be further explored for the antifungal treatment. The *A. flavus* and *C. tropicalis* are few AMB resistant with higher levels of glucan in cell wall. The future studies with analysis of cell wall modifications and AMB resistance are important to further strengthen their relation.

21.8 Need of Drug Resistance Management

The concept of resistance management is limited however the resistance frequencies are still low. Recent research has concluded two approaches to slow down the emergence of resistance, i.e. regarding antibiotic use as cycling and mixing. In cycling approach, a particular class of antibiotics is used for specific period followed by different classes. In mixing, different kinds of antibiotics are prescribed to different types of patients to create a spatial mosaic use of antibiotics. Aim of mixing is to increase the heterogeneity of selective pressure so that microbe resistant antibiotic does not spread rapidly from one patient to another. Cycling dominantly rely on resistance when selection of antibiotic is not present. The resistance will be reduced when the use of antibiotic would be interrupted.

In case of mixing approach, the bacteria resistant to antibiotic A will be treated with antibiotic A. Mixing should consider the transmission network of bacterial targets. The mixing in single ward is less effective, so the transmission be extended to the whole hospital. The separate wards like ICU would be difficult targets to handle if their transmission networks are close and transmission is mainly occurring between workers and patients in single ward. The networks of transmission containing infected bacteria from individuals of unexposed environment would be important in RM mixing.

21.9 Methods to Combat Drug Resistance

Microbial resistance to antibiotics is becoming prevalent with time. Bacterial resistance is common however resistance has been developed against fungi, viruses and protist (Pablos-Mendez, 1998). Resistance is complex phenomenon where microorganisms, environments, and patients are involved either separately or in interactive way. Resistance may develop before exposure of specific drug to microbes or arise because of later exposure. Microbial resistance especially antibiotic resistance in case of flouroquinolone has become a problem (Boehm, 1997). In gram negative bacteria, it is caused by drug effecting the targeted protein. These targets would be DNA gyrase and trypanosomes 4. Additionally, any mutation in targeted genes such as MarA, SoxS and others will result in resistance. MarA is the transcriptional gene involved in activation which is operated by MarRAB promoter causing multiple antibiotic resistance (Tew, 1994). MarRAB locus causes resistance to tetracycline, chloramphenicol, flouroquinolone, nalidixic acid, rifamycin, and penicillin in addition to other such compounds. It doesn't however encode multiple drug efflux system. Instead it controls expression of locus mediating in other drug resistance conditions. For example, the gene omFP is involved in regulating gene for outer membranes in prions (White et al., 1998).

AcrAB is a multidrug pump which in mammals protect against bile salts in small intestine. It is regulated by MarA gene and overexpression results from MarA gene, leading to mutation which in turn causes resistance to antibiotics. The SoxS gene also encodes MarA homolog which regulates ArcAB gene (Li & Nikaido, 2004). Microbes often become resistant to antimicrobial and non-antimicrobial components by acquisition of an enzyme which inactivates genes, modifies drug target or sometimes result in active efflux of genes. Enzyme inactivating antibiotics such as quinolone and chloramphenicol etc. (White, 1999) have yet not found. In this case, where enzymes or other natural products for antibiotics and other such compounds have not been developed, resistance have been generated because of modification of targets. So improved methods for controlling drug resistance in microbes, particularly in microbes having a degree of resistance would offer benefits (Thakur, 2013).

21.10 Clinical Management

Presence of large number of drugs and requirement of medication in initial phase, the current therapy for patients in case of diseases for example, Tb tuberculosis is not only laborious but also resource demanding. Several factors of medical management get specific patient outcomes which include adherence support mechanism i.e. to educate patients about effects resulting from adverse events, counselling, and good structure or management of laboratory with better access to specific susceptibility testing (WHO, 2000). Factors considered include HIV status, CD4 count, age, and persistence of resistance to a specific drug. Suggestions are found on observational examinations, advice of experts, minimum number of medications, and ideal ways of treatment (Kaplan, 2004).

Improved results are related to the utilization of noteworthy medications, antiretroviral treatment (ARV) in HIV effected people and medications with sterilizing activity (Fojo & Bates, 2003). Poor results include the recurrence of resistance from second line drugs, degree of infectious disease, comorbid conditions and high microbial burden (Cannon, 2009).

Flouroquinolone for tuberculosis is effective, but it can also lead to arthralgia like painful conditions. In contrast, aminoglycosides are associated with renal or nephron toxicity. Calofaizine has sterilizing properties however it results in prolongation of QTc and hyper-pigmentation (Sandgren, 2009). High dose of isoniazid is tolerated to the range of 10–15 mg per kilogram. The utility of high dosed ethionamid or isoniazid can be predicted with reference to katG and inhA mutations. Rapid DST and studies need to be developed which can control resistance of antibiotics in case of TB. The studies can use molecular strategies such as XRD and whole genome sequencing for better results (Sandgren, 2009).

Surgery is the oldest clinical method employed with some implications regarding anti-tuberculosis drugs. MDR-XRD has transferred interest towards surgical treatments as adjunctive therapy, mainly for TB, and also for extrapulmonary tuberculosis (EPTB). Pre-operative strategies must confirm cardio-pulmonary rescue after lung resection (Boden, 1999). At least three to a half year of pre-operative chemotherapy is given as treatment. Regardless of the absence of controlled parameters and low methodological abilities, few meta-investigations have improved the treatment results in MDR-tuberculosis patients going through medical procedure (Boden, 1999). Fractional lung resection is related with effective treatment (Wernsdorfer, 1994).

21.11 Vaccines or Immunization

Vaccines can help in reducing drug resistance and spread of antibiotic resistance conditions. It is difficult to recover from antibiotic resistant conditions (Johnson, 2006). Vaccinating humans and animals are the ways to stop intake of antibiotics in considerable amounts which in turn decrease resistance to specific antibiotic. Utilizing existing vaccines and development of new vaccines is effective in tackling antibiotic resistance and reducing morbidity or mortality rates. Expanding the usage of vaccines and limiting antimicrobial substances will eliminate antibiotic resistance (Goldstein & McCusker, 1999). If every child in world suffering from pneumonia receives dosage of vaccine instead of antimicrobial drug, it is estimated that about 1.5 million antibiotic drugs will be saved each year. Vaccines against flu and fever can play role whereas people take antibiotics which are the cause of excessive drug usage and resistance thereof (Townsend & Tew, 2003). In addition, the development, designing and usage of new drugs are present alternative to antibiotic resistance. Antibiotics are considered standard treatment for infections like streptococcus A which cause sore throat and for which no specific vaccine have been discovered yet. Similarly, new vaccines need to be developed targeting klebisela pneumonia which results in pneumonia and blood disorders. Vaccines stop bacterial diseases which later become antibiotic resistant. Multi-drug resistance results in case of MDR-tuberculosis (TB). It has become evident that about 49,000 people in 2015 were infected with MDR-TB (Meads et al., 2009).

A new vaccine against staphylococcus aurous which results in soft tissue and skin infection and clostridium difficile resulting in diarrhea and many others can protect against diseases that are difficult to treat. Getting and developing new disease is lengthy and complex process and drug resistance development needs research (Meads et al., 2009). Vaccines offer solutions which have yet been undermined. Vaccines involve multiple pathways to contract antimicrobial effects which are as follows.

- It reduces direct incidence and sensitivity of infections.
- It reduces appropriate and non-appropriate usage of antibiotics by decreasing disease incidence which also includes virus such as influenza etc.
- The reduced antimicrobial use lowers pressure towards resistance among human flora (Damiano, 1999).

Kennedy and Read pointed out resistance increase in systemic way like that of antibiotic usage. Every used antibiotic acts as an obstacle and hence a new antibiotic will be employed (Schmidt & Löscher, 2005). In contrast, vaccine has a benefit of sustainability and can be used multiple times without generating any resistance. Thus, for several reasons, vaccines overcome resistance as elaborated below;

- First is paraphyletic approach used in vaccine when population of microbes is small, thus reducing chances of generating resistance via mutations which can develop and proliferate (White, 1998).
- Secondly, vaccines can target host in several ways which require multiple mutations to confer resistance.
- Thirdly, there are rare instances where resistance to vaccines have developed or reduction in disease has been attained.
- Fourthly, it is useful for preservation of microbiome as antibiotics can disrupt microbes which can in turn human population, children and other populations (Pfaller, 2012).

Further research on rare cases can create better vaccine targets (Ma et al. 2010) (Fig. 21.1).

21.12 Developing Generation of Antibiotics

Multidrug resistance can turn into morbidity and mortality particularly in the case of organ transplantation or cancer chemotherapy. In generation of antibiotics, generation term is utilized for cephalosporin's and penicillin's based on their action on gram negative and positive bacteria, being classified into two branches as narrow

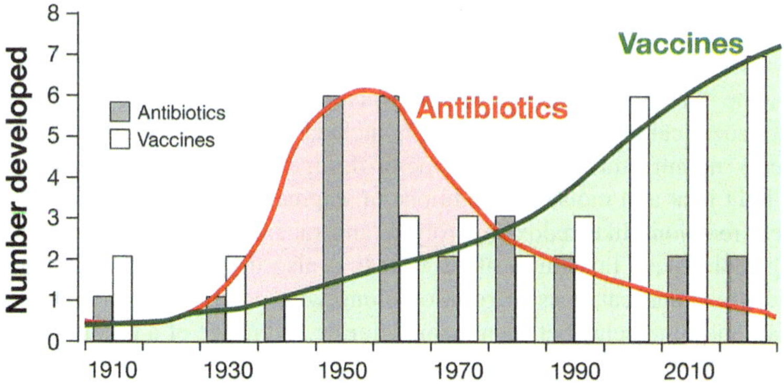

Fig. 21.1 Vaccines and anti-microbial resistant cases developed during century, resulting in new antibiotic pipeline lessens (Massa et al., 2009)

and broad spectrum (Ma et al. 2010). First generation antibiotics have low or narrow spectrum in clinical usage which means there are few organisms which can be treated by this strategy. A lot of organisms lying in this category are resistant to organisms which produce penicillinase or penicillin inactivating enzymes usually termed as beta lactamases. For example, most of gram-positive bacteria causing most types of infections come in this category. Second generation antibiotics have broad spectrum of action on both positive as well as negative strains (Sundar, 2001). In addition to penicillin producing organisms, it also has potential for *E.coli*, hemophilic influenza associated with ear, respiration, and urinary infections. In a list presented by World Health Organization, there are classes of bacteria which can be threatening to humans. These include Acinetobacter, Pseudomonas and many other Enterococcus bacteria or Salmonella. This pressure is formed as a result of un-appropriate usage of antibiotics in agriculture, farming and health care (D'Alessandro & Buttiëns, 2001). An association between host body immune and antimicrobial components is important factor considered in declining or finishing resistance bacteria (Helmbach, 2001). Poor detection of bacteria occurs by lower levels of antibody detection in adaptive immune system by B cells along with CD4 cells and CD8 cells. Persistence for long time in host leads to higher resistance. Antimicrobials clear infection in body by targeting a microorganism. Microbes target organism's immune system and result in generation of host specific antibodies or cell mediated T cell responses (Croft, 2001). Multiple pathways or mechanism of actions are created by vaccines. In relation to anti-microbial drugs, further mutations are needed. Innate molecular pump recognizes pathogen associated molecular patterns (PAMPs) and kill them. Many of such PAMPS are required for growth of bacteria and virulence pathway resulting in targeted ability of PAMP pathways. The immune system can simultaneously kill them resulting in exertion of antibiotic effect (Zheng, 2010).

21.13 Stable Zone Strategy

Stable zone is a complex process which involves variety of factors effecting stability of pharmaceutical product. These factors include interaction of active ingredients, excipients, manufacturing process, type or dosage form, container used for packaging, light heat and moisture conditions in shipment storage and handling. Other chemical reactions like redox, hydrolysis and racemization play role in stability of drug molecules (Takechi et al., 2001). It is also dependent on conditions like pH, radiations and catalysts of reaction along with process of preparation of raw materials and time delay between manufacturing and usage of a product. The pharmaceutical product also encounters changes in its color, density, consistence, and clarity of solution (Thu, 2017). These physical changes also include abrasions and changes in temperature like freezing or warming. The chemical reactions in pharmaceutical drugs can cause loss of ability or potential of activities like anti-oxidant and anti-bacterial (Berens, 2003). The stability of drugs results in microbial changes which include alternation in growth of microbes, non-sterile changes and variation in efficiency (Kamya, 2002). Upon degradation, unstable and inappropriate drug loses activity up to 90%, leading ultimately to death in most cases. It has become crucial for pharmaceutical companies to provide data before any drug approval. Updated database scan be helpful in areas of formulation, drug designing, dosage, and shelf life of a drug (Hastings, 2007). The drug stability testing gives affirmation that the drugs items will stay fit regarding their quality through the period during which they are on shelf. The drugs need to be stable until the patient uses last unit of drug molecule (Bridges et al., 2009).

Real stability involves longer period of test till product degradation under recommended and controlled conditions. The time of test is based on the product which may vary according to the stability and should be long enough to show no measured degradation. During test, data is collected at random frequency and trend analysis is utilized for checking any ambiguity in stability. Sample testing for stability is usual practice for marketed products in batches (Congpuong, 2005). Cyclic temperature testing is designed in a way that conditions such as temperature match with marketing conditions and usually left for 24 h. The maximum and minimum temperatures are selected on the basis of product type, its formation and formulation and other chemical and physical factors. For stability testing, world is divided into four zones based on climatic conditions according to which drugs are stored in proper environmental conditions (Dive & Biot, 2008). The temperatures are decided on the basis of annual mean temperature and levels of humidity in these regions. The conditions are actualized and combined to make them more practical for usage in industry and in generalized applications. The parameters for test would be based on quality, quantity, potency and identity which can change during storage. So, appearance, color, sterility, microbial testing ability, efficiency, dissolution and moisture content are the parameters considered for checking zones (Menegon, 2009) (Table 21.1).

Table 21.1 Climatic zones and the stability distribution (Kühne, 2000)

Area	Countries	Long term conditions
Temperate	UK, Europe etc	21 °C/45% RH
Subtropical and Mediterranean	Japan and Southern Europe	25 °C/60% RH
Hot and dry	Iraq and India	30 °C/35% RH
Hot and humid	Iran and Egypt	30 °C/65% RH
Hot and very humid	Brazil and Singapore	30 °C/75% RH

21.14 Treatment Cycling

Since 1980, diseases are treated using antibiotics for patients having bacterial infections and it eliminates antibiotic resistance. The concept is to limit the usage of antibiotics which in turn reduces selective pressure conferred by resistance, so drug molecules remain effective. The approach used now is termed as treatment cycling and is the conserved sessions of treatment repeated to get rid of resistance. Antibiotic cycling involves usage of antibiotics where specific dosage is given followed by recycling with another drug belonging to different class (Sidhu et al., 2005). Cycle is repeated again. More than two drugs are part of this cycling procedure. Another technique is the antibiotic mixing where each patient is treated with different antibiotic. Numerous techniques have been prescribed to prevent the rise of anti-toxin resistance (Haynes, 2006). A class of antibiotic can be withdrawn from usage for a proper period and again introduced to make bacteria non-resistant to specific antibiotic compound. It is the best method employed to achieve antibiotic heterogeneity where multiple antibiotics are being used in clinical environment (Manohar et al., 2013). Scientists also used mixing of amino glycerides in response to gentamycin among all other negative bacterial species. Experiment also suggest that introduction of antibiotics can reduce drug resistance (Danis & Bricaire, 2003).

21.15 Altering Doses of Drugs or Combinations

Many techniques have combated microbial resistance but the new therapy uses combination of drugs. It results in efficacy of treatment and increases the chances of developing less resistance. It involves combination of two or more antibiotics resulting in the formation of adjuvants which target resistance mechanisms including inhibition of beta lactamase enzymes, removing or targeting resistance genes via interruption of bacterial signal pathway (Weisman, 2006). Other natural substances can be combined with drugs which include nanoparticles, essential oils and plant extracts (Mockenhaupt, 2000). This procedure has its own disadvantages which can further worsen a condition of already developed antibiotic resistance instead of improving it. Identification of patients benefitting this drug therapy is key component in controlling drug or antibiotic resistance. It is mostly used in patients which have suffered from multi-drug resistance. This therapy is preferred because of following reasons;

- It results in broadening of antibiotic spectrum. So choice of primary antibiotics will reduce mortality rate which in resistant patient increases (Bousema, 2003).
- The inter abdominal region leading to gut can result in polymicrobial infections which require one or more than one antibiotic used in combination.
- Antibiotics combinations are preferred because of their synergic actions. Synergy is defined as condition where combined action of two agents is more than that of their individual usage. For example combination of beta lactamase and amino glycerides results in combined effect (Sahu et al., 2008).
- Chances of developing resistance for two or more drugs are lower than that of individual or single drug. Polymyxin and tigecycline are the two regularly utilized anti-microbial agents for carbapenems resistant enterococcus bacteria. As these anti-infection agents are the final solution, chances of developing resistance against these medications should be avoided. There are reports of increase of resistance against tetraglyciene and polymyxin, utilized as immunotherapy. It is estimated that one out of 13 patients under treatment with polymyxin for KPC infection reduce vulnerability to polymyxin B during the treatment (Prentice et al., 2001).

21.16 Health Education

Government's antimicrobial resistance methodology sets out instructions and training people about antimicrobial resistance (Aarestrup, 2005). Antitoxins may be suggested to physicians for comprehending antimicrobial resistance. Educational programs may be launched regarding prevention, executive management, infection control, human medication, nursing, drug stores, dentistry and experts (Aarestrup, 2004). A specific eLearning program or module can be used to support health and care staff for antimicrobial resistance. The public action plan to fight against antimicrobial resistance was provided in 1999 by Task Force controlled by pharmaceutical drug prevention and food administration (Ahmed et al., 2009). This plan involves agencies and drug companies to combat antimicrobial actions. Input data was obtained from states, pharmaceutical agencies, universities, societies, customer population and all other members involved in this health action program against microbial resistance phenomenon. So the need is to:

- Develop a national education plan for antimicrobial resistance.
- Assure availability of specific drugs related to surveillance.
- Ensure monitoring and checking of antimicrobial drug usage.
- Lastly monitoring antibiotic setting to target public health, which ensure safety for plants, foods, animals, and human health.

A well-organized national action plan will help in educating public about antimicrobial resistance in microorganisms, possible risks or threats. The plan should be implemented on all scales or levels in order to define the role of all participants, promoting usage of standard methods for conferring solution to antimicrobial resistance and providing an adequate amount of data to all interested parties or public which include health authorities, officials, authorities for detection of drug resistance in bacteria, fungi and other microorganisms (Morris, 2006). After development of plan for assessing dosage of drugs, it should also be implemented in national action plan (Cokal et al., 2009). It should also include presentations or workshops for patients and prescribers to have knowledge of dealing with infections, sharing pamphlets or leaflets regarding the guidelines, auditing templates, antimicrobial and diagnostic guidelines, educational posters or videos, self-assessment of checklist on prescribed practice work, and other E learning resources (Stonsaovapak & Boonyaratanakornkit, 2010). Delay in antimicrobial prescription forces clinical physician to collect data from patient in which symptoms have not been improved for about a week or have worsened with passage of time. There are more chances to improve antibiotic resistance with delayed or backup prescription. Most of symptoms in people gone through delayed prescription are similar to that of having immediate usage of antibiotics (Ranjbar et al. 2019). Most of people are unaware of backup prescription, therefore before applying back prescription, it is important for patients to have communication with their physician. Rehearsing hand cleanliness is a compelling approach to prevent diseases by stopping spread of germs including those that are anti-microbial resistant. Medical service suppliers do not clean their hands on 60% of the occasions they should. Around one of every 32 emergency clinic patients has medical services related infection (Englen, 2007).

Medical services related diseases are drawing consideration from patients, safety net providers, governments, and administrative bodies. This isn't because of the extent of issue regarding related morbidity, mortality and cost of treatment, yet it is actually preventable (Guven et al., 2010). The adherence to hand cleanliness decreases the danger of cross-transmissions. The clinical network is seeing pathophysiology of irresistible disease conditions and the worldwide spread of multidrug contaminations in medical care set-ups. "Clean Care is Safer Care" is the plan of WHO on understanding security programs (McArthur & Tsang, 2017). Hand cleanliness will result in reducing infections which in turn will decrease usage of antimicrobial agents and the antibiotic resistance (Thyagarajan et al., 2003).

21.17 Stopping Spread of Antibiotic Resistance

Antibiotic resistance is increased by excessive usage of antibiotics as well as by prevention of spread of antibiotics. At individual level, a person should take antibiotics when prescribed by medical expert. A patient should never demand antibiotic dosage if physicians don't prescribe. An individual should follow advice of medical expert to abstain from resistance. One can prevent antibiotic resistance by washing hands, preparing food in hygienic way, and avoiding close contact to infections. Similarly, policy makers can stop the spread of antibiotic resistance by implementing national action plan, improving surveillance of antibiotic resistance, by strengthening policies, programs and plans. It is the need of hour to regulate the usage of bad quality medicines or anti-microbial (Struzycka, 2019). In agricultural sector, it can be prevented by providing animals with antibiotics only under proper conditions. Alternatively animals can be vaccinated instead of antibiotic usage (Bartholomew, 2005). Finally, it can be controlled by biosecurity measures for preventing infections through improvements in hygienic conditions. There are new anti-infection agents being developed, however, not many can be relied against the hazardous types of resistant bacteria. Given the frequency with which individuals currently travel, anti-toxin resistance is a worldwide issue. When diseases cannot be treated by first-line antibiotics, more costly prescriptions are utilized (DO et al., 2006). Anti-microbial resistance is putting the current medication in ambiguity. Organ transplantations, chemotherapy, and medical procedures, for example, cesarean areas become risky without anti-microbial. A global action plan for antimicrobial and resistance was endorsed by world health organization in 2015 (Ohl & Luther, 2011). It aims not only prevention but also treatment of infectious diseases. This global action plan includes following objectives:

- It will improve the understanding of antimicrobial resistance.
- It will embody surveillance and research resulting in lower chances of infections.
- It will optimize the usage of antimicrobial agents or will ensure sustainable investments in case of antimicrobial resistivity.

So, it is not prescribed to save antibiotics for the next time or take medicine without clinical experts. It is also not recommended to take skipped doses of medicine. Another point to ponder is the side effects of medicines. To avoid side effects, the consumption be stopped and treatment be completed with alternative medicine (Lu, 2016). FDA must approve new potential antibiotics which can treat resistant bacteria.

21.18 Environmental and Sanitization Management in Combating Resistance

Anti-infection resistance is a danger to the wellbeing of humans around the world. In numerous world territories, there are no effective anti-toxin treatments thus compromising contaminations. Clinical treatments and medical procedures e.g. gut medical surgery, hip replacement, therapy of leukemia may become untreatable because of chances of infections. WHO in 2013 has declared the development of antimicrobial resistance as an issue where natural pathways of anti-microbial resistance have not yet been discovered (Bloland & WHO, 2001). Ongoing investigations have featured soil and water conditions as source of anti-toxin resistance genes (ARGs) of clinical value. There are three environmental factors contributing to antibiotic resistance genes, i.e., terrestrial agricultures, water from municipal wastes, and pharmaceutical wastes. These water sources do effect as humans are exposed to it for recreational use, bathing, washing, inhaling and breathing. Antibiotic resistant bacterial (ARBs) infections and acquiring antibiotic resistant genes (ARGs) cause morbidity and mortality. Both water and land can be influenced by the modern farming, and wastewaters contribute in the spread of ARBs.

References

Aarestrup, F. M. (2004). Monitoring of Antimicrobial resistance among food animals: Principles and limitations. *Journal of Veterinary Medicine, Series B, 51*(8–9), 380–388.

Aarestrup, F. M. (2005). Veterinary drug usage and Antimicrobial resistance in Bacteria of animal origin. *Basic & Clinical Pharmacology & Toxicology, 96*(4), 271–281.

Ahmed, A. M., Ishida, Y., & Shimamoto, T. (2009). Molecular characterization of antimicrobial resistance in Salmonella isolated from animals in Japan. *Journal of Applied Microbiology, 106*(2), 402–409.

Bartholomew, M. J., et al. (2005). A linear model for managing the risk of antimicrobial resistance originating in food animals. *Risk Analysis, 25*(1), 99–108.

Berens, N., et al. (2003). Plasmodium falciparum: correlation of in vivo resistance to chloroquine and antifolates with genetic polymorphisms in isolates from the south of Lao PDR. *Tropical Medicine & International Health, 8*(9), 775–782.

Bloland, P. B., & WHO (2001). *Drug resistance in malaria*, World Health Organization.

Boden, D., et al. (1999). HIV-1 drug resistance in newly infected individuals. *JAMA, 282*(12), 1135–1141.

Boehm, T., et al. (1997). Antiangiogenic therapy of experimental cancer does not induce acquired drug resistance. *Nature, 390*(6658), 404–407.

Bousema, J. T., et al. (2003). Treatment failure of pyrimethamine-sulphadoxine and induction of Plasmodium falciparum gametocytaemia in children in western Kenya. *Tropical Medicine & International Health, 8*(5), 427–430.

Bridges, D. J., Molyneux, M., & Nkhoma, S. (2009). Low level genotypic chloroquine resistance near Malawi's northern border with Tanzania. *Tropical Medicine & International Health, 14*(9), 1093–1096.

Cannon, R. D., et al. (2009). Efflux-mediated antifungal drug resistance. *Clinical Microbiology Reviews, 22*(2), 291–321.

Clavel, F., & Hance, A. J. (2004). HIV drug resistance. *New England Journal of Medicine, 350*(10), 1023–1035.

Cokal, Y., et al. (2009). Campylobacter spp. and their antimicrobial resistance patterns in poultry. An epidemiological survey study in Turkey. *Zoonoses and Public Health, 56*(3), 105–110.

Congpuong, K., et al. (2005). Molecular epidemiology of drug resistance markers of Plasmodium falciparum malaria in Thailand. *Tropical Medicine & International Health, 10*(8), 717–722.

Croft, S. L. (2001). Monitoring drug resistance in leishmaniasis. *Tropical Medicine & International Health, 6*(11), 899–905.

Croft, S. L., Sundar, S., & Fairlamb, A. H. (2006). Drug resistance in leishmaniasis. *Clinical Microbiology Reviews, 19*(1), 111–126.

D'Alessandro, U., & Buttiëns, H. (2001). History and importance of antimalarial drug resistance. *Tropical Medicine & International Health, 6*(11), 845–848.

Damiano, J. S., et al. (1999). Cell adhesion mediated drug resistance (CAM-DR): Role of integrins and resistance to apoptosis in human myeloma cell lines. *Blood, the Journal of the American Society of Hematology, 93*(5), 1658–1667.

Danis, M., & Bricaire, F. (2003). The new drug combinations: Their place in the treatment of uncomplicated Plasmodium falciparum malaria. *Fundamental & Clinical Pharmacology, 17*(2), 155–160.

Dive, D., & Biot, C. (2008). Ferrocene conjugates of chloroquine and other antimalarials: The development of ferroquine, a new antimalarial. *ChemMedChem, 3*(3), 383–391.

Do, N. T., et al. (2006). Antimicrobial resistance phenotypes of ETEC isolates from Piglets with Diarrhea in North Vietnam. *Annals of the New York Academy of Sciences, 1081*(1), 543–545.

Englen, M. D., et al. (2007). Prevalence and antimicrobial resistance of Campylobacter in US dairy cattle. *Journal of Applied Microbiology, 102*(6), 1570 1577.

Fojo, T., & Bates, S. (2003). Strategies for reversing drug resistance. *Oncogene, 22*(47), 7512–7523.

Goldstein, A. L., & McCusker, J. H. (1999). Three new dominant drug resistance cassettes for gene disruption in Saccharomyces cerevisiae. *Yeast, 15*(14), 1541–1553.

Gottesman, M. M. (2002). Mechanisms of cancer drug resistance. *Annual Review of Medicine, 53*(1), 615–627.

Guven, K., et al. (2010). Occurrence and characterization of Staphylococcus aureus isolated from meat and dairy products consumed in Turkey. *Journal of Food Safety, 30*(1), 196–212.

Hastings, I. M. (2007). Molecular markers as indicators of antimalarial drug failure rates. *Tropical Medicine & International Health, 12*(11), 1298–1301.

Haynes, R. K., et al. (2006). Artemisone—A highly active antimalarial drug of the artemisinin class. *Angewandte Chemie International Edition, 45*(13), 2082–2088.

Helmbach, H., et al. (2001). Drug-resistance in human melanoma. *International Journal of Cancer, 93*(5), 617–622.

Johnson, R., et al. (2006). Drug resistance in Mycobacterium tuberculosis. *Current Issues in Molecular Biology, 8*(2), 97–112.

Kamya, M. R., et al. (2002). Increasing antimalarial drug resistance in Uganda and revision of the national drug policy. *Tropical Medicine & International Health, 7*(12), 1031–1041.

Kaplan, R. M. (2004). Drug resistance in nematodes of veterinary importance: a status report. *Trends in Parasitology, 20*(10), 477–481.

Kühne, M., et al. (2000). Stability of Tetracycline in water and liquid manure. *Journal of Veterinary Medicine Series A, 47*(6), 379–384.

Li, X.-Z., & Nikaido, H. (2004). Efflux-mediated drug resistance in bacteria. *Drugs, 64*(2), 159–204.

Longley, D., & Johnston, P. (2005). Molecular mechanisms of drug resistance. *The Journal of Pathology: A Journal of the Pathological Society of Great Britain and Ireland, 205*(2), 275–292.

Lu, Q., et al. (2016). Pathogen and antimicrobial resistance profiles of culture-proven neonatal sepsis in Southwest China, 1990–2014. *Journal of Paediatrics and Child Health, 52*(10), 939–943.

Ma, J., Dong, C., & Ji, C. (2010). MicroRNA and drug resistance. *Cancer Gene Therapy, 17*(8), 523–531.

Manohar, S., Khan, S. I., & Rawat, D. S. (2013). 4-Aminoquinoline-Triazine-based hybrids with improved in vitro antimalarial activity against CQ-sensitive and CQ-resistant strains of Plasmodium falciparum. *Chemical Biology & Drug Design, 81*(5), 625–630.

Massa, H. M., Cripps, A. W., & Lehmann, D. (2009). Otitis media: Viruses, bacteria, biofilms and vaccines. *Medical Journal of Australia, 191*(S9), S44–S49.

McArthur, A. G., & Tsang, K. K. (2017). Antimicrobial resistance surveillance in the genomic age. *Annals of the New York Academy of Sciences, 1388*(1), 78–91.

Meads, M. B., Gatenby, R. A., & Dalton, W. S. (2009). Environment-mediated drug resistance: A major contributor to minimal residual disease. *Nature Reviews Cancer, 9*(9), 665–674.

Menegon, M., et al. (2009). Monitoring for multidrug-resistant Plasmodium falciparum isolates and analysis of pyrimethamine resistance evolution in Uige province, Angola. *Tropical Medicine & International Health, 14*(10), 1251–1257.

Mockenhaupt, F. P., et al. (2000). Submicroscopic Plasmodium falciparum infections in pregnancy in Ghana. *Tropical Medicine & International Health, 5*(3), 167–173.

Morris, D. O., et al. (2006). Screening of Staphylococcus aureus, Staphylococcus intermedius, and Staphylococcus schleiferi isolates obtained from small companion animals for antimicrobial resistance: A retrospective review of 749 isolates (2003–04). *Veterinary Dermatology, 17*(5), 332–337.

Ohl, C. A., & Luther, V. P. (2011). Antimicrobial stewardship for inpatient facilities. *Journal of Hospital Medicine, 6*(S1), S4–S15.

Pablos-Mendez, A., et al. (1998). Global surveillance for antituberculosis-drug resistance, 1994–1997. *New England Journal of Medicine, 338*(23), 1641–1649.

Pastan, I., & Gottesman, M. (1987). Multiple-drug resistance in human cancer. *New England Journal of Medicine, 316*(22), 1388–1393.

Pfaller, M. A. (2012). Antifungal drug resistance: Mechanisms, epidemiology, and consequences for treatment. *The American Journal of Medicine, 125*(1), S3–S13.

Prentice, H. G., et al. (2001) Oral ciprofloxacin plus colistin: Prophylaxis against bacterial infection in neutropenic patients. A strategy for the prevention of emergence of antimicrobial resistance. *British Journal of Haematology, 115*(1), 46–52.

Ranjbar, R., Yadollahi Farsani, F., & Safarpoor Dehkordi, F. (2019). Antimicrobial resistance and genotyping of vacA, cagA, and iceA alleles of the Helicobacter pylori strains isolated from traditional dairy products. *Journal of Food Safety, 39*(2), e12594.

Sahu, N. K., Sahu, S., & Kohli, D. V. (2008). Novel molecular targets for antimalarial drug development. *Chemical Biology & Drug Design, 71*(4), 287–297.

Sandgren, A., et al. (2009). Tuberculosis drug resistance mutation database. *PLoS Medicine, 6*(2), e1000002.

Schmidt, D., & Löscher, W. (2005). Drug resistance in epilepsy: putative neurobiologic and clinical mechanisms. *Epilepsia, 46*(6), 858–877.

Sidhu, A. B. S., Valderramos, S. G., & Fidock, D. A. (2005). pfmdr1 mutations contribute to quinine resistance and enhance mefloquine and artemisinin sensitivity in Plasmodium falciparum. *Molecular Microbiology, 57*(4), 913–926.

Stonsaovapak, S., & Boonyaratanakornkit, M. (2010). Prevalence and antimicrobial resistance of listeria species in food products in Bangkok Thailand. *Journal of Food Safety, 30*(1), 154–161.

Struzycka, I., et al. (2019). Knowledge of antibiotics and antimicrobial resistance amongst final year dental students of Polish medical schools—A cross-sectional study. *European Journal of Dental Education, 23*(3), 295–303.

Sundar, S. (2001). Drug resistance in Indian visceral leishmaniasis. *Tropical Medicine & International Health, 6*(11), 849–854.

Takechi, M., et al. (2001). Therapeutic efficacy of sulphadoxine/pyrimethamine and susceptibility in vitro of P. falciparum isolates to sulphadoxine-pyremethamine and other antimalarial drugs in Malawian children. *Tropical Medicine & International Health, 6*(6), 429–434.

Tew, K. D. (1994). Glutathione-associated enzymes in anticancer drug resistance. *Cancer Research, 54*(16), 4313–4320.

Thakur, M. D., et al. (2013). Modelling vemurafenib resistance in melanoma reveals a strategy to forestall drug resistance. *Nature, 494*(7436), 251–255.

Thu, A. M., et al. (2017). Combating multidrug-resistant Plasmodium falciparum malaria. *The FEBS Journal, 284*(16), 2569–2578.

Thyagarajan, S., et al. (2003). Geographical difference in antimicrobial resistance pattern of Helicobacter pylori clinical isolates from Indian patients: Multicentric study. *Journal of Gastroenterology and Hepatology, 18*(12), 1373–1378.

Townsend, D. M., & Tew, K. D. (2003). The role of glutathione-S-transferase in anti-cancer drug resistance. *Oncogene, 22*(47), 7369–7375.

Weisman, J. L., et al. (2006). Searching for new antimalarial therapeutics amongst Known Drugs. *Chemical Biology & Drug Design, 67*(6), 409–416.

Wernsdorfer, W. H. (1994). Epidemiology of drug resistance in malaria. *Acta Tropica, 56*(2–3), 143–156.

White, N. (1998). Preventing antimalarial drug resistance through combinations. *Drug Resistance Updates, 1*(1), 3–9.

White, N. (1999). Antimalarial drug resistance and combination chemotherapy. *Philosophical Transactions of the Royal Society of London. Series B: Biological Sciences, 354*(1384), 739–749.
White, N. J. (2004). Antimalarial drug resistance. *The Journal of Clinical Investigation, 113*(8), 1084–1092.
White, T. C., Marr, K. A., & Bowden, R. A. (1998). Clinical, cellular, and molecular factors that contribute to antifungal drug resistance. *Clinical Microbiology Reviews, 11*(2), 382–402.
WHO (2000). *Anti-tuberculosis drug resistance in the world*, World Health Organization.
Zheng, T., et al. (2010). Role of microRNA in anticancer drug resistance. *International Journal of Cancer, 126*(1), 2–10.

The manufacturer's authorised representative in the EU is Springer Nature Customer Service Centre GmbH, Europaplatz 3, 69115 Heidelberg, Germany. If you have any concerns regarding our products, please contact ProductSafety@springernature.com

Printed and bound by CPI Group (UK) Ltd, Croydon, CR0 4YY
28/09/2026
02224065-0013